ESSENTIALS
(of) ELECTRONICS

SECOND EDITION

Frank D. Petruzella

Glencoe
McGraw-Hill

New York, New York   Columbus, Ohio   Woodland Hills, California   Peoria, Illinois

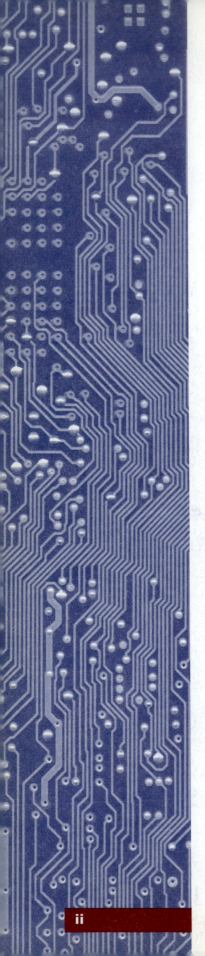

Cover Photo Credits: (center) Bill Frymire/Masterfile, (*others, clockwise from top left*) Chris Sorensen/Stock Market, David Jeff/Stock Market, Fred Eckstein/Photo Take, Tom Ives/Stock Market, Charles O'Rear/Corbis, H. David Seawell/Corbis, Thom Lang/Stock Market, SuperStock, Paul Chauncey/Stock Market, Lester Lefkowitz/Stock Market, D. Boone/Corbis, Chuck Savage/Stock Market, Corbis, Stock/Market.

Acknowledgments

The Author would like to thank the following reviewers, who examined the manuscript to ensure that the information presented is comprehensive, accurate and up to date. Jeff Anderson, Soldier Pond, ME; Bill Hessmiller of Technical Training Associates, Dunmore, PA; William F. Lange of Princeton University, Kirkwood, MO; Dave Longobardi of Antelope Valley College, Lancaster, CA; Gregory T. Shepley, Girard, OH; Dan Siddall of SCP Global Technologies, Boise, ID; and Bill Stevens of Ohio University, Lancaster, OH, all spent a great deal of time reviewing this manuscript and their input is greatly appreciated. Additional thanks go to Rob Ciccotelli, and Linda Jefferson of Glencoe for their hard work and cooperation in the preparation of this text.

Library of Congress Cataloging-in-Publication Data

Petruzella, Frank D.
    Essentials of electronics: Frank D. Petruzella.
      p. cm.
    Includes index.
    ISBN 0-07-821048-8
    1. Electronics. I. Title.
TK7816.P48   1999
621.381 — dc20   99-32921
                                     CIP

Essentials of Electronics, Second Edition   ISBN 0-07-821048-8
Activities Manual for Essentials of Electronics, Second Edition
ISBN 0-07-821049-6
Text with Electronics Experiments Manual   ISBN 0-07-821996-5
Instructor's Manual for Essentials of Electronics, Second Edition with Instructor's Productivity Center CD-ROM   ISBN 0-07-821050-X

Glencoe/McGraw-Hill

A Division of The **McGraw·Hill** Companies

Essentials of Electronics, Second Edition

Send all inquiries to:
Glencoe/McGraw-Hill
936 Eastwind Drive
Westerville, OH 43081

ISBN 0-07-821048-8

Printed in the United States of America.

1 2 3 4 5 6 7 8 9 10 071   05 04 03 02 01 00 99

Brief Contents

Unit 1   DC Circuits

Chapter 1   Safety 1
Chapter 2   Instruments, Tools, and Fasteners 17
Chapter 3   Conductors, Semiconductors, and Insulators 30
Chapter 4   Sources and Characteristics of Electricity 39
Chapter 5   Basic Electrical Units 51
Chapter 6   Electric Connections 62
Chapter 7   Simple, Series, and Parallel Circuits 75
Chapter 8   Measuring Voltage, Current, and Resistance 91
Chapter 9   Circuit Conductors and Wire Sizes 112
Chapter 10   Resistors 123
Chapter 11   Ohm's Law 137
Chapter 12   Solving the Series Circuit 149
Chapter 13   Solving the Parallel Circuit 162
Chapter 14   Solving the Series-Parallel Circuit 172
Chapter 15   Magnetism 187
Chapter 16   Electromagnetism 197
Chapter 17   Batteries 209
Chapter 18   Circuit Protection Devices 225
Chapter 19   Electric Power 239
Chapter 20   Electric Energy 249
Chapter 21   Direct Current and Alternating Current 259

Unit 2   AC Devices

Chapter 22   Low-Voltage Signal Systems 273
Chapter 23   Residential Wiring Requirements and Devices 290
Chapter 24   Residential Branch Circuit Wiring 315
Chapter 25   Appliance Cords and Connections 335
Chapter 26   Lightning Equipment 346
Chapter 27   Electric Motors 367
Chapter 28   Relays 384
Chapter 29   Motor Controls 396
Chapter 30   Inductance and Capacitance 421
Chapter 31   Transformers 443
Chapter 32   Signal Sources 458
Chapter 33   Printed Circuits 474
Chapter 34   Electronic Test Instruments 487
Chapter 35   Semiconductor Diodes 505
Chapter 36   Power Supplies 520
Chapter 37   Transistors 537
Chapter 38   Transistor Switching, Amplification,
             and Oscillation Circuits 555
Chapter 39   Thyristors: The SCR and Triac 579
Chapter 40   Integrated Circuits (ICs) 595
Chapter 41   Digital Fundamentals 613
Chapter 42   The Microcomputer 650
Chapter 43   Network Theorems 674
Chapter 44   Principles of AC Circuits 694

Detailed Contents

Preface   x

Unit 1   DC Circuits

CHAPTER 1   Safety   1

1-1 Safety in the Home   1-2 Safety Outdoors   1-3 Safety in the School Laboratory   1-4 Occupational Safety   1-5 Electric Shock   1-6 First Aid 1-7 Artificial Respiration   1-8 Fire Prevention   1-9 Safety Color Codes 1-10 Hazardous Substances and Waste

CHAPTER 2   Instruments, Tools, and Fasteners   17

2-1 Measuring Devices   2-2 Common Tools   2-3 Organization and Use of Tools

CHAPTER 3   Conductors, Semiconductors, and Insulators   30

3-1 Electron Theory of Matter   3-2 Bohr Model of Atomic Structure   3-3 Ions 3-4 Electricity Defined   3-5 Electric Conductors, Insulators, and Semiconductors 3-6 The Continuity Tester   3-7 Electrical and Electronic Devices

CHAPTER 4   Sources and Characteristics of Electricity   39

4-1 Static Electricity   4-2 Charged Bodies   4-3 Testing for a Static Charge 4-4 Producing a Static Charge   4-5 Current Electricity 4-6 Sources of Electromotive Force [emf]

CHAPTER 5   Basic Electrical Units   51

5-1 Electric Charge   5-2 Current   5-3 Voltage   5-4 Resistance   5-5 Power 5-6 Energy   5-7 Basic Electric Circuit   5-8 Relationship Among Current, Voltage, and Resistance   5-9 Direction of Current Flow

CHAPTER 6   Electric Connections   62

6-1 The Necessity for Proper Electric Connections   6-2 Preparing Wire for Connection   6-3 Terminal-Screw Connections   6-4 Crimp Terminals, Splices, and Mechanical Connectors   6-5 Insulating Repairs   6-6 Soldered Connections 6-7 Electronic Data Communication Interconnection Systems

CHAPTER 7 — Simple, Series, and Parallel Circuits — 75

7-1 Circuit Components   7-2 Circuit Symbols   7-3 Circuit Diagrams
7-4 Simple Circuit   7-5 Series Circuit   7-6 Parallel Circuit
7-7 Constructing Wiring Projects Using Schematic Diagrams
7-8 Breadboarding Circuits   7-9 Computer-Simulated Circuits

CHAPTER 8 — Measuring Voltage, Current, and Resistance — 91

8-1 Analog and Digital Meters   8-2 Multimeter   8-3 Reading Meters
8-4 Measuring Voltage   8-5 Measuring Current   8-6 Measuring Resistance
8-7 Meter Safety   8-8 Multimeter Specifications and Special Features
8-9 Virtual Meters

CHAPTER 9 — Circuit Conductors and Wire Sizes — 112

9-1 Conductor Forms   9-2 Conductor Insulation   9-3 Wire Sizes
9-4 Conductor Ampacity   9-5 Conductor Resistance   9-6 Voltage Drop and Power Loss

CHAPTER 10 — Resistors — 123

10-1 Resistance Wire   10-2 Resistors   10-3 Types of Resistors
10-4 Rheostats and Potentiometers   10-5 Resistor Identification
10-6 Series Connection of Resistors   10-7 Parallel Connection of Resistors
10-8 Series-Parallel Connection of Resistors

CHAPTER 11 — Ohm's Law — 137

11-1 Electrical Units and Prefixes   11-2 Ohm's Law   11-3 Applying Ohm's Law to Calculate Current   11-4 Applying Ohm's Law to Calculate Voltage
11-5 Applying Ohm's Law to Calculate Resistance   11-6 Ohm's Law Triangle
11-7 Power Formulas   11-8 Ohm's Law in Graphical Form

CHAPTER 12 — Solving the Series Circuit — 149

12-1 Series Circuit Connection   12-2 Identifying Circuit Quantities
12-3 Series Circuit Current, Voltage, Resistance, and Power Characteristics
12-4 Solving Series Circuits   12-5 Polarity   12-6 Series-Aiding and Series-Opposing Voltages   12-7 Troubleshooting a Series Circuit

CHAPTER 13 — Solving the Parallel Circuit — 162

13-1 Parallel Circuit Connection   13-2 Parallel Circuit Current, Voltage, Resistance, and Power Characteristics   13-3 Solving Parallel Circuits
13-4 Troubleshooting a Parallel Circuit

CHAPTER 14 — Solving the Series-Parallel Circuit — 172

14-1 Kirchhoff's Voltage Law   14-2 Kirchhoff's Current Law   14-3 Solving Series-Parallel Circuits   14-4 Troubleshooting Series-Parallel Circuits

CHAPTER 15   Magnetism   187

15-1 Properties of Magnets   15-2 Types of Magnets   15-3 Law of Magnetic Poles
15-4 Magnetic Polarity   15-5 The Magnetic Field   15-6 Magnetic Shielding
15-7 Theories of Magnetism   15-8 Uses for Permanent Magnets

CHAPTER 16   Electromagnetism   197

16-1 Magnetic Field Around a Current-Carrying Conductor   16-2 Left-Hand
Conductor Rule   16-3 Magnetic Field About Parallel Conductors   16-4 Magnetic
Field About a Coil   16-5 Left-Hand Coil Rule   16-6 The Electromagnet
16-7 The Magnetic Circuit   16-8 Uses for Electromagnets

CHAPTER 17   Batteries   209

17-1 The Voltaic Cell   17-2 Battery Terminology and Ratings   17-3 Primary Dry
Cells   17-4 Series and Parallel Cell Connections   17-5 Testing Primary Cells and
Batteries   17-6 Lead-Acid Rechargeable Battery   17-7 Testing a Lead-Acid
Battery   17-8 Other Rechargeable Batteries   17-9 Battery Chargers

CHAPTER 18   Circuit Protection Devices   225

18-1 Undesirable Circuit Conditions   18-2 Fuse and Circuit Breaker Ratings
18-3 Types of Fuses   18-4 Testing Fuses   18-5 Circuit Breakers
18-6 Thermal Overload Protection   18-7 Lightning Rods and Arresters

CHAPTER 19   Electric Power   239

19-1 Electric Generating Stations   19-2 Alternate Ways of Generating Electricity
19-3 Transmitting Electricity   19-4 Electric Power   19-5 Calculating Electric
Power   19-6 Measuring Electric Power

CHAPTER 20   Electric Energy   249

20-1 Energy   20-2 Calculating Electric Energy   20-3 The Energy Meter
20-4 Energy Costs   20-5 Energy Management

CHAPTER 21   Direct Current and Alternating Current   259

21-1 Direct Current (DC)   21-2 Alternating Current (AC)   21-3 AC Generation
21-4 The AC Sine Wave   21-5 AC Sine Wave Voltage and Current Values
21-6 Three-Phase AC   21-7 DC Generation   21-8 Types of DC Generators
21-9 Generator Prime Movers

UNIT 2   AC Devices

CHAPTER 22   Low-Voltage Signal Systems   273

22-1 Low-Voltage Signal System   22-2 Door-Chime Circuit   22-3 Annunciator
Circuit   22-4 Electric Door-Lock Circuit   22-5 Telephone Circuits
22-6 Alarm Systems

CHAPTER 23 Residential Wiring Requirements and Devices 290

23-1 Approval of Equipment and Wiring 23-2 The Incoming Service
23-3 The Three-Wire Distribution System 23-4 General Wiring Requirements
23-5 Installing Nonmetallic Sheathed Cable 23-6 Electric-Outlet Boxes
23-7 Electrical Receptacles 23-8 Lampholders 23-9 Switches and Dimmers
23-10 Grounding System 23-11 Overcurrent Protection

CHAPTER 24 Residential Branch Circuit Wiring 315

24-1 National Electrical Code Branch Circuit Requirements 24-2 Pull-Chain
Lampholder and Duplex Receptacle Circuit 24-3 Switch and Lampholder Circuit
24-4 Two Lamps and One Switch Circuit 24-5 Duplex Receptacle, Switch, and
Lamp Circuit 24-6 Switched Split-Duplex Receptacle Circuit 24-7 Three-Way
Switch Circuit 24-8 Four-Way Switch Circuit 24-9 Planning Your Electrical
System

CHAPTER 25 Appliance Cords and Connections 335

25-1 Types of Electric Cords 25-2 Cord Connectors 25-3 Grounding Appliances
25-4 Ground-Fault Circuit Interrupter (GFCI) 25-5 Weatherproof Fixtures

CHAPTER 26 Lighting Equipment 346

26-1 Incandescent Lamps 26-2 Repairing a Table Lamp 26-3 Installing Light
Fixtures 26-4 Fluorescent Lighting 26-5 Fluorescent Light Fixture Circuits
26-6 High-Intensity Discharge (HID) Lamps 26-7 Security Lighting

CHAPTER 27 Electric Motors 367

27-1 Motor Principle 27-2 Direct Current (DC) Motors 27-3 Universal Motors
27-4 Alternating Current (AC) Motors 27-5 Motor Power and Torque
27-6 Troubleshooting Motors

CHAPTER 28 Relays 384

28-1 Electromechanical Relay 28-2 Magnetic-Reed Relay
28-3 Solid-State Relays 28-4 Timing Relays

CHAPTER 29 Motor Controls 396

29-1 Motor Protection 29-2 Motor Starting 29-3 Motor Reversing and Jogging
29-4 Motor Stopping 29-5 Motor Drives 29-6 Motor Pilot Devices

CHAPTER 30 Inductance and Capacitance 421

30-1 Types of Inductors 30-2 Inductance 30-3 Inductive Reactance
30-4 Phase Shift in Inductance 30-5 Capacitance 30-6 Capacitor Ratings
30-7 Types of Capacitors 30-8 RC-Time Constant 30-9 Capacitive Reactance
30-10 Phase Shift in Capacitance 30-11 Impedance 30-12 Power in
AC Circuits 30-13 Troubleshooting Inductors and Capacitors

CHAPTER 31   Transformers   443

31-1 Transformer Action   31-2 Transformer Voltages, Current, and Power Relationships   31-3 Ignition Coil   31-4 Types of Transformers 31-5 Testing Transformers   31-6 Three-Phase Transformer Systems

CHAPTER 32   Signal Sources   458

32-1 Transducers   32-2 Loudspeakers   32-3 Microphones   32-4 Phonograph Cartridges   32-5 Record/Play Tape Head   32-6 Magnetic Disk Drives 32-7 Optical Disk Drives   32-8 Sensors and Detectors

CHAPTER 33   Printed Circuits   474

33-1 PC Board Construction   33-2 Planning and Layout 33-3 Printing the PC Board   33-4 Component Assembly and PC Board Soldering 33-5 Servicing Printed Circuit Boards

CHAPTER 34   Electronic Test Instruments   487

34-1 The Oscilloscope   34-2 Oscilloscope Front-Panel Controls 34-3 Oscilloscope Measurements   34-4 Signal Generators 34-5 Frequency Counters   34-6 DC Voltage Supplies 34-7 Bridge Measuring Circuits

CHAPTER 35   Semiconductor Diodes   505

35-1 Semiconductors   35-2 N-Type and P-Type Material 35-3 PN-Junction Diode   35-4 Diode Operating Characteristics 35-5 Diode Packages and Ratings   35-6 Testing Diodes 35-7 Practical Applications of Diodes   35-8 Special Purpose Diodes

CHAPTER 36   Power Supplies   520

36-1 Power Supplies   36-2 Power Transformer Circuits   36-3 Rectifier Circuits 36-4 Three-Phase Rectifier   36-5 Filter Circuits   36-6 Voltage Regulation 36-7 Voltage Multipliers   36-8 Switching Power Supplies 36-9 Troubleshooting Power Supplies

CHAPTER 37   Transistors   537

37-1 Bipolar-Junction Transistors (BJTs)   37-2 Field-Effect Transistors (FETs) 37-3 Unijunction Transistors (UJTs)   37-4 Phototransistors

CHAPTER 38   Transistor Switching, Amplificaton, and Oscillation Circuits   555

38-1 Transistor-Switching Circuits   38-2 Transistor Amplifiers 38-3 Transistor Oscillators

CHAPTER 39 Thyristors: The SCR and Triac 579

39-1 Thyristors   39-2 Principles of SCR Operation
39-3 DC-Operated SCR Circuits   39-4 AC-Operated SCR Circuits
39-5 Testing the SCR   39-6 The Triac   39-7 Triac Circuit Applications

CHAPTER 40 Integrated Circuits (ICs) 595

40-1 Integrated Circuit (IC) Construction   40-2 Advantages and Limitations of ICs
40-3 IC Symbols and Packages   40-4 Analog and Digital ICs
40-5 Operational Amplifier ICs   40-6 Op-Amp Voltage Amplifiers
40-7 Op-Amp Voltage Comparator   40-8 Linear-Power Amplifier
40-9 Summing and Difference Amplifiers   40-10 The 555 Timer
40-11 Current Sinking and Sourcing of IC Outputs

CHAPTER 41 Digital Fundamentals 613

41-1 Digital Electronics   41-2 Binary Number System   41-3 Logic Gates
41-4 Combinational Logic   41-5 Binary Arithmetic
41-6 Encoders and Decoders   41-7 Multiplexers and Demultiplexers
41-8 Sequential Logic   41-9 Registers   41-10 Binary Counters
41-11 Connecting with Analog Devices   41-12 Digital Logic Probe

CHAPTER 42 The Microcomputer 650

42-1 Microcomputers   42-2 Memory Units   42-3 Input and Output
42-4 Computer Communications   42-5 Computer Software
42-6 Processing Information   42-7 Microprocessor-Based Control Systems
42-8 Personal Computer (PC)   42-9 The Internet

CHAPTER 43 Network Theorems 674

43-1 Voltage and Current Sources   43-2 Superposition Theorem
43-3 Thevenin's Theorem   43-4 Norton's Theorem

CHAPTER 44 Principles of AC Circuits 694

44-1 Impedance in AC Series Circuits   44-2 Current and Voltage in AC
Series Circuits   44-3 Parallel AC Circuits   44-4 Filter Circuits

GLOSSARY 707

INDEX 717

Preface

This second edition of *Essentials of Electronics* provides a strong foundation in the concepts and terminology of both electrical and electronic circuits. The subject material is written in language that is easy to read and understand. Numerous two-color illustrations help students understand the theory and visualize its practical applications. This text is suitable for use in various courses including: Electrical Fundamentals, Electronic Fundamentals, DC Circuits, AC Circuits, Electronic Devices, Digital Fundamentals, Residential Wiring, Electricity/Electronics Tech Prep, and Electricity/Electronics Distance Learning.

Chapters on "Network Theorems," and "Principles of AC Circuits" are new to this edition. "About Electronics" features have also been added throughout the book, which present recent developments in the fields of electricity and electronics. Help Wanted advertisements (edited copies of actual job postings) in each chapter provide insight into the diverse career opportunities in the electricity and electronics industry.

Each chapter opens with an introduction to the topics covered, learning objectives, key terms, and related Internet research sites (suggested organizations to research for additional information on the chapter topics). Key terms are highlighted in the text using ***bold blue italics,*** while terms that are defined in the glossary appear in *red italics*. Each chapter ends with a "Review and Applications" section containing:

- **Related Formulas**—a convenient list of formulas found within the chapter.
- **Review Questions**—to evaluate students' understanding of new concepts.
- **Problems**—which enable students to practice problem-solving techniques.
- **Critical Thinking**—exercises that give students practice with open-ended challenges for which prescribed solutions do not exist.
- **Portfolio Projects**—through which students demonstrate what they have learned.
- **Circuit Challenge**—simulation exercises for use with simulation software packages such as Electronics Workbench.

ANCILLARIES
- *Activities Manual for Essentials of Electronics.* This book contains instructions for hands-on lab work and a multiple-choice test for each chapter.
- *Electronic Experiments Manual.* This optional text with CD-ROM student package adds a Problem Solving and Troubleshooting CD-ROM for use with *Electronics Workbench* to the student text. With over 300 circuit files correlated to every chapter, this CD helps students to apply theory through interactive computer simulations.
- *Instructor's Manual.* This book contains answers to all of the questions and problems in the text and its supplements as well as transparency masters. This package also includes an *Instructor's Productivity Center CD-ROM* which contains a test generator, a curriculum builder program, animated PowerPoint presentations for every chapter, Electronics Workbench files for use with Circuit Challenge simulation exercises in the text, and answer material for the Electronic Experiments Manual.

Frank D. Petruzella

Safety

Working with electricity shouldn't cost an arm, a leg ... or your life. But sad to say for some it has. Many of these casualties are young people just entering the workplace. This chapter covers the basic safety rules and safe practices that apply to the home, outdoors, school lab, and on the job.

Objectives

After studying this chapter, you should be able to:
1. Outline safety rules and safe practices that apply to conditions in the home, outdoors, school lab, and on the job.
2. Explain the factors that determine the severity of an electric shock.
3. Outline the first-aid procedure for bleeding, burns, and electric shock.
4. Describe the mouth-to-mouth method of artificial respiration.
5. List the procedures to be followed in the case of an electrical fire.
6. Identify hazardous materials and describe their characteristics.

Key Terms

- ground-fault circuit interrupter
- lockout
- tagout
- resistance
- voltage

- current
- first aid
- artificial respiration
- hazardous substance
- hazardous waste

- corrosives
- ignitable
- toxic
- reactive

RESEARCH INTERNET SITES

- Occupational Safety and Health Administration (OSHA)
- United States Department of Labor
- Environmental Protection Agency (EPA)

1-1 Safety in the Home

Electrical equipment used in homes today is safe when properly installed, maintained, and used. However, hazards are created when this equipment is improperly used or when suitable safety measures are not employed.

Second only to motor vehicles, the home is the site of most fatal accidents each year in the United States. The best way to reduce accidents at home is to know the potential hazards and take the necessary precautions to eliminate them (Figure 1-1). The following list of electrical safety suggestions is designed to increase your awareness of electrical accidents that can occur in the home.

- Never run extension or power cords under rugs. These cords are not designed for this type of rough service, nor are they a substitute for permanent wiring.
- Do not jerk extension cords from electrical outlets; grasp the plug and pull.
- Cover unused electrical outlets so that children cannot poke pins, or other like objects, into them.

About ◁▭▷ Electronics

In a power outage, do not connect a portable or RV generator directly to your home's electrical system. Electricity can backfeed into the power lines and harm power workers. To prevent this, isolate home wires by using a double-throw double-pole transfer switch when using standby generators at home.

- When using an extension cord, the current rating of the cord should exceed that of the equipment being operated.
- Do not over-fuse a circuit.
- Always turn off the main electrical switch before replacing a blown fuse.
- Keep basement floors around washtubs and machines dry to help eliminate falls and to reduce the hazards of electric shock.
- Replace frayed appliance cords or defective wiring on all appliances as soon as they are discovered.

- Discover the problem's cause if fuses blow or circuit breakers trip often.
- Do not overload electrical outlets by use of multiple tap-off (octopus) devices.
- Do not use electric space heaters, radios, or appliances in the bathroom, laundry room, or near the kitchen sink.
- Turn off the main electrical switch when checking or replacing switches and outlets.
- Do not pry bread from a plugged-in toaster.

1-2 Safety Outdoors

Unfortunately, the human body will conduct electricity. Dry skin offers some insulation, but if the voltage is great enough, the electricity will pass through. When the skin is wet, its resistance is much lower, which explains why shock intensity is greater when hands are wet. A few safety tips (Figure 1-2) to keep in mind outdoors include:

- Never touch a downed wire.
- Stay clear of power lines.
- Do not bring appliances near swimming pools.
- Avoid using power tools on wet grass or other wet surfaces.
- All outdoor electrical outlets should be supplied from circuits equipped with *ground-fault circuit interrupters* to provide extra protection. This is especially important in damp locations where electrical shocks are usually more severe and where a normal fuse or circuit breaker does not provide enough protection.

1-3 Safety in the School Laboratory

Most school accidents are caused because safety rules are not observed or were never explained. The school must ensure safe working conditions and education for the students. However, you must learn how to protect yourself and others working near you. Use your common sense! The school lab, in particular, is not the place for

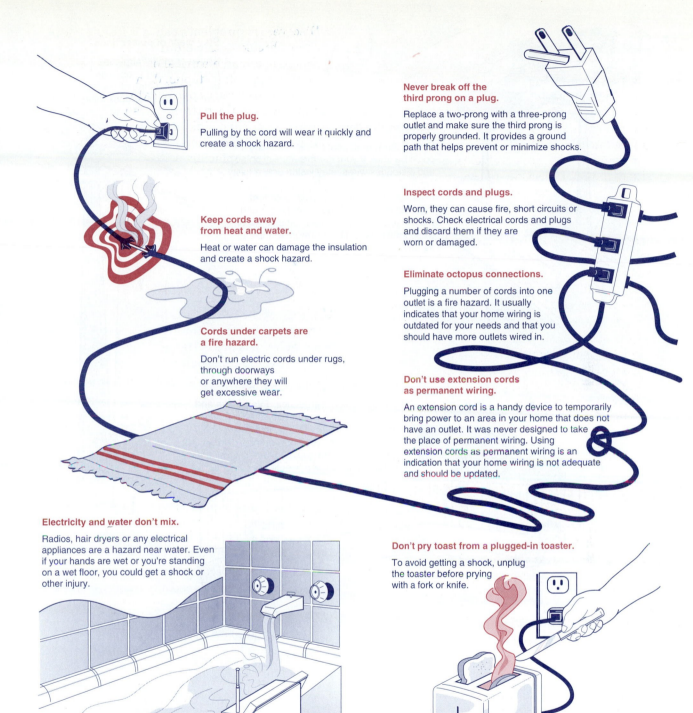

Pull the plug.

Pulling by the cord will wear it quickly and create a shock hazard.

Keep cords away from heat and water.

Heat or water can damage the insulation and create a shock hazard.

Cords under carpets are a fire hazard.

Don't run electric cords under rugs, through doorways or anywhere they will get excessive wear.

Electricity and water don't mix.

Radios, hair dryers or any electrical appliances are a hazard near water. Even if your hands are wet or you're standing on a wet floor, you could get a shock or other injury.

Never break off the third prong on a plug.

Replace a two-prong with a three-prong outlet and make sure the third prong is properly grounded. It provides a ground path that helps prevent or minimize shocks.

Inspect cords and plugs.

Worn, they can cause fire, short circuits or shocks. Check electrical cords and plugs and discard them if they are worn or damaged.

Eliminate octopus connections.

Plugging a number of cords into one outlet is a fire hazard. It usually indicates that your home wiring is outdated for your needs and that you should have more outlets wired in.

Don't use extension cords as permanent wiring.

An extension cord is a handy device to temporarily bring power to an area in your home that does not have an outlet. It was never designed to take the place of permanent wiring. Using extension cords as permanent wiring is an indication that your home wiring is not adequate and should be updated.

Don't pry toast from a plugged-in toaster.

To avoid getting a shock, unplug the toaster before prying with a fork or knife.

Figure 1-1   Know the potential hazards and take the necessary precautions to eliminate them. (*Courtesy Ontario Hydro.*)

Stay clear of power lines.

If kites or model airplanes touch overhead lines, even touching the string or control wire may cause serious injury.

Never touch a downed wire.

Even if the wire appears dead, touching it could be fatal. If your car hits a utility pole and dislodges wires, stay inside until a utility crew removes the wire. If you must get out because of fire, jump free with both feet together and without touching the car and the ground at the same time. Shuffle away using small steps. Do not return to the car for any reason and warn others to keep away.

Don't bring appliances near water.

Electric radios, barbecues, TVs, clocks and other appliances should be kept well away from swimming pools. And swimming pools should never be near power lines. A long-handled skimmer could be fatal if it touches outdoor lighting or power lines.

Figure 1-2 Safety outdoors. (*Courtesy Ontario Hydro.*)

horseplay or carelessness. You should be aware of the following general school safety rules, and your instructor will point out the specific safety rules that apply to the lab you are working in:

1. Inform your instructor immediately when you become aware of a safety hazard.
2. Always notify your instructor when you are injured in the shop. Have proper first aid applied.
3. Do not underestimate the potential danger of a 120-V circuit.
4. Work on live circuits ONLY when absolutely necessary and while under supervision of the instructor.
5. Stand on dry, nonconductive surfaces when working on live circuits.
6. Never bypass an electrical protective device.

7. Arrange your work so that you never have to reach over a hot soldering iron that is on your bench. The iron should be completely shielded when not in use.
8. Keep your work area clean.
9. Always wear safety glasses or goggles when you are operating any kind of power tool or when soldering.
10. Avoid horseplay and practical jokes.
11. Know where the fire extinguisher is and how to use it.
12. Check all "dead" circuits with a meter before you touch them.
13. Do not intentionally touch a circuit that may shock you.
14. Do not touch two pieces of plugged-in equipment at the same time; an equipment defect could cause a shock.
15. Do not open or close any main switch without permission from the instructor.
16. If ever in doubt, ASK your instructor.
17. Make sure all electrical connections are secure before applying a voltage.

1-4 Occupational Safety

Safety has become an increasingly large factor in the working environment. The electrical industry, in particular, regards safety to be unquestionably the single most important priority due to the hazardous nature of the business. *Obey all accident prevention signs!* (See Figure 1-3.)

Personal Safety Attire   The clothing worn at work is important for personal safety. Appropriate attire should be worn for each particular job site and work activity (Figure 1-4). The following points should be observed:

1. Hard hats, safety shoes, and goggles must be worn in areas where they are specified.
2. Safety earmuffs must be worn in noisy areas.
3. Clothing should fit snugly to avoid the danger of becoming entangled in moving machinery.
4. Remove all metal jewelry when working on energized circuits; gold and silver are excellent conductors of electricity.
5. Confine long hair or keep hair trimmed when working around machinery.

Equipment Safety   An ungrounded power tool can kill you! Always use properly grounded power tools. Use only those power tools with three-pronged plugs or double insulated tools with two-pronged plugs (Figure 1-5). Inspect cords and equipment often to make sure ground pins are in safe condition.

Eye Protection Must be Worn | Head Protection Must be Worn | Hearing Protection Must be Worn | Hand Protection Must be Worn | Breathing Protection Must be Worn | Foot Protection Must be Worn

CAUTION! Slippery Floor | CAUTION! Fork Lift | DANGER! Compressed Gases | No Smoking | DANGER! Flammable | DANGER! Poison

Fire Extinguisher | Eyewash | First Aid | Safety Shower

Figure 1-3   Typical warning and caution signs. (*Courtesy Safety Supply Canada.*)

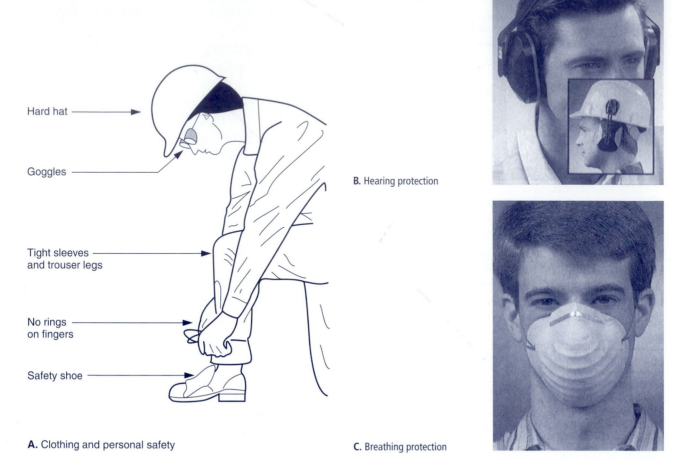

Hard hat

Goggles

Tight sleeves
and trouser legs

No rings
on fingers

Safety shoe

A. Clothing and personal safety

B. Hearing protection

C. Breathing protection

Figure 1-4   Clothing and equipment used for personal safety. (*Courtesy Safety Supply Canada.*)

Electric *lockout* and *tagout* (Figure 1-6) refers to the process of padlocking the power source in the OFF position and indicating, on an appropriate card, the procedure that is taking place. This procedure is necessary so that someone will not inadvertently turn the equipment to the ON position while it is being worked on. Lockout and tagout should be carried out BEFORE any repair is started. When in doubt about this or any other procedure, ask your supervisor! Also, report any unsafe condition, equipment, or work practices as soon as possible.

1-5 Electric Shock

Often we think that serious electric shock can only take place from high-voltage circuits. This is not so! More people are injured or killed by 120-V household voltage every year than in all other electrical-related accidents. If you walked away from your last electric shock, consider yourself lucky. Do not depend on luck. Work safely with electricity and live!

Electric shock occurs when a person's body becomes part of the electric circuit. The three

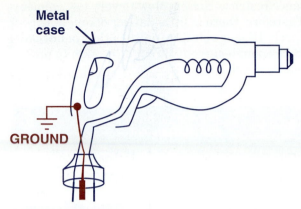

Metal case

GROUND

A. Three-pronged plug type

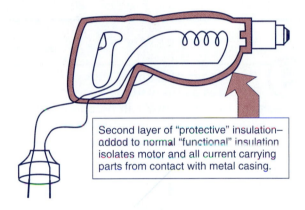

Second layer of "protective" insulation—added to normal "functional" insulation isolates motor and all current carrying parts from contact with metal casing.

B. Double insulated two-pronged plug type

Figure 1-5 Use properly grounded tools.

Figure 1-6 Electrical lockout and tagout. (*Courtesy Safety Supply Canada.*)

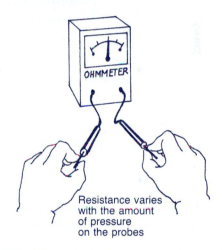

OHMMETER

Resistance varies with the amount of pressure on the probes

Figure 1-7 Measuring body resistance. Increased pressure causes less distance between the outer skin and circulatory system.

electrical factors involved in creating an electric shock are: resistance, voltage, and current.

Resistance　Electrical *resistance* (R) is defined as the opposition to the flow of current in a circuit and is measured in ohms (Ω.) *The lower the body resistance, the greater the potential electric shock hazard.* Body resistance varies with the condition of the skin and the area in contact. Typical body resistance values are listed in Table 1-1. Body resistance can be measured with an instrument called an *ohmmeter* (Figure 1-7).

Example 1-1

A person's finger is placed across the terminals of a 9 volt transistor battery. Assuming a skin resistance of 10,000 Ω the amount of current flow would be:

$$I = \frac{V}{R}$$

$$= \frac{9 \text{ V}}{10,000 \text{ Ω}}$$

$$= 0.0009 \text{ A}$$

$$= 0.9 \text{ mA}$$

This is much less than the 5 mA considered dangerous.

Voltage *Voltage (V)* is defined as the pressure that causes the flow of electric current in a circuit and is measured in units called volts. The amount of voltage that is dangerous to life varies with each individual due to differences in body resistance and heart conditions. **Generally, any voltage above 30 V is considered dangerous.**

Current Electric *current (I)* is defined as the rate of flow of electrons in a circuit and is measured in amperes. The amount of current flowing through a person's body depends on the voltage and resistance. Body current can be calculated using the following formula:

$$\text{Current through body} = \frac{\text{Voltage applied to body}}{\text{Resistance of body}}$$

$$I \text{ (amperes)} = \frac{V \text{ (volts)}}{R \text{ (ohms)}}$$

or

$$I \text{ (milliamperes)} = \frac{V \text{ (volts)}}{R \text{ (kilohms)}}$$

$$\text{where } 1 \text{ ampere} = 1000 \text{ milliamperes}$$
$$1 \text{ kilohm} = 1000 \text{ ohms}$$

It doesn't take much current to cause a painful or even fatal shock. A severe shock can cause the heart and lungs to stop functioning. Also, severe burns may occur where current enters and exits the body. Once current enters the body it follows the circulatory system in preference to the external skin. Figure 1-8 illustrates the relative magnitude and effect of electric current.

> ✋ **Generally, any current flow above 0.005 A (amperes) or 5 mA (milliamperes) is considered dangerous.**

A flashlight cell can deliver more than enough current to kill a human being, yet it is safe to handle. This is because the resistance of human skin is high enough to limit greatly the flow of electric current. In lower voltage circuits, resistance restricts current flow to very low values. Therefore, there is little danger of an electric shock. Higher voltages, on the other hand, can force enough current through the skin to produce a shock.

TABLE **1-1**	Skin Condition or Area and Its Resistance
Skin Condition or Area	**Resistance Value**
Dry skin	100,000 to 600,000 Ω
Wet skin	1,000 Ω
Internal body—hand to foot	400 to 600 Ω
Ear to ear	about 100 Ω

> ✋ **The danger of harmful shock increases as the voltage increases. Voltage as low as 30V can be dangerous.**

1-6 First Aid

First aid is the immediate and temporary care given to the victim of an injury or illness. Its purpose is to preserve life, assist recovery, and prevent aggravation of the condition. A properly stocked first-aid kit should be readily available (Figure 1-9). If someone is hurt, *first send for help immediately*. A few of the basic first-aid procedures are as follows:

BLEEDING To control bleeding, apply direct pressure on the wound using a clean pad or your hand. Raise the arm, leg, or head above heart level.

BURNS For first-degree and minor second-degree burns, immerse the injured area in cold water or apply cold packs to relieve the pain, *do not* break blisters. For second-degree burns with open blisters and all third-degree burns, no water or cold packs should be applied as this increases the likelihood of shock and infection. These serious burns should be treated with thick, clean bandages. No particles of charred clothing should be removed except by skilled medical practitioners. If the victim has suffered facial burns, he or she should be kept propped

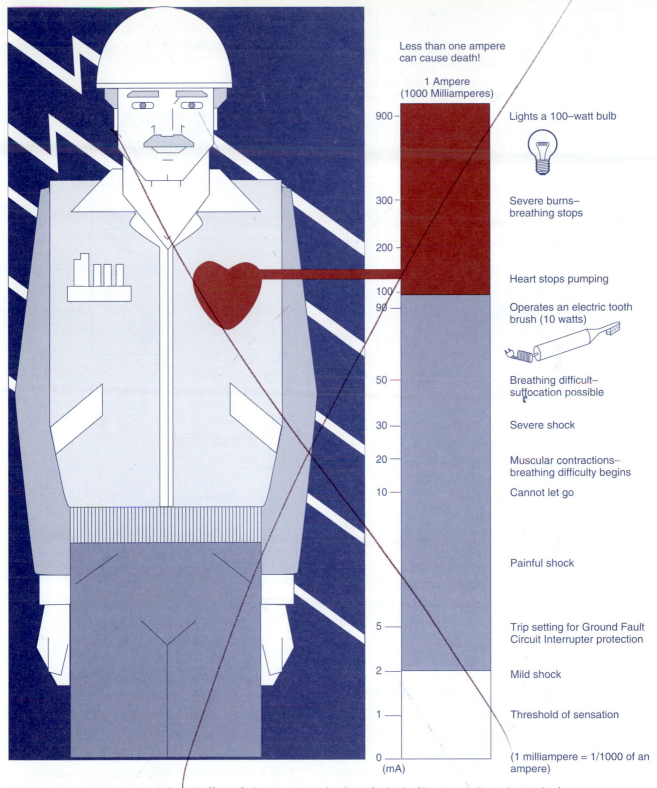

Figure 1-8 Relative magnitude and effect of electric current (mA) on the body. (*Courtesy Ontario Hydro.*)

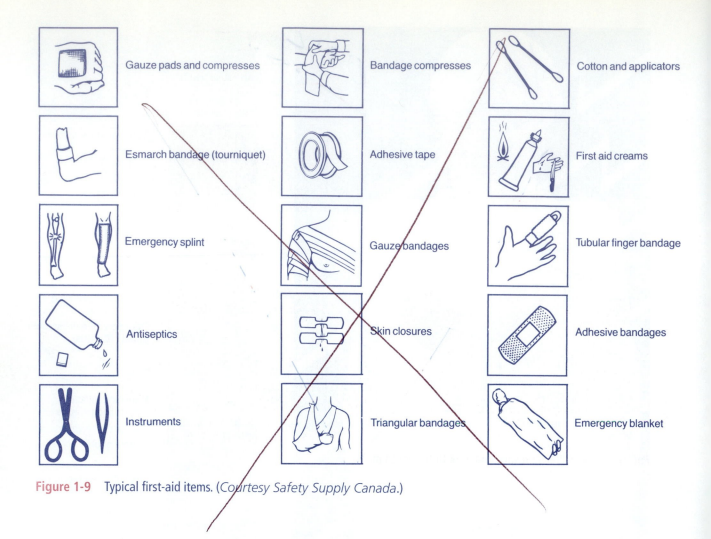

Figure 1-9 Typical first-aid items. (*Courtesy Safety Supply Canada.*)

Items shown: Gauze pads and compresses, Bandage compresses, Cotton and applicators, Esmarch bandage (tourniquet), Adhesive tape, First aid creams, Emergency splint, Gauze bandages, Tubular finger bandage, Antiseptics, Skin closures, Adhesive bandages, Instruments, Triangular bandages, Emergency blanket

up and monitored for breathing difficulty. If only feet, legs, or arms are burned, they should be elevated above the level of the heart. For any serious burn, have medical help on the scene as soon as possible.

ELECTRIC SHOCK   To treat electric shock, turn the power off and remove the electric contact from the victim. *Do not touch the victim until he or she has been separated from the current.* Begin first-aid procedures. Administer artificial respiration if the victim is not breathing. Keep the victim warm; position the victim so that the head is low and turned to one side to encourage flow of blood and to avoid an obstruction to breathing.

1-7 Artificial Respiration

If breathing stops, you can help the victim by knowing how to administer artificial respiration. The basic mouth-to-mouth method of artificial respiration is as follows (Figure 1-10):

1. Place victim on his or her back immediately. *Turn head and clear throat* area of water, mucus, foreign objects, or food.
2. *Tilt victim's head back* to open air passage.
3. *Lift victim's jaw up* to keep the tongue out of the air passage.
4. *Pinch victim's nostrils* closed to prevent air leakage when you blow.

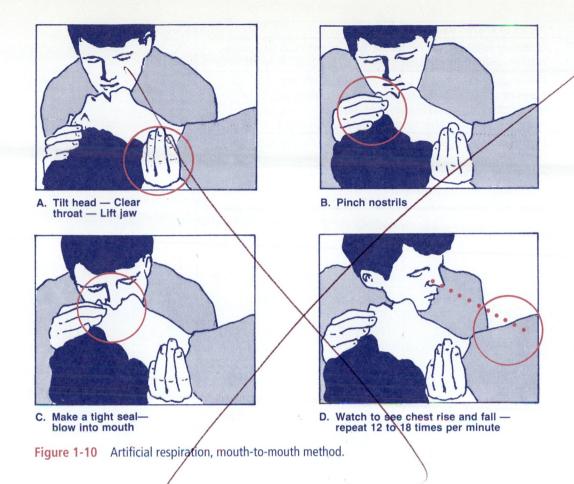

A. Tilt head — Clear throat — Lift jaw

B. Pinch nostrils

C. Make a tight seal— blow into mouth

D. Watch to see chest rise and fall — repeat 12 to 18 times per minute

Figure 1-10   Artificial respiration, mouth-to-mouth method.

5. *Seal your lips* around victim's mouth or use a barrier device.
6. *Blow into the victim's mouth* until you see the chest rise.
7. *Remove your mouth* to allow natural exhalation.
8. *Repeat* 12 to 18 times per minute, watching to see that the chest rises and falls, until natural breathing starts.

1-8 Fire Prevention

Fire prevention is a very important part of any safety program. The chance of a fire occurring can be greatly reduced by good housekeeping. Figure 1-11 illustrates some of the common types of fire extinguishers and their applications. You should know where your fire extinguishers are located and how to use them.

In case of an electrical fire the following procedures should be followed:

1. Trigger the nearest fire alarm to alert all personnel in the workplace as well as the fire department.
2. If possible, disconnect the electric power source.
3. Use a carbon-dioxide or dry-powder fire extinguisher to put out the fire. *Under no circumstances use water,* as the stream of water may conduct electricity through your body and give you a severe shock.
4. Ensure that all persons leave the danger area in an orderly fashion.
5. Do not re-enter the premises unless advised to do so.

Class	Types of Materials Involved
Class A fire	Ordinary combustible materials such as wood, cloth, paper, rubber, and many plastics.
Class B fire	Flammable liquids, gases, and greases. (Only dry-chemical types of extinguishers are effective on pressurized flammable gases and liquids. For deep fat fryers, multi-purpose A/B/C dry chemicals are *not* acceptable.)
Class C fire	Energized electrical equipment. The electrical nonconduc-tivity of the extinguishing media is important.
Class D fire	Combustible metals such as magnesium, titanium, zirco-nium, sodium, and potassium.

A. Classes of fires

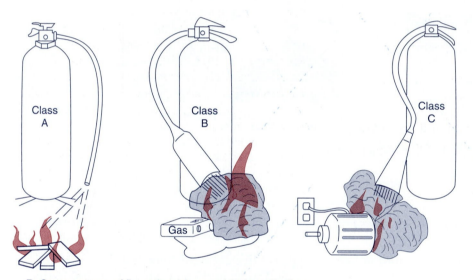

B. Common types of fire extinguishers and their applications

C. Certain multipurpose dry-chemical extinguishers may be used on type A/B/C fires

Figure 1-11   Types of fires and fire extinguishers.

1-9 Safety Color Codes

The Occupational Safety and Health Act (OSHA) lists in detail all the federal legislation covering safety in the workplace. Thanks to OSHA, rarely does anyone today have to work in unsafe conditions. The OSHA has established the following colors to designate certain cautions and dangers:

- *Red* is used to designate:
  Fire protection equipment and apparatus
  Portable containers of flammable liquids
  Emergency stop buttons and switches
- *Yellow* is used to designate:
  Caution and to mark physical hazards
  Waste containers for explosive or combustible materials
  Caution against starting, using, or moving equipment under repair
  Identification of the starting point or power source of machinery
- *Orange* is used to designate:
  Dangerous parts of machines
  Safety starter buttons
  The exposed parts (edges) of pulleys, gears, rollers, cutting devices, and power jaws
- *Purple* is used to designate:
  Radiation hazards
- *Green* is used to designate:
  Safety
  Locations of first-aid equipment (other than fire fighting equipment)

1-10 Hazardous Substances and Waste

Many products contain *hazardous substances,* which if not used and disposed of properly can result in the production of *hazardous waste.* Legally everyone has to properly dispose of hazardous waste. In the United States, the Environmental Protection Agency (EPA) regulates the disposal of hazardous waste.

Recognizing hazardous substances and the type of hazardous waste they produce is the first step in learning how to properly handle and dispose of them. One or more of the following dangerous properties or characteristics identify most hazardous waste: corrosive, ignitable, reactive, or toxic (Figure 1-12).

Corrosives are materials that can attack and destroy human tissue, clothes, and other materials including metals on contact. For example acids found in batteries are corrosive. They can be in the form of gas, liquid, or solid. Most are either acids or bases although some other chemicals are corrosive also.

An *ignitable* material is one that is capable of bursting into flames. For example gasoline, paint, and furniture polish are ignitable substances. Ignitable substances pose a fire hazard; can irritate the skin, eyes, and lungs; and may give off harmful vapors, which can cause explosions.

Toxic materials can poison people and other life. They can cause illness—ranging from severe headaches to cancer—and even death if swallowed or absorbed through the skin. Pesticides, weed killers, and many household cleaners are all examples of toxic materials.

A *reactive* material can explode or create poisonous gas when mixed with another substance or chemical. For example, chlorine bleach and ammonia are reactive. When they come into contact with each other they produce a poisonous gas.

Ideally, hazardous waste is reused or recycled. If this is not possible, hazardous waste

Corrosive   Ignitable

Toxic   Reactive

Figure 1-12 Hazardous properties or characteristics.

must be stored, treated, or disposed of in a way that prevents it from harming people or the environment. Traditional methods include: surface impoundment (storing it in lined ponds), high temperature incineration (controlled burning), landfills (burying it in the ground), and deep well injection (pumping it into underground wells). More acceptable methods focus on minimizing waste, reusing and recycling chemicals, finding less hazardous alternatives, and using more innovative treatments made possible by advances in technology.

Health and Safety Coordinator
Job Description
- Implement company safety/health program.
- Ensure compliance with federal and state laws.
- Train staff to respond to facility emergencies.
- Ensure appropriate corrective action is taken where hazards exist.

Review and Applications

Related Formulas

$$I = \frac{V}{R}$$

$$\text{Amperes} = \frac{\text{Volts}}{\text{Ohms}}$$

$$\text{Milliamperes} = \frac{\text{Volts}}{\text{Kilohms}}$$

1 ampere = 1000 milliamperes   1 kilohm = 1000 ohms

Review Questions

1. Describe three safety rules that deal with the use of extension cords.
2. Describe two safety rules that deal with the replacement of fuses.
3. Since tap water is such a good conductor of electricity, we must observe certain safety rules. Describe two of them.
4. List five electrical safety rules that apply to the outdoors.
5. List five general safety rules that directly apply to your work in the lab.
6. What four special items of safety apparel are a person often required to wear on the job?
7. Why does a double insulated two-pronged plug provide acceptable ground protection for a power tool?
8. Explain what is involved in an electrical lockout and tagout procedure.
9. Does the severity of an electric shock increase or decrease with each of the following changes:
   (a) a decrease in the voltage
   (b) an increase in current
   (c) an increase in body resistance
10. Voltages are generally considered to be dangerous when they are above what value?
11. (a) In most cases, what is the maximum safe current value?
    (b) What is the formula for finding the amount of current flow through a body?
12. Define the term first aid.
13. Outline the basic first-aid procedure for:
    (a) bleeding
    (b) burns
14. What important rescue procedure should be followed in case of an electrical accident involving a live electric circuit?
15. List the important steps to be followed when administering mouth-to-mouth artificial respiration.
16. Why should water not be used to put out an electrical fire?
17. According to OSHA's safety color code, what does the color green designate?

Problem

1. Calculate the body current flow (in amperes and milliamperes) of an electric shock victim who comes in direct contact with a 120-V energy source. Assume a body contact resistance of 1000 ohms.

Critical Thinking

1. You have been asked to dispose of the following components: fluorescent tube, fuse, carbon resistor, and a switch. Which component would be considered hazardous? What precautions should be taken when this item is disposed of?

Portfolio Project

As a student studying electricity and electronics you will be required to experiment with live circuits. In some experiments you may have to use the full 120 V of the power line. Outline what special precautions you should follow to ensure your own safety and that of your fellow students.

Write a short essay on the purposes of legislation concerning safety in the workplace.

A manufacturing company uses battery powered vehicles for transporting components. As the Health and Safety Coordinator, you are asked to write a set of regulations that apply to the safe storage, handling, and disposal of these batteries.

Circuit Challenge

Using a Simulator

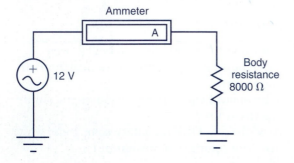

Procedure

(a) Construct the simulated circuit shown, using whatever simulation software package is available to you.

(b) Record the body current flow with the body connected to the 12 V source.

(c) Repeat with the value of the source voltage changed to 120 V.

(d) Repeat with the value of the source voltage changed to 1200 V.

(e) With reference to Figure 1-8, make note of what effect the different current levels would have on a person.

Instruments, Tools, and Fasteners

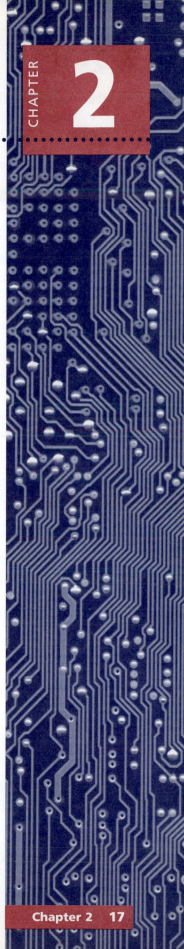

Before you can build, design, or service equipment, you will need to know how to use the tools of the trade. Over the years, a wide variety of instruments as well as hand and power tools, have been developed to help skilled professionals perform their tasks effectively. This chapter will help you select, operate, and maintain basic tools and measuring devices. Typical fastening devices used to secure electrical/electronic devices are also discussed.

Objectives

After studying this chapter, you should be able to:

1. Identify and state the use for common electrical/electronic measuring devices.
2. Identify and state the use for common tools used in the electrical/electronic industry.
3. Outline the procedure to be followed for the proper care and use of tools.
4. Identify and state the feature of common fastening devices.

Key Terms

- **voltmeter**
- **ammeter**
- **ohmmeter**
- **multimeter**
- **logic probe**
- **oscilloscope**
- **hand tool**
- **power tool**
- **machine screw**
- **wood screw**
- **self-tapping screw**

RESEARCH INTERNET SITES

- **Hand Tools Institute (HTI) Organization**
- **BK Precision Instruments**
- **Star Fasteners**

2-1 Measuring Devices

There are many kinds of instruments used for the measurement of electrical/electronic quantities. Some of the more important ones are discussed in more detail in other chapters of this text. Most electrical measuring devices are fairly expensive and delicate. When using any electrical measuring device you should:

- Inspect the device to be sure there are no obvious safety hazards.
- Handle it with care.
- Be sure it is properly connected to the circuit.
- Be sure not to exceed the voltage or current rating of the device.

For experimental work in the lab, individual *voltmeters* to measure voltage, *ammeters* to measure current, and *ohmmeters* to measure resistance are sometimes used. More often, a single *multimeter* is usually used to accurately measure voltage, current, or resistance (Figure 2-1).

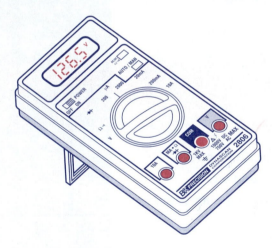

Figure 2-1 Digital multimeter. (*Courtesy B+K Precision.*)

The *voltage tester* (Figure 2-2) is often used by the electrician to measure approximate circuit-operating voltages. Its rugged construction makes it ideally suited for rough on-the-job handling.

The *clip-on ammeter* (sometimes called clamp-on) is used to measure current flow. This device is able to measure current without any direct electrical contact with the circuit. (Figure 2-3).

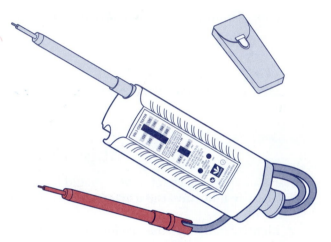

Figure 2-2 Voltage tester. (*Courtesy of Ideal Industries, Inc., Sycamore Illinois, USA; Ajax, Ontario, Canada.*)

Figure 2-3 Clip-on ammeter. (*Courtesy Mercer Electronics.*)

The *neon test light* (Figure 2-4) is an inexpensive device that can be used by the homeowner to indicate the presence of a voltage.

The *logic probe* (Figure 2-5) is designed for quick checking and servicing of digital circuits. It visually displays the presence of correct logic levels by illumination of colored readouts.

The *continuity tester* (Figure 2-6) is used for checking the continuity of dead circuits. This tester is powered by batteries and used to check

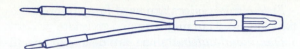

Figure 2-4 Neon test light.

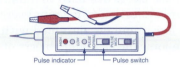

Pulse indicator — — Pulse switch

Figure 2-5 Digital circuit probe. (*Courtesy American Reliance Inc.*)

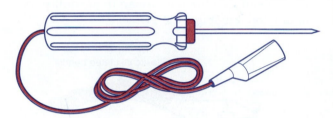

Figure 2-6 Continuity tester. (*Courtesy of Klein Tools, Inc.*)

for defective switches, broken leads, or to identify wires in multiwire cables. A light signal indicates a completed circuit.

The *oscilloscope* (Figure 2-7) is used to troubleshoot and verify inputs and outputs of electronic devices. It can be used to measure voltage like a voltmeter. In addition, it can provide information such as the shape, time period, and frequency of voltage waveforms.

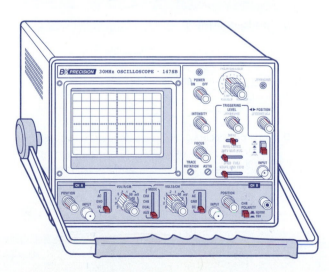

Figure 2-7 Oscilloscope. (*Courtesy B+K Precision.*)

2-2 Common Tools

The skilled technician must be familiar with the proper use of the tools-of-the-trade. As a general rule, higher quality tools tend to be in the higher price ranges but are safer to work with. Cheaper, low-quality tool material and poor design features often put great stress on the tool and the operator.

> Remember, always use the right tool for the job.

Screwdrivers   The screwdriver is a tool designed to loosen or tighten screws. Screwdrivers are identified by the shape of their head (Figure 2-8).

A *slot-head* or *standard* screwdriver is designed for use on screws with slotted heads. This type of screw is often used on the terminals of switches, receptacles, and lampholders.

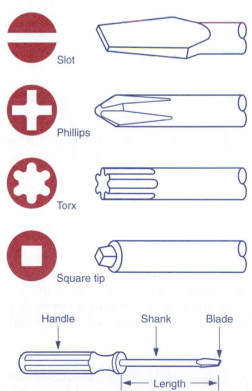

Slot

Phillips

Torx

Square tip

Handle   Shank   Blade

|← Length →|

Figure 2-8   Common screwdriver tip configurations.

The screwdriver's blade should fit the slot of the fastener (Figure 2-9). This prevents damage to the screwdriver's blade and fastener's slot, as well as possible injury to the user's hand or surrounding equipment should the tip slip out of the slot.

The *Phillips* screwdriver is designed for use on screws with an X-shaped insert in their heads. This type of screw is often used on the outside of electrical appliances because there is less likelihood of the screwdriver head slipping out of the slot and damaging the metal finish of the appliance.

The *Torx* screwdriver is designed for use on screws with Torx heads. In recent years, the Torx screwdriver has become particularly popular in the automotive industry for product assembly.

The *square-tip* (also known as the *Robertson* or *Scrulox*) screwdriver is designed for use on screws with square-shaped inserts in their heads. This type of screw creates a snug fit with the screwdriver head, allowing the screw to be easily driven into wooden material. Such screws are sometimes used to secure outlet boxes to joists.

Pliers   The need to cut and shape electric conductors and to grip a variety of objects has caused many types of pliers to be developed (Figure 2-10).

1. This tip is too narrow for the screw slot; it will bend or break under pressure.
2. A rounded or worn tip. Such a tip will ride out of the slot as pressure is applied.
3. This tip is too thick. It will only serve to chew up the slot of the screw.
4. A chisel ground tip will also ride out of the screw slot. It is best to discard it.
5. This tip fits, but it is too wide and will tear the wood as the screw is driven home.
6. The right tip. This tip is a snug fit in the slot and does not project beyond the screw head.

Figure 2-9   Proper and improper use of slot-head (standard) screwdrivers. (*Courtesy Hand Tools Institute,* © *1989.*)

Side-cutting pliers are used for gripping, twisting, and cutting wires.

Diagonal-cutting pliers are designed specifically for cutting wire. They are used for close cutting jobs such as trimming the ends of wire on terminal board connections.

Needle-nose pliers are used to make loop ends on wire for connection to terminal screws.

Curved-jaw pliers are designed with an adjustable joint for gripping objects of various sizes.

Vise Grip® pliers are designed with jaws that can be locked onto the object.

Hammers   Hammers are produced in a variety of head weights and are an important part of any tool kit (Figure 2-11).

A Side-cutting pliers

B. Diagonal-cutting pliers

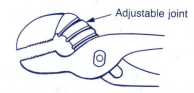

C. Needle-nose pliers

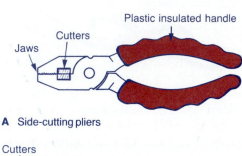

D. Curved-jaw pliers

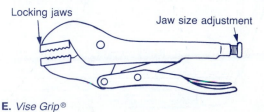

E. *Vise Grip®*

Figure 2-10   Common types of pliers.

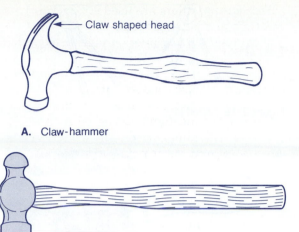

Claw shaped head

A. Claw-hammer

B. Ball peen hammer

Figure 2-11 Common types of hammers.

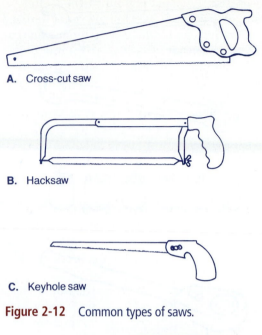

A. Cross-cut saw

B. Hacksaw

C. Keyhole saw

Figure 2-12 Common types of saws.

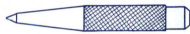

Figure 2-13 Center punch.

Claw hammers are most useful for work on wooden frame structures. The face of the hammer is used to drive nails and staples. The claw is used for removing nails.

Ball peen hammers (also known as machinist's hammers) are ideal for heavy-duty striking operations. This includes such operations as cutting with a cold chisel, producing holes in concrete surfaces, or driving fasteners into place with heavy blows.

Saws A *cross-cut saw* (Figure 2-12A) is usually selected for sawing wood.

The standard *hacksaw* (Figure 2-12B) is required for any metal cutting work. The necessary number of teeth per inch of blade is determined by the thickness and type of metal being cut.

A *keyhole* or *compass saw* (Figure 2-12C) is a narrow saw that is used for cutting holes in finished surfaces or wallboard in order to mount outlet boxes.

Punches A *center punch* (Figure 2-13) is used to mark the location of a hole that is to be drilled and is used most often by electricians. The center punch assists a drill entering exactly at the proper point.

Wrenches Commonly used wrenches include open-end wrenches, box-end wrenches, socket wrenches, and adjustable wrenches (Figure 2-14). The wrench used must correctly fit the nut under pressure, otherwise the nut and the wrench will be damaged.

The *open-end wrench* is designed for use in close quarters. After each short stroke, the wrench can be turned over to fit other flats of the nut.

Box-end wrenches completely surround or "box in" the nut or bolt head when they are being used.

Socket wrenches can be positioned on a nut more quickly. These wrenches use an assortment of handles (e.g., ratchet), which make the technician's work faster and easier.

Adjustable wrenches are especially convenient at times when an odd size of nut is encountered. *When using an adjustable wrench, the pulling force should always be applied to the stationary-jaw side of the handle.*

Nut Drivers *Nut drivers* work like socket wrenches except that they use a straight handle similar to that of a screwdriver (Figure 2-15). The sockets are used to tighten or loosen machine nuts on electrical and electronic equipment. Most nut driver shafts are hollow. This allows them to be used with nuts that are threaded on long bolts.

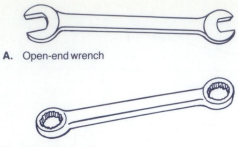

A. Open-end wrench

B. Box-end wrench

Sockets

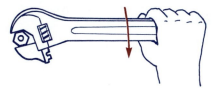

Ratchet handle

C. Ratchet handle

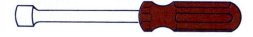

D. Adjustable wrench

Figure 2-14   Common wrenches.

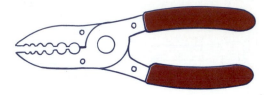

Figure 2-15   Nut driver.

Allen Keys
Some electronic-control knobs are fastened in place by means of a small set screw with a hexagonal socket in the top of it called an Allen head.

Allen keys (sometimes called Allen wrenches) are used for loosening and tightening this type of set screw (Figure 2-16).

Insulation-Removing Devices
Wire and cable preparation requires the removal of a certain amount of insulation (Figure 2-17).

A *wire stripper* is used to remove insulation from small-diameter wire.

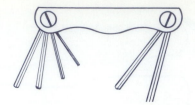

Figure 2-16   Allen keys.

A.   Wire stripper

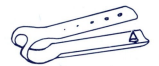

B.   Knife

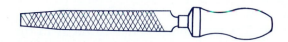

C.   Sheathed cable insulation stripper

Figure 2-17   Insulation-removing devices.

A *knife* is used to remove insulation from cables and larger-diameter wires.

A *cable insulation stripper* is used to remove the insulating sheath from nonmetallic sheath cable.

Files
Both metal and wood files are commonly used.

Metal files are used to remove sharp metal burrs produced as a result of cutting or drilling.

Wood files (Figure 2-18) are used for fitting electrical outlet boxes into finished walls.

Figure 2-18   Wood file.

Chisels

Chisels Two types of chisels are available. They are the *cold chisel* (Figure 2-19A) used for metal work and the *wood chisel* (Figure 2-19B) used for working with wood.

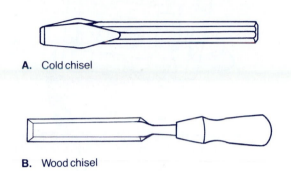

A. Cold chisel

B. Wood chisel

Figure 2-19 Chisels.

Fish Tape A *fish tape and reel* (Figure 2-20) is a tool designed to fish for and pull wire through wall partitions and electrical conduit.

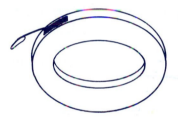

Figure 2-20 Fish tape and reel.

Measuring Tools Different types of measuring tapes and rulers are available (Figure 2-21).

A *steel tape* is used for rapid layout in measurements. *Care must be taken when working near live equipment with a steel tape measure.*

A nonconducting, wooden *folding rule* has sections that pivot to allow it to be opened to whatever length is required.

Electric Drills *Electric drills* (Figure 2-22) are used for drilling holes in wood, metal, and concrete. The size of a drill is determined by the chuck size and the power of the motor. The *chuck* is the part of a drill that holds the twist drill bit. A 3/8-inch drill will hold a bit of any size up to 3/8-inch in diameter. Reversible and battery-powered electric drills are also available.

Auger bits are used in electric drills for drilling holes in wood.

Twist drill bits are used for drilling holes in wood and metal. Twist drill bits are available in carbon-tool steel and in high-speed steel. The more expensive high-speed drill bits are used for the drilling of hard materials because they can withstand a greater heat.

Carbide-tipped masonry drill bits are used for drilling in concrete and masonry material.

A power screwdriver uses a special *screwdriver bit* to install and remove screws.

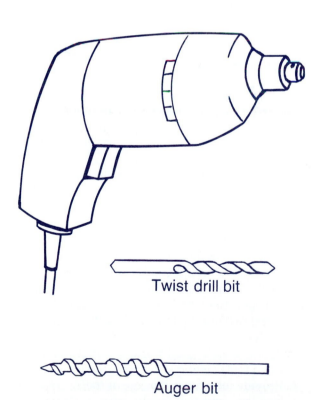

Twist drill bit

Auger bit

Figure 2-22 Electric drill.

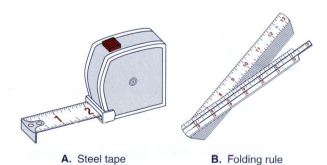

A. Steel tape **B.** Folding rule

Figure 2-21 Measuring tools.

Soldering Equipment The *soldering gun* (Figure 2-23A) is a common soldering tool for general hand-wired circuits.

The *soldering pencil* (Figure 2-23B) is a common soldering tool for printed circuit-board soldering.

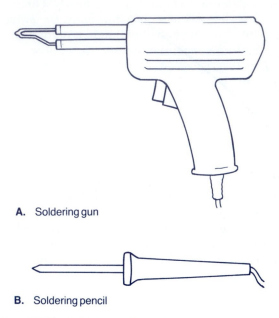

A. Soldering gun

B. Soldering pencil

Figure 2-23 Soldering equipment.

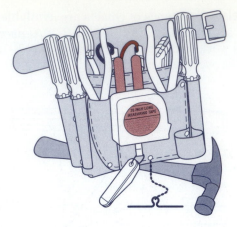

Figure 2-24 Tool pouch with tools. (*Courtesy of Ideal Industries Inc., Sycamore Illinois, USA; Ajax, Ontario, Canada.*)

2-3 Organization and Use of Tools

To be effective, tools must be available when they are needed. Tools can be organized in several ways, depending upon where and how frequently they are used. A leather tool-carry pouch (Figure 2-24) keeps the tools close at hand for the installation and maintenance of equipment. If tools are used at a repair bench, a pegboard arrangement may be appropriate. When the tools will be used both at a bench and on the job site, a portable tool box or case (Figure 2-25) is usually best.

A skilled worker is often judged by the quality and condition of his or her tools. Quality tools that are handled properly will last indefinitely. Some of the steps you can take to help keep your tools in good working condition are:

1. Keep tools cleaned and well oiled.
2. Provide for proper storage of tools.
3. Use the right tool for the job.
4. Use the right size of tool for the job.

5. Keep drills, auger bits, and saw blades sharp.
6. Replace dull hacksaw blades.
7. Never use a file without a firm-fitting handle.
8. Hammers with loose heads should be replaced.
9. Needle-nose pliers must be used on light wires only. The tips will break or bend if abused.
10. Pliers should not be used on nuts, since this will damage both the pliers and the nuts.
11. Never expose pliers to excessive heat. This may draw the temper and ruin the tool.
12. Never use pliers as a hammer nor hammer on the handles.
13. Never use a screwdriver that has a tip that is too large or too small for the screw.
14. Never use a screwdriver as a pry bar or cold chisel.
15. Keep soldering gun and iron tips clean.
16. Whenever possible, pull rather than push on a wrench.
17. Do not use the hammer handle as a driver.
18. Always use an adjustable wrench that is large enough to handle the job. Using a wrench that is too small can cause the movable jaw to break.
19. When replacing a hacksaw blade, be sure to mount the blade with its teeth slanting away from the handle.
20. *Ordinary plastic-dipped handles are designed for comfort—not electrical insulation.* Tools having high dielectric insulation are available and are so identified. Do not confuse the two types.

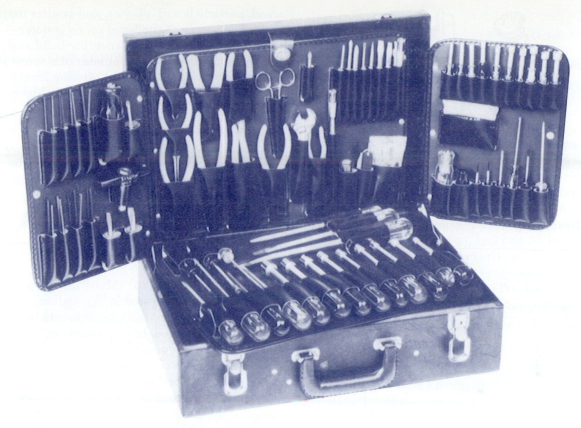

Figure 2-25 Portable tool case with tools. (*Courtesy of The Cooper Tool Group.*)

2-4 Fastening Devices

Fastening devices are available in many forms for the support of electrical/electronic components and equipment. *Permanent fasteners* are used when the parts are not meant to be disassembled and include welding, nailing, gluing, and riveting. *Temporary fasteners* are used when the parts may be disassembled at some future time and include screws, bolts, keys, and pins.

Screw Fasteners   *Machine screw* bolt and nut units (Figure 2-26) are used primarily to join metal to a variety of other materials. They are produced in many thicknesses and thread pitches, depending on the amount of support strength and compression required. The coarse-thread bolt installs faster, since the nut

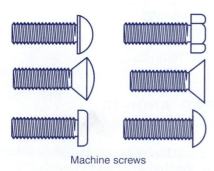

Machine screws

Wing   Hex   Square

Machine screw nuts

Figure 2-26 Machine screw bolt and nut units.

Figure 2-27 Self-tapping screws.

advances along the bolt a greater distance for each complete turn. Fine-threaded units require more turns of the nut to tighten them, but excellent compression is obtained between the surfaces joined.

Self-tapping screws (Figure 2-27) provide an excellent fit and fast assembly when joining metal to metal. They reshape the material in the pilot hole.

Wood screws (Figure 2-28) are produced in various lengths and diameters for fastening to

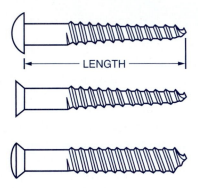

Figure 2-28 Wood screws.

PLASTIC ANCHOR

Figure 2-29 Masonry fasteners.

wood, particle board, plaster, and similar materials. The length of the wood screw is determined by the distance between the head and the tip of the screw. The diameter of a screw is indicated by a gauge number ranging from 0 to 24. The larger the gauge number, the thicker the screw.

Masonry Fasteners   Due to the extensive use of masonry material (concrete and brick), much of the electrical equipment installed must be fastened to masonry surfaces. For this application *screw anchors* available in plastic (Figure 2-29), jute fiber, and lead can be used. The screw anchor is inserted into a predrilled hole in the masonry surface and a screw is then used to complete the fastening system.

Hollow-Wall Fasteners   Many fastening operations must be performed in dry-wall material whose surface is generally too thin and low in density to accommodate any anchor other than the small screw-type anchors. The *spring-wing toggle bolt* fastener (Figure 2-30), is designed to make use of the hollow space behind the dry-wall or similar surface. The steel wings are installed on the machine screw after it has passed through the device being mounted. Then the wings are inserted into a predrilled hole in the mounting surface. Once clear of the back of the hole, the wings spring to an open position. Tightening of the machine screw draws the wings up against the inside surface, thus secur-

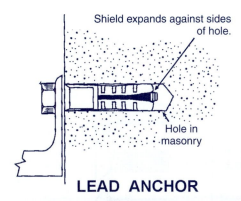

Shield expands against sides of hole.

Hole in masonry

LEAD ANCHOR

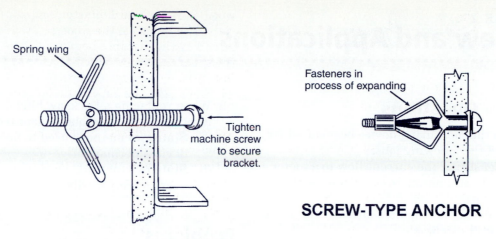

SPRING-WING TOGGLE BOLT

Spring wing

Tighten machine screw to secure bracket.

Fasteners in process of expanding

SCREW-TYPE ANCHOR

Figure 2-30 Hollow-wall fasteners.

ing the mounted device. This type of fastener cannot be reused once installed as it is virtually impossible to remove the open, spring-wing portion from the space behind the wall. Where it is necessary to remove and replace equipment a *hollow-wall screw type anchor,* such as that shown in Figure 2-30, can be used. This type of fastener unit fits into a predrilled hole in the surface, expands, and grips the back of the mounting surface when tightened.

HELP WANTED

Sales Associate Electric Supplies
Job Description
• Extensive training will be provided.
• Two years technical trade school or equivalent required.
• Applicant with an electrical background preferred.
• Knowledge of various hardware components used in the electrical field is a plus.
• Must have good written and verbal communications skills.

Review and Applications

Review Questions

1. Name the measuring device best suited for each of the following jobs:
   (a) measuring the current flow to a motor without making direct contact with the circuit.
   (b) checking for the presence of a voltage at a wall receptacle.
   (c) measuring the voltage, current, or resistance of an electronic circuit.
   (d) quickly checking for the presence of correct digital logic levels.
   (e) checking for broken leads in a cable.
   (f) checking the shape of a voltage waveform.
2. Name the type of screwdriver used to tighten or loosen a screw with:
   (a) a square-shaped insert in its head.
   (b) an X-shaped insert in its head.
   (c) a slot-head.
3. Name the type of plier best suited for each of the following jobs:
   (a) making loop ends on wire for connection to a terminal screw.
   (b) twisting wires together.
   (c) trimming the ends of wire on a terminal board connection.
   (d) locking on to an object.
   (e) gripping a piece of electrical conduit.
4. List three methods by which tools are commonly organized.
5. What two general rules apply to the selection of a tool for a particular job?
6. What two rules apply to the storage of tools?
7. List four rules that pertain to the proper care and use of pliers.
8. State a possible negative effect that can occur as a result of each of the following improper uses of a flathead screwdriver:
   (a) using a tip that is too thick.
   (b) using a tip that is too wide.
   (c) using a tip that is too narrow for the screw slot.
   (d) using a chisel-ground tip.
   (e) exposing the tip to excessive heat.
   (f) using the screwdriver as a pry bar.
9. When replacing a hacksaw blade, in what direction is the new blade mounted?
10. (a) State two methods used to fasten devices permanently.
    (b) State two methods used to fasten devices temporarily.

Problems

1. Name the tool best suited for each of the following jobs:
   (a) cutting a piece of metal or electrical conduit.
   (b) cutting a hole in a finished plastered wall in order to mount an outlet box.
   (c) removing sheathed cable insulation.
   (d) removing metal burrs from a drilled hole.
   (e) soldering a resistor on a printed circuit board.
   (f) marking the location of a hole to be drilled in a metal chassis.
   (g) tightening or loosening a small machine nut that is threaded on a long bolt.
   (h) tightening or loosening a square-head machine nut that is of an odd size.
   (i) removing a large number of standard-size machine nuts with speed and ease.
   (j) notching a wooden wall partition to fit an electrical outlet box.
   (k) removing insulation from small-diameter wire.
   (l) removing nails.
   (m) removing a volume-control knob that is fastened by a set screw with a hexagonal socket.
   (n) pulling wires through electrical conduit.
   (o) drilling a hole in a hard steel plate.
   (p) drilling a hole in a concrete block.
2. State the type of screw fastener most likely used for each of the following applications:
   (a) fastening a metal cover to the metal frame of an appliance.
   (b) fastening an electrical outlet box to particle board.
   (c) fastening the metal frame of a pump motor to a metal base.

(d) mounting a permanent control panel directly to a hollow-wall dry-wall structure.

Critical Thinking

1. As a computer repair technician, you are required to carry a number of hand tools to the job site. Compile a list of the essential hand tools you would require for this task.

Portfolio Project

Using the Internet or any catalogs that are available to you, report on three different makes or models of handheld digital multimeters. Your report should compare the meters:
• Current measurement range.
• Voltage measurement range.
• Resistance measurement range.
• Other features.
Construct a display of the various types of fasteners discussed in this chapter.

Circuit Challenge

Using a Simulator

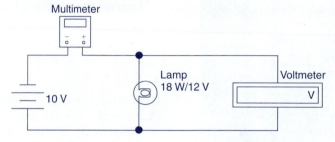

Procedure

(a) Construct the simulated circuit shown, using whatever simulation software package that is available to you. Set the multimeter front panel controls to the settings shown.

(b) Record the current value as indicated by the multimeter.

(c) Record the voltage value as indicated by the voltmeter.

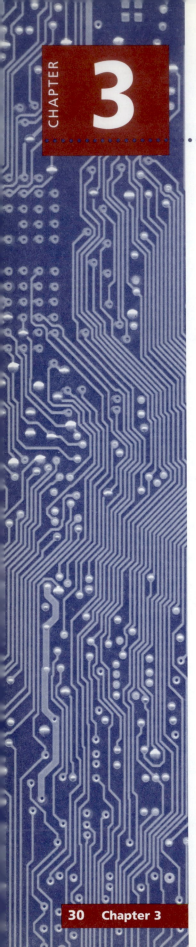

Conductors, Semiconductors, and Insulators

Most materials belong to one of three broad groups: conductors, insulators, and semiconductors. The borderlines are sometimes blurred, though, and many substances cannot easily be put into one group or the other. In this chapter you will learn about the atomic structure of each group.

Objectives

After studying this chapter, you should be able to:
1. Describe the structure of an atom.
2. Recognize, at the atomic level, the characteristics of conductors, semiconductors, and insulators.
3. Explain the ionization process.
4. Recognize common conductor, semiconductor, and insulating materials.
5. Define electricity.

Key Terms

- electron theory of matter
- molecule
- atom
- electron
- proton
- neutron
- nucleus

- element
- atomic number
- atomic mass
- shell
- energy level
- ion
- ionization

- bound electron
- valence electron
- free electron
- electricity
- conductor
- insulator
- semiconductor

RESEARCH **INTERNET** SITES

- Harris Semiconductor
- Dallas Semiconductor
- Fairchild Semiconductor

3-1 Electron Theory of Matter

The *electron theory of matter* helps to explain how electricity works. All matter—solid, liquid, or gas—is made up of tiny particles that are called *molecules*. Molecules are so small that they are invisible to the unaided eye. In fact, millions of molecules can be found on the head of a pin. A molecule is defined as the smallest particle of matter which can exist by itself and still retain all the properties of the original substance.

Molecules are made up of even smaller particles, each of which is called an *atom*. Atoms, in turn, can be broken down into even smaller particles. These smaller subatomic particles are known as *electrons*, *protons*, and *neutrons* (Figure 3-1). The features that make one atom different from another also determine the atom's electrical properties.

There are 92 naturally occurring atoms called *elements*. They are placed in a periodic table in sequence by their atomic number and atomic weight. There are some 14 synthetic elements which do not occur in nature. These two groups comprise the 115 elements known to date. Elements cannot be changed by chemical means, but can be combined to make different types of compounds.

3-2 Bohr Model of Atomic Structure

The model of the atom as proposed by the physicist Niels Bohr gives a concept of its structure, which is helpful in understanding the fundamentals of electricity. According to Bohr, the atom is similar to a miniature solar system. As with the sun in the solar system, the nucleus is located in the center of the atom. Tiny particles called *electrons* rotate in orbit around the nucleus, just as the planets rotate around the sun (Figure 3-2). The electrons are prevented from being pulled into the nucleus by the force of their momentum. They are prevented from flying off into space by an attraction between the electron and the nucleus. This attraction is due to the electric charge on the electron and the nucleus. The electron has a negative (−) charge, while the nucleus has a positive (+) charge. These unlike charges attract each other.

Most of the mass of an atom is found in its nucleus. The particles that can be found in the nucleus are called protons and neutrons. A proton has a positive (+) electrical charge that is exactly equal in strength to that of the negative (−) charge of an electron. The proton is much heavier than the electron. The mass or weight of the neutron is about the same as that of the proton, but it has no electrical charge—hence its name, neutron. Neutrons, as far as is known, do not enter into ordinary electrical

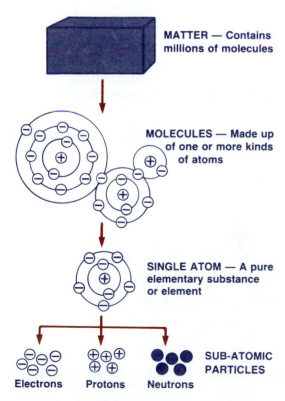

MATTER — Contains millions of molecules

MOLECULES — Made up of one or more kinds of atoms

SINGLE ATOM — A pure elementary substance or element

SUB-ATOMIC PARTICLES

Electrons   Protons   Neutrons

Figure 3-1   A model of the structure of matter.

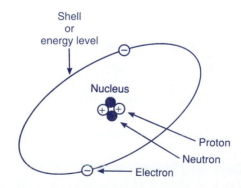

Shell or energy level

Nucleus

Proton

Neutron

Electron

Figure 3-2   Structure of an atom.

activity. Normally, every atom contains an equal number of electrons and protons, making its combined electrical charge *zero* or *neutral*. The total number of protons in the nucleus of an atom is called the *atomic number* of the atom. The total number of both protons and neutrons, is known as the *atomic mass* of the atom. Figure 3-3 shows an aluminum atom according to the Bohr model.

Energy Levels and Shells

According to the Bohr model of the atom, electrons are arranged in *shells* around the nucleus. A shell is an orbiting layer or energy level of one or more electrons. The major shell layers are identified by numbers or by letters, starting with K nearest the nucleus and continuing alphabetically outward. There is a maximum number of electrons that can be contained in each shell. Figure 3-4 illustrates the relationship between the energy shell level and the maximum number of electrons it can contain.

If the total number of electrons for a given atom is known, the placement of electrons in each shell can be determined easily. Each shell layer, beginning with the first and proceeding in sequence, is filled with the maximum number of electrons. For example, a normal copper

Number designation

Letter designation

SHELL LETTER	MAX. ELECTRONS IT CAN HOLD
K	2
L	8
M	18
N	32

Figure 3-4   Electron shells.

atom that has 29 electrons would have the following arrangement of electrons (Figure 3-5):

Shell K (or #1) = 2 (full)
Shell L (or #2) = 8 (full)
Shell M (or #3) = 18 (full)
Shell N (or #4) = 1 (incomplete)

The electrons in any shell of an atom are said to be located at certain *energy levels*. The

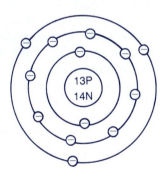

NOTES:
1. No. Protons = No. Electrons
   13 = 13
2. Net electric charge is neutral or zero.
3. Atomic Number = No. Protons
   = 13
4. Atomic Mass = No. Protons + No. Neutrons
   = 13 + 14
   = 27

Figure 3-3   The Bohr model of the aluminum atom.

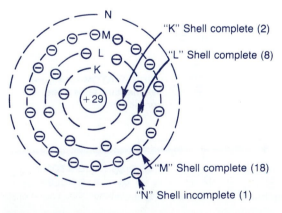

"K" Shell complete (2)
"L" Shell complete (8)
"M" Shell complete (18)
"N" Shell incomplete (1)

Figure 3-5   Placement of electrons in a copper atom.

farther away from the nucleus, the higher the energy level of the electrons. When outside energy such as heat, light, or electricity is applied to certain materials, the electrons within the atoms of these materials gain energy. This may cause the electrons to move to a higher energy level.

3-3 Ions

It is possible, through the action of some outside force, for an atom to lose or acquire electrons (Figure 3-6). The charged atom that results is called an *ion*. A *negatively charged ion* is an atom that has acquired electrons. It has more electrons than protons, and therefore it is negatively (−) charged. A *positively charged ion* is an atom that has lost electrons. It has fewer electrons than protons and therefore is posi-

tively (+) charged. The process by which atoms either gain or lose electrons is called *ionization*.

3-4 Electricity Defined

Bound electrons are electrons in the inner shells of the atom that are bound to that atom because of the strong attraction of the oppositely charged nucleus. The outermost shell of the atom is called the *valence shell*, and its electrons are called *valence electrons*. Because of their greater distance from the nucleus and because of the partial blocking of the electric field by bound electrons in the inner shells, the attracting force exerted on the valence electrons is less. Therefore, valence electrons can be set free easily. Whenever a valence electron is removed from its orbit, it becomes known as a *free electron*. *Electricity* is commonly defined as the flow of these free electrons through a conductor (Figure 3-7).

3-5 Electrical Conductors, Insulators, and Semiconductors

Conductors   Electrons can flow in all matter. However, this flow is much easier through some materials than others. A good *conductor* is a material through which electrons can easily flow with little energy applied. They offer little resistance to current flow. Metals such as silver, copper, gold, aluminum, and iron are considered to be good conductors since they have many free electrons. Copper is the most common metal used as a conductor of electricity because of its relatively low cost and good conducting ability.

The electrical conducting properties of various materials are determined by the number of

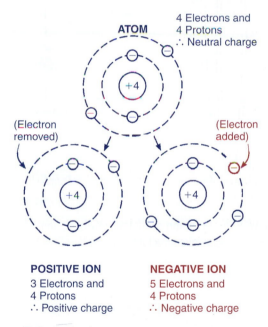

Figure 3-6   Ionization process.

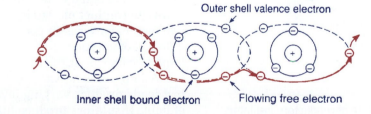

Figure 3-7   Electricity—the flow of free electrons.

electrons in the outer shell of their atoms (Figure 3-8). The outer valence shell never contains more than eight electrons. Generally, a conductor has an incomplete valence shell of one, two, or three electrons. The electrons are held loosely, there is room for more, and a low voltage will cause a flow of free electrons.

Insulators An *insulator* is a material that has few, if any, free electrons and resists the flow of electrons. Generally, insulators have full valence shells of five to eight electrons. The electrons are held tightly, the shell is fairly full, and a very high voltage is needed to cause any electron flow. Some common insulators are air, glass, rubber, plastic, paper, and porcelain. Insulators are used in electric circuits to keep electrons flowing along the intended path of the circuit.

No material has been found to be a perfect insulator. Every material can be forced to permit a small flow of electrons from atom to atom if enough energy in the form of voltage is applied. Whenever a material that is classified as an insulator is forced to pass an electric current, the insulator is said to have been broken down or ruptured.

Semiconductors A *semiconductor* is a material that has some of the characteristics of both a conductor and an insulator. Semiconductors have valence shells containing four electrons. A pure semiconductor may act either as a conductor or an insulator depending upon the temperature at which it is operated. Operated at low temperatures, it is a fairly good insulator. Operated at high temperatures, it is a fairly good conductor. Common examples of pure semiconductor materials are silicon and germanium. Specially treated semiconductors are used to produce modern electronic components such as diodes, transistors, and integrated-circuit chips. These semiconductors are the electronic brains of high-tech machines, driving everything from pocket calculators to powerful computers (Figure 3-9).

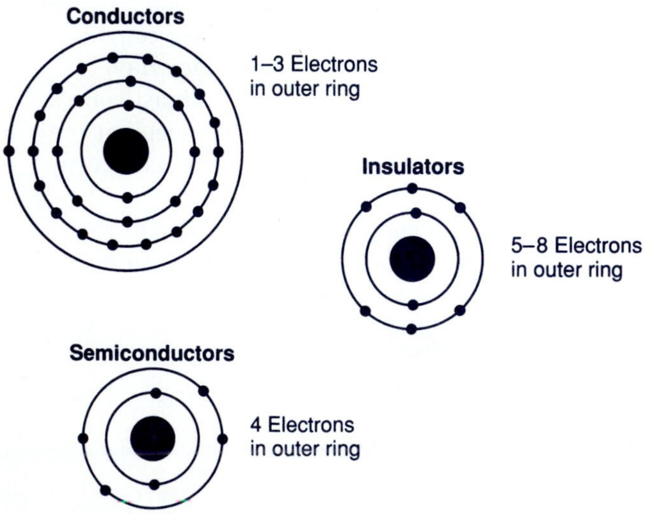

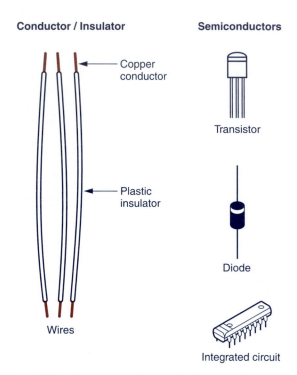

Figure 3-9   Common conductor, insulator, and semiconductors.

3-6 The Continuity Tester

A simple circuit made up of a lamp, battery, and test leads that are connected in a series circuit can be constructed to test materials for their ability to conduct electricity. Connecting the two test leads across a good conductor produces a flow of electrons through it, which causes the lamp to come on at full brightness (Figure 3-10A). Poorer conductors connected across the test leads produce varying levels of brightness of light. Insulators produce little or no flow of electrons or light. (Figure 3-10B).

Figure 3-8   Atomic structure of conductors, insulators, and semiconductors.

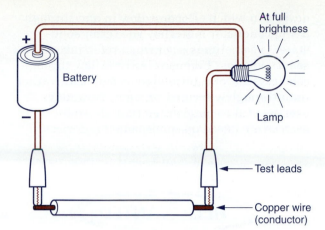

At full
brightness

Battery

Lamp

Test leads

Copper wire
(conductor)

A. Good conductor causes the lamp to come on

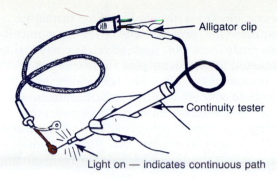

Alligator clip

Continuity tester

Light on — indicates continuous path

Figure 3-11   Continuity tester.

> ✊ **Under no circumstances should the continuity tester be connected to a circuit where other sources of voltage are present, because a serious safety hazard would be created.**

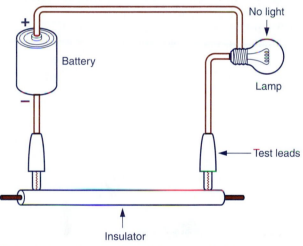

No light

Battery

Lamp

Test leads

Insulator

B. Insulator produces no flow of electrons or light

Figure 3-10   Testing for conductors and insulators.

This same circuit can also be used as a continuity tester to test electrical parts out-of-circuit (Figure 3-11). In this case, a continuous conducting path across the test leads turns the lamp on fully and an open path produces no light at all. Practical applications for this type of testing include: checking extension cords for open leads, checking for blown fuses, and checking for defective switches and pushbuttons. This tester is designed for checking electric components out of their normal circuits.

3-7 Electrical and Electronic Devices

People often refer to a toaster as an electrical device and a radio as an electronic device. Anything that works with electricity is electrical, including both a toaster and a radio, but not all electrical devices are considered electronic. Generally, to be classified as electronic the device must operate using electron tubes or semiconductors.

Electron tubes were the first electronic devices developed (Figure 3-12). In these

Glass
enclosure

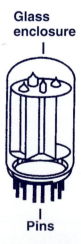

Pins

Figure 3-12   Typical electron tube.

devices, electrons are made to flow through a vacuum or low-pressure gas. They are generally glass enclosures that have physical parts inside connected to conductor pins that come out of the bottom of the tube. The tubes plug into sockets to connect to other circuit components. Before semiconductors were developed, most electronic devices used tubes. Today their use is limited to special applications such as television picture tubes and high-power transmitting tubes for radio and TV signals.

Semiconductors (Figure 3-13) such as transistors and integrated circuits, are more efficient in terms of size and power consumption and have replaced tubes for most electronic applications. Semiconductor devices are solid-state

devices in which electrons flow in and through a semiconductor material. The term "solid-state" is sometimes used when referring to semiconductors because electrons flow through solid crystals of semiconductor materials rather than through a vacuum or a gas. Nowadays, everyone takes for granted that electronic devices are based on semiconductor devices.

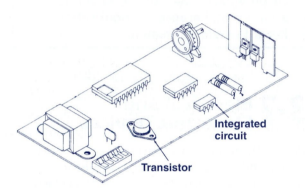

Figure 3-13 Typical semiconductor electronic device.

HELP WANTED

Semiconductor Design Engineer
Job Description
- The company is a leading supplier of integrated circuits.
- Work on all aspects of new product development.
- Must have a basic understanding of digital and analog circuit techniques.
- Training program provided, after which the new engineer works with a design team.
- Requires a BSEE (Bachelor of Science Electronic Engineering) or equivalent (MSEE preferred).

CHAPTER 3
Review and Applications

Review Questions

1. What three particles can be found within an atom?
2. State the type of charge associated with each of the following:
   (a) electron
   (b) proton
   (c) neutron
3. Compare the mass of the proton with that of the electron.
4. How does the number of protons found within an atom compare with the number of electrons?
5. (a) What type of electrical charge is associated with a single atom? Why?
   (b) What type of electrical charge is associated with a negative ion? Why?
   (c) What type of electrical charge is associated with a positive ion? Why?
6. According to Bohr's model of the atom, what is an energy shell?
7. Compare the location of bound electrons and valence electrons.
8. What is a free electron?
9. State a common definition for electricity.
10. Compare electron flow through a conductor to the flow through an insulator.
11. State the number of outer-shell valence electrons usually associated with a conductor, semiconductor, and insulator.
12. (a) What determines whether a pure semiconductor acts as a conductor or as an insulator?
    (b) List three common electronic components produced using specially treated semiconductors.
13. What does the circuit of a simple continuity checker consist of?
14. Explain how a continuity checker is used to test for a blown fuse.
15. Generally, when a device is classified as being electronic, what is the basis for its operation?
16. In what two ways are semiconductors more efficient than tubes?

Problems

1. Classify each of the following as either a conductor or an insulator:

paper	lead	rubber
copper	tap water	plastic
wood	steel	

2. What is the charge of a copper atom if it gains electrons?

Critical Thinking

1. Assuming that you define electricity as the flow of free electrons, the direction of the electron flow in a circuit would have to be from the negative terminal of the voltage source to the positive terminal. Why?

Portfolio Project

Design and construct a simple out-of-circuit continuity tester consisting of a battery, lamp, and test leads. Report on the different electrical components that can be tested using your tester. Report on the limitations of this using your tester to check continuity.

Circuit Challenge

Using a Simulator

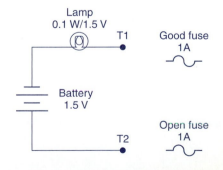

Procedure

(a) Construct the simulated continuity tester circuit shown using whatever simulation software package is available to you. Fault one fuse by double-clicking on it and changing its fault setting to open to simulate a blown fuse.

(b) Test both fuses using the continuity tester circuit. Why does the continuity tester lamp fail to come on when the blown fuse is connected to the circuit?

Sources and Characteristics of Electricity

E lectricity is present in all matter in the form of electrons and protons. Any device that develops and maintains a voltage can be considered a voltage source. To accomplish this, the voltage source must remove electrons from one point and transfer electrons to a second point. In this chapter we will explore the different methods used to produce voltage.

Objectives

After studying this chapter, you should be able to:

1. Define static and current electricity.
2. Explain how static positive and negative charges are produced.
3. State the law of electric charges.
4. Explain the difference between direct current (dc) and alternating current (ac) electricity.
5. List the basic sources of electricity and electrical devices used to convert the various energy forms.

Key Terms

- static electricity
- negative charge
- positive charge
- electrostatic field
- coulomb
- current electricity
- electromotive force
- direct current (dc)
- alternating current (ac)
- solar cell
- battery
- thermocouple
- thermopile
- piezoelectricity
- crystal
- electromagnetic induction
- generator
- alternator

RESEARCH INTERNET SITES

- Savin
- Clark Public Utilities
- Mitsubishi Electric

4-1 Static Electricity

The term static means standing still or at rest. *Static electricity* is an electric charge at rest. You may generate static electricity when you walk across a carpet or run a plastic comb through your hair.

One of the simplest ways to produce static electricity is by *friction*. By rubbing two different materials together, electrons may be forced out of their valence shells in one material and picked up in the shell of the other material. The material that gives up electrons more freely becomes positively charged and the one that gains electrons becomes negatively charged.

For example, a static charge can be produced by rubbing a hard rubber rod with a piece of fur (Figure 4-1). Normally, the atoms of both materials have the same number of electrons and protons and are, therefore, electrically balanced or neutral. When they are rubbed together, electrons move from one material to the other. In this case, the hard rubber rod takes on electrons, giving it an excess of electrons and a *negative charge*. At the same time, the fur loses electrons creating a shortage of electrons and a *positive charge*.

A glass rod and silk cloth can also become charged when rubbed together (Figure 4-2). In this case, the glass rod loses some of its electrons to the silk cloth. This leaves the glass rod with a positive charge and the silk cloth with a negative charge. Note that the charges are produced by removing loosely held electrons from atoms. Since protons are located at the centers of atoms, they are so tightly bound that nor-

mally they do not move. Touching the charged end of the glass rod will immediately discharge or neutralize it. In this instance, electrons flow from your hand to neutralize the charged rod end.

4-2 Charged Bodies

One of the fundamental laws of electricity is: *like charges repel and unlike charges attract.* In Figure 4-3 we see that the balls on the strings were charged as indicated—the unlike charges attract, while the like charges repel.

A force exists in the space between and around charged objects. This region is known as an *electric field of force* or *electrostatic field.* The electrostatic field around two charged objects is represented graphically by lines which are referred to as *electrostatic lines of force* (Figure 4-4). These lines are imaginary and are simply used to represent the direction and relative strength of the field. *The force between the two charged objects varies directly with the quantity of charge on the objects and varies inversely with the square of the distance between them.*

If two strongly charged bodies (one positive and one negative) are moved near to each other, before contact is made, you actually see the equalization of the charges take place in the form of an arc. With very strong charges, static electricity can produce arcs several feet in length. Lightning is a perfect example of the discharge of static electricity, resulting from a strong static charge accumulating in a cloud.

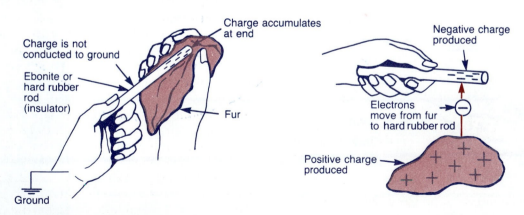

Figure 4-1 Charging a hard rubber rod.

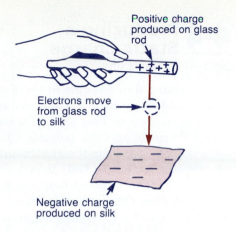

Positive charge produced on glass rod

+ + + +

Electrons move from glass rod to silk

−

Negative charge produced on silk

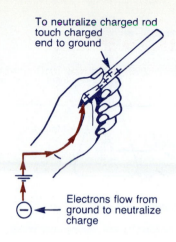

To neutralize charged rod touch charged end to ground

+ + +

Electrons flow from ground to neutralize charge

−

Figure 4-2   Charging and discharging a glass rod.

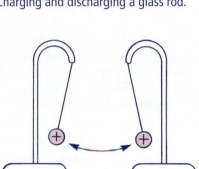

+ ⟷ +

Like charges repel

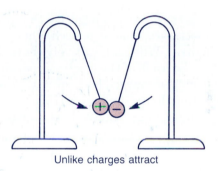

+ −

Unlike charges attract

Figure 4-3   Law of electric charges.

The amount of charge on any given body is expressed in *coulombs.* One coulomb is equal to a charge of approximately 6.25×10^{18} electrons (6,250,000,000,000,000,000 electrons). An object that has gained 6.25×10^{18} electrons has a negative charge of one coulomb. On the other hand,

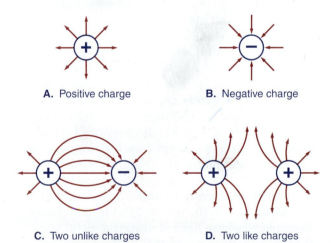

A. Positive charge

B. Negative charge

C. Two unlike charges

D. Two like charges

Figure 4-4   Electrostatic field patterns.

an object that has given up 6.25×10^{18} electrons has a positive charge of one coulomb.

4-3 Testing for a Static Charge

The charge on an object is found by seeing how it affects an object with a known charge. If the two repel, the charges are alike. If they attract, the charges are opposite. If you think the object is neutral, test it with another neutral object. A neutral object does not attract another neutral object. However, a neutral object is attracted by an object with either a positive or negative charge.

The *aluminum-leaf electroscope* is a device for detecting the presence of an electric charge and also for determining whether this charge is positive or negative (Figure 4-5). This device is made up of a glass flask with an insulating stopper. A metal rod passes through the center of the stopper. A metal knob is fastened to the outer end of the rod. Two very thin leaves of

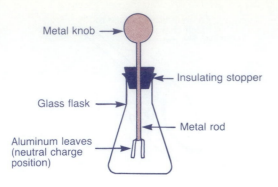

Figure 4-5 Aluminum-leaf electroscope.

aluminum are fastened to the end of the rod inside the flask.

Normally the positive and negative charges within the electroscope balance each other, leaving it neutral. When a negatively charged body touches the knob of the electroscope, electrons flow from the charged body into the knob and down to the aluminum leaves (Figure 4-6). Each leaf then becomes negatively charged. Since like charges repel and both leaves are negative, they will diverge, indicating that the object contained a static charge. The extent to which the leaves diverge is an indication of the quantity or amount of the charge on the object. When the charged body is removed, the electroscope is left with an excess of electrons. The aluminum leaves remain diverged and the electroscope is said to be negatively charged. If a positively charged body touches the knob, electrons flow out of the knob to produce a positive charge on the aluminum leaves.

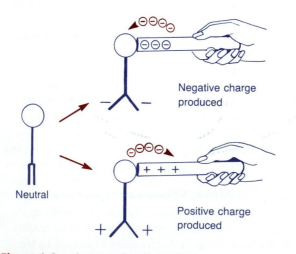

Figure 4-6 Charging the electroscope.

4-4 Producing a Static Charge

There are other ways in which an object can become charged in addition to being charged by friction and by contact. You can also charge an object by *induction*. This method is used to produce a charge of opposite polarity. Charging by induction makes use of the electrostatic field surrounding a charged body in order to charge an object without actually touching it. The principle of charging through induction is illustrated in Figure 4-7. When a negatively charged rod is brought close to the metal sphere the negative charges in the sphere are repelled and drift as far away as possible from the rod. Grounding the sphere by touching it allows electrons to leave the sphere. When the finger is removed in the presence of the negatively charged rod, the electrons cannot return to the sphere and the rod remains positively charged.

For useful applications, static electrical charges are often produced by a *high-voltage dc (direct current) source*. The most effective kind of air cleaner uses positively and negatively charged plates to remove very fine dirt particles from the air in a room. Figure 4-8 illustrates how an electronic air cleaner can be used in a home-heating system to clean the air as it circulates through the furnace. The dirty air passes through a paper prefilter that removes large dust and dirt particles from the air. The air then moves through an electrostatic precipitator consisting of two oppositely charged, high-voltage grids. The precipitator works by giving a positive charge to particles in the air and then attracting them with a negatively charged grid. Finally, the air passes through a carbon filter, which absorbs odors from the air.

Photography and static electricity enable a photocopier to produce almost instant copies of documents. The process is based on the ability of an electrostatically charged drum to attract toner particles in the image of the original document. Figure 4-9 (p. 44) illustrates a typical photocopying cycle. At the heart of the machine is a cylindrical metal drum that is given a charge at the beginning of the copying cycle. The optical system then projects an image of the document on the drum. The electric charge disappears where light strikes the metal surface, so only dark parts of the image remain charged. Oppositely charged particles of black

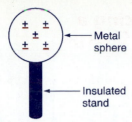

← Metal sphere

← Insulated stand

A. Neutral metal sphere on an insulated stand.

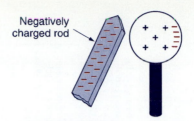

Negatively charged rod

B. Placing the charged rod near the metal sphere causes a redistribution of charges in the sphere.

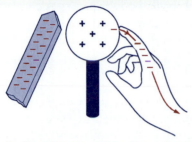

C. Grounding the sphere allows electrons to leave the sphere.

D. When the ground is removed in the presence of the negatively charged rod, the electrons cannot return to the sphere and it remains positively charged.

Figure 4-7   Charging by induction.

toner are then applied to the drum. The charged parts of the drum attract the toner, which is then transferred to a piece of paper. The toner makes the image visible and is fused to the paper with heat. Color is created with colored toners.

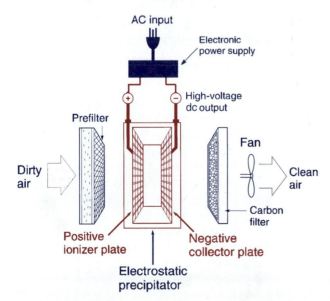

AC input

Electronic power supply

High-voltage dc output

Prefilter

Fan

Dirty air

Clean air

Carbon filter

Positive ionizer plate

Negative collector plate

Electrostatic precipitator

Figure 4-8   Electronic air cleaner.

An electrostatic laser printer operates on information relayed from a computer. This information is directed to the laser, which scans across the drum and creates the charged image area.

4-5 Current Electricity

Current or *dynamic electricity* is defined as an electrical charge in motion (Figure 4-10, p. 45). It consists of a flow of negative electron charges from atom to atom through a conductor. The external force that causes the electron flow is called the *electromotive force* (emf) or *voltage.*

The electromotive force (emf) or voltage is created by a battery which consists of one positive and one negative terminal. The negative terminal has an excess of electrons while the positive terminal has a deficiency of electrons. When a conductor, in this case an electric light bulb, is connected to the two terminals of the battery, a flow of electrons occurs. The positive terminal of the battery has a shortage of electrons and thus attracts electrons from the conductor. The negative terminal has an excess of electrons, which repels electrons into the conductor.

COPY PROCESS

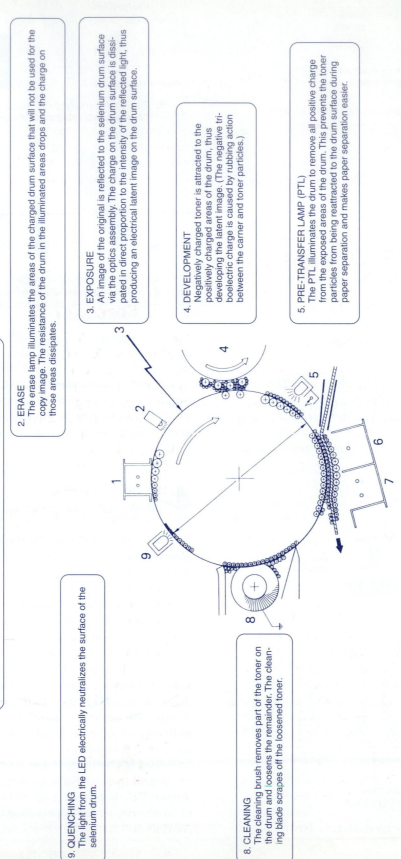

1. DRUM CHARGE
In the dark the charge corona unit gives a uniform positive charge to the selenium drum. The charge remains on the surface of the drum because the photoconductive selenium has high electrical resistance in the dark.

2. ERASE
The erase lamp illuminates the areas of the charged drum surface that will not be used for the copy image. The resistance of the drum in the illuminated areas drops and the charge on those areas dissipates.

3. EXPOSURE
An image of the original is reflected to the selenium drum surface via the optics assembly. The charge on the drum surface is dissipated in direct proportion to the intensity of the reflected light, thus producing an electrical latent image on the drum surface.

4. DEVELOPMENT
Negatively charged toner is attracted to the positively charged areas of the drum, thus developing the latent image. (The negative triboelectric charge is caused by rubbing action between the carrier and toner particles.)

5. PRE-TRANSFER LAMP (PTL)
The PTL illuminates the drum to remove all positive charge from the exposed areas of the drum. This prevents the toner particles from being reattracted to the drum surface during paper separation and makes paper separation easier.

6. IMAGE TRANSFER
Paper is fed to the drum surface at the proper time so as to copy paper and the developed image on the drum surface. Then, a strong positive charge is applied to the reverse side of the copy paper, providing the electrical force to pull the toner particles from the drum surface to the copy paper. At the same time, the copy paper is electrically attracted to the drum surface.

7. PAPER SEPARATION
A strong AC corona discharge is applied to the reverse side of the copy paper, gradually reducing the positive charge on the copy paper and breaking the electrical attraction between the paper and the drum. Then, the stiffness of the copy paper causes it to separate from the drum surface. The pick-off pawls help to separate paper which has low stiffness.

8. CLEANING
The cleaning brush removes part of the toner on the drum and loosens the remainder. The cleaning blade scrapes off the loosened toner.

9. QUENCHING
The light from the LED electrically neutralizes the surface of the selenium drum.

Figure 4-9 Typical photocopying process. (*Courtesy Savin Canada Inc.*)

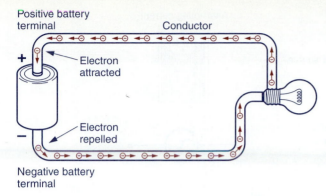

Positive battery terminal

Conductor

+

Electron attracted

Electron repelled

−

Negative battery terminal

Figure 4-10 Current electricity—electrical charge in motion.

Although "static" and "current" electricity may seem different, they are really the same. Both consist of electrical charges. Static electricity consists of electrons at rest on an insulated object and does little work. Current electricity moves and does useful work. When static electricity is discharged, it is no longer static electricity: it is current electricity.

Current electricity may also be classified as *direct current* or *alternating current* based upon the voltage source. Direct current voltage produces a constant flow of electrons in one direction only. Alternating current voltage produces a flow of electrons that changes both in direc-

tion and in magnitude. A battery is a common dc voltage source, while an electrical wall outlet is the most common ac voltage source (Figure 4-11).

Polarity identification is one way to distinguish a voltage source. Polarity can be identified on direct current circuits, but on alternating current circuits the current continuously reverses direction; therefore the polarity cannot be identified. Also it is important to know whether an electrical energy source produces alternating current or direct current. Many control and load components are designed to operate with a specific type of current. Operating the components with the wrong type of current can result in improper operation and/or permanent damage to the component.

4-6 Sources of Electromotive Force (emf)

For electrons to flow there must be a source of electromotive force (emf) or voltage. This voltage source can be produced from a variety of different *primary energy sources*. These primary sources supply energy in one form, which is then converted to electrical energy. Primary sources of electromotive force include friction, light, chemical reaction, heat, pressure, and mechanical-magnetic action.

Light   *Light energy* is directly converted into electric energy by *solar* or *photovoltaic cells*. These are made from a semiconducting, light-sensitive material that makes electrons available when struck by the light energy (Figure 4-12). When light strikes the cell it dislodges electrons from their valence shells, developing an electric charge. A *direct current* output voltage is produced by the cell. The output voltage is directly proportional to the light energy striking the surface of the cell.

When solar cells are combined with a battery, they become a reliable source of electricity. The solar cells provide electricity to use and to charge the battery when there is sunlight. When there is no sunlight available, the battery provides electricity.

One of the best solar cells is the silicon cell. A single cell can produce up to 400 mV (millivolts) with current in the milliampere range. Solar cells are often used as sensing devices in light meters and automatic-lighting circuits. Large

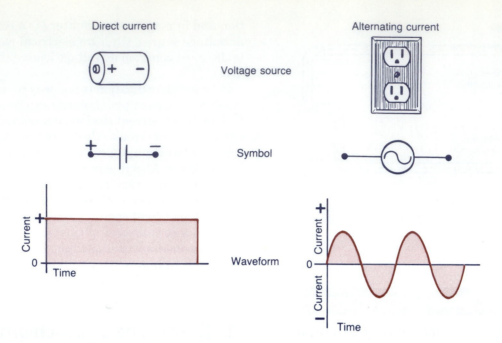

Figure 4-11 DC and ac current electricity.

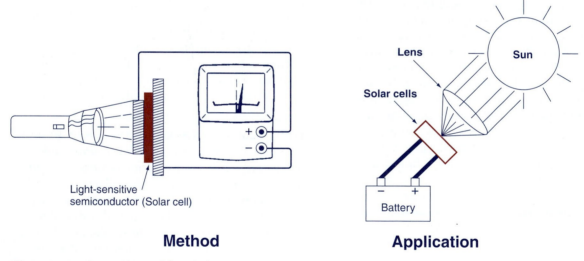

Method

Application

Figure 4-12 Generating emf from light.

panels of solar cells power satellites, while strips of a few cells provide the much smaller current needed to power calculators.

Chemical Reaction The *battery* or *voltaic cell* converts chemical energy directly into electric energy (Figure 4-13). Basically, a battery is made up of *two electrodes* and an *electrolyte solution*.

The chemical action within the cell causes the electrolyte solution to react with the two elec-

trodes. As a result, electrons are transferred from one electrode to the other. This produces a positive and negative charge at the electrode terminals of the cell. When the engine is off, the car's battery provides the voltage to operate its electrical system.

The battery is a popular low-voltage, portable *dc-voltage source*. However, it is a relatively high-cost electric energy source and this limits its applications.

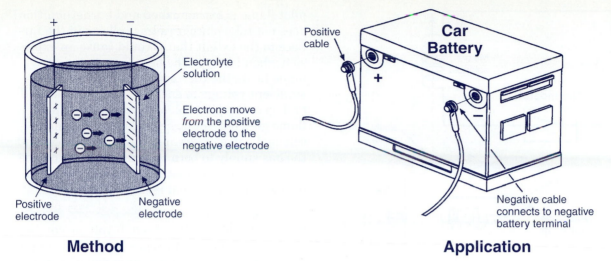

Method

Application

Figure 4-13 Generating emf from chemical reaction.

Heat *Heat energy* can be directly converted into electric energy by a device called a thermo-couple (Figure 4-14). A *thermocouple* is made up of two different types of metals joined at a junction. When heat is applied to the junction, electrons move from one metal to the other. The metal that loses electrons becomes positively charged, while the metal that gains electrons takes on a negative charge. If an external circuit is connected to the thermocouple, a small amount of dc current will flow as a result of the voltage between the two different metals.

One of the most practical applications for the thermocouple is its use as a temperature probe for temperature-measuring devices. Placed inside an industrial furnace it will produce a voltage that is directly proportional to the furnace temperature. A millivoltmeter, calibrated in degrees, is connected across the external thermocouple leads to indicate the temperature. Thermocouples are also used as part of electrical control systems to automatically maintain set-temperature values. The voltage and current produced by a single thermocouple is extremely low. Thermocouples may be arranged in series and in parallel to obtain higher voltage and current capability. An arrangement such as this is called a *thermopile*.

Figure 4-15 shows a thermopile-pilot safety valve circuit. A standing pilot is installed near

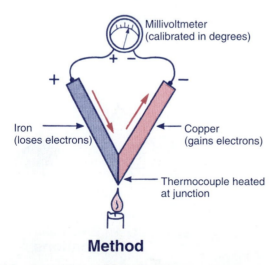

Method

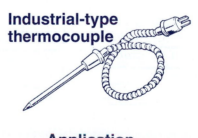

Industrial-type thermocouple

Application

Figure 4-14 Generating emf from heat.

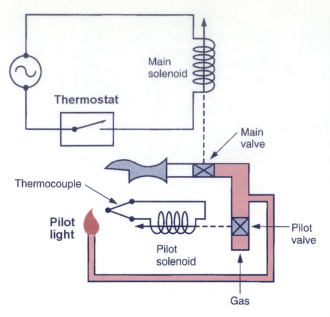

Figure 4-15   Pilot safety valve.

pilot flame is extinguished and burner ignition does not take place, raw gas collects in the furnace to the extent that it could cause an explosion when ignited. The heat from the pilot flame heats the thermopile. This produces sufficient voltage to energize the pilot solenoid and open the pilot solenoid valve. If the pilot flame goes out, all power is lost so the pilot solenoid valve closes and automatically cuts off the gas supply to both the pilot burner and the main burner. The pilot flame can be reignited by operating a manual reset.

Piezoelectric Effect

A small voltage can be produced when certain types of crystals are put under pressure. This effect is called *piezoelectricity,* from the Greek word meaning "to press." If a crystal from one of these substances is placed between two metal plates and a pressure is exerted on the plates, an electric charge will develop as shown in Figure 4-16. The principle illustrated here has many useful applications, even though they are applications with very low power requirements. The applications include use in crystal microphones, phonograph pickups, and automotive engine-knock sensors.

With *crystal microphones,* when the sound waves from your voice reach the microphone,

the burners to provide a safe means of igniting gas. The pilot flame is constantly supplied gas from the combination gas valve, and, once lit, it should remain on. It is absolutely necessary that the pilot flame be burning before the main burners are fed gas from the gas valve. If the

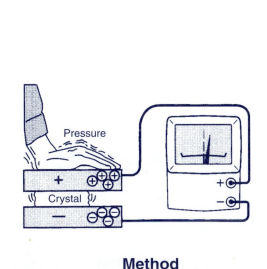

Method

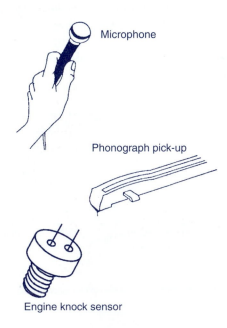

Applications

Figure 4-16   Generating emf from pressure.

pressure is applied to the *crystal* and a small amount of electricity is produced. In a *phonograph pickup,* the needle rides in a groove in the record. The music recorded in the groove causes the needle to vibrate. This vibration applies pressure to the crystal in the pickup, generating a small amount of electricity.

An automotive *engine knock sensor* is sometimes located on the engine block. Vibrations produced by engine knock apply pressure to the crystal producing an output voltage which is monitored by the engine computer.

Mechanical-Magnetic Most of the electricity used is produced by converting mechanical and magnetic energy into electric energy. If a conductor is moved through a magnetic field, a voltage is developed in the conductor (Figure 4-17). This is called *electromagnetic induction* and is the principle used in powering *generators.* The mechanical-power source spins a coil between the poles of a magnet or electromagnet. As it cuts through the magnetic lines of force, an electric current flows through the coil (Figure 4-18). Enormous generators driven by water or steam turbines are used to supply our communities with electricity. The generator may be designed to produce ac or dc electricity. Automobiles use ac generators, called *alternators,* as the main source of voltage when the engine is running. This is rectified or changed into dc for use in the automobile.

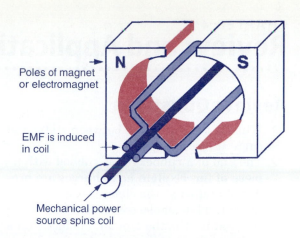

Poles of magnet or electromagnet

EMF is induced in coil

Mechanical power source spins coil

Figure 4-18 Generator principle.

HELP WANTED

Business-Machine Service Technician
Job Description
- Diagnose and repair electromechanical business machines.
- Product line includes computers, printers, fax machines, and copiers.
- Technical competence to diagnose problems to root cause level and effect repairs.
- College degree in a related field (Business, Computer Science, Electronics).
- Valid driver's license required.

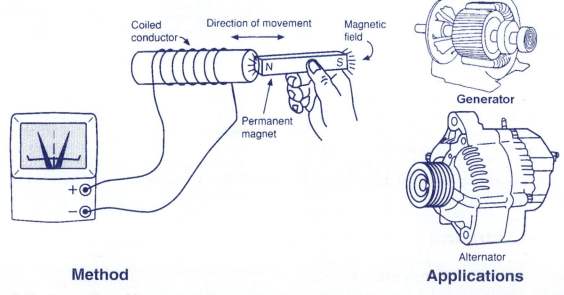

Coiled conductor   Direction of movement   Magnetic field

Permanent magnet

Method

Generator

Alternator

Applications

Figure 4-17 Generating emf from mechanical-magnetic action.

Review and Applications

Review Questions

1. What is the difference between static and current electricity?
2. The end of a rubber rod is rubbed with a piece of fur. Explain how charges are produced and what the charges are.
3. State the law of electrostatic charges.
4. In what unit is the amount of charge on a body expressed?
5. A positively charged glass rod is placed in contact with the knob of an electroscope. Explain the reaction that takes place.
6. Name three ways in which an object can become charged.
7. Explain how an electronic air cleaner precipitator works.
8. Upon what process is the operation of a photocopier based?
9. Explain the difference between *direct-current* (dc) and *alternating-current* (ac) electricity.
10. List six primary energy sources that can be used to produce an electromotive force (emf) or voltage.
11. What is the primary energy source used in each of the following devices:
    (a) generator   (b) thermocouple
    (c) solar cell   (d) engine-knock sensor
    (e) battery

Problems

1. Under what condition will a static charge change to current electricity?
2. What is the purpose of the photovoltaic cell in an automatic door opener device?
3. The battery operates the car's electrical system when the engine is off. What operates the system with the engine running?
4. How is electricity produced in most electric power stations?

Critical Thinking

1. The thermocouple operated pilot solenoid shown in Figure 4-15 fails to function normally. List four possible reasons for this.

Portfolio Project

Plan, organize, and deliver a class presentation on any one of the voltage sources discussed in this chapter. Use computer software (e.g. Power Point) and other technology to produce an effective presentation.

Circuit Challenge

Using a Simulator

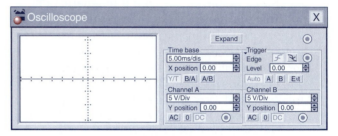

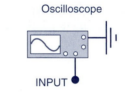

6-V ac
Voltage source

6-V dc
Voltage source

Procedure

(a) Construct the simulated circuit shown using whatever simulation software package is available to you. Set all front panel oscilloscope controls to the settings shown.
(b) Which voltage source waveform is a straight line?

CHAPTER 5

Basic Electrical Units

I n practical situations, you must be able to measure electricity if you want to work with it. Just as you determine the water pressure in a tank with a pressure gauge, so you measure electrical pressure with a voltmeter. In each case a unit of measurement is used, and the meter must be calibrated in that unit. This chapter will focus on the measuring of voltage, current, and resistance.

Objectives

After studying this chapter, you should be able to:

1. Define electric current, voltage, resistance, power, energy, and list the unit of measurement of each.
2. Identify the essential parts of a circuit and state the purpose of each.
3. Explain the relationship among current, voltage, and resistance.
4. State the difference between electron flow and conventional current flow.
5. Make measurements of current, voltage, and resistance.

Key Terms

- ampere (A)
- volt (V, emf, or E)
- ohm (Ω)
- milli (m)
- micro (μ)
- kilo (k)
- mega (M)
- electric power
- watt (W)
- joule (J)
- watthour (Wh)
- load
- control device
- protection device
- Ohm's law
- electron flow
- conventional current flow

RESEARCH INTERNET SITES

- Iguana Labs
- BC Hydro
- International Rectifier (IRF)

5-1 Electric Charge

As noted in Chapter 4, there is a tiny negative charge on each electron. For practical purposes, scientists decided to combine many of these charges so they would be able to measure them. The practical unit of electrical charge is the *coulomb.* One coulomb of charge is the total charge on 6.25×10^{18} electrons. This is an important unit that is used to describe or define other electrical units.

5-2 Current

The rate of flow of electrons through a conductor is called *current*. The letter I (which stands for intensity) is the symbol used to represent current. Current is measured in *amperes* (A). The term ampere refers to the number of electrons passing a given point in one second. If we could count the individual electrons, we would discover that approximately 6.25×10^{18} (6,250,000,000,000,000,000) electrons go by a given point in the circuit during 1 second for a current flow of 1 ampere (Figure 5-1). Flow of electric current may be compared to the flow of water through a pipe.

When the electrons begin to flow, the effect is felt instantly all along the conductor, much like the force that can be transmitted through a row of billiard balls (Figure 5-2). Although the individual electrons do not travel more than a few

Figure 5-2   Transmission of an impulse.

inches per second, the current effectively travels through the conductor nearly at the speed of light (186,000 miles/second).

About ⬅🔳➡ Electronics

The unit for current, the ampere (A), was named for French physicist André Marie Ampere, who discovered that two parallel wires attract each other when currents flow through them in the same direction and repel each other when currents are made to flow in opposite directions.

One *ampere is equal to one coulomb of charge moving past a given point in one second.* An instrument called an *ammeter* is used to measure current flow in a circuit (Figure 5-3). The ammeter is inserted into the path of the current flow, or in *series*, to measure current. This means the circuit must be opened and the meter leads placed between the two open points. Although the ammeter measures electron flow in coulombs per second, it is calibrated or marked in *amperes*. For most practical applications, the term amperes is used instead of coloumbs per second when referring to the amount of current flow.

5-3 Voltage

Voltage (V, emf, or E) is electrical pressure, a potential force or difference in electrical charge between two points. Voltage pushes current through a wire similar to water pressure pushing water through a pipe. The voltage level or value is proportional to the difference in the electrical potential energy between two points. Voltage is measured in *volts* (V). A voltage of one volt is required to force one ampere of current through one ohm of resistance (Figure 5-4).

Voltage, or a difference in potential, exists between any two charges that are not exactly

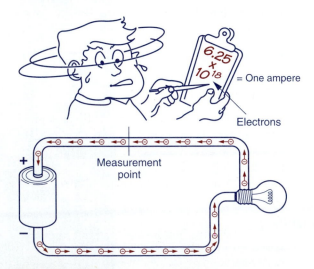

6.25 × 10¹⁸ = One ampere

Electrons

Measurement point

Figure 5-1   Current flow in a conductor.

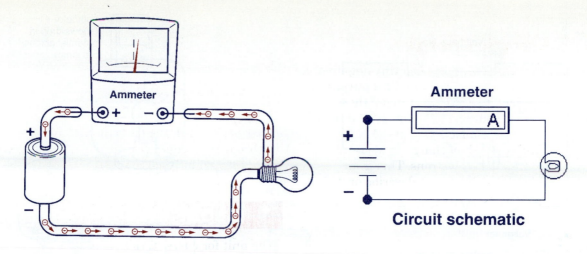

Ammeter

Ammeter

Circuit schematic

Current	Base Unit	Units for Very Small Amounts		Units for Very Large Amounts	
Symbol	A	μA	mA	kA	MA
Pronounced As	Ampere (Amp)	Microampere	Milliampere	Kiloampere	Megampere
Multiplier	1	0.000001	0.001	1000	1,000,000

Figure 5-3   Current measurement.

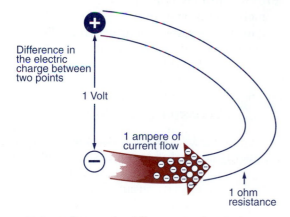

Difference in the electric charge between two points

1 Volt

1 ampere of current flow

1 ohm resistance

Figure 5-4   Voltage—the difference in electric charge between two points.

equal to each other. Even an uncharged body has a potential difference with respect to a charged body; it is positive with respect to a negative charge and negative with respect to a positive charge. Voltage also exists between two unequal positive charges or between two unequal negative charges. Therefore, voltage is purely relative and is not used to express the actual amount of charge, but rather to compare one charge to another and indicate the electro-

motive force between the two charges being compared.

A *voltmeter* is used to measure the voltage, or potential energy difference of a load or source. Voltage exists between two points and does not flow through a circuit as current does. As a result, a voltmeter is connected *across*, or in *parallel*, with the two points. Very small amounts of voltage are measured in millivolts and micro-volts, whereas high-voltage values are expressed in kilovolts and megavolts (Figure 5-5).

5-4 Resistance

Resistance (*R*) is the opposition to the flow of electrons (current). It is like electric friction that slows the flow of current. Resistance is measured in *ohms*. The Greek symbol Ω (omega) is often used to represent ohms.

Resistance is due, in part, to each atom resisting the removal of an electron by its attraction to the positive nucleus. Collisions of countless electrons and atoms, as the electrons move through the conductor, create additional resis-

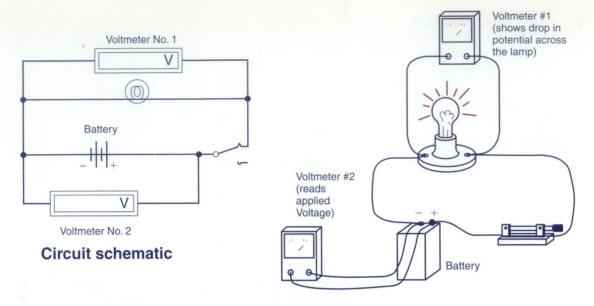

Voltage	Base Unit	Units for Very Small Amounts		Units for Very Large Amounts	
Symbol	V	mV	μV	kV	MV
Pronounced As	Volt	Millivolt	Microvolt	Kilovolt	Megavolt
Multiplier	1	0.001	0.000001	1000	1,000,000

Figure 5-5   Voltage measurement.

tance also. The resistance created causes heat in the conductor when current flows through it.

Every electrical component has resistance and this resistance changes electric energy into another form of energy—heat, light, or motion. A special meter called an *ohmmeter* can measure the resistance of a device in ohms when no current is flowing (Figure 5-6).

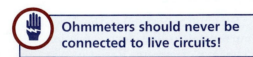 **Ohmmeters should never be connected to live circuits!**

The ease with which electric current flows through a material depends on whether there are relatively large numbers of *free electrons*. A material with few electrons would have a substantial resistance or opposition to current flow.

5-5 Power

Electric power (*P*) refers to the amount of electric energy converted to another form of energy in a given length of time. Power is the work performed by an electric circuit and is measured in *watts* (*W*). Power in an electric circuit is equal to:

$$\text{Power} = \text{Voltage} \times \text{Current}$$
$$\text{Watts} = \text{Volts} \times \text{Amperes}$$
$$P = VI$$

where:

P is the power in watts
V is the voltage in volts
I is the current in amperes

Electric irons, toasters, lamps, and radios are examples of electrical devices that are rated in watts (Figure 5-7). This rating is not always

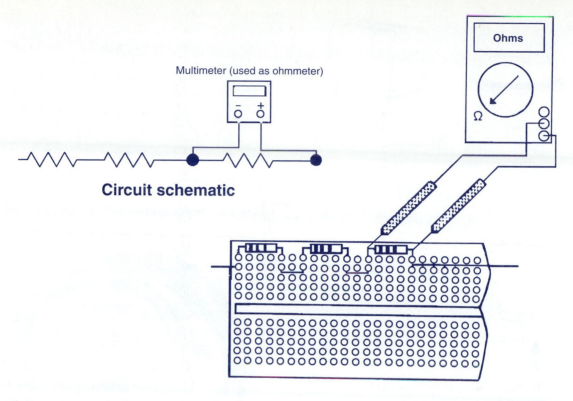

Circuit schematic

Multimeter (used as ohmmeter)

Ohms

Resistance	Base Unit	Units for Very Small Amounts		Units for Very Large Amounts	
Symbol	Ω	μΩ	mΩ	kΩ	MΩ
Pronounced As	Ohm	Microhm	Milliohm	Kilohm	Megohm
Multiplier	1	0.000001	0.001	1000	1,000,000

Figure 5-6   Resistance measurement.

specified simply in terms of watts. Ratings may give voltage and current or voltage and watts.

The wattage rating of a lamp indicates the rate at which the device can convert electrical energy into light. The faster a lamp converts electrical energy to light, the brighter the lamp will be. For example, a 100-watt lamp will give off more light than a 40-watt lamp (Figure 5-8). Similarly, electric soldering irons are made with various wattage ratings. Higher-wattage irons change electric energy to heat faster and, therefore, operate at higher temperatures than those of low-wattage rating.

5-6 Energy

Electric energy refers to the energy of *moving electrons*. In a complete electrical circuit, the voltage pushes and pulls electrons into motion. When electrons are forced into motion, they have kinetic energy, or the energy of motion.

A unit of energy, called the *joule (J)*, is used in scientific work to measure electric energy. By definition, a joule is the amount of energy carried by 1 coulomb of charge propelled by an electromotive force of 1 volt.

1200 W
120 V
10 A

1440 W
120 V
12 A

60 W
120 V
0.5 A

240 W
120 V
2 A

Figure 5-7 Power rating of electrical devices.

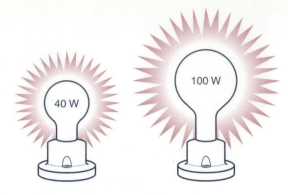

40 W

100 W

Figure 5-8 Higher wattage bulb produces more light.

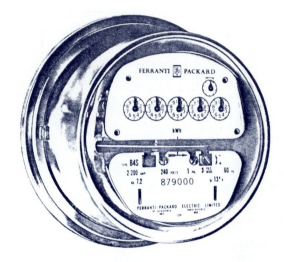

Figure 5-9 Kilowatthour energy meter. (*Courtesy Ferranti-Packard Ltd.*)

The *watthour* (Wh) is the more practical unit of measurement of electric energy. *Power* and *time* are factors that must be considered in determining the amount of energy used. This is usually done by multiplying watts by hours. The result is watthours, abbreviated Wh. If power is measured in kilowatts and multiplied by hours, the result is *kilowatthours*, abbreviated kWh. Energy measurements are used in calculating the cost of electric energy. A kilowatthour meter connected to the home electrical system is used to measure the amount of energy used (Figure 5-9).

5-7 Basic Electric Circuit

An electric circuit can be compared to a race track, with free electrons racing around the circuit instead of cars, and copper wires serving as the track. There are differences, too. Cars are self-propelled, but free electrons must be pushed through a circuit by the electromotive force (voltage) source or power supply. Also, a race track generally has the same road surface around the whole track; while an electric circuit contains a restriction, the load, similar to a

narrow tunnel. The load is the place where the free electrons are put to work! Finally, the race track has no gates to interrupt its roadway, but the electric circuit contains a start-stop gate (i.e., switch) that controls the electron flow.

All devices and machines powered by current electricity contain an electric circuit (Figure 5-10). A *closed electric circuit* can be defined as a complete electric path from one side of a voltage source to the other. Its essential parts consist of a power source, conductors, loads, control devices, and protection devices.

A battery is an example of a *power source*. The battery creates a potential energy difference or voltage across its two terminals. The electric energy from the battery is transported through the circuit by moving electrons. These moving electrons drift from atom to atom in the direction of the positive terminal of the battery.

Closed circuit

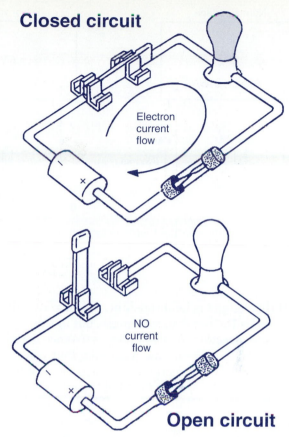

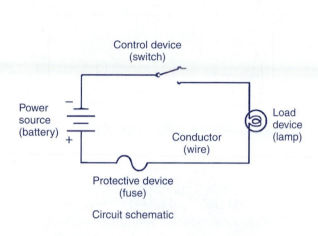

Electron current flow

NO current flow

Open circuit

Figure 5-10 Basic electric circuit.

Conductors provide a low-resistance path from the source to the load. In an ideal circuit, electrons lose all of their available energy while going through the load. In reality, a slight energy loss occurs in the wires as electrons flow through the circuit.

A *load* is any device that uses electric energy or changes it into other forms. A lamp is an example of a load in a circuit. Electrons flow through the lamp converting the electric energy of the source into light and heat energy.

Control devices vary current flow, or turn it ON and OFF. A switch is an example of a common control device. When the switch is in the "on" position, it acts as a conductor to keep electrons moving through the circuit and the circuit is said to be closed. When the switch is in the "off" position, the circuit path is interrupted and the circuit is said to be open.

Protection devices open the current path if too much current flows. Too much current can cause damage to conductors and load devices. Fuses and circuit breakers are common exam-

ples of protection devices. If the current in any part of the circuit increases to a dangerously high level, the fuse will melt and automatically open the circuit.

5-8 Relationship Among Current, Voltage, and Resistance

The amount of current (electron) flow in a circuit is determined by the voltage and resistance. As you know, voltage is the force that causes current to flow. Therefore, the higher the voltage applied to a circuit, the higher the current through the circuit. On the other hand, a decrease in the applied voltage will result in a decrease in circuit current. This assumes that the circuit resistance, or opposition to current flow remains constant (Figure 5-11). The amount of source voltage normally is not affected by either current or resistance.

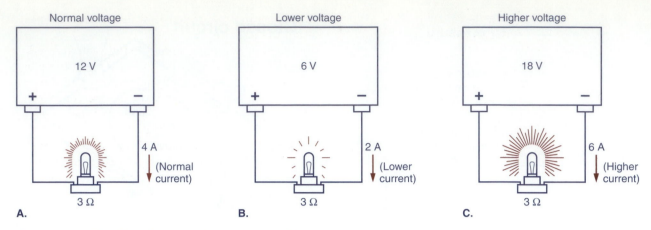

Figure 5-11   The effect of source voltage on current flow.

If the voltage is held constant, the current will change as the resistance changes, but in the opposite direction. As you know, resistance is the opposition to current flow. Assuming that voltage is constant, an increase in resistance results in a decrease in current flow. On the other hand, lowering the resistance causes an increase in current (Figure 5-12). Resistance is not normally affected by either voltage or current.

To better understand the relationship of voltage, current, and resistance, compare it to a house watering system (Figure 5-13). Voltage is like water pressure and current in a wire or circuit is like the water flow in the hose. The valve offers resistance to the water flow and can be

adjusted to control the amount of water flow. The amount of water flow depends on the setting of the valve and the amount of water pressure present.

The relationship between current, voltage, and resistance is defined by *Ohm's law*. Ohm's law is the most important and basic law of electricity and electronics and is usually stated as follows: *the current flowing in a circuit is directly proportional to the applied voltage and inversely proportional to the resistance.* In equation format, this is:

$$I(\text{current}) = \frac{V(\text{voltage})}{R(\text{resistance})}$$

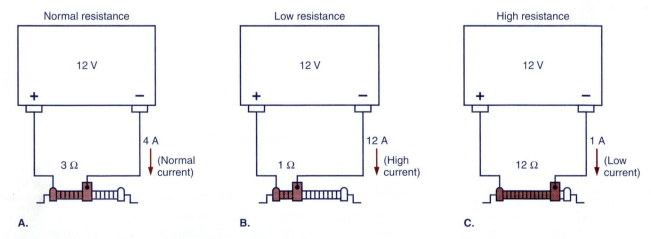

Figure 5-12   The effect of circuit resistance on current flow.

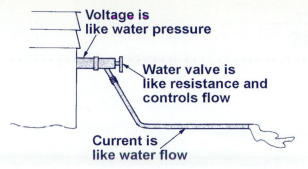

Figure 5-13 Watering system analogy of an electric circuit.

and assumes a current-flow direction from positive to negative.

Because the electron is the lightest charged particle, it would stand to reason that this particle could be most easily forced into directed motion as an electric current. If electrons flow one way in a material, the conventional electric current is in the opposite direction.

Both conventional current flow and electron flow are acceptable and are used for different applications. Regardless of whether the electric current is considered to be electron flow or conventional flow, the operation of the circuit is basically the same. It is important to understand and be able to think in terms of both conventional current flow and in terms of electron flow.

5-9 Direction of Current Flow

The direction of current flow in a circuit can be designated either as *electron flow* or *conventional current flow* (Figure 5-14). Electron flow is based on the electron theory of matter and, therefore, indicates the flow of current from negative to positive. Conventional current flow is based on an older fluid theory of electricity

Unless otherwise specified, the direction of current flow assumed in this text will be according to the electron flow (negative to positive). All arrows used to indicate the direction of current in a circuit shall indicate electron flow.

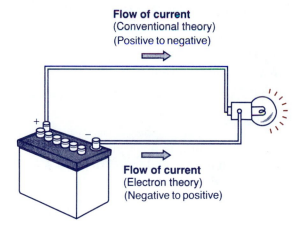

Flow of current
(Conventional theory)
(Positive to negative)

Flow of current
(Electron theory)
(Negative to positive)

Figure 5-14 Direction of current flow.

HELP WANTED

Electronics Instructor
Job Description
- Teach electronics at a community college.
- Bachelor's degree required.
- At least 3 years of experience in an electronics related field.
- Teaching background is desirable.
- Must have good written and verbal communication skills.

Review and Applications

Related Formulas

1 ampere = 1 coulomb per second	Watts = Volts × Amperes
Power = Voltage × Current	$P = VI$
milli (m) = prefix for 10^{-3}	micro (μ) = prefix for 10^{-6}
kilo (k) = prefix for 10^3	mega (M) = prefix for 10^6

Review Questions

1. What is the practical unit of electrical charge?
2. Define electric current.
3. State the basic unit used to measure current.
4. The current flow in a circuit is measured with an ammeter and found to be 10 A. Express this in terms of coulombs per second.
5. How is an ammeter inserted into a circuit to measure current flow?
6. Define voltage.
7. State the basic unit used to measure voltage.
8. How is a voltmeter connected into a circuit to measure voltage?
9. Define electrical resistance.
10. State the basic unit used to measure resistance.
11. When using an ohmmeter to measure resistance, what precaution must be observed?
12. Define electric power.
13. State the basic unit used to measure power.
14. A lamp draws a current of 0.5 A when 120 V is applied across it. What is the power rating of the lamp?
15. Why does a 140-W soldering iron produce more heat than a 20-W soldering iron?
16. Define electric energy.
17. Define 1 joule of electric energy.
18. What is the more practical unit of measurement of electric energy?
19. For what practical applications are electric energy measurements most often used?
20. Define a closed electric circuit.
21. Name five basic parts of any electric circuit and state the basic purpose of each part.
22. Explain how the amount of current flowing in a circuit is related to the voltage of the circuit.
23. Explain how the amount of current flowing in a circuit is related to the resistance of the circuit.
24. What is the direction of current flow according to electron flow?
25. What is the direction of current flow according to conventional current flow?

Problems

1. The current flow in a circuit is measured with an ammeter and found to be 10 A. Express this in terms of coulombs per second.
2. A lamp draws a current of 0.5 A when 120 V is applied across it. What is the power rating of the lamp?
3. Convert the following data to prefix units:
   (a) 0.034 V   (b) 12,000 A
   (c) 3,600 Ω   (d) 5,600,000 Ω
   (e) 0.000009 A   (f) 0.00004 V
4. Convert these prefix units back to basic units:
   (a) 3 μ V   (b) 6.2 kΩ
   (c) 5 mA   (d) 1.8 MΩ
   (e) 8 kV   (f) 4.7 mA

Critical Thinking

1. Assume applying 8 V to a resistor load produces a current of 4 mA.
   (a) What would the value of the current flow be if the voltage were increased to 16 V?
   (b) What would the value of the current flow be if the voltage were decreased to 4 V?
   (c) What would the value of the current flow be if the load resistance were doubled?
   (d) What would the value of the current flow be if the load resistance were halved?

Portfolio Project

As a middle school teacher, you are required to present a lesson on the basic electric circuit to a grade 7 science class. Terms to be taught include: power source, conductors, load, control device, and protective device. Prepare a packet of material to be used in your presentation. Include a multiple-choice test for feedback on how well the information presented was understood by the students.

Circuit Challenge

Using a Simulator

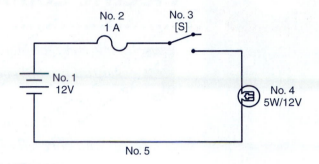

Procedure

(a) Construct the simulated basic circuit shown, using whatever simulation software package is available to you.
(b) Identify the component parts (No. 1 through No. 5) and state the function of each.

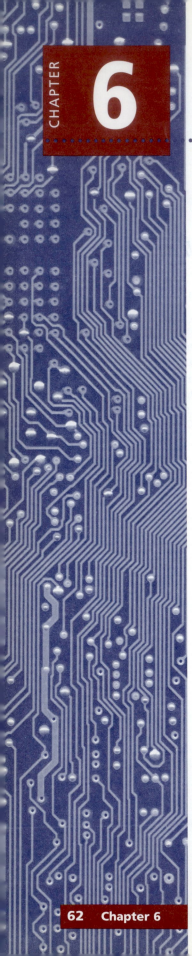

6

Electric Connections

*E*lectric connections that provide maximum conductivity are an essential part of every circuit. Manufacturers have recognized this fact and have developed a variety of wiring devices that ensure safe and reliable installations. This chapter will help you select and correctly install various kinds of connectors.

Objectives

After studying this chapter, you should be able to:
1. Describe the negative effects of a poorly formed electric connection.
2. Properly remove insulation from a wire.
3. Correctly install terminal-screw, crimp-on, and mechanical connectors.
4. Explain the principles of soldering.
5. Properly insulate wire repairs.
6. Assemble, test, and install cables used as part of electronic interconnection systems.

Key Terms

- wire stripper
- terminal screw
- crimp-on connector
- compression connector
- twist-on connector
- set-screw connector

- wire splicing
- heat shrink tubing
- plastic tape
- soldering
- solder
- solder flux

- cable connectors
- coaxial cable
- fiber optics
- wire-wrapping

- AMP Incorporated
- Weidmuller
- 3M Corporation

6-1 The Necessity for Proper Electric Connections

Almost all electrical installations and repairs consist of connecting wires to terminals or to other wires. Cutting, splicing, and connecting electrical wires must be done properly or problems will result. *A poorly made electrical connection will have a much higher than normal resistance.* In high-current electric circuits, this results in an excessive amount of heat being produced at the connection when normal *current* flows through the circuit (Figure 6-1). The poor electric connection will also reduce the total energy normally available for the load. This is due to the fact that a portion of the energy supplied is used to produce unwanted heat at the faulty connection point.

Figure 6-2 Electronic printed circuit-board connections. (*Photograph courtesy of Gail Motion Control, Sunnyvale, CA.*)

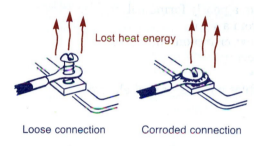

Figure 6-1 High-resistance connections.

Electronic devices, such as computers, operate at very low voltage and current levels. *High-resistance* connections in these circuits can cause weakened or complete loss of control signals. It is important to maintain good electric contact with this type of connection in order to minimize resistance losses. A common pencil eraser makes a very good abrasive in cleaning contacts, especially the edge connectors on printed circuit boards (Figure 6-2).

6-2 Preparing Wire for Connection

Electric wire is usually fully insulated with a covering of rubber or plastic. This covering must be removed (stripped) before the wire can be properly connected to anything. The amount of *insulation* to be stripped from the end of the wire depends on the kind of connection you are going to make.

Wire strippers are special tools used to remove insulation from wires that are to be joined or connected. One type (Figure 6-3) has a series of sharp openings in its scissor blade to allow stripping of wire of different gauge sizes or diameters. The gauge size of the wire must be matched to the opening in the wire stripper to prevent cutting into the wire and weakening it.

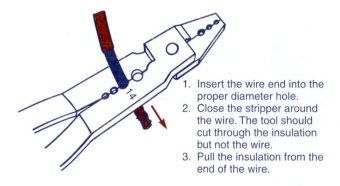

1. Insert the wire end into the proper diameter hole.
2. Close the stripper around the wire. The tool should cut through the insulation but not the wire.
3. Pull the insulation from the end of the wire.

Figure 6-3 Using a wire stripper.

If you are stripping wire with a knife, you must be careful not to nick the wire and weaken it. Do not cut straight into the insulation. Slant your knife blade at an angle toward the end of the wire as you would if you were sharpening a pencil (Figure 6-4, p. 64). Once cut, pull the insulation off the wire using your fingers or a pair of pliers.

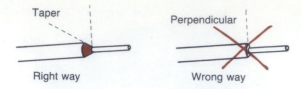

Figure 6-4 Using a knife to remove insulation.

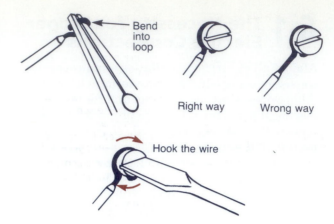

Right way   Wrong way

Hook the wire

The insulation of some wires, including some magnet wires, can be removed by the heat of a soldering iron. Wires with enamel or varnish insulation are often stripped chemically with a paint solvent.

6-3 Terminal-Screw Connections

The simplest and most common type of electrical connection is the **terminal-screw** type. When you are wiring electrical devices such as switches or lampholders, this is the type of connection most often used. Connections to screw-type terminals are made by forming the end of the wire into a loop that is fitted around the head of the screw (Figure 6-5).

Wire markers (Figure 6-6) are often used to identify wires and matching terminal ends. They make it possible to quickly find and trace wires in a circuit. The marking of wires and terminals is also very useful in testing certain circuits and in replacing wires and components in them.

6-4 Crimp Terminals, Splices, and Mechanical Connectors

Broken wires are often joined by using **crimp-on** or **compression connectors**. A **butt-type connector** is used to join two wires. Wires are inserted into a special insulated or noninsulated conductor-retaining sleeve. A *crimping tool* is then used to crimp the wires in the sleeve (Figure 6-7). Different shapes of *crimp-terminal lugs* are used for connecting wires to *terminals* (Figure 6-8).

Electric connections made by means of **wire splicing** involve twisting two wires together, soldering them, and taping the splices. Figure 6-9 illustrates two methods of splicing a wire. The joint must be as strong as if the wire were unbroken. Contact points between the two conductors must be such that current will pass

1. Remove only enough insulation to make a loop of bare wire around the screw terminal (about ³/₄ in.). If the wire is stranded, the bare strands should be tightly twisted together, and then tinned.
2. Use needle-nose pliers to bend the bared wire into a loop.
3. Loosen the screw terminal with a screwdriver, but do not remove it from its hole.
4. Hook the wire over the top of the screw and tighten in a **clockwise** direction. In this way the wire will be drawn around the screw and not pushed away from it when the screw is tightened.
5. Close the loop around the screw with the pliers and tighten the screw.
6. Almost no bare wire should extend beyond the screw head. If it does, you have stripped away too much insulation and the bare wire should be shortened.

Figure 6-5 Connecting to terminal screws.

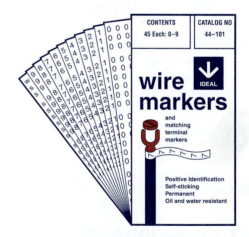

Figure 6-6 Wire markers. (*Courtesy of Ideal Industries Inc., Sycamore, Illinois, USA; Ajax, Ontario, Canada.*)

from one wire to the other as if the wires were in one piece. Scraping the conductor clean with the back of a knife before forming your splice helps to improve the electrical contact between the two conductors.

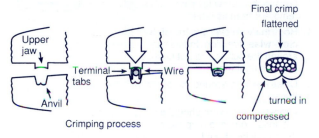

Noninsulated
butt connector

Pre-insulated
butt connector

Stripping area

Crimping
area

Cutting area

Crimping tool

1. Select the proper-size butt connector to fit the wires.
2. Strip a small amount of insulation from the ends of the wires to be joined.
3. Insert one of the wires halfway into the conector and crimp the connector to hold the wire.
4. Insert the other wire into the other end of the butt connector and crimp the connector to hold the wire.
5. Pull the wires to be certain they are held securely.
6. For a noninsulated type of butt connector, wrap the connection with plastic tape or use a section of heat shrink tubing to seal it from dirt and moisture.

Upper jaw

Terminal tabs

Wire

Anvil

Crimping process

Final crimp flattened

turned in

compressed

Figure 6-7 Installing a crimp-on butt connector.

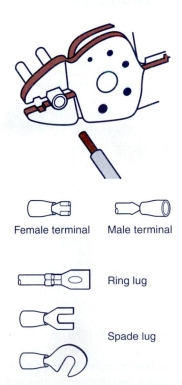

Female terminal   Male terminal

Ring lug

Spade lug

Figure 6-8 Crimp-terminal lugs.

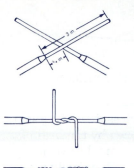

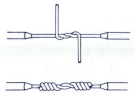

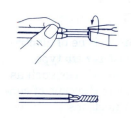

How to make a Western Union splice

1. Remove about 3 in. of insulation from each conductor.
2. Cross the conductors about ¾ in. from the insulation and bend their ends so that one wire wraps around the other.
3. Wrap one wire in a clockwise direction until it reaches the insulation. Wrap the other in a counter-clockwise direction for the same number of turns.
4. Cut off any excess wire and bend the ends down.
5. Solder the entire splice and tape.

How to make a pigtail splice

1. Remove about 1½ in. of insulation from each conductor.
2. Hold the wires tightly parallel with your left hand and begin twisting the free ends.
3. Twist the free ends in a clockwise direction forming about 6 turns and cut off the wire ends.
4. Solder the entire splice and tape.

Figure 6-9 Splicing wires.

Mechanical connectors that require no soldering or taping are used by electricians for making most electrical connections. These devices save both time and labor and, as a result, are used extensively. The *twist-on connector* uses a metal spring that threads itself around the conductors. As the connector is rotated it holds the conductors in place (Figure 6-10A-C, p. 66). The internal spring design takes advantage of leverage and vise action to multiply the strength of a person's hand. It is available in a range of sizes for splicing conductors from No. 18 gauge up to No. 8 gauge. *When using any type of mechanical connector, it is very important to match the size of the connector with the gauge size of the wire to make a proper connection.*

The *set-screw connector* is a two-piece connector that uses a set screw to hold the conductors in place (Figure 6-11, p. 67). This design allows conductor connections to be interchanged easily. They are used mainly in commercial and industrial circuits where equipment must be changed frequently for maintenance purposes.

Mechanical-lug cable connectors (Figure 6-12 p. 67) are often used in conjunction with larger electric-cable installations. The connec-

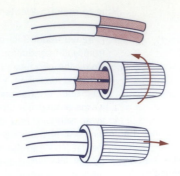

1. Remove about ¾ in. of insulation from both conductors and scrape clean.
2. Hold the two conductor ends even and insert into connector shell.
3. Twist the connector clockwise onto the wires until it is tight.
4. Be certain no bare wire is visible when the cap is in place. Test the connection by trying to pull the connector away from the wires.

A. Installing a twist-on connector

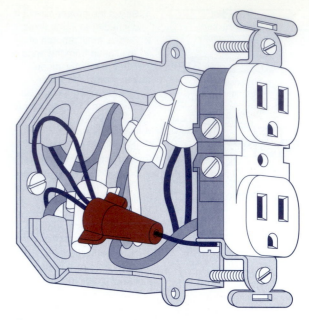

C. Installed in an electrical outlet box

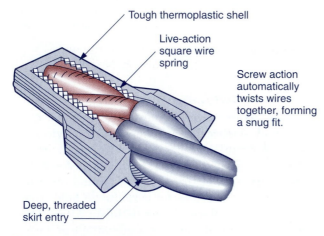

Tough thermoplastic shell

Live-action square wire spring

Screw action automatically twists wires together, forming a snug fit.

Deep, threaded skirt entry

B. Conductors when installed

Figure 6-10 Twist-on connector. (*Courtesy of Ideal Industries Inc., Sycamore, Illinois, USA; Ajax, Ontario, Canada.*)

tor holds all strands of the cable securely without damage to any of the strands. The connector is made of a metal that resists electrolysis between itself and the cable. When installing a conductor in a lug, tighten the holding screw once the cable has been fully inserted into the lug. Allow several minutes to elapse, and then retighten the holding screw. The strands will have settled in place, making a second tightening necessary for a secure connection.

6-5 Insulating Repairs

The insulation used to cover wire repairs must be equivalent to the amount of insulation that was removed from the wire. *Plastic tape* is preferred for taping wire. To do the actual taping, begin on the insulation of the conductors and wrap the plastic tape tightly around the entire splice, ending back on the insulation (Figure 6-13).

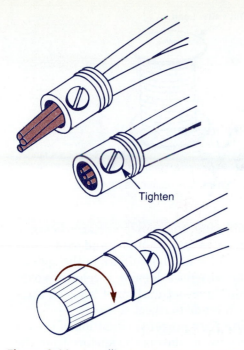

1. Remove about ¾ in. of insulation from conductors and scrape clean.
2. Remove the brass fitting from the connector and loosen the set screw.
3. Hold the two conductor ends even and insert them into the brass fitting so that the threaded shoulder is next to the insulation.
4. Tighten the set screw and cut off any excess wire that extends beyond the end of the brass fitting.
5. Thread the plastic cap onto the brass fitting until snug.
6. Be certain no bare wire is visible when the cap is in place. Test the connection by trying to pull the connector away from the wires.

Tighten

Figure 6-11   Installing a set-screw connector.

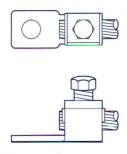

A. Single conductor lug cable connector

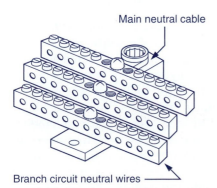

Main neutral cable

Branch circuit neutral wires

B. Neutral block for a distribution panel

Figure 6-12   Mechanical-lug cable connector.

The turns should overlap by half the width of the tape to produce a double layer of insulation.

Heat shrink tubing provides an easy and highly effective means of insulating and protecting terminal connections and splices against moisture, dirt, and corrosion (Figure 6-14). The tubing is first slipped over the wire and then over the connection once the wires have been secured. Brief exposure to heat causes the tubing to shrink to half its original size and to force its way into voids and around the terminal connection or splice.

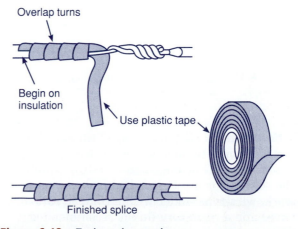

Overlap turns

Begin on insulation

Use plastic tape

Finished splice

Figure 6-13   Taping wire repairs.

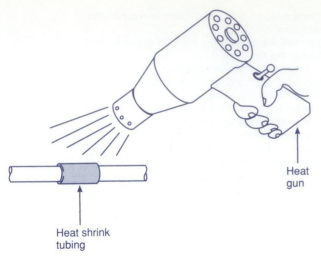

Heat
gun

Heat shrink
tubing

Figure 6-14   Heat shrink tubing.

6-6 Soldered Connections

Soldering is a quick, efficient method of joining metals permanently. Soldering an electric-terminal connection or splice improves its mechanical strength, helps to lower its electrical resistance, and prevents any corrosion or oxidation of the copper. Some manufacturers require all wiring, including crimp-on terminals, to be soldered to ensure proper contact (Figure 6-15).

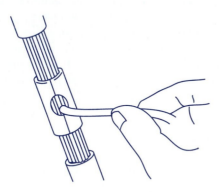

Figure 6-15   Soldering a crimp-on terminal to ensure proper contact.

Solder is an alloy made up of tin and lead; it has a low melting point. The tin/lead ratio determines the strength and melting point of the solder. For most electrical and electronic work, wire-type solder with a tin/lead ratio of 60/40 and a resin core flux is recommended (Figure 6-16).

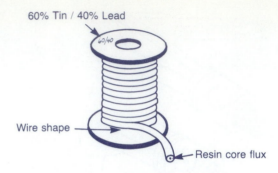

60% Tin / 40% Lead

Wire shape

Resin core flux

Figure 6-16   Wire-type resin core flux.

When soldering, the copper surfaces being soldered must be free from dirt and oxide, otherwise the solder will not adhere to the splice. Heating the wire for soldering accelerates oxidation. This leaves a thin film of oxide on the wires, which tends to reject solder.

Soldering flux prevents oxidation of the copper surfaces by insulating the surface from the air. Both acid- and resin-based fluxes are available. Acid-based fluxes should not be used in electrical work as they tend to corrode the copper wires. Resin flux is available in paste form or as a continuous core inside solder wire (Figure 6-17).

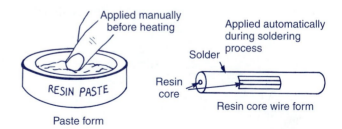

Applied manually
before heating

Applied automatically
during soldering
process

Solder

RESIN PASTE

Resin
core

Resin core wire form

Paste form

Figure 6-17   Resin-based flux.

The most common method of applying heat for soldering is by means of a *soldering gun* or *soldering iron* (Figure 6-18). Heat in a soldering gun is produced by means of transformer action. The soldering gun heats up very quickly but has a short duty cycle. Heat in a soldering iron is produced by means of a heating element similar to that found in a toaster. The soldering iron heats up at a slower rate but has a longer duty cycle.

For the best soldering results, the copper heating tip of the soldering iron or gun must be kept clean or well *tinned*. New tips must be tinned before they can be used. This can be

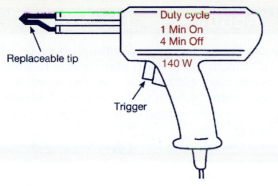

A. Soldering gun

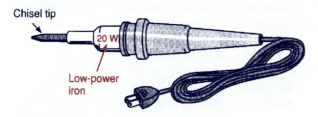

B. Soldering iron

Figure 6-18   Soldering heat sources.

accomplished by applying solder to the heated tip and wiping it clean. A well-tinned tip will conduct the maximum amount of heat from the tip to the surface being soldered (Figure 6-19).

> ✋ Safety glasses should be worn during both the tinning and soldering processes to provide adequate eye protection.

To solder a wire splice, position the heated copper tip below the splice and establish a good contact between the tip and splice (Figure 6-20). Wait until the splice is hotter than the

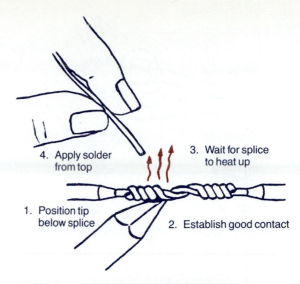

4. Apply solder from top

3. Wait for splice to heat up

1. Position tip below splice

2. Establish good contact

Figure 6-20   Soldering a splice.

melting point of the solder before beginning to solder, then apply the solder from the top. Gravity and adhesion will cause the flux and solder to reach all parts of the splice. A good soldered splice should be covered entirely with a thin coat of glossy solder.

Some electronic installations make use of soldered terminal connections for connecting component parts and wiring. Connections to soldered terminals are made by forming a tight mechanical connection that is soldered in place (Figure 6-21).

A good soldered connection has just enough solder to bond the wire to the terminal. It should have a smooth, semi-gloss appearance, and edges that seem to blend cleanly and smoothly into the terminal and wire. Any other appearance is a sign of improper soldering. A large glob on the terminal indicates too much solder was applied. If the soldered connection has a dull and pitted appearance, or has a ball of solder that does not blend into the metal, then the temperature of the metal was too low.

6-7 Electronic Data Communication Interconnection Systems

Cable connectors are designed to be fitted together and used in matching pairs (male and female). They are most often used for connecting components that are some distance apart. The connector is the device that holds the

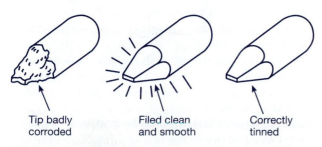

Tip badly corroded

Filed clean and smooth

Correctly tinned

Figure 6-19   Tinning the tip.

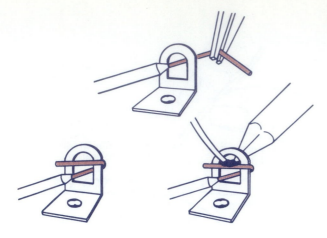

1. Remove only enough insulation to make a loop of bare wire around the terminal.
2. Use needle-nose pliers to bend the bare wire through and around the terminal hole.
3. Apply heat from a soldering iron or gun to the back side of the terminal. After a few seconds, touch the resin-core solder to the wire at the front of the terminal. If the solder melts, the connection is sufficiently heated and the solder will flow. (Do not touch the solder to the iron.)
4. When the connection has been coated with solder, remove the length of solder. Leave the iron on the connection to boil away the resin. If any resin remains under the solder, an insulation barrier could be formed. Now remove the iron and let the connection cool. Do not move the connection while it is cooling.

Figure 6-21 Soldering to a terminal.

terminals in position. Some connectors use a locking device to prevent the connector from separating and to ensure good electric connection (Figure 6-22). The locking device may be of the squeeze-to-unlock type. Always be sure to unlock the connector before pulling it apart and

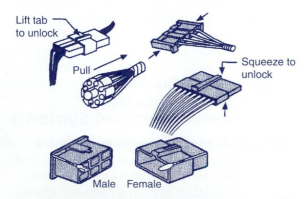

Lift tab to unlock

Pull

Squeeze to unlock

Male   Female

Figure 6-22 Typical locking-type connectors.

to lock the connector after completing the connection. Connectors make excellent test points because the circuit can be opened without need for wire repairs after testing (Figure 6-23).

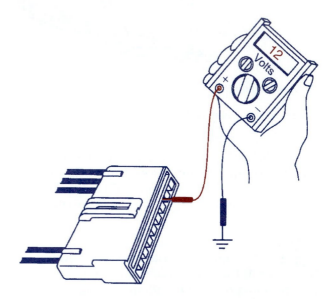

Figure 6-23 Using a connector for test point access.

Computers make extensive use of cables and connectors for connection to monitors, printers, disk drives, and modems. These *peripheral* (external) *devices* are often connected by cable to make up the total computer system. The **D-type** connector is one of the most common types of computer connectors (Figure 6-24). Cable wires can be connected to a connector by soldering or crimping.

Cables constructed with metallic shields are manufactured to reduce the effects of electromagnetic and radio interference. Shielded **coaxial cable** consists of a solid center copper-wire conductor surrounded by a plastic insulator. Over the insulator is a second conductor, a braided-copper shield. An outer plastic sheath

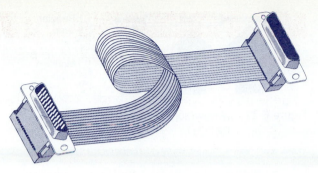

A. Ribbon cable assembly

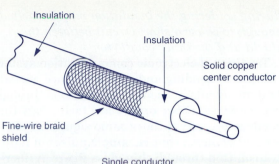
Insulation
Insulation
Solid copper center conductor
Fine-wire braid shield

Single conductor

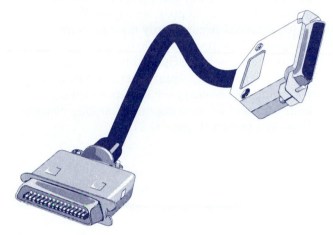

B. D-type male and female cable assembly

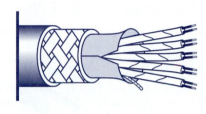

Multiconductor

A. Cable construction

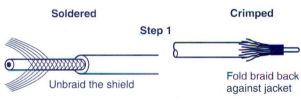

Soldered	Crimped
Step 1	

Unbraid the shield

Fold braid back against jacket

Step 2

Twist unbraided portion together

Insert cable into connector

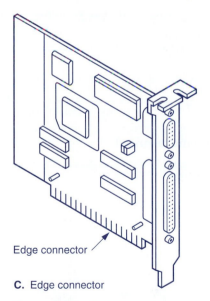

Edge connector

C. Edge connector

Figure 6-24 Typical computer connectors. (*Courtesy Promark Electronics.*)

protects and insulates the braid (Figure 6-25). When installing coaxial cable, be careful not to damage the insulation between the shield and the inner conductor. *If too much heat is applied*

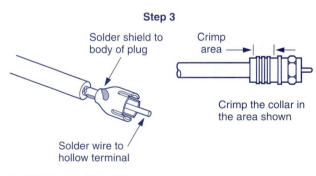

Step 3

Solder shield to body of plug

Crimp area

Solder wire to hollow terminal

Crimp the collar in the area shown

B. Cable installation

Figure 6-25 Shielded coaxial cable.

during soldering the insulation may be melted enough to cause a short circuit between the shield and the inner conductor.

Traditional electronic communication systems operate by sending and receiving voice, visual, or data signals in the flow of electrons through copper wires. **Fiber optics** is a technology that sends and receives those same signals in the form of *photons*, that is, simple pulses of light transmitted through thin glass fibers. A fiber-optic cable consists of a core, cladding, and a protective jacket (Figure 6-26). Special connectors are used for connection to fiber-optic cables.

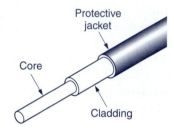

A. Cable construction

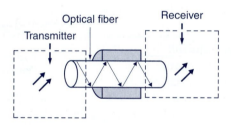

B. Transmission of signal

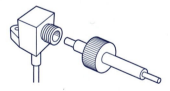

C. Connection to cable

Figure 6-26 Fiber-optic cable.

Figure 6-27 Wire-wrapping. (Courtesy OK Machine and Tool Co.)

Wire-wrapping involves making an electric connection by tightly coiling a wire around a metal terminal. Copper wire from No. 26 to No. 32 AWG is commonly used for wire-wrapping. This process is faster than soldering and requires no heat. Special terminal posts, which are generally square, are used in wire-wrapping. Wire-wrapping and unwrapping tools are used for making and changing connections in a circuit (Figure 6-27).

HELP WANTED

Fiber-Optics Technician
Job Description
- Install, test, and maintain fiber-optic cable networks.
- Experience with fusion splicing of fiber-optic cable desirable.
- Considerable climbing and physical mobility required.
- Training provided before joining installation crew.
- High school diploma required.

CHAPTER 6

Review and Applications

Review Questions

1. Describe three negative effects of a poor electric connection.
2. What precaution must be observed when using wire strippers in order to prevent damage to the wire?
3. When removing insulation from a wire with a knife, in what position is the knife blade held?
4. State two important features of a properly made terminal-screw connection.
5. What special tool is required to install a compression connector?
6. What is involved in making an electric connection by splicing wires?
7. State the main advantages that solderless mechanical connectors have over the soldered splice type of connection.
8. What precaution must be observed when using any type of mechanical connector in order to ensure a good connection?
9. When installing a set-screw connector, into which end of the brass fitting are the bare-wire ends inserted?
10. What final check can be made on mechanical connectors to test the mechanical strength of the connection?
11. State two methods used to insulate wire repairs.
12. State two positive effects of soldering an electric connection.
13. (a) What characteristic of solder is determined by its tin/lead ratio?
    (b) What tin/lead ratio is most often recommended for electrical work?
14. When soldering copper surfaces, what three things could cause the solder not to adhere to the copper surface?
15. What type of flux should not be used for electrical work? Why?
16. Compare the soldering gun and iron with regard to:
    (a) how heat is produced.
    (b) duty cycle.
    (c) heat-up time.

17. What is the reason for keeping the soldering tip well cleaned or tinned?
18. What safety precaution should be observed when tinning or soldering?
19. Describe two signs of an improperly soldered terminal connection.
20. Describe two methods commonly used to unlock connectors that use a locking device.
21. Name a commonly used type of computer connector.
22. What is the purpose of the metallic shield used in coaxial cable?
23. Explain how a fiber-optic cable sends and receives signals.
24. What is the advantage of using a wire-wrapped connection over a soldered connection?

Problems

1. Name the type of connection or connector best suited for each of the following applications:
   (a) joining broken wires
   (b) joining wires in an electrical outlet box
   (c) permanent wire connection to a printed circuit board
   (d) connecting wires to light switch
2. A computer cable used for the printer connection is suspected of having one or more open wires in the cable. How could you verify this using a multimeter?

Critical Thinking

1. Assume in the process of stripping insulation from a wire the copper is nicked. How will this effect the current carrying capacity of the wire? Why?

Portfolio Project

Collect examples of as many of the different kinds of connection devices studied in this chapter as you can. Identify each and give a specific example of their application.

Circuit Challenge

Using a Simulator

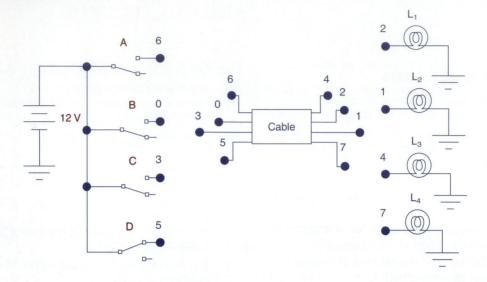

Procedure

(a) Construct the simulated cable circuit shown, using whatever simulation software package is available to you.

(b) Complete the wiring from the switches through the cable to the lights, according to wire number markings.

(c) Operate the circuit and identify which lamp is operated by which switch.

Simple, Series, and Parallel Circuits

Electric circuits can be as simple as a flashlight that uses but a single circuit, or as complex as a computer that uses thousands of circuits working together. In this chapter you will study the three basic circuit types: simple, series, and parallel.

Objectives

After studying this chapter, you should be able to:

1. Identify the basic components of a circuit and the symbols used to represent them.
2. Compare pictorial, schematic, wiring, and block diagrams.
3. Explain the operation of series and parallel connected loads and controls.
4. Construct a circuit schematic from a set of written instructions.
5. Read and construct circuits from schematic diagrams.

Key Terms

- energy source
- protective device
- conductors
- control device
- load device
- fuse
- circuit breaker
- symbols

- pictorial diagram
- schematic diagram
- wiring diagram
- block diagram
- simple circuit
- series connection
- parallel connection

- wiring number sequence
- common points
- breadboarding
- solderless breadboard socket
- computer simulated circuits

RESEARCH INTERNET SITES

- Radio Shack
- Electronics Workbench (interactiv.com)
- Warner Electric (warernet.com)

7-1 Circuit Components

Three fundamental quantities—voltage, current, and resistance—are present in every electric circuit (Figure 7-1). These quantities are determined by the proper arrangement of component parts to produce the desired function of the circuit.

The component parts that make up any circuit are as follows: the energy source, the protective device, conductors, the control device, and the load device.

The **energy source** supplies the voltage required to move the free electrons along the conducting path of the circuit. It is also referred to as the power supply. Two types of **sources**, direct current (dc) and alternating current (ac), are used (Figure 7-2).

The purpose of the **protective device** is to protect circuit wiring and equipment. It is designed to allow only currents within safe limits

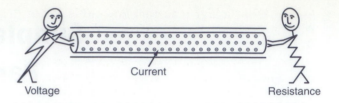

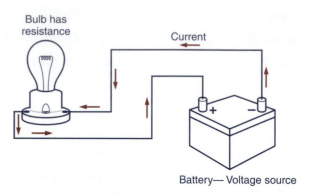

Figure 7-1 Voltage, current, and resistance are present in every circuit.

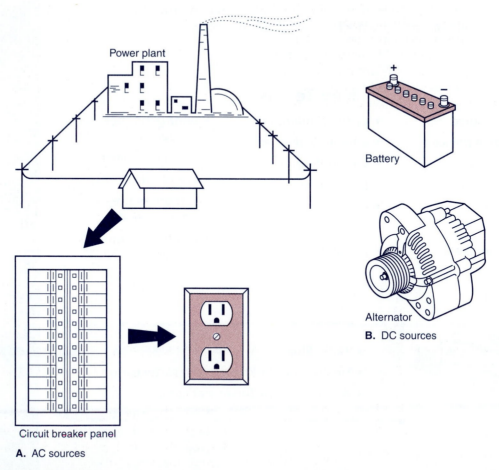

Figure 7-2 Power sources.

to flow. When a higher current flows, this device will automatically open the circuit. This effectively shuts off current until the problem is corrected. Two types of protective devices, namely *fuses* and *circuit breakers* are normally used. Often the protection device is a part of the voltage source or power supply device (Figure 7-3).

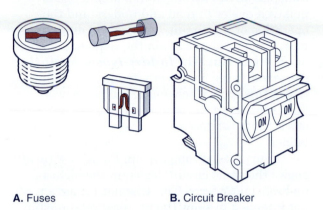

A. Fuses   **B.** Circuit Breaker

Figure 7-3   Protective devices.

Conductors or wires are used to complete the path from component to component. Conductors provide a low-resistance path for electrons. They are usually insulated to protect against accidental contact with the circuit. The most common conductor type is rubber- or plastic-insulated copper wire (Figure 7-4).

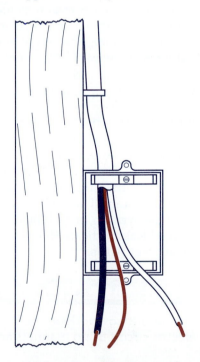

Figure 7-4   Conductors used for house wiring.

A *control device* is usually included in the circuit to allow you to easily start, stop, or vary the electron flow. Common control devices include switches, thermostats, and lamp dimmers (Figure 7-5).

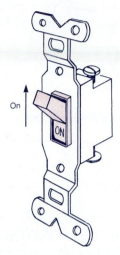

A. A switch manually starts and stops electron flow.

B. A thermostat automatically starts and stops electron flow.

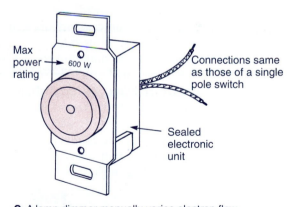

C. A lamp dimmer manually varies electron flow.

Figure 7-5   Control devices.

The *load* is the part of the circuit that converts the electric energy so it can produce the desired function or useful work of the circuit. Lamps, motors, heaters, and resistors are just a few common load devices (Figure 7-6). For

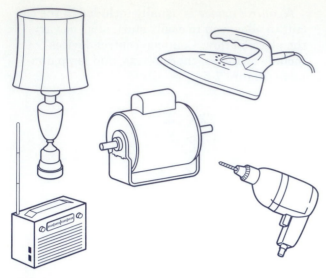

Figure 7-6 Load devices.

purposes of calculation, all of the circuit resistance is considered to be contained in the load device.

7-2 Circuit Symbols

The use of **symbols** to represent electrical and electronic components can be considered to be a form of technical shorthand. The use of these symbols tends to make circuit diagrams less complicated and easier to read and understand. In electricity and electronics, symbols and related lines are used to show how the parts of a circuit are connected to one another. Unfortunately, not all electrical and electronic symbols are standardized. You will find slightly different symbols used by different manufacturers. Also, symbols sometimes look nothing like the real thing and we have to learn what they mean. The symbols we will be using in this chapter are illustrated in Figure 7-7.

7-3 Circuit Diagrams

There are four major types of diagrams used to show the layout of circuits: *pictorial*, *schematic*, *wiring*, and *block*.

A *pictorial **diagram*** is used to show the physical details of the circuit visually. The advantage here is that a person can simply take

a group of parts, compare them with the pictures in the diagram, and wire the circuit as shown. The main disadvantage is that many circuits are so complex that this method is impractical.

A *schematic diagram* uses symbols to represent the various components and, as a result, is not as cluttered as a pictorial diagram. The components are arranged in a manner that makes it easier to read and understand the operation of the circuit. This type of diagram is most often used to explain the sequence of operation of a circuit. The ***ladder-type schematic diagram*** is the one most often used in industry. In this type of schematic the two power lines connect to the power source and the various circuits connect across them like rungs in a ladder.

For purposes of comparison, Figure 7-8 (p. 80) shows the same circuit drawn in the pictorial and schematic form. The schematic diagram is not intended to show the physical relationship of the various components of the circuit. Rather, it leans toward simplicity, emphasizing only the operation of the circuit.

A ***wiring diagram*** is intended to show, as closely as possible, the actual *connection* and *placement* of all component parts in a circuit. Unlike the schematic, the components are shown in their relative physical position. All connections are included to show the actual routing of the wires. A color code may identify certain wires. Such diagrams show the necessary information for doing the actual wiring, or for physically tracing wires in the event of troubleshooting, or for making changes to the circuit (Figure 7-9 p. 80).

A ***block diagram*** (Figure 7-10 p. 80) is a method of representing the major functional parts of complex electronic and electrical systems by blocks. Individual components and wires are *not* shown. Instead, each block represents electrical circuits that perform specific functions in the system. The function the circuits perform are written in each block. Arrows connecting the blocks indicate the general direction of current paths.

A. Direct current sources

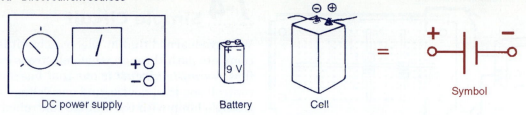

DC power supply   Battery   Cell   Symbol

B. Alternating current sources

Receptacle   Transformer   Symbol

C. Switches

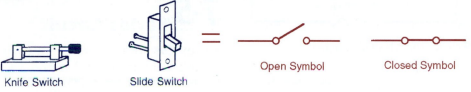

Knife Switch   Slide Switch   Open Symbol   Closed Symbol

D. Push button

Symbol

E. Buzzer

Symbol

F. Bell

Symbol

G. Lamp

Symbol

H. Conductors

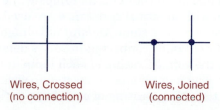

Wires, Crossed
(no connection)   Wires, Joined
(connected)

I. Resistor

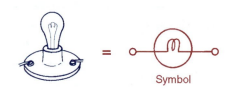

Symbol

Figure 7-7   Circuit symbols.

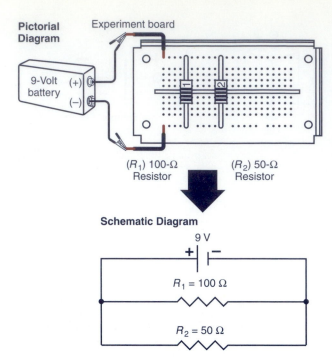

Figure 7-8   Pictorial and schematic diagrams.

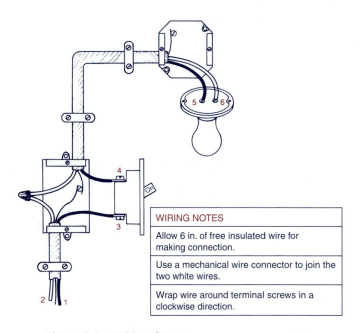

WIRING NOTES

Allow 6 in. of free insulated wire for making connection.

Use a mechanical wire connector to join the two white wires.

Wrap wire around terminal screws in a clockwise direction.

Figure 7-9   Wiring diagram.

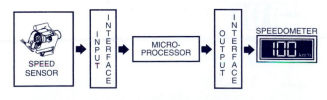

Figure 7-10   Block diagram.

7-4 Simple Circuit

We have learned that an electric circuit is a complete path through which electrons can flow. A *simple circuit* is one that has only one control, one load device, and one voltage source. A single lamp with the voltage controlled by a single switch is one example of a simple circuit. Each component is connected or wired "end-to-end." This simple circuit is controlled by opening and closing the *switch*. When the switch is closed, electrons flow to turn the lamp ON (Figure 7-11A). When the switch is open, the flow is interrupted and the lamp turns OFF (Figure 7-11B). When on, the voltage across the lamp is the same as the voltage of the source.

7-5 Series Circuit

Series-Connected Loads   If two or more loads are connected end-to-end, they are said to be connected *in series* (Figure 7-12). This is called a *series circuit*. The *same amount* of current flows through each of them. Also, there is only *one* possible path through which the current flows. The current stops if this path is opened or the circuit is broken at any point. For example, if two lamps are connected in series and one burns out, *both* lamps will go out.

When loads are connected in series, each receives part of the applied source voltage. For example, if three *identical* lamps are connected in series, each will receive one-third of the applied source voltage (Figure 7-13). The amount of voltage each load in series receives is directly proportional to its electrical resistance. The higher the resistance of the load connected in series, the more voltage it receives.

The filaments of some types of holiday tree lamps illustrate loads that are connected in series (Figure 7-14, p. 82). These are of the type that if one lamp goes out, all lamps will go out. Lamps with identical resistance are used so each receives the same amount of voltage. The number of lamps connected in series determines the voltage rating of each lamp in the string. The more lamps connected in series the lower the voltage rating of each lamp.

Series-Connected Control Devices   Two or more control devices may also be connected in

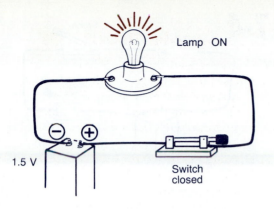

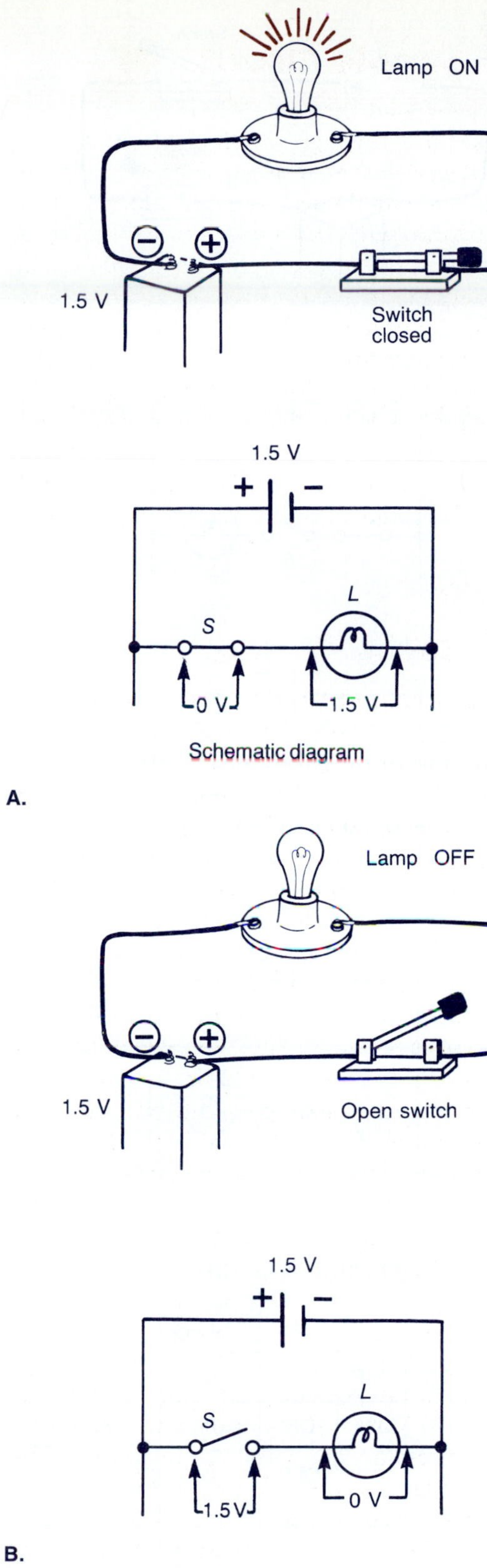

Lamp ON

Switch closed

1.5 V

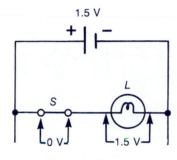

1.5 V

+ | −

S

L

0 V

1.5 V

Schematic diagram

A.

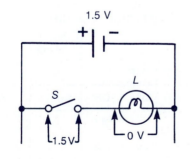

Lamp OFF

1.5 V

Open switch

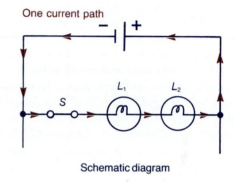

1.5 V

+ | −

S

L

1.5 V

0 V

Schematic diagram

B.

Figure 7-11   Simple circuit.

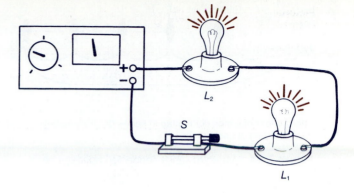

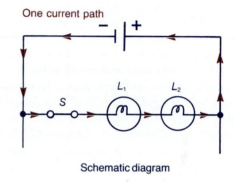

One current path

− +

S

L_1 L_2

Schematic diagram

Figure 7-12   Two lamps connected in series.

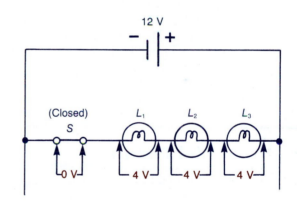

12 V

− | +

(Closed)
S

L_1 L_2 L_3

0 V   4 V   4 V   4 V

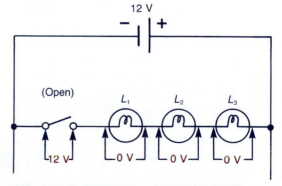

12 V

− | +

(Open)

L_1 L_2 L_3

12 V   0 V   0 V   0 V

Figure 7-13   Voltage drop across identical series-connected lamps.

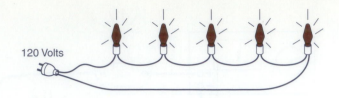

120 Volts

The number of lamps connected in series determines the voltage rating of each lamp.

Figure 7-14   String of series-connected holiday tree lamps.

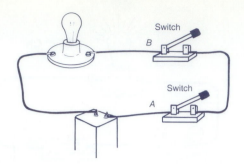

series. The connection is the same as that used for loads, which is end-to-end. Connecting control devices in series results in what can be called an AND type of control circuit. Take the example of two switches, A and B, connected in series to a lamp (Figure 7-15). In order for the lamp to turn ON both switch A AND switch B would have to be closed. A *truth table* is a way of showing how the switches in the circuit work the lamps: "0" means the lamp is OFF, "1" means the lamp is ON.

Series connection of control devices is used in electrical and electronic control systems. Large workshop machines are often controlled by two series-connected switches for safety reasons. To start the machine, both switches must be turned ON, but to stop it, only one of the switches has to be turned OFF.

7-6 Parallel Circuit

Parallel-Connected Loads   If two or more loads are connected across the two voltage source leads they are said to be connected in parallel. This is called a *parallel circuit*. The parallel connection of load devices is used for circuits where each device is designed to operate at the same voltage as the *power supply* (Figure 7-16). This is the case in wiring lights and small appliances in the home. Here, with the source voltage being 120 V, all appliances and lights connected in parallel to it must be rated for 120 V. The use of lower voltage devices, such as 12-V devices would cause such units to burn out.

A parallel circuit has the same amount of voltage applied to each device. However, the current to each device will vary with the resis-

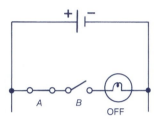

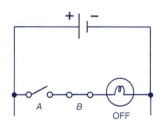

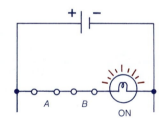

TRUTH TABLE

SWITCHES		LAMP
A	B	
OFF	OFF	0
OFF	ON	0
ON	OFF	0
ON	ON	1

"AND" Type control circuit A *"AND"* B are closed to switch lamp **ON**

Figure 7-15   Two switches connected in series.

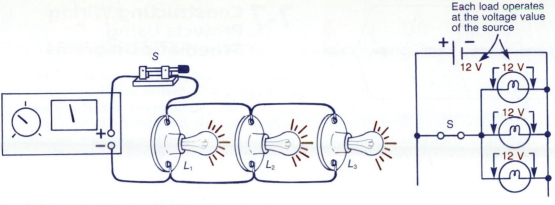

A. Three lamps connected in parallel

Schematic diagram

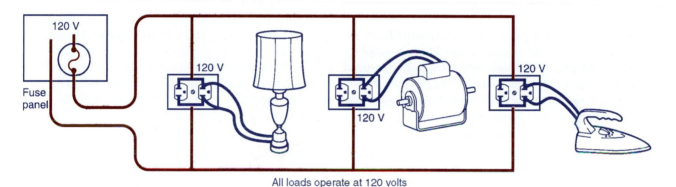

All loads operate at 120 volts

B. Household appliances connected in parallel

Figure 7-16 Loads connected in parallel.

tance of the device. The amount of current each in-parallel load passes is *inversely* proportional to its resistance value. The higher the resistance of the load connected in parallel, the less current it passes.

When load devices are operated in parallel, each load operates *independently* of the others. This is because there are as many current paths as there are loads. For example, if two lamps are connected in parallel, there will be two paths created. If one lamp burns out, the other will not be affected (Figure 7-17).

Holiday tree lamp strings wired in parallel (Figure 7-18) have the advantage that if any one lamp goes out, the other lamps in the string will continue to operate. All lamps in a parallel string are rated for 120 V regardless of the number of lamps used.

Parallel-Connected Control Devices When two or more control devices are connected across each other, they are said to be connected

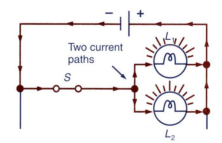

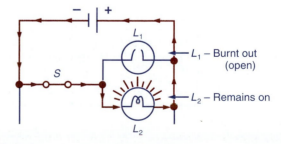

Figure 7-17 Current paths for two lamps connected in parallel.

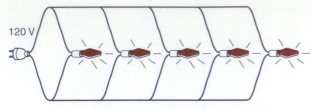

120 V

All lamps rated for 120 volts

Figure 7-18 String of parallel-connected holiday tree lamps.

in parallel. Connecting control devices in parallel results in what can be called an *OR type of control circuit*. Take the example of two pushbuttons, *A* and *B,* connected in parallel to a lamp. In order for the lamp to turn ON, either Push button *A* OR Push button *B* or both would have to be pressed (Figure 7-19). Parallel connection of control devices is also used in electrical and electronic control systems.

7-7 Constructing Wiring Projects Using Schematic Diagrams

Planning is an important part of an electrical wiring project. It involves thinking through the job to be done. What do you want the circuit to do? What type of electrical diagrams are required? What should be done first, second, and so on? All of these questions should be answered before you begin to construct the project.

Take, for example, a circuit consisting of two lamps connected in parallel and controlled by a single pushbutton. Assume a 12-V dc source and two 12-V lamps are to be used. The best way to begin is to draw a schematic diagram based on the set of instructions given (Figure 7-20A).

Your next step is to make a *wiring number-sequence* chart for the circuit. This will help you make the right wire connections. Begin by assigning a number (start with 1) to each

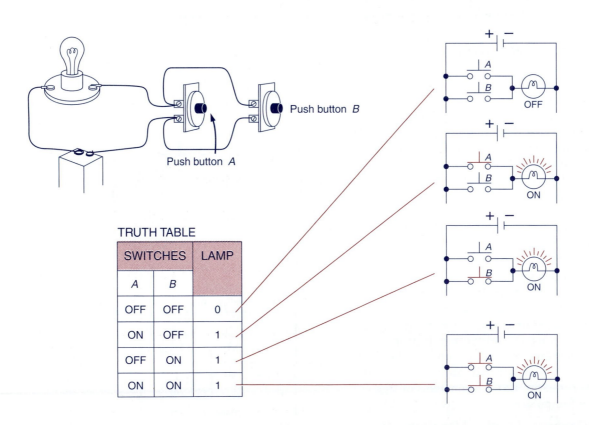

Push button *B*

Push button *A*

TRUTH TABLE

SWITCHES		LAMP
A	*B*	
OFF	OFF	0
ON	OFF	1
OFF	ON	1
ON	ON	1

OR type control circuit *A* or *B* are pushed to switch lamp ON

Figure 7-19 Two push buttons connected in parallel.

Figure 7-20 illustrates the following steps:

Step 1: Draw a schematic diagram based on the set of instructions given.

Step 2: Assign a number to each terminal of each component part.

Group all common terminal numbers together.

Step 3: Number the terminals of each component on the wiring board (numbers assigned are the same as those used on the schematic).

Step 4: Interconnect all common sets of terminals using the minimum amount of wire.

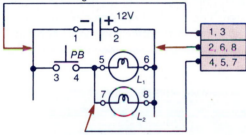

A. Circuit schematic diagram

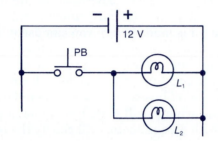

B. Wiring number sequence chart

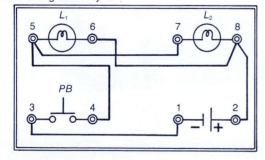

C. Circuit wiring board layout

D. Circuit wiring diagram

Figure 7-20   Planning a wiring project.

terminal of each component part. Next, all common terminal numbers are grouped together. Common terminals represent *common points* in the circuit that are connected together. Record these common number groups in chart form as illustrated in Figure 7-20B.

The final step is to complete a wiring diagram of the circuit. This diagram shows the different components and wires as they would be located in the actual circuit. Note that the component parts in the wiring layout do not appear in the same position as they did in the schematic (Figure 7-20C). The first step in completing the wiring diagram is to number the terminals of each component. The numbers assigned are the same as those used on the schematic.

To complete the wiring diagram, all common sets of terminals are interconnected. The wiring number-sequence chart is used to determine which terminals connect together. Draw wires neatly from point to point using straight lines with right angle turns. Make all connections at terminal screws. Plan the "wire runs" from point to point so that you use the minimum amount of wire (Figure 7-20D).

Once the wiring diagram is completed, the project is ready to be wired. The type of experimental wiring board used to wire the actual project varies from school to school (Figure 7-21). You may want to construct your own wiring board. In general, the actual components are fastened to the top or the bottom of the board, while their schematic symbols are printed on top. Terminal connections can be made by way of binding ports, spring terminals, or clips. The actual component layout on the top of the board can be reproduced for use by students in making wiring diagrams.

7-8 Breadboarding Circuits

Breadboarding refers to the building of a temporary version of a circuit. This technique is used extensively for electronic lab experiments. At times it is also a good idea to build a breadboard version of a circuit before assembling it in permanent form. This gives you a chance to find out how well the circuit works and allows you to make changes before assembling it in permanent form.

The most popular device for the assembly of temporary electronic circuits is the **solderless**

modular-breadboard socket (Figure 7-22). It consists of a plastic block with common multiple-conductor terminal rows. Electronic components such as resistors, capacitors, diodes, transistors, and integrated circuits simply plug in as needed. Jumper wires are used to interconnect parts whose leads are not inserted in the same row of terminals. Special components such as lamps, switches, transformers, potentiometers, and the like can be used with sockets by simply soldering short lengths of wire to their terminals and then inserting them into the board.

If properly used, the breadboard socket can make the assembly of experimental circuits simple and quick. You must understand how your board is made before you can use it. Both horizontal and vertical electrical buses (common connection points) are provided on most boards. Depending on the model of board, both the number of buses and the number of contacts on each bus will vary. If in doubt, a quick check with an ohmmeter will clarify the bus scheme for you. The top and bottom horizontal rows can be used for common ground and ± voltage. The vertical columns of contacts are for component interconnection. Axial component

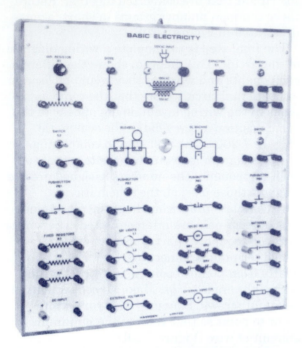

A. Electricity experimental wiring board. (*Courtesy Hampden Ltd.*)

B. Electronics experimental wiring board. (*Photograph supplied by L. J. Technical Systems.*)

Figure 7-21 Typical experimental wiring boards.

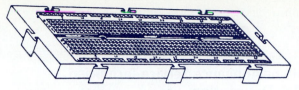

A. Solderless breadboard

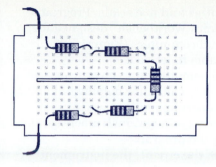

B. Wired board

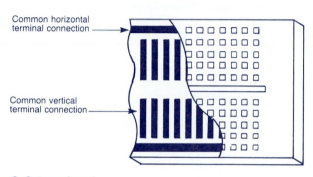

C. Cutaway board

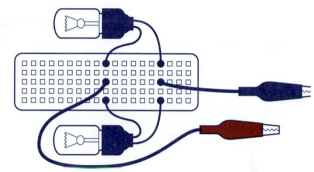

D. Special components added

Figure 7-22 Solderless breadboard socket.

leads and jumper wires should be bent at right angles before being inserted into the socket. Keep jumper wires as short as possible and mounted flush with the top of the board. It is sometimes a good idea to make a pictorial layout diagram of the circuit, from the schematic, prior to wiring the board.

7-9 Computer Simulated Circuits

The computer has fast become a powerful learning tool. Computer programs allow you to experiment with electrical and electronic circuits without using scarce and expensive

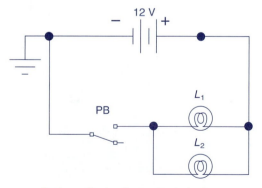

A schematic circuit, constructed using Electronics Workbench software, can be used to simulate real-world circuits.

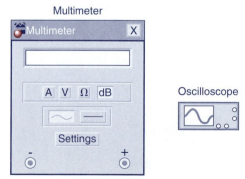

Circuit measurements can be made using simulated test instruments.

Figure 7-23 Computer simulated circuit.

laboratory facilities and materials. Programs such as Electronics Workbench® designed for use with personal computers (Figure 7-23, p. 87) are available to let you:

- Construct a schematic for an electric or electronic circuit on a computer display;
- Simulate the operation of that circuit;
- Display its activity on test instruments contained within the program;
- Print a copy of the circuit, the instrument readings, and a parts list; and
- Troubleshoot the circuit.

HELP WANTED

Home Appliance Service Technician

Job Description

- Diagnosis/repair of electrical cooking, laundry, and refrigeration products.
- Must possess excellent customer relation skills.
- Vocational/technical diploma desired.
- Position requires a minimum of 3 years experience in major home appliance repair.

Review and Applications

Review Questions

1. Name the five component parts of any basic circuit.
2. Draw an acceptable symbol that can be used on a schematic diagram to represent:
   (a) a dc power supply.   (e) a bell.
   (b) an ac voltage source. (f) a buzzer.
   (c) a lamp.   (g) a switch.
   (d) a resistor.   (h) a pushbutton.
3. Describe one advantage and one limitation of the pictorial type of electrical diagram.
4. (a) How are components normally represented in a schematic diagram?
   (b) What makes a schematic diagram easier to read?
5. What is a wiring diagram intended to show?
6. (a) What type of circuits often require the use of a block diagram?
   (b) What does each block in a block diagram represent?
   (c) What do the arrows connecting the blocks indicate?
7. A buzzer is to be operated when either one or the other of two pushbuttons is pressed. What type of electrical connection should be used to connect the two pushbuttons?
8. What connection of two or more pushbuttons can be considered to be an *AND* type of control circuit?
9. Draw the schematic diagram for circuits a–e:
   (a) one lamp controlled by one pushbutton and operated from an ac source.
   (b) two lamps connected in series, controlled by one switch, and operated from a dc source.
   (c) one lamp controlled by two pushbuttons connected in series and operated from an ac source.
   (d) one lamp controlled by two switches connected in parallel and operated from a dc source.
   (e) two lamps connected in parallel, controlled by one pushbutton, and operated from an ac source.

Problems

1. Two identical 12-V lamps are connected in series to a 12-V dc source.
   (a) How many current paths are produced?
   (b) What is the value of the voltage drop across each lamp?
   (c) Comment on the brightness of each lamp (full or dim)? Why?
   (d) Assume that one lamp burns open (or out). What happens to the other lamp as a result of this? Why?
2. Two identical 12-V lamps are connected in parallel to a 12-V dc source.
   (a) How many current paths are produced?
   (b) What is the value of the voltage drop across each lamp?
   (c) Comment on the brightness of each lamp (full or dim)? Why?
   (d) Assume that one lamp burns open (or out). What happens to the other lamp as a result of this? Why?

Critical Thinking

1. A control system calls for a light to come on when switch A and either switch B or switch C, is closed. State the connection of the switches that will accomplish this. Draw a schematic diagram of the circuit.
2. A 12-V pilot light is to be operated from a 120-V source by using a resistor to lower the voltage applied to it. How would the resistor be connected relative to the light?
3. One bulb in a 20-string series connected holiday tree lamp set needs to be replaced. If the lamp set is rated for an input voltage of 120-V ac, what voltage rating of bulb would be required?
4. Why are fuses always connected in series with the circuit they protect?

Portfolio Project

Obtain a wiring diagram of a major electric home appliance such as a washer, dryer, or stove. Produce a schematic diagram using this wiring diagram. With reference to your schematic, prepare a written report on how the different circuits of the appliance operate.

Circuit Challenge

Using a Simulator

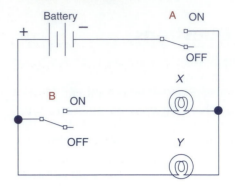

Procedure

(a) Construct the simulated circuit shown, using whatever simulation software package is available to you.

(b) Operate the circuit and construct a truth table showing how the two switches control the two lamps.

Measuring Voltage, Current, and Resistance

Electricity cannot be seen. However, making circuit measurements of voltage, current, or resistance, using a meter, can help in understanding an electric circuit and determining if it is operating properly. This chapter will help you to understand how meters work and how to use them to make basic electrical measurements.

Objectives

After studying this chapter, you should be able to:
1. Compare the operation of analog and digital meters.
2. Correctly read an analog scale and digital meter display.
3. Use a multimeter to measure voltage, current, and resistance.
4. List the safety precautions to be observed when using multimeters.
5. Explain multimeter specifications and special features.

Key Terms

- analog
- digital
- multimeter
- multiplier
- rectified
- signal conditioner
- ac-to-dc converter
- analog-to-digital converter

- digital display
- voltage drop
- shunt
- clamp-on ammeter
- zero-adjust knob
- continuity test
- impedance

- resolution
- LED
- LCD
- auto ranging
- auto polarity
- hold feature
- response time

- Knight Kits
- Elenco Kits
- Marcraft Kits

8-1 Analog and Digital Meters

The three basic electrical test instruments are the voltmeter, ammeter, and ohmmeter. They are used to obtain accurate information about the voltage, current, and resistance of a circuit. Both *analog* and *digital* types are available.

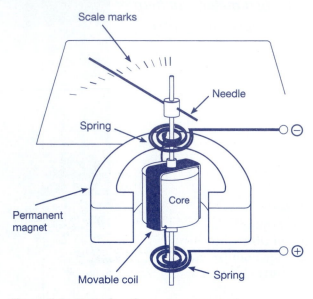

Figure 8-1 Typical analog meter movement.

Analog meters use a needle and mechanical type of meter movement to indicate the measurement. The permanent magnet, moving-coil **galvanometer** is the basis of most analog meters (Figure 8-1). It consists of a moving coil suspended between the poles of a horseshoe magnet. With no current in the coil, no magnetic field is set up and the needle balances, so there is no tension on the spring. This gives a reading of 0. With current applied, the coil becomes an electromagnet. Magnetic forces cause the coil to turn until the forces are balanced with the spring force. This results in a reading other than 0.

The major difference between analog and digital meters is the type of display used. In a digital meter, the meter movement is replaced by an electronic digital display. The internal circuitry of a digital meter is made up of electronic circuits (Figure 8-2). Any quantity that is measured appears as a number on the digital display.

8-2 Multimeter

Often, the voltmeter, ammeter, and ohmmeter are combined in a single instrument called a **multimeter**. The multimeter has become the

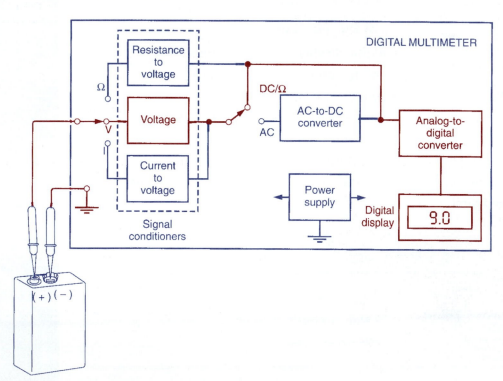

Figure 8-2 Typical internal electronic circuitry of a digital meter.

basic measuring and electronic tool of the electrician and technician. The reasons for this are that it is much easier to carry a single instrument than three separate meters, and it is much cheaper to buy a single multimeter than three single-function meters.

An **analog multimeter** consists of a single meter movement and the associated meter circuitry required of a voltmeter, ammeter, and ohmmeter (Figure 8-3). The analog multimeter is often referred to as a **Volt-Ohm-Milliammeter (VOM).** The technician must make the decision of what measurement is to be taken. A **function** switch selects the type of measurement: voltage, current, or resistance. A **range switch** selects the full-scale range of the measurement. This permits proper selection of the internal circuits so that only one range, of one type of measurement, is selected at any one time. Depending on the design, multimeters often combine the function and range selection into one switch.

The **test leads** and **input jacks** connect the multimeter to the circuit or component you want to measure. Test leads are usually red and black. Normally you put the red test lead in the POS (+) input jack and the black test lead in the NEG (−) or common input jack. Multimeters usually have more than these two input jacks. You use other jacks primarily for measuring higher voltages and current and other special functions.

A typical **digital multimeter (DMM)** is shown in Figure 8-4. The digital readout is the main meter output, indicating the numerical value of the measurement. Like analog

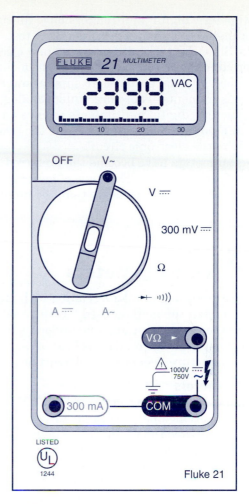

A. Digital multimeter

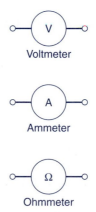

B. Typical meter symbols

Figure 8-4 Digital multimeter. (*Reproduced with permission from the John Fluke Mfg. Co., Inc.*)

Figure 8-3 Analog multimeter.

multimeters, digital multimeters have function and range switches and jacks to accept test leads. Since digital multimeters contain electronic circuits to produce their measurements, they need internal batteries to supply power for all measurements. Therefore, unlike analog multimeters, digital multimeters require an ON/OFF power switch that connects the power supply to the electronic circuits.

Digital meters have become the preferred choice for technicians. This type of meter is easier to read and more accurate than the analog type.

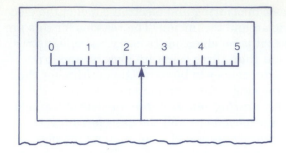

Figure 8-5 Reading a single-range analog meter scale.

8-3 Reading Meters

Before attempting to take a reading with any type of meter, the method used for range selection and measurement must be understood. The format used varies, so it is best to refer to the manufacturer's operating instructions.

Analog Meters   The most accurate reading of an analog meter scale is obtained when the head is positioned perpendicular to the scale and directly over the needle. Some analog meters use a mirror on the scale. Mirror-scale meters are used to prevent errors due to parallax. Parallax occurs when the reader's line of sight on the needle is not exactly at right angles to the plane of the scale. A meter zero-adjust screw on the front of the meter is used to set the meter needle to zero on the scale when no current is flowing.

Reading a *single-range* meter scale is similar to reading a ruler scale (Figure 8-5). Generally, the major divisions are marked and the value of the minor divisions can be easily calculated. The scale in Figure 8-5 is read as follows:

Value of each major division = 1
Value of each minor division = 0.2
∴ Reading = 2.4

Multirange meters are harder to read because one scale is often used with two or more ranges. To read this type of meter, first determine the reading on the scale and then apply the appropriate multiplier or divider as indicated by the range switch (Figure 8-6).

Analog ohmmeter scales are not marked off evenly. Such a meter scale is said to be nonlinear.

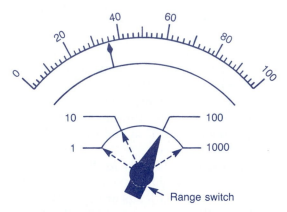

RANGE SWITCH	SCALE READING	X or ÷	CORRECT READING
100	36	X 1	36
1 000	36	X 10	360
10	36	÷ 10	3.6
1	36	÷ 100	0.36

Figure 8-6 Reading a multirange analog meter scale.

In the analog ohmmeter example shown in Figure 8-7, the range-selector switch gives you the multiplier, in this case, $R \times 100$. To obtain the reading of 150 Ω, first accurately read the scale as 1.5 Ω. Then multiply 1.5 by 100, which gives $1.5 \times 100 = 150$ Ω.

The face of an analog VOM has a combination of scales. These scales are generally used to measure a wide range of values of dc voltage and current and ac voltage and resistance. In the analog multimeter example shown in Figure 8-8, the range switch is set to take dc-voltage readings. Assuming that four different voltage measurements were taken (switch positions 1, 2, 3, and 4) and the needle moved to the

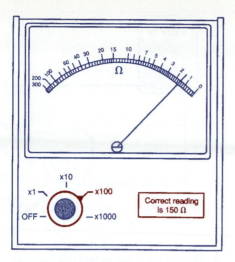

Figure 8-7 Reading an analog ohmmeter.

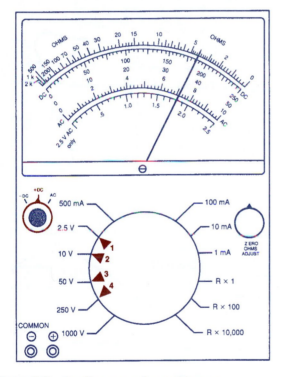

Figure 8-8 Reading an analog multimeter.

same position for each measurement, the voltage readings would be as follows:

Setting 1:   1.85 V
Setting 2:   7.4 V
Setting 3:   37 V
Setting 4:   185 V

Digital Meters   Once you become familiar with them, digital meters are easier to read

than analog meters. Many digital meters are **autoranging**, that is, the meter itself adjusts to the range needed for the specific measurement.

The digital readouts shown in Figure 8-9 are typical readings you might see, depending on the meter being used. Note that the symbol for ohms (Ω) is used for units below 1000, kΩ is used for units above 1000, and MΩ is used for units above 1,000,000. The chart (Figure 8-9B) shows how the digits are arranged for each range. The position of the decimal point and the suffix (m, k, or M) are most important.

The following examples may help your understanding:

125.5 mV = 125.5 millivolts

1.255 V = 1.255 volts
      or 1255 millivolts (multiply by 1000)

9.75 A = 9.75 amps
      or 9750 milliamps (multiply by 1000)

A. Typical resistance readings

Functions	Range	Digital (d) Display
DCV/ACV	300 mV (DCV only) 3 V 30 V 300 V 3000 V	ddd.d mV d.ddd V dd.dd V ddd.d V dddd V
DCA/ACA	300 mA 10 A	ddd.d mA dd.dd A
kΩ	300 ohm 3 kilohm 30 kilohm 300 kilohm 3000 kilohm 30 megohm	ddd.d Ω d.ddd kΩ dd.dd kΩ ddd.d kΩ dddd kΩ dd.dd MΩ

B. Typical method of displaying digits for each range

Figure 8-9 Reading a digital meter.

220.6 Ω = 220.6 ohms
      or 0.2206 kilohms (divide by 1000)

30.5 k Ω = 30.5 kilohms
      or 30,500 ohms (multiply by 1000)

0.750 kΩ = 0.750 kilohms
      or 750 ohms (multiply by 1000)

8-4 Measuring Voltage

A *voltmeter* is used to measure electromotive force (emf) or voltage (potential difference) in an electric circuit. Voltmeters can be used for checking the available voltage at an outlet receptacle, across battery terminals, and between two points in a circuit (Figure 8-10).

The voltage measuring range of the basic analog meter movement or *galvanometer* is limited to the millivolt range. This is due to the delicate nature of the coil and springs. To extend the voltage range, a resistor with a *high-resistance value* is connected in series with the meter movement. This resistor is called a **multiplier** because it multiplies the range of the meter movement. The voltage range can be changed by changing the value of the multiplier resistor (Figure 8-11). The higher the resistance value of the multiplier, the higher the voltage range of the meter.

In order to measure an ac voltage with an analog meter movement, the incoming voltage first must be **rectified** or converted to direct current. A semiconductor diode is usually used for this purpose. The diode is connected in series with the ac voltage. It allows the current to pass through it in only one direction, thereby converting the ac voltage to be measured into pulsating dc voltage that can be measured by the meter movement (Figure 8-12).

A block diagram of an ac digital voltmeter is shown in Figure 8-13. The test probes measure the ac voltage and bring it into the *voltage-conditioner* circuit. The voltage conditioner provides attenuation or amplification of the voltage signal to reduce it or to increase it to a level that the measuring circuits are designed to handle. The signal is then fed to an **ac-to-dc converter** circuit which converts the voltage signal from ac to dc. The *analog-to-digital (A/D) converter* circuit then accepts this voltage and changes it to a digital code that represents

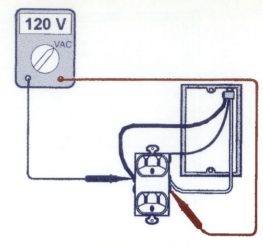

A. Measuring outlet receptacle voltage

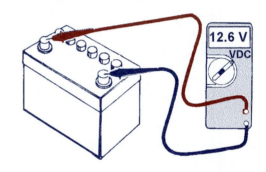

B. Measuring battery voltage

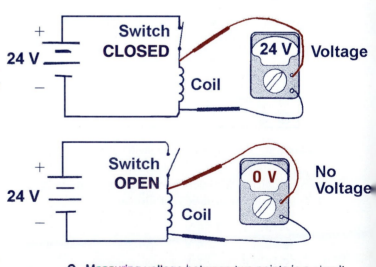

C. Measuring voltage between two points in a circuit

Figure 8-10   Checking available voltage with a voltmeter.

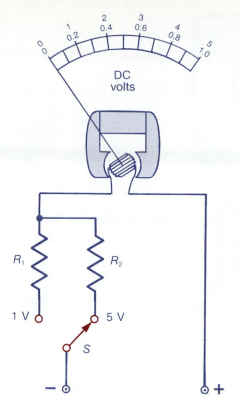

Figure 8-11 Multirange analog dc voltmeter.

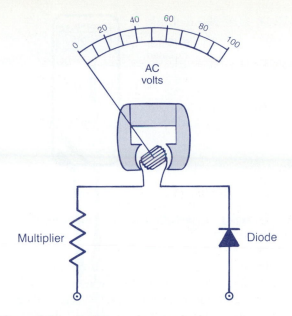

Figure 8-12 Analog ac voltmeter circuit.

the magnitude of the voltage. The digital code is then used to generate the numerical digits that show the measured value in the *digital display.* If the input voltage to be measured is dc, the ac-to-dc converter circuit is bypassed

and the signal is fed directly from the voltage conditioner to the analog-to-digital converter.

The voltmeter must be connected in parallel, or across the load or circuit. It has a high resistance and taps off a small amount of current to operate the meter circuit. If it were connected in series, this high resistance would reduce circuit current, and the meter would give an incorrect reading.

Either dc or ac types of voltmeters are selected according to the type of voltage to be

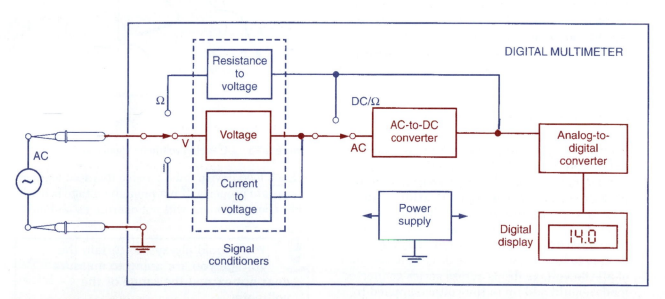

Figure 8-13 AC digital voltmeter.

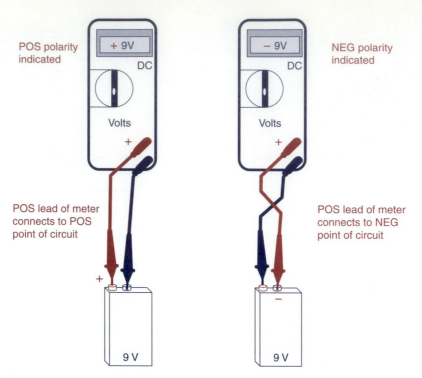

POS polarity indicated

NEG polarity indicated

POS lead of meter connects to POS point of circuit

POS lead of meter connects to NEG point of circuit

Figure 8-14   DC digital voltmeter polarity identification.

measured. Digital voltmeters will automatically indicate the correct polarity of a dc voltage measurement (Figure 8-14). When the positive lead of the meter is connected to the positive point of the circuit, the meter will indicate a positive (+) polarity on the digital display. When the positive lead of the meter is connected to the negative point of the circuit, the meter will indicate a negative (−) polarity on the digital display.

Analog voltmeters must be connected with the correct polarity. The negative (−) voltmeter lead connects to the negative (−) side of the circuit and the positive (+) voltmeter lead connects to the positive (+) side of the circuit (Figure 8-15). If the leads are reversed, the needle will move off scale to the left of zero and the movement may be damaged.

Voltage drop is the loss of voltage caused by the flow of current through a resistance. Increases in resistance increase the voltage drop. To check the voltage drop, use a voltmeter connected between the points where the voltage drop is to be measured. In a dc circuit the total of all the voltage drops across series-connected loads should add up to the voltage applied to the circuit (Figure 8-16).

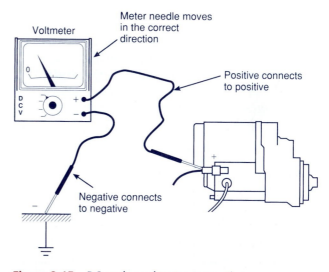

Figure 8-15   DC analog voltmeter connection.

Each load device must receive its rated voltage to operate properly. If not enough voltage is available, the component will not operate as it should.

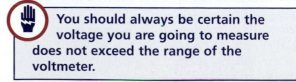

> You should always be certain the voltage you are going to measure does not exceed the range of the voltmeter.

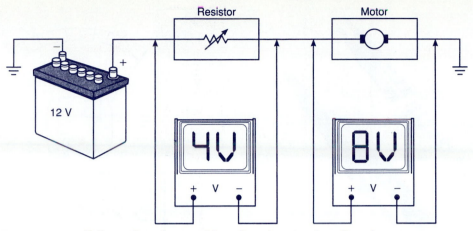

Battery voltage rise = resistor voltage drop + motor voltage drop

Figure 8-16 Measuring voltage drops across loads.

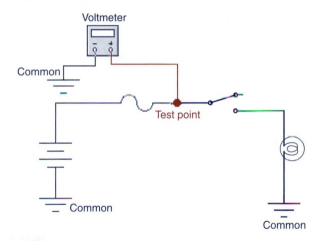

Figure 8-17 Measuring a voltage to common or circuit ground.

This may be difficult if the voltage is unknown. If such is the case, you should always start with the highest range. Attempting to measure a voltage higher than the voltmeter can handle may cause damage to the voltmeter.

At times you may be required to measure a voltage from a specific point in the circuit to ground or a common reference point (Figure 8-17). To do this, first connect the black, common test probe of the voltmeter to the circuit ground or common. Then connect the red test probe to whatever point in the circuit you want to measure.

The voltage tester is a special type of voltmeter commonly used by electricians (Figure 8-18). Its rugged construction makes it ideally

suited for on-the-job voltage measurements. The voltage tester indicates the approximate level of the voltage present and not the exact value. It is primarily used to test for the presence or absence of a voltage.

8-5 Measuring Current

An **ammeter** is used to measure the amount of current flowing in a circuit. Ammeters measure current flow in amperes. For ranges less than 1 ampere, milliammeters and microammeters are used to measure current (Figure 8-19).

To extend the current range of a galvonometer, a resistor with a low resistance value is connected in parallel with the meter movement.

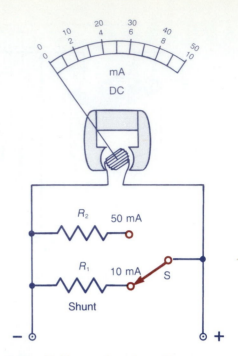

Figure 8-20 Multirange dc analog milliammeter.

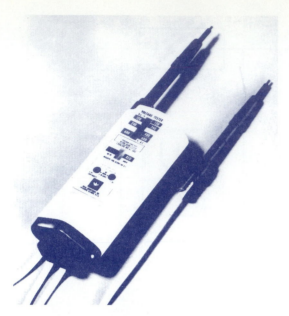

Figure 8-18 Voltage tester. (*Courtesy of IDI Electric (Canada) Ltd. and Idea Industries Inc.*)

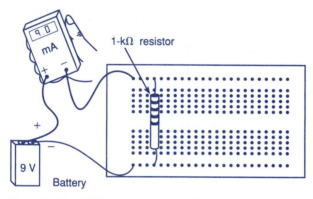

Figure 8-19 Millammeter connected to measure current.

This resistor is called a *shunt*. Its purpose is to provide an alternate low resistance path around the meter movement. This results in most of the current being measured flowing through the shunt and only a small amount through the actual meter movement. The lower the resistance value of the shunt, the higher the current range of the meter (Figure 8-20).

A block diagram of a dc digital milliammeter is shown in Figure 8-21. The test probes are connected in series with the circuit so that the circuit is completed through the meter. The current-conditioner circuit changes the current to be measured into a voltage by passing it through a shunt resistor, of low resistance value, and measuring the voltage drop across the resistor. The range switch determines the resistor used for each range. When the proper current range is selected, the proper resistance is selected so that the voltage out of the current conditioner will be within the range required by the A/D converter. The A/D converter then accepts this voltage and changes it to a digital code that is used to drive the digital display and to show the measured current value. A high current range, such as 10 amperes, usually is measured using a special input jack to which a special high-power resistor is connected.

Current measurements are more difficult to make than voltage measurements because to measure current, you must physically break into the current path. Once you break into the current path, you can insert your ammeter (Figure 8-22). The resistance of the circuit controls the current flow through the meter. The ammeter itself has a very low internal resistance value to allow normal current to flow during the metering process. As a result, accidentally connecting it in parallel will cause the meter to draw a high current, which could ruin the meter.

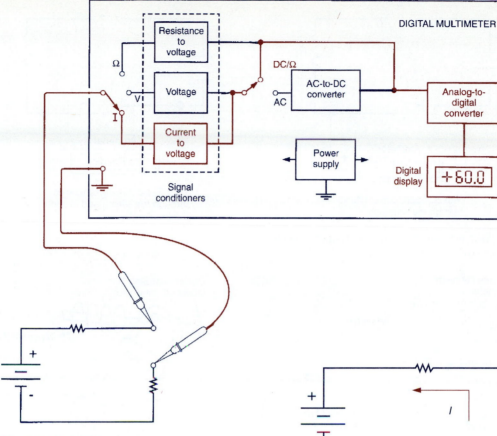

Figure 8-21 DC digital milliammeter.

> ✋ **The ammeter must always be connected in series with the circuit; the circuit current must flow through the meter.**

A switch can be used to measure current without disturbing the circuit. Figure 8-23 illustrates how this is accomplished. When the switch is closed, the ammeter registers zero current because the current flows through the switch instead of the ammeter. When you open the switch, the current flows through the ammeter. You can now measure the value of the current. As soon as you complete your measurement, you can close the switch and remove your ammeter and the circuit will continue to operate normally.

Analog dc ammeters must be connected into the circuit, making sure you observe the correct polarity. That is, the positive and negative leads of the meter must connect to the positive

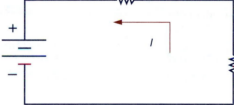

A. Original circuit

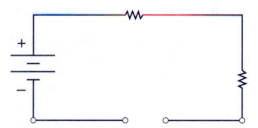

B. Circuit with a break

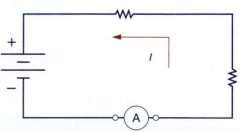

C. Circuit with the ammeter connected

Figure 8-22 Current measurement.

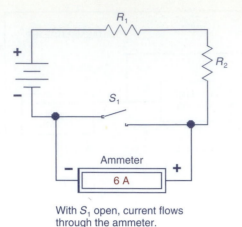

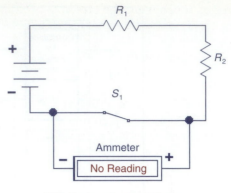

With S_1 open, current flows through the ammeter.

With S_1 closed, current flows through the switch and around the ammeter.

Figure 8-23   Switching an ammeter in and out of a circuit.

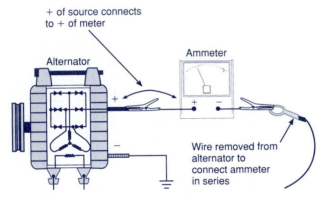

Figure 8-24   DC analog ammeter connected to measure current flow from an automobile alternator.

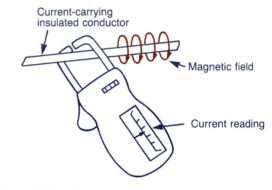

Figure 8-25   Clamp-on ammeter.

and negative leads of the circuit, respectively (Figure 8-24). Digital ammeters, like digital voltmeters, automatically give the correct polarity of the dc current source. When taking an ammeter reading, make sure the meter-range switch is set high enough for the current being measured. Never place an ammeter in a circuit that has already caused a fuse to blow, as the resulting high current flow may damage the meter.

To avoid the necessity of breaking the circuit, technicians often use a ***clamp-on ammeter*** (Figure 8-25) to measure higher currents in the ampere range. It is more convenient to use than is a standard ammeter because the circuit does not have to be opened to take a current reading. This instrument clamps around the circuit conductor and indicates the current flow in the

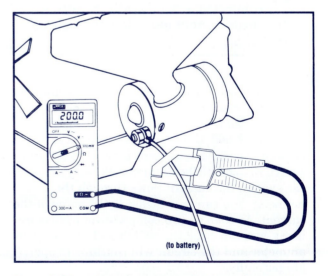

Figure 8-26   Clamp-on current probe. (Reproduced with permission from the John Fluke Mfg. Co., Inc.)

conductor by measuring the strength of the magnetic field about the current-carrying conductor.

Accessories to some digital multimeters include a clamp-on **Hall-effect probe** that can be plugged into the meter to read ac or dc currents. Figure 8-26, shows a digital multimeter used in conjunction with a clamp-on current probe to measure the starting current of an automobile starter motor.

8-6 Measuring Resistance

The ohmmeter is used to measure the amount of electrical resistance offered by a complete circuit or a circuit component. Ohmmeters measure resistance in ohms. For ranges greater than 1000 ohms, kilohms (kΩ) and megohms (MΩ) are used to measure resistance.

The basic analog ohmmeter circuit is made up of a meter movement, a dry cell, a fixed resistor, and a variable resistor, which are all connected in series (Figure 8-27). The principle of its operation is a simple one. Current is made to flow through an unknown resistance. Its resistance is then determined by measuring the amount of resultant current flow. According to

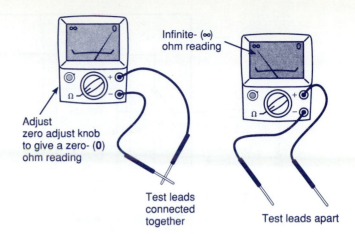

Figure 8-28 Zero-setting an analog ohmmeter.

Ohm's law, the amount of current flow will be inversely proportional to the resistance value. The amount of current measured by the meter is an indication of the unknown resistance. Therefore, the scale of the meter movement can be marked off in ohms.

Before taking a reading with an analog-type ohmmeter, the meter scale must be set for zero. To set the pointer for a reading of 0 Ω, connect the two test leads of the ohmmeter together and adjust the **zero-adjust knob** either way (Figure 8-28). This procedure is not required with most digital-type ohmmeters, as they usually have automatic zeroing.

The block diagram of a digital ohmmeter is shown in Figure 8-29. Typically, the resistance to a voltage signal-conditioner circuit uses a ratio method to measure the value of the unknown resistor. This voltage ratio is made by placing the unknown resistor in series with an internal known reference resistor and voltage source. This voltage ratio is then applied to the A/D converter circuit. Extra circuits are designed into the A/D converter so it can be used to measure the voltage ratio and calculate the unknown resistance. The reference resistance and voltage value are changed for different resistance ranges.

Both digital and analog ohmmeters are self-powered by a battery located within the meter itself and can be damaged if connected to a live circuit. To make an out-of-circuit resistance measurement (Figure 8-30) with an ohmmeter, simply connect the leads of the ohmmeter

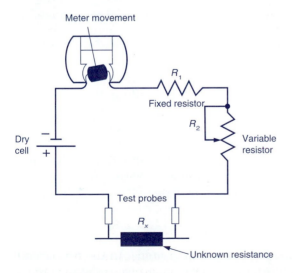

Figure 8-27 Basic analog ohmmeter circuit.

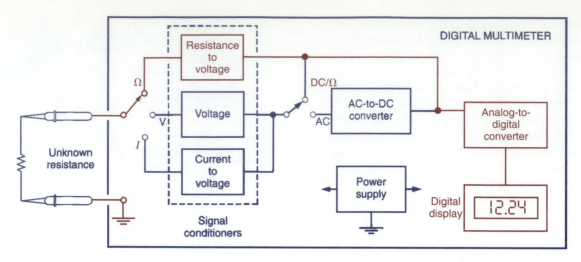

Figure 8-29 Digital ohmmeter.

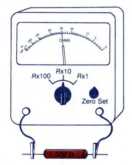

A. Out-of-circuit measurement

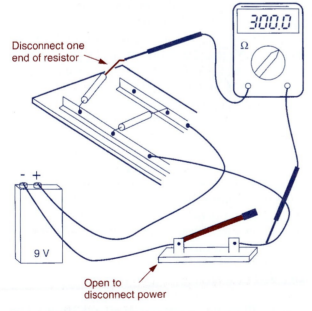

B. In-circuit measurement

Figure 8-30 Measuring resistance.

across the component (similar to a voltmeter connection) and set the meter to the appropriate resistance range. When using an ohmmeter to measure a component in a circuit, two precautions must be observed:

1. Turn the power source OFF. Disconnect the equipment from the power source if possible.
2. Disconnect, if possible, one lead end of the component to open any parallel paths so only the resistance of the single component is measured.

In addition to measuring resistance, an ohmmeter is used to make ***continuity tests***. These show whether or not there is a closed electrical path from one test point to another (Figure 8-31). When making a continuity test, the ohmmeter is set to its lowest resistance range. A complete conducting path is indicated by a low resistance reading. The value of the reading is not important as long as it is low. An open, or incomplete, path in the circuit is indicated by an infinite resistance reading. If you have continuity, but the resistance reading is high, you know that an excessive amount of resistance exists in the circuit.

Some digital multimeters indicate continuity with an audible tone or beeper. A continuous tone sounds if the resistance between the terminals is less than approximately 150 ohms (Figure 8-32).

A continuity test is also very useful in checking for shorts and grounds. In the test circuit shown in Figure 8-33, a zero-resistance reading is indicated across the two terminals of the coil.

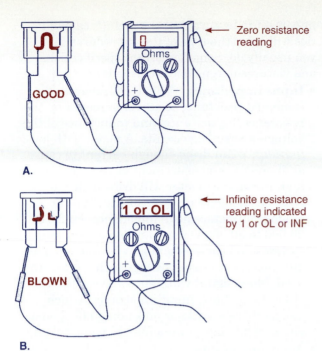

A.

Zero resistance reading

GOOD

Infinite resistance reading indicated by 1 or OL or INF

1 or OL

B.

BLOWN

Figure 8-31   Using an ohmmeter to test a fuse for continuity.

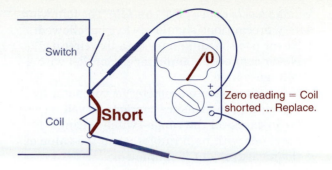

Switch

Coil

Short

Zero reading = Coil shorted ... Replace.

Figure 8-33   Continuity check for short circuit..

This means the wire insulation has become defective, causing a **short circuit** between the wire turns of the coil, and the coil should be replaced.

In the test circuit shown in Figure 8-34, a zero-resistance reading is indicated between the transformer winding and the core. This means the metal frame of the transformer will be at **ground potential (grounded)**. The wire insulation has become defective at some point, and since this condition might be dangerous or cause the circuit to work improperly, the transformer should be replaced.

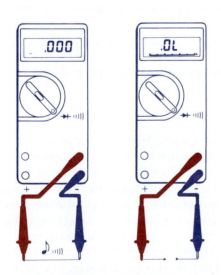

.000   **.OL**

Figure 8-32   Continuity beeper. (*Reproduced with permission from the John Fluke Mfg. Co., Inc.*)

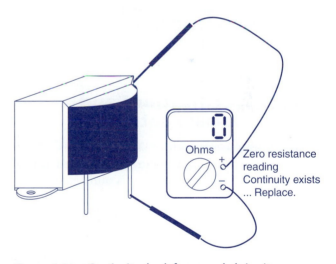

Ohms

Zero resistance reading Continuity exists ... Replace.

Figure 8-34   Continuity check for grounded circuit.

8-7 Meter Safety

Using a meter safely is an important technical skill. When measuring electricity, you are dealing with an invisible and often lethal force.

Voltage levels above 30 V can kill! The following safety precautions should always be followed:

- Never use the ohmmeter on a live circuit.
- Never connect an ammeter in parallel with a voltage source.
- Never overload an ammeter or voltmeter by attempting to measure currents or voltages far in excess of the range-switch setting. Start out with high ranges when the value of the measurement is unknown.
- Make certain that any terminals you are measuring across are not accidentally shorted together, or to ground, by the test leads (Figure 8-35).
- Never measure unknown voltages in HIGH VOLTAGE circuits. Refer to the technical information or the manufacturer for a reference voltage before proceeding.
- Check meter test leads for frayed or broken insulation before working with them.
- Avoid touching the bare metal clips or tips of test probes.
- Whenever possible, remove voltage before connecting meter test leads into the circuit.
- When connecting meter test leads to live circuits, work with one hand at your side to lessen the danger of accidental shock.
- To lessen the danger of accidental shock, disconnect meter test leads immediately after the test is completed.

8-8 Multimeter Specifications and Special Features

Some multimeters feature other functions beyond the basic voltage, current, and resistance measurement. When selecting a multimeter it is important to be sure that the meter's capabilities will cover the type of test procedures that you usually do. Some important specifications and features to consider are as follows:

- **Input impedance** This term refers to the combined resistance to current created by the resistance, capacitance, and inductance of the voltage-measuring circuits. A meter with low *impedance* can draw enough current to cause an inaccurate measurement of voltage drop. A high-impedance meter will draw little current, ensuring accurate readings. Typical analog voltmeters have impedance values of 20,000 to 30,000 Ω per volt. Some electronic systems and components may be damaged or give inaccurate results when such meters are used. Most digital voltmeters have a 10 MΩ (10 million ohms) input impedance, which means there is little or no loading effect on a circuit while measurements are being taken.

- **Accuracy and resolution** Simple accuracy specifications are given as a plus/minus percentage of full scale. *Resolution* refers to the smallest numerical value that can be read on the display of a digital meter. Factors that determine resolution are the number of digits displayed and the number of ranges available for each function. Digital meter error is less than $\pm1\%$ of the actual count, plus one or more counts of the least-significant digit. Analog meter error is in the range of ±1–3% of the actual count. The most common DMM display is called a $3\frac{1}{2}$-digit display, because it shows three full digits (0–9), preceded by a so-called half-digit. At this position, the meter can only display 0 or 1. Thus, the largest number that can be displayed on a $3\frac{1}{2}$-digit meter is 1999. This also is called a 2000-count display, because the actual count may be 0000 to 1999.

- **Battery life** Whereas the analog multimeter simply draws power from the circuit under test for voltage and current measurements, the digital multimeter requires a battery to operate. Battery life (rated in hours) is a major consideration in selecting a digital multimeter. Either a *light-emitting diode (LED)* or *liquid-crystal display (LCD)* can be used in a digital multimeter. The LCD-type digital meter requires less power and, therefore, has a much longer battery-life rating.

- **Protection** Meter-protection circuits prevent damage to the meter in cases of acciden-

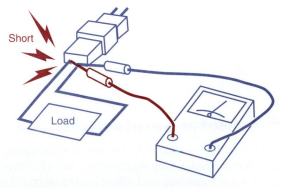

Figure 8-35   Do not short-circuit test leads.

tal overloading. The resistance and current measuring circuitry is usually protected by a fuse connected in series with the input lead. Safety precautions, involving the possible misuse of meters, have lead to the use of a second high-energy fuse within the meter circuit. This fuse is a high-voltage rated element (around 600 V) and is usually two or three times higher in fusing current than the standard user-replaceable fuse.

- **Combination digital and analog display** Analog meters are particularly suited for trend observation, as in slowly changing voltage levels. Some digital multimeters use a combination display that includes a bar graph to provide simulation of an analog needle (Figure 8-36) for watching changing signals or for adjusting circuits.
- **Auto ranging** *Automatic ranging* automatically adjusts the meter's measuring circuits to the correct voltage, current, or resistance ranges.
- **Auto polarity** With the *automatic polarity* feature a + or − activated on the digital display indicates the polarity of dc measurements and eliminates the need for reversing leads.
- **Hold feature** Many digital multimeters have a **HOLD** button that captures a reading and displays it from memory even after the probe has been removed from the circuit. This is particularly useful when making measurements in a confined area where you cannot quite read the meter.
- **Response time** Response time is the number of seconds a digital multimeter requires for its electronic circuits to settle to their rated accuracy.

- **Diode test** This test is used to check the forward and reverse bias of a semiconductor junction. Typically, when the diode is connected in *forward bias* the meter displays the *forward voltage drop* and beeps briefly (Figure 8-37). When connected in *reverse bias* or open circuit the meter displays *OL*. If the diode is shorted, the meter displays zero and emits a continuous tone.
- **Averaging or true rms** A *true rms* meter responds to the effective heating value of an

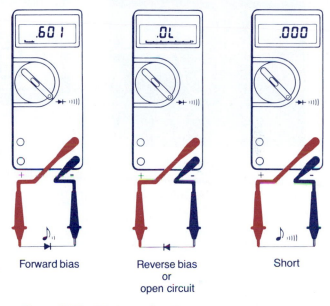

Figure 8-37 Diode test function.

ac waveform. Meters that have rectifier-type circuits have scales that are calibrated in RMS values for ac measurements, but actually are measuring the average value of the input voltage and are depending on the voltage to be a sine wave. Where the ac signal approximates a pure sine wave there will be little or no difference in the two readings (Figure 8-38).

- **Multifunction digital meters** The integrated-circuit (IC) chip revolution has helped to combine the capabilities of other test instruments into a multifunction digital meter. Volts, ohms, and amps are the most often used functions; however, the multifunction digital meter allows the reading of dBm, frequency, capacitance, logic level, and temperature measurements, as well.

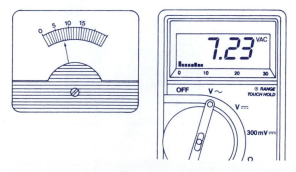

Figure 8-36 Bar graph display used to simulate an analog needle. (*Reproduced with permission from the John Fluke Mfg. Co., Inc.*)

A. Sine wave: true rms meter and averaging meter give same reading

B. Non-sine wave: true rms meter reads higher than the averaging meter

Figure 8-38 Averaging or true rms. (*Reproduced with permission from the John Fluke Mfg. Co., Inc. © John Fluke Mfg. Co., Inc.*)

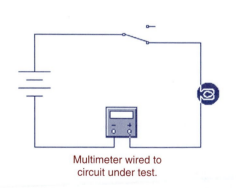

Multimeter wired to
circuit under test.

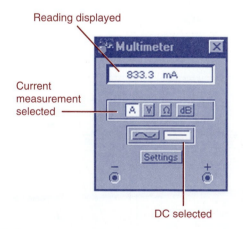

Multimeter set to measure
dc current and display reading.

Figure 8-39 Virtual multimeter.

8-9 Virtual Meters

Virtual instrumentation is based on using the computer and associated software to create the measuring instrument. In principle, this allows users to create instruments at lower costs that are capable of higher performance.

Figure 8-39 shows a typical virtual multimeter used as one of the simulated instruments in the Electronics Workbench® simulation package. This multimeter can be used to measure ac or dc voltage or current, or resistance, between any two points in the circuit. The multimeter is auto-ranging, so a measurement range does not need to be specified. Its internal resistance and current are preset to near-ideal values, which can be changed by adjusting the settings.

HELP WANTED

Meter Reader
Job Description
- Recording the amount of electricity used by residential and commercial customers.
- Taking readings of meter dials in route books or handheld microcomputers.
- Inspecting meters and connections for defects, damages, or unauthorized connections.
- Electrical background is desirable.

CHAPTER 8
Review and Applications

$\cdots\cdots\cdots\cdots\cdots\cdots\cdots\cdots\cdots\cdots\cdots\cdots\cdots\cdots\cdots\cdots\cdots\cdots\cdots\cdots\cdots\cdots\cdots\cdots\cdots\cdots$

Related Formulas

1V = 1000 mV	1kΩ = 1000 Ω
1A = 1000 mA	1MΩ = 1,000,000 Ω

Review Questions

1. Compare the way measurements are indicated on analog and digital meters.
2. What are the three basic metering functions that can be performed using a multimeter?
3. State the purpose of each of the following parts of a multimeter front-panel switching arrangement:
   (a) function switch
   (b) range switch
   (c) jacks
4. Give two reasons why digital meters are preferred over analog types.
5. In what way is an analog ohmmeter scale different from that of a voltmeter or ammeter scale?
6. How is the voltage measuring range of the basic analog meter movement extended?
7. What is the purpose of the diode rectifier used with ac analog voltmeters?
8. Explain the function of each of the following circuit blocks of an ac digital voltmeter:
   (a) voltage conditioner circuit
   (b) ac-to-dc converter circuit
   (c) A/D converter circuit
   (d) digital display
9. Define what is meant by voltage drop in a circuit.
10. State one advantage and one limitation of a voltage tester.
11. How is the current-measuring range of the basic analog meter movement extended?
12. Explain the function of the current-conditioner circuit of a digital milliammeter.

13. Compare the internal resistance value and circuit connection of a voltmeter and ammeter.
14. The voltage and current of a lamp connected to a dc source are to be measured. Draw the schematic diagram of this circuit showing the normal connection of an ammeter and voltmeter.
15. When measuring voltages and currents of unknown levels, what meter range switch setting should be used?
16. Why do technicians prefer to use the clamp-on type of ammeter when checking circuit currents?
17. Explain the principle of operation for the basic analog ohmmeter circuit.
18. How is the pointer of an analog-type ohmmeter set for zero?
19. Explain the operation of the resistance to voltage signal-conditioner circuit of a typical digital ohmmeter.
20. When using an ohmmeter to measure a component in a circuit, what two precautions must be observed?
21. Outline five safety precautions that should be observed when operating multimeters.
22. Give a brief explanation of each of the following multimeter specifications or functions:
    (a) input impedance
    (b) accuracy
    (c) resolution
    (d) auto ranging
    (e) auto polarity
    (f) hold
    (g) response time
    (h) diode test

Problems

1. Convert each of the following multimeter readings:
   (a) 340 mV to volts
   (b) 0.75 V to millivolts
   (c) 2 A to milliamps
   (d) 1950 mA to amps
   (e) 7.5 Ω to kilohms
   (f) 2.2 kΩ to ohms
   (g) 1.5 kΩ to ohms

2. (a) A dc analog voltmeter is connected incorrectly, so that the positive lead of the meter connects to the negative side of the circuit. What will happen?
   (b) If a dc digital multimeter were connected in the same manner, what would happen?

3. An ohmmeter is connected across the two leads of a switch to check its continuity out of circuit. If the switch is turned on and off manually, what ohmmeter readings should be indicated for each of the following conditions: switch in "ON" and switch in "OFF" position?
   (a) Switch operating properly.
   (b) Switch is open circuited.
   (c) Switch is short-circuited.

Critical Thinking

1. When troubleshooting a circuit you measure a dc voltage of 4.7 V at a test point where the schematic indicates it should be 5 V. What should you infer from this reading? Why?

2. While measuring current in a circuit the two test leads are accidentally shorted together. What effect will this have on the circuit under test? Why?

Portfolio Project

Purchase a multimeter in kit form. Assemble the meter and test it for proper operation. Give a demonstration showing how the meter is used to measure voltage, current, and resistance.

Circuit Challenge

Using a Simulator

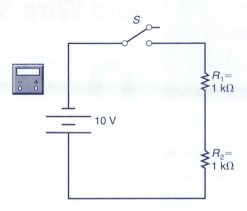

Procedure

(a) Construct the simulated circuit shown, using whatever simulation software package is available to you.
(b) How can the current be measured without disconnecting any wires?
(c) Use the multimeter to measure the value of the current flow.
(d) Use the multimeter to measure the value of the voltage across R_1.
(e) Use the multimeter to measure the value of the voltage across R_2.
(f) Use the multimeter to measure the total resistance of the circuit.

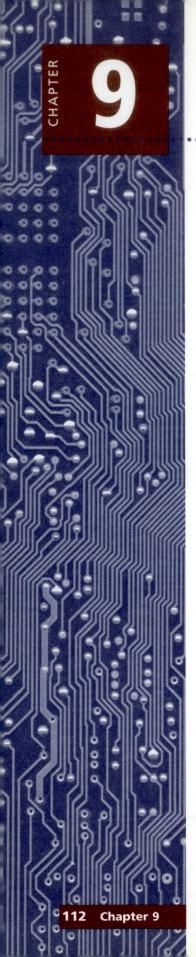

Circuit Conductors and Wire Sizes

O *ne of the most important parts of any electrical system is the conductor that connects all the components. The conductors must be able to deliver the necessary energy on a continuous basis, without overheating or causing unacceptable voltage drops. In this chapter we will study the different conductor forms and their application.*

Objectives

After studying this chapter, you should be able to:
1. Identify uses for different conductor forms.
2. Properly select wire-insulating materials.
3. Compare the AWG size and diameter of conductors.
4. List the factors that determine a conductor's ampacity rating.
5. Identify the factors that contribute to the resistance value of a conductor.
6. Calculate line-voltage drop and line power loss.

Key Terms

- ground
- positive terminal
- negative terminal
- cable
- cord
- printed circuit board

- thermoplastic
- neoprene
- enamel insulation
- American Wire Gauge (AWG)
- circular mil (CM) area

- ampacity
- National Electrical Code (NEC)
- ambient temperature
- line voltage drop
- line power loss

RESEARCH INTERNET SITES

- Interstate Wire
- AVO (megger)
- General Cable

Circuit Conductors and Wire Sizes

9-1 Conductor Forms

Conductors carry current to and from each component that is being operated on the circuit. The best electric conductors are metals. Automotive circuits are often referred to as **single-wire**, or ground-return, systems (Figure 9-1). This is because the metal frame of the vehicle is used as one of the wires in the circuit. The hot, or insulated, side of the circuit connects to the positive of the power source. The other half of the path for current flow is the vehicle's engine, frame, and body. This is called the *ground* side of the circuit. An insulated cable connects the battery's *positive (+) terminal* to the vehicle's loads. A second insulated cable connects the battery's *negative (−) terminal* to the engine or frame.

The most popular metal used as a conductor is copper. In addition to its low resistance, copper is easy to work with and makes excellent electric connections to terminal screws. In electronic circuits it can be easily soldered, ensuring a secure electric connection.

Copper conductors used for wiring circuits can be made in the form of wire, cable, cord, or printed circuit boards. A solid wire is a single conductor covered by some form of insulation. A stranded wire is a single conductor made up of many small-diameter wires running alongside each other. The purpose of stranding conductors is to provide increased flexibility. Both solid and stranded **hook-up wire** is used to complete circuits within electrical and electronic equipment (Figure 9-2).

The term *cable* can refer to a larger stranded, insulated wire or two or more separately insulated wires that are assembled within a com-

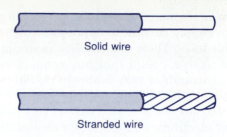

Figure 9-2   Solid and stranded hook-up wire.

mon covering (Figure 9-3). Large single-conductor cables are used for high-current car battery terminal connections. Multi-conductor, non-metallic sheathed cable is used for permanent wiring circuits in a house.

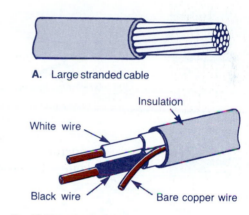

A.   Large stranded cable

B.   Multiconductor cable

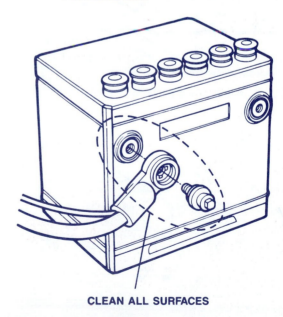

CLEAN ALL SURFACES

C.   Battery cable terminal connection

Figure 9-3   Cable forms.

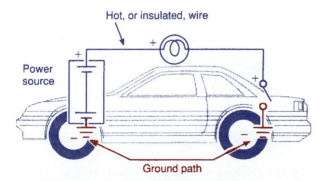

Figure 9-1   Single-wire automotive circuit.

Cord is the name given to very flexible cables used to supply current to appliances and portable tools. Their construction is similar to that of a cable, except that cord conductors are made of strands of very fine wire that are twisted together (Figure 9-4).

Printed circuits are used extensively in low-current electronic equipment for connecting components together. A *printed circuit board* consists of conducting paths of thin copper strips etched or printed on a flat, insulated plate (Figure 9-5). Printed circuit boards also provide a convenient base for mounting small electronic components along with the means for interconnecting them.

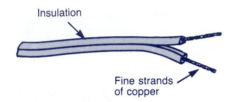

Figure 9-4   Lamp cord.

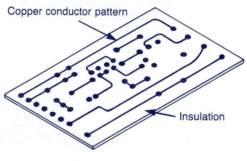

A. Conducting paths

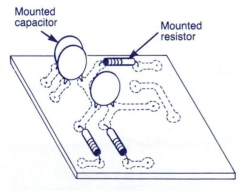

B. Mounted components

Figure 9-5   Printed circuit board.

9-2 Conductor Insulation

Conductors must normally be insulated to make them useful and safe. Insulation of conductors is accomplished by coating or wrapping them with various materials having high resistance. An ordinary ohmmeter cannot be used for measuring insulation resistance. To test adequately for insulation breakdown, it is necessary to use a much higher voltage than is furnished by an ohmmeter's battery. Special high-voltage insulation resistance testers such as a "megger" or megohmmeter are used (Figure 9-6) to measure insulation resistance.

Some of the factors that must be considered when selecting a wire insulation are circuit voltages, surrounding temperature conditions, moisture, and conductor flexibility. Since insulation is expensive, it is desirable to use only the minimum required by the application, and therefore, a wide variety of insulated conductors are available.

Thermoplastic is one of the most commonly used insulators for residential and industrial wiring. Regular thermoplastic is an excellent insulator, but it is sensitive to extremes of temperature. Type TW thermoplastic is a weatherproof type while type THW is weatherproof and heat resistant.

Neoprene is a special rubber-type insulation used for power line cords on heat-producing applications such as kettles and frypans. The abbreviation for heat-proof neoprene insulations is HPN.

Enamel insulation is used on magnetic wire. It is applied by repeatedly passing the bare wire through a solution of hot enamel to

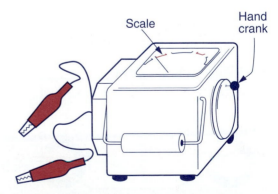

Figure 9-6   High-voltage megger used to check insulation resistance.

form a thin, tough coating that is a varnish-like plastic compound. It is called magnetic wire because it is used for winding the coils in electromagnets, solenoids, transformers, motors, and generators (Figure 9-7).

Conductors are often identified by a color code that is part of the insulation. Wire colors refer to the base color of the insulation and sometimes to a stripe, hash mark, or dot in a contrasting color (Figure 9-8).

9-3 Wire Sizes

The size of a solid wire is determined by its diameter. For convenience, wire sizes are usually referred to by an equivalent gauge number rather than by the actual diameter.

The *American Wire Gauge (AWG) table* is the standard used and consists of forty wire sizes

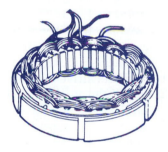

Figure 9-7   Magnetic wire—used for winding electromagnets. (*Courtesy Chrysler Corporation.*)

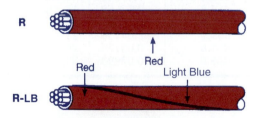

WIRING COLOR CODE

Wire colors are indicated by an alphabetical code.

B = Black	LB = Light Blue	R = Red
BR = Brown	LG = Light Green	V = Violet
G = Green	O = Orange	W = White
GR = Gray	P = Pink	Y = Yellow

The first letter indicates the basic wire color and the second letter indicates the color of the stripe.

Figure 9-8   Typical wire insulation color code.

ranging from AWG 40 (the smallest) all the way up to AWG 0000 (Figure 9-9). Note that the larger the gauge number, the smaller the actual diameter of the conductor.

In the AWG system, the cross-sectional area of wire is measured in circular mils. A mil is 0.001 in. A circular mil is the area of a circle 1 mil (0.001 in.) in diameter. The *circular mil (CM)* area of a round wire is found simply by squaring the diameter of the wire expressed in mils. In the metric system, wire sizes are based on cross-sectional area in square millimeters (mm^2). For stranded wire, the size in circular mils is the total cross-sectional area, and is equal to the area (in circular mils) of one strand multiplied by the number of strands.

When comparing conductor sizes, *remember that the external diameter of a wire, including its insulation, has nothing to do with its wire size or current capacity.* Thick insulation will make a small-gauge wire look much larger. Heavy insulation is required for high-voltage wires and lighter insulation is used for low-voltage circuits.

The AWG number for an unknown solid conductor can be determined by measuring it with a wire gauge (Figure 9-10). To gauge the wire, the insulation is carefully removed from one end. Then the bare end is inserted into the smallest slot in which it will fit without using force. The number stamped below the slot is the AWG of the wire.

9-4 Conductor Ampacity

The *ampacity* of a conductor refers to the maximum amount of current it can safely carry without becoming overheated. This current rating, or ampacity, is determined by its material,

AWG (B and S) Gauge	Standard Metric Size (mm)	Diameter in mils	Cross-sectional Area		Ohms per 1000 ft at 20°C (68°F)	Lbs per 1000 ft	Ft per lb
			Circular mils	Square inches			
0000	11.8	460.0	211,600	0.1662	0.04901	640.5	1.561
000	10.0	409.6	167,800	0.1318	0.06180	507.9	1.968
00	9.0	364.8	133,100	0.1045	0.07793	402.8	2.482
0	8.0	324.9	105,500	0.08289	0.09827	319.5	3.130
1	7.1	289.3	83,690	0.06573	0.1239	253.3	3.947
2	6.3	257.6	66,370	0.05213	0.1563	200.9	4.977
3	5.6	229.4	52,640	0.04134	0.1970	159.3	6.276
4	5.0	204.3	41,740	0.03278	0.2485	126.4	7.914
5	4.5	181.9	33,100	0.02600	0.3133	100.2	9.980
6	4.0	162.0	26,250	0.02062	0.3951	79.46	12.58
7	3.55	144.3	20,820	0.01635	0.4982	63.02	15.87
8	3.15	128.5	16,510	0.01297	0.6282	49.98	20.01
9	2.80	114.4	13,090	0.01028	0.7921	39.63	25.23
10	2.50	101.9	10,380	0.008155	0.9989	31.43	31.82
11	2.24	90.74	8234	0.006467	1.260	24.92	40.12
12	2.00	80.81	6530	0.005129	1.588	19.77	50.59
13	1.80	71.96	5178	0.004067	2.003	15.68	63.80
14	1.60	64.08	4107	0.003225	2.525	12.43	80.44
15	1.40	57.07	3257	0.002558	3.184	9.858	101.4
16	1.25	50.82	2583	0.002028	4.016	7.818	127.9
17	1.12	45.26	2048	0.001609	5.064	6.200	161.3
18	1.00	40.30	1624	0.001276	6.385	4.917	203.4
19	0.90	35.89	1288	0.001012	8.051	3.899	256.5
20	0.80	31.96	1022	0.0008023	10.15	3.092	323.4
21	0.71	28.46	810.1	0.0006363	12.80	2.452	407.8
22	0.63	25.35	642.4	0.0005046	16.14	1.945	514.2
23	0.56	22.57	509.5	0.0004002	20.36	1.542	648.4
24	0.50	20.10	404.0	0.0003173	25.67	1.223	817.7
25	0.45	17.90	320.4	0.0002517	32.37	0.9699	1031.0
26	0.40	15.94	254.1	0.0001996	40.81	0.7692	1300
27	0.355	14.20	201.5	0.0001583	51.47	0.6100	1639
28	0.315	12.64	159.8	0.0001255	64.90	0.4837	2067
29	0.280	11.26	126.7	0.00009953	81.83	0.3836	2607
30	0.250	10.03	100.5	0.00007894	103.2	0.3042	3287
31	0.224	8.928	79.70	0.00006260	130.1	0.2413	4145
32	0.200	7.950	63.21	0.00004964	164.1	0.1913	5227
33	0.180	7.080	50.13	0.00003937	206.9	0.1517	6591
34	0.160	6.305	39.75	0.00003122	260.9	0.1203	8310
35	0.140	5.615	31.52	0.00002476	329.0	0.09542	10,480
36	0.125	5.000	25.00	0.00001964	414.8	0.07568	13,210
37	0.112	4.453	19.83	0.00001557	523.1	0.06001	16,660
38	0.100	3.965	15.72	0.00001235	659.6	0.04759	21,010
39	0.090	3.531	12.47	0.000009793	831.8	0.03774	26,500
40	0.080	3.145	9.888	0.000007766	1049.0	0.02993	33,410

A. AWG table for standard annealed copper

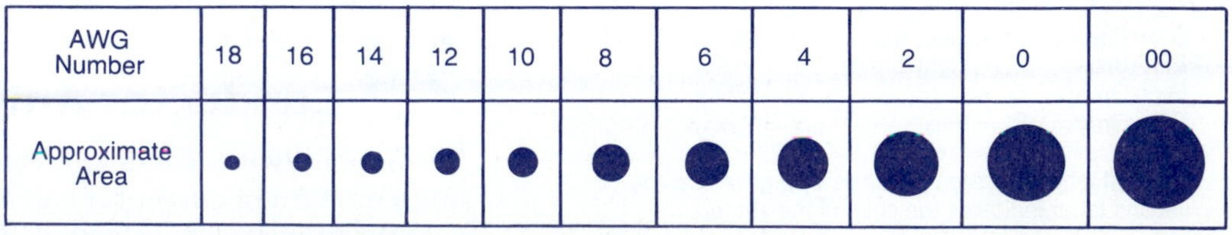

AWG Number	18	16	14	12	10	8	6	4	2	0	00
Approximate Area	•	•	•	•	•	•	•	•	•	•	•

B. Wire sizes

Figure 9-9 AWG wire sizes

A. Wire gauge

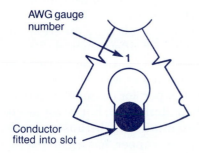

AWG gauge number

Conductor fitted into slot

B. Using the gauge

Figure 9-10   AWG wire gauge.

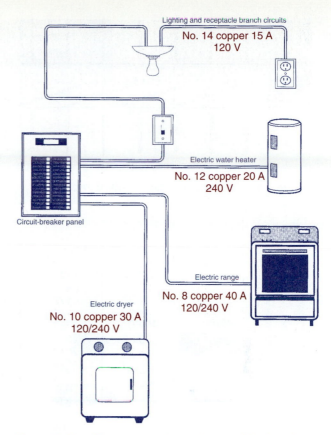

Lighting and receptacle branch circuits
No. 14 copper 15 A
120 V

Circuit-breaker panel

Electric water heater
No. 12 copper 20 A
240 V

Electric range
No. 8 copper 40 A
120/240 V

Electric dryer
No. 10 copper 30 A
120/240 V

Figure 9-11   Common conductor sizes used for branch circuits in a home.

gauge size, type of insulation, and conditions under which it is installed. Copper is a better conductor than aluminum and so it can carry more current for a given gauge. Similarly, the smaller the gauge number, the larger the conductor and the more current it can carry. A conductor with a heat-resistant insulation will have a higher ampacity rating than one of equivalent size with a lower insulator temperature rating. In addition, conductors that are run singly in free air will have a higher ampacity rating than a similar conductor that is enclosed with other conductors in a cable or conduit.

The **National Electrical Code** contains tables that list the ampacity for the approved types of conductor size, insulation, and operating conditions. These tables are a practical source of information that should be referred to for specific circuit installations. Figure 9-11 shows some of the common copper conductor sizes used for branch circuits in a home.

Manufacturers usually list the correct wire and fuse sizes in the installation instructions. However, in many cases, the person responsible for the installation must calculate the wire and fuse size. The National Electrical Code governs the types and sizes of wire that can be used for a particular application and a certain amper-

age. The correct wire and fuse size is important to the safety, life, and efficiency of any equipment.

9-5 Conductor Resistance

When a conductor carries a current, the conductor's resistance causes the conversion, into heat, of a portion of the electric energy being transmitted (Figure 9-12). The resistance of a length of wire is determined by its length, diameter, operating temperature, physical condition, and type of material used (Figure 9-13).

The resistance of a wire varies directly with its length. If two wires are of the same material and diameter, the longer wire has more resistance than the shorter wire. The greater the diameter of the wire, the lower its resistance. Large conductors allow more current flow with less voltage. If two wires are of the same material and length, the thinner wire will have more resistance than the thicker wire.

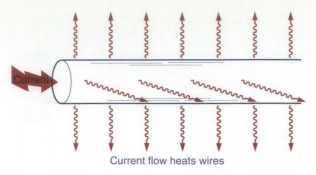

Figure 9-12 Conductor resistance heats wires.

The resistance of a conductor varies with temperature. For copper, the higher the temperature, the higher the resistance. The two factors that determine the operating temperature of a conductor are the temperature of the surrounding air space (*ambient temperature*) and the amount of current flow through the conductor.

Partially cut or nicked wire will act like smaller gauge wire, with high resistance in the damaged area. Broken strands in the wire, poor splices, and loose or corroded connections also increase resistance.

Different materials have different atomic structures, which affect their ability to conduct electrons. Materials with many free electrons are good conductors with low resistance to current flow. Copper, for example, is a better conductor than aluminum, but is not as good as silver. Aluminum wire, because it is not as good a conductor as copper, has an ampacity approximately equal to that of copper wire two sizes smaller. For example, aluminum wire No. 12 has about the same ampacity as copper wire No. 14.

9-6 Voltage Drop and Power Loss

The *resistance* of the wire conductors of a circuit is low when compared to that of the load. In most circuits, the conductors are treated as being ideal or perfect conductors of electricity. As a result, they are said to have zero resistance (Figure 9-14). In this case, the voltage value of the source is the same as that across the load. In other words, no voltage is lost in the line.

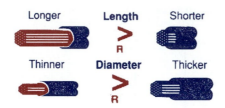

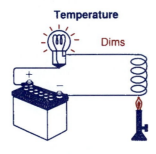

Figure 9-13 Factors affecting wire resistance.

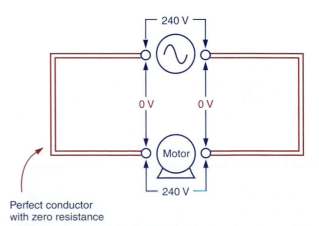

Figure 9-14 Zero-line voltage drop.

In some circuits, the resistance of the conductors is important and must be taken into account. This is often the case where the load is located some distance from the voltage source. In this type of circuit, the voltage at the load can be much less than what appears at the energy source. The **line voltage drop** can be easily measured when the equipment is operating by reading the voltage at the supply and subtracting it from that of the voltage read at the equipment. If we read 240 V at the supply and 230 V at the equipment when it is operating, then there is a voltage drop of 10 V in the circuit (Figure 9-15).

Circuit faults such as broken strands in a cable or poor high-resistance connections also cause higher than normal drops in a line. To check for excessive line voltage drop, the voltage across the conductor is measured when the equipment is operating (Figure 9-16).

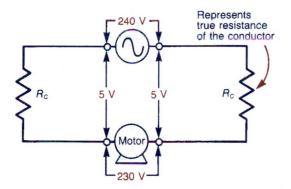

Figure 9-15 Measuring line voltage drop.

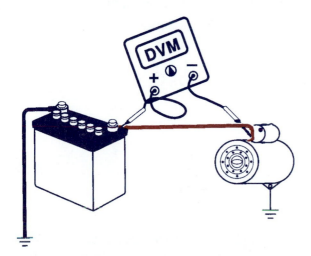

Figure 9-16 Testing for excessive conductor voltage drop.

It is always desirable to keep the line voltage drop as low as possible. Any voltage that drops between the supply and the equipment is lost to the equipment. If the voltage drop is large enough, it will seriously affect the operation of the equipment.

The actual voltage drop in a length of wire is equal to the product of the current and resistance of the wire:

$$V_{\text{drop}} = I \times R_{\text{wire}}$$

where:

V = voltage drop in volts (V)
I = current in amperes (A)
R = resistance in ohms (Ω)

Line voltage drops are kept low by keeping the resistance of the line wires low. The resistance of a length of wire decreases as its diameter increases and increases as its length increases.

Excessive line voltage drop can be caused by using wires that are too small in diameter for the circuit's current requirements and length. When sizing circuit conductors for extremely long circuits the expected voltage drops are estimated before the installation. If necessary, wire diameter sizes are increased above the rated current capacity that is required for the circuit to keep the line voltage loss within acceptable limits.

Current flow through a wire also causes a **power loss** due to the conductor's resistance. Power loss in a wire is equal to the square of the current multiplied by the resistance of the wire:

$$P = I^2 \times R_{\text{wire}}$$

where:

P = power in watts (W)
I = current in amperes (A)
R = resistance in ohms (Ω)

Naturally, it is desirable to hold line power loss to a minimum. The larger the conductors are, the smaller the power or I^2R losses will be.

A compromise is generally reached in which the voltage drop will be held within acceptable limits and in which there is an economic balance between the cost of the conductors and the cost of the power lost.

HELP WANTED

Electrical Line Worker Apprentice

Job Description

- Build and maintain power lines.
- Apprenticeship lasts four years and includes 700 hours of paid related classroom instruction.
- High school diploma with credits in math, physics, and electricity preferred.
- Good physical condition and no fear of heights.
- Must be willing to work out of doors in inclement weather when repairs are required.

Review and Applications

Related Formulas

Line voltage drop = Supply voltage − Load voltage

$$V_{drop} = I_{wire} \times R_{wire}$$

$$P_{loss} = I^2_{wire} \times R_{wire}$$

Review Questions

1. Why are automotive circuits often referred to as single-wire systems?
2. Name four common conductor forms and state one practical application for each.
3. What are four important factors to be considered when selecting the type of wire insulation to be used for a particular job?
4. What is the relationship between the AWG number and the diameter of the wire?
5. (a) Define the term *conductor ampacity*.
   (b) List four factors that determine the ampacity of a conductor.
6. Compare the ampacity of a copper and aluminum conductor of the same gauge size.
7. State the AWG size of copper wire commonly used for each of the following house circuits:
   (a) 120/240 V, 30 A electric dryer circuit
   (b) 120/240 V, 40 A electric range circuit
   (c) 120 V, 15 A electric lighting and receptacle branch circuit
   (d) 240 V, 20 A electric water heater circuit
8. State the effect (increase or decrease) of each of the following on the resistance value of a circuit conductor:
   (a) increasing the length of the conductor
   (b) decreasing the diameter of the conductor
   (c) increasing the operating temperature of the conductor
   (d) using an aluminum conductor in place of a copper one
   (e) using a partially cut, damaged conductor
9. (a) What is meant by the term *line voltage drop?*
   (b) Under what condition is the line voltage drop considered to be zero?

Problems

1. As a rule, the voltage drop in a wire or cable should not exceed 5% of the supply voltage. Assuming the supply voltage feeding a load is 120 V and the voltage measured at the load is 116 V calculate:
   (a) the amount of line voltage drop.
   (b) the maximum line voltage drop acceptable based on the 5% rule.
   (c) the percent of the supply voltage loss in this circuit.
2. (a) What is the resistance of 500 ft. of No. 10 AWG solid copper wire? (Refer to the AWG table.)
   (b) Calculate the line voltage drop if the overall length of wire in (a) is used in a circuit that draws 25 A of current.
   (c) Calculate the line power loss of the circuit.

Critical Thinking

1. The cable running from the battery to the starter motor of a car is incorrectly replaced with one that is $\frac{1}{2}$ the diameter called for. Explain how this could cause each of the following to occur:
   (a) Excessive heat generated in the cable on starting.
   (b) Failures, at times, to turn engine over at a fast enough speed for starting.

Portfolio Project

Construct an experimental circuit containing a 12-V dc source and light, with two No. 18 copper line wires, one 1 ft. in length and the other 2 ft. in length.

(i) Using a digital multimeter, record the mV voltage drop across each line wire. Comment on how length effects the amount of line voltage drop.

(ii) Increase the line current by connecting a second light in parallel. Repeat the measurements of the line voltage drops. Comment on how the increased current effects the amount of line voltage drop.

(iii) Change the size of the line wires to No. 12 copper and repeat the measurements of line voltage drop. Comment on how the increase in the diameter of the line wires effects the amount of line voltage drop.

Circuit Challenge

Using a Simulator

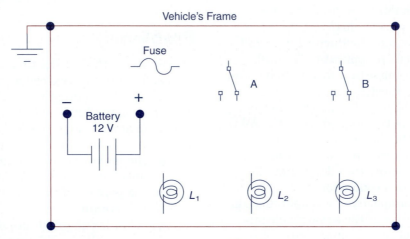

Procedure

Complete the wiring for the automotive circuit shown using whatever simulation software package is available to you. The circuit is to be wired so that:

- The fuse is connected to protect all components of the circuit.
- Switch A is connected to control L_1 and L_2.
- Switch B is connected to control L_3.

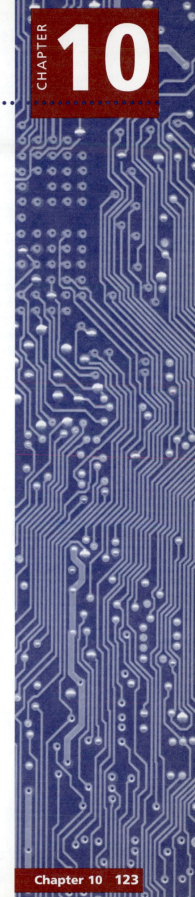

Resistors

esistors are components that are specifically designed to have a certain amount of resistance. The principal applications of resistors are to generate heat, limit current, and divide voltage. This chapter discusses the different types of resistors and the ways in which they are used.

Objectives

After studying this chapter, you should be able to:

1. Explain the different ways in which resistors are used.
2. State the ways in which resistors are rated.
3. Use the resistor color-code to determine resistance.
4. Calculate the total resistance of different resistor configurations.
5. Make resistance measurements.
6. Discuss how resistors are used as voltage and current dividers.

Key Terms

- resistance wire
- resistor
- percentage tolerance
- wire-wound resistor
- power resistor
- fusible resistor
- carbon-composition resistor
- film-type resistor

- resistor network
- fixed resistor
- precision resistor
- adjustable resistor
- variable resistor
- rheostat
- potentiometer
- trimmer potentiometer

- linear taper
- nonlinear taper
- resistor color code
- four-band resistor
- five-band resistor
- resistor failure rate
- voltage divider
- current divider

RESEARCH INTERNET SITES

- **Ohmite**
- **Nicrom**
- **Precision Resistor**

10-1 Resistance Wire

A *resistance wire* is used to produce heat for heating with electricity (Figure 10-1). The most popular type of resistance wire is made of a high-resistance alloy ckel and chromium and is referred to a *ome wire*. This wire is used for the hea nents in stoves, dryers, toasters, and o ating appliances.

When a voltage is ed to the heating element, the high resistance of the wire converts most of the electrical energy into heat energy. In tubular elements, the wire carrying current is enclosed in a tubing with a powdered mineral insulation. The powder insulation insulates the wire from the tubing as well as seals the wire off from contact with the air. This seal prevents oxidation and prolongs the life of the element.

A heated automotive rear-window system uses an electric-resistor grid baked on to the inside surface of the glass to form a heating element (Figure 10-2). Current flowing through the grid produces heat, which is used to clear fog, ice, and snow from the window. Cleaning the inside rear glass should be done carefully to avoid scratching the grid material and causing an open in the circuit.

10-2 Resistors

No electronic system could function without resistors! *Resistors* are used in a circuit to adjust and set voltage and current levels. You can make a simple resistor by drawing a line with a lead pencil on a sheet of paper (Figure 10-3). The resistance of the line or points along it can be measured using an ohmmeter. Set the ohmmeter to its highest resistance scale to measure this resistance. You will find that the resistance varies directly with the length of the path and inversely with its cross-sectional area.

Resistors are rated in three ways (Figure 10-4). The first rating is **resistance**, measured in ohms (Ω). It is hard to manufacture a resistor with an exact number of ohms of resistance; therefore, most resistors carry a **percentage tolerance** or **accuracy rating**. Electric current passing through a resistor causes it to heat and if the temperature rises too high, the material of the resistor may burn out. Resistors are, therefore, rated for power in watts (W), and

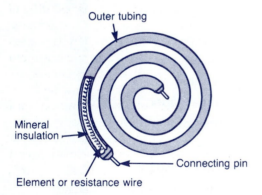

Figure 10-1   Resistance wire heating element for a stove.

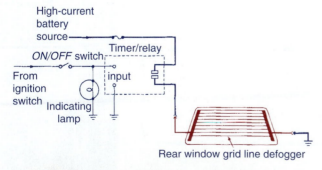

Figure 10-2   Typical automotive rear-defogger circuit.

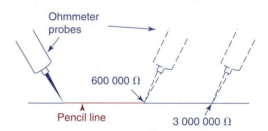

Figure 10-3   Pencil-lead resistor.

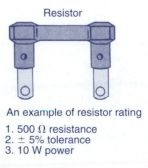

An example of resistor rating

1. 500 Ω resistance
2. ± 5% tolerance
3. 10 W power

Figure 10-4   Resistor ratings

resistors of any one ohm rating can usually be obtained in various wattage ratings. The product of the voltage across a resistor multiplied by the current through it must not exceed the wattage rating of the resistor, or the resistor will overheat and probably be damaged. The larger the physical size of the resistor, the more heat it can safely dissipate, thus, the greater the power rating.

All electrical devices can be separated into two groups: active and passive. **Active** devices are capable of delivering power, while **passive** devices are capable of receiving and possibly storing power. An ordinary resistor is classified as a passive device because all it does is dissipate power; it cannot generate power. A battery, on the other hand, is an active device because it can generate power.

10-3 Types of Resistors

Resistors can be classified according to their construction. **Wire-wound resistors** are made by wrapping high-resistance wire around an insulated cylinder (Figure 10-5). The smaller the wire diameter and the longer the wire, the higher the resistance. This type of resistor is expensive to manufacture. They are generally used in circuits that carry high currents or in circuits where accurate resistance values are required. Large wire-wound resistors are called **power resistors** and range in size from $1/2$ watt to tens or even hundreds of watts. Special wire wound *fusible resistors* are designed to burn open easily when their power rating is exceeded. They serve the dual functions of a fuse and resistor to limit the current.

Carbon-composition resistors are made from a paste consisting of carbon and a filler material (Figure 10-6). The resistance of a carbon resistor is determined by the amount of

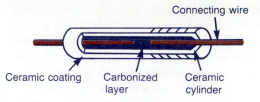

Figure 10-6 Carbon-composition resistor.

carbon used in making the resistor. The resistor element is enclosed in a plastic case for insulation and mechanical strength. Joined to the two ends of the carbon resistance element are metal caps with leads for soldering the connections into a circuit. Carbon-composition resistors are inexpensive and, at one time, were the most common type used in electronics. Generally, they cannot handle large currents and their actual value of resistance can vary as much as 20 percent from their rated value.

Presently, the most popular type of resistor is the **film-type resistor** (Figure 10-7). In these devices, a resistance film is deposited on a nonconductive rod. Then the value of resistance is set by cutting a spiral groove through the film. The length and width of the groove determines the resistance value. These resistors are not cylindrical. Instead, they look like tiny bones. There are two kinds: **carbon-film** types and **metal-film** types. Their advantages are more precise resistance values and lower cost.

A **chip resistor** is a tiny ceramic block-shaped resistor. These have a thick carbon film that is deposited on the ceramic chip. Wraparound metal end terminals are attached for easy **surface mounting** to a printed circuit board. Power dissipation is typically $1/8$ to $1/4$ W with tolerance rating of $\pm 1\%$ or $\pm 5\%$.

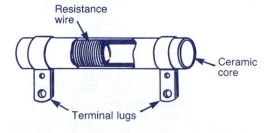

Figure 10-5 Wire-wound resistor.

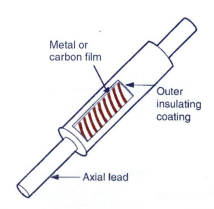

Figure 10-7 Film resistor.

Resistor networks consist of several resistors integrated into a single IC. This reduces the time required to assemble resistors on a printed circuit board as well as reducing the space these resistors take up. Resistor networks are widely used in circuits in which there are numerous requirements for large numbers of identical resistors. They are packaged in a single in-line package (SIP) or dual in-line packages (DIP) (Figure 10-8).

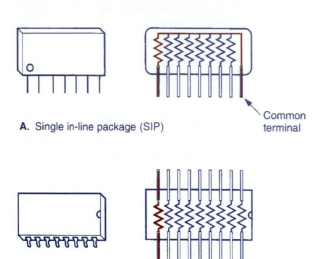

A. Single in-line package (SIP)

Common terminal

B. Dual in-line package (DIP)

Figure 10-8 Resistor networks.

A second way in which resistors are classified is in terms of how they function. A **fixed resistor** has a single value of resistance (Figure 10-9). The three types of fixed resistors are: general purpose resistors, power resistors, and precision resistors. **Precision resistors** are usually made of metal film material and have a ±1% percent or better tolerance.

Sometimes requirements dictate that the resistor value be fixed after you assemble the circuit. An **adjustable resistor** (Figure 10-10)

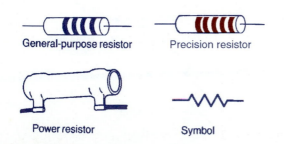

General-purpose resistor

Precision resistor

Power resistor

Symbol

Figure 10-9 Fixed resistors.

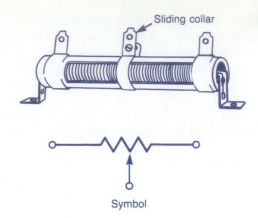

Sliding collar

Symbol

Figure 10-10 Adjustable resistor.

is one designed to provide for a range of different resistance values. This resistor contains a sliding contact that can be positioned and secured to provide different resistance values up to the maximum value of the resistor. It is, however, **not** designed to be continuously variable.

A **variable resistor** (Figure 10-11) is one designed to provide for continuous adjustment of resistance. Variable resistors have a resistive body and a wiper. The wiper slides on the resistive body, changing the length of the resistive material between one end of the device and the wiper. Since resistance depends directly on

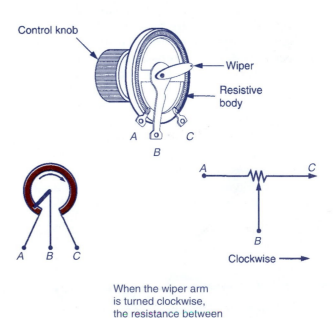

Control knob

Wiper

Resistive body

A   C

B

A   C

B

Clockwise ⟶

When the wiper arm is turned clockwise, the resistance between B and A increases and the resistance between B and C decreases.

Figure 10-11 Variable resistor.

length, increasing the length of the resistive material between the end of the resistor and the wiper makes the resistance higher.

10-4 Rheostats and Potentiometers

Variable resistors are of two different types: *rheostat* and *potentiometer (pot)*. A rheostat is a variable resistor connected using only two of its terminals. The rheostat is used to control current by varying the resistance in a circuit. Common examples of rheostats are auto dashboard light dimmer, model car speed controller, and video game paddle control (Figure 10-12).

Rheostats are usually used in higher power-level applications to control current levels.

A *potentiometer (pot)* is a variable resistor that makes use of all three of its terminals. The potentiometer is used to control *voltage*. Potentiometers are generally low-power variable resistors used to adjust the level of an ac or dc voltage. The two fixed maximum-resistance leads are connected across the voltage source and the variable wiper arm lead provides a voltage that varies from zero to maximum (Figure 10-13). Potentiometers also are used as volume and tone controls on audio equipment and are available in different sizes and shapes (Figure 10-14, p. 128).

Trimmer (trim) potentiometers are used when the ohmic value of a resistor is set at the time a circuit is manufactured and tested. They are usually a miniature size and mounted on a printed circuit board (Figure 10-15). Often they are used to fine-tune or calibrate a circuit. Unlike a typical potentiometer that rotates just short of one complete revolution, some **trim** pots are what we call *multiturn pots*. A ten-turn trim pot, for example, must be rotated completely ten times in order for the wiper to move from one end of the resistive element

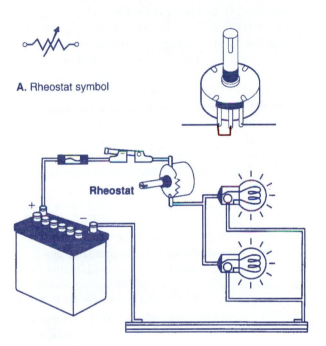

A. Rheostat symbol

B. Auto dashboard light dimmer

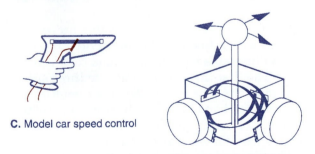

C. Model car speed control

D. Video game paddle control

Figure 10-12 Rheostat as a variable current control.

A. Potentiometer symbol

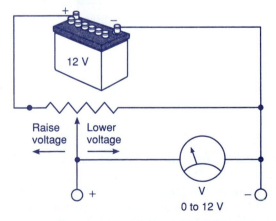

B. Variable dc control circuit

Figure 10-13 Potentiometer as a variable dc voltage control.

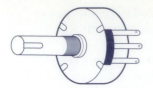

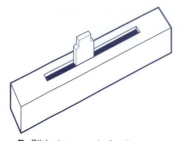

A. Rotary-type control pot

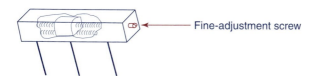

B. Slide-type control pot

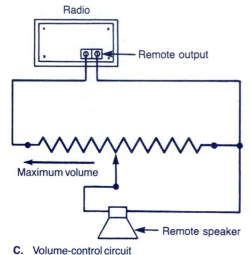

Radio

Remote output

Maximum volume

Remote speaker

C. Volume-control circuit

Figure 10-14 Potentiometer as a volume control.

Fine-adjustment screw

Figure 10-15 Trimmer potentiometer.

to the other. This allows for very precise adjustments.

The *taper* of a potentiometer refers to the way its resistance varies with shaft rotation. With a *linear taper*, the resistance changes in direct proportion to rotation. The resistance for a *nonlinear taper* changes more gradually at one end, with bigger changes at the opposite

end. The effect is accomplished by different densities of the resistance element in one half than in the other. Audio volume controls are of the nonlinear type, allowing greater control of loudness at normal or low listening levels.

10-5 Resistor Identification

Some resistors are large enough to have their resistance value, tolerance, and power rating stamped on them. For small fixed resistors, a system of *color coding* is often used to identify the resistance value and tolerance.

Figure 10-16 illustrates how a *four-band resistor color* code is read. Each color has the numerical value as indicated. Color bands are always read from the end that has the band closest to it. The first two bands identify the first and second digits of the resistance value, and the third band indicates the number of zeros. An exception to this is when the third band is either silver or gold, which indicates a 0.01 to 0.1 multiplier, respectively. The fourth band is always either silver or gold, and in this position silver indicates a ±10% tolerance and gold indicates ±5% tolerance. Where no fourth band is present, the resistor to tolerance is ±20%.

Some carbon composition resistors have a fifth color band, usually separated from the others, which indicates the *failure rate*. Figure 10-17 lists the colors and corresponding percentage of failures or changes in resistance that can be expected during a 1000-hour period of normal operation. If, for example, the fifth band is yellow, the reliability rate is 0.001% per 1000-hour period. This means that 1 out of a group of 100,000 resistors that all have that color band is likely to fail in the first 1000-hour period.

Precision 1% and 2% film resistors are labeled with a *five-band color code*. Figure 10-18 illustrates how the five-band color code is read. The first three bands indicate three significant digits. The fourth band is the multiplier and the fifth band indicates the percentage of tolerance. Color and multiplier values are the same as that used for the four-band color code.

The physical size of a resistor has nothing to do with its resistance. A very small resistor can have a very low or a very high resistance. The physical size of a resistor is, however, an indication of its wattage rating. For a given value of resistance, the physical size of a resistor increases as the wattage rating increases (Figure 10-19). Fixed resistors typically come in five sizes, ranging from $\frac{1}{8}$ W to 2 W. With experience, you can soon learn to tell the wattage ratings of resistors by looking at their physical sizes.

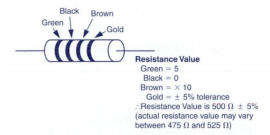

Resistance Value
Red = 2
Blue = 6
Orange = × 1 000
Silver = ± 10% tolerance
∴Resistance Value is 26 000 Ω ± 10%
(actual resistance value may vary between 23 400 Ω and 28 600 Ω)

10-6 Series Connection of Resistors

Often more than one resistor is used in a circuit. Resistors are connected in **series** by connecting them end-to-end together in a line (Figure 10-20). The total resistance of the circuit formed is simply the sum of the individual resistances. If these are numbered R_1, R_2, and R_3, then the total resistance or R_T is calculated using the formula:

$$R_T = R_1 + R_2 + R_3 \ldots$$

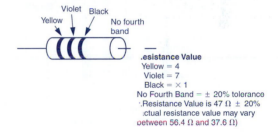

Resistance Value
Green = 5
Black = 0
Brown = × 10
Gold = ± 5% tolerance
∴Resistance Value is 500 Ω ± 5%
(actual resistance value may vary between 475 Ω and 525 Ω)

Resistance Value
Yellow = 4
Violet = 7
Black = × 1
No Fourth Band = ± 20% tolerance
∴Resistance Value is 47 Ω ± 20%
(actual resistance value may vary between 56.4 Ω and 37.6 Ω)

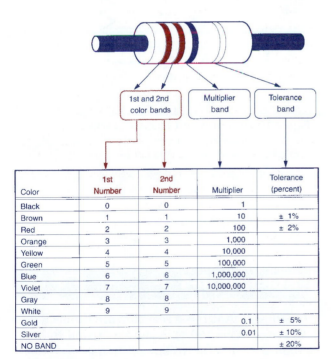

Color	1st Number	2nd Number	Multiplier	Tolerance (percent)
Black	0	0	1	
Brown	1	1	10	± 1%
Red	2	2	100	± 2%
Orange	3	3	1,000	
Yellow	4	4	10,000	
Green	5	5	100,000	
Blue	6	6	1,000,000	
Violet	7	7	10,000,000	
Gray	8	8		
White	9	9		
Gold			0.1	± 5%
Silver			0.01	± 10%
NO BAND				± 20%

A. Color values

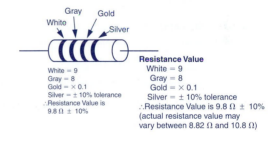

Resistance Value
White = 9
Gray = 8
Gold = × 0.1
Silver = ± 10% tolerance
∴Resistance Value is 9.8 Ω ± 10%

White = 9
Gray = 8
Gold = × 0.1
Silver = ± 10% tolerance
∴Resistance Value is 9.8 Ω ± 10%
(actual resistance value may vary between 8.82 Ω and 10.8 Ω)

GIVEN RESISTANCE VALUE	COLOR CODE			
	1st Band	2nd Band	3rd Band	4th Band
360 Ω ± 5%	Orange	Blue	Brown	Gold
10 Ω ± 10%	Brown	Black	Black	Silver
4 700 Ω ± 20%	Yellow	Violet	Red	No Fourth Band
5 Ω ± 10%	Green	Black	Gold	Silver
8 000 Ω ± 5%	Gray		Red	Gold
3 300 000 Ω ± 20%	Orange		Green	No Fourth Band

B. Reading resistor values

Figure 10-16 Four-band resistor color code.

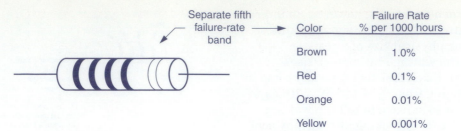

Color	Failure Rate % per 1000 hours
Brown	1.0%
Red	0.1%
Orange	0.01%
Yellow	0.001%

Figure 10-17   Resistor failure-rate code.

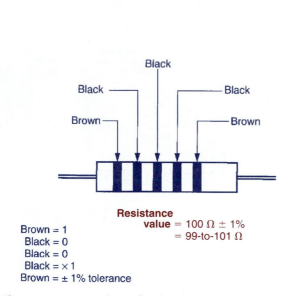

Brown = 1
Black = 0
Black = 0
Black = × 1
Brown = ± 1% tolerance

Figure 10-18   Reading a five-band precision resistor.

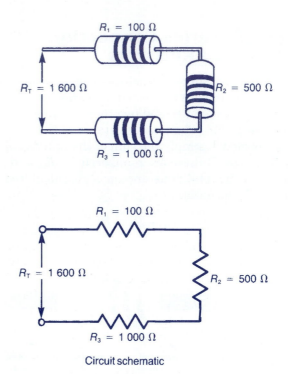

Circuit schematic

Figure 10-20   Resistors connected in series.

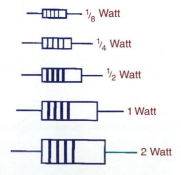

Figure 10-19   Relationship between resistor wattage
and physical size.

Example 10-1

   Suppose that three resistors are connected in series. R_1 is 100 Ω, R_2 is 500 Ω, and R_3 is 1000 Ω. The total resistance then is

$$R_T = R_1 + R_2 + R_3$$
$$R_T = 100\ \Omega + 500\ \Omega + 1000\ \Omega$$
$$R_T = 1600\ \Omega$$

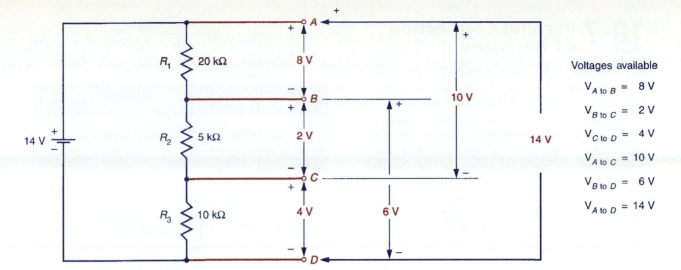

Figure 10-21 Voltage-divider circuit.

Resistors connected in series are used as *voltage dividers*. The voltage drops are proportional to the series resistances. A higher resistance has a greater voltage drop than a smaller resistance in the same series circuit; equal resistances have the same amount of voltage drop. The voltage-divider principle is widely used in circuits where one voltage source must supply several different voltages for different parts of a circuit. Figure 10-21 shows the schematic of a typical voltage-divider circuit. In this circuit six different voltage levels are available from the single 14-V source.

When resistors are connected in series, with the same current flowing through each, the voltage at the point that the resistors are connected is proportional to the ratio of the value of one of the resistors to the sum of the resistor values. Stated as a formula:

$$V_1 = \frac{R_1}{R_T} \times V_T$$

Example 10-2

Resistors R_1, R_2, and R_3 (3 kΩ, 6 kΩ, and 1 kΩ, respectively) are in series across an applied voltage of 24 V (Figure 10-22). The voltage drop across each resistor is:

$$V_1 = \frac{R_1}{R_T} \times V_T$$

$$= \frac{3 \text{ k}\Omega}{10 \text{ k}\Omega} \times 24 \text{ V}$$

$$= 7.2 \text{ V}$$

and

$$V_2 = \frac{R_2}{R_T} \times V_T$$

$$= \frac{6 \text{ k}\Omega}{10 \text{ k}\Omega} \times 24 \text{ V}$$

$$= 14.4 \text{ V}$$

and

$$V_3 = \frac{R_3}{R_T} \times V_T$$

$$= \frac{1 \text{ k}\Omega}{10 \text{ k}\Omega} \times 24 \text{ V}$$

$$= 2.4 \text{ V}$$

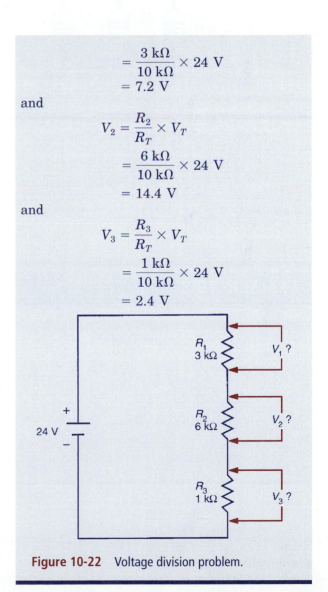

Figure 10-22 Voltage division problem.

10-7 Parallel Connection of Resistors

Resistors are connected in parallel by connecting them *across* a common set of line wires. The total resistance of the circuit formed is less than that of the lowest value of resistance present. This is because each resistor provides a separate parallel path for the current to flow. Assume all the resistors connected in parallel have the same value of resistance. The total resistance is then found by dividing the common resistance value by the total number of resistors connected (Figure 10-23).

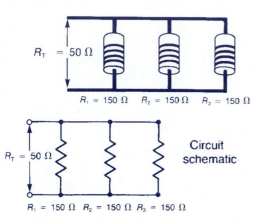

Figure 10-23 Resistors of the same value connected in parallel.

Example 10-3

Suppose that three 150-Ω resistors are connected in parallel. The total resistance is then:

$$R_T = \frac{R(\text{common value})}{\text{number of resistors}}$$

$$R_T = \frac{150\ \Omega}{3}$$

$$R_T = 50\ \Omega$$

To find the total resistance of two unequal values of resistors connected in parallel (a very common use) the product over sum formula is used. This formula is:

$$R_T = \frac{R_1 \times R_2}{R_1 + R_2}$$

Example 10-4

Suppose that a 60-Ω resistor is connected in parallel with one of 40-Ω. The total resistance is:

$$R_T = \frac{R_1 \times R_2}{R_1 + R_2}$$

$$R_T = \frac{60\ \Omega \times 40\ \Omega}{60\ \Omega + 40\ \Omega}$$

$$R_T = \frac{2400\ \Omega}{100\ \Omega}$$

$$R_T = 24\ \Omega$$

For more than two unequal resistors connected in parallel, the general formula for total resistance of a parallel circuit is used. This formula is:

$$R_T = \frac{1}{\dfrac{1}{R_1} + \dfrac{1}{R_2} + \dfrac{1}{R_3}}, \text{etc.}$$

Example 10-5

Suppose that a 120-Ω, 60-Ω, and 40-Ω resistor are all connected in parallel. The total resistance is:

$$R_T = \frac{1}{\dfrac{1}{R_1} + \dfrac{1}{R_2} + \dfrac{1}{R_3}}$$

$$R_T = \frac{1}{\dfrac{1}{120\ \Omega} + \dfrac{1}{60\ \Omega} + \dfrac{1}{40\ \Omega}}$$

$$R_T = \frac{1}{0.00833 + 0.0167 + 0.025}$$

$$R_T = \frac{1}{0.050}$$

$$R_T = 20\ \Omega$$

Resistors connected in parallel are used as *current dividers*. If two resistors are connected in parallel, with the same voltage across each, the current in one resistor is proportional to the ratio of the value of the other resistor to the sum of the resistor values. Stated as a formula:

$$I_1 = \frac{R_2}{R_1 + R_2} \times I_T \qquad I_2 = \frac{R_1}{R_1 + R_2} \times I_T$$

(This current-divider formula can be used only for two branch resistances. For more branches, it is possible to combine the branches in order to work with only two divided currents at a time.)

Example 10-6

Suppose the total current to two parallel-connected resistors is 30 mA. The resistance of R_1 is 200 Ω and the resistance of R_2 is 400 Ω (Figure 10-24). The current flow through each resistor would be:

$$I_1 = \frac{R_2}{R_1 + R_2} \times I_T \qquad I_2 = \frac{R_1}{R_1 + R_2} \times I_T$$

$$I_1 = \frac{400\ \Omega}{600\ \Omega} \times 30\ \text{mA} \qquad I_2 = \frac{200\ \Omega}{600\ \Omega} \times 30\ \text{mA}$$

$$I_1 = 20\ \text{mA} \qquad I_2 = 10\ \text{mA}$$

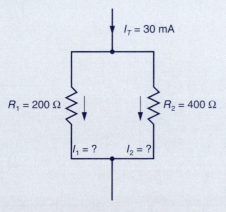

Figure 10-24 Current division problem.

10-8 Series-Parallel Connection of Resistors

A *series-parallel* circuit contains both series and parallel resistors. The rules governing these circuits are the same as those developed for series circuits and for parallel circuits. First, the resistance of the parallel portion is found. Then the total resistance of the parallel section is added to any series resistance to find the total resistance of the series-parallel circuit.

Example 10-7

A 30-Ω resistor, R_1, and a 60-Ω resistor, R_2, are connected in parallel with each other and in series with a 40-Ω resistor, and R_3, (Figure 10-25). The total resistance of the series-parallel circuit would be:

$$R_1 \| R_2 = \frac{R_1 \times R_2}{R_1 + R_2} \qquad R_T = R_1 \| R_2 + R_3$$

$$R_1 \| R_2 = \frac{30\ \Omega \times 60\ \Omega}{30\ \Omega + 60\ \Omega} \qquad R_T = 20\ \Omega + 40\ \Omega$$

$$R_1 \| R_2 = \frac{1800}{90} \qquad R_T = 60\ \Omega$$

$$R_1 \| R_2 = 20\ \Omega$$

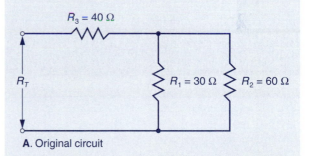

A. Original circuit

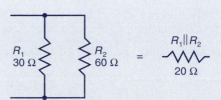

B. Equivalent resistance of the parallel portion

Figure 10-25 Series-parallel circuit problem.

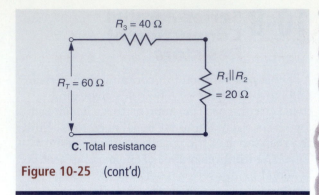

C. Total resistance

Figure 10-25   (cont'd)

Review and Applications

Related Formulas

$$R_T(\text{series circuit}) = R_1 + R_2 + R_3 \ldots$$

$$V_X(\text{voltage divider}) = \frac{R_X}{R_T} \times V_T$$

$$R_T(\text{parallel circuit}) = \frac{R_1 \times R_2}{R_1 + R_2}$$

$$R_T(\text{parallel circuit}) = \frac{1}{\dfrac{1}{R_1} + \dfrac{1}{R_2} + \dfrac{1}{R_3}}$$

$$I_1(\text{current divider}) = \frac{R_2}{R_1 + R_2} \times I_T$$

Review Questions

1. (a) What is a common, practical application for resistance wire?
   (b) What type of wire is most often used for this application?
2. Explain the function of a resistor in a circuit.
3. Name the three ways by which resistors are rated.
4. What type of circuits generally require the use of wire-wound resistors?
5. Explain the function of a fusible resistor.
6. Why have film resistors become more popular than the carbon-composition types?
7. Name the two kinds of film-type resistors.
8. Describe the construction of a chip resistor.
9. (a) Describe the construction of a resistor network.
   (b) In what type of circuits are resistor networks widely used?
10. In terms of how they function, name the three ways resistors are classified.
11. What is the tolerance rating of a precision resistor?
12. Compare the connection and control function of a rheostat with that of a potentiometer.
13. If the wiper arm of a linear potentiometer is one-quarter of the way around the contact surface, what is the resistance between the wiper arm and each terminal if the total resistance is 25 kΩ?
14. (a) When are trimmer potentiometers used?
    (b) Describe the operation of a ten-turn trim pot.

15. Compare the way the resistance varies in a linear and nonlinear potentiometer.
16. Identify the color bands for each of the following four-band color-coded resistors:
    (a) 100 Ω ± 10%
    (b) 2200 Ω ± 10%
    (c) 47,000 Ω ± 20%
    (d) 1,000,000 Ω ± 10%

Problems

1. A 680-Ω resistor has a rated tolerance of 10 percent. What is the rated resistance range for this resistor?
2. A separate fifth orange stripe is present on a color-coded carbon-composition resistor. If 100,000 of these were made, how many of these would likely fail in the first 1000-hour period?
3. What would be the color code for a five-band 909 Ω precision resistor with a 1% tolerance?
4. State the resistance value and percentage of tolerance for each of the four-band color-coded resistors shown in the table.

	1st Band	2nd Band	3rd Band	4th Band
(a)	Red	Green	Yellow	Silver
(b)	Orange	Blue	Brown	Gold
(c)	White	Brown	Red	none
(d)	Gray	Black	Blue	Gold
(e)	Violet	Green	Gold	Silver
(f)	Blue	Red	Black	Gold

5. Calculate the total resistance for each of the following circuit connections:
   (a) series circuit: $R_1 = 40\ \Omega$, $R_2 = 75\ \Omega$
   (b) parallel circuit: $R_1 = 200\ \Omega$, $R_2 = 200\ \Omega$, $R_3 = 200\ \Omega$
   (c) series circuit: $R_1 = 2000\ \Omega$, $R_2 = 6000\ \Omega$, $R_3 = 2200\ \Omega$
   (d) parallel circuit: $R_1 = 14\ \Omega$, $R_2 = 32\ \Omega$
   (e) series circuit: $R_1 = 4700\ \Omega$, $R_2 = 800\ \Omega$, $R_3 = 200\ \Omega$
   (f) parallel circuit: $R_1 = 60\ \Omega$, $R_2 = 30\ \Omega$, $R_3 = 15\ \Omega$

6. Resistors R_1, R_2, and R_3 ($50\,k\Omega$, $30\,k\Omega$, and $20\,k\Omega$, respectively) are connected in series across an applied voltage of 200 V. Calculate voltages V_1, V_2, and V_3 for this voltage-divider network.

7. The total current to two parallel-connected resistors is 3 A. The resistance of R_1 is 10 Ω and the resistance of R_2 is 40 Ω. Calculate currents I_1 and I_2 for this current-divider circuit.

8. A 5-kΩ resistor, R_1, and a 20-kΩ resistor, R_2, are connected in parallel with each other and in series with a 6-kΩ resistor, R_3. Calculate the total resistance of this series-parallel circuit.

Critical Thinking

1. You have been given three 330 Ω resistors to connect together. Three circuit configurations are possible using all three resistors. Describe each configuration and the total resistance value you would expect to measure for each configuration.

2. State the resistance value and the percentage tolerance for each of the five-band color-coded resistors shown in the table.

	1st Band	2nd Band	3rd Band	4th Band	5th Band
(a)	Green	Blue	Red	Red	Brown
(b)	Violet	Gray	Violet	Silver	Red
(c)	Orange	Blue	Green	Black	Brown
(d)	Brown	Black	Green	Brown	Red

Portfolio Project

Obtain 10 color-coded resistors. Record the resistance and percent tolerance of each according to their color code. Calculate the acceptable resistance range of each resistor, according to its tolerance rating. Using a digital multimeter record the actual resistance value of each resistor.

Circuit Challenge

Using a Simulator

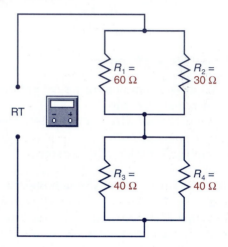

Procedure

(a) Calculate the total resistance of the resistor network shown. Complete the wiring for this circuit, using whatever simulation software package is available to you. Measure the total circuit resistance and compare this reading to your calculated value.

(b) Assume resistor R_1 burnt open (infinite resistance). Calculate the new value of the total resistance. Disconnect R_1 from the circuit and measure the total resistance. Compare this reading to your calculated value.

(c) Assume resistor R_3 of the original circuit is short-circuited (zero resistance across it). Calculate the new value of the total resistance. Connect a wire across R_3 to simulate a short in it and measure the total resistance. Compare this reading to your calculated value.

Ohm's Law

Ohm's law is easily the most important formula in electricity and electronics. A simple equation $I = V/R$, summarizes the relationship between the values of current, voltage, and resistance in an electric circuit. Before George Simon Ohm discovered his now famous law in 1827, work with electric circuits was a hit and miss affair. People simply guessed at what would happen in a circuit; they were not able to plan and then build a specific circuit as we do today.

Objectives

After studying this chapter, you should be able to:
1. Use prefixes to convert electrical quantities.
2. State Ohm's law and define the relationship between current, voltage, and resistance.
3. Use Ohm's law to solve for unknown quantities of current, resistance, or voltage.
4. Apply the power formula to calculate the power in a circuit.

Key Terms

- metric prefixes
- micro
- milli
- kilo
- mega

- voltage
- current
- resistance
- Ohm's law
- Ohm's law triangle

- power
- wattage
- linear
- nonlinear

RESEARCH INTERNET SITES

- Fluke
- Toshiba
- Hewlett Packard

11-1 Electrical Units and Prefixes

There are three basic measurements made on electrical circuits: (1) voltage, (2) current, and (3) resistance. Figure 11-1 lists these basic electrical quantities and the symbols that identify them. The table also explains the function of the quantity in an electrical circuit.

In certain circuit applications the basic units—volt, ampere, and ohm—are either too *small* or too *big* to work with. For example, in a television set, the signal for the antenna may have a strength of 0.00000125 V, while the voltage applied directly to the picture tube may be in the range of 27,000 V. In such cases *metric prefixes* are used. Using prefixes, these values would be expressed as 1.25 μV (microvolts) and 27 kV (kilovolts).

We base metric prefixes on powers of ten. Figure 11-2 lists the metric prefixes commonly used in electrical and electronic measurement values. The table shows the symbols for the metric prefixes and their equivalent decimal and powers of ten numbers.

Knowing how to convert metric prefixes back to base units is needed when reading a digital multimeter or using electrical circuit formulas. The prefix chart in Figure 11-3 shows how many positions the decimal point is moved to get from a base unit to a multiple or a fraction of a base unit, or to get back to a base unit.

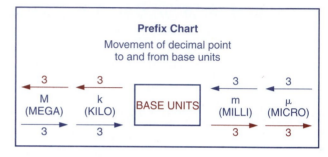

Figure 11-3　Prefix chart.

Quantity		Unit of Measure		Function
Name	**Symbol**	**Name**	**Symbol**	
Voltage	V, emf, or E	Volt	V	Voltage is the electromotive force or pressure which makes current flow in a circuit.
Current	I	Ampere	A	Current is the flow of electrons through a circuit.
Resistance	R	Ohm	Ω	Resistance is the opposition to current flow offered by electric devices in a current.

Figure 11-1　Voltage, current, and resistance.

Number		Power of Ten	Prefix	Symbol
One billion	1,000,000,000	10^9	giga	G
One million	1,000,000	10^6	mega	M
One thousand	1,000	10^3	kilo	k
One	1	10^0	——	——
One thousandth	0.001	10^{-3}	milli	m
One millionth	0.000001	10^{-6}	micro	μ
One billionth	0.000000001	10^{-9}	nano	n
One trillionth	0.000000000001	10^{-12}	pico	p

Figure 11-2　Table of common metric prefixes.

Example 11-1

To convert amperes (A) to milliamperes (mA), it is necessary to move the decimal point three places to the right (this is the same as multiplying the number by 1000).

$$0.012 \text{ A} = ? \text{ mA}$$

$$0.012 \text{ A} = 0.012$$

$$0.012 \text{ A} = 12 \text{ mA}$$

Example 11-2

To convert milliamperes (mA) to amperes (A), it is necessary to move the decimal point three places to the left (this is the same as multiplying by 0.001).

$$450.0 \text{ mA} = ? \text{ A}$$

$$450.0 \text{ mA} = 450.0$$

$$450.0 \text{ mA} = 0.45 \text{ A}$$

Example 11-3

To convert ohms (Ω) to kilohms (kΩ), it is necessary to move the decimal point three places to the left.

$$47000.0 \ \Omega = ? \text{ k}\Omega$$

$$47000.0 \ \Omega = 47000.0$$

$$47000.0 \ \Omega = 47.0 \text{ k}\Omega$$

Example 11-4

To convert from megohms (MΩ) to ohms (Ω), it is necessary to move the decimal point six places to the right.

$$2.2 \text{ M}\Omega = ? \ \Omega$$

$$2.2 \text{ M}\Omega = 2.200000$$

$$2.2 \text{ M}\Omega = 2,200,000 \ \Omega$$

Example 11-5

To convert from microamperes (μA) to amperes (A), it is necessary to move the decimal point six places to the left.

$$500 \ \mu\text{A} = ? \text{ A}$$

$$500 \ \mu\text{A} = 000500.$$

$$500 \ \mu\text{A} = 0.0005 \text{ A}$$

11-2 Ohm's Law

Electricity always acts in a predictable manner. By using different laws for electric circuits, we can predict what should happen in a circuit or diagnose why things are not operating as they should.

Ohm's law expresses the relationship between the *voltage* (V), the *current* (I), and the *resistance* (R) in a circuit. Ohm's law can be stated as follows:

 The current (*I*) in a circuit is *directly* proportional to the applied voltage (*V*) and *inversely* proportional to the circuit resistance (*R*).

In other words, Ohm's law states that the current in an electrical circuit depends on two things:
- The voltage applied to the circuit
- The resistance in the circuit

The relationship of voltage and current can be more easily understood if you compare the circuits of Figure 11-4. All three circuits have the same fixed resistance. Note that when the voltage increases, the current also increases. The current, therefore, is directly proportional to the voltage.

If the voltage is held constant, the current will change as the resistance changes, but in the opposite direction. This is illustrated in the circuits of Figure 11-5. All three circuits have the same fixed voltage. Note that when the resistance increases, the current decreases;

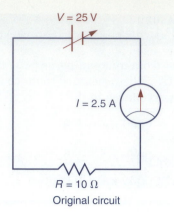

V = 25 V

I = 2.5 A

R = 10 Ω
Original circuit

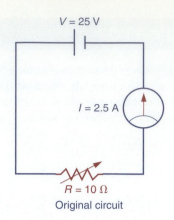

V = 25 V

I = 2.5 A

R = 10 Ω
Original circuit

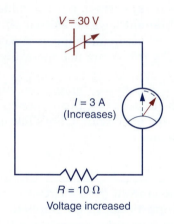

V = 30 V

I = 3 A
(Increases)

R = 10 Ω
Voltage increased

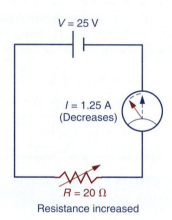

V = 25 V

I = 1.25 A
(Decreases)

R = 20 Ω
Resistance increased

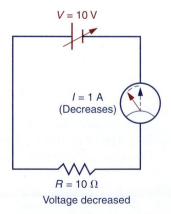

V = 10 V

I = 1 A
(Decreases)

R = 10 Ω
Voltage decreased

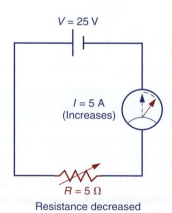

V = 25 V

I = 5 A
(Increases)

R = 5 Ω
Resistance decreased

Figure 11-4   The effect of changes in voltage on current.

Figure 11-5   The effect of changes in resistance on current.

when the resistance decreases, the current increases. Thus, the current is inversely proportional to the resistance.

Mathematically, Ohm's law can be expressed in the form of three formulas: one basic formula and two others derived from it. Using these three formulas, and knowing any two of the values for voltage, current, or resistance, it is possible to find the third value.

Ohm's Law Formulas

Find Current	Find Voltage	Find Resistance
$I = \dfrac{V}{R}$	$V = I \times R$	$R = \dfrac{V}{I}$
Current equals voltage divided by resistance.	Voltage equals current multiplied by resistance.	Resistance equals voltage divided by current.

When applying the mathematical equations of Ohm's law to a circuit, it is important to use the correct units of measurement. *Improperly mixing units will result in incorrect answers.* Common combinations of units that can be used for different types of circuits include:

Electrical Circuits
I = current in amperes (A)
V = voltage in volts (V)
R = resistance in ohms (Ω)

Electronic Circuits
I = current in milliamperes (mA)
V = voltage in volts (V)
R = resistance in kilohms (kΩ)

Microelectric Circuits
I = current in microamperes (μA)
V = voltage in volts (V)
R = resistance in megohms (MΩ)

11-3 Applying Ohm's Law to Calculate Current

By using Ohm's law, we can predict what is going to happen in a circuit before we apply power. When any two of the three quantities (V, I, or R) are known, the third can be calcu-

lated. For example, if the voltage and resistance are known, the current can be calculated. The formula used is:

$$I = \frac{V}{R}$$

Example 11-6

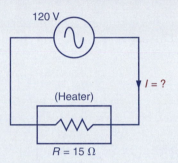

Suppose that a portable electric heater with a resistance of 15 Ω is directly connected to a 120-V ac electric outlet as shown. The current flow in this circuit is:

$$\text{Current} = \frac{\text{Voltage}}{\text{Resistance}}$$
$$I = \frac{V}{R}$$
$$I = \frac{(120\ V)}{(15\ \Omega)}$$
$$I = 8\ A$$

Electronic and microelectronic circuits operate at much lower current values than electric circuits. This is mainly because they usually contain much higher resistance values. If the resistance of these circuits is expressed in kilohms or megohms, the current can be calculated directly in milliamperes or microamperes:

$$mA = \frac{V}{k\Omega}$$

or

$$\mu A = \frac{V}{M\Omega}$$

Example 11-7

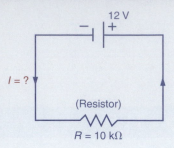

12 V

$I = ?$

(Resistor)

$R = 10\ k\Omega$

Suppose a 10-kΩ carbon resistor is connected to a 12-V battery as shown. The current flow in this circuit is:

$$I = \frac{V}{R}$$

$$I = \frac{(12\ V)}{(10\ k\Omega)}$$

$$I = 1.2\ mA$$

11-4 Applying Ohm's Law to Calculate Voltage

When the current flow and resistance of a circuit are known, the voltage being applied can be calculated. The formula is: $V = I \times R$.

Example 11-8

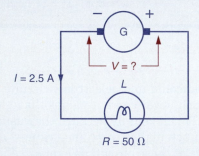

$-$ $+$

G

$V = ?$

$I = 2.5\ A$

L

$R = 50\ \Omega$

Suppose a dc generator delivering a current of 2.5 A to a lamp bank that has a combined resistance of 50 Ω, as shown. The voltage output of the generator is:

$$\text{Voltage} = \text{Current} \times \text{Resistance}$$

$$V = I \times R$$

$$V = (2.5\ A)(50\ \Omega)$$

$$V = 125\ V$$

For low current electronic and microelectronic circuits, current and resistance may be expressed in either milliamperes and kilohms or microamperes and megohms. Then the voltage can be calculated directly using the common combination of:

$$V = mA \times k\Omega \qquad \textbf{or} \qquad V = \mu A \times M\Omega$$

Example 11-9

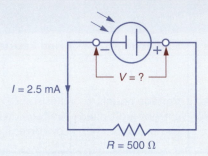

$V = ?$

$I = 2.5\ mA$

$R = 500\ \Omega$

Suppose a solar cell provides a current of 2.5 mA to a 500-Ω (0.5-kΩ) load. The voltage output of the solar cell is:

$$V = I \times R$$

$$V = (2.5\ mA)(0.5\ k\Omega)$$

$$V = 1.25\ V$$

11-5 Applying Ohm's Law to Calculate Resistance

The resistance of a load can be calculated when the applied voltage across it and the current flow through it are known. The formula used is:

$$R = \frac{V}{I}$$

Example 11-10

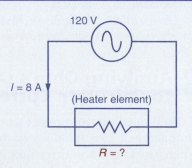

120 V

$I = 8\ A$

(Heater element)

$R = ?$

Suppose an electric kettle draws a current of 8 A when connected to a 120-V ac electric outlet as shown. The resistance of the kettle heating element is:

$$\text{Resistance} = \frac{\text{Voltage}}{\text{Current}}$$

$$R = \frac{V}{I}$$

$$R = \frac{(120 \text{ V})}{(8 \text{ A})}$$

$$R = 15 \text{ } \Omega$$

In electronic circuits, it is often more convenient to calculate the resistance values of resistors rather than measuring them with an ohmmeter. If voltage and current values are known, it is faster to calculate the resistance value than to measure it. The current may be expressed in milliamperes or microamperes. When this is the case, the resistance can be found by using the common combination of:

$$k\Omega = \frac{V}{mA} \quad \textbf{or} \quad M\Omega = \frac{V}{\mu A}$$

Example 11-11

9 V

$I = 2 \text{ } \mu\text{A}$ $R = ?$

Circuit Schematic

Suppose the current flow through a resistor is known to be 2 μA, as shown. The voltage across the resistor is measured and found to be 9 V. The resistance value of the resistor is:

$$R = \frac{V}{I}$$

$$R = \frac{(9 \text{ V})}{(2 \text{ } \mu\text{A})}$$

$$R = 4.5 \text{ M}\Omega$$

11-6 Ohm's Law Triangle

The three variations of **Ohm's law** formulas can be easily remembered by arranging the three quantities within a **triangle** as shown in Figure 11-6. A finger placed over the symbol standing for the unknown quantity leaves the remaining two symbols in the correct relationship to solve for the unknown value.

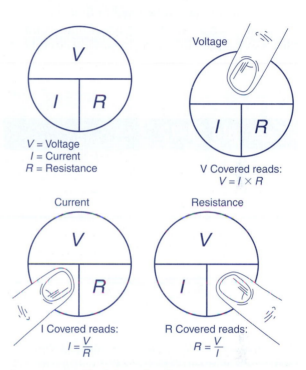

V = Voltage
I = Current
R = Resistance

V Covered reads:
$V = I \times R$

I Covered reads:
$I = \frac{V}{R}$

R Covered reads:
$R = \frac{V}{I}$

Figure 11-6   Ohm's law triangle.

11-7 Power Formulas

Power (P) is the amount of work performed by an electric circuit when the voltage forces current to flow through the resistance. The base unit used to measure power is the *watt* (W). The power formulas show the relationships among electric power and voltage, current, and resistance in a dc circuit. The basic power formula is:

$$\text{Power in watts} = \text{volts} \times \text{amperes}$$
$$P = V \times I$$

From this formula, it is possible to get two other commonly used power formulas. For example, Ohm's law states that $V = I \times R$.

By substituting $I \times R$ for V in the basic power formula, we have

$$P = (I \times R) \times I$$
$$P = I^2R$$

Also, from Ohm's law we know that $I = V/R$. By substituting V/R for I in the basic power formula we have,

$$P = V \times \left(\frac{V}{R}\right) \quad \textbf{or} \quad P = \frac{V^2}{R}$$

Power formulas can be used to find the **wattage** ratings of circuit parts. Some examples using power formulas in solving circuit problems follow:

Example 11-12

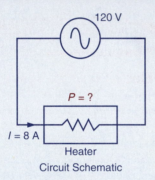

120 V

$P = ?$

$I = 8\ A$

Heater

Circuit Schematic

Suppose an electric heater draws a current of 8 A when connected to its rated voltage of 120 V. The power rating of the heater is:

$$\text{Power} = \text{Voltage} \times \text{Current}$$
$$P = V \times I$$
$$P = (120\ \text{V})(8\ \text{A})$$
$$P = 960\ \text{W}$$

Example 11-13

Suppose a current of 30 A is being supplied to an electric range. The total resistance of the wire used to supply this current is 0.1 Ω. The power that is lost in the wire is:

$$\text{Power} = \text{Current}^2 \times \text{Resistance}$$
$$P = I^2R$$

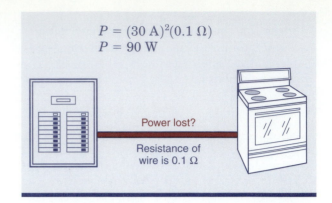

$$P = (30\ \text{A})^2(0.1\ \Omega)$$
$$P = 90\ \text{W}$$

Power lost?

Resistance of wire is 0.1 Ω

Example 11-14

6 V

48 Ω
$P = ?$

Suppose a 48-Ω resistor is to be connected to a 6-V source. The wattage that must be dissipated by the resistor is:

$$\text{Power} = \frac{\text{Voltage}^2}{\text{Resistance}}$$
$$P = \frac{V^2}{R}$$
$$P = \frac{(6\ \text{V})^2}{48\ \Omega}$$
$$P = 0.75\ \text{W}$$

To keep a resistor from becoming overheated, its wattage rating should be about twice the wattage rating computed from a power formula. Thus, the resistor used in this circuit should have a wattage rating of about 2 W.

11-8 Ohm's Law in Graphical Form

The relationship between current and voltage may be shown graphically as in Figure 11-7. When the voltage is varied, the meters show that the current values are directly proportional to the voltage values. For instance, with 10 V applied the current equals 1 A; for 20 V the current is 2 A.

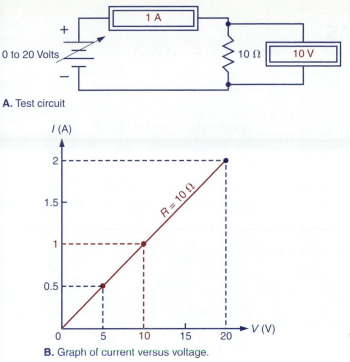

A. Test circuit

B. Graph of current versus voltage.

Figure 11-7 The relationship between current and voltage shown graphically.

An ordinary resistor is often called a *linear* device because a graph of its current versus voltage is a straight line. A linear resistance has a constant value of ohms. Its resistance does not change with the applied voltage. On the other hand, the resistance of the tungsten filament in a light bulb is *nonlinear*. This is because the resistance of the filament increases as the filament becomes hotter. Increasing the

applied voltage does produce more current, but this current does not increase in the same proportion as the increase in voltage (Figure 11-8).

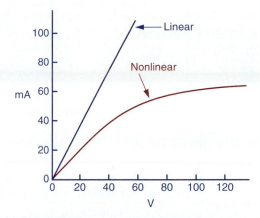

Figure 11-8 Linear and nonlinear devices.

Industrial Electrical Technician
Job Description
• Troubleshoot machine controls and equipment.
• Mechanical aptitude desirable.
• Must be able to troubleshoot using electrical schematics, multimeters, and oscilloscopes.
• High school diploma and two years of college or technical school required.

Review and Applications

· ·

Related Formulas

milli (m) = prefix for 10^{-3}

micro (μ) = prefix for 10^{-6}

kilo (k) = prefix for 10^3

$$\% \text{ of error} = \frac{\text{measured reading} - \text{calculated value}}{\text{calculated value}} \times 100$$

$I = V/R$

$V = I \times R$

$R = V/I$

mega (M) = prefix for 10^6

$$k\Omega = \frac{V}{mA}$$

$$M\Omega = \frac{V}{\mu A}$$

$$\text{Ohms} = \frac{\text{Volts}}{\text{Amps}}$$

$P = V \times I$

$P = I^2 R$

$$P = \frac{V^2}{R}$$

Review Questions

1. What is the base unit and symbol used for electric:
   (a) current
   (b) voltage
   (c) resistance
   (d) power
2. Write the metric prefix and symbol used to represent each of the following:
   (a) one thousandth
   (b) one million
   (c) one millionth
   (d) one thousand
3. State Ohm's law.
4. Write the voltage, current, and resistance formulas for Ohm's law.

Problems

1. Convert each of the following:
   (a) 2500 Ω to kilohms
   (b) 120 kΩ to ohms
   (c) 1,500,000 Ω to megohms
   (d) 2.03 MΩ to ohms
   (e) 0.000466 A to microamps
   (f) 0.000466 A to milliamps
   (g) 378 mV to volts
   (h) 475 Ω to kilohms
   (i) 28 μA to amps
   (j) 5 kΩ + 850 Ω to kilohms
   (k) 40,000 kV to megavolts
   (l) 4,600,000 μA to amps
   (m) 2.2 kΩ to ohms

2. In your notes, make a copy of the following chart. Calculate and record the missing values.

	Current	Resistance	Voltage
(a)	I = ?	R = 6 Ω	V = 12 V
(b)	I = 0.1 A	R = 120 Ω	V = ?
(c)	I = ?	R = 10 Ω	V = 120 V
(d)	I = 20 A	R = ?	V = 24 V
(e)	I = 0.001 A	R = 30 Ω	V = ?
(f)	I = 0.0005 A	R = ?	V = 40 V
(g)	I = ?	R = 1.5 kΩ	V = 12 V
(h)	I = 24 mA	R = ?	V = 12 V
(i)	I = 1.2 mA	R = 12 kΩ	V = ?
(j)	I = ?	R = 1/2 Ω	V = 12 V
(k)	I = 40 μA	R = ?	V = 3 V
(l)	I = 0.02 mA	R = 0.5 MΩ	V = ?

3. A headlamp bulb working off a 12-V battery takes a current of 3 A. What is the hot resistance of the bulb filament? (Hot resistance refers to the resistance of the bulb filament under normal operating condition and is much higher than its cold or out-of-circuit resistance.)

4. A resistor has a resistance of 220 Ω. The current flow through the resistor is measured and found to be 30 mA. What is the value of the voltage across the resistor?

5. An electric soldering iron with a 40-Ω heating element is plugged into a 120-V outlet. How much current will be drawn by the iron?

6. A baseboard heater draws a current of 8 A when connected to a 240-V source. What is the resistance value of the heater element?

7. A computer control module has an internal resistance of 50,000 ohms. What is its normal operating current, in microamperes, if the applied working voltage is 6 V?

8. The current through and the voltage drop across a precision resistor is measured and found to be 8 mA and 1.5 V, respectively. What is the resistance value of the resistor?

9. A window defogger heater draws 15 A when connected to a 12-V source. What is the amount of power dissipated by the heater?

10. How much power is lost in the form of heat when 25 A of current flows through a conductor that has 0.02 Ω of resistance?

11. The voltage drop across a 330-Ω resistor is measured and found to be 9 V. How much wattage is being dissipated by the resistor?

12. In your notes, make a copy of the following chart. Calculate and record the missing values.

	Current	Resistance	Voltage	Power
(a)	100 mA	?	250 V	?
(b)	?	4.7 kΩ	24 V	?
(c)	3 A	40 Ω	?	?

Critical Thinking

1. (a) A 12-Ω resistor is connected across a 60-V dc source. Calculate the current flow through the resistor.
   (b) If the value of the applied voltage is reduced to 48 V, does the current flow increase or decrease?
   (c) Calculate the amount of change in current flow.

2. In an electric circuit consisting of a battery and a fixed resistor, if the resistor value is doubled, what happens to the current?

3. What is the percentage of error if the measured value of voltage in a circuit is 8.2 V and the calculated value is 8 V?

Portfolio Project

A job interviewer can often tell how well you really understand the concept. Often, the questions are general in nature to see how well you can express your understanding of the topic. With this in mind, prepare an answer to each of the following job interview questions:

• I have a simple circuit breadboarded on my lab bench. Tell me how I can determine the amount of current flow using Ohm's law.

• Show me, by drawing a graph, the relationship between voltage and current for a 10 Ω and 20 Ω resistor.

• How would you determine the correct wattage size for a resistor?

Circuit Challenge

Using a Simulator

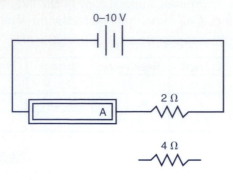

Procedure

(a) Construct the circuit shown, using whatever simulation software package is available to you.

(b) Change the supply voltage (in 2 V steps) from 0 to 10 V and record the amount of current flow for each step. Plot a graph of current versus voltage.

(c) Insert the 4 Ω resistor in place of the 2 Ω and repeat the process. On your original graph paper plot a second graph using values for the 4 Ω resistor.

Solving the Series Circuit

The main feature of a series circuit is the way its parts are connected: they form a single loop, beginning and ending at the power supply. Series circuits obey a specific set of rules that apply only to them. This chapter examines these special characteristics of a series circuit.

Objectives

After studying this chapter, you should be able to:
1. State the voltage, current, resistance, and power characteristics of a series circuit.
2. Solve for unknown circuit values in a series circuit.
3. Measure current, voltage drops, and resistance values in a series circuit.
4. Apply the concepts of relative polarity, series-aiding, and series-opposing voltages.
5. Troubleshoot a series dc resistive circuit.

Key Terms

- series circuit
- $R_1 V_1 I_1$
- $R_T V_T I_T$
- $P_1 P_T$
- voltage drop
- polarity
- common ground
- series-aiding
- series-opposing
- open circuit
- short circuit

RESEARCH INTERNET SITES

- Turck
- Amprobe
- C&K Components

12-1 Series Circuit Connection

A *series circuit* is a circuit that has only one path for current flow. All the components in the circuit are connected in a single line end-to-end. See Figure 12-1, for an example of a series circuit (pictorial and schematic). Only one pathway can be traced for the flow of electrons from one side of the voltage source to the other. If this path is broken at any point, all current flow in the circuit stops.

12-2 Identifying Circuit Quantities

The symbols used when referring to voltage, current, resistance, and power are *V, I, R,* and *P,* respectively. In circuits that contain more than one load resistor, it is necessary to use a system of letter and number subscripts to correctly identify the different circuit quantities.

The total resistance of the circuit can be represented by the symbol R_T and the individual resistors by the symbols $R_1, R_2, R_3,$ etc. The applied source voltage can be represented by the symbol V_T and the voltage drop across the individual resistors by the symbols $V_1, V_2, V_3,$ etc. The total or source current can be represented by the symbol I_T and the current flow through the individual resistors is represented by the symbols $I_1, I_2, I_3,$ etc. (Figure 12-2). The total power dissipated can be represented by the symbol $P_T,$ while the power dissipated by the individual resistors is given by the symbols $P_1, P_2, P_3,$ etc.

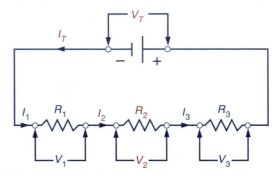

Figure 12-2 Identifying quantities in a series circuit.

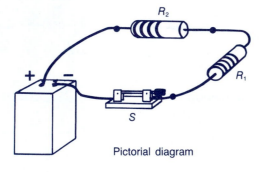

Pictorial diagram

12-3 Series Circuit Current, Voltage, Resistance, and Power Characteristics

Current The current is the same value throughout a series circuit (Figure 12-3). This is because there is only one current path. There is

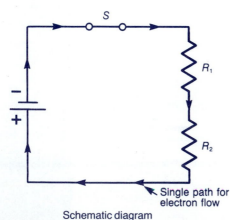

Schematic diagram

Figure 12-1 Series circuit.

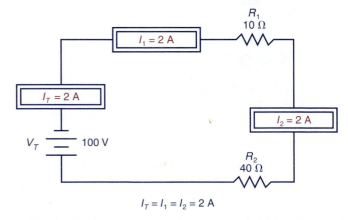

$$I_T = I_1 = I_2 = 2 \text{ A}$$

Figure 12-3 The current is the same at all points in a series circuit.

no loss or gain of current in a circuit. All of the electrons that flow out of the negative side of the voltage source must flow through each component and return to the positive side of the voltage source. Although the load resistors may be different in value, they all carry the same amount of current when connected in series.

Voltage The voltage applied to a series circuit is divided among each of the loads (Figure 12-4). The voltage across each load resistor is called the *voltage drop*. The amount of voltage each load resistor receives is directly proportional to the resistance value of the load. The higher the resistance value, the greater the voltage drop. Total source voltage is then the sum of the voltage drops across each of the individual load resistors.

Resistance The total resistance (R_T) of a series circuit is equal to the sum of the individual load resistances (Figure 12-5). Since there is only one path for electrons to follow, they must flow through each of the load resistors in their journey from one side of the voltage source to the other. Therefore, all individual load resistance values are added together to determine the total circuit resistance.

Power The power dissipation of a resistor in a circuit is determined by the amount of current flowing through the resistor and the amount of voltage across the resistor. The total power dissipated (P_T) in a series circuit is always the sum of the power dissipated by the individual resistors or the product of total current (I_T) and source voltage (V_T) (Figure 12-6).

In the construction of any circuit, it is important to know the power dissipation of the components. Heat is one of the worst enemies of electric and electronic circuits. Overdissipation of components with too much electric power will cause them to overheat and burn out (Figure 12-7).

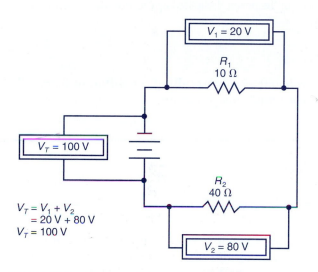

$$V_T = V_1 + V_2$$
$$= 20\ V + 80\ V$$
$$V_T = 100\ V$$

Figure 12-4 The sum of the voltage drops across the individual components in a series circuit is equal to the applied voltage.

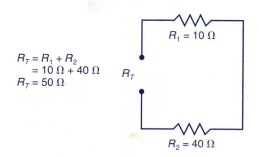

$$R_T = R_1 + R_2$$
$$= 10\ \Omega + 40\ \Omega$$
$$R_T = 50\ \Omega$$

A. Finding total resistance

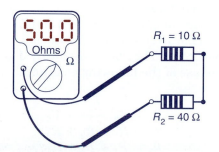

B. Reading total resistance

Figure 12-5 The total resistance in a series circuit is equal to the sum of the individual resistances.

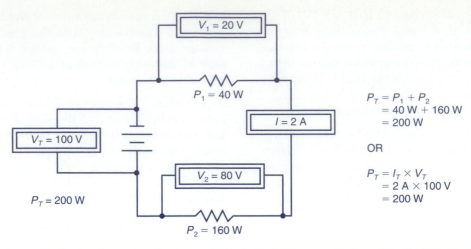

$$P_T = P_1 + P_2$$
$$= 40\ W + 160\ W$$
$$= 200\ W$$

OR

$$P_T = I_T \times V_T$$
$$= 2\ A \times 100\ V$$
$$= 200\ W$$

Figure 12-6 Power dissipation in a series circuit equals the sum of the power dissipated by the individual resistors, or the total current multiplied by the total voltage.

Overheated resistor

Figure 12-7 Overheated resistor.

12-4 Solving Series Circuits

Given series circuit values of voltage, current, and resistance, it is possible to calculate any unknown values of voltage, current, resistance, and power. The ability to make these calculations is important if you are truly to understand the operation of a series circuit.

In solving a series circuit problem, you must know the voltage, current, resistance, and power characteristics (as previously presented) of the circuit. They are most easily determined by the use of the following equations:

$$I_T = I_1 = I_2 = I_3 \ldots$$
$$V_T = V_1 + V_2 + V_3 \ldots$$
$$R_T = R_1 + R_2 + R_3 \ldots$$
$$P_T = P_1 + P_2 + P_3 \ldots$$

Solving the series circuit will also require the use of Ohm's law. Ohm's law is most easily remembered by recalling the different forms of the Ohm's law equation:

$$V = I \times R$$
$$I = \frac{V}{R}$$
$$R = \frac{V}{I}$$
$$P = V \times I$$

Ohm's law is applied to the total circuit or to the individual loads. When applying it to the total circuit, it becomes:

$$V_T = I_T \times R_T$$
$$I_T = \frac{V_T}{R_T}$$
$$R_T = \frac{V_T}{I_T}$$
$$P_T = V_T \times I_T$$

When applying it to the individual load it becomes:

$$V_1 = I_1 \times R_1$$
$$I_1 = \frac{V_1}{R_1}$$
$$R_1 = \frac{V_1}{I_1}$$
$$P_1 = V_1 \times I_1$$

One helpful method of solving series circuits is to use a chart to record all given and calculated values of voltage, current, resistance, and power. We use this method when solving the following typical series circuit problems.

Example 12-1

Problem   Find all the unknown values of V, I, R, and P for the series circuit of Figure 12-8.

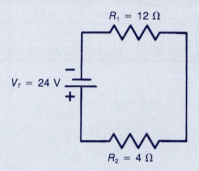

Figure 12-8   Circuit for Example 12-1.

Solution

STEP 1   Make a chart and record all known values.

	Voltage	Current	Resistance	Power
R_1			12 Ω	
R_2			4 Ω	
Total	24 V			

STEP 2   Calculate R_T and enter in the chart.

$$R_T = R_1 + R_2$$
$$= (12\ \Omega) + (4\ \Omega)$$
$$R_T = 16\ \Omega$$

	Voltage	Current	Resistance	Power
R_1			12 Ω	
R_2			4 Ω	
Total	24 V		16 Ω	

STEP 3   Calculate I_T, I_1, and I_2 and enter in the chart.

$$I_T = V_T/R_T$$
$$= (24\ \text{V}) \div (16\ \Omega)$$
$$I_T = 1.5\ \text{A}$$
$$I_1 = I_T$$
$$I_1 = 1.5\ \text{A}$$

$$I_2 = I_T$$
$$I_2 = 1.5\ \text{A}$$

	Voltage	Current	Resistance	Power
R_1		1.5 A	12 Ω	
R_2		1.5 A	4 Ω	
Total	24 V	1.5 A	16 Ω	

STEP 4   Calculate V_1 and V_2 and enter the values in the chart.

$$V_1 = I_1 \times R_1$$
$$= (1.5\ \text{A})(12\ \Omega)$$
$$V_1 = 18\ \text{V}$$
$$V_2 = I_2 \times R_2$$
$$= (1.5\ \text{A})(4\ \Omega)$$
$$V_2 = 6\ \text{V}$$

	Voltage	Current	Resistance	Power
R_1	18 V	1.5 A	12 Ω	
R_2	6 V	1.5 A	4 Ω	
Total	24 V	1.5 A	16 Ω	

STEP 5   Calculate P_T, P_1, and P_2 and enter the values in the chart.

$$P_T = V_T \times I_T$$
$$= (24\ \text{V})(1.5\ \text{A})$$
$$P_T = 36\ \text{W}$$
$$P_1 = V_1 \times I_1$$
$$= (18\ \text{V})(1.5\ \text{A})$$
$$P_1 = 27\ \text{W}$$
$$P_2 = V_2 \times I_2$$
$$= (6\ \text{V})(1.5\ \text{A})$$
$$P_2 = 9\ \text{W}$$

	Voltage	Current	Resistance	Power
R_1	18 V	1.5 A	12 Ω	27 W
R_2	6 V	1.5 A	4 Ω	9 W
Total	24 V	1.5 A	16 Ω	36 W

Example 12-2

Problem

Find all the unknown values of V, I, R, and P for the series circuit of Figure 12-9.

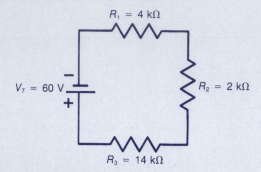

Figure 12-9   Circuit for Example 12-2.

Solution

STEP 1   Make a chart and record all known values.

	Voltage	Current	Resistance	Power
R_1			4 kΩ	
R_2			2 kΩ	
R_3			14 kΩ	
Total	60 V			

STEP 2   Calculate R_T and enter the value in the chart.

$$R_T = R_1 + R_2 + R_3$$
$$= (4\text{ k}\Omega) + (2\text{ k}\Omega) + (14\text{ k}\Omega)$$
$$= 20\text{ k}\Omega$$

	Voltage	Current	Resistance	Power
R_1			4 kΩ	
R_2			2 kΩ	
R_3			14 kΩ	
Total	60 V		20 kΩ	

STEP 3   Calculate I_T, I_1, I_2, and I_3 and enter the values in the chart.

$$I_T = \frac{V_T}{R_T}$$
$$= (60\text{ V}) \div (20\text{ k}\Omega)$$
$$I_1 = 3\text{ mA}$$
$$I_2 = I_T$$

$$I_1 = 3\text{ mA}$$
$$I_2 = I_T$$
$$I_1 = 3\text{ mA}$$
$$I_2 = I_T$$
$$I_3 = 3\text{ mA}$$

	Voltage	Current	Resistance	Power
R_1		3 mA	4 kΩ	
R_2		3 mA	2 kΩ	
R_3		3 mA	14 kΩ	
Total	60 V	3 mA	20 kΩ	

STEP 4   Calculate V_1, V_2, and V_3 and enter the values in the chart.

$$V_1 = I_1 \times R_1$$
$$= (3\text{ mA})(4\text{ k}\Omega)$$
$$V_1 = 12\text{ V}$$
$$V_2 = I_2 \times R_2$$
$$= (3\text{ mA})(2\text{ k}\Omega)$$
$$V_2 = 6\text{ V}$$
$$V_3 = I_3 \times R_3$$
$$= (3\text{ mA})(14\text{ k}\Omega)$$
$$V_3 = 42\text{ V}$$

	Voltage	Current	Resistance	Power
R_1	12 V	3 mA	4 kΩ	
R_2	6 V	3 mA	2 kΩ	
R_3	42 V	3 mA	14 kΩ	
Total	60 V	3 mA	20 kΩ	

STEP 5   Calculate P_T, P_1, P_2, and P_3 and enter the values in the chart.

$$P_T = V_T \times I_T$$
$$= (60\text{ V})(3\text{ mA})$$
$$P_T = 180\text{ mW}$$
$$P_1 = V_1 \times I_1$$
$$= (12\text{ V})(3\text{ mA})$$
$$P_1 = 36\text{ mW}$$
$$P_2 = V_2 \times I_2$$
$$= (6\text{ V})(3\text{ mA})$$
$$P_2 = 18\text{ mW}$$
$$P_3 = V_3 \times I_3$$

$$= (42 \text{ V})(3 \text{ mA})$$
$$P_3 = 126 \text{ mW}$$

	Voltage	Current	Resistance	Power
R_1	12 V	3 mA	4 kΩ	36 mW
R_2	6 V	3 mA	2 kΩ	18 mW
R_3	42 V	3 mA	14 kΩ	126 mW
Total	60 V	3 mA	20 kΩ	180 mW

Example 12-3

Problem   Find all the unknown values of V, I, R, and P for the series circuit in Figure 12-10.

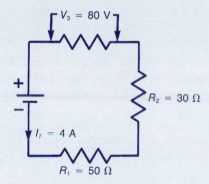

Figure 12-10   Circuit for Example 12-3.

Solution
STEP 1   Make a chart and record all known values.

	Voltage	Current	Resistance	Power
R_1			50 Ω	
R_2			30 Ω	
R_3	80 V			
Total		4 A		

STEP 2   Calculate I_1, I_2, and I_3 and enter the values in the chart.

$$I_T = I_1 = I_2 = I_3 = 4 \text{ A}$$

	Voltage	Current	Resistance	Power
R_1		4A	50 Ω	
R_2		4A	30 Ω	
R_3	80 V	4A		
Total		4A		

STEP 3   Calculate V_1, V_2, and V_T and enter the values in the chart.

$$V_1 = I_1 \times R_1$$
$$= (4 \text{ A})(50 \text{ Ω})$$
$$V_1 = 200 \text{ V}$$
$$V_2 = I_2 \times R_2$$
$$= (4 \text{ A})(30 \text{ Ω})$$
$$V_2 = 120 \text{ V}$$
$$V_T = V_1 + V_2 + V_3$$
$$= (200 \text{ V}) + (120 \text{ V}) + (80 \text{ V})$$
$$V_T = 400 \text{ V}$$

	Voltage	Current	Resistance	Power
R_1	200 V	4A	50 Ω	
R_2	120 V	4A	30 Ω	
R_3	80 V	4A		
Total	400 V	4A		

STEP 4   Calculate R_3 and R_T and enter the values in the chart.

$$R_3 = \frac{V_3}{I_3}$$
$$= (80 \text{ V}) \div (4 \text{ A})$$
$$R_3 = 20 \text{ Ω}$$
$$R_T = \frac{V_T}{I_T}$$
$$= (400 \text{ V}) \div (4 \text{ A})$$
$$R_T = 100 \text{ Ω}$$

	Voltage	Current	Resistance	Power
R_1	200 V	4A	50 Ω	
R_2	120 V	4A	30 Ω	
R_3	80 V	4A	20 Ω	
Total	400 V	4A	100 Ω	

STEP 5   Calculate P_T, P_1, P_2, and P_3 and enter the values in the chart.

$$P_T = V_T \times I_T$$
$$= (400 \text{ V})(4 \text{ A})$$
$$P_T = 1600 \text{ W}$$
$$P_1 = V_1 \times I_1$$
$$= (200 \text{ V})(4 \text{ A})$$
$$P_1 = 800 \text{ W}$$

$$P_2 = V_2 \times I_2$$
$$= (120 \text{ V})(4 \text{ A})$$
$$P_2 = 480 \text{ W}$$
$$P_3 = V_3 \times I_3$$
$$= (80 \text{ V})(4 \text{ A})$$
$$P_3 = 320 \text{ W}$$

	Voltage	Current	Resistance	Power
R_1	200 V	4A	50 Ω	800 W
R_2	120 V	4A	30 Ω	480 W
R_3	80 V	4A	20 Ω	320 W
Total	400 V	4A	100 Ω	1600 W

be obtained, depending on what point in the circuit is common ground (Figure 12-12).

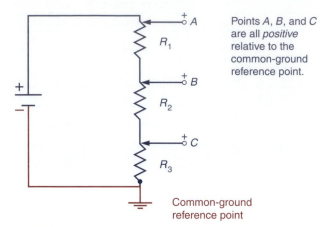

Points *A*, *B*, and *C* are all *positive* relative to the common-ground reference point.

A. Positive voltages

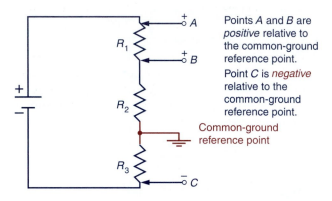

Points *A* and *B* are *positive* relative to the common-ground reference point.

Point *C* is *negative* relative to the common-ground reference point.

B. Positive and negative voltages

Figure 12-12   Circuit with common ground reference point.

12-5 Polarity

The *polarity* (− or +) of the voltage drop across a resistor depends on the direction of current flow through the resistor. Current enters at the negative polarity and leaves at the positive polarity. Figure 12-11 illustrates the concept of relative polarity:

Point *A* is negative relative to Point *B*.
Point *B* is negative relative to Point *C*.
Point *C* is negative relative to Point *D*.

In practical circuits, often some point in the circuit is designated as *common ground* and all voltages are measured relative to that point. Both negative and positive output voltages can

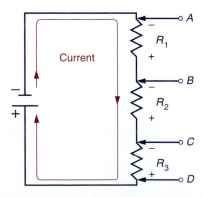

Figure 12-11   Voltage drop polarity.

12-6 Series-Aiding and Series-Opposing Voltages

Some electric and electronic control circuits may contain more than one voltage source within the same series loop. When this is the case, the voltage sources may be connected as series-aiding or series-opposing.

Series-aiding voltage sources are connected with polarities that allow the current to flow in the same direction. The positive terminal of one voltage source is connected to the negative terminal of the other as shown in Figure 12-13. Series-aiding voltages are added to obtain the total equivalent voltage.

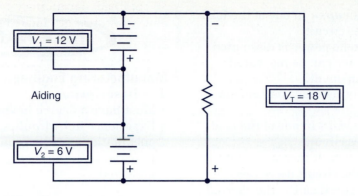

Figure 12-13 Series-aiding voltages.

Series-opposing voltages are connected with polarities that allow the current to flow in opposite directions. The positive terminal of one voltage source is connected to the positive terminal of the other as shown in Figure 12-14. Series-opposing voltages are subtracted to obtain the total equivalent voltage. Subtract the smaller voltage valve from the larger value and give the equivalent voltage the polarity of the larger voltage. If the two voltages are equal, the net voltage and current will be zero.

12-7 Troubleshooting a Series Circuit

Using Ohm's law, you can predict what changes will occur in a circuit under different fault conditions. This information often can be very useful in pinpointing the cause of a problem. As an example, we shall use one of the original series circuits studied and examine how different faults can affect voltage, current, resistance, and power.

Example 12-4

Problem *Open* in component R_1 of the series circuit.

Solution The table shows the circuit conditions with R_1 open.

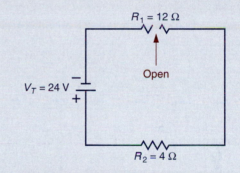

	Voltage	Current	Resistance	Power
R_1	24 V	0	Infinite	0
R_2	0 V	0	4 Ω	0
Total	24 V	0	Infinite	0

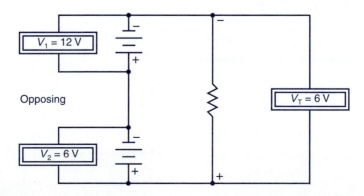

Figure 12-14 Series-opposing voltages.

In summary, with an *open* in one of the load components of a series circuit:

- No current flows, so no power is dissipated.
- The full source voltage can be measured across the open component.
- All voltages across other load components drop to 0 V.

In summary, with a *short* in one of the load components of a series circuit:

- Current flow in the circuit increases and the total resistance of the circuit decreases.
- Zero voltage is measured across the shorted component and the voltage across the other load components increases.
- The total power dissipated by the circuit and the other load components increases.

Example 12-5

Problem   Short across component R_3 of the series circuit.

Solution   The table shows the circuit conditions with R_3 shorted.

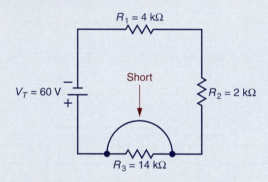

	Voltage	Current	Resistance	Power
R_1	40 V	10 mA	4 kΩ	400 mW
R_2	20 V	10 mA	2 kΩ	200 mW
R_3	0 V	10 mA	0	0
Total	60 V	10 mA	6 kΩ	600 mW

Review and Applications

. . . .

Related Formulas

$I_T = I_1 = I_2 = I_3 \ldots$

$R_T = R_1 + R_2 + R_3 \ldots$

$V_T = V_1 + V_2 + V_3 \ldots$

$P_T = P_1 + P_2 + P_3 \ldots$

$I = V/R$

$V = I \times R$

$R = V/I$

$P = V \times I$

Review Questions

1. (a) State the current characteristic of a series circuit.
   (b) Express this characteristic in the form of an equation.
2. (a) State the voltage characteristic of a series circuit.
   (b) Express this characteristic in the form of an equation.
3. (a) State the resistance characteristic of a series circuit.
   (b) Express this characteristic in the form of an equation.
4. (a) State the power dissipation characteristic of a series circuit.
   (b) Express this characteristic in the form of an equation.

Problems

1. (a) Find all unknown values for V, I, R, and P for the series circuit shown.
   (b) Make a chart in your notebook and record the given and calculated circuit values.
   (c) Show all steps and equations used to arrive at your answers.

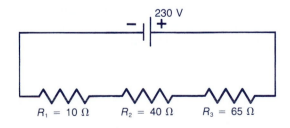

2. (a) Find all unknown values of V, I, R, and P for the series circuit shown.
   (b) Make a chart in your notebook and record the given and calculated circuit values.
   (c) Show all steps and equations used to arrive at your answers.

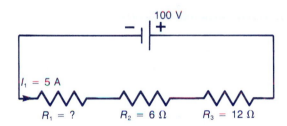

3. (a) Find all unknown values for V, I, R, and P for the series circuit shown.
   (b) Make a chart in your notebook and record the given and calculated circuit values.
   (c) Show all steps and equations used to arrive at your answers.

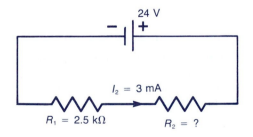

4. (a) Find all unknown values for V, I, R, and P for the series circuit shown.
   (b) Make a chart in your notebook and record the given and calculated circuit values.
   (c) Show all steps and equations used to arrive at your answers.

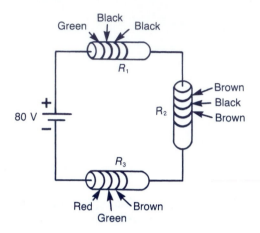

5. What is the polarity of:
   (a) point A relative to the common ground?
   (b) point B relative to the common ground?
   (c) point C relative to the common ground?

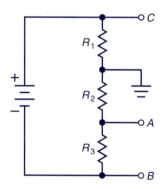

6. (a) Are the two voltage sources connected series-aiding or series-opposing?
   (b) What is the value of V_T applied to the circuit?
   (c) What is the polarity of point A relative to point B?

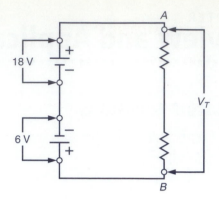

Critical Thinking

1. When an open fault occurs in one of the load components of a series circuit:
   (a) what happens to the current flow?
   (b) how much power is dissipated by the circuit?
   (c) what value of voltage can be measured across the open component?
   (d) what value of voltage can be measured across the other load components?
   (e) what value of total resistance can be measured?

2. When a short occurs in one of the load components of a series circuit:
   (a) what happens to the current flow?
   (b) what happens to the total resistance of the circuit?
   (c) what value of voltage can be measured across the shorted component?
   (d) what happens to the voltage across the other load components?
   (e) what happens to the total power dissipated by the circuit and the other load components?

Portfolio Project

Acquire a 6 V lamp and measure its normal operating current. You are required to have this lamp indicate when power to a 12 V battery is on. Calculate the value of resistor (resistance and wattage) you need to wire in series with the lamp to drop the 12 volts to 6 volts. Connect the circuit and record the value of the current flow and the voltage drop across the resistor and lamp.

Circuit Challenge

Using a Simulator

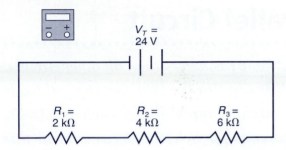

$V_T = 24\ V$

$R_1 = 2\ k\Omega$ $R_2 = 4\ k\Omega$ $R_3 = 6\ k\Omega$

Procedure

(a) Construct the circuit shown, using whatever simulation software package is available to you.

(b) Compute all values of voltage, current, and resistance and record in chart form.

(c) Use the multimeter to measure all values of voltage, current, and resistance.

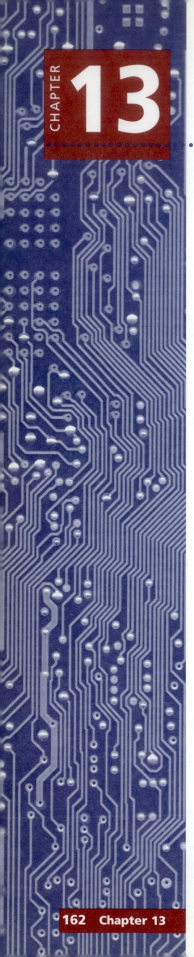

Solving the Parallel Circuit

Parallel circuits are used more frequently than all other circuits. The main reason for their popularity is the fact that all loads can be designed for a common supply voltage such as 12 volts for automobiles and 120 volts for household circuits. In this chapter, we will study the unique characteristics of the parallel circuit.

Objectives

After studying this chapter, you should be able to:
1. State the voltage, current, resistance, and power characteristics of a parallel circuit.
2. Solve for unknown circuit values in a parallel circuit.
3. Measure current paths, voltages, and resistance values in a parallel circuit.
4. Troubleshoot a parallel dc resistive circuit.

Key Terms

- acts independently
- parallel circuit
- branch
- open
- short circuit

- IEEE Organization
- Global Specialties
- Johnson Controls

13-1 Parallel Circuit Connection

A *parallel circuit* is a circuit that has two or more paths (or branches) for current flow. In parallel circuits, the loads connect across each other, like rungs on a ladder, with the voltage source across each load (Figure 13-1). Each load **acts independently** of the other. This connection results in as many pathways or *branches* for the current as there are load components connected in parallel.

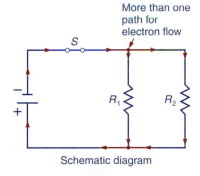

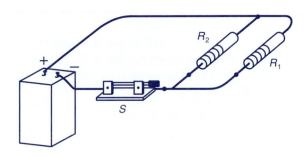

Figure 13-1 Parallel circuit.

13-2 Parallel Circuit Current, Voltage, Resistance, and Power Characteristics

Current   The total current in a parallel circuit is equal to the sum of all the branch currents (Figure 13-2). Each load is connected directly across the supply lines and acts independently of the other loads as far as current is concerned. The amount of current drawn by each load is determined by the resistance value of each load. The higher the resistance of the load, the lower the amount of current in each load.

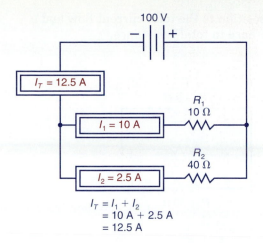

$$I_T = I_1 + I_2$$
$$= 10\ A + 2.5\ A$$
$$= 12.5\ A$$

Figure 13-2   The total current in a parallel circuit is equal to the sum of the branch currents.

Voltage   The voltage across each branch of a parallel circuit is the same and is equal in value to that of the voltage source (Figure 13-3). This is due to the fact that each load is connected directly across the two line wires.

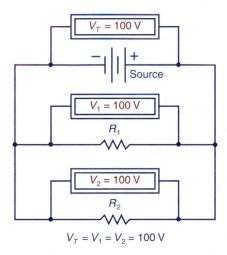

$$V_T = V_1 = V_2 = 100\ V$$

Figure 13-3   The voltage across each branch of a parallel circuit is the same as the applied voltage.

Resistance   The total resistance of a parallel circuit is always lower than the value of any of the resistors in the branches of the circuit (Figure 13-4). When you connect loads in parallel, you have more paths through which the current can flow. This results in less total opposition to the total current flow and a decrease in total resistance.

A.

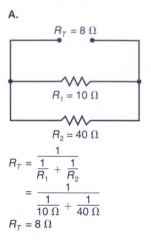

$$R_T = \cfrac{1}{\cfrac{1}{R_1} + \cfrac{1}{R_2}}$$

$$= \cfrac{1}{\cfrac{1}{10\ \Omega} + \cfrac{1}{40\ \Omega}}$$

$$R_T = 8\ \Omega$$

B.

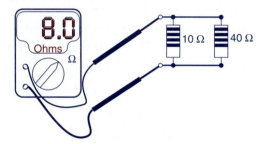

Figure 13-4 The total resistance in a parallel circuit is less than the resistance of any branch.

Power Power dissipation in a parallel circuit is the same as that of a series circuit. Therefore, the total power dissipated (P_T) in a parallel circuit equals the sum of the power dissipated by the individual branch resistors or the product of total current (I_T) and the source voltage (V_T) (Figure 13-5).

13-3 Solving Parallel Circuits

The procedure for solving parallel-circuit values of voltage, current, resistance, and power is similar to that used for solving values for series circuits. Ohm's law as it applies to the circuit as a whole and to individual loads, as well, is used. In addition, the parallel-circuit characteristics of voltage, current, resistance, and power must also be used. The parallel-circuit charac-

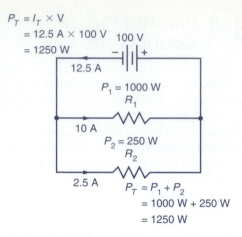

$$P_T = I_T \times V$$
$$= 12.5\ \text{A} \times 100\ \text{V}$$
$$= 1250\ \text{W}$$

$$P_T = P_1 + P_2$$
$$= 1000\ \text{W} + 250\ \text{W}$$
$$= 1250\ \text{W}$$

Figure 13-5 Power dissipation in a parallel circuit equals the sum of the power dissipated by individual branch resistors or the total current multiplied by the total voltage.

teristics expressed in the form of equations are as follows:

$$I_T = I_1 + I_2 + I_3 \ldots$$

$$V_T = V_1 = V_2 = V_3 \ldots$$

$$R_T = \frac{R_1 \times R_2}{R_1 + R_2} \quad \text{(for 2 loads)}$$

$$R_T = \cfrac{1}{\cfrac{1}{R_1} + \cfrac{1}{R_2} + \cfrac{1}{R_3}} \ldots \quad \text{(for 2 or more loads)}$$

$$P_T = P_1 + P_2 + P_3 \ldots$$

Example 13-1

Problem Find all the unknown values for V, I, R, and P for the parallel circuit of Figure 13-6.

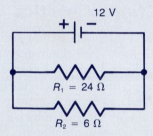

Figure 13-6 Circuit for Example 13-1.

Solution

STEP 1 Make a chart and record all known values.

	Voltage	Current	Resistance	Power
R_1			24 Ω	
R_2			6 Ω	
Total	12 V			

STEP 2 Calculate V_1 and V_2 and enter the values in the chart.

$$V_T = V_1 = V_2 = 12 \text{ V}$$

	Voltage	Current	Resistance	Power
R_1	12 V		24 Ω	
R_2	12 V		6 Ω	
Total	12 V			

STEP 3 Calculate I_1, I_2, and I_T and enter the values in the chart.

$$I_1 = \frac{V_1}{R_1}$$
$$= (12 \text{ V}) \div (24 \text{ Ω})$$
$$I_1 = 0.5 \text{ A}$$
$$I_2 = \frac{V_2}{R_2}$$
$$= (12 \text{ V}) \div (6 \text{ Ω})$$
$$I_2 = 2 \text{ A}$$
$$I_T = I_1 + I_2$$
$$= (0.5 \text{ A}) + (2 \text{ A})$$
$$I_T = 2.5 \text{ A}$$

	Voltage	Current	Resistance	Power
R_1	12 V	0.5 A	24 Ω	
R_2	12 V	2 A	6 Ω	
Total	12 V	2.5 A		

STEP 4 Calculate R_T and enter the value in the chart.

$$R_T = \frac{V_T}{I_T}$$
$$= (12 \text{ V}) \div (2.5 \text{ A})$$
$$R_T = 4.8 \text{ Ω}$$

or

$$R_T = \frac{R_1 \times R_2}{R_1 + R_2}$$
$$= \frac{(24 \text{ Ω})(6 \text{ Ω})}{(24 \text{ Ω}) + (6 \text{ Ω})}$$
$$= \frac{144}{30}$$
$$R_T = 4.8 \text{ Ω}$$

	Voltage	Current	Resistance	Power
R_1	12 V	0.5 A	24 Ω	
R_2	12 V	2 A	6 Ω	
Total	12 V	2.5 A	4.8 Ω	

STEP 5 Calculate P_T, P_1, and P_2 and enter in chart.

$$P_T = V_T \times I_T$$
$$= (12 \text{ V})(2.5 \text{ A})$$
$$P_T = 30 \text{ W}$$
$$P_1 = V_1 \times I_1$$
$$= (12 \text{ V})(0.5 \text{ A})$$
$$P_1 = 6 \text{ W}$$
$$P_2 = V_2 \times I_2$$
$$= (12 \text{ V})(2 \text{ A})$$
$$P_2 = 24 \text{ W}$$

	Voltage	Current	Resistance	Power
R_1	12 V	0.5 A	24 Ω	6 W
R_2	12 V	2 A	6 Ω	24 W
Total	12 V	2.5 A	4.8 Ω	30 W

Example 13-2

Problem Find all the unknown values of V, I, R, and P for the parallel circuit of Figure 13-7.

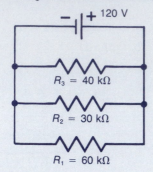

Figure 13-7 Circuit for Example 13-2.

Solution

STEP 1 Make a chart and record all known values.

	Voltage	Current	Resistance	Power
R_1	120 V		60 kΩ	
R_2	120 V		30 kΩ	
R_3	120 V		40 kΩ	
Total	120 V			

STEP 2 Calculate I_1, I_2, I_3, I_T, and R_T and enter in the chart.

$$I_1 = \frac{V_1}{R_1}$$
$$= (120 \text{ V}) \div (60 \text{ k}\Omega)$$
$$I_1 = 2 \text{ mA}$$
$$I_2 = \frac{V_2}{R_2}$$
$$= (120 \text{ V}) \div (30 \text{ k}\Omega)$$
$$I_2 = 4 \text{ mA}$$
$$I_3 = \frac{V_3}{R_3}$$
$$= (120 \text{ V}) \div (40 \text{ k}\Omega)$$
$$I_3 = 3 \text{ mA}$$
$$I_T = I_1 + I_2 + I_3$$
$$= (2 \text{ mA}) + (4 \text{ mA}) + (3 \text{ mA})$$
$$I_T = 9 \text{ mA}$$

$$R_T = \frac{V_T}{I_T}$$
$$= (120 \text{ V}) \div (9 \text{ mA})$$
$$R_T = 13.3 \text{ k}\Omega$$

	Voltage	Current	Resistance	Power
R_1	120 V	2 mA	60 kΩ	
R_2	120 V	4 mA	30 kΩ	
R_3	120 V	3 mA	40 kΩ	
Total	120 V	9 mA	13.3 kΩ	

STEP 3 Calculate P_T, P_1, P_2, and P_3 and enter in the chart.

$$P_T = V_T \times I_T$$
$$= (120 \text{ V})(0.009 \text{ A})$$
$$P_T = 1.08 \text{ W}$$
$$P_1 = V_1 \times I_1$$
$$= (120 \text{ V})(0.002 \text{ A})$$
$$P_1 = 0.24 \text{ W}$$
$$P_2 = V_2 \times I_2$$
$$= (120 \text{ V})(0.004 \text{ A})$$
$$P_2 = 0.48 \text{ W}$$
$$P_3 = V_3 \times I_3$$
$$= (120 \text{ V})(0.003 \text{ A})$$
$$P_3 = 0.36 \text{ W}$$

	Voltage	Current	Resistance	Power
R_1	120 V	2 mA	60 Ω	0.24 W
R_2	120 V	4 mA	30 Ω	0.48 W
R_3	120 V	3 mA	40 Ω	0.36 W
Total	120 V	9 mA	13.3 Ω	1.08 W

Example 13-3

Problem Find all the unknown values of V, I, R, and P for the parallel circuit of Figure 13-8.

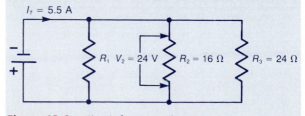

Figure 13-8 Circuit for Example 13-3.

Solution

STEP 1 Make a chart and record all known values.

	Voltage	Current	Resistance	Power
R_1	24 V			
R_2	24 V		16 Ω	
R_3	24 V		24 Ω	
Total	24 V	5.5 A		

STEP 2 Calculate I_2, I_3, I_1, R_1, and R_T and enter in the chart.

$$I_2 = \frac{V_2}{R_2}$$
$$= (24 \text{ V}) \div (16 \text{ Ω})$$
$$I_2 = 1.5 \text{ A}$$

$$I_3 = \frac{V_3}{R_3}$$
$$= (24 \text{ V}) \div (24 \text{ Ω})$$
$$I_3 = 1 \text{ A}$$

$$I_1 = I_T - (I_2 + I_3)$$
$$= (5.5 \text{ A}) - (2.5 \text{ A})$$
$$I_1 = 3 \text{ A}$$

$$R_1 = \frac{V_1}{I_1}$$
$$= (24 \text{ V}) \div (3 \text{ A})$$
$$R_1 = 8 \text{ Ω}$$

$$R_T = \frac{V_T}{I_T}$$
$$= (24 \text{ V}) \div (5.5 \text{ A})$$
$$R_T = 4.36 \text{ Ω}$$

	Voltage	Current	Resistance	Power
R_1	24 V	3 A	8 Ω	
R_2	24 V	1.5 A	16 Ω	
R_3	24 V	1 A	24 Ω	
Total	24 V	5.5 A	4.36 Ω	

STEP 3 Calculate P_T, P_1, P_2, and P_3 and enter the values in the chart.

$$P_T = V_T \times I_T$$
$$= (24 \text{ V})(5.5 \text{ A})$$
$$P_T = 132 \text{ W}$$
$$P_1 = V_1 \times I_1$$

$$= (24 \text{ V})(3 \text{ A})$$
$$P_1 = 72 \text{ W}$$

$$P_2 = V_2 \times I_2$$
$$= (24 \text{ V})(1.5 \text{ A})$$
$$P_2 = 36 \text{ W}$$

$$P_3 = V_3 \times I_3$$
$$= (24 \text{ V})(1 \text{ A})$$
$$P_3 = 24 \text{ W}$$

	Voltage	Current	Resistance	Power
R_1	24 V	3 A	8 Ω	72 W
R_2	24 V	1.5 A	16 Ω	36 W
R_3	24 V	1 A	24 Ω	24 W
Total	24 V	5.5 A	4.36 Ω	132 W

13-4 Troubleshooting a Parallel Circuit

Troubleshooting parallel circuits for open or shorted resistors is not much different from troubleshooting a series circuit. Again, using Ohm's law, you can predict what changes will occur under different fault conditions. Using one of the original parallel circuits studied, we shall examine how different faults can affect voltage, current, resistance, and power.

Example 13-4

Problem OPEN in component R_1 of the parallel circuit.

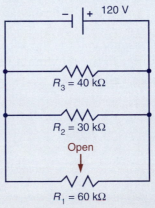

Figure 13-9 Circuit for Example 13-4.

Solution The table shows the circuit conditions with R_1 open.

	Voltage	Current	Resistance	Power
R_1	120 V	0	Infinite	0
R_2	120 V	4 mA	30 kΩ	0.48 W
R_3	120 V	3 mA	40 kΩ	0.36 W
Total	120 V	7 mA	17.1 kΩ	0.84 W

In summary, with an **open** in one of the load components of a parallel circuit:

• The total resistance increases, resulting in less total current flow and total power dissipation.

• The open branch will have normal voltage, but infinite resistance, which will result in zero current flow and power dissipation.

• All other branches will remain unchanged with normal voltage, resistance, current, and power dissipation.

Troubleshooting for a *shorted* resistor in a parallel circuit is not quite so easy. This is because a shorted resistor shorts out the voltage source. A shorted branch has practically zero resistance. With no opposition, the voltage source will deliver its maximum current flow. Depending on the voltage source, the amount of current can be dangerously high. Normally the circuit-protective device (fuse or circuit breaker) would open the circuit in the event of a short across any of the branches.

When troubleshooting for a shorted resistor, open all branch current paths in the circuit. Then, metering power supply voltage, close each branch, one at a time. Consider Figure 13-10. When you close the branch of the shorted resistor:

• The power supply voltage drops to near zero

• Current jumps to the maximum value the power supply can provide

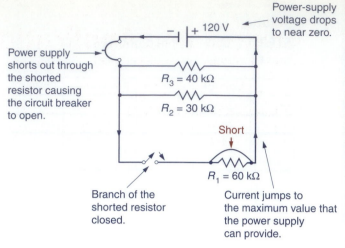

Power supply shorts out through the shorted resistor causing the circuit breaker to open.

Power-supply voltage drops to near zero.

Branch of the shorted resistor closed.

Current jumps to the maximum value that the power supply can provide.

Figure 13-10 Effect of a short circuit across parallel branches.

• The power supply shorts out through the shorted resistor and the circuit breaker opens to prevent burnout of the power supply and wiring.

HELP WANTED

HVAC Apprentice
Job Description
• Maintain and install heating, ventilation, and air conditioning systems.
• Multiyear apprenticeship program for high school graduate.
• High school courses in electricity/electronics, sheet metal, air conditioning, CAD, computer applications desirable.
• Gain hands-on skills by working in the field, earn money, and learn at the same time.

Review and Applications

Related Formulas

$$I_T = I_1 + I_2 + I_3 \ldots$$

$$R_T = \cfrac{1}{\dfrac{1}{R_1} + \dfrac{1}{R_2} + \dfrac{1}{R_3} \ldots}$$

$$V_T = V_1 = V_2 = V_3 \ldots$$

$$P_T = P_1 + P_2 + P_3 \ldots$$

$$I = V/R$$

$$V = I \times R$$

$$R = V/I$$

$$P = V \times I$$

Review Questions

1. (a) State the current characteristic of a parallel circuit.
   (b) Express this characteristic in the form of an equation.
2. (a) State the voltage characteristic of a parallel circuit.
   (b) Express this characteristic in the form of an equation.
3. (a) State the resistance characteristic of a parallel circuit.
   (b) Express this characteristic in the form of an equation.
4. (a) State the power dissipation characteristic of a parallel circuit.
   (b) Express this characteristic in the form of an equation.

Problems

1. (a) Find all unknown values of V, I, R, and P for the parallel circuit shown.
   (b) Make a chart in your notebook and record given and calculated circuit values.
   (c) Show all steps and equations used to arrive at your answers.

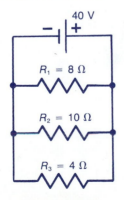

40 V

$R_1 = 8 \ \Omega$

$R_2 = 10 \ \Omega$

$R_3 = 4 \ \Omega$

	Voltage	Current	Resistance	Power
R_1	__ V	__ A	__ Ω	__ W
R_2	__ V	__ A	__ Ω	__ W
R_3	__ V	__ A	__ Ω	__ W
Total	__ V	__ A	__ Ω	__ W

2. (a) Find all unknown values of V, I, R, and P for the parallel circuit shown.
   (b) Make a chart in your notebook and record given and calculated circuit values.

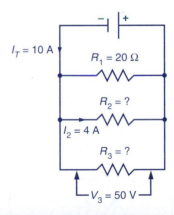

$I_T = 10 \ A$

$R_1 = 20 \ \Omega$

$R_2 = ?$

$I_2 = 4 \ A$

$R_3 = ?$

$V_3 = 50 \ V$

	Voltage	Current	Resistance	Power
R_1	__ V	__ A	__ Ω	__ W
R_2	__ V	__ A	__ Ω	__ W
R_3	__ V	__ A	__ Ω	__ W
Total	__ V	__ A	__ Ω	__ W

   (c) Show all steps and equations used to arrive at your answers.

3. (a) Find all unknown values of V, I, R, and P for the parallel circuit shown.
(b) Make a chart in your notebook and record given and calculated circuit values.
(c) Show all steps and equations used to arrive at your answers.

	Voltage	Current	Resistance
R_1	__ V	__ mA	__ kΩ
R_2	__ V	__ mA	__ kΩ
Total	__ V	__ mA	__ kΩ

R_1 = 4.7 kΩ
I_1 = 22 mA
R_2 = 3.3 kΩ

	Voltage	Current	Resistance	Power
R_1	__ V	__ mA	__ kΩ	__ mW
R_2	__ V	__ mA	__ kΩ	__ mW
Total	__ V	__ mA	__ kΩ	__ mW

4. (a) Find all unknown values of V, I, R, and P for the parallel circuit shown.
(b) Make a chart in your notebook and record given and calculated circuit values.
(c) Show all steps and equations used to arrive at your answers.

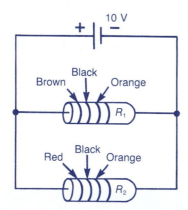

(Assume the tolerance value of each resistor is ±0%)

Critical Thinking

1. When an open fault occurs in one of the load components of a parallel circuit what changes, if any, occur in the:
   (a) total circuit resistance?
   (b) total current flow?
   (c) total power dissipation?
   (d) voltage across the open branch?
   (e) resistance of the open branch?
   (f) current of the open branch?
   (g) power dissipation of the open branch?
   (h) voltage across all other branches?
   (i) resistance of all other branches?
   (j) current through all other branches?
   (k) power dissipation of all other branches?
2. When a short occurs in one of the branches of a parallel circuit what happens to the:
   (a) power supply voltage?
   (b) total current flow?

Portfolio Project

Breadboard a circuit consisting of a battery, fuse, switch, and three identical lamps connected in parallel. Simulate each of the following faults and report on the test procedure used in locating the problem:
(1) weak or dead battery
(2) blown fuse
(3) shorted switch
(4) open lamp filament
(5) shorted lamp socket

Circuit Challenge

Using a Simulator

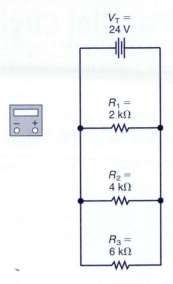

$V_T = 24$ V

$R_1 = 2$ kΩ

$R_2 = 4$ kΩ

$R_3 = 6$ kΩ

Procedure

(a) Construct the circuit shown, using whatever simulation software package is available to you.

(b) Compute all values of voltage, current, and resistance and record in chart form.

(c) Use the multimeter to measure all values of voltage, current, and resistance.

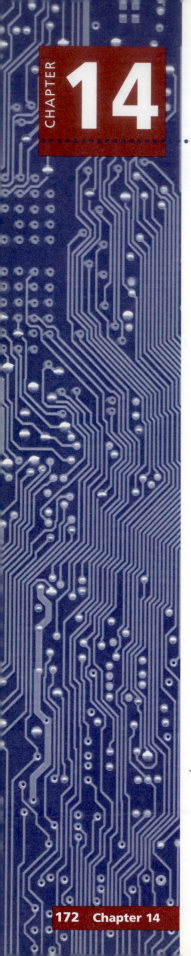

Solving the Series-Parallel Circuit

*T*he two previous chapters discussed circuit loads that were connected in series or in parallel. A series-parallel circuit consists of a group of series and parallel loads in which the total current flows through at least one of the loads. Most electrical and electronic equipment require the use of combination series-parallel circuits.

Objectives

After studying this chapter, you should be able to:
1. State and correctly apply Kirchhoff's voltage and current laws.
2. Recognize the series and parallel branches in series-parallel circuits.
3. Determine equivalent resistance in series-parallel circuits.
4. Solve a series-parallel circuit for unknown quantities of voltage, current, and resistance.
5. Troubleshoot series-parallel dc resistive circuits.

Key Terms

- Kirchhoff's voltage law
- closed loop
- Kirchhoff's current law
- node
- combination circuit
- equivalent resistance
- parallel component
- series component

- Arcade-Electronics
- Aromat
- Omron

14-1 Kirchhoff's Voltage Law

Kirchhoff's laws are an extension of Ohm's law. They can be considered as additional tools for solving values for electric circuits. Used along with Ohm's law, these laws allow you to analyze complex series-parallel circuit networks.

Kirchhoff's voltage law is used to describe the voltage characteristics of a series dc circuit. It states the relationship between the voltage drops and the applied voltage of a series circuit. Recall that the sum of all the voltage drops in a series circuit is equal to the applied voltage (Figure 14-1). This fact is referred to as Kirchhoff's voltage law.

$$V_T = V_1 + V_2 + V_3$$

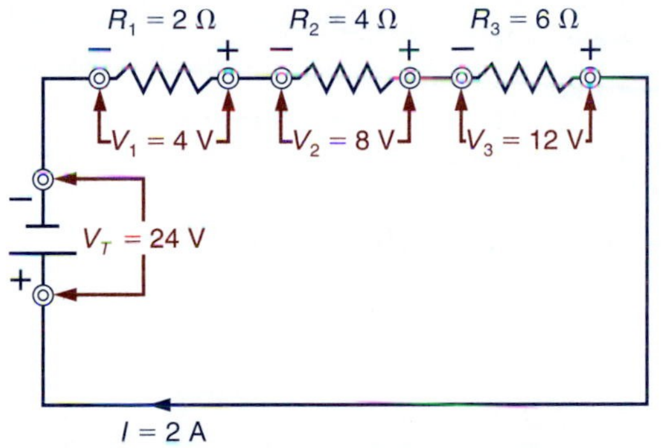

R_1 = 2 Ω   R_2 = 4 Ω   R_3 = 6 Ω

$V_1 = 4\ V$ $V_2 = 8\ V$ $V_3 = 12\ V$

$V_T = 24\ V$

$I = 2\ A$

Figure 14-1   Kirchhoff's voltage law applied to the series circuit.

This voltage law can also be worded as "around a *closed loop,* the algebraic sum of all the voltages is zero." It becomes apparent that this statement is true when we trace the volt-

age around the loop of a series circuit. The polarity of the voltage drop across each load is opposite to that of the applied voltage (Figure 14-2). Thus, the applied voltage is a voltage rise or positive voltage. The voltage across each load is a voltage drop or negative voltage.

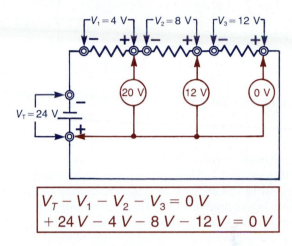

$V_1 = 4\ V$ $V_2 = 8\ V$ $V_3 = 12\ V$

20 V   12 V   0 V

$V_T = 24\ V$

$$V_T - V_1 - V_2 - V_3 = 0\ V$$
$$+ 24\ V - 4\ V - 8\ V - 12\ V = 0\ V$$

Figure 14-2   Tracing the voltages in a series circuit.

14-2 Kirchhoff's Current Law

Kirchhoff's current law is used to describe the current characteristics of a parallel dc circuit. The law refers to the relationship between the total current flow in a parallel circuit and the current flow through each of its branches. Recall that the sum of the branch currents is equal to the current flowing from the voltage source (Figure 14-3). This is referred to as Kirchhoff's current law.

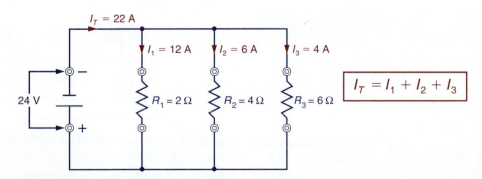

$I_T = 22\ A$

$I_1 = 12\ A$ $I_2 = 6\ A$ $I_3 = 4\ A$

24 V

$R_1 = 2\ Ω$ $R_2 = 4\ Ω$ $R_3 = 6\ Ω$

$$I_T = I_1 + I_2 + I_3$$

Figure 14-3   Kirchhoff's current law applied to the parallel circuit.

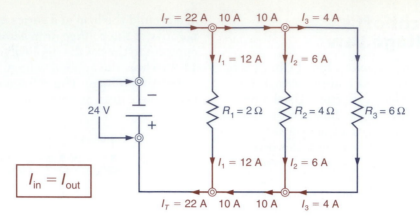

Figure 14-4 Tracing the currents in a parallel circuit.

Stated simply, Kirchhoff's current law reads, "the current entering any point in a circuit is equal to the current leaving that same point." It becomes apparent that this statement is true when you analyze the currents in and out of different points in a parallel circuit. No matter what point is selected, the total current flow into the point is always equal to the current flowing out of the point (Figure 14-4). A connecting point or junction for two or more components is often referred to as a *node*.

14-3 Solving Series-Parallel Circuits

A *series-parallel* or *combination circuit* is one that is made up of both series and parallel paths. Most electric and electronic devices require the use of series-parallel circuits. You can apply your knowledge of how series and parallel circuits operate to simplify and analyze series-parallel circuits.

Example 14-1

Problem Both Ohm's law and Kirchhoff's laws are used in solving a series-parallel circuit. The general procedure followed is to find the *equivalent resistance* of each parallel group of components and then treat the result as a series circuit. Find all unknown values of V, I, R, and P for the series-parallel circuit of Figure 14-5.

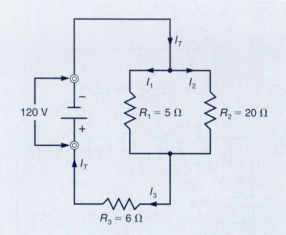

Circuit

According to Kirchhoff's laws

$$I_T = I_1 + I_2$$
$$I_T = I_3$$

$$V_T = V_1 + V_3$$
$$V_T = V_2 + V_3$$

Figure 14-5 Circuit for Example 14-1.

Solution

STEP 1 Make a chart and record all known values.

	Voltage	Current	Resistance	Power
R_1			5 Ω	
R_2			20 Ω	
R_3			6 Ω	
Total	120 V			

STEP 2 Calculate R_T and enter in chart. First calculate the equivalent resistance of R_1 and R_2 in parallel.

$$R_{eq} = \frac{R_1 \times R_2}{R_1 + R_2}$$

$$R_{eq} = \frac{(5\ \Omega)(20\ \Omega)}{(5\ \Omega) + (20\ \Omega)}$$

$$R_{eq} = 4\ \Omega$$

Replacing parallel resistors R_1 and R_2 with one of an equivalent value produces an equivalent series circuit (Figure 14-6). The total resistance then becomes:

$$R_T = R_{eq} + R_3$$

$$R_T = (4\ \Omega) + (6\ \Omega)$$

$$R_T = 10\ \Omega$$

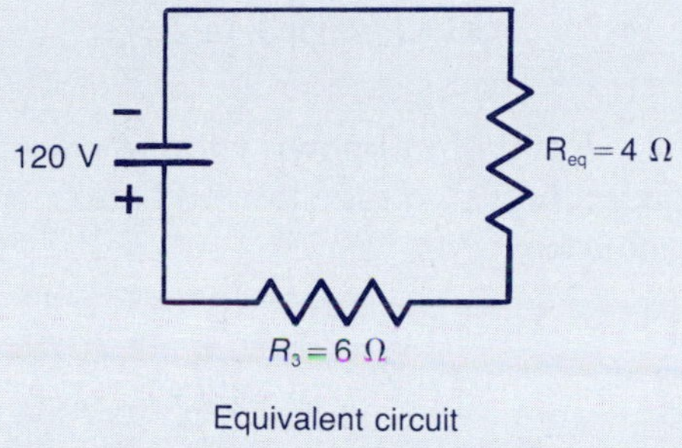

Equivalent circuit

Figure 14-6   Simplifying the series-parallel circuit.

	Voltage	Current	Resistance	Power
R_1			5 Ω	
R_2			20 Ω	
R_3			6 Ω	
Total	120 V		10 Ω	

STEP 3   Calculate I_T and I_3 and enter the values in the chart. Total current is calculated by applying Ohm's law.

$$I_T = \frac{V_T}{R_T}$$

$$I_T = \frac{(120\ \text{V})}{(10\ \Omega)}$$

$$I_T = 12\ \text{A}$$

According to Kirchhoff's law, I_3 is equal to I_T or 12 A.

	Voltage	Current	Resistance	Power
R_1			5 Ω	
R_2			20 Ω	
R_3		12 A	6 Ω	
Total	120 V	12 A	10 Ω	

STEP 4   Calculate V_3, V_2, and V_1 and enter in chart.

Using Ohm's law to calculate

$$V_3 = I_3 \times R_3$$

$$V_3 = (12\ \text{A})(6\ \Omega)$$

$$V_3 = 72\ \text{V}$$

Using Kirchhoff's laws to calculate

$$V_2 = V_T - V_3$$

$$V_2 = (120\ \text{V}) - (72\ \text{V})$$

$$V_2 = 48\ \text{V}$$

$$V_1 = V_T - V_3$$

$$V_1 = (120\ \text{V}) - (72\ \text{V})$$

$$V_1 = 48\ \text{V}$$

	Voltage	Current	Resistance	Power
R_1	48 V		5 Ω	
R_2	48 V		20 Ω	
R_3	72 V	12 A	6 Ω	
Total	120 V	12 A	10 Ω	

STEP 5   Calculate I_1 and I_2 and enter in the chart.

Using Ohm's law to calculate:

$$I_1 = \frac{V_1}{R_1}$$

$$= \frac{(48\ \text{V})}{(5\ \Omega)}$$

$$I_1 = 9.6\ \text{A}$$

$$I_2 = \frac{V_2}{R_2}$$

$$= \frac{(48\ \text{V})}{(20\ \Omega)}$$

$$I_2 = 2.4\ \text{A}$$

Or using Kirchhoff's laws to calculate:

$$I_2 = I_T - I_1$$

$$I_2 = (12\ \text{A}) - (9.6\ \text{A})$$

$$I_2 = 2.4\ \text{A}$$

	Voltage	Current	Resistance	Power
R_1	48 V	9.6 A	5 Ω	
R_2	48 V	2.4 A	20 Ω	
R_3	72 V	12 A	6 Ω	
Total	120 V	12 A	10 Ω	

STEP 6 Calculate P_T, P_1, P_2, and P_3 and enter in the chart.

$$P_T = V_T \times I_T$$
$$= (120 \text{ V})(12 \text{ A})$$
$$P_T = 1440 \text{ W}$$
$$P_1 = V_1 \times I_1$$
$$= (48 \text{ V})(9.6 \text{ A})$$
$$P_1 = 461 \text{ W}$$
$$P_2 = V_2 \times I_2$$
$$= (48 \text{ V})(2.4 \text{ A})$$
$$P_2 = 115 \text{ W}$$
$$P_3 = V_3 \times I_3$$
$$= (72 \text{ V})(12 \text{ A})$$
$$P_3 = 864 \text{ W}$$

	Voltage	Current	Resistance	Power
R_1	48 V	9.6 A	5 Ω	461 W
R_2	48 V	2.4 A	20 Ω	115 W
R_3	72 V	12 A	6 Ω	864 W
Total	120 V	12 A	10 Ω	1440 W

Example 14-2

Problem If certain key voltage and current measurements of a circuit are known, all unknown voltages and currents can be found by using only Kirchhoff's laws. Given the circuit in Figure 14-7 with voltages and currents as indicated, find all unknown values of V, I, R, and P.

Solution
STEP 1 Make a chart and record all known values.

	Voltage	Current	Resistance	Power
R_1	24 V			
R_2		3 mA		
R_3		1.5 mA		
R_4	12 V			
Total		5 mA		

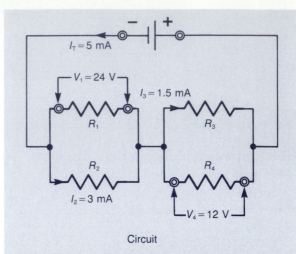

Circuit

Figure 14-7 Circuit for Example 14-2.

STEP 2 Find all unknown voltage values using Kirchhoff's voltage law and enter them in the chart.

$$V_T = V_1 + V_4$$
$$= (24 \text{ V}) + (12 \text{ V})$$
$$V_T = 36 \text{ V}$$
$$V_2 = V_T - V_4$$
$$= (36 \text{ V}) - (12 \text{ V})$$
$$V_2 = 24 \text{ V}$$
$$V_3 = V_T - V_1$$
$$= (36 \text{ V}) - (24 \text{ V})$$
$$V_3 = 12 \text{ V}$$

	Voltage	Current	Resistance	Power
R_1	24 V			
R_2	24 V	3 mA		
R_3	12 V	1.5 mA		
R_4	12 V			
Total	36 V	5 mA		

STEP 3 Find all unknown current values using Kirchhoff's current law and enter in chart.

$$I_1 = I_T - I_2$$
$$= (5 \text{ mA}) - (3 \text{ mA})$$
$$I_1 = 2 \text{ mA}$$
$$I_4 = I_T - I_3$$
$$= (5 \text{ mA}) - (1.5 \text{ mA})$$
$$I_4 = 3.5 \text{ mA}$$

	Voltage	Current	Resistance	Power
R_1	24 V	2 mA		
R_2	24 V	3 mA		
R_3	12 V	1.5 mA		
R_4	12 V	3.5 mA		
Total	36 V	5 mA		

STEP 4 Calculate all resistance values and enter them in the chart.

$$R_T = \frac{V_T}{I_T}$$
$$= \frac{(36\ \text{V})}{(5\ \text{mA})}$$
$$R_T = 7.2\ \text{k}\Omega$$
$$R_1 = \frac{V_1}{I_1}$$
$$= \frac{(24\ \text{V})}{(2\ \text{mA})}$$
$$R_1 = 12\ \text{k}\Omega$$
$$R_2 = \frac{V_2}{I_2}$$
$$= \frac{(24\ \text{V})}{(3\ \text{mA})}$$
$$R_2 = 8\ \text{k}\Omega$$
$$R_3 = \frac{V_3}{I_3}$$
$$= \frac{(12\ \text{V})}{(1.5\ \text{mA})}$$
$$R_3 = 8\ \text{k}\Omega$$
$$R_4 = \frac{V_4}{I_4}$$
$$= \frac{(12\ \text{V})}{(3.5\ \text{mA})}$$
$$R_4 = 3.43\ \text{k}\Omega$$

	Voltage	Current	Resistance	Power
R_1	24 V	2 mA	12 kΩ	
R_2	24 V	3 mA	8 kΩ	
R_3	12 V	1.5 mA	8 kΩ	
R_4	12 V	3.5 mA	3.43 kΩ	
Total	36 V	5 mA	7.2 kΩ	

STEP 5 Calculate all power dissipation values and enter them in the chart.

$$P_T = V_T \times I_T$$
$$= (36\ \text{V})(5\ \text{mA})$$
$$P_T = 180\ \text{mW}$$
$$P_1 = V_1 \times I_1$$
$$= (24\ \text{V})(2\ \text{mA})$$
$$P_1 = 48\ \text{mW}$$
$$P_2 = V_2 \times I_2$$
$$= (24\ \text{V})(3\ \text{mA})$$
$$P_2 = 72\ \text{mW}$$
$$P_3 = V_3 \times I_3$$
$$= (12\ \text{V})(1.5\ \text{mA})$$
$$P_3 = 18\ \text{mW}$$
$$P_4 = V_4 \times I_4$$
$$= (12\ \text{V})(3.5\ \text{mA})$$
$$P_4 = 42\ \text{mW}$$

	Voltage	Current	Resistance	Power
R_1	24 V	2 mA	12 kΩ	48 mW
R_2	24 V	3 mA	8 kΩ	72 mW
R_3	12 V	1.5 mA	8 kΩ	18 mW
R_4	12 V	3.5 mA	3.43 kΩ	42 mW
Total	36 V	5 mA	7.2 kΩ	180 mW

Example 14-3

There are two basic types of series-parallel circuit connections. One involves *parallel components* connected in series. Examples 14-1 and 14-2 dealt with this type. The other type involves *series components* connected in parallel. An example of this type is illustrated in the following problem.

Problem Find all unknown values of V, I, R, and P for the series-parallel circuit of Figure 14-8.

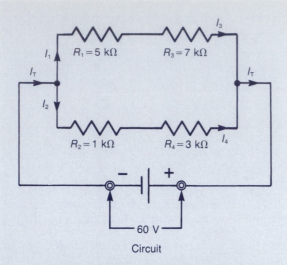

According to Kirchhoff's laws:

$$I_T = I_1 + I_2$$
$$I_1 = I_3$$
$$I_2 = I_4$$
$$I_T = I_3 + I_4$$

$$V_T = V_1 + V_3$$
$$V_T = V_2 + V_4$$

Figure 14-8 Circuit for Example 14-3.

STEP 1 Make a chart and record all known values.

	Voltage	Current	Resistance	Power
R_1			5 kΩ	
R_2			1 kΩ	
R_3			7 kΩ	
R_4			3 kΩ	
Total	60 V			

STEP 2 Calculate R_T and enter in chart. R_1 and R_3 in series become:

$$R_{eq} = R_1 + R_3$$
$$= (5 \text{ k}\Omega) + (7 \text{ k}\Omega)$$
$$R_{eq} = 12 \text{ k}\Omega$$

R_2 and R_4 in series become:

$$R_{eq} = R_2 + R_4$$
$$= (1 \text{ k}\Omega) + (3 \text{ k}\Omega)$$
$$R_{eq} = 4 \text{ k}\Omega$$

Replace series resistors R_1 and R_3 with one of an equivalent value (R_A), and replace series resistors R_2 and R_4 with one of an equivalent size (R_B). The circuit is thus reduced to the equivalent parallel circuit of Figure 14-9. The total resistance then becomes:

$$R_T = \frac{R_A \times R_B}{R_A + R_B}$$
$$= \frac{(12 \text{ k}\Omega)(4 \text{ k}\Omega)}{(12 \text{ k}\Omega) + (4 \text{ k}\Omega)}$$
$$R_T = 3 \text{ k}\Omega$$

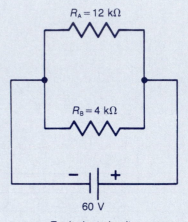

Equivalent circuit

Figure 14-9 Equivalent circuit for Example 14-3.

	Voltage	Current	Resistance	Power
R_1			5 kΩ	
R_2			1 kΩ	
R_3			7 kΩ	
R_4			3 kΩ	
Total	60 V		3 kΩ	

STEP 3 Calculate all current values and enter them in the chart.

Using Ohm's law to calculate

$$I_T = \frac{V_T}{R_T}$$
$$= \frac{(60 \text{ V})}{(3 \text{ k}\Omega)}$$
$$I_T = 20 \text{ mA}$$

The equivalent current flow, I_A and I_B, can be calculated using their equivalent series resistance as follows:

$$I_A = \frac{V_T}{R_A}$$

$$= \frac{(60 \text{ V})}{(12 \text{ k}\Omega)}$$

$$I_A = 5 \text{ mA}$$

$$I_B = \frac{V_T}{R_B}$$

$$= \frac{(60 \text{ V})}{(4 \text{ k}\Omega)}$$

$$I_B = 15 \text{ mA}$$

Using Kirchhoff's laws to calculate

$$I_A = I_1 = I_3 = 5 \text{ mA}$$
$$I_B = I_2 = I_4 = 15 \text{ mA}$$

	Voltage	Current	Resistance	Power
R_1		5 mA	5 kΩ	
R_2		15 mA	1 kΩ	
R_3		5 mA	7 kΩ	
R_4		15 mA	3 kΩ	
Total	60 V	20 mA	3 kΩ	

STEP 4 Calculate all unknown voltage values and enter in chart.

Using Ohm's law to calculate

$$V_1 = I_1 \times R_1$$
$$= (5 \text{ mA})(5 \text{ k}\Omega)$$
$$V_1 = 25 \text{ V}$$
$$V_2 = I_2 \times R_2$$
$$= (15 \text{ mA})(1 \text{ k}\Omega)$$
$$V_2 = 15 \text{ V}$$
$$V_3 = I_3 \times R_3$$
$$= (5 \text{ mA})(7 \text{ k}\Omega)$$
$$V_3 = 35 \text{ V}$$
$$V_4 = I_4 \times R_4$$
$$= (15 \text{ mA})(3 \text{ k}\Omega)$$
$$V_4 = 45 \text{ V}$$

	Voltage	Current	Resistance	Power
R_1	25 V	5 mA	5 kΩ	
R_2	15 V	15 mA	1 kΩ	
R_3	35 V	5 mA	7 kΩ	
R_4	45 V	15 mA	3 kΩ	
Total	60 V	20 mA	3 kΩ	

STEP 5 Calculate all power dissipation values and enter in chart.

$$P_T = V_T \times I_T$$
$$= (60 \text{ V})(20 \text{ mA})$$
$$P_T = 1.2 \text{ W}$$
$$P_1 = V_1 \times I_1$$
$$= (25 \text{ V})(5 \text{ mA})$$
$$P_1 = 125 \text{ mW}$$
$$P_2 = V_2 \times I_2$$
$$= (15 \text{ V})(15 \text{ mA})$$
$$P_2 = 225 \text{ mW}$$
$$P_3 = V_3 \times I_3$$
$$= (35 \text{ V})(5 \text{ mA})$$
$$P_3 = 175 \text{ mW}$$
$$P_4 = V_4 \times I_4$$
$$= (45 \text{ V})(15 \text{ mA})$$
$$P_4 = 675 \text{ mW}$$

	Voltage	Current	Resistance	Power
R_1	25 V	5 mA	5 kΩ	125 mW
R_2	15 V	15 mA	1 kΩ	225 mW
R_3	35 V	5 mA	7 kΩ	175 mW
R_4	45 V	15 mA	3 kΩ	675 mW
Total	60 V	20 mA	3 kΩ	1.2 W

Example 14-4

Some series-parallel circuits are connected so that it is difficult to identify which components are in parallel and which are in series. The most difficult part of solving this type of circuit may be in identifying the series and parallel parts. In this case, it is best to redraw the circuit in a simpler schematic form before starting any calculations. An example of this type is illustrated in the following problem.

Problem Find all unknown values of V, I, R, and P for the series-parallel circuit shown in Figure 14-10.

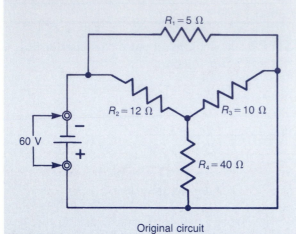

Original circuit

Figure 14-10 Circuit for Example 14-4.

STEP 1 Simplify the circuit as shown in Figure 14-11.

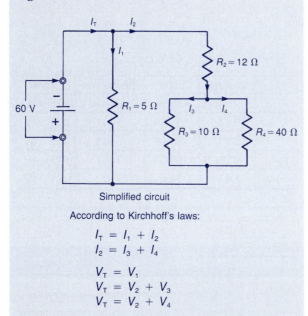

Simplified circuit

According to Kirchhoff's laws:

$$I_T = I_1 + I_2$$
$$I_2 = I_3 + I_4$$

$$V_T = V_1$$
$$V_T = V_2 + V_3$$
$$V_T = V_2 + V_4$$

Figure 14-11 Simplified circuit for Example 14-4.

STEP 2 Make a chart and record all known values.

	Voltage	Current	Resistance	Power
R_1	60 V		5 Ω	
R_2			12 Ω	
R_3			10 Ω	
R_4			40 Ω	
Total	60 V			

STEP 3 Calculate R_T and enter the value in the chart.

Solution R_3 and R_4 in parallel become:

$$R_{eq} = \frac{R_3 \times R_4}{R_3 + R_4}$$
$$= \frac{(10\ \Omega)(40\ \Omega)}{(10\ \Omega) + (40\ \Omega)}$$
$$R_{eq} = 8\ \Omega = R_A$$

Replace parallel resistors R_3 and R_4 with one of equivalent size (R_A). R_A is connected in series with R_2. The equivalent resistance of this branch (R_B) then becomes

$$R_{eq} = R_2 + R_A$$
$$= (12\ \Omega) + (8\ \Omega)$$
$$R_{eq} = 20\ \Omega = R_B$$

This is shown in Figure 14-12.

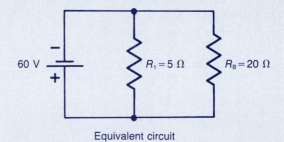

Equivalent circuit

Figure 14-12 Equivalent circuit for Example 14-4.

The total resistance is then found by solving the equivalent parallel circuit as follows:

$$R_T = \frac{R_1 \times R_B}{R_1 + R_B}$$

$$R_T = \frac{(5\ \Omega)(20\ \Omega)}{(5\ \Omega) + (20\ \Omega)}$$

$$R_T = 4\ \Omega$$

	Voltage	Current	Resistance	Power
R_1	60 V		5 Ω	
R_2			12 Ω	
R_3			10 Ω	
R_4			40 Ω	
Total	60 V		4 Ω	

STEP 4   Calculate all unknown current and voltage values and enter in the chart.

$$I_T = \frac{V_T}{R_T}$$

$$= \frac{(60\ \text{V})}{(4\ \Omega)}$$

$$I_T = 15\ \text{A}$$

$$I_1 = \frac{V_1}{R_1}$$

$$= \frac{(60\ \text{V})}{(5\ \Omega)}$$

$$I_1 = 12\ \text{A}$$

$$I_2 = I_T - I_1$$

$$= (15\ \text{A}) - (12\ \text{A})$$

$$I_2 = 3\ \text{A}$$

$$V_2 = I_2 \times R_2$$

$$= (3\ \text{A})(12\ \Omega)$$

$$V_2 = 36\ \text{V}$$

$$V_3 = V_T - V_2$$

$$= (60\ \Omega) - (36\ \text{V})$$

$$V_3 = 24\ \text{V}$$

$$I_3 = \frac{V_3}{R_3}$$

$$= \frac{(24\ \text{V})}{(10\ \Omega)}$$

$$I_3 = 2.4\ \text{A}$$

$$V_4 = V_3 = 24\ \text{V}$$

$$I_4 = \frac{V_4}{R_4}$$

$$= \frac{(24\ \text{V})}{(40\ \Omega)}$$

$$I_4 = 0.6\ \text{A}$$

	Voltage	Current	Resistance	Power
R_1	60 V	12 A	5 Ω	
R_2	36 V	3 A	12 Ω	
R_3	24 V	2.4 A	10 Ω	
R_4	24 V	0.6 A	40 Ω	
Total	60 V	15 A	4 Ω	

STEP 5   Calculate all power dissipation values and enter in chart.

$$P_T = V_T \times I_T$$

$$= (60\ \text{V})(15\ \text{A})$$

$$P_T = 900\ \text{W}$$

$$P_1 = V_1 \times I_1$$

$$= (60\ \text{V})(12\ \text{A})$$

$$P_1 = 720\ \text{W}$$

$$P_2 = V_2 \times I_2$$

$$= (36\ \text{V})(3\ \text{A})$$

$$P_2 = 108\ \text{W}$$

$$P_3 = V_3 \times I_3$$

$$= (24\ \text{V})(2.4\ \text{A})$$

$$P_3 = 57.6\ \text{W}$$

$$P_4 = V_4 \times I_4$$

$$= (24\ \text{V})(0.6\ \text{A})$$

$$P_4 = 14.4\ \text{W}$$

	Voltage	Current	Resistance	Power
R_1	60 V	12 A	5 Ω	720 W
R_2	36 V	3 A	12 Ω	108 W
R_3	24 V	2.4 A	10 Ω	57.6 W
R_4	24 V	0.6 A	40 Ω	14.4 W
Total	60 V	15 A	4 Ω	900 W

14-4 Troubleshooting Series-Parallel Circuits

To troubleshoot a series-parallel circuit, the same method you used to troubleshoot series circuits and parallel circuits is used. First, solve the circuit for all normal operating voltage, current, and resistance values. Next, measure the actual circuit values and compare these values to your calculated values. You usually measure voltages first, because these measurements are easy to make and because you don't have to disturb the circuit. Often voltage measurements give you enough information to determine the faulty component.

Generally, if your measured and calculated values disagree by more than 10 percent, you can assume the circuit is defective. This percent of error allowable can vary. With very sensitive circuits, a smaller percentage difference can indicate a faulty circuit.

Normal-Operating Circuit

	Voltage	Current	Resistance	Power
R_1	48 V	9.6 A	5 Ω	461 W
R_2	48 V	2.4 A	20 Ω	115 W
R_3	72 V	12 A	6 Ω	864 W
Total	120 V	12 A	10 Ω	1440 W

Faulty Circuit—Opened R_1 Resistor

	Voltage	Current	Resistance	Power
R_1	92.4 V	0	Infinite	0
R_2	92.4 V	4.62 A	20 Ω	427 W
R_3	27.7 V	4.62 A	6 Ω	128 W
Total	120 V	4.62 A	26 Ω	555 W

Example 14-5

Problem   Opened R_1 resistor in the series-parallel circuit.

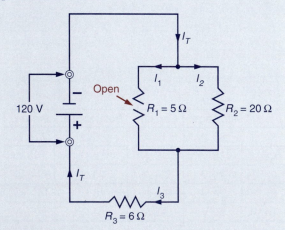

Figure 14-13   Circuit for Example 14-5.

Solution   The tables show a comparison between the normal-operating circuit and the faulty circuit.

Example 14-6

Problem   Shorted R_2 resistor in the series-parallel circuit.

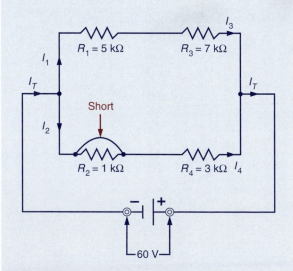

Figure 14-14   Circuit for Example 14-6.

Solution   The tables show a comparison between the normal-operating circuit and the faulty circuit.

Normal-Operating Circuit

	Voltage	Current	Resistance	Power
R_1	25 V	5 mA	5 kΩ	125 mW
R_2	15 V	15 mA	1 kΩ	225 mW
R_3	35 V	5 mA	7 kΩ	175 mW
R_4	45 V	15 mA	3 kΩ	675 mW
Total	60 V	20 mA	3 kΩ	1.2 W

Faulty Circuit—Shorted R_2 Resistor

	Voltage	Current	Resistance	Power
R_1	25 V	5 mA	5 kΩ	125 mW
R_2	0	20 mA	0	0
R_3	35 V	5 mA	7 kΩ	175 mW
R_4	60 V	20 mA	3 kΩ	1.2 W
Total	60 V	25 mA	2.4 kΩ	1.5 W

HELP WANTED

Consumer Electronics Technician
Job Description
- Repair televisions, stereos, VCRs, and CD players.
- Broad knowledge of electronic circuits and mechanical devices required.
- High school diploma plus 3 to 5 years experience.
- Good attendance record.

Review and Applications

Related Formulas

$$V_T = V_1 + V_2 + V_3 \ldots \qquad I = V/R$$

$$I_T = I_1 + I_2 + I_3 \ldots \qquad V = I \times R$$

$$R_{eq} = \frac{R_1 \times R_2}{R_1 + R_2} \qquad R = V/I$$

$$P_T = P_1 + P_2 + P_3 \ldots \qquad P = V \times I$$

Review Questions

1. State Kirchhoff's voltage law.
2. State Kirchhoff's current law.
3. Determine, using Kirchhoff's laws, V_T, V_1, I_1, and I_3 for the circuit shown in Figure 14-15.

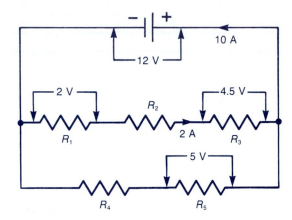

Figure 14-16 Circuit for question 4.

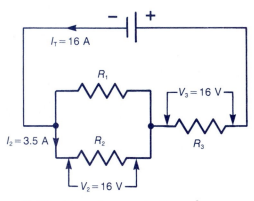

Figure 14-15 Circuit for question 3.

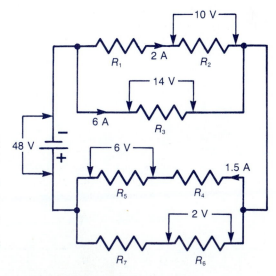

4. Determine, using Kirchhoff's laws, I_1, V_2, I_4, and V_4 for the circuit shown in Figure 14-16.
5. Determine, using Kirchhoff's laws, V_1, V_4, V_7, I_7, and I_T for the circuit shown in Figure 14-17.

Figure 14-17 Circuit for question 5.

Problems

1. (a) Find all unknown values of V, I, R, and P for the series-parallel circuit shown in Figure 14-18. Make a chart in your notebook and record given and calculated circuit values. Show all steps and equations used to arrive at your answers.

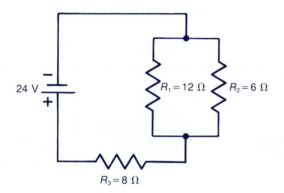

Figure 14-18   Circuit for problem 1.

(b) Repeat for an opened R_3 resistor fault.
(c) Repeat for a shorted R_2 resistor fault.

2. Find all unknown values of V, I, R, and P for the series-parallel circuit shown in Figure 14-19. Make a chart in your notebook and record given and calculated circuit values. Show all steps and equations used to arrive at your answers.

3. (a) Find all unknown values of V, I, R, and P for the series-parallel circuit shown in Figure 14-20. Make a chart in your notebook and record given and calculated values. Show all steps and equations used to arrive at your answers.

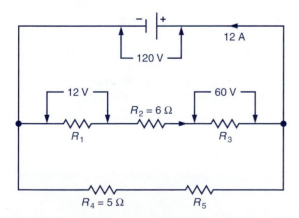

Figure 14-19   Circuit for problem 2.

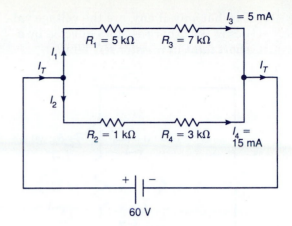

Figure 14-20   Circuit for problem 3.

(b) Repeat for an opened R_1 resistor fault.
(c) Repeat for a shorted R_4 resistor fault.

4. (a) Find all unknown values of V, I, R, and P for the series-parallel circuit (Figure 14-21). Make a chart in your notebook and record given and calculated values. Show all steps and equations used to arrive at your answers.

(b) Repeat for an opened R_3 resistor fault.
(c) Repeat for a shorted R_1 resistor fault.

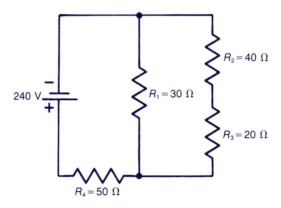

Figure 14-21   Circuit for problem 4.

Critical Thinking

1. (a) Find all unknown values of V, I, R, and P for the series-parallel circuit shown in Figure 14-22. Make a chart in your notebook and record given and calculated values. Show all steps and equations used to arrive at your answers.

(b) In what way, if any, are the voltage values of the other resistors affected by an open fault in resistor R_5? Why?

(c) In what way, if any, are the voltage values of the other resistors affected by a short fault in resistor R_5? Why?

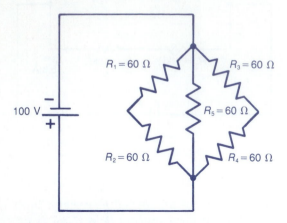

Figure 14-22 Circuit for Critical Thinking question 1.

Portfolio Project

Breadboard a series-parallel circuit consisting of a single resistor in series with two other resistors connected in parallel with each other. Calculate all values of voltage, current, and resistance when the circuit is connected to a 9 V-battery source. Operate the circuit and measure all values of voltage, current, and resistance. Explain any minor differences between calculated and measured values.

Circuit Challenge

Using a Simulator

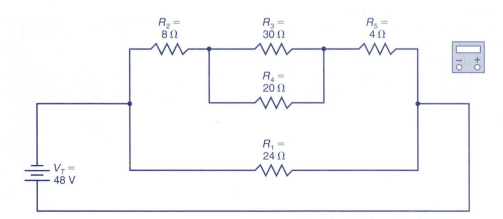

Procedure

(a) Construct the circuit shown, using whatever simulation software package is available to you.

(b) Compute all values of voltage, current, and resistance and record in chart form.

(c) Use the multimeter to measure all values of voltage, current, and resistance.

Magnetism

A magnet *is a piece of iron oxide or special alloy that exerts an invisible force of attraction on objects made of iron, nickel, or cobalt. The invisible force itself is called magnetism or magnetic force. This chapter discusses the characteristics of magnetic fields, along with their field patterns. Permanent and temporary magnets are introduced and the differences between them examined.*

Objectives

After studying this chapter, you should be able to:
1. Define common magnetic terms.
2. Identify magnetic materials, nonmagnetic materials, and magnetic alloys.
3. State the law of magnetic poles.
4. Describe the characteristics of magnetic lines of force.
5. Explain the theory of magnetism.

Key Terms

- magnet
- magnetism
- magnetic material
- nonmagnetic material
- lodestone
- north pole
- south pole
- ring magnet
- magnetic polarity

- north-seeking pole
- magnetic shielding
- molecular theory of magnetism
- temporary magnet
- permanent magnet
- retentivity
- Alnico
- magnetic poles

- law of magnetic poles
- horseshoe magnet
- magnetic field
- magnetic flux
- magnetic lines of force
- electron theory of magnetism
- magnetic saturation

RESEARCH INTERNET SITES

- Magnet Sales
- A & A Magnetics
- Dexter Magnetic Technologies

15-1 Properties of Magnets

The ability of certain materials to attract objects made of iron or iron alloys is the most familiar of all magnetic effects. This property of a material to attract pieces of iron or steel is called *magnetism.*

Magnetic materials are those materials that magnets attract. Some common magnetic materials are iron, steel, nickel, and cobalt (Figure 15-1). You can magnetize any magnetic material.

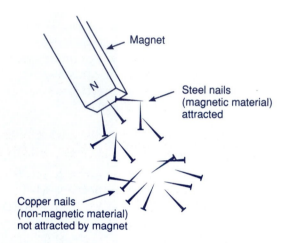

Figure 15-1 Types of magnets.

Nonmagnetic materials are those materials that magnets do not attract. Examples of nonmagnetic materials are copper, aluminum, lead, silver, brass, wood, glass, liquids, and gases. You cannot magnetize any nonmagnetic material.

15-2 Types of Magnets

Natural and Artificial Magnets   The effects of magnetism were first observed in pieces of iron ore called *lodestone* or magnetite. Lodestone is said to be a natural magnet because it possesses magnetic qualities when found in its natural form. Natural magnets have very little practical use because it is possible to produce much stronger magnets by artificial means. Artificial magnets are those made from ordinary unmagnetized magnetic materi-

als. The bar magnet, horseshoe magnet, and compass needle are all examples of artificial magnets (Figure 15-2).

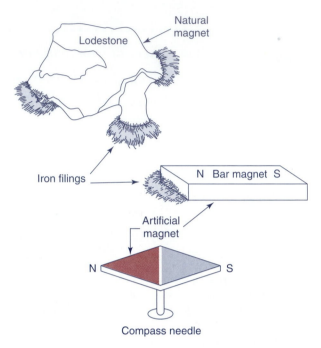

Figure 15-2   Natural and artificial magnets.

Most artificial magnets are produced electrically. The process used is a simple one. To magnetize a magnetic material using electricity, the material to be magnetized is first placed in a coil of insulated wire. A direct current voltage source is then momentarily applied to the coil leads (Figure 15-3A). To demagnetize an artificial magnet, the same process is repeated but the voltage source used is alternating current (Figure 15-3B).

When a material is easy to magnetize, it is said to have high **permeability**.

Temporary and Permanent Magnets

Different magnetic materials have different abilities to retain their magnetism once they are magnetized. The ability of a material to retain its magnetism is determined by the **retentivity** of the material. **Temporary magnets** have low retentivity (Figure 15-4A). They lose most of their magnetic power once the magnetizing force is removed. Soft irons have a low retentivity factor and as such make good temporary magnets. **Permanent magnets** are made from hard iron and steel (Figure 15-4B).

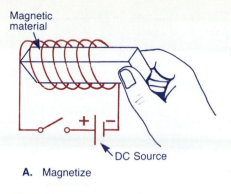

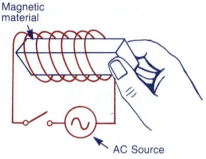

A. Magnetize

B. Demagnetize

Figure 15-3 The magnetizing and demagnetizing processes.

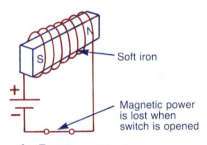

A. Temporary magnet

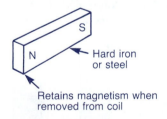

B. Permanent magnet

Figure 15-4 Retaining magnetism.

It requires more energy to magnetize them; however, once magnetized, they will retain their magnetism for a long period of time.

Magnetic alloys are a combination of certain magnetic and nonmagnetic materials. **Alnico**, for example, is a combination of one nonmagnetic metal (aluminum), two weakly magnetic metals (nickel and cobalt), and a good magnetic metal (iron). You can magnetize any magnetic alloy.

One special category of permanent magnets is that of **ceramic magnets**, often referred to as ferrites. Ceramic magnets are made by combining iron-oxide particles with a ceramic compound. Ceramic magnets can be molded to any shape and have very high electrical resistance.

The magnetism that remains in a magnetic material, once the magnetizing force is removed, is called residual magnetism. This term is usually only applied to **temporary magnets**. Residual magnetism is of importance in certain types of generators, because it provides the initial voltage required for the generator to build up to its rated voltage.

About ⬛ Electronics

The ocean floor is magnetized north to south except around underwater mountains, where lava flow makes magnetic spokes. Hammerhead sharks use the spokes as roads to relocate feeding grounds; they produce a current and then sense the magnetic differentials.

15-3 Law of Magnetic Poles

The effects of magnetism are strong at the ends of the magnet and weak in the middle. The ends of the magnet, where the attractive forces are the greatest, are called the *poles of the magnet*. Every magnet has two such poles. These poles are identified as the **north** and **south poles** of the magnet (Figure 15-5).

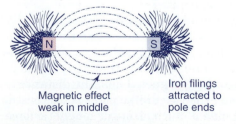

Figure 15-5 Magnetic poles.

The *law of magnetic poles* states that like poles repel and unlike poles attract. Placing the north pole of a suspended magnet near the south pole of a second magnet will cause the two pole ends to come together, or to be attracted to each other (Figure 15-6A). Repeating this experiment using the two north pole ends will cause the two pole ends to move apart, or to produce a repelling effect (Figure 15-6B).

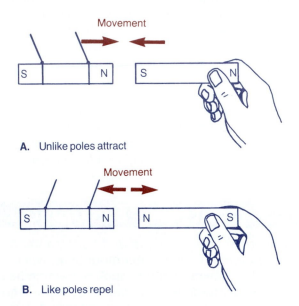

A. Unlike poles attract

B. Like poles repel

Figure 15-6   Law of magnetic poles.

The attraction or repulsion between magnets varies directly with the product of their strengths.

If one bar magnet is placed on a desk and a second magnet is moved slowly toward it, you will observe that the attracting or repelling force will increase as the distance between the poles of the magnet decreases. Actually, this magnetic force varies inversely as the square of the distance between poles changes. For example, if the distance between two unlike poles is increased to twice the distance, the attracting force will be reduced to one-fourth of its former value (Figure 15-7).

The *horseshoe magnet* is actually a bar magnet that is bent around in the shape of a horseshoe. This brings the two poles of the magnet closer together than they are in a straight bar magnet. Thus, the distance between the two unlike poles is decreased, producing a much stronger magnetic force (Figure 15-8).

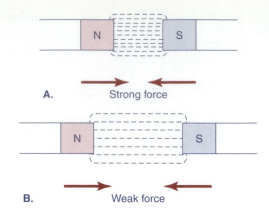

A.   Strong force

B.   Weak force

Figure 15-7   Distance and force between poles.

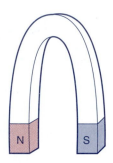

Figure 15-8   Horseshoe magnet.

A *ring magnet* (Figure 15-9) is actually like two horseshoe magnets placed together with opposite poles touching. They form a closed loop with a hole in the center. Since the loop has no

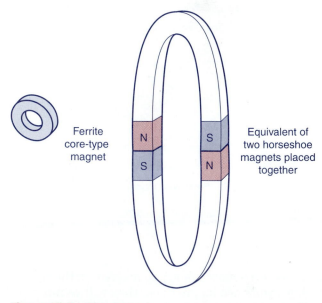

Ferrite core-type magnet

Equivalent of two horseshoe magnets placed together

Figure 15-9   Ring magnet.

open ends, there can be no air gap and no designated poles.

15-4 Magnetic Polarity

Just as a dc-voltage source has negative (−) and positive (+) terminals representing electrical polarity, a magnetic source has north (N) and south (S) poles representing **magnetic polarity**.

The earth itself is a natural magnet with magnetic poles located near the north and south geographic poles (Figure 15-10). A compass is simply a permanent magnet pivoted at its midpoint so that it is free to move in a horizontal plane. Due to the magnetic attraction between poles the compass will always come to rest with the same end pointing toward the north. The end of the compass that points to the geographic north was established as being the *north-seeking pole* of the compass. Thus, the north-seeking end of the compass is considered to be the north pole of the compass. The opposite end is the south pole of the compass.

If a small compass is placed near one end of a bar magnet, the force between the poles of the magnet and the poles of the compass-needle magnet will move the compass needle away from its usual north-south direction. As long as the compass needle is small in comparison with the bar magnet, the compass will point in the direction of force exerted by the bar magnet on the poles of the compass. The compass can be used to identify the polarity of the poles of a magnet (Figure 15-11). First, identify the north and south poles of the compass. Remember the north pole of the compass points to the geographic north pole. Next, place the compass at one of the pole ends of the magnet. Apply the law of magnetic poles to identify the unmarked pole of the magnet. If the north-seeking pole of the compass is attracted to the pole, then it is a south-magnetic pole. If the south pole of the compass is attracted to the pole then it is a north-magnetic pole.

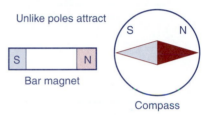

Figure 15-11 Using a compass to mark the polarity of a magnet.

15-5 The Magnetic Field

The area surrounding a magnet, in which the invisible magnetic force is evident, is called the **magnetic field** of the magnet. A representation of this magnetic field pattern can be observed through the use of iron filings sprinkled in the area around the magnet (Figure 15-12). When you bring unlike poles together, the lines of force unite to produce one magnetic field equal to the sum of the two separate magnetic fields.

At times, it is necessary to illustrate the direction and intensity of magnetic field patterns. A commonly used method of representing the forces in a magnetic field is by use of lines called **magnetic lines of force**. An entire group of magnetic field lines is called **magnetic flux** or simply flux.

Although lines of force are invisible, they are assumed to have certain characteristics. These characteristics are summarized as follows:
- Lines of force never cross one another.
- Lines of force form closed loops.

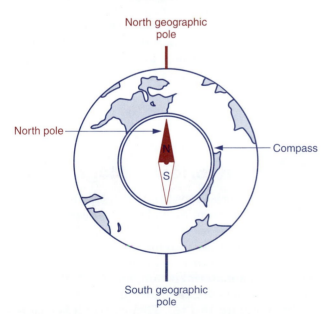

Figure 15-10 The earth is a natural magnet.

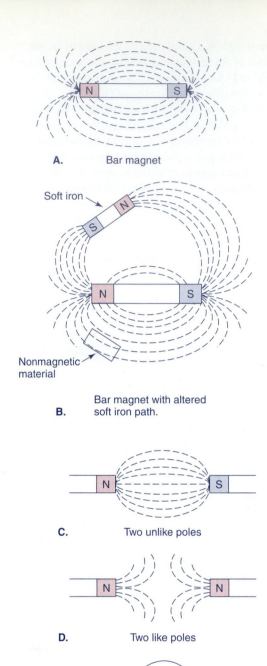

A.   Bar magnet

Soft iron

Nonmagnetic material

B.   Bar magnet with altered soft iron path.

C.   Two unlike poles

D.   Two like poles

E.   Horseshoe magnet

Figure 15-12   Magnetic field patterns.

- Lines of force travel from the north pole to the south pole outside the magnet and from the south pole to north pole inside the magnet.
- Lines of force follow the easiest path, passing most easily through soft iron.
- The stronger the magnet, the greater the flux density (lines per square inch).
- Lines of force repel each other.
- There is no known insulator for lines of force.

Rather than using iron filings, the magnetic field can be investigated more accurately using a compass. When a compass is placed in a magnetic field, the north pole of the compass will point in the direction of the lines of force.

15-6 Magnetic Shielding

Certain types of electric and electronic equipment are affected in their operation and accuracy by stray magnetic lines of force. As mentioned, one of the characteristics of lines of force is that there is no known insulator for them and this presents a problem in protecting devices from stray magnetic fields (Figure 15-13A). The problem is overcome by making use of another characteristic. That characteristic is that lines of force travel most easily through soft iron. As an example, meters that need to be protected are surrounded by a low resistance soft iron magnetic path so that any stray magnetic lines of force travel around rather than through the meter (Figure 15-13B). The same principle of design is applied in motors and transformers to minimize the radiation of lines of force from the magnetic fields of these devices.

15-7 Theories of Magnetism

Different theories have been developed over the years in an attempt to explain what causes magnetism. The ***molecular theory of magnetism*** assumes that each molecule (group of atoms) of a substance is, in fact, a small magnet. When a material is unmagnetized, its molecular magnets are arranged in a random fashion (Figure 15-14A). The net result is a cancellation of the magnetic effect. In a magnetized

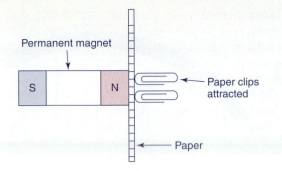

A. No insulator for lines of force

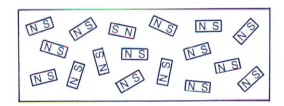

B. Protecting against lines of force

Figure 15-13 Shielding against magnetic lines of force.

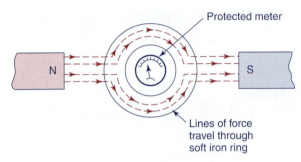

A. Unmagnetized bar

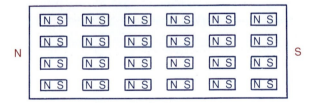

B. Magnetized bar

Figure 15-14 Molecular theory of magnetism.

bar the molecular magnets are arranged so that their magnetic fields are aligned in the same direction (Figure 15-14B).

If a magnet is split in half, the molecular theory also applies, since each half will possess both a north and south pole.

The ***electron theory of magnetism*** is a more modern theory of magnetism. Electrons

are believed to spin on their axes, in the same way that the earth turns on its axis, as they orbit around the nucleus. The spinning effect of the electron creates a magnetic field. The polarity of this magnetic field is determined by the direction the electron is spinning.

Unmagnetized materials have electrons spinning in different directions causing the cancellation of the magnetic effect (Figure 15-15A). Magnetized materials tend to have most or all of their electrons spinning in the same direction (Figure 15-15B).

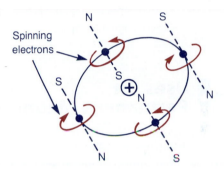

A. Unmagnetized material

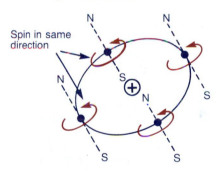

B. Magnetized material

Figure 15-15 Electron theory of magnetism.

There is a definite limit to the amount of magnetism a material can have. This limit is reached when all the molecular magnets are aligned or all the electrons are spinning in the same direction. When maximum magnetic strength is reached the material is said to be *magnetically saturated*.

Proper handling of permanent magnets is important. Any striking or dropping of a magnet can disturb the alignment of the molecular magnets. Heat can also cause the molecular magnets to move; this disturbs their alignment and might cause them to demagnetize. Proper storage of permanent magnets is important to prolong their strength. Always store magnets so they form closed magnetic loops with opposite poles touching (Figure 15-16).

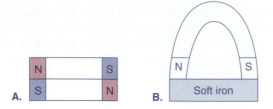

Figure 15-16 Storing permanent magnets to form closed magnetic loops.

15-8 Uses for Permanent Magnets

Permanent magnets of various shapes are used extensively in electric and electronic equipment. Horseshoe magnets are often used in the construction of analog-type deflection meters (Figure 15-17).

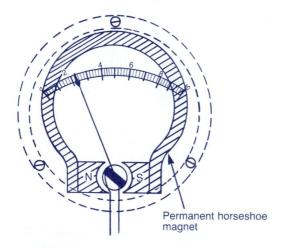

Figure 15-17 Permanent-magnet meter.

Both small wheel-operated bicycle generators and hand-operated flashlight generators use permanent magnets to generate the voltage used to operate lights (Figure 15-18). Rotating the shaft of the generator provides the mechanical force or motion required for generator action, while the internal permanent magnet provides the required magnetism. The faster

A. Bicycle generator

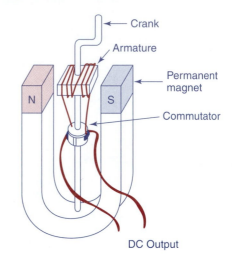

Internal Construction

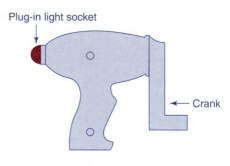

B. Flashlight generator

Figure 15-18 Permanent-magnet generator.

the shaft is turned, the more voltage is generated and the more brightly the lamp will glow.

Permanent-magnet dc motors are used to convert electric energy into mechanical energy. Their operation depends on the interaction of two magnetic fields. One magnetic field is produced by a fixed permanent magnet and the other by an electromagnet wound on a movable armature (Figure 15-19).

Permanent-magnet loudspeakers are the most common of all speakers. They are

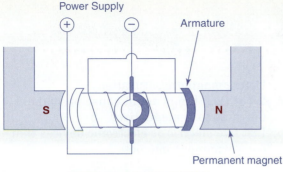

INTERNAL CONSTRUCTION

MOTOR

Figure 15-19   Permanent-magnet dc motor.

designed to convert electric energy into sound energy. The voice coil is suspended in the air gap of a permanent-magnet arrangement. When current flows through the coil, a second magnetic field is established, which causes the coil to vibrate (Figure 15-20).

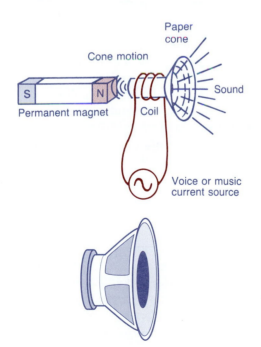

Figure 15-20   Permanent-magnet speaker.

Magnetic switches are used in alarm systems to detect the opening of a door or window (Figure 15-21). A permanent magnet is mounted on the window or door and a special switch is mounted on the frame. When the window or door is closed, the two units are aligned and the magnetic field attracts a metal bar, keeping the switch contacts closed. If the window or door is opened, the magnet moves and the switch contacts open to activate a circuit that rings the alarm.

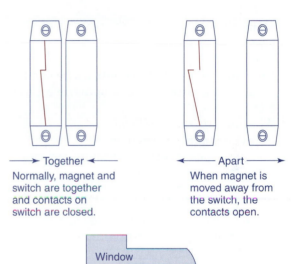

→ Together ←
Normally, magnet and switch are together and contacts on switch are closed.

← Apart →
When magnet is moved away from the switch, the contacts open.

Figure 15-21   Permanent-magnet switch.

Review and Applications

Review Questions

1. Define magnetism.
2. Classify each of the following as being a magnetic or nonmagnetic material:
   copper
   iron
   nickel
   aluminum
   brass
   steel
3. Explain how a coil of insulated wire can be used to magnetize and demagnetize an iron bar.
4. What are magnetic alloys?
5. Define residual magnetism.
6. Compare the retentivity of temporary and permanent magnets.
7. State the law of magnetic poles.
8. What is the relationship between the distance between two unlike poles and the amount of magnetic attraction between them?
9. The polarity of one end of a bar magnet is to be determined using a compass. If the north pole of the compass needle is attracted to this end, what is its polarity?
10. List five characteristics of magnetic lines of force.
11. Explain how instruments are shielded against stray magnetic fields.
12. Compare the way the magnetic effect is explained according to the molecular theory of magnetism and the electron theory of magnetism.

13. Define magnetic saturation.
14. What two things might cause permanent magnets to demagnetize?
15. List five electric or electronic devices that use permanent magnets.

Problems

1. What happens when you bring the north pole of one magnet near to the south pole of another magnet?
2. What type of magnet ceases to be a magnet after you remove the magnetizing force?

Critical Thinking

1. A horseshoe magnet is often stored with a soft iron bar across the space between its poles. Why?

Portfolio Project

Design and construct a magnetism demonstration board showing each of the following:
• Identifying the poles of a magnet showing the path of magnetic lines of force.
• Proving the law of magnetic poles.
• Identifying magnetic and nonmagnetic materials.
• Inducing temporary magnetism in soft iron.
• Inducing permanent magnetism in steel.
• Testing various materials as magnetic shields.

Electromagnetism

Many electrical/electronic devices depend on electromagnetism for their operation. Electromagnetism deflects an electron beam to create a picture on your TV screen, generates the high-voltage spark in your car's ignition system, and turns the shaft of an electric motor. Electromagnetism is produced as free electrons move through a conductor in the form of current. In this chapter we will study this important relationship between electricity and magnetism.

Objectives

After studying this chapter, you should be able to:

1. Correctly apply the conductor and coil left-hand rules.
2. Sketch common electromagnetic field patterns.
3. State the factors that determine the strength of an electromagnet.
4. Explain Ohm's law for the magnetic circuit.
5. Describe the basic operation of practical electromagnetic devices.

Key Terms

- left-hand conductor rule
- left-hand coil rule
- toroid core
- permeability
- magnetic circuit
- magnetomotive force
- ampere turns
- reluctance
- solenoid
- transformer

RESEARCH INTERNET SITES

- Takaha America
- C R Magnetics
- Pinnacle-Speakers

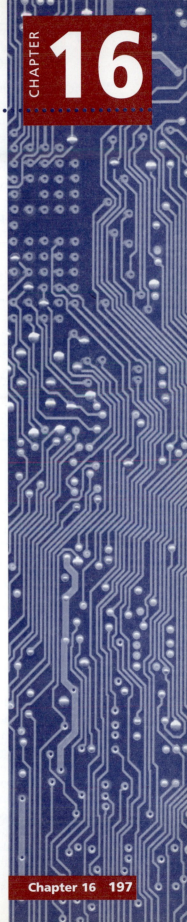

16-1 Magnetic Field Around a Current-Carrying Conductor

Whenever electrons flow through a conductor, a magnetic field is created around the conductor (Figure 16-1). This important relationship between electricity and magnetism is known as *electromagnetism*, or the magnetic effect of current. If dc current flows, the magnetic field will act in one direction, either clockwise or counterclockwise, around the conductor. An ac-current flow will produce a magnetic field that varies in direction with the direction of the electron flow.

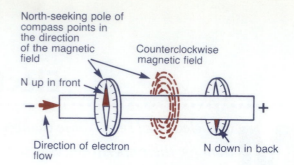

Figure 16-2 Using a compass to indicate the magnetic field.

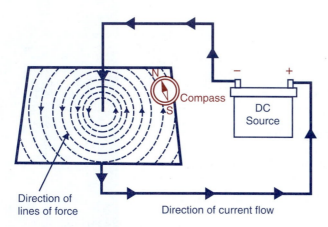

Figure 16-1 Magnetic field around a current-carrying conductor.

The strength of the magnetic field around a single conductor is usually weak and therefore goes undetected. A compass can be used to reveal both the presence and direction of this magnetic field (Figure 16-2). When the compass is brought close to a conductor carrying dc current, the *north-seeking pole* of the compass needle will point in the direction the magnetic lines of force are traveling. As the compass is rotated around the conductor, a definite circular pattern will be indicated.

The amount of current flowing through a single conductor determines the strength of the magnetic field produced around the conductor. *The greater the current flow, the stronger the magnetic field produced.* Currents of 2 to 3 A can be produced by momentarily shorting a single piece of wire across an ordinary D cell. The presence of the magnetic field around the

shorted conductor can be detected by dipping the wire into a pile of iron filings (Figure 16-3). The filings will be attracted to the wire and will cling to it as long as a complete circuit is maintained to create an electron flow.

An ordinary compass can be used to trace a short-to-ground fault in automotive wiring (Figure 16-4). The compass makes use of the fact that a wire-carrying current creates a magnetic field. To locate a short-to-ground in circuits that are protected by a circuit breaker, turn the circuit

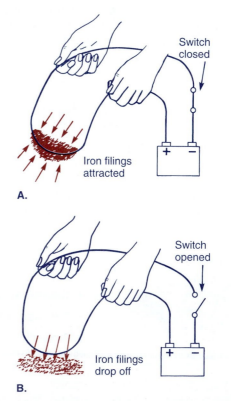

Figure 16-3 Using iron filings to detect the presence of a magnetic field.

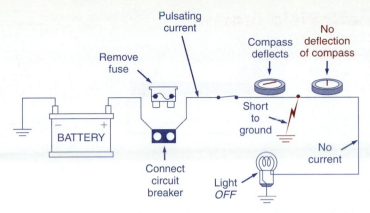

Figure 16-4 Using a compass to locate a short-to-ground fault.

ON and follow the wiring with a compass. The circuit breaker will open and close the circuit as it heats up and cools down. The compass needle will "kick" each time the breaker closes. As the compass passes the point of the short-to-ground, it will stop deflecting because the current is not flowing in this portion of the circuit. Since there is no *insulator* for lines of force, the compass will show a deflection even through trim panels. Thus, the problem can be pinpointed without removing trim. If the circuit is fused, the problem can be located in the same manner by substituting a circuit breaker for the fuse. Short detectors that operate on the same principle also are commercially available.

16-2 Left-Hand Conductor Rule

A definite relationship exists between the direction of current flow through a conductor and the direction of the magnetic field created. A simple rule has been established for determining the direction of the magnetic field when the direction of the electron flow is known (Figure 16-5). This rule is known as the ***left-hand conductor rule***, and it uses electron flow from negative to positive to determine the current direction. The rule is stated as follows:

> ✋ WHEN YOU PLACE YOUR *LEFT* HAND SO YOUR THUMB POINTS IN THE DIRECTION OF THE ELECTRON FLOW, YOUR CURLED FINGERS POINT IN THE DIRECTION OF THE MAGNETIC LINES OF FORCE THAT CIRCLE THE CONDUCTOR.

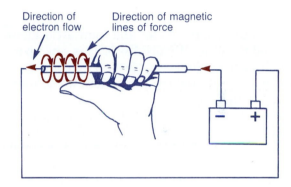

Figure 16-5 Left-hand conductor rule.

Using this rule, if either the direction of the lines of force or the current is known, the other factor can be determined.

An end view of the wire is often used to simplify the drawing of a conductor that is carrying current (Figure 16-6). The end of the wire is represented by a circle. Current flow into the conductor is represented by a *cross* to indicate the feathers on the back of the arrow going away from you (i.e., into page). Current flow out of the conductor is represented by a *dot* to indicate the point of the arrow coming toward you (i.e., out of page).

16-3 Magnetic Field About Parallel Conductors

The resulting magnetic field produced by current flow in two adjacent conductors tends to cause the attraction or repulsion of the two conductors. If the two parallel conductors are carrying current in opposite directions, the direction of the magnetic field is clockwise

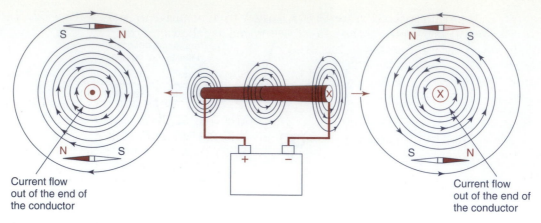

Current flow
out of the end of
the conductor

Current flow
out of the end of
the conductor

Figure 16-6   End view of conductor and magnetic field.

around the one conductor and counterclockwise around the other (Figure 16-7). This sets up a repelling action between the two individual magnetic fields and the conductors would tend to move apart.

When two parallel conductors are carrying current in the same direction, the direction of the magnetic field is the same around each field (Figure 16-8). The magnetic lines of force between the conductors oppose each other leaving essentially no magnetic field in this area. At the top and bottom of the conductors, the lines of force act in the same direction and link together and act around both conductors. This sets up an attracting action between the two individual magnetic fields and the conductors will tend to move together. The two conductors under this condition will create a magnetic field equivalent to one conductor carrying twice the current.

The interaction of magnetic fields, by creating forces that repel or attract, is a means of con-

Two magnetic fields attract

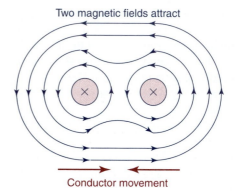

Conductor movement

Figure 16-8   Parallel conductors with current flows in the same direction.

verting electric energy to motion or mechanical work, and this makes the operation of electric motors possible.

The magnetic force between parallel conductors must be taken into consideration when designing large pieces of electrical equipment that handle very high current flows. In this situation, increased stress can result in damage to the conductors if they are not properly secured.

16-4 Magnetic Field About a Coil

As mentioned previously, two conductors lying alongside each other carrying currents in the same direction create a *magnetic field* twice as strong as that of the single conductor. If we take a single piece of wire and wind it into a number of loops to form a coil we create the equivalent of several parallel conductors carry-

Two magnetic fields repel

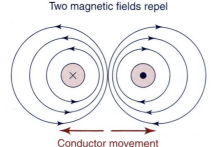

Conductor movement

Figure 16-7   Parallel conductors with current flows in opposite directions.

ing current in the same direction (Figure 16-9). The total resulting magnetic field is the sum of all the single-loop magnetic fields. A coil so formed will have a magnetic-field pattern similar to that of a bar magnet with a definite north and south pole.

flow from negative to positive. The rule is stated as follows:

> IF THE COIL IS GRASPED IN THE LEFT HAND, WITH FINGERS POINTING IN THE DIRECTION OF THE CURRENT FLOW, THE THUMB WILL POINT IN THE DIRECTION OF THE NORTH POLE OF THE MAGNET (FIGURE 16-10).

Figure 16-9 Magnetic field produced by a current-carrying coil.

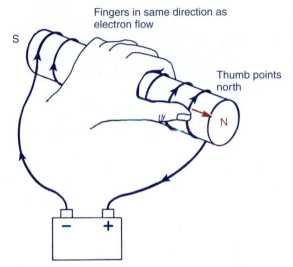

A.

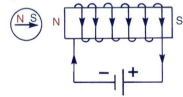

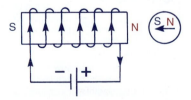

B.

Figure 16-10 Left-hand coil rule.

16-5 Left-Hand Coil Rule

A definite relationship exists between the direction of current flow through a coil, the direction in which the wire is wound to form the coil, and the location of the north and south poles. If you reverse the direction of current flow through the coil, you reverse the north and south poles. You can also reverse the poles in a coil by reversing the direction of the winding of the coil.

If the electromagnet is operated with direct current, the polarity of its magnetic poles remains fixed. If the electromagnet is operated with alternating current, its magnetic polarity reverses with each reversal of the direction of the current.

The *left-hand coil rule* has been established to determine any one of these three factors (polarity, current direction, and winding direction) when the other two factors are known. The current flow direction used is the electron

16-6 The Electromagnet

When a coil of insulated wire is wound over a core of magnetic material such as soft iron, the device becomes a practical electromagnet (Figure 16-11). The strength of the magnetic field is greatly increased by adding the iron core. This increase in magnetic strength is the result of magnetism induced into the core. When current flows through the coil, the core becomes magnetized by induction. The magnetic lines of force produced by the magnetized core align themselves with those of the coil to produce a much stronger magnetic field. Once the current stops flowing in the coil, both the coil and the soft iron core lose their magnetism.

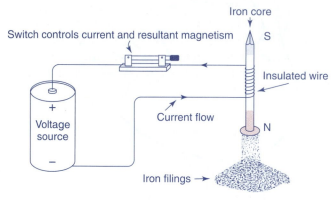

Figure 16-11 Basic electromagnet.

The magnetic field will have the same polarity regardless of whether the iron core is present or not. If the current direction is reversed through the coil, the polarity of both the coil and iron core are reversed.

Several factors affect the magnetic-field strength of an electromagnet formed by a coil (Figure 16-12). These include
- Core material, length, and area.
- Number of turns on the coil and the spacing of the turns.
- Amount of current flowing through each turn.

A *toroid-core* electromagnet (Figure 16-13) has a magnetic-field pattern similar to that of a ring magnet. The lines of force produced by the coil are totally contained in the toroidal core instead of being allowed to escape into the air. For this reason, toroid-core electromagnets are said to be self shielding.

The *permeability* of a material is a measure of the ease with which magnetic lines of force pass through a material. Iron and steel have much greater possibilities than air and other non-magnetic materials.

16-7 The Magnetic Circuit

The *magnetic circuit* is similar to the electric circuit. Basically, a magnetic circuit is a closed-loop path for magnetic lines of force, much like an electric circuit is a closed-loop path for the flow of electrons. In the electric circuit, electrons travel from the negative to the positive terminal of the voltage source (Figure 16-14A). In the magnetic circuit, lines of force travel from the north pole to the south pole of the electromagnet (Figure 16-14B). The rate of flow of electrons in the electric circuit is called *current* (I) and is measured in *amperes* (A). The total number of lines of force in the magnetic circuit is called *magnetic flux* (Φ). The unit commonly used to measure magnetic flux is the *weber* (Wb).

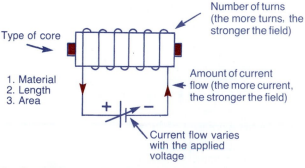

A. Current, core, and turns

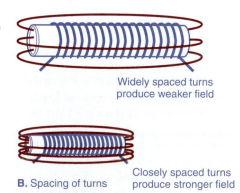

B. Spacing of turns

Figure 16-12 Factors which determine the strength of an electromagnet.

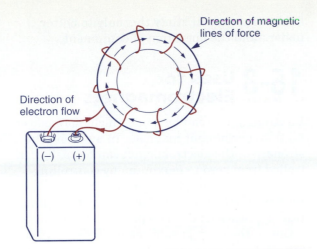

Figure 16-13   Toroid-core electromagnet.

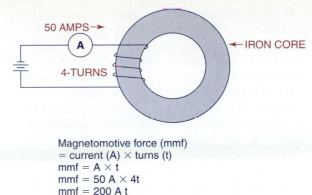

Magnetomotive force (mmf)
= current (A) × turns (t)
mmf = A × t
mmf = 50 A × 4t
mmf = 200 A t

Figure 16-15   Magnetomotive force (mmf) measured in ampere turns.

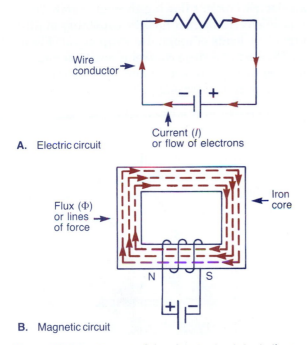

A.   Electric circuit

B.   Magnetic circuit

Figure 16-14   Current of the electric circuit is similar to magnetic flux of the magnetic circuit.

Just as an electric current is the result of an electromotive force (emf), or voltage, acting on the electric circuit, so a magnetic flux is produced by a **magnetomotive force** (mmf) acting upon the magnetic circuit (Figure 16-15). The magnetomotive force is the product of the current in amperes (A) and the number of turns (*t*) on the coil. The unit commonly used to measure mmf is the **ampere turn (At).**

The magnetic circuit counterpart of resistance in an electric circuit is called **reluc-**

tance ($\mathcal{R}$) (Figure 16-16). Therefore, reluctance is a measure of the opposition offered by a magnetic circuit to the setting up of flux, just as resistance is the opposition to current flow in an electric circuit. The reluctance of a magnetic circuit depends upon the type of material(s) used in the circuit, the circuit's length, and its cross-sectional area.

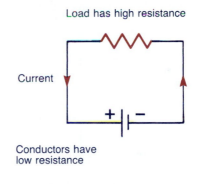

A.   Electrical resistance (R)

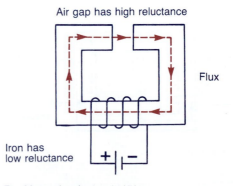

B.   Magnetic reluctance ($\mathcal{R}$)

Figure 16-16   Resistance of the electric circuit is similar to reluctance of the magnetic circuit.

In some applications the magnetic core is not continuous. For example, an air gap may exist in the circuit. Air gaps are often deliberately placed in magnetic circuits to increase reluctance. By increasing the total reluctance, saturation of the core may be prevented.

The similarity between magnetic and electric circuits extends to *Ohm's law*. Just as electromotive force (*V*) must work against resistance (*R*) to produce current (*I*), magnetomotive force (mmf) must work against reluctance ($\mathcal{R}$) to produce flux (Φ) (Figure 16-17). The *magnetic Ohm's law formula* states that the flux produced by a magnetic circuit is directly proportional to the magnetomotive force and inversely proportional to the reluctance.

$$\Phi = \frac{\text{mmf}}{\mathcal{R}}$$

Calculation and measurement of voltage, current, and resistance in the electric circuit is relatively easy to do and useful for troubleshooting. The same is *not* true for quantities in the magnetic circuit. A thorough knowledge of the laws of magnetic circuits is important to the designer of equipment, but for practical use in field work, we will study this only to better understand the operation of equipment.

16-8 Uses for Electromagnets

Electromagnets can be made much more powerful than permanent magnets. In addition, the strength of the electromagnet can be easily controlled from zero to maximum by controlling the current flowing through the coil. For these reasons, electromagnets have many more practical applications than do permanent magnets.

One of the most graphic examples of a working electromagnet is the one used in cranes that are used to move scrap iron. The crane magnet is a big block of soft iron that is magnetized by an electric current flowing through a coil. This type of electromagnet has the capability of lifting heavy loads of magnetic scrap metal (Figure 16-18). Lift-and-drop control is easily accomplished by the connection and disconnection of voltage to the electromagnet.

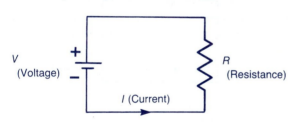

A. Electric circuit

$$I = \frac{V}{R}$$

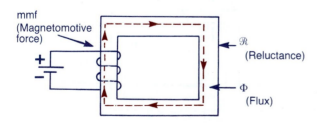

B. Magnetic circuit

$$\Phi = \frac{\text{mmf}}{\mathcal{R}}$$

Figure 16-17 Ohm's law for the electric and magnetic circuit.

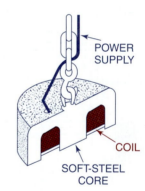

A. Cross-section of lifting magnet.

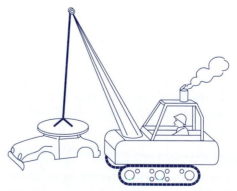

B. Lifting scrap metal

Figure 16-18 Crane magnet.

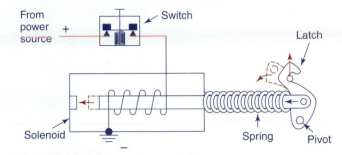

A. Solenoid

B. Solenoid-operated latch

Figure 16-20 Solenoid.

All motors and *generators* make use of electromagnets. In these machines, the strength of the electromagnet can be varied to change the generated voltage or the speed of the motor. In a typical generator circuit, the current flow through the field coils is adjusted by means of a variable resistor or rheostat connected in series with the coils and the dc-voltage source (Figure 16-19). Varying the current causes the strength of the magnetic field to vary.

A *solenoid* is an electromagnet with a moveable iron core or plunger. When power is applied, the magnetic field that is produced pulls the plunger into the coil. Figure 16-20 shows a typical automobile solenoid-operated trunk-lid release circuit. When the circuit to the coil is completed, the moveable core is drawn into the coil, and the latch moves to release the trunk lid. Solenoids are used in a variety of applications, for example, as solenoid valves, solenoid clutches, and solenoid switches.

Transformers are electric devices that are used to raise or lower ac voltages (Figure 16-21). This device uses two electromagnetic coils to transform or change ac-voltage levels.

The input current goes to a primary coil wound around an iron core. The output current emerges from a secondary coil also wound around the core. The alternating input current produces a magnetic field that continually switches ON and OFF. The core transfers this field to the secondary coil where it induces an output current. The change in voltage depends on the ratio of turns in the primary and secondary coils.

The electric bell is an example of electromagnets used to produce sound through simple

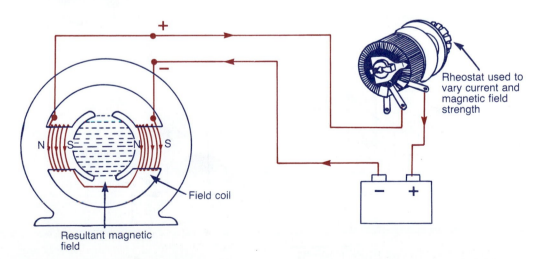

Figure 16-19 Generator magnetic field circuit.

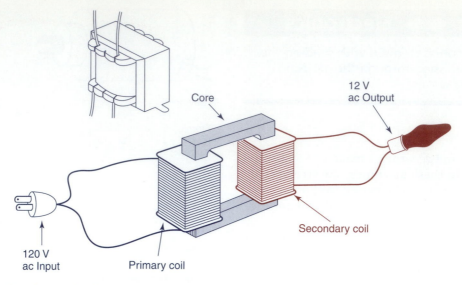

Figure 16-21 Transformer circuit.

vibration (Figure 16-22). It uses two small coils wound on a core to form a horseshoe magnet. When the button is pressed, current passes through the coils and magnetizes the core, thus attracting the moveable armature to it and causing the hammer to strike the bell. As the hammer strikes the bell, the movement of the armature opens the contacts. The current stops

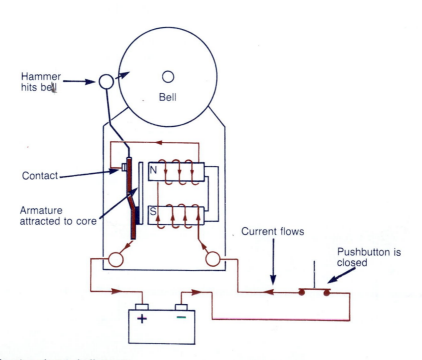

Figure 16-22 Vibrating electric bell circuit.

flowing to the electromagnet, which then loses its magnetism. A spring pulls the armature back, and the hammer moves away from the bell. The contacts then close again, and the cycle repeats itself for as long as the button is pressed.

HELP WANTED

Electric Power Plant Operator
Job Description
- Required by a utility company.
- Entry-level position.
- Includes extensive training in the production and distribution of electric power.
- College diploma in Electrical Power Systems desirable but not mandatory.
- Your school record of attendance, reliability, effort, and cooperation is most important.

Review and Applications

···

Related Formula

mmf = A × t

Review Questions

1. Explain the relationship between electricity and magnetism.
2. What two methods can be used to show the presence of a magnetic field around a conductor that carries a dc current?
3. What is the difference in the direction of the magnetic field produced around a conductor by dc flow and by ac flow?
4. What determines the strength of the magnetic field that is produced around a single conductor?
5. If two parallel conductors are carrying the same amount of current in the same direction:
   (a) in what direction will the magnetic forces tend to move the conductors?
   (b) what is the equivalent strength of the magnetic field created?
6. Describe the construction of a practical electromagnet.
7. What two factors determine the location of the north and south poles of an electromagnet?
8. List the three main factors that determine the strength of an electromagnet.
9. What is meant by the term permeability?
10. Define each of the following terms in relation to a magnetic circuit:
    (a) magnetic flux
    (b) magnetomotive force (mmf)
    (c) reluctance
11. State the magnetic Ohm's law formula.
12. Name two advantages that electromagnets have over permanent magnets.
13. (a) How is lift-and-drop control of a crane-operated lifting magnet accomplished?
    (b) Why is this type of crane application useless for moving scrap copper material?

14. What effect does varying the strength of the electromagnets have on the operation of an electric:
    (a) generator?
    (b) motor?
15. (a) Describe the basic construction of a solenoid.
    (b) List three common applications for solenoid-operated devices.
16. In the operation of a transformer, what does the degree of change in voltage depend on?
17. In the operation of a vibrating electric bell, what part of the bell is responsible for making and breaking the circuit current flow once the button is pressed?

Problems

1. Calculate the magnetomotive force of the following sources:
   (a) a 500-turn coil with an electron flow of 24 mA.
   (b) a 10-turn coil with an electron current of 25 A.
   (c) a current of 1.5 amperes flowing through a 75-turn coil.

Critical Thinking

1. The conducting medium in most magnetic circuits is iron or an alloy of iron. Give a reason for this preference.
2. Why will a transformer not operate from a constant dc source?

Portfolio Project

Prepare a written résumé and cover letter for the help wanted position listed for this chapter.

Batteries

*T*his chapter is a practical guide for the selection, use, and testing of batteries. A battery is a combination of materials used to change chemical energy into electric energy in the form of voltage. The emphasis is not on the internal design of these devices, but on their application and performance.

Objectives

After studying this chapter, you should be able to:
1. Compare the characteristics and application of different types of cells.
2. Explain the ways in which cells and batteries are rated.
3. Design a suitable connection of cells to obtain a desired voltage and current capacity.
4. Explain how cells and batteries are tested.
5. Properly charge a rechargeable battery.

Key Terms

- positive electrode
- negative electrode
- electrolyte
- primary cell
- secondary cell
- ampere hours (Ah)
- internal resistance
- energy density
- shelf life
- dry cell

- wet cell
- carbon-zinc cell
- zinc-chloride cell
- alkaline cell
- mercury cell
- lithium cell
- lead-acid battery
- charging
- discharging
- dry-charged battery

- maintenance-free battery
- cold-cranking amperes
- reserve capacity
- hydrometer
- specific gravity
- heavy-load test
- nickel-cadmium (Ni-Cd) battery
- battery memory

RESEARCH INTERNET SITES

- Panasonic
- Raychem
- Eveready

17-1 The Voltaic Cell

A *voltaic cell* is the basic device for converting chemical energy into electric energy. It consists of two different metal plates immersed in a solution. The metal plates are called *positive* and *negative electrodes* and the solution is called the *electrolyte*.

One type of voltaic cell uses copper and zinc as the two electrodes and sulfuric acid as the electrolyte (Figure 17-1). When placed together, a chemical reaction occurs between the electrodes and the sulfuric acid. This reaction produces a negative charge on the zinc (surplus of electrons) and a positive charge on the copper (deficiency of electrons). If an external circuit is connected across the two electrodes, electrons will flow from the negative zinc electrode to the positive copper electrode. The electric current will flow as long as the chemical action continues. In this type of cell, the zinc electrode is eventually consumed as part of the chemical reaction.

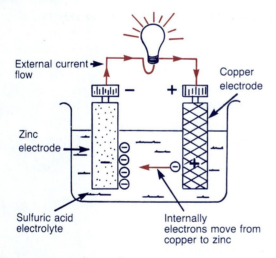

Figure 17-1   Voltaic cell.

Frequently, a single cell is called a battery. By strict definition, a *battery* consists of two or more cells connected together. These cells are usually enclosed in one case.

Primary cell is the name given to any cell in which an electrode is consumed gradually during normal use. It cannot be restored to its original useful state by electric recharging. A typical primary cell is the kind used in flashlights.

The *secondary cell* has the ability to be used over a longer period of time. Its chemical energy is replenished by recharging it at specific intervals. A typical example of the use of a secondary cell is the battery used in automobiles.

17-2 Battery Terminology and Ratings

Batteries were in use before incandescent lamps. More widespread today than ever, they can be found in everything from cellular phones to vehicles, enabling the use of electronic devices where an electrical power supply is either unfeasible or unavailable. Among the most desirable qualities of a battery are its compact size, light weight, and long operating life.

Voltage   All cells and batteries are rated for their normal output voltage. The voltage output from a single cell is 1–3 V and depends on the material used for the electrodes and the type of electrolyte. Batteries with higher output voltages contain cells connected in series (Figure 17-2).

Most battery-driven products will operate until the battery voltage diminishes to a certain cutoff voltage and then they cease to operate.

Energy Capacity   The battery energy capacity is rated in *ampere hours* (Ah). This capacity is determined by multiplying the amperes of current that the battery will deliver by the number of hours the battery will deliver it. For example, if a battery is rated at 16 Ah, it simply means that the battery can deliver 16 A of current to a load for 1 hour, or it could deliver 8 A for 2 hours, or 1 A for 16 hours, and so on (Figure 17-3). *For any given cell or battery type, the energy capacity of the battery is directly proportional to its physical size* (Figure 17-4).

Internal Resistance   The *internal resistance* of a battery cell is mainly the resistance of the electrolyte and limits the maximum amount of current the cell can produce. Internal resistance of a good cell is very low, with typical values of less than 1 Ω. The internal resistance of a battery varies with battery design, state of charge, temperature, and age.

Often the terminal voltage of a battery drops when the load is connected. The amount of volt-

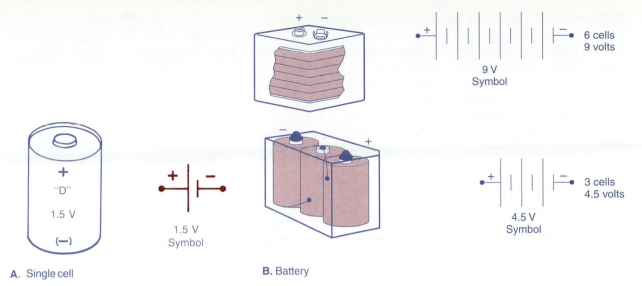

6 cells
9 volts

9 V
Symbol

+ −

3 cells
4.5 volts

4.5 V
Symbol

"D"

+

1.5 V

(—)

1.5 V
Symbol

+   −

A. Single cell

B. Battery

Figure 17-2   Typical single cell and battery voltage ratings.

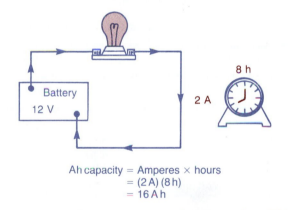

Battery
12 V

2 A

8 h

Ah capacity = Amperes × hours
= (2 A) (8 h)
= 16 Ah

Figure 17-3   Battery energy capacity, ampere-hour (Ah) rating.

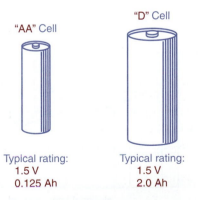

"AA" Cell

"D" Cell

Typical rating:
1.5 V
0.125 Ah

Typical rating:
1.5 V
2.0 Ah

Figure 17-4   For any given cell or battery type, physical size determines battery capacity.

age drop is directly related to the internal resistance of the battery. The higher the internal resistance, the greater the voltage drops under load conditions.

Energy Density   The *energy density* of a battery is the ratio of its stored energy to its weight or volume. High-energy density batteries provide more capacity in the same size or even in smaller packages.

Shelf Life   Batteries are also rated for *shelf life* that is given in years. Even if a cell is not being used, an internal chemical reaction takes place within the cell. As a battery ages, it very slowly discharges itself internally. Also, the electrolyte may dry out very slowly. This process will eventually render the battery useless. Shelf life is defined as the time in years a stored battery will produce at least 75 percent of its initial capacity.

About ◁▷ Electronics

Not all electric vehicles are designed to stick to the road. Among those with broader purposes are airport baggage shuttles, forklifts, lawn mowers, vehicles for mining and farming, and military electric vehicles.

Temperature   Batteries normally operate best near room temperature and are usually rated for a specific output capacity at room temperature or 70°F. Operating them above and below this temperature will reduce their service life. For example, an automobile battery output drops on cold days making it more difficult to turn the engine over.

17-3 Primary Dry Cells

In **dry cells** the electrolyte is a paste of powdered chemicals. **Wet cells**, like those used in car batteries, contain a liquid electrolyte.

Carbon-Zinc Cells
The most common general purpose and least expensive type of dry cell battery is the **carbon-zinc** type (Figure 17-5). This cell consists of a zinc container which acts as the negative electrode. In the center is a carbon rod, which is the positive electrode. The electrolyte takes the form of a moist paste made up of a solution containing ammonium chloride. As with all primary cells, one of the electrodes becomes decomposed as part of the chemical reaction. In this cell the negative zinc container electrode is the one that is used up. As a result, cells left in equipment for long periods of time can rupture, spilling electrolyte and causing damage to the electronic parts.

Carbon-zinc cells are produced in a range of common standard sizes (Figure 17-6). These include 1.5-V, AAA, AA, C, and D cells and 9-V rectangular batteries. The service life of a carbon-zinc battery is higher with *lower* current drains. Usually you triple the service life of the

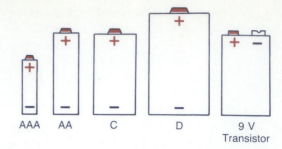

Figure 17-6   Relative sizes of standard carbon-zinc batteries.

battery if you decrease the current drain by half. Also, an intermittent duty cycle allows access to more of the battery capacity.

Zinc-Chloride Cells
The **zinc-chloride cell** is an improved version of the carbon-zinc cell. Variation in the chemical mix increases its energy capacity by approximately 50 percent.

Alkaline Cells
Alkaline cells offer higher energy capacity, better high- and low-temperature performance, and longer shelf life than do carbon-zinc or zinc-chloride cells.

An alkaline cell has a different electrolyte chemistry and is constructed differently than a carbon-zinc cell. It is this difference that provides higher electrochemical efficiency.

Voltage output from the alkaline cell is 1.5 V. Like the carbon-zinc cell this voltage is *sloping*, because the voltage falls with usage (Figure 17-7).

Alkaline batteries are more expensive than carbon-zinc types. They are perfect for products with high current demand such as cassette recorders, camera flash units, battery-driven toys, and electronic games.

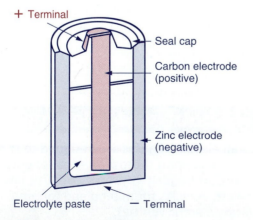

Figure 17-5   Carbon-zinc dry cell.

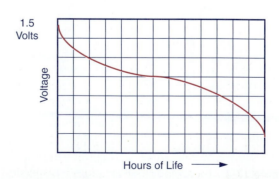

Figure 17-7   Sloping cell voltage-discharge curve.

Mercury Cells A *mercury cell* will develop a voltage of 1.34 V by the chemical action between zinc and mercuric oxide. It is expensive to manufacture, but has a decided advantage of producing about five times the ampere-hour output compared to that of a carbon-zinc cell of equivalent size. In addition to a longer shelf life (approximately three and one half years), it has a flat cell voltage-discharge curve, which results in a constant terminal voltage even as it approaches the end of its usefulness. This characteristic is important for electronic devices that require constant voltage level for proper and reliable operation. Figure 17-8 shows a button-type mercury battery used in watches and calculators.

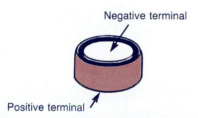

Figure 17-8   Button-type mercury battery.

Lithium Cells *Lithium primary cells* are relatively new and offer a big advantage over alkaline cells. Characterized by high-output voltage (3 V per cell), long shelf life (over 10 years), high energy density (low weight and small volume), and high energy capacity, they are an excellent choice for special applications.

Lithium primary cells are available in a wide variety of electrochemical and physical configurations. They find use in cardiac pacemakers, smoke detectors, computer memory power backup, watches, calculators, and in other military and medical applications.

17-4 Series and Parallel Cell Connections

All batteries consist of one or more electrochemical cells that are electrically connected to provide the correct amount of power for a specific application.

Series Connection Often an electric or electronic circuit requires a voltage or current that a single cell alone is not capable of supplying. In this case, it is necessary to connect groups of cells in various series and parallel arrangements.

Cells are connected in series by connecting the positive terminal of one cell to the negative terminal of the next cell (Figure 17-9).

A. Pictorial diagram

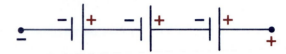

B. Schematic diagram

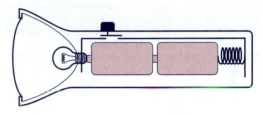

C. In a flashlight

Figure 17-9   Series connection of cells.

Example 17-1

Suppose three D flashlight cells are connected in series. Each cell has a rating of 1.5 V and 2 Ah. The voltage and ampere hour rating of this battery would be:

$$V_{Battery} = \text{Volts per cell} \times \text{Number of cells}$$
$$= (1.5 \text{ V})(3)$$
$$= 4.5 \text{ V}$$
$$\text{Ah Battery Rating} = \text{Ah rating of 1 cell}$$
$$= 2 \text{ Ah}$$

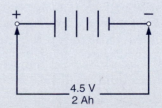

4.5 V
2 Ah

Figure 17-10   Circuit schematic

Identical cells are connected in series to obtain a higher voltage than is available from a single cell. With this connection of cells, the output voltage is equal to the sum of the voltages of all cells. However, the ampere-hour or energy-capacity rating remains the same and is equal to that of a single cell.

If, by error, one cell connection is reversed in a series group, its voltage will oppose that of the other cells. This will produce a lower than expected battery output voltage.

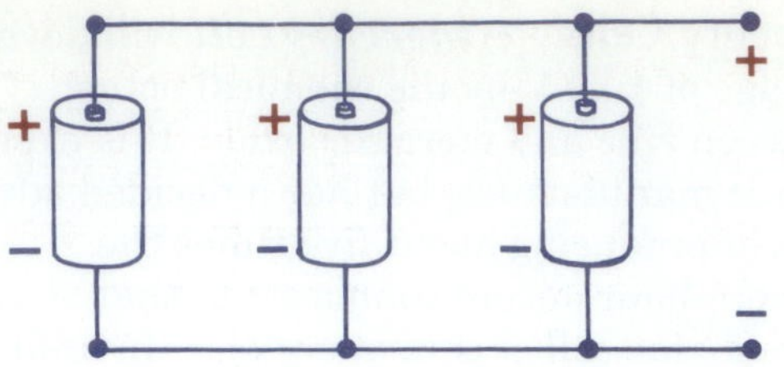

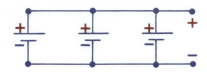

A. Pictorial diagram

B. Schematic diagram

Figure 17-12   Parallel connection of cells.

Example 17-2

Suppose that one of the three D flashlight cells from Example 17-1 is connected in reverse. The output voltage then would be:

$$V_{\text{Battery}} = (1.5 \text{ V}) + (1.5 \text{ V}) - (1.5 \text{ V})$$
$$= (3 \text{ V}) - (1.5 \text{ V})$$
$$= 1.5 \text{ V}$$

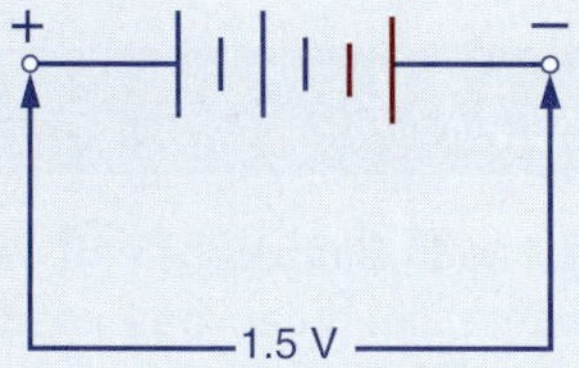

Figure 17-11   Circuit schematic

Example 17-3

Suppose four alkaline cells are connected in parallel. Each cell has a rating of 1.5 V and 8 Ah. The voltage and ampere-hour rating of this battery would be:

$$V_{\text{Battery}} = \text{Voltage rating of 1 cell}$$
$$= 1.5 \text{ V}$$
$$\text{Ah Battery Rating} = \text{Ah rating per cell} \times \text{Number of cells}$$
$$= (8 \text{ Ah})(4)$$
$$= 32 \text{ Ah}$$

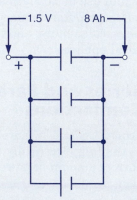

Figure 17-13   Circuit schematic

When connecting a number of cells in series, all cells must be equally matched in terms of voltage and current capacity. *The weakest cells in series will determine the overall energy capacity of your battery.*

Parallel Connection   Cells are connected in parallel by connecting all the positive terminals together and all the negative terminals together (Figure 17-12).

Identical cells are connected in parallel to obtain a higher output current or ampere-hour rating. With this connection of cells, the energy capacity in terms of ampere-hour rating is equal to the sum of the ampere-hour ratings of all cells. However, the output voltage remains the same as the voltage of a single cell. Again, as in series, the individual cells must be matched as closely as possible so the output current can be shared equally among them.

If, by error, a cell connection is reversed in a parallel group, it will act as a short circuit. All cells will discharge their energy through this short-circuit path. Maximum current will flow and the cells may be permanently damaged.

In emergency situations, it is sometimes necessary to jump start a car by connecting a charged booster battery in parallel with the discharged battery of the vehicle.

> ✋ **WHEN JUMP STARTING A VEHICLE, REMEMBER TO ALWAYS CONNECT THE *POSITIVE TERMINAL* OF THE *DISCHARGED BATTERY* TO THE *POSITIVE TERMINAL* OF THE *BOOSTER BATTERY*. WRONG POLARITY HOOKUP, EVEN MOMENTARILY, CAN DO EXTENSIVE DAMAGE!**

Figure 17-14 shows the correct jumper hookup procedure for jump starting a vehicle.

Series-Parallel Connection Sometimes the requirements of a piece of equipment exceed both voltage and energy capacity rating of a single cell. In this case, a series-parallel grouping of cells must be used. This is not as common as straight series or parallel, but it can be done. The number of cells that must be connected in series is calculated first. Then, the number of parallel rows of series-connected cells is calculated.

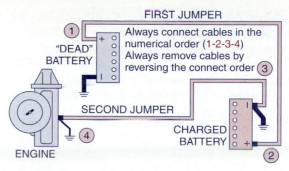

FIRST JUMPER
Always connect cables in the numerical order (1-2-3-4)
Always remove cables by reversing the connect order

Figure 17-14 Cable hookup for jump starting a car.

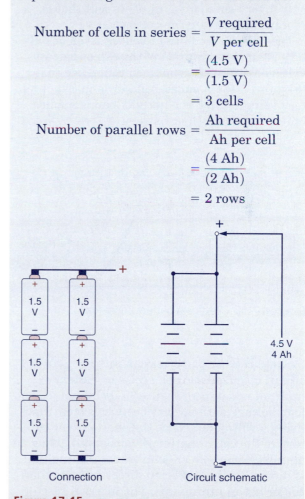

Example 17-4

Suppose a battery-operated circuit requires 4.5 V and a capacity of 4 Ah. Cells rated at 1.5 V and 2 Ah are available to do the job. The required arrangement of cells would then be:

$$\text{Number of cells in series} = \frac{V \text{ required}}{V \text{ per cell}}$$

$$= \frac{(4.5 \text{ V})}{(1.5 \text{ V})}$$

$$= 3 \text{ cells}$$

$$\text{Number of parallel rows} = \frac{Ah \text{ required}}{Ah \text{ per cell}}$$

$$= \frac{(4 \text{ Ah})}{(2 \text{ Ah})}$$

$$= 2 \text{ rows}$$

Connection   Circuit schematic

Figure 17-15

When connecting groups of cells or batteries in parallel, each group must be at the same voltage level. Paralleling two batteries of unequal voltage levels sets up a difference of potential energy between the two. As a result, the higher voltage battery will discharge its current into the other battery until both are at equal voltage value.

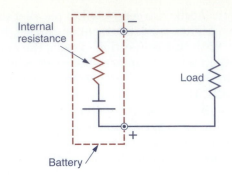

Battery

17-5 Testing Primary Cells and Batteries

A visual inspection will tell you little about the useful life of a cell or battery unless it has deteriorated to the point where the electrolyte is spilling from the case.

A no-load voltage test of the cell or battery is another poor indication of cell or battery life. This test requires the cell or battery to deliver only a very small amount of current that is required to operate the voltmeter.

The best method used to check a cell or battery is an in-circuit test of the cell or battery voltage with the normal load connected to it. (Figure 17-16). A substantial drop in cell or battery voltage when the normal load is applied indicates a bad cell or battery.

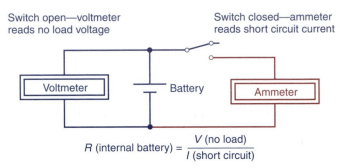

Switch open—voltmeter reads no load voltage

Switch closed—ammeter reads short circuit current

$$R \text{ (internal battery)} = \frac{V \text{ (no load)}}{I \text{ (short circuit)}}$$

Figure 17-17 Calculating the internal resistance of a cell.

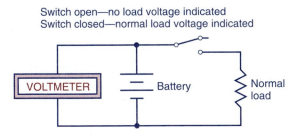

Switch open—no load voltage indicated
Switch closed—normal load voltage indicated

Figure 17-16 Testing a battery under load.

Small cells or batteries can be tested out-of-circuit by calculating the cell's or battery's internal resistance (Figure 17-17). First, the no-load voltage of the cell or battery is measured using a voltmeter. Next, the short-circuit current of the cell or battery is measured, using an ammeter of a high enough current rating that is momentarily connected directly across the cell or battery terminals. The internal resistance is then calculated using Ohm's law. A high value of internal cell or battery resistance indicates a bad cell or battery.

Since the materials from which cells are made are not perfect conductors, they have resistance. Current flowing through the external circuit also flows through the internal resistance of the cell. Therefore, if a short circuit is connected across the cell, the internal resistance of the cell prevents the current from becoming infinitely high. Typically, the internal resistance of an alkaline 1.5 V D size cell is approximately 0.2 Ω.

17-6 Lead-Acid Rechargeable Battery

Where high values of load current are required, the lead-acid wet cell is the type most commonly used. The *lead-acid cell* consists of two different types of lead plates immersed in a liquid electrolyte (Figure 17-18). The chemical action of the electrolyte removes electrons from one type of plate and adds them to the other, causing the plates—and the terminals connected to them—to become oppositely charged. The positive plate is brown-colored lead peroxide (PbO_2) and the negative plate is gray-colored sponge lead (Pb). The electrolyte is a mixture of 36 percent sulfuric acid (H_2SO_4) and 64 percent water (H_2O). This mixture results in

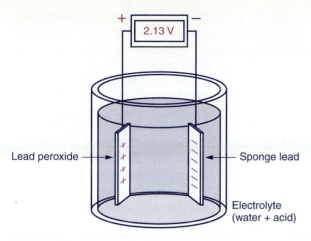

Figure 17-18 Lead-acid cell.

a voltage of 2.13 V per cell. The water (H_2O) is not the same as your tap water. If a battery should need water added, it should be distilled water. Regular tap water has impurities which will cause a chemical reaction on the plates and cause them to break down prematurely.

As the cell supplies voltage to an electric circuit producing current flow, an electrochemical reaction causes the lead-acid cell to *discharge* (Figure 17-19). As current flows, the acid content of the electrolyte diminishes and the active metals on the plates are replaced with lead sulfate. The discharging continues until the two plates become alike in chemical composition and the acid is used up.

The lead-acid cell is a secondary cell or storage cell that can be recharged. The discharging action of a lead-acid cell can be reversed by **charging**, or by sending current

into the cell in the reverse direction (Figure 17-20). During charging, the chemical reaction of discharging is reversed. The accumulated sulfate on the positive and negative plates is forced back into the electrolyte, and water in the electrolyte is forced to split into hydrogen and oxygen, which are given off as gases. The battery can be restored to a fully charged state by applying a dc current to the battery at a slightly higher voltage than the battery. Gasing occurs as the battery nears full charge, and hydrogen bubbles out at the negative plate and oxygen appears at the positive plate.

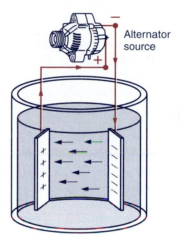

Figure 17-20 Charging action of a lead-acid cell.

> ✋ **HYDROGEN GAS IS EXPLOSIVE! ANY FLAME OR SPARK CAN IGNITE IT. IF THE FLAME TRAVELS INTO THE CELLS, THE BATTERY MAY EXPLODE!**

The most common application for the lead-acid storage battery (Figure 17-21) is to supply power to start the engine in an automobile. The voltage of the battery depends on the number of individual cells connected together in series. A standard 12-V battery has six cells connected in series to produce an open circuit voltage of 12.6 V, commonly referred to as 12 V. The current capacity of the battery is determined by the product of the plate surface area multiplied by the number of plates and the volume of electrolyte. Batteries with larger plates, or many plates, produce more current and can maintain their voltage output at a higher current-discharge rate.

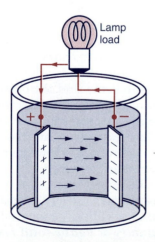

Figure 17-19 Discharging action of a lead-acid cell.

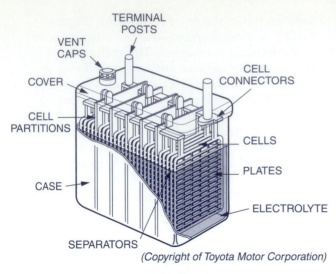

VENT CAPS
TERMINAL POSTS
COVER
CELL CONNECTORS
CELL PARTITIONS
CELLS
PLATES
CASE
ELECTROLYTE
SEPARATORS

(Copyright of Toyota Motor Corporation)

Figure 17-21   Components of a lead-acid storage battery.

Dry-charged lead-acid batteries are built, charged, washed and dried, sealed, and shipped *without* any electrolyte. They can then be stored longer than wet batteries. When a dry-charge battery is put in use, it is filled to the correct level with electrolyte and activated by a short period of charging before installation.

Maintenance-free lead-acid batteries are made with *calcium* in the grids. The calcium reduces the production of battery gases. As a result, the loss of electrolyte is reduced. The term maintenance-free is attached to the calcium-grid battery because, under normal charging conditions, water does not have to be added to the electrolyte periodically. Some are completely sealed units except for small vent holes, through which the battery breathes and gases are emitted. On others, the top of the battery cells is covered with a large snap-in vented cover.

The battery's primary function of providing energy to crank the car's engine requires a large discharge of current in a short time. The *cold-cranking amperes* (CCAs) rating specifies how much current (in amperes) the battery can deliver for 30 seconds at 0° F while maintaining a terminal voltage of 7.2 V, or 1.2 V per cell. CCAs usually range from 300 to 650 A for passenger vehicles.

The battery must provide emergency energy for ignition, lights, and accessories if the vehicle's charging system fails. The *reserve-capacity* (RC) rating specifies, in units of minutes, the length of time a fully charged battery

at 80° F can be discharged at 25 A while maintaining a terminal voltage of at least 10.5 V, or 1.75 V per cell.

Sealed-lead batteries are also used for stand-by applications (those that rely on a battery when electric power fails). These applications include emergency lighting, alarm systems, computer and medical system backup, and telecommunications. The sealed-lead battery is a clean, sealed, high-performance battery that is economical and easy to charge. It does not need maintenance, can be used in any position, and does not cause the corrosion that is traditionally associated with lead-acid batteries.

17-7 Testing a Lead-Acid Battery

A *state-of-charge* test checks the condition of a lead-acid battery. The most accurate method uses a *hydrometer* to measure the *specific gravity* (density or strength) of the electrolyte. Specific gravity means exact weight. The hydrometer compares the exact weight of the electrolyte with that of water. Strong electrolyte solution in a charged battery is heavier than a weak electrolyte solution in a discharged battery. The specific gravity of a fully charged battery is about 1270 at 80° F. By measuring the specific gravity of the electrolyte, you can tell if the battery is fully charged or requires charging (Figure 17-22). A specific gravity difference of 0.050 or more between one or more cells indicates a defective battery.

Only batteries with removable vent caps can be tested with a hydrometer. Sealed maintenance-free batteries usually have a built-in *charge-indicator eye* located on the top of the battery case that indicates the general charge of the battery (Figure 17-23). The charge indicator changes color with changes in the state of charge of the battery. For example, a bright green eye may indicate that the battery is at least 75 percent charged. A dark or black eye may indicate the battery is in need of a charge. A light or yellow eye may mean that the electrolyte level has dropped below the plates and the battery must be discarded.

Although a hydrometer test determines a battery's state of charge, it does not measure the battery's ability to deliver adequate cranking power. A *capacity* or *heavy-load test* is designed to simulate starter motor current

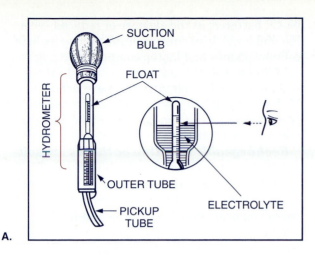

A.

B.

36% ACID	64% WATER	ELECTROLYTE
S.G. = 1.835	S.G. = 1.000	S.G. = 1.270

STATE OF CHARGE	SPECIFIC GRAVITY*	OPEN-CIRCUIT VOLTAGE
100%	1.265	12.6
75%	1.225	12.4
50%	1.190	12.2
25%	1.155	12.0
DEAD	1.120	11.9

C.

*Difference between cells should not vary more than 50 points (0.050). If it does, replace the battery.
(Copyright of Toyota Motor Corporation)

Figure 17-22 Testing specific gravity with a hydrometer.

drawn on a fully charged battery to determine if the battery is capable of meeting the current demand required to start the engine. To make a simple light-load voltage test of a car battery, check the value of the battery output voltage with and without the headlights on. A maxi-

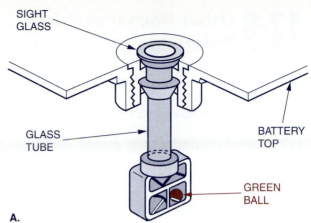

A.

BATTERY TOP	BATTERY TOP	BATTERY TOP
DARK GREEN DOT VISIBLE CHARGED	DARK GREEN DOT NOT VISIBLE DISCHARGED	LIGHT OR YELLOW ELECTROLYTE LEVEL LOW

B.
(Courtesy Chrysler Corporation)

Figure 17-23 Built-in eye hydrometer in a battery.

mum-load voltage test can be made by metering the battery voltage while operating the starting motor (Figure 17-24). If the voltmeter reads 9.6 V or more, the battery power is acceptable.

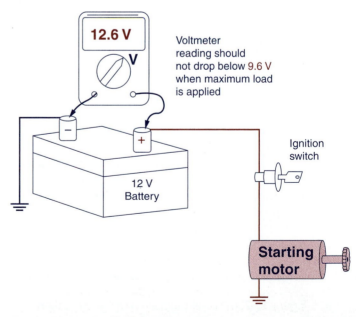

Voltmeter reading should not drop below 9.6 V when maximum load is applied

Figure 17-24 Testing a battery under load.

17-8 Other Rechargeable Batteries

Sealed *nickel-cadmium (Ni-Cd) batteries* consist of wound cylindrical cells that can be assembled in many different configurations. Because of their high energy density, high current capability, and fast recharge times (as low as 15 minutes to return to 90 percent charge), sealed nickel-cadmium batteries are regularly chosen to power portable products. The Ni-Cd cell is a durable secondary cell that can be recharged up to 1000 times.

The chemical reaction that occurs within the Ni-Cd cell is similar to that of the lead-acid battery. At full charge, the output voltage per cell is 1.2 V compared to a common primary battery cell voltage of 1.5 V. Unlike standard primary batteries, the Ni-Cd batteries' terminal voltage remains flat or constant until the cell is almost completely drained. Also, the Ni-Cd cell can be stored for a long time, even when discharged, without damage.

Ni-Cd cells and batteries are available in all standard sizes. They are used extensively with portable power tools and electronic equipment (Figure 17-25). Quite often, both the cell and matching charger unit are supplied with the piece of equipment it operates.

Nickel metal hydride (Ni-MH) rechargeable batteries provide improvements in capacity over Ni-Cd batteries at the expense of reduced cycle life and lower load current. Applications include cellular phones and laptop computers.

Lithium ion (Li-ion) rechargeable batteries require protective circuitry and are used where high-energy density is needed and cost is secondary. Applications include notebook computers and video cameras.

Rechargeable alkaline batteries are suitable for low-power, low-cost applications. Their limited cycle life is compensated for by low self-discharge, making this battery useful in portable entertainment devices and flashlights.

When considering battery maintenance, note that Ni-Cd has the shortest charge time, delivers the highest load current, and offers the lowest cost-per-cycle. However, it is the most demanding in exercise (discharge-charge cycle) requirements in preventing *memory*. The word memory was originally derived from cyclic memory, meaning that a Ni-Cd battery can remember how much discharge was required on previous discharges. Improvements in technology have virtually eliminated this phenomenon. Memory of a modern Ni-Cd battery is the effect of crystalline formation on cell operation. The active materials of a Ni-Cd battery (nickel and cadmium) are present in finely divided crystals. In a good cell, these crystals remain small, obtaining maximum surface area. When the memory phenomenon occurs, the crystals grow up to 150 times their original size, drastically reducing the surface area. The result is voltage depression which leads to loss of capacity.

The effects of crystalline formation are most prominent if a Ni-Cd battery is left in the charger for days or if repeatedly recharged without a periodic full discharge. It is not necessary, nor is it advised, to discharge a Ni-Cd before each charge because excessive cycling puts extra strain on the battery. An analogy can be drawn with a hand towel. A hand towel, if washed after each use, wears out its fabric too quickly. It is, however, necessary to clean the towel on a periodic basis. A full discharge to one volt per cell once a month is sufficient to keep crystal formation under control.

Another form of memory that occurs on some Ni-Cd cells is the formation of a metallic compound of nickel and cadmium which develops extra internal resistance in the cell. Reconditioning by deep discharge helps to break up this compound and reverse the capacity loss.

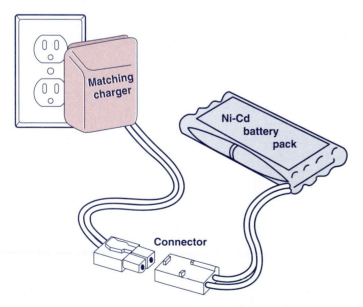

Figure 17-25 Nickel-cadmium rechargeable battery system.

17-9 Battery Chargers

When the chemical reaction in a rechargeable battery has ended, the battery is said to be **discharged** and can no longer produce the rated flow of electric current. This battery can be recharged, however, by causing direct current from an outside source to flow through it in a direction opposite to that in which it flowed out of the battery. When charging a battery, the negative lead of the charger must connect to the negative terminal of the battery and the positive lead of the charger to the positive lead of the battery (Figure 17-26). A reversal of these connections will produce a short circuit and may damage both the charger and battery.

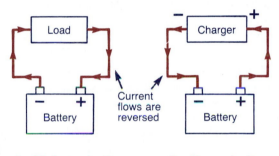

A. Discharge circuit   **B.** Charge circuit

Figure 17-26 Battery charging and discharging currents.

The automobile uses an automatic-charging circuit as part of the car's electrical system. It is designed to recharge the battery as required. Out-of-car battery charging is done using large commercial type battery chargers (Figure 17-27). The two basic charging concepts are **slow charging** and **fast charging**. Basically, slow charging involves a low current applied over a long period of time. Fast charging is just the opposite: a higher current applied over a shorter period of time. Slow charging is the preferred method of charging a battery. Since the charging current is relatively low, the chances of overcharge damage are minimized. In addition, it is the only method that can restore a battery to full charge.

When charging any battery, it is important to set the charging rate to a value recommended by the manufacturer. The charging rate is simply the amount of reverse current applied to the

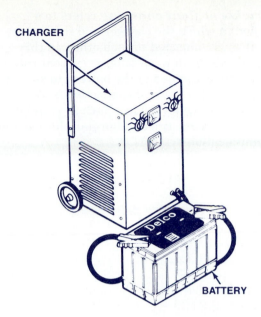

(Courtesy General Motor)

Figure 17-27 Typical automotive battery charger.

battery over a period of time. For example, a nickel-cadmium cell might be charged at the rate of 200 mA for 8 hours. The current is set by adjustment of the output voltage on the charger and read by an ammeter that is connected in series with the charger and battery (Figure 17-28). When the battery and charger are at the same voltage, no current flows. The charger voltage is set to a value slightly higher than that of the battery to produce a current flow. Undercharging or overcharging a battery can permanently damage it.

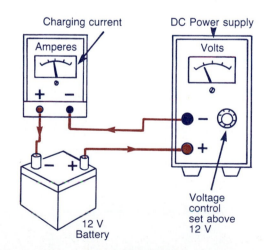

Figure 17-28 Metering the charging current.

Trickle or float charging refers to a method in which the charger and the battery are always connected to each other. In this arrangement, a low overcharge current rate is continuously applied to the battery to keep it at full charge. This type of connection is often used for emergency devices such as standby lighting systems, fire and burglar alarms, and communication systems.

Cable TV Service Technician

Job Description
- Install equipment at customer sites.
- Repair underground lines.
- Monitor signal leakage.
- Keep accurate records of service calls.
- Strong mechanical aptitude.
- An electronics degree or diploma required.

CHAPTER 17
Review and Applications

Related Formulas

Battery voltage rating (series connection) = Volts per cell × number of cells

Battery Ah rating (parallel connection) = Ah rating of 1 cell × number of cells

$$\text{Internal resistance} = \frac{V\,(\text{no-load})}{I\,(\text{short circuit})}$$

1. Name the three main parts of a voltaic cell.
2. By strict definition, what is the difference between a cell and a battery?
3. What is the main difference between a primary and secondary cell?
4. (a) What is the approximate voltage-output range for a single cell?
   (b) What determines the value of this output voltage?
5. (a) In what unit is the energy capacity of a cell or battery rated?
   (b) For a given cell or battery type, what determines the output capacity?
6. (a) What accounts for most of the internal resistance of a battery cell?
   (b) How does the internal resistance of a cell affect its terminal voltage when a load is applied?
7. Define energy density as it applies to a battery.
8. Define the term shelf life as it applies to cell or battery ratings.
9. Explain how cells or batteries are rated with regard to operating temperature.
10. What is the main difference between a wet and dry cell?
11. How does the carbon-zinc cell compare with the equivalent alkaline cell in price, energy capacity, and shelf life?
12. Button-type mercury cells are often used in watches and calculators. What voltage characteristic of this type of cell makes it suited for this application?
13. List four advantages that lithium cells have over alkaline cells.

14. What battery rating is increased when you connect cells in:
    (a) series
    (b) parallel
    (c) series-parallel
15. When jump starting a vehicle with a booster battery what polarities are connected together?
16. Why is a no-load test of any battery a poor indication of battery life?
17. Outline the procedure that is followed to make an internal resistance check of a small primary cell.
18. What is a lead-acid battery's electrolyte solution made up of?
19. How can a lead-acid battery be restored to a fully charged state?
20. (a) What connection of cells is used in a standard 12-V car battery?
    (b) What open-circuit voltage is produced by this connection?
    (c) What factors determine the current-capacity rating of the battery?
21. In what ways are maintenance-free, lead-acid batteries different from conventional types?
22. Define what is specified by a lead-acid battery cold-cranking amperes (CCAs) rating.
23. Explain how a battery hydrometer is able to test the state of charge of a lead-acid battery.
24. How do built-in eye hydrometers usually indicate the general state of charge of a maintenance-free, lead-acid battery?

25. Outline the procedure that is followed to make a maximum-load voltage test of a lead-acid automobile battery.
26. Compare the open circuit terminal voltage of each of the following cells:
    (a) carbon zinc
    (b) alkaline
    (c) mercury
    (d) lithium
    (e) lead-acid
    (f) nickel-cadmium
27. Explain the difference between slow charging and fast charging.
28. Define the term charging rate as it applies to recharging of a battery.
29. How is the amount of charging current controlled in a battery charger?

Problems

1. A battery with a rated output of 12 V and 16 Ah supplies an emergency lighting system. If the lighting system draws 2.5 A at 12 V, calculate rated length of time that the lights will be operative in case of a power failure.
2. Design a battery arrangement that can supply 4.5 V and 8 Ah using cells rated at 1.5 V and 2 Ah each. Draw a schematic diagram of the cell arrangement.
3. A 1.5 V cell is temporarily short-circuited and produces a current of 5 A. Calculate the internal resistance of the cell.

Critical Thinking

1. A 12-V lead acid battery has an internal resistance of 0.01 Ω. How much current will flow if the battery terminals are short-circuited?
2. Although low-voltage, high-current sources such as lead-acid batteries cannot deliver an electric shock they can still be dangerous. Why?

Portfolio Project

Conduct a longevity comparison test of an AA carbon-zinc and alkaline cell. Connect each continuously to a 10-Ω load and record the time it takes for the cell voltage to drop below a 0.9-volt cutoff point. Compare cost and longevity to determine which is the better investment.

Circuit Challenge

Using a Simulator

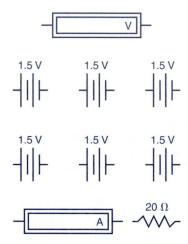

Procedure

(a) Wire the six cells for an output voltage of 9 volts, using whatever simulation software package is available to you.
(b) Calculate the amount of current that would flow with this battery connected to a 20 Ω load.
(c) Connect the battery to the 20 Ω load and record the actual current flow as indicated by the ammeter.

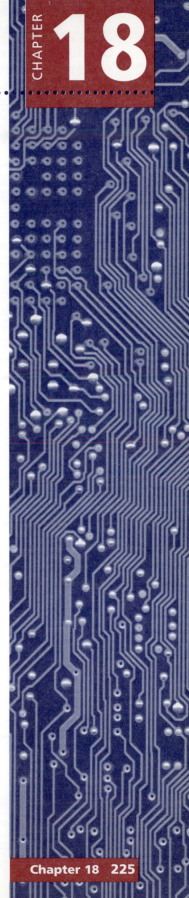

Circuit Protection Devices

Electric circuits can be damaged or even destroyed if their voltage and current levels exceed those for which they are designed. Fuses and circuit breakers are used to prevent this from occurring. Both operate on the same principle: to interrupt or open the circuit as quickly as possible before damage can occur.

Objectives

After studying this chapter, you should be able to:
1. Define the terms overload and short circuit.
2. Compare the basic principle of operation of a fuse and a circuit breaker.
3. State how fuses and circuit breakers are rated.
4. Identify basic fuse types and typical applications.
5. Test fuses and circuit breakers in and out of circuits.
6. Explain how lightning rods and arresters protect electrical equipment.

Key Terms

- overload
- short circuit
- fuse
- circuit breaker
- continuous-current rating
- voltage rating
- interrupting current rating

- fast-acting fuse
- slow-blow fuse
- plug fuse
- cartridge fuse
- time-delay fuse
- renewable fuse
- fusible link
- thermal circuit breaker

- magnetic circuit breaker
- tripped breaker
- self-resetting circuit breaker
- thermal overload protection
- lightning rod
- lightning arrester

RESEARCH INTERNET SITES

- Wickmann
- Bussmann
- Eaton

18-1 Undesirable Circuit Conditions

Overloads   An electric circuit is limited in the amount of current it can handle. The **current capacity** of the circuit is determined by the size of the wire conductors used. An electric circuit is said to be *overloaded* when the amount of current flowing through it is more than its rated current capacity.

Overloads can occur in the home when too many lamps and appliances are plugged into the same circuit (Figure 18-1). For example, a general-purpose house branch circuit is usually wired for a maximum capacity of 15 A. If the sum of the parallel load currents connected to it exceeds 15 A, the circuit is said to be overloaded. The solution to this problem is to remove some of the loads.

In an overloaded circuit, the conductors are required to carry more current than they are safely rated to carry. *This results in excessive heat being produced by the conductors, creating a potential fire hazard.*

Short Circuits   The term *short circuit* refers to a circuit that is completed in the *wrong way.* Sometimes a direct path is accidentally created from one side of the voltage source to the other, without passing through a load. This situation is called a *short circuit* (Figure 18-2). When a complete circuit is shorted, its total resistance is reduced to near *zero* value. As a result, lots of current will flow! Actual value of current flow will be limited only by the capacity of the source.

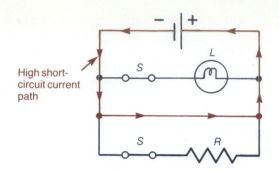

Figure 18-2   Shorted circuit.

A short circuit usually occurs when wire insulation wears away and a wire touches another bare wire. This allows current to flow into another circuit. This type of short can result in two circuits operating when only one is turned ON (Figure 18-3).

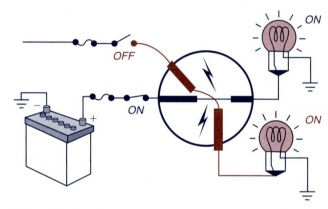

Figure 18-3   Short circuit resulting in two circuits operating when only one is turned ON.

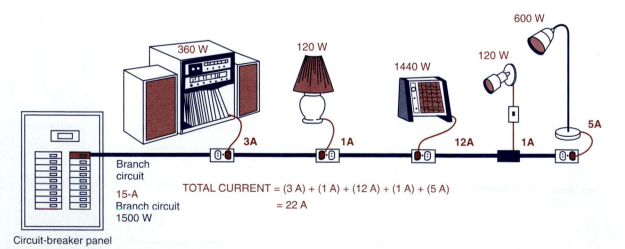

Branch circuit

15-A Branch circuit 1500 W

Circuit-breaker panel

TOTAL CURRENT = (3 A) + (1 A) + (12 A) + (1 A) + (5 A)
= 22 A

Figure 18-1   Overloaded circuit.

Short circuits can be wired by accident when the circuit is first constructed. When constructing any type of circuit, it is wise to check for shorts before applying voltage.

Any bare or exposed wire spells trouble (Figure 18-4). Many short circuits occur in flexible cords, plugs, or appliances. Look for black smudge marks on faceplates or frayed or charred cords connected to dead circuits. To correct this problem, simply replace the damaged cord or plug.

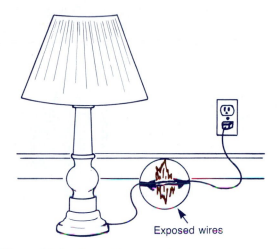

Figure 18-4 Shorted lamp cord.

A short circuit across the voltage source is one of the most dangerous circuit faults, because of the amount of current that flows through it. Most voltage sources are capable of producing hundreds of amperes of short-circuit current that can instantly melt wires and cause fires.

18-2 Fuse and Circuit Breaker Ratings

Circuit protection devices are sensitive to current. *Fuses* and *circuit breakers* are commonly referred to as **overcurrent-protection devices** (Figure 18-5). Their purpose is to protect electric circuits from damage by too much current flow. They are both *low-resistance* (usually less than 1 ohm) devices and are connected in *series* with the circuit they protect. Whenever wiring is forced to carry more current than it can safely handle, fuses will blow or circuit breakers will trip. These actions open the circuit, disconnecting the supply of electricity.

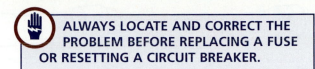

ALWAYS LOCATE AND CORRECT THE PROBLEM BEFORE REPLACING A FUSE OR RESETTING A CIRCUIT BREAKER.

Fuses and circuit breakers are rated for both current and voltage. The **continuous-current rating** marked on the fuse or circuit breaker (Figure 18-6) represents the maximum amount of current the device will carry without blowing or tripping open the circuit. The continuous-current rating is matched to the current rating of the load device and ampacity of the circuit conductors it protects. For example, a circuit using a No. 14 AWG copper conductor has a rated capacity of 15 A.

NEVER REPLACE ANY FUSE OR CIRCUIT BREAKER WITH ONE OF A HIGHER CURRENT RATING!

The **voltage rating**, as marked on a fuse or circuit breaker, is the *highest* voltage at which it is designed to safely interrupt the current. It may be used at any voltage at or below the rated voltage without affecting its operating characteristics. Common voltage ratings used are 32 V, 125 V, 250 V, and 600 V.

The **interrupting-current rating** (also known as short-circuit rating) of a fuse or circuit breaker is the maximum current it can safely interrupt at rated voltage. Safe operation requires protective devices to remain intact (no explosion or body rupture) during high short-circuit current conditions. The interrupting-current rating is also greater than the continuous-current rating and should be far in excess of the maximum current the power source can deliver. Typical interrupting ratings are as follows:

Ampere Rating of Fuse	Interrupting Rating in Amperes
0 to 1	35
1.1 to 3.5	100
3.6 to 10	200
10.1 to 15	750
15.1 to 30	1500

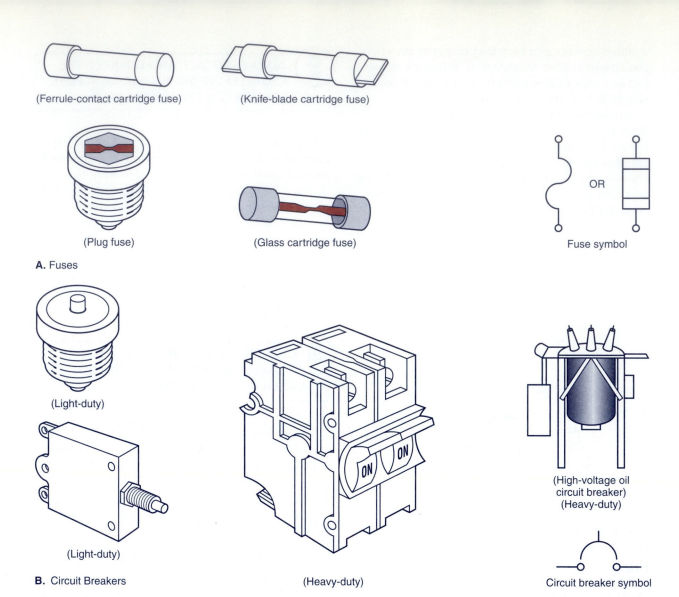

(Ferrule-contact cartridge fuse)   (Knife-blade cartridge fuse)

(Plug fuse)   (Glass cartridge fuse)   OR   Fuse symbol

A. Fuses

(Light-duty)

(Light-duty)   (Heavy-duty)   (High-voltage oil circuit breaker) (Heavy-duty)

B. Circuit Breakers   Circuit breaker symbol

Figure 18-5   Overcurrent-protection devices.

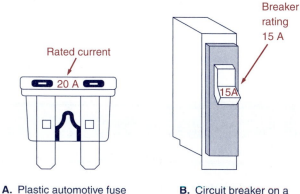

Rated current

20 A

A. Plastic automotive fuse

Breaker rating 15 A

15A

B. Circuit breaker on a household breaker-panel box

Figure 18-6   Fuse and circuit breaker current rating.

The ***time-current*** characteristics or *response time* of a protective device refers to the length of time it takes for the device to operate under difficult current overload conditions. Fast-acting rated protective devices may respond to an overload in a fraction of a second, while slow-acting types may take one to thirty seconds, depending on the amount of the current overload. ***Fast-acting*** fuses are used to protect exceptionally delicate electronic circuits that have a steady flow of current through them. ***Slow-blow*** fuses are used on circuits with normally high inrush or surge currents. This type of circuit would cause a fast- or medium-acting fuse to blow prematurely when power is first applied or during transient peak current conditions.

The *resistance* of a fuse or circuit breaker is low and usually an insignificant part of the total circuit resistance.

18-3 Types of Fuses

Screw-base or Plug Fuse All fuses contain a short metal strip made of an alloy with a *low melting point*. When installed in a socket or fuse-holder, the metal strip becomes a link in the circuit. When the current flow through this link is greater than the rating of the fuse, the metal strip will melt opening the circuit.

The **screw-base or plug fuse** is used to protect branch circuits in a home. It is constructed with a fusible link enclosed in a glass housing (Figure 18-7). This prevents the melted metal link from splattering when the fuse blows. It uses a screw-in base for connection into the fuseholder socket.

Plug fuses have a maximum voltage rating of 125 V. They are available in a number of common current sizes up to a maximum of 30 A. They are used most often in 120-V general house lighting and receptacle circuits. The current rating of the fuse is matched to the maximum-current rating of the circuit conductors.

Noninterchangeable plug fuses are designed to prevent a fuse from being replaced with one of a higher current rating. Nontampering adapters are installed in the fuse socket so the socket will not accept a fuse with a higher current rating.

Fuse links melt for two reasons: either the circuit has developed a short, or it has been overloaded with too many appliances. The see-through glass body of the plug fuse is a great help for finding what caused a fuse to blow (Figure 18-8). If the glass front is black, it indicates that there has been a short circuit. A careful check of the circuit should be made before replacing the fuse. If the glass front is clean and clear, it indicates that the circuit is overloaded. In this case some of the appliances should be removed from the circuit before replacing the fuse.

Cartridge Fuses There are two basic types of **cartridge fuses**. They are the *ferrule-contact* type (Figure 18-9) and the *knife-blade* type (Figure 18-10).

The ferrule-contact type is used in house wiring to protect 120/240-V heavy-duty appliance circuits. These include stoves, clothes dryers, and water heaters. They are rated for 250 V and are available in different current sizes up to 60 A. Again, the fuse current rating selected must be matched to the conductor ampacity rating.

The knife-blade cartridge fuse is used for circuit-current ratings in excess of 60 A. The contact points of this fuse are larger and more rugged. This allows it to handle higher current flows. Available in current ratings from 65 A to 600 A, they are also suitable for 240-V circuits. Knife-blade fuses are sometimes used in fused main disconnect switches in the home. They

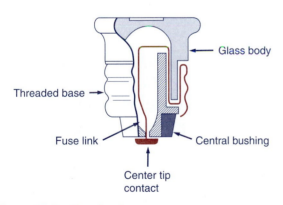

Glass body

Threaded base

Fuse link

Central bushing

Center tip contact

Figure 18-7 The plug fuse.

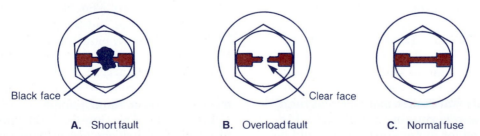

Black face

Clear face

A. Short fault **B.** Overload fault **C.** Normal fuse

Figure 18-8 Fuse fault indicators.

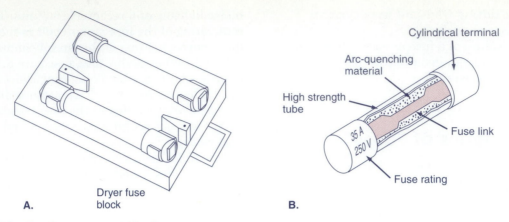

A. Dryer fuse block

B.

Figure 18-9 Ferrule-contact cartridge fuse.

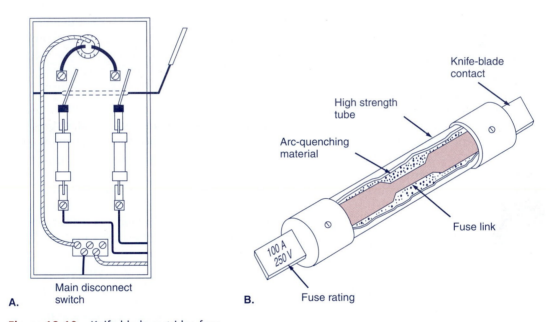

A. Main disconnect switch

B. Fuse rating

Figure 18-10 Knife-blade cartridge fuse.

provide protection for the consumer's main service wiring.

Whether cartridge or plug fuses are used, it takes one fuse (in the "hot" line) to protect a 120-V circuit, but two fuses are required (one in each "hot" line) to protect a 240-V circuit.

Time-Delay Fuses   *Time-delay fuses* are designed to withstand the high surge of current required to get a motor started. A motor will take several seconds to come up to speed on starting. This start takes more current than when the motor is running at full speed. If a fuse was sized for motor start-up, it would be too large to protect the wire in the circuit. If the fuse was sized properly for the wiring in the circuit, it would blow each time the motor started. Special *time-delay* fuses are designed to overcome this problem.

Time-delay fuses are available in both plug and cartridge types. They do not blow like standard fuses on large, but temporary, overloads. They will, however, blow like standard fuses on small continuous overloads, and instantly on short circuits. This type of fuse is used in the home to protect motor circuits such as the furnace, freezers, or water pumps.

A time-delay fuse (Figure 18-11) has a metal fuse strip that has one end attached to the case of the fuse. The other end is attached to a pin

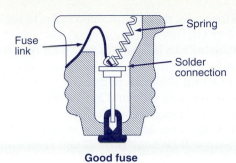

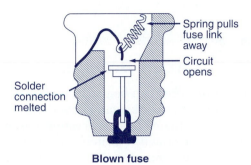

A. Plug type

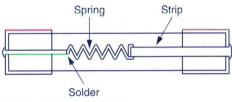

B. Slow-blow cartridge type

Figure 18-11 Time-delay fuses.

held under spring tension. The end of the pin is embedded in solder. If the solder gets hot enough to melt, the pin pulls out of the solder, breaking the circuit. This solder will withstand a momentary overload (i.e., motor start-up) without melting. If the heat from the overload is continuous, the solder melts and breaks the circuit. This might take several seconds. If a direct short should occur, the metal fuse strip melts instantly and opens the circuit.

Renewable Fuses Only nonrenewable types of fuses are approved by the National Electrical Code for residential use. Once a fuse of this type blows, it must be replaced by a new one. This feature prevents people from tampering with the fuse link.

Renewable cartridge fuses are approved for industrial applications. Unlike the nonrenewable type used in the home, these cartridge fuses contain a fuse link that can be

replaced once blown (Figure 18-12). Although initially more costly, this fuse will reduce maintenance costs over a long period of time.

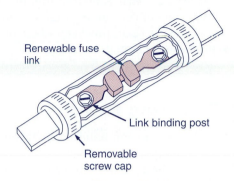

Figure 18-12 Renewable cartridge fuse.

Fusible Links Certain automobile wiring harnesses are equipped with *fusible links* to protect against harness damage in the event of a short in a main feed circuit. A fusible link (Figure 18-13) is a short length of wire, *smaller* in size than the wire in the circuit that it protects. The fusible-link wire is covered with a

Wire to be protected	Link to be used	Color code
10 gauge	14 gauge	Brown
12 gauge	16 gauge	Black
14 gauge	18 gauge	Green
16 gauge	20 gauge	Orange

Figure 18-13 Fusible links.

thick nonflammable (Hypalon) insulation. An overload causes the link to heat and the insulation to blister. If the overload remains, the link will melt, causing an open circuit. The links are color-coded for wire size. A melted link must be replaced with one of the same gauge size after the cause of the overload has been identified and the problem corrected.

18-4 Testing Fuses

With glass fuses, you can usually see if the fuse link inside is burned open. Fuses may be tested out-of-circuit by using a simple *continuity lamp tester* (Figure 18-14). A good fuse connected in series with the test leads will complete the circuit and cause the lamp to turn on.

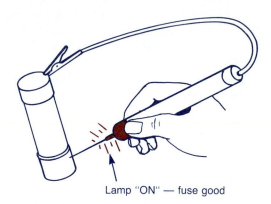

Lamp "ON" — fuse good

Figure 18-14   Simple continuity fuse tester.

The ohmmeter can also be used to make an out-of-circuit test of a fuse (Figure 18-15). A good fuse should indicate a near-zero resistance reading on the meter. An open fuse is indicated by an infinite (∞) reading on an analog ohmmeter. Depending on the manufacturer, digital

ohmmeters may display an open infinite reading as 1 or OL (overload).

A voltmeter can be used to make an in-circuit test of a fuse (Figure 18-16). Voltage is checked on the line and the load side of the fuse. Full voltage on the line side and zero voltage on the load side indicates a blown fuse.

The voltage drop across the two terminals of a good fuse will be near zero. If you read appreciable voltage across the fuse, this means it is an open circuit.

18-5 Circuit Breakers

A **circuit breaker** can be used (in place of a fuse) to protect circuits against overloads and short circuits. Like a fuse, it is connected in series with the circuit it protects. Circuit breakers are rated in a manner similar to the one used for fuses. As with fuses, the ampere rating of a breaker must match the ampacity of the circuit it protects.

A *manually reset*, toggle-type circuit breaker serves both as a switch and a nondestructible fuse (Figure 18-17). As a switch, a circuit breaker lets you open a circuit (turn the switch to OFF) whenever you want to do work on it. As a fuse, it provides overcurrent protection that can be reset.

Circuit breakers can be classified according to tripping operation as being thermal, magnetic, or thermal-magnetic. **Thermal circuit breakers** consist of a heating element and mechanical latching mechanism. The heating element is usually a bimetallic strip that heats up when current flows through it. A thermal circuit breaker must cool off before you can reset it. In addition, ambient temperatures affect the trip point. Thus, a thermal circuit

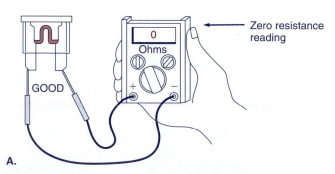

A.

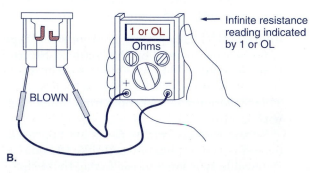

B.

Figure 18-15   Using an ohmmeter to test fuses.

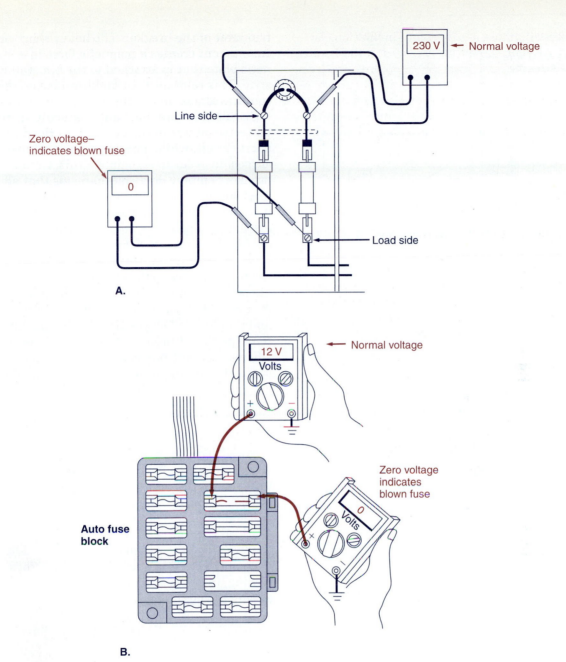

Figure 18-16 In-circuit test of fuses using a voltmeter.

breaker will require more current and take a longer time to trip in a very cold environment than it will in a hot environment. Applications include motors, transformers, and solenoid protection.

Magnetic circuit breakers work on the principle of magnetism. The current flowing through the circuit passes through a coil in the circuit breaker housing. If the current exceeds the rating of the breaker, the magnetic field becomes strong enough to produce a force to trip the breaker action. Typical applications include printed-circuit board and power semiconductor protection.

Thermal-magnetic circuit breakers combine heat and magnetic sensing (Figure 18-18). An overload causes the bimetallic strip to heat and bend. This action releases the trip lever and opens the breaker's contacts. For short circuits a quicker method is used to release the

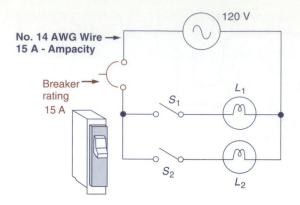

Figure 18-17 Circuit breaker connection.

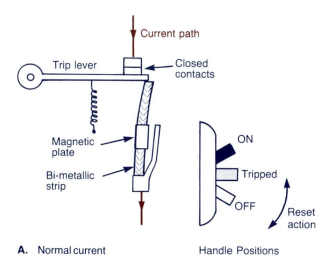

A. Normal current

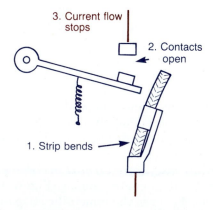

B. Overload current

Figure 18-18 Action of a manually reset circuit breaker.

trip lever of the breaker. The heavy short circuit current creates a magnetic force in a magnetic plate that is attached to the bimetallic strip. This releases the trip lever to open the contacts almost instantly.

Circuit breakers must automatically open the protected circuit in the event of overload or short circuit, while ignoring transient current surges. In order to accomplish this, circuit breakers employ delay mechanisms that allow brief surges to occur without tripping. The magnetic solenoid used in most breakers is an extremely sensitive system that can be set to operate with precise delay responses.

When certain types of circuit breakers trip, the toggle moves to the **tripped** or intermediate position. To reset this type of breaker, it is first moved all the way to the full OFF position and then to the ON position. With a pushbutton-type circuit breaker, the button pops out when the circuit breaker is tripped, which opens the circuit. To reset the circuit breaker, you push the button back in.

Self-resetting circuit breakers are used for protecting circuits where temporary overloads may occur and where power must be quickly restored. The automobile headlamp, windshield wiper, and air conditioner circuits are often protected with self-resetting circuit breakers. The breaker consists of a bimetallic strip made up of two bonded strips of metal (Figure 18-19). An overload current generates heat, which causes the two metals to expand differently and bend to open the circuit. When the current stops flowing, the bimetallic strip cools, once again closing the circuit.

Although circuit breakers are initially more expensive to install than fuses, they have the following advantages:

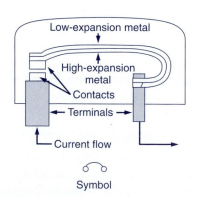

Figure 18-19 Self-resetting type circuit breaker.

- They can be used both as a protective device and an on/off control switch.
- No cost is involved in replacing them when they operate due to overcurrent.
- It is more convenient and safer to reset them than to replace fuses.
- They can be accurately produced for different time-delay tripping action.

18-6 Thermal Overload Protection

A motor is considered to be *overloaded* when it draws too much current. Overloading slows down the motor which results in an increase of input current. A large amount of current may be drawn by a motor at starting but is permissible for a short time. If an overload current continues for any length of time the motor will overheat and burn out. Most small motors are equipped with a manual or automatic reset **thermal overload**. This overload may be built into the motor or mounted externally on the motor housing (Figure 18-20) and senses motor temperature, not current. High motor winding temperatures that result from an overload cause the thermal overload to operate to automatically open the circuit.

18-7 Lightning Rods and Arresters

Lightning Rods A lightning bolt is the result of an electrical path set up between a charged cloud and the earth. The cloud and earth behave as two voltage-supply leads with a voltage potential energy in the million-volt range. The resistance path between the two can result in short bursts of current in the thousand-ampere range. This amount of current flow directed through electric lines and equipment can do extensive damage to the system.

While lightning-caused damage to electrical installations is quite rare in larger cities, it is rather frequent in rural areas. One protection against a lightning discharge is to provide an alternate, direct low-resistance path to ground. A *lightning rod* works on this principle. A sharply pointed metal rod is mounted on the highest part of a house or barn and a heavy copper wire is run from it to the ground (Figure 18-21). Any lightning discharge will be directed through this low resistance path, thereby protecting the structure and its contents.

A television antenna mast or other metal support acts the same as a lightning rod and must, therefore, be well grounded. The steel frame structures of tall commercial and industrial buildings also act as unintended lightning rods

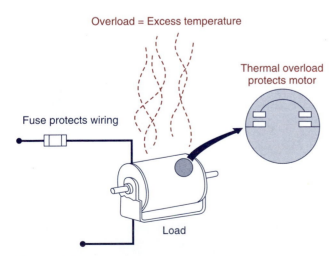

Figure 18-20 Thermal overload protection.

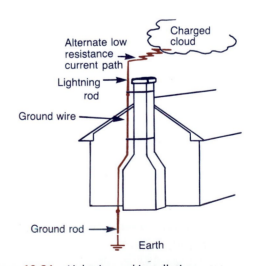

Figure 18-21 Lightning rod installation.

and must be grounded to provide the same protection.

Lightning Arresters   Lightning arresters are used to drain the energy from power lines struck by lightning. The *lightning arrester* works on the spark gap principle like the spark plug in a car. One side of the arrester is connected to ground and the other is connected to the wire to be protected (Figure 18-22). Under normal circuit voltage conditions, these two points are insulated by the air gap between them. When lightning strikes the line, the resulting high voltage ionizes the air and produces a low resistance discharge path to ground. Specially designed arresters are available for use on overhead power lines as well as on signal circuits such as telephone circuits and antenna lead-in wires.

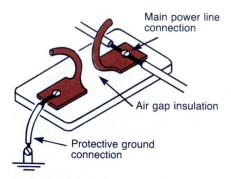

Figure 18-22   Lightning arrester.

Control System Technician
Job Description
- Install, repair, and maintain electrical control systems.
- Systems include programmable logic controllers, motor drives, instrumentation, and generators.
- Graduate of a college program in Electrical, Electronics, or Instrumentation Technology.
- Possess excellent troubleshooting skills and function well in a semiautonomous team.

Review and Applications

Related Formula

$$I = \frac{P}{V}$$

Review Questions

1. When is an electric circuit considered to be overloaded?
2. What is the most common cause of electric circuit overloads in a home?
3. Define the term short circuit.
4. Why is a short circuit, created directly across the voltage source, the most dangerous of all circuit faults?
5. What happens when conductors are required to carry more current than they are safely rated to handle?
6. How are fuses and circuit breakers connected with respect to the circuits they protect?
7. How does the resistance of a fuse or circuit breaker compare with that of a load device?
8. Explain what each of these fuse and circuit breaker ratings specify:
   (a) continuous-current rating
   (b) voltage rating
   (c) short-circuit current rating
   (d) response time
9. To what is the continuous-current rating of a fuse or circuit breaker matched?
10. Explain how a fuse operates to protect a circuit.
11. Name the type of fuse most likely used in each of the following circuits:
    (a) 35-A dryer circuit
    (b) 15-A lighting circuit
    (c) 100-A main disconnect switch
12. What is a noninterchangeable plug fuse designed to protect against?
13. (a) Describe the special feature of time-delay fuses.
    (b) In what type of circuit should time-delay fuses be used?
14. In what way is a fusible-link wire different from the wire it protects?
15. State two methods that can be used to check fuses out-of-circuit.
16. Explain how a circuit breaker trips when the circuit becomes overloaded.
17. List three advantages that circuit breakers have over fuses.
18. Explain how a thermal-overload protection device prevents motor burn out.
19. Explain how a lightning rod circuit protects an electrical installation.
20. Explain the principle of operation for a lightning arrester.

Problems

1. A 100-W accessory device is to be installed in the electrical system of an automobile. What size fuse would be required to protect it?
2. A slow-blow fuse is replaced with a fast-acting fuse in a circuit that calls for a slow-blow fuse. What is most likely to happen? Why?

Critical Thinking

1. A good fuse operating in a normal circuit will have practically zero voltage across it. Why?

Portfolio Project

As a control systems technician, you have been asked to investigate the frequent tripping of a circuit breaker. Report on how you would proceed to remedy the problem.

Circuit Challenge

Using a Simulator

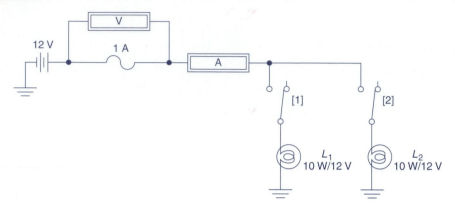

Procedure

(a) Wire the fused circuit shown, using what-ever simulation software package is available to you.

(b) Calculate the amount of current that would normally flow through the fuse with:
   (i) switch 1 closed
   (ii) both switches closed

(c) Operate the circuit and record the actual amount of current flow through the fuse and the value of the voltage drop across the fuse, with only switch 1 closed.

(d) Operate the circuit with both switches closed. What happens? Why?

(e) Record the value of the voltage drop across the blown fuse.

Electric Power

Early electric power plants were small and generated direct current. Today's plants are huge by comparison and use alternating current generators or alternators. In this chapter we will study the different types of power plants and the transmission systems used in conjunction with them. Also discussed is the electric power formula as it applies to rating of devices and the measurement of power.

Objectives

After studying this chapter, you should be able to:
1. Describe the various methods of power generation.
2. Outline the method used in transmitting power over long distances.
3. Define electric power.
4. Perform calculations using the electric power formula.
5. Properly connect a wattmeter into a circuit.

Key Terms

- hydroelectric
- thermal-electric
- nuclear
- turbine
- generator
- solar cell

- high voltage
- transmission network
- transformer
- brownout
- electric energy

- electric power
- watt
- kilowatt
- power rating
- power formula
- wattmeter

- Pacific Gas and Electric
- Arizona Electric Power Cooperative
- Orlando Utilities Commission

19-1 Electric Generating Stations

The electric energy supplied to our home is produced at a central location called a **generating station**. This station is usually equipped with several large generators each of which is driven separately. Most of the electric power generated today is three-phase, 60-cycle alternating current (ac). Three-phase power permits more economical and efficient distribution systems.

Hydroelectric Generating stations are classified according to the method used to drive the generator. **Hydroelectric** generating stations use water power in the form of waterfalls and water under pressure from giant dams or reservoirs to drive the generators (Figure 19-1). The generating station, where the turbines and generators are installed, is a building located at the base of the dam. A large pipe, the *penstock*, carries water from the reservoir to the turbines.

The water flows into the penstock and races down the long slope. At the bottom, it meets the blades of the *turbine* with great force and makes the turbine spin. A *generator* is attached to the turbine, and it is also set turning.

The flow of water is adjustable. A control valve on the penstock can be partly closed, just like a water tap. This slows down the turbine and the generator, so less electricity is produced.

A hydroelectric generating station is the cheapest to operate. Unlike a factory that has to pay for its raw materials, the hydroelectric plant gets its raw material—flowing water— free. It uses a renewable energy resource (energy provided by the source can be renewed as it is used) and is the most environmentally safe method of producing electricity. The water that flows through the turbine is not changed or polluted. After doing its work, the river continues on its way downstream.

Thermal-electric Power There are not enough suitable places for hydroelectric installations to meet the increasing demand for electric energy. As a result, steam power is also used to drive generators. Stations that use generators driven by steam turbines are called **thermal-electric** generating stations.

Thermal-electric stations use heat to convert water to steam, which then forces a turbine wheel to turn. A conventional (fossil fuel) thermal-electric generating station may burn coal, oil, or natural gas to release heat energy (Figure 19-2). After the water in the boiler turns to steam, it is fed to a steam turbine. The steam rushes into the turbine under great pressure and it forces the blades around very rapidly. As the shaft holding the blades revolves, it spins the generator and electric energy is produced.

Nuclear Power The heart of a **nuclear power** generating station is its nuclear reactor (Figure 19-3). Here, immense heat is generated by the splitting of uranium atoms or nuclear fission. The heat is transferred from the reactor to a boiler, where it boils water to steam. The turbine-generator part of a nuclear plant is the same as a thermal-electric plant, and the prod-

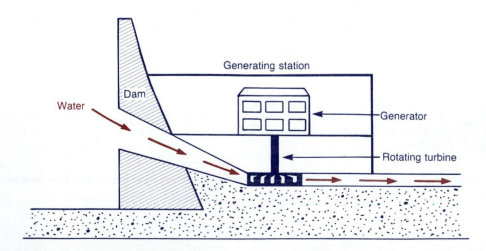

Figure 19-1 Hydroelectric generating station.

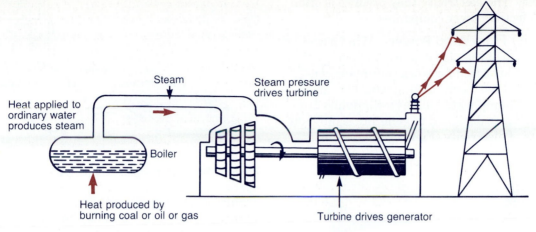

Figure 19-2   Thermal-electric, conventional (fossil fuel) generating station.

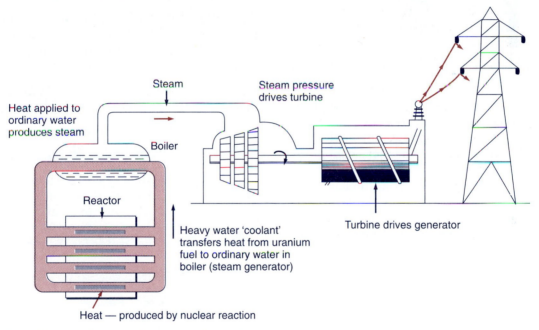

Figure 19-3   Thermal-electric nuclear generating station.

uct is identical—electricity. In addition, both kinds of power use nonrenewable resources that can adversely affect the environment.

19-2 Alternate Ways of Generating Electricity

Other ways of generating electricity, using renewable resources, are constantly being investigated. At the present time, we are seeing a number of small energy suppliers using a variety of renewable energy sources to produce

About ◄▆► Electronics

Hawaiians are generating electricity profitably by using the temperature differences between warm surface water and deep cold water to move turbines. This provides 255 kW of power.

electricity. The electricity they produce is often sold to electric power companies.

Electricity From Wind Power
Wind power must be one of the oldest sources of energy that has been harnessed for work. For many centuries, wind has been turning wheels to grind grain and pump water. Today's windmills are wind turbines, designed to generate significant amounts of electricity (Figure 19-4).

Figure 19-4 Wind turbines used to generate electricity.

To extract as much energy from the wind as possible, the wind turbine blades are huge—up to 330 feet apart. Wind sensors enable the turbine's computer to control the movement of the rotor and to produce optimum power in all wind conditions. If one is very sure of a steady supply of wind where one lives, a wind ac alternator can be used. This will produce the alternating current needed for appliances. If one lives where the wind is not always blowing, a wind dc generator is needed to produce direct current for storage batteries. An inverter is also needed to convert the direct current into alternating current, so that regular electric appliances can be used.

Electricity From Solar Energy
Solar energy, that is, energy harnessed from the sun, is one of the most important alternate sources for producing electricity. Using *solar* or *photovoltaic cells* we can convert the sun's energy directly into electric energy. These cells are silicon wafers (Figure 19-5). Solar energy dislodges electrons from the wafer to create an electric current.

Photovoltaic cells are still too costly compared to the cost of fossil fuels as a source of electricity. However, in the future, as more and more of them are sold, and as we learn to make them

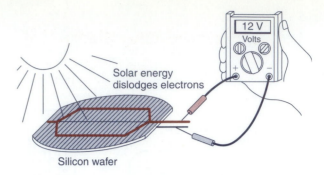

Figure 19-5 Solar or photovoltaic cell.

more efficiently, silicon cells should become cheaper. Solar power plants require considerable space. They must be located in places that have mostly days of sunshine. Desert and other arid locations are ideal for solar power plants.

Co-generation Power Station
Co-generation is a process in which electricity and thermal energy are generated from a single fuel source. The fact that both the electricity and heat produced by the system are used brings overall efficiency up to 80 percent or higher. This is in contrast to the approximately 30 percent or so efficiency of a traditional coal fired plant generating electricity where much of the thermal energy is lost as exhaust gases.

Figure 19-6 shows a typical co-generation system. In this application, a gasoline engine is used to turn the generator to create electricity with the waste exhaust, in turn, funneled into a boiler to produce hot water for the plant.

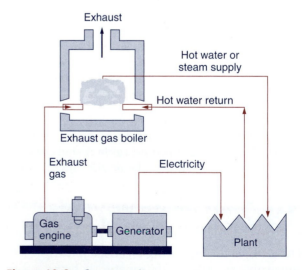

Figure 19-6 Co-generation.

19-3 Transmitting Electricity

It is not always possible or practical to locate a generating station close to where the electric energy will be put to use. In most instances, the electricity must travel hundreds of miles from the source to the load. Transmitting large amounts of electric energy over fairly long distances is accomplished most efficiently by using **high voltages**.

High voltages are used in transmission lines to reduce the required amount of current flow. The power transmitted in a system is proportional to the voltage multiplied by the current. If the voltage is raised, the current can be reduced to a small value, while still transmitting the same amount of power.

Example 19-1

If 10,000 W are to be transmitted, a current of 100 A would be required if the voltage used were only 100 V.

$$100 \text{ V} \times 100 \text{ A} = 10,000 \text{ W}$$

On the other hand, if the transmission voltage is stepped up to 10,000 V, a current flow of only 1 A would be needed to transmit the same amount of power.

$$10,000 \text{ V} \times 1 \text{ A} = 10,000 \text{ W}$$

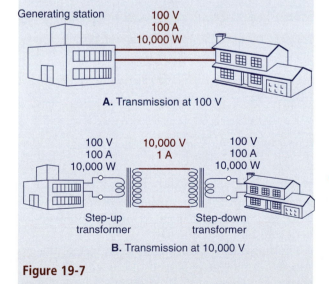

Generating station
100 V
100 A
10,000 W

A. Transmission at 100 V

100 V · 100 A · 10,000 W
10,000 V · 1 A
100 V · 100 A · 10,000 W

Step-up transformer · Step-down transformer

B. Transmission at 10,000 V

Figure 19-7

Due to the reduction of necessary current flow at high voltage, the size and cost of conductor wires required to handle the entire output of a generating plant may be no larger than the lower voltage leads serving industrial customers. Reducing the current also minimizes the amount of power lost in the lines.

There are certain limitations to the use of high-voltage transmission systems. The higher the transmitting voltage, the more difficult and expensive it becomes to safely insulate between line wires, as well as from line wires to ground. For this reason, the voltages in a typical high-voltage grid system are reduced in stages as they approach the area of final use (Figure 19-8). When the power reaches a city, transformer stations lower the voltage—first to 27,600 V, then to 4000 V. Some heavy machinery uses power at this voltage, but home appliances operate at still lower voltages—240 V or 120 V. The *transformers* in which this last voltage conversion takes place are often visible on street poles.

A **network of transmission lines** and generating stations are connected together. In this way, excess power from one region can be fed to another region in response to demand. During the periods when the demand for electric power drops, power stations shut down some generators. At times of peak demand, auxiliary equipment is set in operation. Networking results in a decrease in the electric-generating capacity needed by any one area. This produces lower costs for all users.

In a typical power distribution system, the flow of current is not a constant. Loads vary with the hour of the day and the day of the week. Since the voltage drop in parts of the system is determined by the amount of current flow, a continuous change in the voltage measured at the customer's home could occur. Automatic voltage regulators are used to try to maintain constant voltage levels. Electric power companies generally allow variations of plus or minus 10 percent from a nominally stated voltage.

A significant reduction in power line voltage for extended periods is commonly referred to as a **brownout**. Intentional brownouts are sometimes used by the power utility to force a reduction in power consumption. For example, during a heat wave in a large city, the demand for power to operate air-conditioning equipment

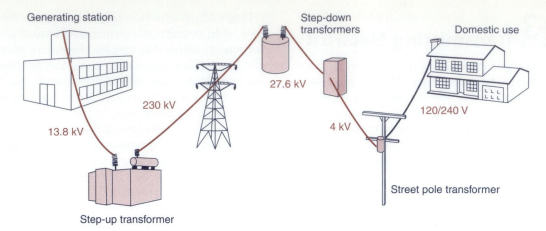

Generating station

Step-down transformers

Domestic use

230 kV

27.6 kV

13.8 kV

4 kV

120/240 V

Step-up transformer

Street pole transformer

Figure 19-8   Typical high-voltage transmission system.

may tax the capacity of the electrical system to deliver additional load. Since there is no easy way to prevent additional load from coming on-line, the power company may choose to lower voltage to all customers. The effect of lower voltage is an immediate reduction in power demand. For example, with a 10 percent reduction in voltage, the power demand will automatically be lowered by 19 percent.

19-4 Electric Power

Electric energy is simply the work performed by an electric current. Whenever current exists in a circuit, there is a conversion of electric energy into other forms of energy. For example, current flow through a lamp filament converts electric energy into light and heat energy. *Electric power* is defined as the rate at which electric energy is converted. The *watt* (W) is the unit used to measure electric power.

Most electric devices carry an electrical wattage or power rating. The wattage rating indicates the rate at which the device can convert electric energy into another form of energy, such as light, heat, or motion. The faster a device converts electric energy, the higher its wattage rating. The **power rating** of a device can be read directly from its nameplate or calculated from other information given (Figure 19-9).

If the normal wattage rating is exceeded, the equipment or device will overheat and perhaps be damaged. For example, if a lamp is rated at 100 W at 120 V and is connected to a source of

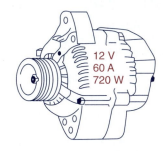

120 V
1050 W

Nameplate

A. Appliance power rating

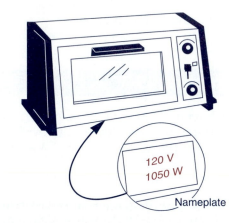

12 V
60 A
720 W

B. Car alternator power-output rating

Figure 19-9   Power ratings for electric devices.

240 V, the current through the lamp will double. The lamp will then use four times the wattage for which it is rated, and it will overheat and burn out quickly.

There is a direct relationship between measurement of electric power and mechanical power. By definition 746 W equals 1 horsepower (hp). This relation can be remembered more easily as 1 hp equals approximately 3/4 kilowatt (kW).

19-5 Calculating Electric Power

The watt, which is the basic unit of electric power, is equal to the voltage multiplied by the current. This represents the rate at which work is being done in moving electrons through the circuit. Electric power can be calculated directly using the measurements of voltage and current. To calculate the power in watts, you simply multiply the voltage in volts by the current in amperes:

$$P = V \times I$$

where

P = power in watts (W)
V = voltage in volts (V)
I = current in amperes (A)

Electrical-load devices such as lights and appliances are designed to operate at a specific voltage. They may be marked with both their operating voltage and power output at that voltage. If the current rating is not marked, it can be calculated by using the rated power and voltage values. To calculate the current, the power is divided by the rated voltage:

$$I = \frac{P}{V}$$

In some cases it is more convenient to calculate power by using the resistance of the material rather than the voltage applied to it. This occurs quite often when calculating the power dissipated in a resistor or lost as heat in a conductor. By substituting the product IR for V in the basic power formula, we get:

$$P = V \times I$$
$$P = (IR)(I)$$
$$P = I^2R$$

When large amounts of power are being considered, it is more convenient to use a larger unit than a watt, and thus avoid large numbers. Such a larger unit is a **kilowatt** (kW), which is equal to 1000 watts.

Example 19-2

Suppose a portable electric heater draws a current of 8 A when connected to its rated voltage of 120 V (Figure 19-10). The power rating of the heater is then:

$$\text{Power} = \text{Voltage} \times \text{Current}$$
$$P = V \times I$$
$$P = (120\ \text{V})(8\ \text{A})$$
$$P = 960\ \text{W}$$

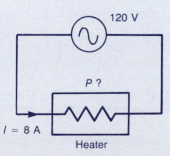

Figure 19-10   Circuit schematic

Example 19-3

Suppose a 150-W light bulb has a voltage rating of 120 V (Figure 19-11). When connected to a 120-V source, the current flow through the bulb would be:

$$I = \frac{P}{V}$$
$$I = \frac{(150\ \text{W})}{(120\ \text{V})}$$
$$I = 1.25\ \text{A}$$

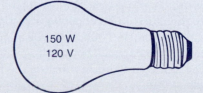

Figure 19-11

Example 19-4

Suppose the current flow through a 100-Ω resistor is measured and found to be 0.5 A.
(a) How much power is being dissipated as heat?
(b) How much power would be dissipated if the current is doubled?

$$\text{(a)} \quad P = I^2 \times R$$
$$P = (0.5 \text{ A})^2 (100 \ \Omega)$$
$$P = 25 \text{ W}$$
$$\text{(b)} \quad P = I^2 R$$
$$P = (1 \text{ A})^2 (100 \ \Omega)$$
$$P = 100 \text{ W}$$
$$(4 \text{ times the power})$$

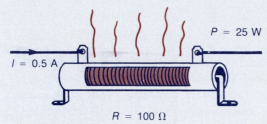

$P = 25$ W

$I = 0.5$ A

$R = 100 \ \Omega$

A. Circuit for Part a

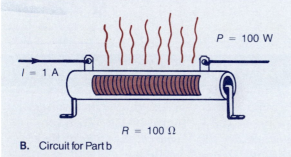

$P = 100$ W

$I = 1$ A

$R = 100 \ \Omega$

B. Circuit for Part b

Figure 19-12

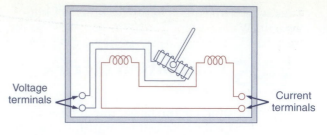

Voltage terminals — Current terminals

A. Internal construction

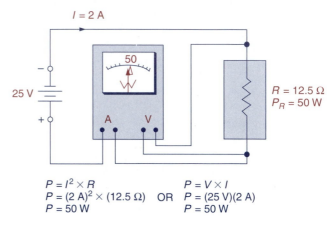

$I = 2$ A

25 V

50

A   V

$R = 12.5 \ \Omega$
$P_R = 50$ W

$$P = I^2 \times R \qquad\qquad P = V \times I$$
$$P = (2 \text{ A})^2 \times (12.5 \ \Omega) \quad \text{OR} \quad P = (25 \text{ V})(2 \text{ A})$$
$$P = 50 \text{ W} \qquad\qquad P = 50 \text{ W}$$

B. Inserted into circuit

Figure 19-13   Wattmeter.

section is connected the same as a regular voltmeter is, in parallel or across the load. The ammeter section is connected like a regular ammeter is, in series with the load. With an analog wattmeter, a reverse pointer movement is corrected by reversing the two voltmeter leads or the two ammeter leads, but not both.

19-6 Measuring Electric Power

A *wattmeter* is an electrical instrument which is used to measure electric power directly. This meter is a combination of both a voltmeter and an ammeter. It measures both voltage and current at the same time and indicates the resultant power value.

The wattmeter has four terminal connections: two for the voltmeter section and two for the ammeter section (Figure 19-13). The voltmeter

HELP WANTED

Power Distribution System Operator
Job Description
- Control the electric grid that delivers power to customers.
- Monitor and direct all activity on power lines and substations at voltages up to 60,000 volts.
- Requires sound judgment, patience, and good communications skills in highly stressful environments.
- Experience with operating diagrams and schematic drawings.

Review and Applications

∙∙

Related Formulas

$P = V \times I$

$I = \dfrac{P}{V}$

$P = I^2 \times R$

Review Questions

1. Explain how electricity is generated using the hydroelectric process.
2. (a) Explain how electricity is generated using the thermal-electric process.
   (b) State the two primary sources of heat used in the thermal-electric process.
3. State two alternative ways of generating electricity using renewable resources.
4. (a) Why are high voltages used when transmitting electric power over long distances?
   (b) What limitation is there to the use of high-voltage transmission systems?
5. Define electric power.
6. What is the basic unit used to measure electric power?
7. (a) A wattmeter is a combination of what two types of meters?
   (b) When connecting a wattmeter into a circuit, how is each meter section connected with regard to the load?
   (c) With an analog wattmeter, how is a reversed pointer movement corrected?

Problems

1. Calculate the power rating of a range element that draws 7.5 A when connected to its rated voltage of 230 V.
2. How many amperes will a 1250-W toaster draw when connected to its rated voltage of 120 V?
3. How much power is dissipated in the form of heat when 0.5 A of current flows through a 50-Ω resistor?

Critical Thinking

1. With reference to Example 19-1, determine how much the transmission line current would be if 10 megawatts of power were transmitted over a 60,000 volt transmission line.

Portfolio Project

Report on how electric power is produced and distributed in your geographic area. Include location and type of power plant, transformers, transmission lines, and substations that make up the system.

Circuit Challenge

Using a Simulator

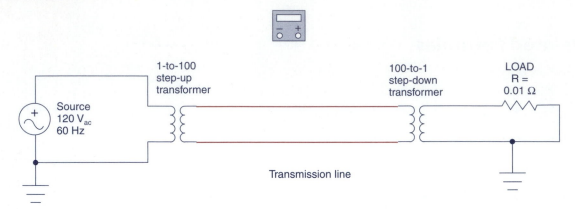

Procedure

(a) Wire the simulated transmission circuit shown, using whatever simulation software package is available to you.

(b) Calculate the value of the transmission line voltage.

(c) Calculate the value of the load voltage.

(d) Calculate the value of the load current.

(e) Calculate the value of the transmission line current.

(f) Use the multimeter to measure each of the calculated values.

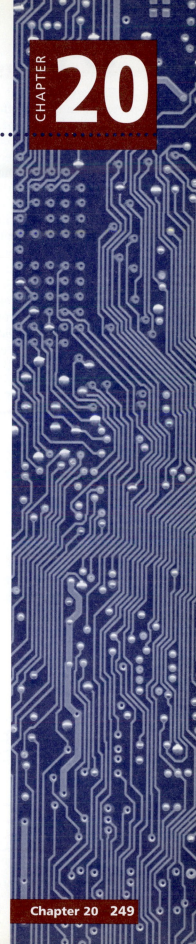

Electric Energy

The main purpose of all electrical installations is to deliver electric energy to a load. The load, in turn, uses this energy to develop power and do work for us. In this chapter we study the measurement of energy and its efficient use.

Objectives

After studying this chapter, you should be able to:
1. Define electric energy.
2. Calculate energy consumption.
3. Correctly read a dial-type kilowatthour meter.
4. Calculate energy costs.
5. Outline a strategy for more efficient use of electric energy.

Key Terms

- electric energy
- joule
- kilowatthour

- kWh meter
- digital meter
- dial meter

- ENERGY GUIDE label
- intelligent building

RESEARCH **INTERNET** SITES

- Duke Energy
- Niagara Mohawk
- Ohio Edison

20-1 Energy

Energy is defined as the ability to do work. It exists in many different forms. These include electric, heat, light, mechanical, chemical, and sound energy. *Electric energy* is the energy carried by moving electric charges.

Every kind of energy can be measured in *joules* (J). One joule of electric energy is equivalent to the energy carried by 1 coulomb of electric charge being propelled by a force of 1 volt.

One of the most important laws of science states that energy can neither be created, nor be destroyed; it can only be changed from one form to another. For example, an incandescent lamp transforms electric energy into useful light energy. However, not all the electric energy is converted into light energy. About 95 percent of it is converted into wasted heat (Figure 20-1). In this process the important thing to note is that while the total energy may take on different forms, in each case the total amount of energy converted remains constant.

20-2 Calculating Electric Energy

In scientific work, the unit used for measuring electric energy is the joule (J). One joule represents a very small amount of energy. The energy used by a device is most easily calculated from its power rating. To calculate energy (E) in joules, we multiply the power in watts (W) by the time the device is in use in seconds (s):

$$E = P \times t$$

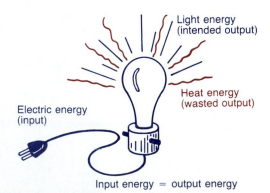

Figure 20-1 Energy conversion in a light bulb.

Light energy (intended output)

Heat energy (wasted output)

Electric energy (input)

Input energy = output energy

Example 20-1

Suppose a 100-W light bulb is operated for 5 minutes (Figure 20-2). The amount of energy converted in joules is then:

$$\text{Energy} = \text{Power} \times \text{Time}$$
$$E = (100 \text{ W})(5 \times 60 \text{ s})$$
$$E = 30,000 \text{ J}$$
$$E = 30 \text{ kJ}$$

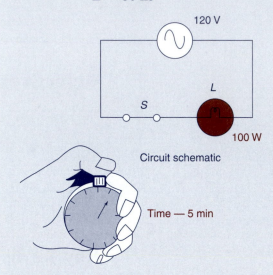

120 V

L

S

100 W

Circuit schematic

Time — 5 min

Figure 20-2

where

$$E = \text{energy in joules (J)}$$
$$P = \text{power in watts (W)}$$
$$t = \text{time in seconds (s)}$$

The practical unit for measuring electric energy is the *kilowatthour* (kWh). To find the energy used by a device in kilowatthours, the same basic energy formula is used. However, in this case, the power rating is expressed in kilowatts (kW) and the time is given in hours (h):

$$E = P \times t$$

where

$$E = \text{energy in kilowatthours (kWh)}$$
$$P = \text{power in kilowatts (kW)}$$
$$t = \text{time in hours (h)}$$

Example 20-2

Suppose an electric coffee maker is rated for 900 W of power (Figure 20-3). It is used an average of 6 hours per month. The average monthly energy conversion in kilowatthours is then:

$$Energy = Power \times Time$$
$$E = Pt$$
$$E = \left(\frac{900}{1000}\,kW\right)(6\,h)$$
$$E = 5.4\,kWh$$

900 W

Monthly energy input
5.4 kW h

Figure 20-3

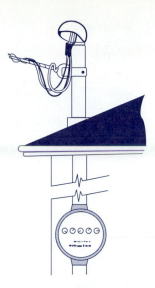

Figure 20-4 Kilowatthour energy meter attached to a home.

20-3 The Energy Meter

The **kilowatthour (kWh) meter** is used to measure the amount of electric energy used in a house, building, or factory (Figure 20-4). The meter records how many watts of power are used over a period of time. At regular intervals, a power company employee comes to read a customer's meter. The power company calculates the amount of electricity used and charges the customer for it.

The watt is a relatively small unit, so the basic unit used by the power company is the kilowatt. A kilowatthour (kWh) is 1000 watts of electricity used for one hour. To give you some comparison, one kilowatthour is equivalent to the energy required to burn a 100-W light bulb for 10 hours or to operate a 5000-W electric clothes dryer for approximately 12 minutes.

The kilowatthour meter records the energy used in the same way as the odometer of a car records the distance travelled. By means of a small motor that runs faster or slower depending on how much current passes through it, the meter registers the amount of electricity that is consumed. As the motor runs, it turns numbered indicators to register the cumulative total energy used to the time of a reading.

Some meters will give a direct reading on a **digital** display, while others use a series of **dials** that must be read in order to determine the value recorded. More common, however, is the dial-type meter with four or five separate dials, each of which supplies one digit for the reading. The dials are linked together by gears. The left dial registers tens of thousands; the next, thousands; and so on to the right dial, which registers units, that is, single kilowatthours. The numbers on the dial faces are arranged alternately clockwise and counterclockwise, and the pointer moves accordingly, but no matter which direction the pointer follows, it always goes from 0 to 9 (Figure 20-5).

The kilowatthour meter gives a total cumulative reading. To determine the amount of energy used in a specific time period you must take two consecutive readings and subtract the earlier one from the later one.

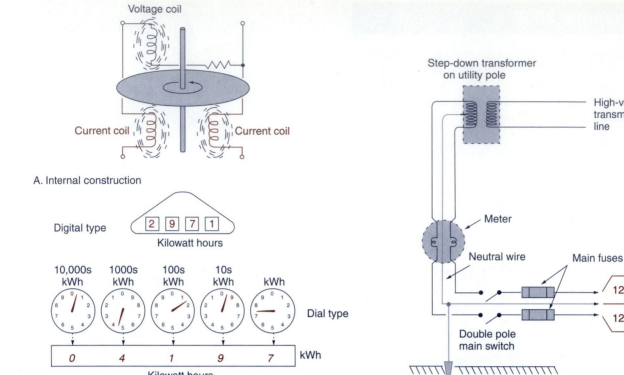

A. Internal construction

B. Types of meters

C. Connection

Figure 20-5 Kilowatthour meter construction and connection.

Example 20-3

Steps to follow when reading a dial-type meter (Figure 20-6):

1. Read the order of the dials from left to right. Note that some dials increase their readings in a counterclockwise direction while others increase in a clockwise direction.
2. When the dial pointer is between two numbers, read the smaller of the two numbers.
3. When the pointer rests almost squarely on a number, refer to the dial to the immedi-

ate right to determine which number you record. If it is 0 or 1, the number in question has been passed. If it is 8 or 9, the number has not been passed.

4. Your meter may have four or five dials. The kilowatthour reading can be obtained by reading the dials and by using the multiplier (if applicable) shown on the meter.

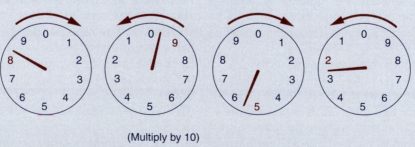

(Multiply by 10)

Figure 20-6   Read: 89 520 kW h (89 525 is also correct)

Example 20-4

Suppose two consecutive kWh meter readings appear as shown in Figure 20-7. The amount of electricity used in the time period between the two readings is:

1st reading = 17,650 kWh

2nd reading = 18,349 kWh

kWh used = 18,349 kWh − 17,650 kWh

kWh used = 699 kWh

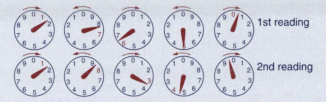

1st reading

2nd reading

Figure 20-7

20-4 Energy Costs

Electric energy is sold by the kilowatthour (kWh). The bill from the electric utility company always gives the previous and present meter readings. The difference is the amount used. The company also issues a rate card with the charge per kilowatthour. To calculate your bill you simply multiply the kWh used by the cost per kWh.

Example 20-5

The billing from an electric utility company indicates a present meter reading of 6060 kWh and a previous reading of 3140 kWh. If the rate is 5 cents per kWh the cost is:

present reading = 6060 kWh

previous reading = 3140 kWh

kWh used = 6060 kWh − 3140 kWh

kWh used = 2920 kWh

Cost = kWh × rate

Cost = 2920 kWh × \$0.05

Cost = \$146.00

The amount of electric energy used for electric appliances depends on two factors. One is the length of time they are used. The other fac-

tor is the amount of electric power required to make each appliance operate. These two factors determine the electric energy we use each month. The meter keeps a record of the total electric energy used, but does not tell us about individual appliances.

20-5 Energy Management

In the past, a typical approach to inefficient use of power was to supply larger quantities and an increasing number of outlets. This would be similar to taking a gas-guzzling automobile, upgrading the tank, and only driving on streets with many service stations.

More efficient use of electric energy can reduce demand and offset rising costs. Have you ever wondered how much electricity your appliances consume and what it costs when you use them? Table 20-1 lists some appliances, the wattage for each, and an estimate of the time each is in use during a one-month period. This listing of how much electricity your appliances use should help you manage your electricity

About ⊏▬▶ Electronics

Because our industries are so dependent on machines controlled by sensitive electronic devices, blackouts cost 3 to 5 billion dollars per year in production down-time and lost data.

Example 20-6

An electric clothes dryer rated for 4.2 kW is used an average of 20 hours per month (Figure 20-8). The rate for the energy used is 4.5 cents per kWh. The cost of operating it for a one-month period is:

$$Energy = Power \times Time$$
$$E = (4.2 \text{ kW})(20 \text{ h})$$
$$E = 84 \text{ kWh}$$
$$Cost = Energy \times rate \text{ per kWh}$$
$$Cost = (84 \text{ kWh})(4.5¢)$$
$$Cost = \$3.78$$

Figure 20-8

more efficiently. Monthly kWh consumption will vary depending on the size of your family, appliance use, and living habits. The first step to energy conservation is to be aware of the big energy users.

Effective energy management involves the wise use of electricity, rather than doing without. There are many ways to conserve electric energy in the home. The following list describes a few ways to trim ever-increasing electricity bills:

- Use a single bulb of higher wattage rather than several low-watt bulbs. Two 60-W incandescent bulbs produce less light than one 100-W bulb while they consume about 20 percent more energy.
- Make regular incandescent lamps more energy efficient by taking control of their energy consumption with dimmer switches. (Remember that electronic dimmer switches do not work with fluorescent lights.)
- Use bulbs with a long-lasting rating only when the extended life is an advantage that outweighs the fact they put out 30 percent less light and still use the same amount of power.
- Halogen flood lights provide more light than incandescent bulbs and consume less energy. Halogen bulbs are compatible with dimmer switches.
- Use the lowest power rated bulb required to do the job.
- Run a dishwasher, washing machine, and clothes dryer only when there is a full load.
- Set the thermostat on the hot water tank as low as possible.
- Take showers rather than baths.
- Use a microwave oven instead of a regular oven. The microwave is more efficient.
- Adjust the air conditioner and the furnace at night. Set the air conditioner to a higher temperature and the furnace to a lower temperature.

Energy is often wasted when it is converted from one form to another. In many energy conversions, there is more energy lost than there is energy doing useful work. For example, of the total electric energy supplied to an incandescent lamp only about 5 percent is converted to useful light energy. Fluorescent lamps are much more energy efficient. A standard 40-W fluorescent tube, for example, produces as much as six times more light than a 40-W incandescent bulb. Some conversions, however, are highly efficient. The conversion of electric energy into mechanical energy with electric motors is one example. Only about 10 percent of the electric energy is wasted.

Manufacturers are encouraged to construct energy-efficient appliances. Most major appliances carry an ***ENERGY GUIDE label*** which gives the *energy-efficiency rating* (EER) of the appliance. The label can provide the following information:

- the energy-efficiency rating (EER) of the appliance; the higher the rating, the more efficient the appliance
- the efficiency range or energy consumption of similar products
- the estimated cost of operating the appliance for one year

Appliance	Wattage (Approximate Average)	Hours of Use (Monthly)	kWh (Monthly)
TABLE 20-1 Power Rating of Electric Appliances and Average Monthly Use			
Air cleaner (room and furnace)	40	250–720	10–29
Air conditioner (room)			
6000 BTU	750	120–720	90–540
9000 BTU	1050	120–720	126–756
Clothes dryer	5000	10–25	50–125
Clothes washer, automatic (excluding hot water)	500	10–40	5–20
Coffee maker	900	4–30	4–27
Computer (monitor and printer)	200	50–160	10–32
Dehumidifier	350	185–720	65–252
Dishwasher (excluding hot water)	1300	10–23	13–30
Electric blanket	180	30–90	5–16
Electric heater (portable)	1000	30–90	30–90
Fan (portable)	115	18–52	2–6
Food blender	390	3–5	1–2
Food freezer (15 cu ft)	335	180–420	60–140
Electric frying pan	1150	10–20	12–23
Furnace fan motor			
intermittent	350	160–415	56–145
continuous	350	720	252
Hair dryer (portable)	1000	5–10	5–10
Humidifier (portable)	100	80–540	8–54
Iron (hand-held)	1000	5–10	5–10
Kettle	1500	7–10	10–15
Microwave oven	1000	15–30	15–30
Oil furnace (burner)	260	96–288	25–75
Range	12,500	4–6	50–80
self-cleaning cycle only	3200	1/2–1 1/2	2–5
Refrigerator-freezer			
frost free (17 cu ft)	500	200–300	100–150
non-frost free (13 cu ft)	300	200–300	60–90
Sewing machine	75	4–14	0.3–1
Swimming pool			
filter motor 1/2 hp	900	720	648
3/4 hp	1200	720	864
1.0 hp	1500	720	1080
1.5 hp	2100	720	1512
Television	200	60–200	12–40
Toaster	1150	1–3	1–3
Vacuum cleaner			
portable	800	4–6	3–5
central	1600	4–6	6–10

	Power Rating of Electric Appliances and Average Monthly Use			
TABLE 20-1 **(CONTINUED)**				
Appliance	**Wattage (Approximate Average)**	**Hours of Use (Monthly)**	**kWh (Monthly)**	
Video cassette recorder	40	50–200	1–8	
Water bed heater	400	180–300	72–120	
Water heater				
typical family of 4	3800	98–138	375–525	
typical family of 2	3800	66–92	250–350	

WATTAGE × HOURS USED ÷ 1000 = kWh

kWh × COST PER kWh = TOTAL COST

Always compare the energy efficiency ratings of similar appliances. If all other factors are equal, choose the most energy-efficient unit.

The greatest savings of energy must come from the largest users—business, industry, and government. Electricity is not a natural resource. It is difficult to store large amounts of energy until it is wanted. For large-scale use, it must be manufactured as it is needed. This characteristic creates many problems for the power-supply authority. They must be able to generate electric energy to meet varying demands. *Time-of-day* metering allows customers to be charged at different rates depending on the time of day that they are consuming power. A higher on-peak rate is charged during the daylight hours of Monday to Friday, and a lower off-peak rate is charged overnight and on the weekends. With this incentive, medium and large energy users need to monitor their total power needs closely and to know exactly where the energy is going.

The use of electronic control devices to improve the energy efficiency of electrical systems is increasing. Conventional buildings use many separate wiring systems that are large consumers of electricity. These include heating, ventilating, air-conditioning (HVAC), as well as lighting systems. Energy-efficient *"intelligent" or "smart" buildings* use a single, integrated computerized control system. Various input sensors keep the computer constantly informed of the operating conditions of the building. The computer processes the information supplied by the sensors and makes decisions according to the energy-efficient program that is stored in its memory. The computer outputs signals that bring actuators into operation by means of pressure or electric or hydraulic power to operate the system automatically. For maximum energy efficiency, the system can be programmed to shut everything down if the occupants leave for more than a few minutes and to restart when they return (Figure 20-9).

HELP WANTED

Certified Energy Manager
Job Description
- Required for a large office facility.
- Carry out energy audits.
- Develop energy management strategies.
- Evaluate energy rate options.
- Monitor energy usage.

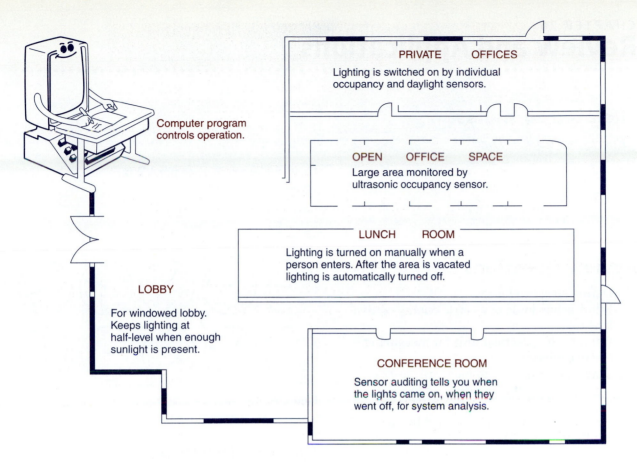

PRIVATE OFFICES
Lighting is switched on by individual occupancy and daylight sensors.

Computer program controls operation.

OPEN OFFICE SPACE
Large area monitored by ultrasonic occupancy sensor.

LUNCH ROOM
Lighting is turned on manually when a person enters. After the area is vacated lighting is automatically turned off.

LOBBY
For windowed lobby. Keeps lighting at half-level when enough sunlight is present.

CONFERENCE ROOM
Sensor auditing tells you when the lights came on, when they went off, for system analysis.

Figure 20-9 Energy-efficient computer controlled lighting system.

Review and Applications

Related Formulas

$Energy = Power \times Time$

$E = P \times t$

$J = W \times s$

$kWh = kW \times h$

$Cost = Energy \times rate\ per\ kWh$

Review Questions

1. Define electric energy.
2. What is one joule of electric energy equivalent to?
3. What is the practical unit for measuring electric energy?
4. What meter is used to measure electric energy in a house, building, or factory?
5. State the two factors that determine the amount of electric energy used by an appliance.
6. List five ways you can increase the energy efficiency of home lighting circuits.
7. What information is found on an ENERGY GUIDE major appliance label?

Problems

1. A 150-W light bulb is on for a total of 16 hours. How many kilowatthours of energy are converted?
2. (a) A 2.5-kW water heater is operated for a total of 42 hours over a one-month period. Calculate the kilowatthours of energy converted.
   (b) Calculate the cost of operating the heater for the same time period at an average rate of 8 cents per kWh.
3. Refer to Figure 20-10 to answer these questions:
   (a) Two successive readings of a kilowatthour meter are as shown. What are the readings?
   (b) How much energy was used?
   (c) Calculate the cost of the energy converted using an average rate of 6 cents per kWh.

Figure 20-10

Critical Thinking

1. If electric energy costs 6.5 cents per kilowatthour, how much would it cost to use a 100-W lamp for 4 hours each evening for 30 days?
2. A 1.5-kW portable heater is to be used 5 hours a day for 60 days. What will the operating cost be if the electricity costs 8 cents per kWh?
3. An electric baseboard heater draws 8 A from a 240-V source and is operated on an average of 8 hours per day.
   (a) What is the power rating of the heater?
   (b) What will the operating cost be for 30 days if the electricity costs 7 cents per kWh?

Portfolio Project

As an energy manager of a hotel chain you have been asked to identify energy savings opportunities. Report on how you might use each of the following control devices to reduce electric energy consumption: photocells, time clocks, occupancy sensors, programmable thermostats, and spring-wound timers.

Direct Current and Alternating Current

Practical generators of electricity fall into two general groups: direct current (dc) and alternating current (ac). Direct current is produced by those generators that do not change the polarity of their terminals. Because the polarity of their terminals is constantly reversing, ac generators produce an alternating current flow. In this chapter, we will examine the ways in which ac and dc voltages are generated and measured.

Objectives

After studying this chapter, you should be able to:
1. Define direct and alternating current.
2. Explain the basic generator principle.
3. Calculate effective, peak, and peak-to-peak values of a sine wave.
4. Compare single-phase and three-phase ac systems.

Key Terms

- direct current
- alternating current
- generator principle
- left-hand generator rule
- armature
- slip rings
- stator
- field
- single-phase

- three-phase
- wye connection
- delta connection
- commutator
- diode
- rectifier
- sine wave
- cycle
- period
- frequency

- peak value
- peak-to-peak value
- effective value
- rms
- separately-excited
- self-excited
- shunt generator
- prime mover
- motor-generator set
- alternator

RESEARCH INTERNET SITES

- ZZZAP Power Corporation
- Power Quality
- Magnetek

21-1 Direct Current (DC)

Electric current can be defined as the flow of electrons in a circuit. Based on the electron theory, electrons flow from the negative (−) polarity to the positive (+) polarity of a voltage source.

Direct current (*dc*) is current that flows *only in one direction* in a circuit (Figure 21-1). Current, in this type of circuit, is supplied from a dc voltage source. Since the *polarity* (negative and positive terminals) of a dc source is fixed, the current produced flows in one direction only.

Batteries are commonly used as a dc voltage source. Both the voltage and polarity of the battery are fixed. When connected to a load, the current produced flows in one direction at some steady or constant value.

Direct current flow need not necessarily be constant, but it must travel in the same direction at all times. There are several types of direct current and all depend upon the value of the current in relation to time (Figure 21-2). *Pure* or *constant* direct current shows no variations in value over a period of time and is the type associated with a battery. Both *varying* and *pulsating* direct currents have a changing

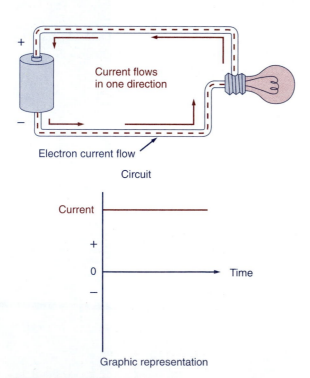

Figure 21-1   Direct current.

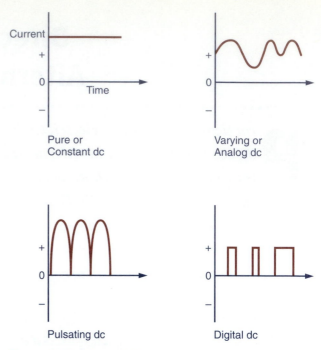

Figure 21-2   Types of direct current.

value when plotted against time. *Varying* or *analog* direct current variations have no definite pattern and are associated with certain types of electrical sensor signals. *Pulsating* or *digital* direct current variations are uniform and repeat at regular intervals. These are associated with electronic dc-power supplies and digital computer signals.

21-2 Alternating Current (AC)

An *alternating current* (*ac*) circuit is one in which the direction and amplitude of the current flow changes at regular intervals. Current in this type of circuit is supplied from an ac voltage source. The polarity (negative and positive terminals) of an ac source changes at regular intervals resulting in a reversal of the circuit current flow.

Alternating current usually changes in both value and direction. The current increases from zero to some maximum value and then drops back to zero as it flows in one direction. This same pattern is then repeated as it flows in the opposite direction. The waveform or exact manner in which the current increases and decreases is determined by the type of ac voltage source used (Figure 21-3).

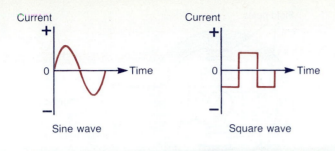

Sine wave

Square wave

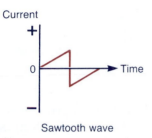

Sawtooth wave

Figure 21-3   AC waveforms.

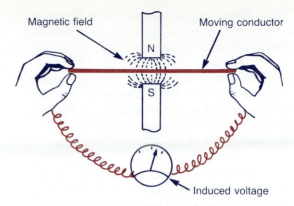

Figure 21-4   Generator principle.

21-3 AC Generation

Almost all of the electric energy supplied for domestic and commercial purposes is ac. Alternating current, generated and transmitted at high voltage, is easily reduced by the use of transformers to the relatively low voltages suitable for use. Since there is no dc transformer, direct current must be generated and transmitted at practically the same voltage at which it is used. In order to avoid unreasonably large and expensive transmission lines and power losses, dc transmission is limited to comparatively short distances.

The basic method of obtaining ac is by use of an ac generator (also called an alternator). A *generator* is a machine that uses magnetism to convert mechanical energy into electric energy. The **generator principle,** simply stated, is that a voltage is induced in a conductor whenever the conductor is moved through a magnetic field so as to cut lines of force.

Figure 21-4 shows the basic *generator principle.* As we have learned previously, electrons that are in motion set up a magnetic field around a conductor. Conversely, a change in a magnetic field around a conductor tends to set electrons in motion. The mere existence of a magnetic field is not enough; there must be some form of change in the field. If we move the

conductor through the magnetic field, a force is exerted by the magnetic field on each of the free electrons within the conductor. These forces add together and the effect is that voltage is generated or induced into the conductor.

The **left-hand rule for a generator** (Figure 21-5) shows the relationship between the direction the conductor is moving, the direction of the magnetic field, and the resultant direction of the induced current flow. When the thumb is pointed in the direction of the conductor's motion, and the index finger in the direction of the flux, the middle finger will point in the direction of the induced electron flow. The rule is also applicable when the magnet is moved, instead of the conductor. In this case, however, the thumb must point in the direction of relative conductor motion.

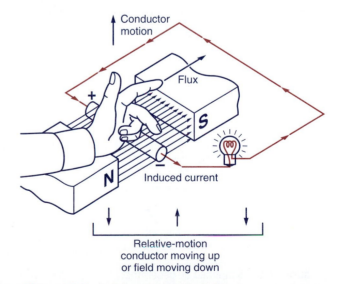

Figure 21-5   Left-hand generator rule.

The amount of voltage induced in a conductor as it moves through a magnetic field depends upon:

- The strength of the magnetic field; the stronger the field the more voltage induced.
- The speed at which the conductor cuts through the flux; increasing the conductor speed increases the amount of voltage induced.
- The angle at which the conductor cuts the flux; maximum voltage is induced when the conductor cuts the flux at 90°, and less voltage is induced when the angle is less than 90°.
- The length of the conductor in the magnetic field; if the conductor is wound into a coil of several turns, its effective length increases, and so the induced voltage will increase.

An ac generator or alternator produces an ac voltage by causing a loop of wire to turn within a magnetic field. This relative motion between the wire and the magnetic field causes a voltage to be induced between the ends of the wire. This voltage changes in magnitude and polarity as the loop is rotated within the magnetic field (Figure 21-6). The mechanical force required to turn the loop can be obtained from various sources. For example, the very large ac generators are turned by steam turbines or by the movement of water.

The voltage produced by a single-loop generator is too weak to be of much practical value. A practical ac generator has many more turns of wire wound on an *armature* (Figure 21-7). The armature is the movable part of the generator and is made up of a number of coils wound on an iron core. The ac voltage induced in the armature coils is connected to a set of **slip rings**

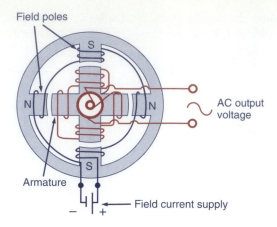

Figure 21-7 Practical ac generator.

from which the external circuit receives the voltage through a set of brushes. Electromagnets are used to produce a stronger magnetic field. The coil windings of the electromagnets are fixed or stationary and are, therefore, called **stators** or stator windings. Since these stator windings provide the magnetic field for the generator, they are often referred to as **field coils** or windings.

21-4 The AC Sine Wave

As the armature of the alternator is rotated through one complete revolution, a *sine wave* voltage is produced at its output terminals. This generated sine wave voltage varies in both voltage value and polarity.

If the coil is rotated at a constant speed, the number of magnetic lines of force that are cut per second varies with the position of the coil.

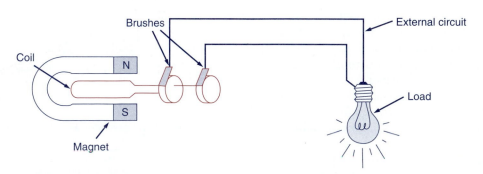

Figure 21-6 Simplified ac generator.

When the coil is moving parallel to the magnetic field, it cuts no lines of force. Therefore, no voltage is generated at this instant. When the coil is moving at right angles to the magnetic field it cuts the maximum number of lines of force. Therefore, maximum or peak voltage is generated at this instant. Between these two points the voltage varies in accordance with the sine of the angle at which the coil cuts the lines of force.

The coil is shown in four specific positions in Figure 21-8. These are intermediate positions that occur during one complete revolution of the coil position. The graph shows how the voltage increases and decreases in amount during one rotation of the loop. Note that the direction of the voltage reverses each half-cycle. This is because for each revolution of the coil, each side must first move down and then up through the field.

The sine wave is the most basic and widely used ac waveform. Some of the important electrical characteristics and terms used when referring to ac sine wave voltages or currents are cycle, period, and frequency.

Cycle

One *cycle* is one complete wave of alternating voltage or current (Figure 21-9). During the generation of one cycle of output voltage, there are two changes or alternations in the polarity of the voltage. These equal, but opposite, halves of a complete cycle are referred to as alternations. The terms positive and negative are used to distinguish one alternation from the other.

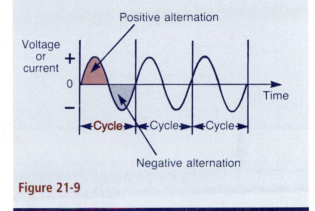

Figure 21-9

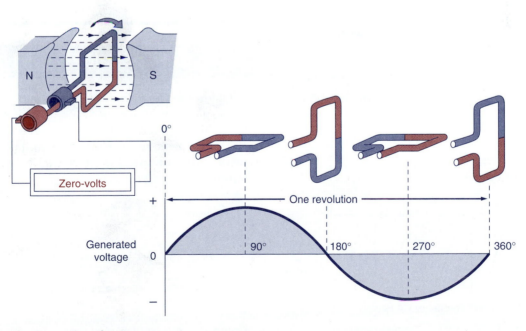

Figure 21-8 Generation of a sine wave.

Period

The time required to produce one complete cycle is called the *period* of the waveform (Figure 21-10). The period is usually measured in seconds (s) or smaller units of time such as milliseconds (ms) or microseconds (μs).

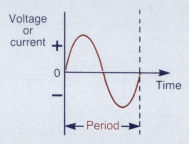

Figure 21-10

Frequency

The *frequency* of an ac sine wave is the number of cycles produced per second (Figure 21-11). The unit of frequency is the hertz (Hz). One hertz equals one cycle per second. For example, the 120-V ac electric outlet in your home has a frequency of 60 Hz.

In many electronic applications, the frequencies involved are much higher. An example is your favorite radio station with a frequency of 1400 kHz. Every second that station transmits 1,400,000 complete cycles. When dealing with large quantities of cycles, prefixes are used. The more common prefixes are:

K for kilo or thousand (10^3)
M for Mega or million (10^6)
G for Giga or billion (10^9)

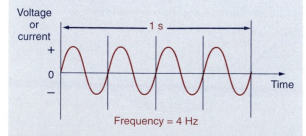

Figure 21-11

21-5 AC Sine Wave Voltage and Current Values

Since the value of a sine wave of voltage or current continually changes, you must be specific when referring to the value of the waveform you are describing. There are several ways of expressing the value of a sine wave.

Peak Value

Each alternation of the sine wave is made up of a number of instantaneous values. These values are plotted at various heights above and below the horizontal line to form a continuous waveform. The *peak value* of a sine wave refers to the maximum voltage or current value (Figure 21-12). Notice that two equal peak values occur during the cycle. One is during the positive alternation and the other is during the negative alternation.

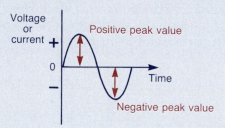

Figure 21-12

Peak-to-Peak Value

The *peak-to-peak value* of a sine wave refers to its total overall value from one peak to the other (Figure 21-13). It is equal to two times the peak value. Both peak and peak-to-peak values are used when taking ac measurements using an oscilloscope.

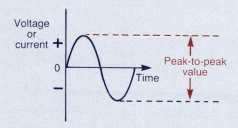

Figure 21-13

Effective Value

The *effective* or *rms value* of a sine wave is the one most extensively used when referring to an ac voltage or current. It is common practice to assume that all ac voltage and current readings are effective values unless otherwise stated. Likewise, voltmeters and ammeters are calibrated to read effective (rms) values unless otherwise stated.

When 1 A (ampere) of dc current flows through a resistor a certain amount of energy is dissipated by the resistor in the form of heat. An ac current that will produce the same amount of heat in the resistor is considered to have an effective current value of 1 A. Of course, the ac current must have a peak value that is higher than 1 A in order to be equivalent to a constant dc current of 1 A. In this case (Figure 21-14), the effective value of the ac current would be 1 A, while the peak value would be approximately 1.414 A.

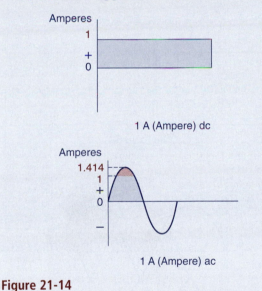

Figure 21-14

The effective value of a sine wave of current can be determined by a mathematical process known as the *root-mean-square,* or rms method. This is the reason why the effective value is often referred to as the rms value. By using this process, it can be proven that the effective value of a sine wave of current is always equal to 0.707 times its peak value.

Example 21-1

Assume a current sine wave has a peak value of 10 A (Figure 21-15). The effective or rms value of this current would be:

$$I_{rms} = I_{peak} \times 0.707$$
$$I_{rms} = (10 \text{ A})(0.707)$$
$$I_{rms} = 7.07 \text{ A}$$

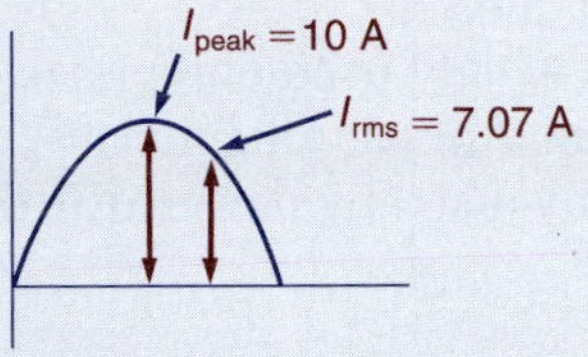

Figure 21-15

Since an alternating current is produced by an alternating voltage, the same rule applies in terms of its effective value. The effective value of a voltage sine wave is equal to 0.707 times its peak value. This equation can be transposed so that the peak value of a sine wave can be determined if its effective value is known. Accordingly, the peak value of a sine wave is equal to 1.414 times its effective value.

Example 21-2

The voltage that is used in your home has an effective or rms value of 120 V (Figure 21-16). Its peak and peak-to-peak values are:

$$V_{peak} = V_{rms} \times 1.414$$
$$V_{peak} = (120 \text{ V})(1.414)$$
$$V_{peak} = 170 \text{ V}$$
$$V_{peak\text{-}to\text{-}peak} = V_{peak} \times 2$$
$$V_{peak\text{-}to\text{-}peak} = (170)(2)$$
$$V_{peak\text{-}to\text{-}peak} = 340 \text{ V}$$

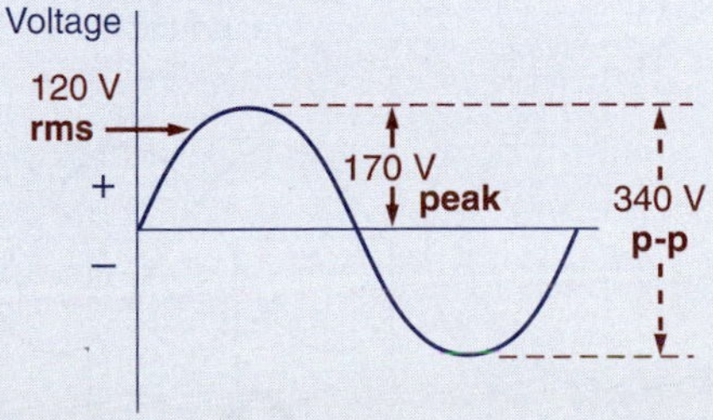

Figure 21-16

21-6 Three-Phase AC

Figure 21-17 shows the fundamental arrangement used to generate single-phase (1ϕ), two-phase (2ϕ), and three-phase (3ϕ) ac voltages. In this alternator design, the stator coil(s) provides the output voltage and current. The armature, or rotor, is actually a rotating electromagnet providing both the magnetic field and relative motion.

The most widely used polyphase system is *three-phase*. Almost all generation and distribution systems used by power utilities are three-phase.

The three sets of stator coils of the three-phase alternator may be connected in *wye* (also known as star) or *delta*, as illustrated in Figure 21-18. The three-phase four-wire wye system is very common and is the standard system supplied by many power utilities to commercial and industrial customers. It is very versatile, since within the four-wire system shown in Figure 21-19 it can deliver both single-phase and three-phase and both 208 V and 120 V without the use of a transformer (Figure 21-19).

As load requirements increase, the use of single-phase power is no longer practical or economical. In terms of the conductors necessary to deliver a given amount of power, three-phase systems can deliver 1.73 times more power for the same amount of wire than a single-phase system. In addition, three-phase systems permit the design of more efficient electric equipment. Three-phase motors, for example, are more efficient than comparable single-phase

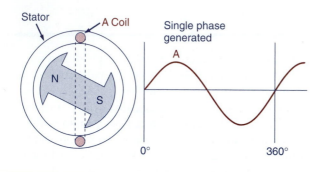

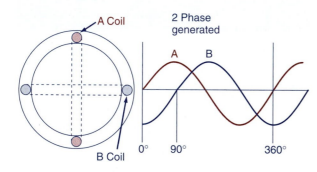

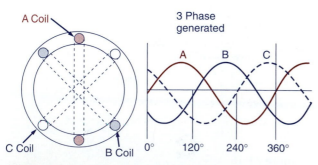

Figure 21-17   Stator-generated voltages.

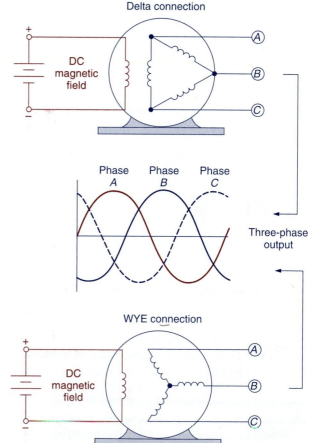

Figure 21-18   Wye- and delta-stator connections.

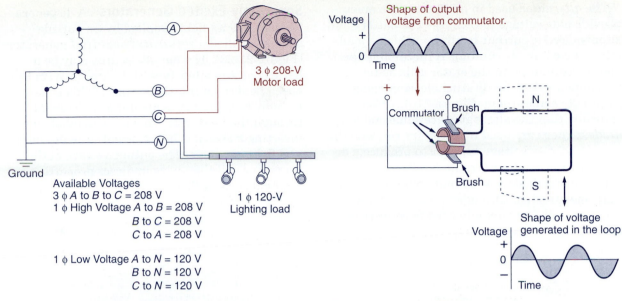

Figure 21-19 Voltages available from a four-wire wye-connected alternator.

Available Voltages
3 ɸ A to B to C = 208 V
1 ɸ High Voltage A to B = 208 V
                 B to C = 208 V
                 C to A = 208 V

1 ɸ Low Voltage A to N = 120 V
                B to N = 120 V
                C to N = 120 V

3 ɸ 208-V Motor load

1 ɸ 120-V Lighting load

Ground

Shape of output voltage from commutator.

Commutator   Brush

Brush

Shape of voltage generated in the loop

Figure 21-20 Simplified dc generator.

motors. For that reason, larger size ac motors are available only in three-phase.

21-7 DC Generation

Generating dc is basically the same as generating ac. The only difference is the manner in which the generated voltage is supplied to the output terminals.

Figure 21-20 shows a simplified dc generator. The shape of the voltage generated in the loop is still that of an ac sine wave. Notice, however, that the two slip rings of the ac generator have been replaced by a single segmented slip ring called a ***commutator***. This commutator acts like a mechanical switch or rectifier to automatically convert the generated ac voltage into a dc voltage. As the armature starts to develop a negative alternation, the commutator switches the polarity of the output terminals via the brushes. This keeps all positive alternations on one terminal and all negative alternations on the other. Then the only essential difference between an ac and a dc generator is the use of slip rings on the one and a commutator on the other.

A single-loop generator produces a direct current output that is pulsating. Most applications

require a dc-generated voltage that has a steady value with minimum ripple or variation. This is accomplished by adding more coils and commutator segments in the armature (Figure 21-21). All the coils are connected in series between the two brushes. The voltage appearing at the brushes is the sum of the voltages in the separate coils. The result is a nearly steady dc output.

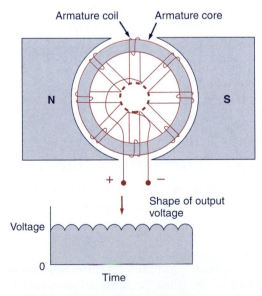

Armature coil   Armature core

N   S

Shape of output voltage

Figure 21-21 Practical dc generator.

The *alternator* used in an automobile must provide a dc voltage to recharge the battery and supply electric current to all the vehicle's loads when the engine is running. A three-phase ac voltage is induced in the stator coils as the magnetic field of the rotating electromagnet rotor cuts across them (Figure 21-22). Contained within the alternator are semiconductor *diodes*. These are used to *rectify* (or convert) the three-phase ac voltage in order to produce a dc voltage at the output terminals. Basically, the function of a diode is to permit current flow in only one direction through a circuit. A six-diode rectifier pack is commonly used to convert the three-phase ac stator voltage to dc.

Separately-Excited Generators A dc generator that has its field supplied by an outside source is called a **separately-excited** generator (Figure 21-23). The outside source may be a battery or any other type of dc supply. With the speed held constant, the output of this generator may be varied by controlling the current through the field coils. This is accomplished by inserting a rheostat in series with the dc source and field windings. The output voltage of the generator will then vary in direct proportion to the field current flow.

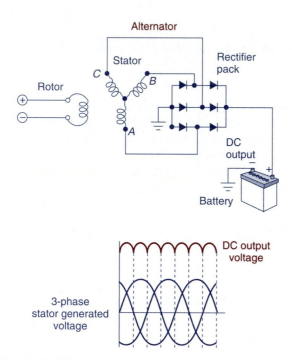

Figure 21-22 Automobile alternator.

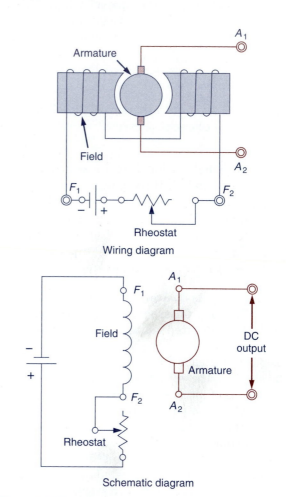

Figure 21-23 Separately-excited generator.

21-8 Types of DC Generators

Practical dc generators use electromagnetic field coils rather than permanent magnets. The direct current used to energize the field windings is called the exciting current. DC generators are classified according to the method by which current is supplied to these field coils. The two major classifications are separately-excited and self-excited generator types.

Self-Excited Generators The inconvenience of a separate dc source for field excitation led to the development of the self-excited generators. **Self-excited generators** use part of the generated current to excite the field. They are classified according to the method by which the field coils are connected. Self-excited generators may be series-connected, shunt-connected, or compound-connected.

In a **shunt generator** the field windings are connected in parallel with the armature (Figure 21-24). The field windings consist of many turns of relatively small wire and actually use only a small part of the generated current. The initial voltage the generator is required to build up is produced by the residual magnetism in the iron of the field poles. Residual magnetism is that magnetism which remains in the poles when no current flows through the field coils. As the generated voltage increases, the current through the field coils also increases. This strengthens the magnetic field and allows the generator to build up to its rated output voltage. A rheostat connected in series with the field coils is used to vary the field current, which in turn controls the generator output voltage.

21-9 Generator Prime Movers

The output voltage of any generator is directly proportional to the field strength multiplied by the speed of rotation. Mechanical energy must be applied to the generator armature shaft in order to spin it around and produce electricity. The source of the mechanical energy is called the **prime mover**. With small generators, the armature can be turned by hand or hooked up to a bicycle wheel and spun by pedal power. The electricity produced would be enough to operate a small light bulb.

Larger generators need more mechanical energy to generate greater amounts of electricity. For example, the gasoline engine in the automobile is used to drive the car's alternator, which supplies voltage to the electrical system (Figure 21-25). Electricity can only be generated when the engine is running, and therefore a battery is required to provide the electric energy required to start the engine.

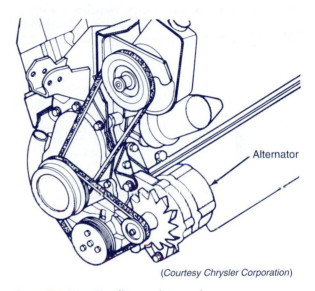

(Courtesy Chrysler Corporation)

Figure 21-25 Gasoline engine—prime mover.

Industrial plants sometimes use **motor-generator (M-G) sets** for converting ac to dc. In this application an ac motor is used to drive a dc generator. AC voltage is applied to the motor and dc voltage is obtained from the commutator of the generator (Figure 21-26).

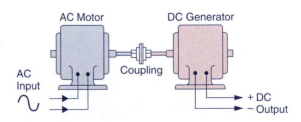

Figure 21-26 Motor-generator set.

When power in an electrical system is disrupted for an extended period, life-threatening conditions may occur. Continuous lighting and

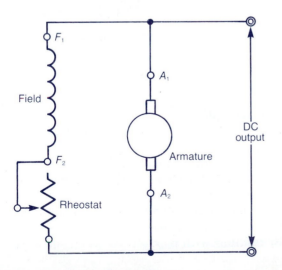

Figure 21-24 Shunt generator.

power are essential in places of public assembly, theaters, hotels, sports arenas, health-care facilities, and the like. Engine-driven alternators are commonly used to provide emergency power in the event of a power failure. Prime movers for these on-site alternators may be powered by gasoline, diesel, or gaseous fuel engines. Figure 21-27 shows the block diagram of a typical transfer arrangement used to automatically energize the emergency system upon failure of the normal current supply.

High-voltage generating stations, which are our main source of electricity, require a tremendous amount of mechanical energy to generate their electric voltage requirements. For exam-

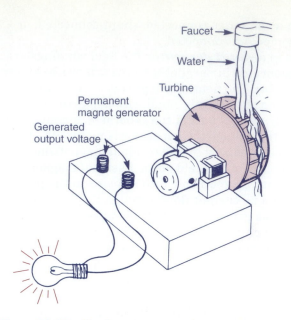

Figure 21-28 Simple water-powered generator.

ple, hydroelectric generating stations use the force of falling water as the prime mover (Figure 21-28). In these plants, large turbine generator units are made to revolve under the force of the falling water. Most power company alternators have their dc field in the rotor. The current ratings of the alternators can be relatively high, for the output of the alternator is taken directly from the stator windings to the external circuit without the use of slip rings.

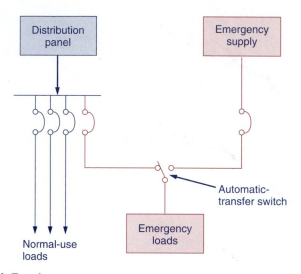

A. Transfer arrangement

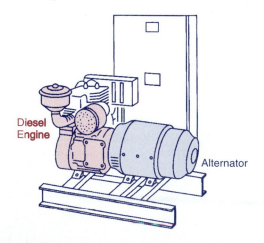

B. On-site alternator

Figure 21-27 Typical emergency power system.

HELP WANTED

Technical Sales/Support Representative
Job Description
- Required by a supplier of portable and on-site alternators.
- Build upon an established client base.
- Prepare quotations and presentations for site installations.
- Must have completed a post-secondary electrical or electronic program.
- Be familiar with popular PC based software.

Review and Applications

Related Formulas

peak-to-peak = 2 × peak

peak = rms × 1.414

rms = peak × 0.707

Review Questions

1. Define a direct current (dc) circuit.
2. Define an alternating current (ac) circuit.
3. Compare the polarity of an ac and a dc voltage source.
4. What is the basic operating principle of a generator?
5. List four factors that determine the amount of voltage induced in a conductor as it moves through a magnetic field.
6. With reference to the generation of a sine wave voltage in a coil:
   (a) why is the voltage zero two times during each cycle of rotation?
   (b) why does the induced voltage reverse in polarity during each cycle of rotation?
7. Define each of the following as they apply to a sine wave:
   (a) cycle
   (b) alternation
   (c) period
   (d) frequency
8. What is the standard frequency of the ac voltage supplied to our homes?
9. Sketch an ac sine voltage waveform. On this sketch indicate each of the following:
   (a) positive-peak voltage value
   (b) peak-to-peak voltage value
   (c) effective, or rms, voltage value
10. List the different 1ϕ and 3ϕ voltages available from a standard four-wire wye-connected ac system.
11. As load requirements increase, the use of single-phase power is no longer practical or economical. What are two reasons for this?

12. Explain in what way the armature circuit of an ac generator is different from that of a dc generator.
13. With reference to an automobile alternator:
   (a) what type of voltage is induced in the stator windings?
   (b) what is the purpose of the dc voltage applied to the rotor coil?
   (c) what is used to convert the voltage from ac to dc?
14. In what way is a self-excited dc generator different from the separately-excited type?
15. How will each of the following affect the voltage output of a dc shunt generator?
   (a) a decrease in speed of the prime mover
   (b) an increase in the field current
   (c) a decrease in the resistance of the field rheostat
16. Explain the function of the prime mover in the generating process.

Problems

1. (a) A steady dc current of 20 A flows through a given resistor producing a certain amount of heat. What peak value of ac current is required to produce the same amount of heat?
   (b) What rms, or effective value, of ac current is required to produce the same amount of heat?
2. (a) An ac sine wave voltage is measured using an ac voltmeter and found to be 10 V. What is the peak value of this voltage?

(b) What is the effective, or rms, value of this voltage?

(c) What is the peak-to-peak value of this voltage?

Critical Thinking

1. The output voltage fails to build up on a self-excited generator that has not been operated for a long period of time. It is suspected that the magnetic field may have lost all its residual magnetism. How could the residual magnetism be restored?

2. With reference to the transfer arrangement of the emergency power supply system of Figure 21-27, why aren't all loads normally operated by the emergency supply?

Portfolio Project

Construct a miniature motor-generator set by coupling together the shafts of two hobby type permanent magnet dc machines with a piece of plastic tubing. Report on the generator output for different rotational speeds, direction of rotation conditions, and load conditions.

Circuit Challenge

Using a Simulator

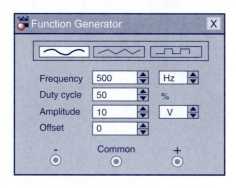
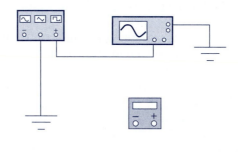

Procedure

(a) Wire the signal generator circuit shown, using whatever simulation software package is available to you.

(b) With the sine wave signal set to 10 volts peak, record the peak-to-peak value of the waveform as displayed on the scope.

(c) Calculate the rms value of the voltage waveform.

(d) Use the multimeter to measure the rms value of the voltage waveform.

Low-Voltage Signal Systems

Low-voltage signal systems generally operate from transformers that can not deliver more than about 30 volts and 5 watts. Because of the low voltage and power level used, there is little danger of shock or fire even in the case of an accidental short circuit. The signal systems covered in this chapter include the door chime, annunciator, door lock, telephone, and alarm circuits.

Objectives

After studying this chapter, you should be able to:

1. Read and construct electrical diagrams for typical signal systems.
2. Wire typical signal systems in accordance with a planned wiring layout.
3. Explain the basic operation of a door-chime circuit, annunciator circuit, door-lock circuit, telephone system, and alarm system.

Key Terms

- energy-limiting circuit transformer
- bell transformer
- solenoid
- door chime
- annunciator
- electric door lock
- card access system
- reader

- cable pair
- protector
- 42A block
- transmitter
- receiver
- handset
- tip wire
- ring wire

- on hook
- off hook
- pulse dialing
- tone dialing
- perimeter protection
- area protection
- hard-wired
- wireless

(continued)

- Chubb
- Honeywell
- National Burglar and Fire Alarm Association (NBFAA)

- magnetic switch
- ultrasonic detector
- microwave detector
- infrared detector
- microprocessor-based control

- normally-open loop
- normally-closed loop
- end-of-line resistor
- instant loop
- delayed loop
- zone

- armed
- disarmed
- LEDs
- automatic telephone dialer
- battery backup

22-1 Low-Voltage Signal System

Low-voltage signal circuits operate at less than 120 V. The maximum voltage in such systems is 30 V, thus reducing the danger of electric shock. Low-voltage signal systems are used for such applications as door chimes, annunciators, door locks, telephones, and security systems.

Installing low-voltage signal circuits is, in most respects, cheaper and simpler than other wiring jobs. Power is often supplied by a *transformer* that is approved for this purpose. Approved transformers have built-in overcurrent protection and are purposely limited to very low power values. For this reason, these circuits are referred to as **energy-limiting circuits.** It is assumed that a short circuit in this type of circuit will not start a fire or pose any other threat to life or property. Article 725 of the National Electrical Code outlines the rules that apply to these circuits.

A *bell transformer* permanently connected to the 120-V ac line is often used to step the voltage down to a lower level (Figure 22-1). Standard bell transformer secondary voltages are 6–10 Vac and 12–18 Vac.

Smaller diameter wires are approved for use on energy-limited signal circuits. Because of the lower current levels a No. 18 gauge copper conductor with thermoplastic insulation is generally used. The conductors can be enclosed in a jacket that forms a cable (Figure 22-2), or twisted together without an overall jacket. Two-, three-, and four-conductor cables are commonly used. Each conductor is color coded for identification. Cables are commonly supported with special insulated staples. They must be run entirely separate from electric lighting or power circuits to prevent accidental transfer of voltage to the signal circuit.

Thermoplastic insulation   No. 18 AWG

Figure 22-2   Two-conductor cable.

22-2 Door-Chime Circuit

The **door chime** is a very popular signal device that is used in homes today. A typical two-tone door chime allows you to identify signals from two locations. It is made up of two 16-V electric solenoids and two tone bars. A *solenoid*, as you may recall, is an electromagnet that has a movable core or plunger. When voltage is momentarily applied to the *front* solenoid, its plunger strikes both tone bars. When voltage is momentarily applied to the *back* solenoid, its plunger strikes only one tone bar. Thus a double tone (ding-dong) is produced by a signal from the front solenoid and a single tone (dong) from the back solenoid (Figure 22-3).

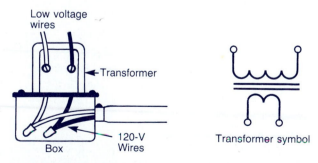

Figure 22-1   Transformer connection.

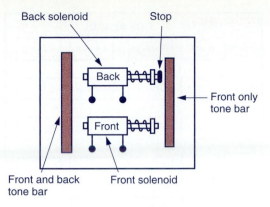

Back solenoid   Stop

Back

Front only
tone bar

Front

Front and back
tone bar   Front solenoid

Figure 22-3   Two-tone door chime.

The terminal board of the chime unit usually has three terminal screws (Figure 22-4). The one marked "F" (front) connects to one side of the front door solenoid. The one marked "B" (back) connects to one side of the back door solenoid. The one marked "T" (transformer) connects to both remaining leads of the solenoids. This makes the "T" terminal common to both solenoids.

The complete schematic and wiring number-sequence chart for a door-chime circuit is drawn in Figure 22-5. A 120-V/16-V bell transformer is used as the power supply unit. The schematic

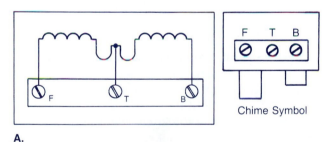

F   T   B

Chime Symbol

A.

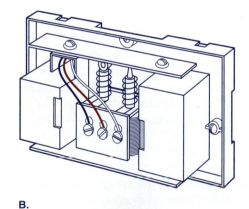

B.

Figure 22-4   Door-chime terminal board.

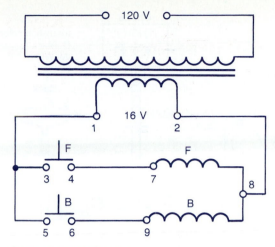

A.   Chime schematic diagram

1, 3, 5
2, 8
4, 7
6, 9

B.   Wiring number sequence chart

Figure 22-5   Door-chime schematic diagram.

circuit can be read easily to show how the circuit operates. Pushing the appropriate push button will complete the circuit to the front or back solenoid. Push buttons are used instead of switches so that the circuit will remain operative only as long as the button is being pressed. A two-tone chime indicates the signal is from the front door location. A single-tone chime indicates the signal is from the back door location.

When this circuit is wired in a house there can be any number of different types of wiring layouts and cable runs for the same schematic. The one drawn in Figure 22-6 is designed to simulate one that is typical of a house layout. The transformer is usually located in the basement of the house with its 120-V primary permanently wired into the house's electrical system. Three cable runs are used. A single two-conductor cable is installed from the transformer to each of the doors and a three-conductor cable runs from the transformer to the chime. The chime is located in a central location on the first floor. Note that the components have been numbered according to the numbering sequence used on the schematic. The buttons and transformer are represented

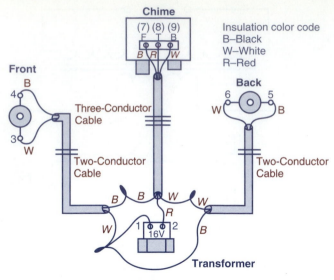

Figure 22-6 Typical door-chime wiring diagram.

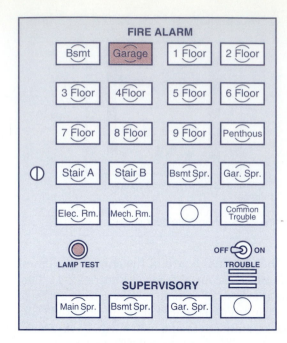

Figure 22-7 Typical back-light annunciator.

pictorially. The wiring diagram is completed by connecting the terminals according to the wiring number-sequence chart. Use the wire insulation color code to properly identify cable wire ends. A two-conductor cable usually contains a white and black wire. The three-conductor cable wires are usually color coded white, black, and red. Low-voltage wiring for items such as door chimes does not require outlet boxes.

22-3 Annunciator Circuit

An *annunciator* is used to identify signals from several different locations. In large buildings where a fire alarm system is required, an annunciator is used to provide fire location information to fire fighting personnel. In a back-light annunciator (Figure 22-7), lamps are mounted inside individual egg-crate housings located behind engraved name plates. A lamp, under alarm conditions, will illuminate the lettering on the respective nameplate.

Figure 22-8 shows the schematic for a simple four-point electronic annunciator circuit. Silicon-controlled rectifiers (SCRs) are used to switch the lamps ON and OFF on command from the remote push buttons and reset push buttons. For example, momentarily pressing remote PB_1 switches SCR_1 into conduction. The circuit to L_1 is completed, and this lamp on the

annunciator unit turns ON. This lamp remains ON until the reset PB_1 is momentarily pressed.

22-4 Electric Door-Lock Circuit

An *electric door-lock circuit* is commonly used in large apartment buildings to allow each apartment to unlock the main entrance by remote control. A typical electric door latch (Figure 22-9) consists of an electromagnet with an armature that serves as a latch-release plate. When current flows through the electromagnet, it attracts the latch-release and allows the door to be pushed open.

The schematic and wiring number-sequence chart for a simple two apartment electric door-lock circuit is shown in Figure 22-10. Current flow to the door-lock electromagnet is controlled by two push buttons connected in parallel. These push buttons (A_1 and B_1) are located in their respective apartments. Pressing one or the other of the apartment push buttons completes the circuit to the door-lock electromagnet. Current flow to the buzzer located in Apartment A is controlled by push button A_2, which is located in the lobby. Similarly, current to the buzzer located in Apartment B is controlled by push button B_2, which is located in the lobby. Used in conjunction with an intercom

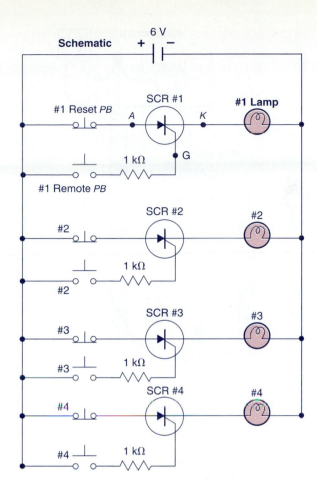

Schematic

6 V

#1 Reset *PB* SCR #1 #1 Lamp

A K

G

1 kΩ

#1 Remote *PB*

#2 SCR #2 #2

1 kΩ

#2

#3 SCR #3 #3

1 kΩ

#3

#4 SCR #4 #4

1 kΩ

#4

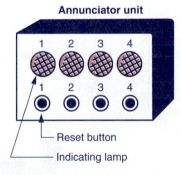

Annunciator unit

1 2 3 4

1 2 3 4

— Reset button

— Indicating lamp

Figure 22-8 Simple four-point electronic annunciator circuit.

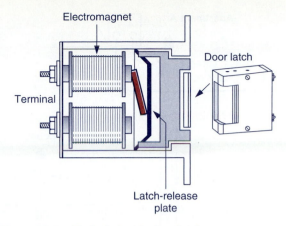

Electromagnet

Door latch

Terminal

Latch-release plate

Figure 22-9 Typical electric door latch.

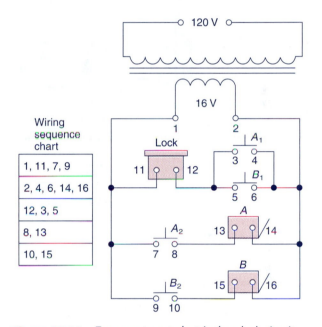

120 V

16 V

Wiring sequence chart

| 1, 11, 7, 9 |
| 2, 4, 6, 14, 16 |
| 12, 3, 5 |
| 8, 13 |
| 10, 15 |

Lock

1 2 A_1

3 4

11 12 B_1

5 6

A_2 A 13 14

7 8

B_2 B 15 16

9 10

Figure 22-10 Two apartment electric door-lock circuit schematic.

a table lookup is performed by the microprocessor to determine if the card is authorized. If the card is authorized, the microprocessor outputs a signal to unlock the door. When a card is lost or stolen, its number is simply deleted from the

system, visitors can identify themselves and the door can be unlocked from the apartment being rung. A typical wiring diagram for the circuits is shown in Figure 22-11.

Electric door locks are used as part of electronic **card access systems** (Figure 22-12). Unlike keys used in a mechanical lock, access-control cards each contain a unique encoded number. When a card is presented at a **reader**,

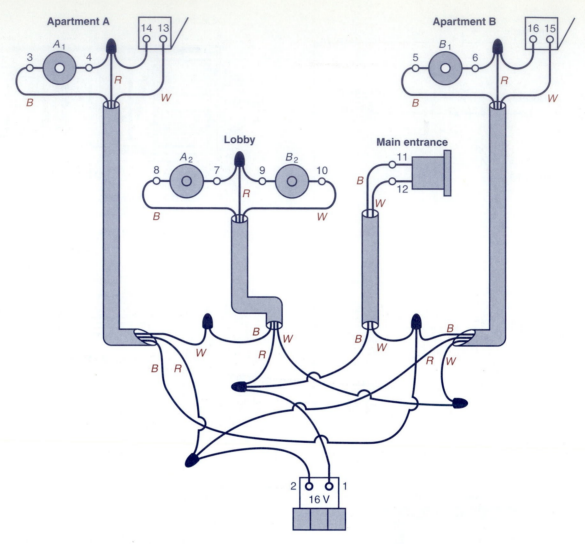

Figure 22-11 Typical wiring diagram for a two apartment electric door-lock circuit.

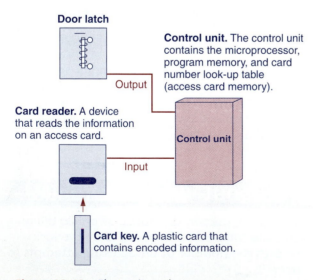

Figure 22-12 Electronic card access system.

lookup table(s). Security is not compromised, and the only cost incurred is for the replacement card.

22-5 Telephone Circuits

Telephone service enters the house by one of two methods. One is an overhead service connected to a nearby telephone pole. The other is an underground service connected to a cable-termination box (Figure 22-13). Actually, the telephone line is two wires called a ***cable pair***. Whether the telephone line comes to your location overhead or underground, the pair of wires terminate at a ***protector***. The protector serves as a lightning arrestor. It protects the pair of

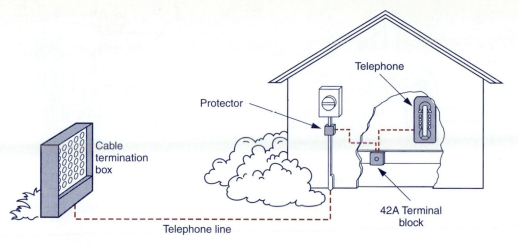

Figure 22-13   Typical underground telephone service connection.

telephone wires into your building from high voltage spikes that might occur on the wires due to lightning bolts. The most common place for the initial telephone line connection as the line comes in from the outside protector is to a terminal block called the **42A block**.

A telephone system basically consists of a transmitter, a battery, and a receiver (Figure 22-14). A **transmitter** is used to change sound energy into electric energy. This electric energy flows through the wires to the **receiver**, which changes it back to sound energy. Both the transmitter and receiver are housed in one unit called the **handset**.

The telephone requires only two wires to operate: the **tip wire** and the **ring wire**. The tip wire is identified by the letter "T" and is usually the green wire. The ring wire is identified by the letter "R" and is usually the red wire. A standard telephone cable will have four wires colored yellow, black, green, and red. These wires are connected at the terminal box and run to the individual telephones (Figure 22-15).

All telephones that are installed in a home are connected electrically in parallel across the ring and tip wires. All connections originate at the terminal block and the wire pair connection is extended as each telephone is added (Figure 22-16).

Each telephone line is connected to a central office that contains switching equipment, signalling equipment, and batteries that supply direct current to operate the system. Switches

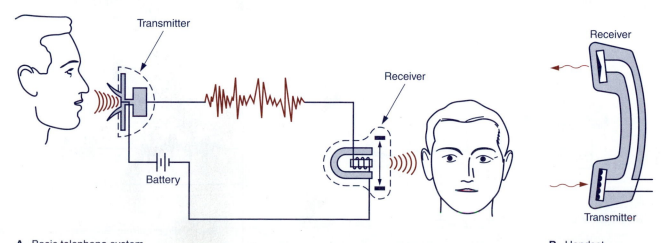

A. Basic telephone system

B. Handset

Figure 22-14   Telephone system.

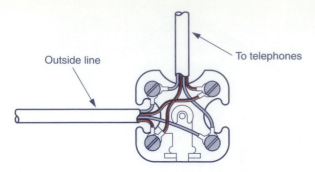

Figure 22-15 Connection of incoming telephone line.

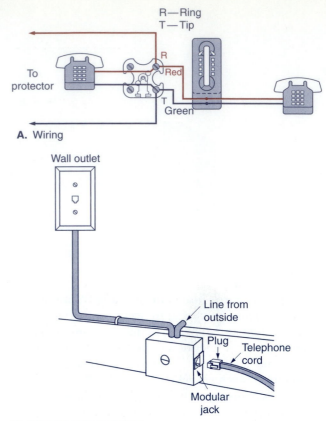

A. Wiring

B. Modular jacks and plugs

Figure 22-16 Typical interconnection of telephones.

in the central office respond to the dial pulses or tones from the calling telephone to connect the calling phone to the called phone. When connection is established, the two telephones communicate using the direct current (dc) supplied by the central office batteries. The dc voltage between the red (ring) and green (tip) wires should be about 50 V with all of the telephones either **on hook** (hung up) or unplugged from their jacks. The voltage that will be measured across the line in the **off hook** condition (handset lifted and ready for use) is typically in the 5–10-Vdc range (Figure 22-17). An open or short on the line can result in a zero voltage reading at a jack.

The central office notifies you that a call is waiting by ringing your telephone. It does this by placing an alternating current (ac) voltage signal of about 90-V ac on the tip (green) and ring (red) wires to your telephone (Figure 22-18). This voltage is used to operate the ringer in your telephone. The capacitor (*C*) passes the ac for ringing, but blocks the flow of

any dc. The ac ringing voltage can shock you. Normally telephone voltages are not dangerous, but you should take some precautions.

Some telephone sets send the telephone number by dial pulses, while others send it by audio tones. ***Pulse dialing*** (known as rotary dialing) provides the telephone system with the tele-

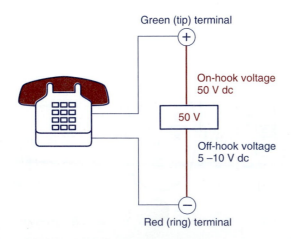

Figure 22-17 Telephone dc voltages and currents.

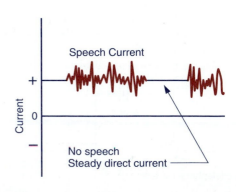

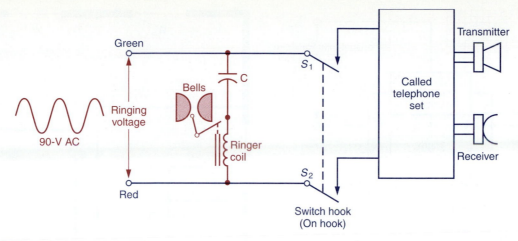

Figure 22-18 Telephone ringing voltage.

phone number called by switching the dc current in the local loop ON and OFF (Figure 22-19). A finger is placed in the hole for the digit to be dialed and rotated clockwise to the stop. When the dial is released, it opens and closes switches coupled to the dial that break the circuit and then make it again. The breaking and making of the circuit causes the current in the circuit to stop flowing and then to begin flowing again. As a result, ON-OFF current pulses occur in the local loop to the central office. The number of pulses correspond with the digit dialed.

Tone dialing uses audio tones to send the telephone number. It is much faster than pulse dialing and uses a push button keypad (Figure 22-20) instead of a rotary dial. Pressing one of

the keys causes an electronic circuit in the keypad to generate combinations of audio frequency tones that are sent to the central office to identify the digits of the telephone number. The name given to the combination tone dialing is Dual-Tone MultiFrequency (DTMF) dialing. The frequency of the audio tone generated corresponds with the number pressed.

If your existing telephones have a rotary dial and you want to replace them with tone-dialing telephones, you must call the local telephone company and ask them to connect you for tone-dialing service. If you already have push button tone-dialing telephones and install the electronic telephones that have push buttons, but are also able to do pulse dialing, everything will work properly. Pulse dialing works on tone-

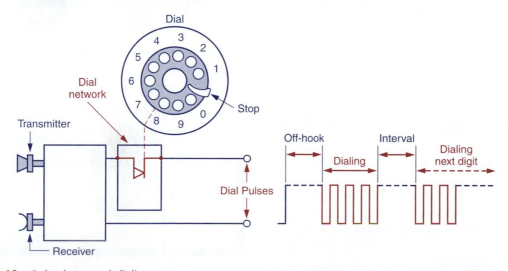

Figure 22-19 Pulse (or rotary) dialing.

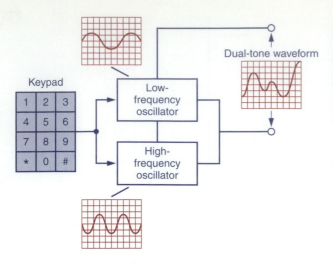

Figure 22-20 Tone dialing.

dialing circuits, but tone dialing does *not* work on pulse-dialing circuits.

22-6 Alarm Systems

Security alarm systems use two basic types of protection to detect burglars and intruders: perimeter protection and area protection. A *perimeter protection* system surrounds your home with security by protecting every point where an intruder could gain entry. For perimeter protection to be effective, every potential entry point should be protected with a sensor or switch. This includes all doors and windows (Figure 22-21).

Area protection is the second line of defense against burglars and intruders. Instead of detecting a door or window opening, area protection systems detect the presence of an intruder after the intruder is inside the house.

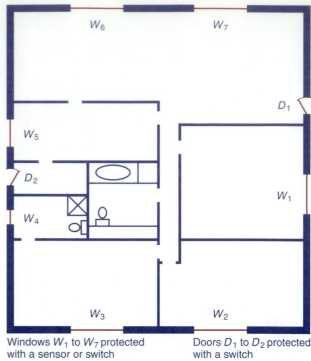

Windows W_1 to W_7 protected with a sensor or switch

Doors D_1 to D_2 protected with a switch

Figure 22-21 Perimeter protection.

Area protection sensors and detectors are more sophisticated and usually more expensive than their perimeter protection counterparts. They are usually placed where a burglar is most likely to pass through as he or she searches for valuables (Figure 22-22).

The best security systems use both perimeter and area protection. The idea is that one system backs up the other if the equipment fails or if an intruder defeats part of the alarm system.

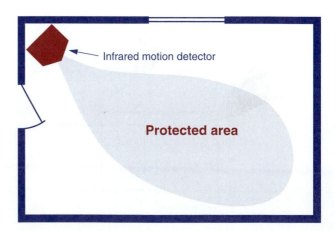

Infrared motion detector

Protected area

Figure 22-22 Area protection.

There are two kinds of burglar alarms: wireless and hard-wired systems. With a **wireless** alarm system, you don't use wire to connect the detection devices to the control panel. In a wireless system, the control panel is basically a radio receiver and the detection devices are transmitters. Wireless systems are extremely easy to install, but they are relatively expensive and prone to radio frequency interference. *Hard-wired alarm* systems are more difficult to install, but they are usually cheaper and more reliable than wireless types (Figure 22-23).

Detection devices are the eyes and ears of an alarm system. The **magnetic switch** (Figure 22-24) is the most popular for perimeter protection of doors and windows. It consists of a switch and a magnet. In a typical installation, you mount the magnet on the door or window

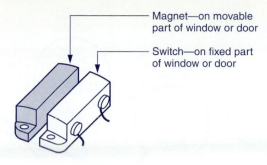

Figure 22-24 Magnetic switch.

sash and align the switch on the door or window frame. Moving the magnet out of alignment with the switch—by opening the door or window—opens or closes the switch contacts and activates the alarm.

For area protection, there are three types of motion detectors: ultrasonic, infrared, and microwave. **Microwave** and **ultrasonic** alarm units work on a radar-like principle. They emit waves of energy. The alarm is activated when movement disturbs those waves. A passive **infrared motion detector** (Figure 22-25) works like a thermometer. It detects changes in infrared (or heat) energy. In a residential situation, a passive infrared device is usually the best kind of motion detector. It consumes less

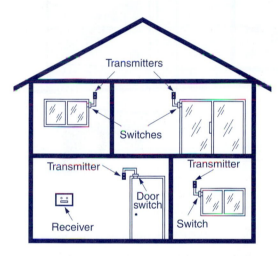

A. Wireless system

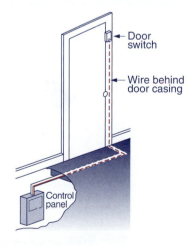

B. Hard-wired

Figure 22-23 Types of alarm systems.

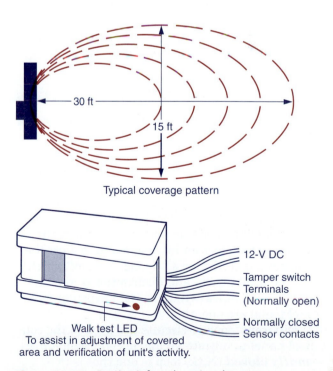

Typical coverage pattern

12-V DC

Tamper switch Terminals (Normally open)

Normally closed Sensor contacts

Walk test LED
To assist in adjustment of covered area and verification of unit's activity.

Figure 22-25 Passive infrared motion detector.

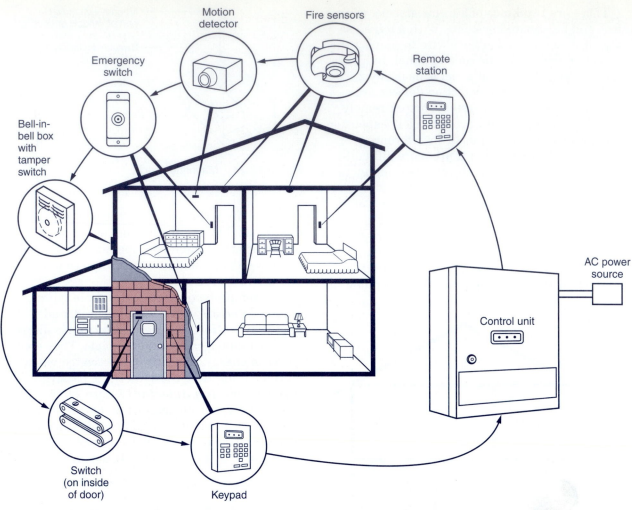

Figure 22-26 Basic parts of a security alarm system.

power than the others and is less prone to false alarms.

All alarm systems consist of a control panel, a keypad or key switch, a sounding device such as a bell or siren, and detection devices (Figure 22-26). The **microprocessor-based control panel** is the brain of the system; it holds programmed information that tells the alarm system how to function. The detection devices are positioned in strategic places throughout the building to detect an intrusion and to alert the control panel. Either a keypad or a key switch can be used to arm and disarm the system. In some small alarm systems, the keypad and control panel are one unit.

Different loops or circuits are used in the control panel to initiate an alarm condition. A **normally closed** (N.C.) loop is one that is in a no-fault or normally-operating state when the

circuit is complete and current is flowing (Figure 22-27). It consists of sensors or switches

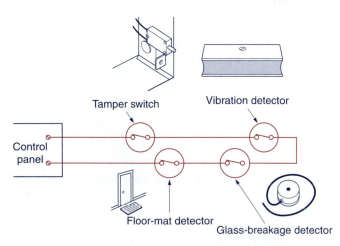

Figure 22-27 Typical normally closed (N.C.) circuit loop.

connected in series. Any break in the circuit causes an alarm condition. Normally closed loops are used most often because they are considered supervised. This means that if the loop circuit is broken or cut, the alarm will be activated.

A *normally open* (N.O.) loop is one that is in a no-fault or normally-operating state when the circuit is open and current is unable to flow (Figure 22-28). It consists of sensors or switches connected in parallel. Closing the circuit completes a current path and an alarm condition is indicated. An *end-of-line resistor* can be used to make a normally open loop be self supervised. This ensures that a signal is given by the system to warn that a problem exists if a conductor is cut or broken or if connections come loose. A small predetermined current constantly flows through the wiring during the normal state. If this current falls below a certain level, the problem signal is triggered. However, if the current increases greatly because the sensors closed their contacts, the alarm will be sounded. The current to supervise the system is

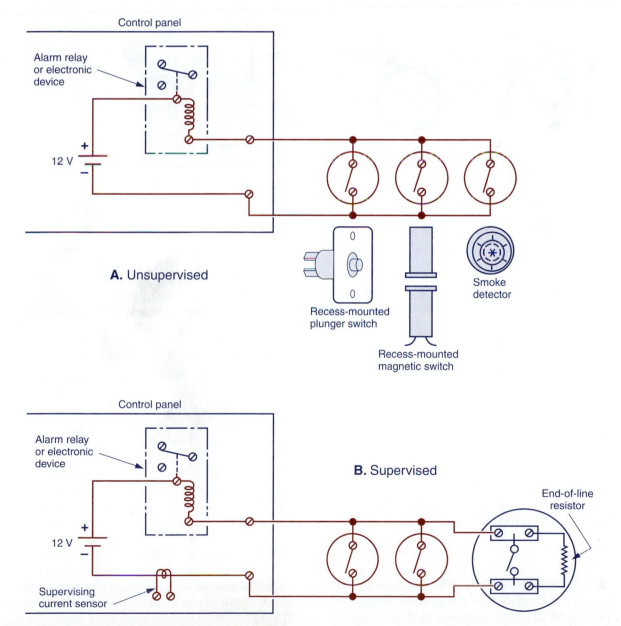

Figure 22-28 Typical normally open (N.O.) circuit loop.

dependent on the resistance of the end-of-line resistor.

Instant loops cause an instant alarm condition when a violation occurs. *Delayed loops* cause an alarm condition to sound after a fixed period of time when a violation occurs. The exit delay gives you time to get out the door and return the loop to its normal state before the alarm goes OFF. The entry delay gives you time to come in and disarm the control panel before the alarm goes OFF. Alarm time-delay adjustment controls are used to set the delay time (Figure 22-29).

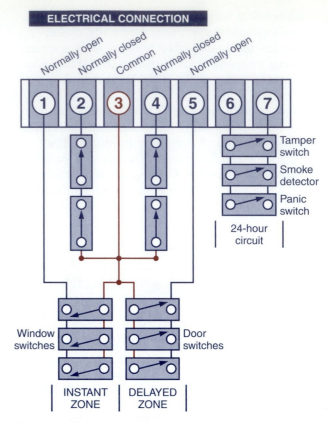

Figure 22-30 Typical control-panel circuit connections.

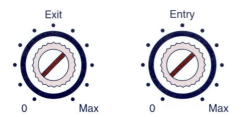

Figure 22-29 Alarm time-delay adjustment controls.

The 24-hour protection loop is designed to activate the alarm system instantly anytime, whether the system is armed or not. Applications for this protection loop include monitoring of fire-alarm detectors, panic buttons, and tamper switches.

A *zone* is a detecting circuit. Generally, the more zones a control panel has, the better (Figure 22-30). Zoned circuits let you close down your system in all those areas where you want to move about freely, while still keeping all the other zones on alert. A loop can be representative of a zone; or a zone might be all the windows or all the doors. In a home a zone may be the upstairs area as opposed to the downstairs area, with each zone on a separate alarm loop.

Arming an alarm system pertains to the method used to turn the system ON. Typical arming devices include key-lock switch and digital key pad (Figure 22-31). *Disarming* the system pertains to the method used to turn the alarm OFF. In most cases, the same device is used to arm and disarm the system. Often an array of *LEDs* (light-emitting diodes) of differ-

ent colors are used to indicate the state of the system. They make operating and servicing the alarm system easy by lighting up and blinking in ways that let you know how the system is functioning.

Alarm output devices include bells, sirens, and strobe lights. For the sake of your neighbors, the system should shut off a wailing siren within about 15 minutes. Bylaws in some communities limit the length of time an alarm can sound. After that, the alarm should automatically reset in case a burglar returns.

The alarm system can also be monitored using an *automatic telephone dialer* connected to the control panel. When the alarm is activated, the dialer calls one or more programmed numbers and leaves a recorded message. The idea is that someone, a relative or friend, will get the message and call the police.

You can also use the services of a central station to monitor your alarm. If the alarm goes OFF, the central station calls back to ask for a

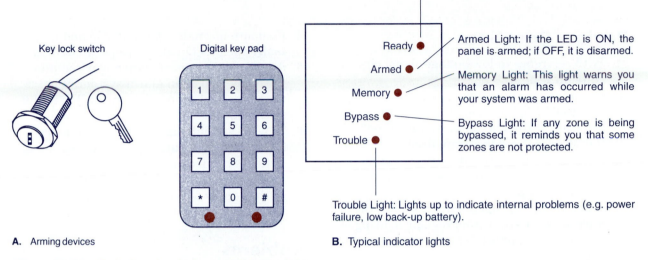

Ready Light: This LED lights when all loops are in their normal condition, whether open or closed. If one of the loops is violated, the LED will not light, and the unit will not arm. This usually means that a door or window has been left open.

Key lock switch

Digital key pad

Ready ●

Armed ●

Memory ●

Bypass ●

Trouble ●

Armed Light: If the LED is ON, the panel is armed; if OFF, it is disarmed.

Memory Light: This light warns you that an alarm has occurred while your system was armed.

Bypass Light: If any zone is being bypassed, it reminds you that some zones are not protected.

Trouble Light: Lights up to indicate internal problems (e.g. power failure, low back-up battery).

A. Arming devices

B. Typical indicator lights

Figure 22-31 Typical arming devices and indicator lights.

code word. If you don't give the right code word, they immediately call the police.

A *battery backup* enables the alarm system to continue to operate during an ac power failure. Generally, it consists of rechargeable batteries and charging circuits.

HELP WANTED

Telephone Installer
Job Description
- Install and maintain telephones and related equipment in homes and offices.
- Training is available.
- Electrical/electronic background would be a definite asset.
- Technical skills are important but the following are more important: hard working; self starting; customer oriented; team player; quick learner.

Review and Applications

Review Questions

1. With reference to low-voltage signal circuits, in general:
   (a) what type of voltage-supply unit is used?
   (b) what is the maximum voltage used?
   (c) what is the size of wire used?
2. Why can wire approved for a door-chime circuit not be used for home lighting or power circuits?
3. When installing signal-circuit cables, they must be kept apart from regular lighting or power circuit cables. What is the reason for this rule?
4. Explain the construction and operation of a two-tone door chime.
5. In general, what are annunciators used for?
6. Explain the construction and operation of a typical electric door latch.
7. What is the function of each of the following in the operation of a card access system?
   (a) card key
   (b) card reader
   (c) microprocessor
8. With reference to a telephone system:
   (a) what is the purpose of the protector?
   (b) what is the purpose of the transmitter?
   (c) what is the purpose of the receiver?
   (d) how are all telephones in a home connected in relation to the telephone line?
   (e) what are the approximate "on hook" and "off hook" dc voltage values of the telephone line?
   (f) what is the approximate ac voltage value of the ringing voltage?
9. Compare the way the telephone number is generated in pulse dialing and tone dialing telephones.
10. Compare perimeter- and area-alarm detection.
11. Explain the construction and operation of a magnetic switch.
12. Compare the operation of microwave and ultrasonic motion detectors with the passive infrared type.
13. Explain the function of the control panel as part of an alarm system.

14. Compare normally closed (N.C.) and normally open (N.O.) alarm loops or circuits.
15. An entry door sensor switch is wired into the delay loop of an alarm system. What is the purpose of connecting the switch to this special loop?
16. To what special alarm loop are fire alarm detectors, panic buttons, and tamper switches connected? Why?
17. Explain the terms arming and disarming as they apply to an alarm system.

Problems

1. Refer to Figure 22-5. Redraw the schematic with a second remote door chime added to the original circuit.
2. Refer to Figure 22-10. A key operated normally-open push button is to be installed in the lobby for entrance into the building. Redraw the schematic with this push button properly connected.

Critical Thinking

1. Refer to Figure 22-6. The front chime does not operate when the front push button is pressed. The rear chime operates fine. What components would normally be eliminated as a possible cause of the problem? Why?
2. Refer to Figure 22-8. Assume No. 1 remote PB is stuck closed. In what way would this effect the operation of the circuit?
3. Refer to Figure 22-10. The buzzer in apartment A will not operate when the A_2 push button is pressed. The buzzer is suspected of being bad. How could a voltmeter be used to verify this?
4. Refer to Figure 22-24. Explain how you would use an ohmmeter to test the operation of the magnetic switch.
5. Refer to Figure 22-28. Assume the end-of-line resistor of the supervised circuit becomes short circuited. Will a problem signal or alarm signal be triggered? Why?

Portfolio Project

Complete a study of the fire alarm system used in the school or similar institute. Make a diagram of the main annunciator panel, showing all controls and zones monitored. Include the location and type of sensors and actuators used.

Circuit Challenge

Using a Simulator

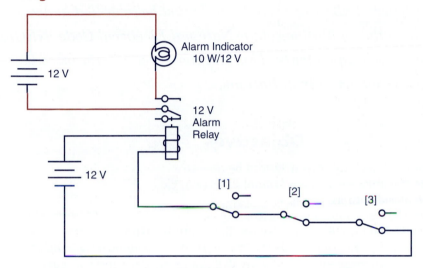

Procedure

(a) Wire the three-sensor normally-closed alarm control loop circuit shown, using whatever simulation software package is available to you.

(b) Draw a schematic of the circuit rewired to simulate a three-sensor normally-open alarm control loop. Wire the simulated circuit to verify its operation.

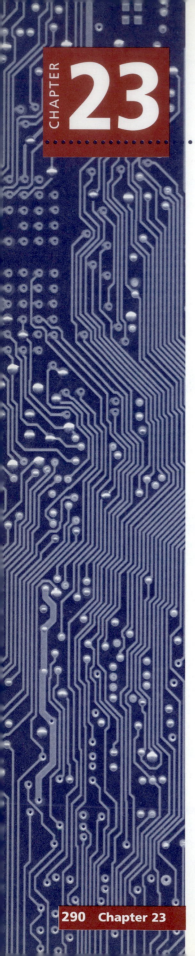

23

Residential Wiring Requirements and Devices

Boxes, switches, receptacles, cables, and numerous other hardware items are essential parts of all residential electrical installations. This hardware should be approved by the Underwriters' Laboratories (UL) and must be installed in accordance with the regulations of the National Electrical Code (NEC) and its local supplements. This chapter serves as an introduction to house wiring installations.

Objectives

After studying this chapter, you should be able to:
1. Outline the purpose of the National Electrical Code (NEC) and the method of code enforcement.
2. Identify the components of an incoming service.
3. Draw the schematic for a three-wire distribution system.
4. State common NEC requirements that pertain to residential wiring.
5. Explain the purpose and process involved in grounding and overcurrent protection.

Key Terms

- **Underwriters' Laboratories (UL)**
- **National Electrical Code (NEC)**
- **incoming service**
- **step-down transformer**
- **service mast**
- **service box**

- **distribution panel**
- **panelboard**
- **branch circuit**
- **underground service**
- **three-wire distribution system**
- **neutral wire**
- **live wire**

- **hot wire**
- **ground wire**
- **double-pole circuit breaker**
- **general-purpose circuits**

(continued)

RESEARCH INTERNET SITES

- **Underwriters' Laboratories (UL)**
- **National Fire Protection Association (NFPA)**
- **Leviton**

- small-appliance circuits
- individual circuits
- metallic conduit
- PVC conduit
- nonmetallic sheathed cable
- armored cable
- circuit breaker
- loop wiring system
- cable strap
- cable staple
- cable clamp

- surface-type installation
- flush-type installation
- octagon box
- switch box
- square box
- utility box
- knockout
- ganged outlet box
- hanger bar
- duplex receptacle

- keyless lampholder
- pull-chain lampholder
- single-pole switch
- double-pole switch
- three-way switch
- four-way switch
- dimmer switch
- grounding
- fuse
- bonding

23-1 Approval of Equipment and Wiring

Approval Authorities   Electric products for sale in the United States are generally required to pass standardized tests for safe usage. One of the best-known testing organizations is the *Underwriters' Laboratories*, identified with the UL symbol shown in Figure 23-1. The Underwriters' Laboratories places its UL symbol on electric products to attest to the fact that the item has passed certain safety tests. *The tests are only the minimum safety requirements, and the label does not mean that all listed items are of equal quality.* The various types of material used in electric wiring should be of a type listed by UL to assure that an acceptable safety level is being maintained. Most stores will only sell merchandise that is UL listed. The important thing is to shop for quality and do quality work.

National Electrical Code   Electrical installations in the United States must be made in accordance with the specific electrical codes of the locale in which the installation is being made. Most local governments adopt the National Electrical Code as the legal code in their area, with minor changes made to deal with particular conditions of the area. Thus, the *National Electrical Code (known as the NEC)* becomes the legally enforceable electrical code of the area. The National Electrical Code consists of a set of rules and guidelines designed to safeguard persons and property from the dangers arising from the presence and use of electricity. The Code is not a textbook. Rather, it is a set of rules, developed over many years, that have been found to provide safe and practical electrical installations.

The Code book is revised and updated every three years. Chapters 1, 2, 3, and 4 contain general requirements that cover most situations. Chapters 5, 6, and 7 contain information that pertains to special occupancies, equipment, and conditions. Chapter 8 deals with communications systems such as telephone, fire and burglar alarm circuits. Chapter 9 contains examples and tables. Chapters are divided into Articles. Articles are divided into Sections. Sections are divided into subsections and exceptions.

Approval of the Electrical Installation
Electric wiring should be installed only by a qualified person, and the installation must be performed according to the codes and laws governing the locale in which the work is done. All work must be inspected and approved by the

Figure 23-1   The Underwriters' Laboratories label.

local electrical inspector before voltage is applied. The local inspector not only sees that the governing rules are followed but also that, in his or her judgement, the quality of the work is acceptable. Normally, the electrical utility company will not provide power until an inspection has been completed satisfactorily.

The first inspection, called the rough-in inspection, is done after the cables are run between the boxes, but before the insulation, wallboard, switches, and fixtures are installed. The second inspection, called the final inspection, is done after the walls and ceilings are finished and all electrical connections are made.

23-2 The Incoming Service

The electrical installation in a home includes all the equipment and wiring beyond the point where the power company makes connection. The *incoming service* specifically refers to that portion of the installation from the service box up to and including the point at which the power company makes connection.

With an *overhead incoming service* (Figure 23-2) voltage is applied to the house from a *step-down transformer* located on a nearby pole. Most supply lines consist of two insulated cables supported by a bare neutral cable. The supply lines run overhead from the pole trans-

former and connect to the wires coming from the *service mast*. A variety of service entrances are available. The choice of entrance depends upon the style of home, desired entrance location, and specific electrical code requirements. The most common entrance is the riser type where the conduit is mounted to the outside wall and extended up through the eave, with the service head positioned above the roof line. Depending upon local codes, conduit can be used for all electrical installations and cable installation where flat-wall mounting is possible without going underground or through the eave.

A kilowatthour meter is installed outdoors at the end of the service mast. This meter records the amount of electric energy used in the home.

Service conduit is used to protect the wires that run from the meter to the *service box*, which is usually located in the basement of the house. The service box contains the main service switch and fuses, or the main service circuit breaker. It provides protection for the service conductors and a means for disconnecting voltage to all circuits in the house.

The *distribution panel* or *panelboard* is located next to the main switch. The various branch circuits in the house are supplied from this panelboard. *Branch circuits* carry electricity to points of use throughout the house. In addition to providing terminating points for branch circuits, the distribution panel contains fuses or circuit breakers that provide overcurrent protection for the branch circuits. A combination panelboard (Figure 23-3), which has both the main switch or breaker and distribution panel built into one unit, reduces installation time and cost.

In recent years the *underground incoming service* (Figure 23-4) has become very popular. It is used in residential areas where local regulations ban overhead distribution systems. Not only does it eliminate unsightly poles and overhead wires, it gives the system added protection against unfavorable weather conditions.

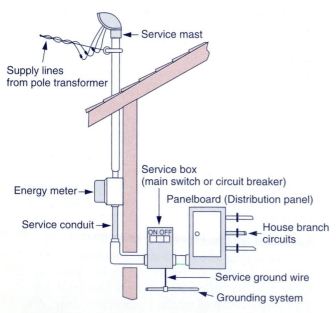

Figure 23-2 Overhead incoming service.

Supply lines from pole transformer

Service mast

Energy meter

Service box (main switch or circuit breaker)

Panelboard (Distribution panel)

Service conduit

ON OFF

House branch circuits

Service ground wire

Grounding system

23-3 The Three-Wire Distribution System

The *three-wire distribution system* refers to the wiring method used to operate the electrical system in a home. This is a single-phase 60-Hz (cycles) system consisting of three wires, which

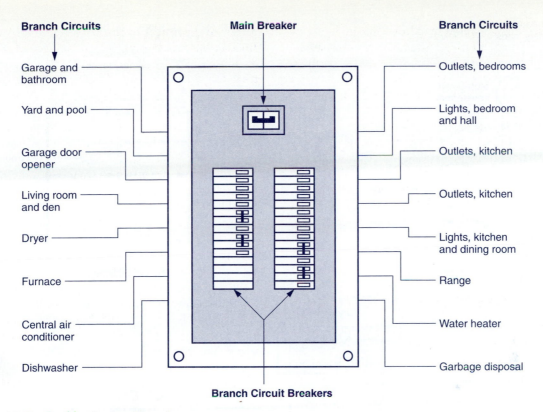

Branch Circuits

- Garage and bathroom
- Yard and pool
- Garage door opener
- Living room and den
- Dryer
- Furnace
- Central air conditioner
- Dishwasher

Main Breaker

Branch Circuits

- Outlets, bedrooms
- Lights, bedroom and hall
- Outlets, kitchen
- Outlets, kitchen
- Lights, kitchen and dining room
- Range
- Water heater
- Garbage disposal

Branch Circuit Breakers

Figure 23-3 Combination panelboard.

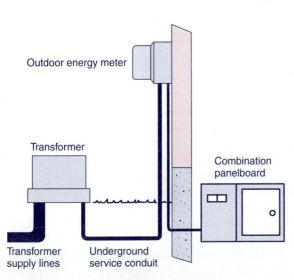

Outdoor energy meter

Transformer

Combination panelboard

Transformer supply lines

Underground service conduit

Figure 23-4 Underground incoming service.

provides one 240-V circuit and two 120-V circuits. The system provides 120-Vac for operating lights and small portable appliances. In addition, it provides 240-Vac for operating heavy-duty appliances such as the electric range, clothes dryer, and hot water heater.

The diagram for a three-wire distribution system is shown in Figure 23-5. The primary voltage is in the kilovolt range. This voltage is stepped down to 240 V, which appears across the two outside leads of the transformer secondary winding. A center-tap wire on the transformer divides this voltage in half, giving 120 V between the center-tap connection and the outer leads. The two outside wires are called the **hot** or **live wires** and have a *black* or *red* insulation covering. The center-tap wire is *grounded* (connected to earth) at the base of the transformer and is known as the **neutral wire**. The neutral wire is color coded with a *white* insulation covering. For safety purposes, the *live* wires are switch controlled and have fuses or circuit breakers connected in series with them. The *neutral* wire is grounded (connected to the earth) at the transformer and in the residential main switch box.

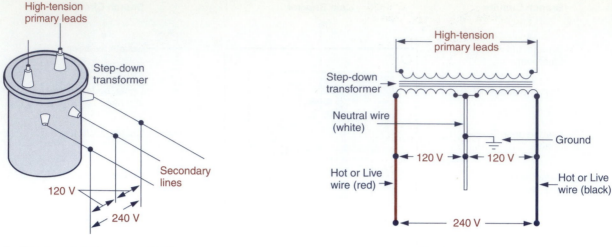

A. Step-down transformer connection

B. Circuit schematic

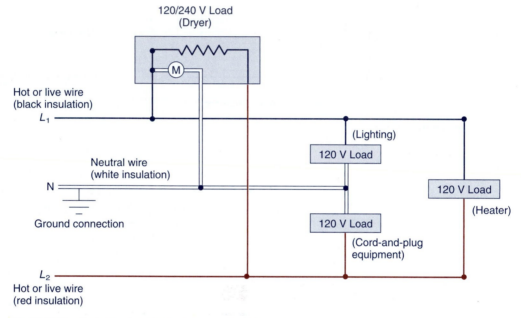

C. 120/240 volt, three-wire, single phase, residential service.

Figure 23-5 Residential three-wire distribution system.

23-4 General Wiring Requirements

Service Size Requirements The size of an incoming residential service is rated in terms of the maximum current the system can handle. Articles 220 and 230 of the Code outline the minimum service requirements. Generally, homes are required to have a minimum 100-A 3-wire distribution system. The current rating of the service is determined by the size of the conductors used from the utility lines down to the distribution panel. If properly protected, the current rating of the service is the same as the main fuse or circuit breaker rating. A 100-A service might have been adequate a few years ago, but if you consider future needs, a 200-A service might be necessary.

A typical wiring diagram of a complete incoming service is shown in Figure 23-6. The white neutral wire is never switched or fused. It connects to the neutral terminal blocks of both the main switch and distribution panel. The neutral wire is grounded at both the transformer

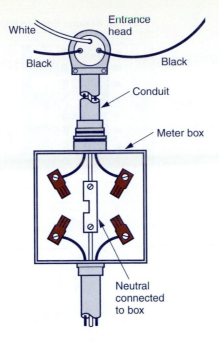

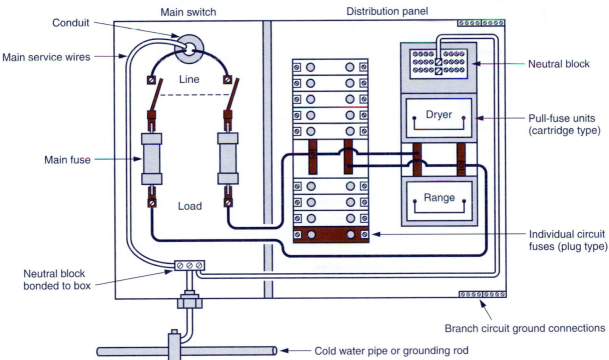

Figure 23-6 Typical residential service wiring diagram.

and to an approved grounding system in the house. The two black live wires are connected in series with the kilowatthour meter coils, main switch, and main fuses. They supply voltage to a series of secondary fuses that protect the branch circuits of the house.

The wiring of a circuit breaker panel is basically the same as that of a fuse panel. A circuit breaker panel offers additional features such as tamper-proof current ratings, ease of resetting (instead of replacing), and absence of exposed live parts.

Connection of branch circuits to the distribution panel is really quite simple (Figure 23-7). To connect a 120-V circuit, the hot wire may be fed from any single-pole breaker, and the neutral wire is fed from the neutral terminal block. The National Electrical Code requires the use of a **double-pole circuit breaker** for 240-V circuits, so that both hot wires are protected and are opened and closed together.

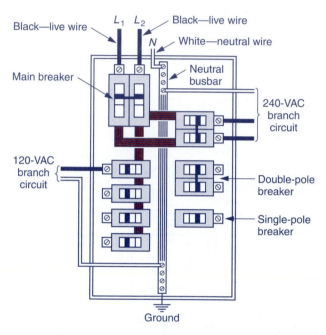

Figure 23-7 Branch circuit connections to the distribution panel.

Most *circuit breakers* plug into their terminals (Figure 23-8). Usually, you insert the breaker in an anchor clip at one end, then push the connector clip into a snap-in hole at the other end. The cover for the panel will have knockout blanks to accommodate each circuit breaker. Only remove the number of spaces necessary for the number of breakers installed. If any blanks are left, they may be removed later if another circuit is installed.

Conductor Size Requirements

Rules regarding the size of conductors used in house wiring are outlined in Article 220 of the NEC. The most important factor in determining conductor size is the maximum amount of current that the circuit can be expected to handle. Other factors considered in determining conductor size include the number of conductors in

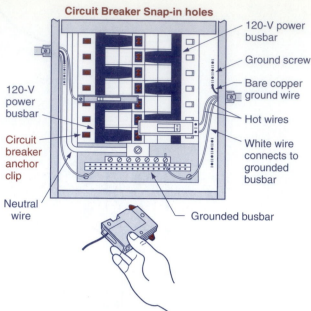

Figure 23-8 Installing a circuit breaker.

the cable, type of insulation used, and the local temperature conditions.

Depending on the local regulations, a service-entrance cable or conduit with separate wires may be used for the main feed wires of a service. In general, No. 2 gauge copper wires are used for 100-A services and No. 3/0 gauge copper wires for 200-A services.

The branch circuits used in homes can be divided into three general types: general-purpose circuits, small appliance circuits, and individual circuits. **General-purpose circuits** carry the smallest loads in the house and use a No. 14 or No. 12 AWG copper wire. The Code does not permit a wire smaller than No. 14 gauge copper wire protected by a 15-A fuse or circuit breaker for permanently installed house wiring. There is a trend toward using a No. 12 gauge copper wire, with 20-A protection, and in some localities it is required as a minimum. These circuits provided power for lights and for convenience receptacles. One of these circuits should be planned for every 500 square feet of floor space.

The **small-appliance circuit** handles a larger load and requires at least a No. 12 AWG copper wire with 20-A protection. The Code requires that at least *two* of these circuits be provided for the receptacles in the kitchen, breakfast room, dining room, and pantry. No lighting is connected to small appliance circuits. These two circuits can be merged into one three-wire circuit.

Individual circuits would include larger appliances such as water heaters and ranges. These individual circuits run directly from the distribution panel to only one outlet or appliance.

Device	Minimum Copper Conductor Size
Refrigerator receptacle	No. 12 AWG, protected for 20 A
Laundry room or area receptacle	No. 12 AWG, protected for 20 A
Electric range	No. 8 AWG, protected for 40 A
Central air/heat	Varies with the size of the unit
Electric dryer	No. 10 AWG, protected for 35 A
Electric water heater	No. 12 AWG, protected for 20 A

Rigid Conduit Wiring

In many communities, local Codes specify the use of conduit for wiring the service entrance equipment in a house. A conduit system offers the best mechanical protection for conductors, but is also the most expensive to install. Both rigid *metallic (steel or aluminum)* and *PVC (polyvinylchloride) conduit* are approved for house service installations (Figure 23-9). The Code requires conductors be drawn into the conduit system after it has been completely assembled.

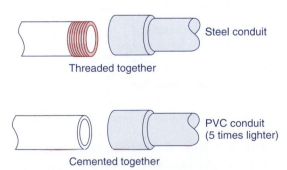

Figure 23-9   Rigid conduit.

The PVC plastic-type rigid conduit is being used extensively for residential service entrance conduit. This conduit is approximately five times lighter than steel conduit and twice as light as aluminum conduit of the same size, making it easier to handle. Also, the conduit does not have to be threaded, because lengths of PVC conduit and fittings are assembled using a solvent cement.

Rules regarding the installation of rigid conduit are found in Articles 346 and 347 of the Code. This type of wiring system provides the best protection for conductors, but it is expensive.

Nonmetallic Sheathed-Cable Wiring

Nonmetallic sheathed cable is the most common type of cable used for branch-circuit wiring inside a house. It is made up of two or three insulated wires, along with a bare copper ground wire, and all are enclosed in a thermoplastic sheath. The cable is marked to show the type of wire, the wire size, and the maximum working voltage (Figure 23-10).

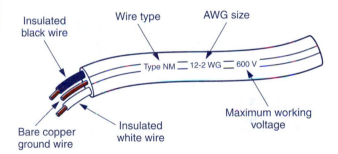

Figure 23-10   Nonmetallic sheathed cable.

Rules regarding the use and installation of nonmetallic sheathed cable are found in Article 336 of the NEC. Of all the wiring systems, this is by far the easiest to install. Nonmetallic sheathed cable is available in three types:

1. NM is available in cable assemblies of two or three insulated conductors in sizes No. 14 to No. 2 AWG. It is acceptable for most interior wiring, ceiling outlet boxes, and the like, which are not subject to dampness.
2. NMC is similar to type NM, except that the outer covering is also fungus resistant and corrosion resistant. It is used in damp indoor locations and for outdoor wiring that is above ground.

3. UF cable has the wires embedded in a solid sheath. It can be used for direct burial underground as well as in areas where NM and NMC cable is used.

Armored-Cable Wiring

Armored cable is made up of two, three, or four insulated wires and a bare copper wire, all of which are enclosed in a protective metal covering of aluminum or galvanized steel (Figure 23-11). It combines the flexibility of nonmetallic sheathed cable with the metal protection of rigid conduit.

Steel or aluminum armor

Figure 23-11 Armored cable.

Rules regarding the use and installation of armored cable are found in Article 333 of the Code. Armored cable is widely used in industrial plants as well as public and commercial buildings. Its use in a house is limited to installations where the possibility of physical damage to the cable makes nonmetallic sheathed cable unsuitable.

Loop-Wiring System

Years ago, homes were wired using single wires that were run separately and supported on porcelain knobs. The Code now requires that all cables must be run in continuous lengths between electrical boxes (Figure 23-12). All connections must be made in electrical boxes and these boxes must be accessible for inspection and troubleshooting after the installation is completed.

23-5 Installing Nonmetallic Sheathed Cable

The proper handling and installation of nonmetallic sheathed cable is important if the electrical installation is to pass inspection and give years of safe and trouble-free service.

Rules regarding the installation of nonmetallic sheathed cable are found in Article 336 of the Code. When any doubt exists about these rules, the local inspector should be consulted.

The following are a number of important rules and tips to remember when installing nonmetallic sheathed cable:

1. The cable must be secured by a *strap* or *staple* within 12 in. of every box (Figure 23-13).

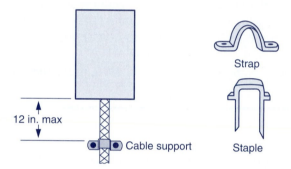
Strap

12 in. max

Cable support

Staple

Figure 23-13 Cable must be supported by a strap or staple.

2. The cable should be supported at intervals of not more than 5 ft in open runs between boxes (Figure 23-14).

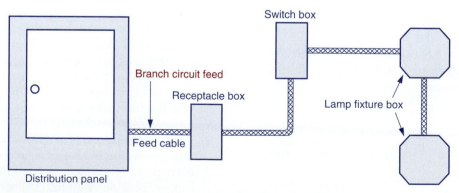
Switch box

Branch circuit feed

Receptacle box

Lamp fixture box

Feed cable

Distribution panel

Figure 23-12 Loop-wiring system.

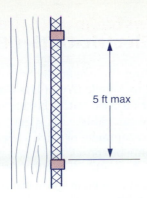

Figure 23-14   Maximum interval between staples.

3. Use the correct size of cable strap or staple for the size of cable being installed. Also, cable can be damaged if the staples are driven too deeply into the wood (Figure 23-15).

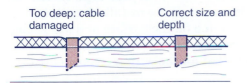

Figure 23-15   Use of the correct size of staple.

4. The cable must be held securely by a **clamp** or a box connector at the point of entry into the box (Figure 23-16).
5. The outer sheath of the cable should extend beyond the clamp by a minimum of 1/4 inch to protect the conductors from being shorted together by the clamp. The cable clamp should not be over-tightened or it will crush the cable.

6. A minimum of 6 in. of free insulated wire must be provided in each box for making connections to electrical devices (Figure 23-17).

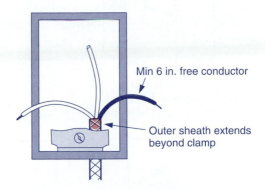

Figure 23-17   Free wire in box.

7. To protect the cable from accidental puncture by nails it should be kept at least 1-1/4 in. from the outer edge of any wooden stud. This is done by drilling the holes for the cable in the center of the stud (Figure 23-18). The hole should be large enough for the cable to pass through it without damaging the insulation.
8. If the distance from the outer edge of the stud to the cable is less than 1-1/4 in., the cable must be protected by a steel plate that is fastened to the wooden stud in front of the cable to be protected (Figure 23-19).
9. Nonmetallic sheathed cable is not designed to be: (a) buried in plaster, (b) located near hot-water pipes, or (c) located near hot-air ducts.

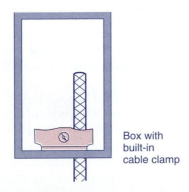

A. Cable clamp

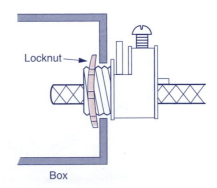

B. Box connector

Figure 23-16   Cable supports in a box.

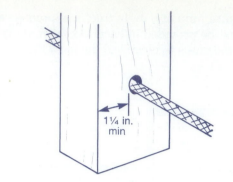

Figure 23-18 Protecting the cable.

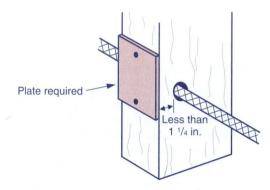

Plate required

Less than 1 ¼ in.

Figure 23-19 Using a plate to protect the cable.

10. The cable should not be bent sharply or handled so that the outer covering is damaged.

11. When removing the outer covering of the cable in order to make connections in outlet boxes, the wires must not be damaged. Correct removal is done by slitting the outer covering of the cable as far back as necessary with a cable stripping tool (Figure 23-20).

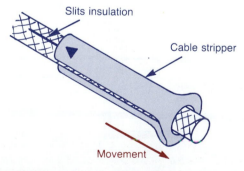

Slits insulation

Cable stripper

Movement

Figure 23-20 Cable-stripping tool.

23-6 Electric-Outlet Boxes

Rules regarding the use of electric outlet boxes can be found in Article 370 of the Code. The Code calls for electric-outlet boxes to be installed between cable runs where a switch, lampholder, receptacle, or wire splice is located. They serve four purposes:
- Reduce fire hazards
- Contain all the electric connections
- Support the wiring devices
- Provide grounding continuity

Electrical boxes can be obtained in two forms—metallic and nonmetallic. Galvanized steel boxes are the most popular type approved for use with nonmetallic sheath cable.

There are two types of electric-outlet box installations that can be used for the loop system. One is called **surface-type installation**, and it means all the installation components are mounted on the surface and are visible. The other is called **flush-type installation** and it means that all the wiring components are concealed behind the finished surface and only the opening in the electric-outlet box is visible through the finished surface.

Outlet boxes come in various shapes. The **octagon-shaped box** is used to support light fixtures, or is used as a junction point for splicing wires from different cables (Figure 23-21A). The device box (also referred to as a **switch box**) is used to house switches and receptacles (Figure 23-21B). The **square box** is used to house electric range and dryer receptacles and as a junction box for surface and concealed wiring systems (Figure 23-21C).

The cable enters the box through one of the **knockouts** in the box. A knockout is a partially punched hole that can be removed easily with a sharp blow. If a knockout is removed, the space provided must be taken up by the cable, or, in case of accidental removal, with a hole closure washer.

On some types of boxes, the pry-out type of knockout is used. This type of knockout has a small slot in the knockout that is designed to be pried out with a screwdriver (Figure 23-22).

The sides of metal sectional switch boxes are easily removed for **grouping** or **ganging** a series of boxes together (Figure 23-23). This feature allows you to quickly assemble a box

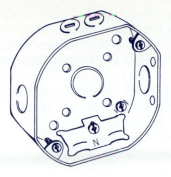

A. Octagon box

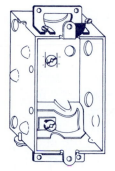

B. Device or switch box

Utility type

C. Square box

Figure 23-21   Outlet boxes.

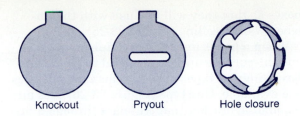

Knockout   Pryout   Hole closure

Figure 23-22   Types of knockouts and closures.

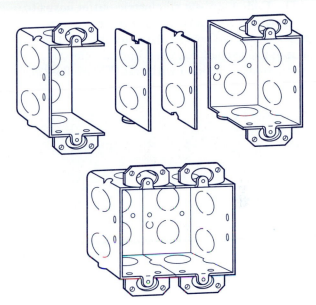

Figure 23-23   Ganged outlet box.

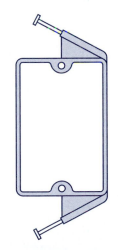

Figure 23-24   Nonmetallic outlet box.

capable of supporting any number of switches or receptacles.

Nonmetallic boxes are widely used with nonmetallic-sheathed cable because they are inexpensive. These plastic boxes even come equipped with the nails for quick mounting to studs (Figure 23-24). Nonmetallic boxes are also available in two-gang sizes or round-shaped sizes.

Various methods are used to mount or secure outlet boxes. For simple surface wiring, the box is fastened to the surface of the wall or ceiling. Concealed wiring systems often make use of **_hanger bars_**, brackets, or nails to position boxes so that they will be flush with the finished wall or ceiling (Figure 23-25).

When a box is installed in a finished wall, a special-purpose unit can be used to mount the box. Different types of special fastening devices are available. One type (Figure 23-26) uses an expanding bracket that will pass through a precut hole in the wall, expand, and grip the plaster when a tension bolt is tightened.

The Code requires all electric-outlet boxes to be provided with a cover for protective purposes. Different covers are manufactured to match the type of box used and the electric component contained within it (Figure 23-27).

There is a limit to the number of conductors the Code will allow per outlet box. Besides being unsafe because of possible damage to the insulation, crowding a box with wires makes it difficult to install devices in the box. The factors that determine the number of insulated conductors allowed per given box include:
- Inside volume of the box
- Wire gauge size of conductors
- The use of mechanical wire connectors, mounting nails, and fixture studs
- The housing of receptacles, switches, or other devices inside the box

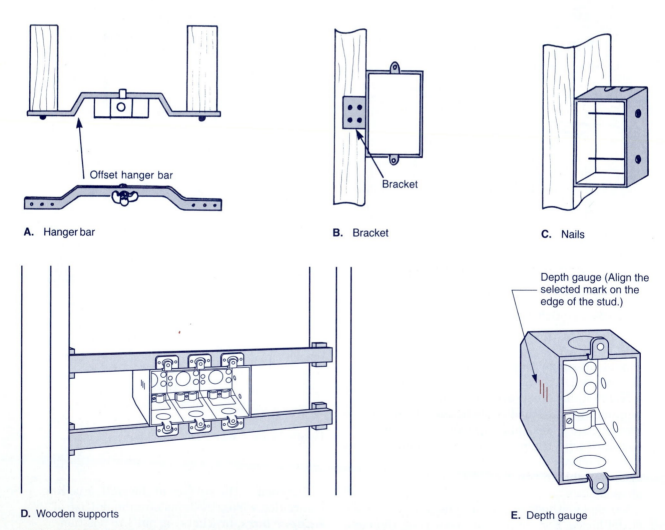

Offset hanger bar

A. Hanger bar

Bracket

B. Bracket

C. Nails

Depth gauge (Align the selected mark on the edge of the stud.)

D. Wooden supports

E. Depth gauge

Figure 23-25 Methods for securing outlet boxes.

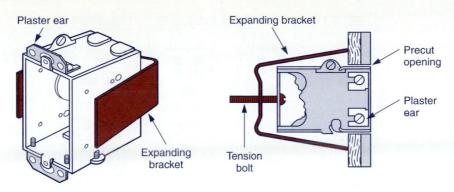

Figure 23-26   Mounting a box in a finished wall.

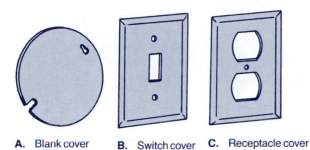

A. Blank cover   **B.** Switch cover   **C.** Receptacle cover

Figure 23-27   Types of box covers.

Article 370 and Table 370-6(a) of the Code should be referred to in order to select boxes of approved size.

Example 23-1

A 4 in. × $2^1/_8$ in. octagonal ceiling box is mounted using a hanger bar and fixture stud. If the box is to contain four insulated mechanical wire connectors, the maximum number of No. 14 AWG conductors permitted in the box would be:

10	maximum allowed according to Table 370-6(a) of the Code
−2	space for each pair of insulated wire connectors
−1	space for the stud
7	maximum number of No. 14 AWG conductors that can be installed within the box

---

23-7 Electrical Receptacles

A *receptacle* is used to supply power to portable plug-in devices such as lamps and appliances. Figure 23-28 shows some of the different types of receptacles found in the typical home. Each has a unique arrangement of slots and is designed for a specific application:

• The polarized two-slot receptacle is common in homes built before 1960. Slots are of different sizes to accept polarized plugs.
• The polarized three-slot grounded receptacle has two different size slots and a U-shaped hole for grounding. It is required in all new wiring installations.
• The 20 A three-slot grounded receptacle features a special T-shaped slot. It is installed on circuits wired with conductors rated for 20 A, for use with large appliances or portable tools.
• The 250 V three slot receptacle is used primarily for 250 V window air conditioners. It is available as a single unit, or as half of a duplex receptacle, with the other half wired for 125 V.

The three-wire-ground *duplex receptacle* is the most common electrical device used to supply current to lamps, appliances, or portable plug-in devices in a home (Figure 23-29). It is designed to receive two electric plugs and delivers a voltage of approximately 120 V between its two parallel slots. The ground connection provides a safety ground connection for those electrical devices that require it.

The terminal screws of the duplex receptacle are color coded to assist with the proper

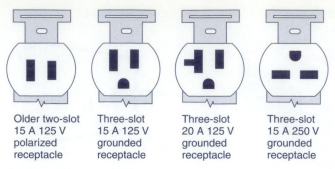

Older two-slot
15 A 125 V
polarized
receptacle

Three-slot
15 A 125 V
grounded
receptacle

Three-slot
20 A 125 V
grounded
receptacle

Three-slot
15 A 250 V
grounded
receptacle

Figure 23-28 Basic types of receptacles.

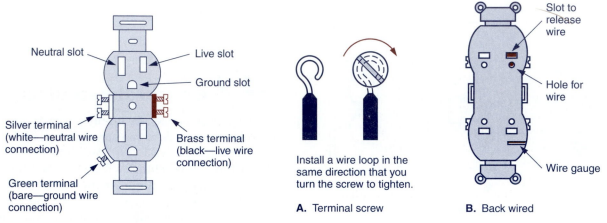

Neutral slot

Live slot

Ground slot

Silver terminal
(white—neutral wire
connection)

Brass terminal
(black—live wire
connection)

Green terminal
(bare—ground wire
connection)

Figure 23-29 Duplex receptacle connection.

Install a wire loop in the
same direction that you
turn the screw to tighten.

A. Terminal screw

Slot to
release
wire

Hole for
wire

Wire gauge

B. Back wired

Figure 23-30 Making connections to receptacles.

connection of the device. When connecting a
duplex receptacle, the following rules apply:

1. White neutral wire connects to silver
   terminal,
2. Black live wire connects to brass terminal,
   and
3. Bare copper ground wire connects to green
   terminal.

To terminate the end of an insulated wire
around a terminal screw on a receptacle,
approximately 7/8 in. of the insulation around
the wire must be removed and a terminal eye
formed. If you prefer back wiring, use the strip
gauge on the back of the receptacle for removing the insulation, then simply insert the end in
the proper hole. To remove the wire, press the
end of a small screwdriver into the slot next to
the hole to release the wire (Figure 23-30).

To determine whether a receptacle is properly
polarized, use an ac voltmeter and proceed as
follows (Figure 23-31).

1. Check the voltage between the two parallel
   slots on the receptacle. Line voltage should
   be read.

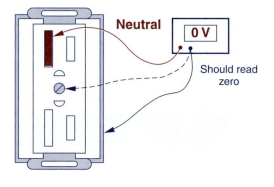

Neutral   0 V

Should read
zero

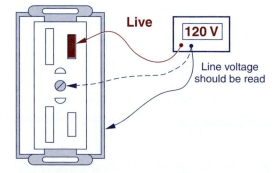

Live   120 V

Line voltage
should be read

Figure 23-31 Checking for correct polarization of a
receptacle.

2. Check the voltage between the wide neutral slot and the electric-outlet box or the machine screw securing the cover plate. Voltage should be zero.
3. Check the voltage between the narrow live slot and the electric-outlet box or the machine screw securing the cover plate. Line voltage should be read.

23-8 Lampholders

Incandescent lamps operate in what the Code calls lampholders, or what many people refer to as lamp sockets. Lampholders are available in a large variety of types. One of the simplest types is the *keyless ceiling lampholder* (Figure 23-32A). The body of this lampholder is made of porcelain or bakelite and the two connection terminals are color coded to assist with the safe connection of the device.

The Code requires the live black wire to be connected to the brass-colored terminal screw (inside, this brass terminal joins with the center contact). The white neutral wire must be connected to the silver-colored terminal screw (inside, this silver terminal joins with the threaded shell of the lampholder). This connection will prevent an electric shock in the event that a person replacing a lamp makes electric connection with the shell of the lampholder via the threaded base of the lamp (Figure 23-32B).

A lampholder with a built-in switch is known as a key-type lampholder. One of the most popular key-type lampholders used in house wiring is the *pull-chain lampholder* (Figure 23-33). The Code requires that the pull-chain switch mechanism have an insulating link in the chain to prevent connection with the 120-V circuit should a defect occur in the switch. Internally, the switch is connected in series with the brass-colored terminal. Wiring is similar to that of the keyless lampholder with the white neutral wire connected to the silver terminal and the live hot wire connected to the brass terminal.

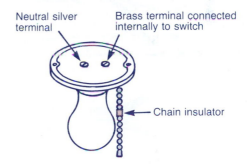

Figure 23-33   Pull-chain lampholder.

Light fixtures are made already wired (Figure 23-34). Solderless connectors are used to connect the outlet wires to the fixture wires. The wires connect black-to-black and white-to-white. In case the identification of the fixture wires is not clear, trace the wires into the fixture: the one that connects to the screw shell of the socket is for connection to the white neutral wire.

23-9 Switches and Dimmers

Switches approved for use in house-wiring circuits are rated for the maximum number of amperes the switch is designed to interrupt along with its maximum-voltage rating. For example, a switch may be stamped "15-A 125-V" (Figure 23-35). This means that the switch may be used to interrupt a circuit that has a maximum of 15 A of current at 125 V

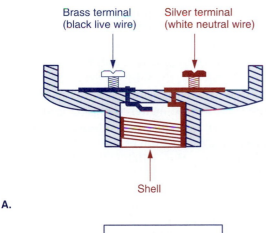

A.

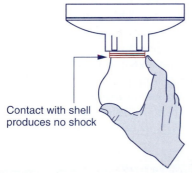

B.

Figure 23-32   Keyless lampholder connection.

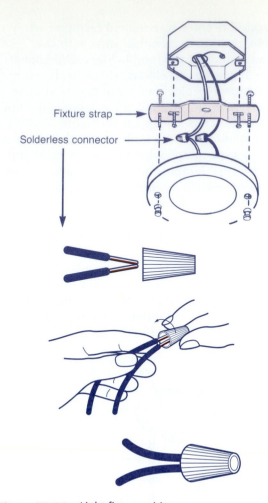

Figure 23-34   Light fixture wiring.

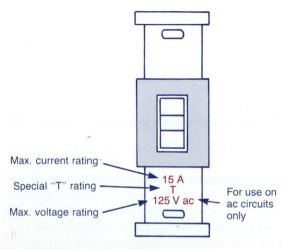

Figure 23-35   Switch ratings.

flowing through the switch contacts. The letters "AC" stamped on the switch mean that it can be used only on ac circuits. Switches marked with the letter "T" (tungsten) are designed to handle the heavy surge of current that occurs when voltage is first applied to a circuit containing an incandescent lamp.

When aluminum wire was first used, it was connected to ordinary terminals that were quite suitable for copper, but it soon became evident that those terminals were not suitable for aluminum; the connections heated badly and led to loose connections, excessive heating, and even fires. If your wire is made of pure aluminum, check receptacles and switches to make sure that their terminals are marked CO-ALR for 15-A and 20-A devices or CU-AL for higher ratings. If they are not, replace the outlets with the correct type. Also, see that terminal screws are screwed tightly—aluminum wires expand and contract a good deal, and thus, they can work loose. Tight attachment is necessary so that the terminal screws penetrate the oxidation to permit the smooth flow of electricity.

Single-Pole Switch   A *single-pole, single-throw (SPST) switch* is used in house wiring to control light(s) from one point. When this switch is pushed to the ON position, the circuit is completed between the two terminal points, allowing current to flow through the switch (Figure 23-36A). In the OFF position, the contact between the two terminal points is broken, opening the internal-switch circuit (Figure 23-36B). These switches are fastened to the outlet box so that the handle of the switch is moved up to turn the switch ON and down to turn if OFF.

Double-Pole Switch   A *double-pole, single-throw (DPST) switch* is used on 240-V heavy-duty house circuits. This voltage requires two live wires and a switch capable of opening both wires at the same time.

The double-pole switch is similar in construction to a pair of single-pole switches. Two sets of terminal screws marked "line" and "load" are provided, and the handle of the switch is identified as being ON or OFF (Figure 23-37).

Three-Way Switch   A pair of *single-pole, double-throw (SPDT) switches,* commonly referred to in house wiring as *three-way switches*, must be used to control light(s) from

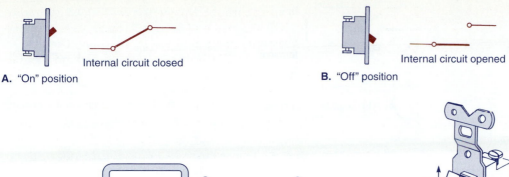

A. "On" position
Internal circuit closed

B. "Off" position
Internal circuit opened

(Note: Ground wires are not shown.)

120 V ◁
To distribution panel

On

C. Typical connection

Figure 23-36 Single-pole, single-throw switch.

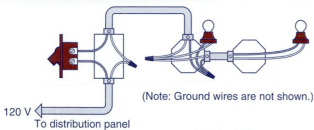

A. "On" position internal contacts closed

Line

B. "Off" position internal contacts open

Load

Figure 23-37 Double-pole, single-throw switch.

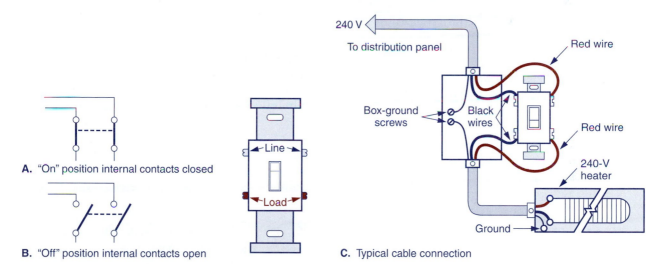

240 V ◁
To distribution panel

Red wire

Box-ground screws

Black wires

Red wire

240-V heater

Ground

C. Typical cable connection

two positions. Common examples include control of hall or stairway lights.

The internal circuit of this device is such that it allows current to flow through the switch in either of its two positions (Figure 23-38). For this reason there are no ON/OFF marks on the operating handle. A three-way switch is recognized by its three terminals. The common terminal is identified by the word "line," or by a differently colored terminal screw. *When*

replacing this type of switch be careful to connect the line wire to the proper terminal screw.

Four-Way Switch To control light(s) from three different positions you need a pair of three-way switches and a ***four-way switch***. The four-way switch, like the three-way switch, allows current to flow through the switch in either of its two positions (Figure 23-39). For this reason there are no ON/OFF marks on the operating handle.

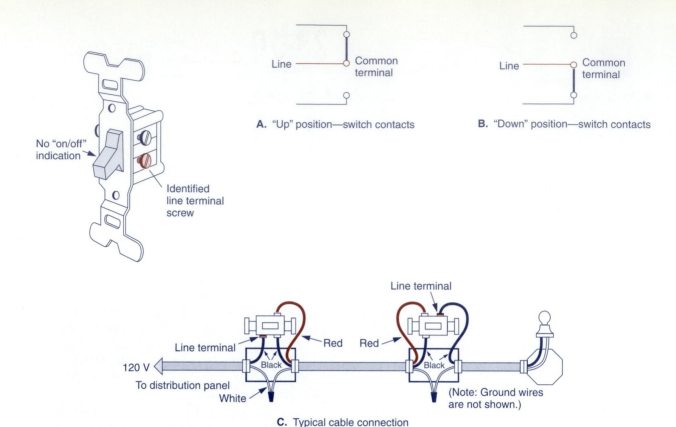

A. "Up" position—switch contacts

B. "Down" position—switch contacts

C. Typical cable connection

120 V ◁

To distribution panel

Line terminal

Line terminal

Red Red

Black Black

White

(Note: Ground wires are not shown.)

No "on/off" indication

Identified line terminal screw

Figure 23-38 Three-way switch.

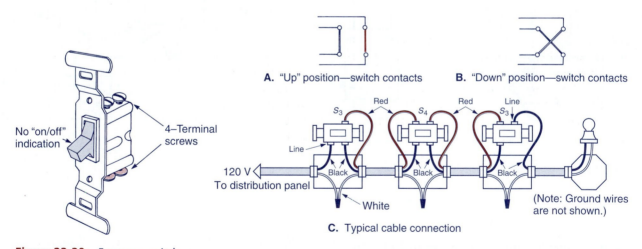

A. "Up" position—switch contacts

B. "Down" position—switch contacts

S_3 S_4 S_3

Red Red Line

Line

Black Black Black

White

(Note: Ground wires are not shown.)

120 V ◁

To distribution panel

C. Typical cable connection

No "on/off" indication

4–Terminal screws

Figure 23-39 Four-way switch.

Lamp Dimmers   Lamp dimmers are used to control the brightness of lights in addition to being used to turn them ON and OFF (Figure 23-40). They are sealed, electronic-controlled units that vary the amount of voltage applied to the light to control brightness.

Dimmers designed for use with incandescent lights are quite inexpensive and, if used properly, conserve energy. They are rated as to the maximum-power rating of load(s) that can be connected to them and are available in types designed to replace both single-pole and three-

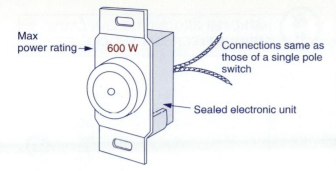

Figure 23-40 Single-pole dimmer switch.

Max power rating → 600 W

Connections same as those of a single pole switch

Sealed electronic unit

way switches. A dimmer that is made for incandescent lighting should never be used with ceiling fans or fluorescent lighting. There are special dimmers that are used with these types of circuits.

Combination Switches A pilot-light switch (Figure 23-41) has a built-in bulb that glows when power flows through the switch to a load. Pilot-light switches are often installed for convenience when the device being operated cannot be seen from the switch location. Attic lights and fans are often controlled by pilot-light switches. This switch requires a neutral wire connection.

Also available are combination switch/receptacle units that allow an installer to locate the switch and receptacle in a single gang box. The switch can be wired to control the receptacle portion of the unit or another remote light or device. Like the pilot light switch, the switch/receptacle requires a neutral wire connection. A switch box that contains a single two-wire cable has only hot wires, and cannot be fitted with a switch/receptacle.

23-10 Grounding System

The purpose of *grounding* is safety. In a normal house electric circuit, current flows to an appliance or light through a live wire and flows back to the distribution panel through a white neutral wire. The live wire is charged with voltage. The neutral one is at zero, the voltage of the earth—in fact, the neutral wire is connected to the earth. Any deviation from this normal path is dangerous, and to protect you and your home against these hazards, electrical codes require a safety system called grounding, which keeps every outlet box and cover plate at zero voltage.

In general, grounding guards against two hazards—*fire* and *shock*. A fire hazard can occur when current leaks from a broken live wire or connection and reaches a point of zero voltage by some path other than the normal one. Such a path offers high resistance, so that the current can generate enough heat to start a fire.

A shock hazard generally arises when there is little or no leaking current, but the potential for abnormal current flow exists. If a bare live wire touched the cover plate of a switch or a receptacle and the cover plate was not grounded, the voltage of the live wire would charge the plate. If you then touched the charged plate, your body could provide a current path to zero voltage, and you could suffer a serious shock.

Grounding refers to the deliberate connection of parts of the house wiring installation to a common earth connection. In order for this protection system to work, *both* the electric current-carrying conductor *system* and the circuit *hardware* (or boxes) must be grounded. In a properly grounded system, a direct ground

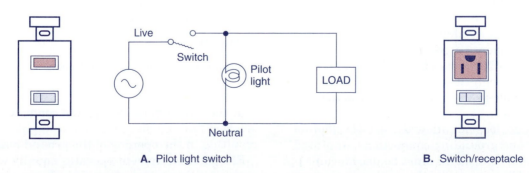

A. Pilot light switch

B. Switch/receptacle

Figure 23-41 Combination switches.

fault produces a short-circuit current surge. This current blows a fuse or trips a circuit breaker to immediately open the circuit (Figure 23-42).

The *neutral white wire* is used to ground the electric *current-carrying system*. This wire is connected to ground at the main service entrance switch box by a ground cable that connects to the house grounding system (Figure 23-43). The grounding wire is usually a No. 6 or larger. Article 250 of the Code deals with grounding requirements. One important Code requirement is that *the neutral wire never be fused or switched*. This wire must run to all electric outlets without interruption to ensure that the ground circuit is complete regardless of any other circuit operating conditions.

The bare ground wire in the nonmetallic sheath cable is used to ground the normal non-current-carrying electrical hardware of the system. This includes all metal boxes and receptacles! The ground wire must be run without interruption to all outlet boxes and be securely connected to the ground-screw terminal of the box. In duplex receptacle circuits, the bare ground must first connect to the ground-screw terminal of the box and then on to the green hexagon-head ground terminal on the receptacle (Figure 23-44).

Bonding refers to the permanent joining of metallic parts to form an electrically conductive path. This is accomplished by installing bonding jumper wires. Bonding ties equipment together to keep voltage buildup on the equipment the same, so there is no difference in potential between the various pieces of equipment. Bonding equipment together does not necessarily mean that this equipment is properly grounded. A grounding conductor must be installed as part of the circuit to connect the equipment to a grounding electrode.

At the main service, the grounded neutral conductor must be bonded to the metal enclosure. For most residential panels, this main bonding jumper is a *green* bonding screw that is furnished with the panel. This bonding screw is inserted through the neutral bar into a threaded hole in the back of the panel itself (Figure 23-45).

The Code does not allow the use of solder for bonding and grounding connections. Approved clamps and connectors must be used for this purpose. The reason for this is that in the event

IMPROPER GROUNDING CAN CAUSE SERIOUS ELECTRICAL SHOCK!

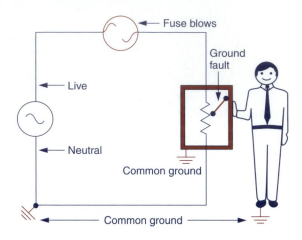

- Ground fault produces a short circuit which blows the fuse
- No shock is received from touching the metal frame

A. Properly grounded circuit

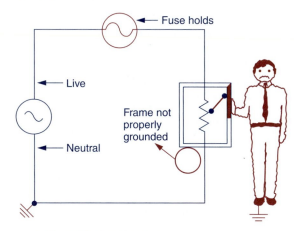

- Ground fault produces no abnormal current flow
- Fuse holds
- Shock is received by touching the metal frame and ground

B. Improperly grounded circuit

Figure 23-42 Grounding for protection.

that the circuit is called upon to conduct high levels of fault current, the solder might melt, resulting in the opening of the ground path.

Electrical equipment operates equally well with or without grounding. This fact sometimes

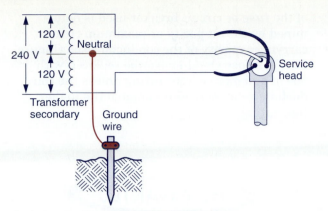

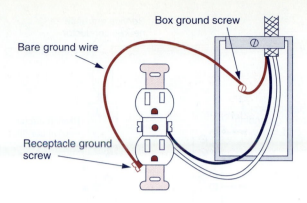

A. End of the run

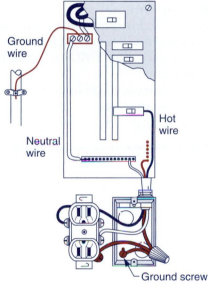

B. Main service and outlet box

A. At the transformer on the utility pole

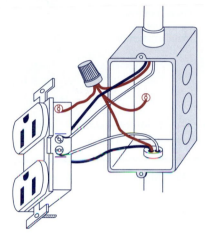

B. Middle of the run

Figure 23-44 Outlet box ground connection.

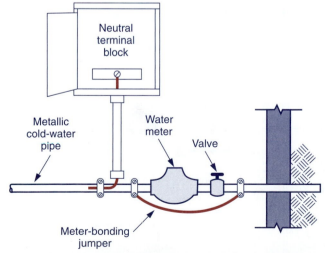

C. Grounding to a water pipe

Figure 23-43 Grounding the electrical system.

leads to carelessness or neglect of the all-important matter of proper and sufficient grounding. Always remember, the purpose of grounding is *safety*.

23-11 Overcurrent Protection

The purpose of fuses or circuit breakers used in house-wiring systems is to limit the amount of electric current (amperes) that can be passed through a given conductor. Overloading a wire beyond its rated current capacity will cause the wire to overheat, creating a potential fire hazard.

These overcurrent protective devices are connected in *series* with the *live* conductors and are designed to open the circuit when their rated current capacity is exceeded. The current rating

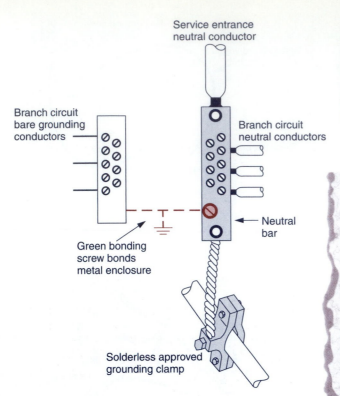

Service entrance
neutral conductor

Branch circuit
bare grounding
conductors

Branch circuit
neutral conductors

Neutral
bar

Green bonding
screw bonds
metal enclosure

Solderless approved
grounding clamp

Figure 23-45   Main service grounding connections.

of the *fuse* or circuit breaker used is determined by the ampacity or maximum current-carrying capacity of the conductor used (Figure 23-46). A fuse should never be replaced with one of a higher current rating unless the conductor size is increased to match the higher fuse rating.

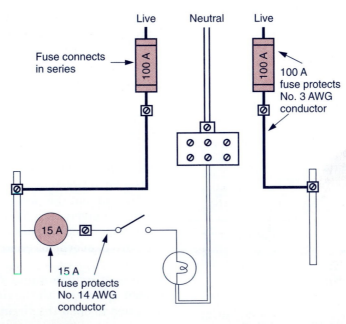

Live   Neutral   Live

Fuse connects
in series

100 A

100 A

100 A
fuse protects
No. 3 AWG
conductor

15 A

15 A
fuse protects
No. 14 AWG
conductor

Figure 23-46   Fuse connection and rating.

CHAPTER 23

Review and Applications

Review Questions

1. What does the Underwriters' Laboratories label on electrical products indicate?
2. Explain what the National Electrical Code is and how it applies to house wiring.
3. What is the role of the local electrical inspector with regard to house wiring?
4. Explain the purpose of each of the following parts of a consumer's service:
   (a) kilowatthour meter
   (b) main-service switch or circuit breaker
   (c) distribution panel
5. State the two different voltage levels that can be obtained from a three-wire distribution system and the application for each level.
6. In general, what is the minimum service size required for a one-family home?
7. List three advantages of a circuit breaker distribution panel over the fused type of panel.
8. Why is a double-pole circuit breaker required for 240-V branch circuits?
9. Rank rigid conduit, nonmetallic sheathed cable, and armored cable wiring with regard to:
   (a) protection for conductors
   (b) costs of installation
10. Name the three types of nonmetallic sheathed cable and the application for each.
11. Describe the "loop" system of wiring used in homes today.
12. At what points should nonmetallic sheathed cable be supported or secured?
13. How much free insulated wire must be provided in each outlet box for making electric connections?
14. Why should nonmetallic sheathed cable be kept at least 1-1/4 in. from the outer edge of wooden studs?
15. In general, list purposes of outlet boxes as part of an electrical system.
16. State two uses for octagonal outlet boxes.
17. What two electrical devices are housed in device boxes?
18. How does a cable normally enter an outlet box?
19. What factors determine the number of insulated conductors allowed per outlet box?
20. State the color-code rules that apply when connecting a three-wire ground duplex receptacle.
21. To what part of a lampholder does the white neutral wire connect?
22. Why does the Code require an insulating link on the chain of a pull-chain lampholder?
23. What does 15-A 125-Vac that is stamped on a switch mean?
24. State the type of switch(es) that would most likely be used for each of the following control applications:
   (a) two lamps controlled from one position
   (b) switching a 240-V baseboard heater ON and OFF
   (c) one lamp controlled from two positions
   (d) one lamp controlled from three positions
25. In general, what two types of hazards does grounding guard against?
26. State the function of the ground wire and neutral wire with regard to the grounding protective system.
27. What does the Code have to say about the running, fusing, and switching of the neutral wire?
28. (a) What is the purpose of fusing electric circuits in houses?
   (b) What factor determines the size of circuit breaker or fuse that can be used in a particular circuit?
   (c) How are circuit breakers and fuses connected with regard to the circuits they are protecting?

Problems

1. Refer to Figure 23-3.
   (a) How many 120 V circuit breakers could be installed in the panelboard?
   (b) How many 120 V circuit breakers are installed?
   (c) How many 240 V circuit breakers are installed?

2. Refer to Figure 23-10. What connection of lamps is used?

Critical Thinking

1. Refer to Figure 23-5. Under normal conditions what is the value of the voltage between the neutral wire and ground? Why?
2. Refer to Figure 23-6. Assume one of the two main fuses is blown. With the main switch closed, how much voltage would be measured on the load size:
   (a) across the two fuses?
   (b) from the blown fuse to the neutral block?
   (c) from the good fuse to the neutral block?
3. Refer to Figure 23-31. Assume the voltmeter indicates a 120 Vac reading when connected between the neutral slot and the machine screw securing the cover plate. What wiring error does this indicate?
4. Refer to Figure 23-36. Assume a three-way switch is to be used to replace the single-pole switch shown in the circuit. What two terminals of the three-way switch would have to be used?
5. Refer to Figure 23-40. Assume the dimmer switch shown is to be connected to control a 1000 W incandescent lighting load. What problem could develop? Why?

Portfolio Project

Prepare a branch circuit directory of your home's electrical system. List which rooms and appliances are protected by which fuses or circuit breakers.

Circuit Challenge

Using a Simulator

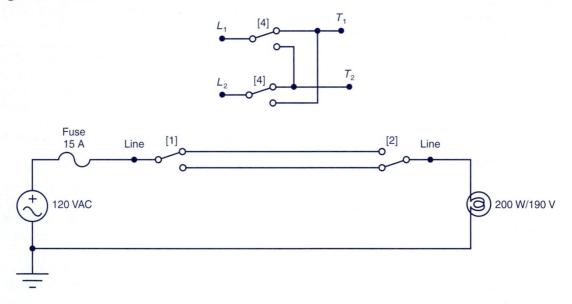

Procedure

(a) Wire the three-way switch circuit shown, using whatever simulation software package is available to you.

(b) Connect the four-way switch into the circuit to control the lamp from three locations.

Residential Branch Circuit Wiring

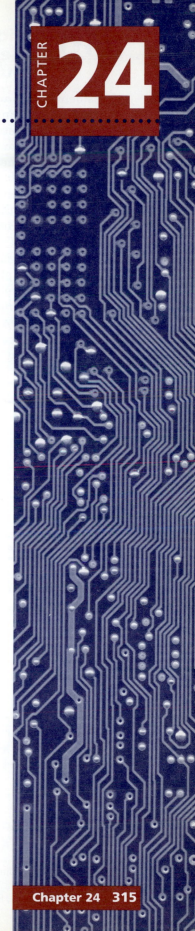

*T*he residential wiring techniques described in this chapter follow the guidelines of the National Electric Code (NEC) and common trade practices. Always be careful when working with electric circuits. In careless hands, the 120 volts used in residential circuits can be a swift killer. Always treat every circuit as live until proven otherwise.

Objectives

After studying this chapter, you should be able to:
1. Define the term branch circuit as it applies to the NEC.
2. Outline NEC branch circuit requirements for different areas of a home.
3. Describe the physical features of a split receptacle.
4. Complete typical branch circuit schematic and wiring diagrams.
5. Wire typical branch circuits in accordance with the guidelines of the NEC.

Key Terms

- branch circuit
- duplex receptacle
- GFCI
- split-link receptacle
- dryer receptacle

- range receptacle
- lampholder
- schematic diagram
- wiring diagram
- single-line diagram

- number sequence chart
- three-way switch
- four-way switch

- Hubbell
- National Electrical Manufacturers Association (NEMA)
- Paragon Electric

24-1 National Electrical Code Branch Circuit Requirements

A residential **branch circuit** consists of all the parts of a circuit in the distribution panel that lie beyond the overcurrent device (fuse or circuit breaker) up to the last outlet on the circuit. The majority of these branch circuits supply voltage to the lights and small appliances in the home. These circuits are generally wired using nonmetallic sheathed *cables* with a No. 12 *AWG* wire with 20-A protection or a No. 14 AWG with 15-A protection. There are different codes used for older installations that used aluminum wiring. For example, No. 12 AWG aluminum wiring would have a 15-A protection.

A maximum of twelve outlets may be wired on a single branch circuit as illustrated in Figure 24-1, except in locations such as the kitchen, laundry, and utility room. Each light fixture and each **duplex receptacle** is counted as an outlet (switches are not counted). Since a 15-A branch circuit cannot safely supply loads where the total power rating exceeds 1500 W, it is advisable to calculate the power rating of all appliances likely to be used in the branch circuit. If the total rating is more than 1500 W, add more branch circuits and/or decrease the number of outlets per circuit. In general, no more than 8 to 10 outlets (depending on the load of each) should be included in one branch circuit.

According to Section 210-23 of the Code, the permissible loads on branch circuits can be summarized as follows:
- The load shall not exceed the branch-circuit rating or 80 percent if a continuous load (3 hours or more).
- The branch-circuit must be rated 15, 20, 30, 40, or 50 amperes when serving two or more outlets.

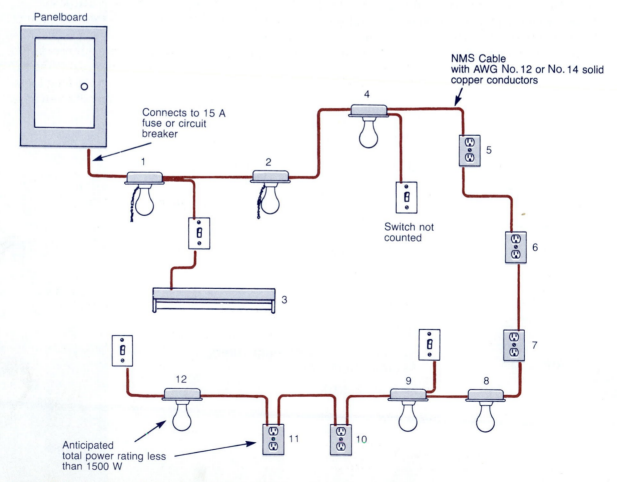

Figure 24-1   Maximum of 12 outlets are allowed per branch circuit.

- An individual branch circuit can supply any size load.
- Both 15 and 20 ampere branch-circuits:
  a. Can supply lighting, other equipment, or both types of loads.
  b. Cord-and-plug connected equipment shall not exceed 80 percent of the branch-circuit rating.
  c. Equipment fastened in place shall not exceed 50 percent of the branch-circuit rating if the branch-circuit also supplies lighting, other cord-and-plug connected equipment, or other types of loads.

A *separate circuit* for one duplex receptacle must be installed in the laundry or utility area. This receptacle must be located within 6 feet of the intended location of the washing machine. The circuit must be a 20-ampere circuit.

Except in kitchens, bathrooms, hallways, utility areas, and unfinished basements, duplex receptacles are required so that there is no more than 6 ft of usable wall space between receptacles. In other words, if a lamp with a 6-ft cord is placed anywhere in the room, the cord must be able to reach a receptacle.

A duplex receptacle installed in each washroom and at counter height in each bathroom must be protected by a **ground-fault circuit interrupter (GFCI).**

At least one duplex receptacle in an outdoor location must be installed and be readily accessible for appliances used outdoors. A separate circuit is required for receptacles installed outdoors and the receptacle also must be protected by a GFCI. Outlet boxes and covers for receptacles must be weatherproof.

At least one duplex receptacle shall be provided in a garage. Receptacles in garages shall be supplied from a separate circuit or may be connected to the circuit that feeds the outdoor receptacles, since it is on a GFCI.

Rooms in which food is prepared or served require receptacles for small appliances such as mixers and toasters. The Code calls for two or more 20-A small appliance branch circuits to serve all receptacles in a kitchen, pantry, and dining room, and in other eating or food preparation areas. Kitchen countertop areas must have receptacles on two separate circuits with no point along the counter space being more than 24 in. from a receptacle outlet. **Split-link receptacles** can be used for this purpose. These receptacles allow wiring two separate hot wires in the duplex unit. This is made possible by breaking the metal link that connects the two parts of the receptacle. Figure 24-2 shows this type of receptacle. The separate feed to each outlet consists of two live conductors (one

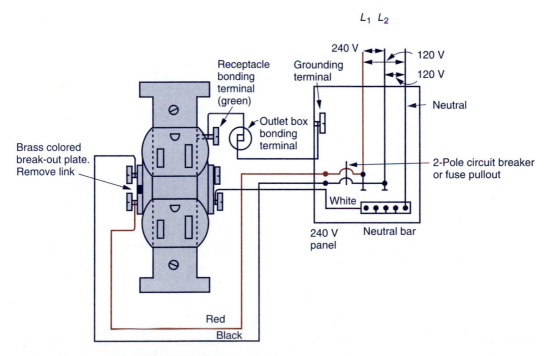

Figure 24-2 Kitchen split-receptacle connection.

red, one black), and a neutral conductor (white). Three-wire circuits must be protected by fuses in pull-out blocks or circuit breakers. When manually operated, these fuses or *circuit breakers* disconnect the two live conductors at the same time.

The kitchen small appliance circuits are provided to serve cord-and-plug appliances. They are not permitted by the Code to serve lighting fixtures or appliances such as dishwashers, garbage disposals, or exhaust fans. Also, all 120 V receptacles that serve kitchen countertops, must have GFCI protection.

An approved 50-A, 120/240-V, three- or four-wire receptacle must be installed to supply electricity to a free-standing electric range that is connected by an approved cord and cap. Wiring installed for an electric clothes dryer must terminate in an approved receptacle (most dryers require a 30-A, 120/240-V, three- or four-wire receptacle). (See Figure 24-3.)

Individual *range* and *dryer circuits* are designed to supply both 120-V and 240-V circuits. Generally, for a range, the heating elements operate at 240 V, while the clock, timer, lights, and so on operate at 120 V. Likewise, the

Surface mount range receptacle   Wall mount range receptacle

Typical 30-A, 240-V Three-Wire Dryer Receptacle

Typical 50-A, 240-V Three-Wire Range Receptacle

Typical 50-A, 240-V Four-Wire Range Receptacle

Range plug

A. Range and dryer receptacles have different current ratings as well as plug configurations

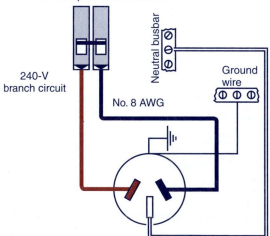

B. Typical range circuit schematic

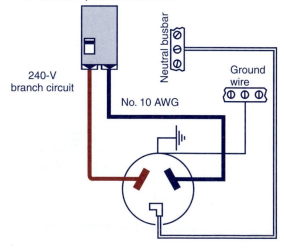

C. Typical dryer branch-circuit schematic

Figure 24-3 Individual range and dryer circuits.

heating elements in the dryer require 240 V, while the drum motor, timer, lights, and so on operate at 120 V.

Branch circuits for 240-V devices (e.g., ranges, dryers, water heaters, split receptacles, permanently installed heaters, and 240-V pump motors) require fuse pullouts or circuit breakers that simultaneously disconnect the live conductors in the circuit. Disconnection is accomplished by a two-pole circuit breaker, a handle-tie on two single-pole circuit breakers, or a two-pole fuse pullout. All branch circuits must be identified in the panelboard with the name of the device supplied and the permitted maximum-current rating of the overcurrent protection.

In order to maintain the identity of insulated wires, the Code has established a color code for these conductors. The color code is as shown in Table 24-1 and applies to all branch circuit wiring in a home.

TABLE 24-1	Branch Circuit Color-Code Chart
Identified Conductor	**Color**
Neutral wire	White or natural gray
Live or hot wire	Black or red
Insulated ground wire	Green

24-2 Pull-Chain Lampholder and Duplex Receptacle Circuit

The first branch circuit we will discuss consists of a pull-chain *lampholder* and duplex receptacle. Each device is connected directly across the two supply-line wires and they operate independently of each other. The circuit breaker is connected in series with the live wire and the neutral wire is connected directly to the receptacle and lampholder. The *schematic diagram* for this circuit is shown in Figure 24-4A.

When this branch circuit is wired in a home, the circuit layout is not as it is represented in the schematic diagram. Two- or three-conductor cable is used, instead of single conductors. Also, the different devices are not located in the positions shown in the schematic diagram. As a result, many different wiring diagram layouts are possible for any given schematic layout.

Before making a wiring diagram of the circuit, it is helpful to make a wiring *number-sequence chart* for the circuit that includes a wire color-code chart. The procedure followed is similar to that used for signal circuit wiring projects. First, the terminal of each component in the circuit is assigned a number (Figure 24-4B). Then, the common terminals that are to be connected together are recorded in chart form. Each group of terminal connections is then assigned a color in accordance with Code requirements.

The wiring diagram shows the different components and cable runs as they are located in the actual circuit (Figure 24-4C). In this particular layout, the voltage-source cable runs from the distribution panel to the receptacle-outlet box and from there to the lampholder box. When simulating the wiring of this circuit in the classroom, a U-ground plug cap may be connected to the end of the voltage-source cable to act as the supply feed.

The wiring diagram can be completed with the aid of the schematic diagram and the wiring number-sequence chart. Each component is numbered according to the schematic diagram and connected according to the wiring number-sequence chart. The bare-ground wire, while not part of the circuit schematic, must be connected to each outlet box and convenience receptacle according to NEC requirements.

24-3 Switch and Lampholder Circuit

One of the most common house light-control circuits is that of a single-pole switch that controls a single lampholder. This circuit is often used in bedrooms, kitchens, and bathrooms. According to NEC requirements, the switch must be connected in series with the live or hot wire, while the neutral wire is run directly to the lampholder. The schematic diagram and typical wiring number-sequence chart for the circuit is shown in Figure 24-5A.

The first wiring diagram layout of this circuit has the voltage-source cable run into the switch outlet box and then on to the lampholder box (Figure 24-5B). With this arrangement of cables, the circuit is easily wired following the schematic diagram and the wiring-sequence chart.

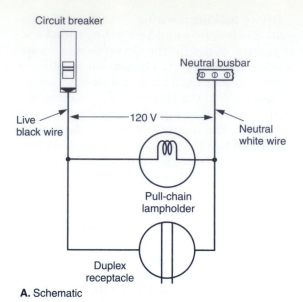

Circuit breaker

Neutral busbar

Live
black wire

120 V

Neutral
white wire

Pull-chain
lampholder

Duplex
receptacle

A. Schematic

Circuit breaker

Neutral busbar

1 ◄—— 120 V ——► 2

3   4

P.C.

5   6

B. Wiring number-sequence chart

NUMBER SEQUENCE CHART	COLOR CODE*
1, 3, 5	Black
2, 4, 6	White

WIRING NOTES
Receptacle Wiring — White to silver terminal — Black to brass terminal — Bare copper to green terminal
Lampholder Wiring — White to silver shell terminal — Black to brass switch terminal

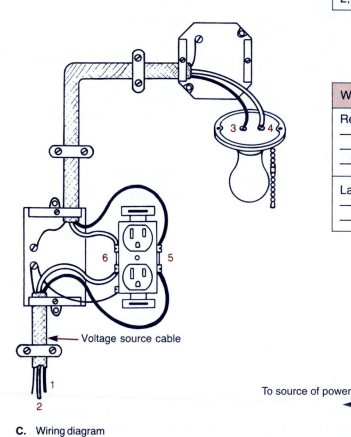

Voltage source cable

C. Wiring diagram

To source of power

Single-line diagram

* The black wires noted appear as solid blue wires in this text.

Figure 24-4   Wiring for a pull-chain lampholder and duplex receptacle circuit.

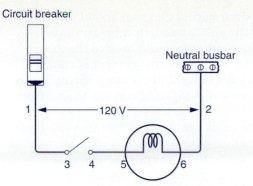

Circuit breaker

Neutral busbar

1 ← 120 V → 2

3 4 5 6

NUMBER SEQUENCE CHART	COLOR CODE *
1, 3	Black
4, 5	Black
2, 6	White

A. Schematic and wiring number-sequence chart

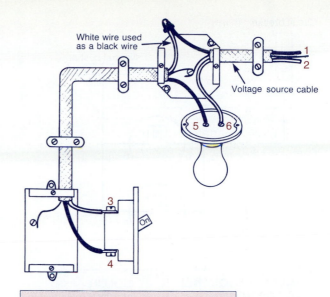

White wire used as a black wire

Voltage source cable

1
2

5   6

3

On

4

WIRING NOTES

Arrange connections so you end up with a black live and white neutral wire at the lampholder.
Mount switch so that the handle is "on" in the up position.
Extend the outer sheath of the cable beyond the cable clamp.

Single-line diagram

C. Wiring diagram—source cable run to lampholder-outlet box

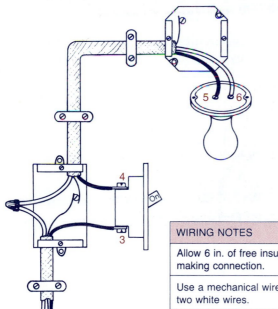

5   6

4

On

3

2   1

WIRING NOTES

Allow 6 in. of free insulated wire for making connection.
Use a mechanical wire connector to join the two white wires.
Wrap wire around terminal screws in a clockwise direction

Single-line diagram

B. Wiring diagram—source cable run to switch-outlet box

* The black wires noted appear as solid blue wires in this text.

Figure 24-5   Wiring for a switch and lampholder circuit.

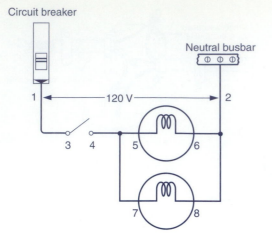

Circuit breaker

Neutral busbar

120 V

NUMBER SEQUENCE CHART	COLOR CODE*
1, 3	Black
4, 5, 7	Black
2, 6, 8	White

A. Schematic and wiring number-sequence chart

* The black wires noted appear as solid blue wires in this text.

The second wiring diagram layout of this circuit has the voltage-source cable run into the lampholder outlet box and then to the switch (Figure 24-5C). This arrangement of cables requires a deviation from normal wiring procedures. The wire supplying current to and from the switch should be black. But the two-conductor cable that is used contains one black and one white wire. Under this arrangement of cables, the Code will permit the use of the white wire as a black wire. The accepted procedure to follow is to connect the white wire from the source and the black wire from the switch outlet directly to the lampholder. The black wire from the source is then connected to the white wire from the switch to complete the connection. This connection of wires assures that you end up with a black live wire and white neutral wire at the lampholder. The same procedure is acceptable when using three-wire cable for three- or four-way switches.

24-4 Two Lamps and One Switch Circuit

Some house circuits require two lamps to be operated by a single switch. This circuit is often

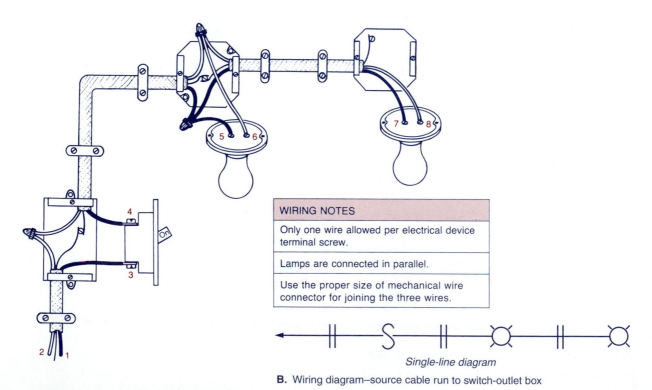

WIRING NOTES

Only one wire allowed per electrical device terminal screw.

Lamps are connected in parallel.

Use the proper size of mechanical wire connector for joining the three wires.

Single-line diagram

B. Wiring diagram—source cable run to switch-outlet box

Figure 24-6 Wiring for two lamps controlled from a single-switch circuit.

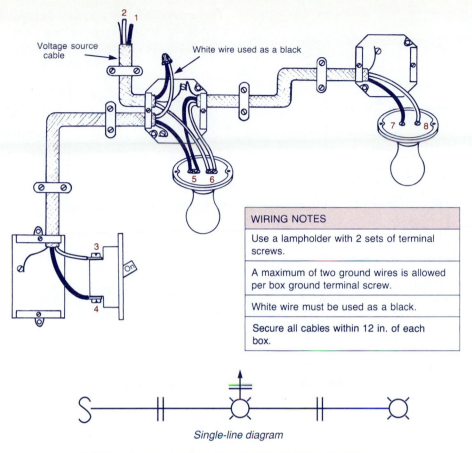

WIRING NOTES
Use a lampholder with 2 sets of terminal screws.
A maximum of two ground wires is allowed per box ground terminal screw.
White wire must be used as a black.
Secure all cables within 12 in. of each box.

Single-line diagram

C. Wiring diagram–source cable run to lampholder-outlet box

Figure 24-6   Continued.

used in large open areas of the home such as in a basement or a double garage.

When connecting two or more lamps controlled by a single switch, the parallel connection of lamps is used (Figure 24-6A). Connecting the lamps in parallel provides the full 120 V across each lamp when the switch is closed. Connecting the two lamps in series would result in less than the normal 120-V operating voltage across each light. A second advantage of the parallel connection of lamps is that they operate independently of each other so that if one burns out, the operation of the other is not affected.

The first wiring diagram layout of this circuit shows the voltage-source cable run into the switch-outlet box (Figure 24-6B). The wiring of the switch and first lampholder is the same as that of the previous branch circuit. A two-conductor cable is run between the two lampholders in order to connect the lamps in parallel.

The second wiring diagram layout of this circuit has the voltage-source cable run into the first lampholder-outlet box as in the previous branch circuit (Figure 24-6C). Again, the wiring is the same and a two-conductor cable is added to connect the two lamps in parallel.

24-5 Duplex Receptacle, Switch, and Lamp Circuit

The next circuit discussed calls for two lamps controlled by one switch and one uncontrolled duplex receptacle. This circuit is often used in a room where the lights are to be controlled by a switch but the receptacle is to be "live" at all times. The lamps and switch are connected as in the previous circuit, while the duplex receptacle is connected across the two supply-line wires. A schematic and wiring sequence chart for the circuit are shown in Figure 24-7A.

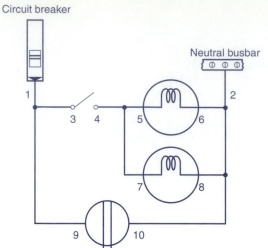

A. Schematic and wiring number-sequence chart

NUMBER SEQUENCE CHART	COLOR CODE*
1, 3, 9	Black and Red
4, 5, 7	Black
2, 6, 8, 10	White

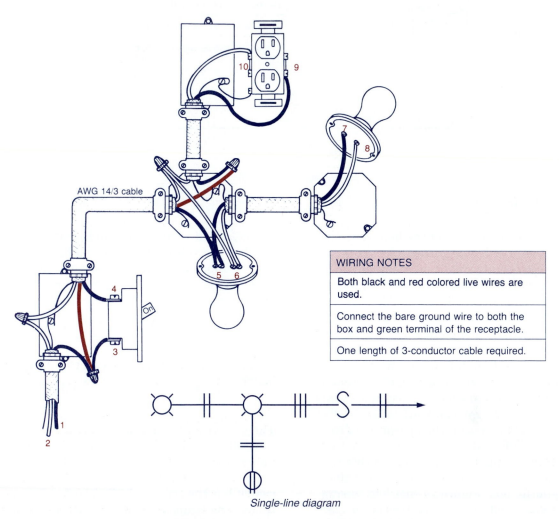

AWG 14/3 cable

WIRING NOTES

Both black and red colored live wires are used.

Connect the bare ground wire to both the box and green terminal of the receptacle.

One length of 3-conductor cable required.

Single-line diagram

B. Wiring diagram–source cable run to switch-outlet box

Figure 24-7   Wiring for duplex receptacle, two lamps, and switch circuit (continued on p. 329).

*The black wires noted appear as solid blue wires in this text.

C. Wiring diagram—source cable
run to lampholder-outlet box

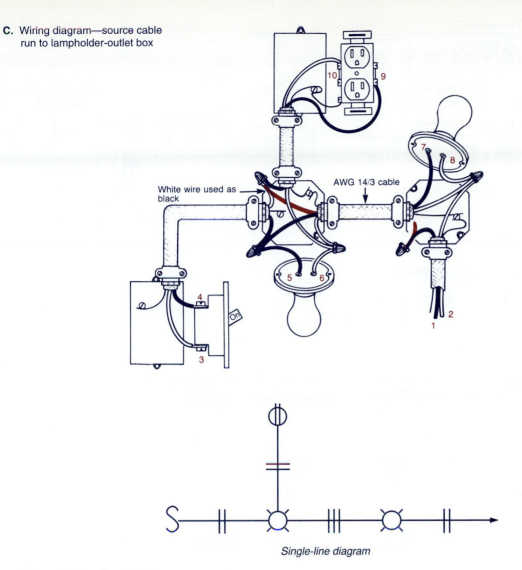

White wire used as black

AWG 14/3 cable

Single-line diagram

Figure 24-7 Continued.

Wiring this circuit requires the use of a three-conductor cable. This cable contains a black, white, and red colored conductor. According to the wiring layout shown in Figure 24-7B, the voltage-source cable is run into the switch-outlet box. The white neutral wire runs uninterrupted to both lamps and to the duplex receptacle. The live black wire is switched on its way to the lights and also runs directly to the receptacle by means of the red wire.

The wiring diagram in Figure 24-7C, is of the same circuit, but in this case, the voltage-source cable is run into the lampholder-outlet box. This arrangement of cables requires the

use of one length of three-conductor cable run between the two lampholder outlets.

24-6 Switched Split-Duplex Receptacle Circuit

Often it is desirable to have a switch that controls one half of a duplex receptacle. To do this, remove the link between the two live brass terminal screws of the receptacle. On the neutral side, the link is left intact to connect the two halves of the receptacle (Figure 24-8).

The wiring of a split receptacle, lamp, and switch circuit is shown in Figure 24-9. For this

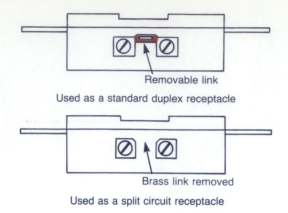

Removable link

Used as a standard duplex receptacle

Brass link removed

Used as a split circuit receptacle

A. Receptacle is "split" by removing the link between the two live brass terminal screws

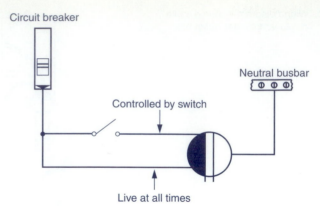

Circuit breaker

Neutral busbar

Controlled by switch

Live at all times

B. Circuit schematic

Figure 24-8 Switched split-duplex receptacle.

application, one half of the receptacle is wired so that it is alive at all times, while the other half is controlled by a switch. A portable lamp plugged into one half of the receptacle is controlled by the switch. At the same time, the other half is alive at all times for use with appliances such as vacuum cleaners, radios, or televisions.

24-7 Three-Way Switch Circuit

To turn lamps ON and OFF from two different locations requires the use of two ***three-way switches.*** The schematic circuit in Figure 24-10A is that of one lamp that is controlled from two positions. Note that the three-way switch has three terminal connections. The common terminal is shown as the center terminal on the schematic and is the identified terminal on the actual switch.

A typical wiring diagram of this circuit is shown in Figure 24-10B. A three-conductor cable must be run between the two switches, and, because of color limitations, the Code permits the use of the white wire as part of the live-wire switching circuit. However, it is important that you arrange your connections so that you end up with a white neutral and black (or red) live-colored wire at the lampholder terminals.

24-8 Four-Way Switch Circuit

To control a light from more than two locations, you must use a ***four-way switch*** with two three-way switches. The schematic circuit of Figure 24-11A is that of one lamp that is controlled from three positions. Note that the four-way switch has four terminal connections and no ON/OFF markings on the handle.

A typical wiring diagram of this circuit is shown in Figure 24-11B. The four-way switch is installed between the two three-way switches. Moving the handle of the four-way switch reverses the connections from one pair of traveler wires to the other. Traveler wires are those that start at one switch and end at the other. Different brands of four-way switches have different switching arrangements. If you are not sure of the type, use an ohmmeter to verify the switching arrangement. To control a light from more locations, install another four-way switch between the two three-way switches for each additional location.

24-9 Planning Your Electrical System

In planning an electrical installation it is recommended that a circuit layout of the type shown in Figure 24-12, be made. This will

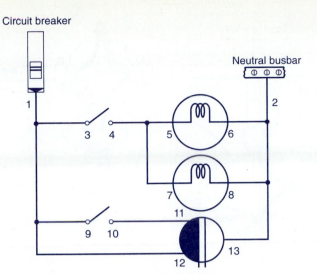

Circuit breaker

Neutral busbar

NUMBER SEQUENCE CHART	COLOR CODE*
1, 3, 9, 12	Black or Red
4, 5, 7	Black or Red
10, 11	Black or Red
2, 6, 8, 13	White

A. Schematic and wiring number-sequence chart

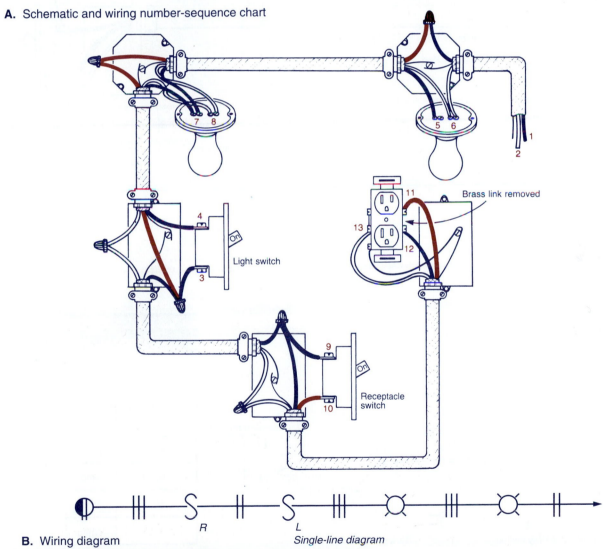

Brass link removed

Light switch

Receptacle switch

B. Wiring diagram

R

L

Single-line diagram

*The black wires noted appear as solid blue wires in this text.

Figure 24-9 Wiring for a split receptacle, lamp, and switch circuit.

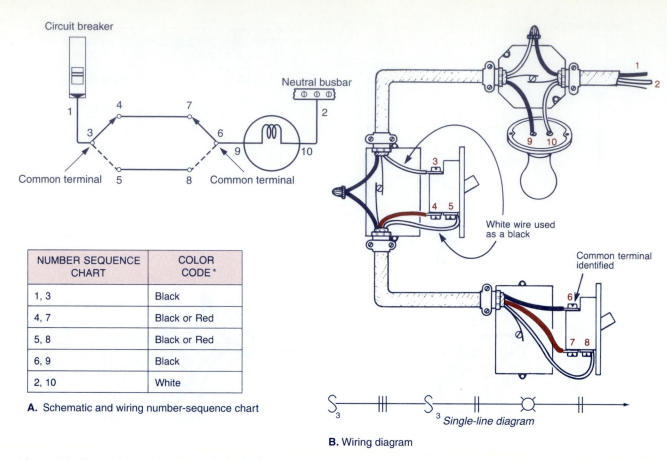

NUMBER SEQUENCE CHART	COLOR CODE*
1, 3	Black
4, 7	Black or Red
5, 8	Black or Red
6, 9	Black
2, 10	White

A. Schematic and wiring number-sequence chart

B. Wiring diagram

Figure 24-10 Wiring a three-way switch circuit.

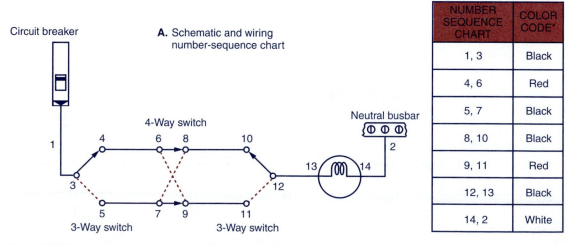

A. Schematic and wiring number-sequence chart

NUMBER SEQUENCE CHART	COLOR CODE*
1, 3	Black
4, 6	Red
5, 7	Black
8, 10	Black
9, 11	Red
12, 13	Black
14, 2	White

Figure 24-11 Wiring a four-way switch circuit (continued on p. 329).

*The black wires noted appear as solid blue wires in this text.

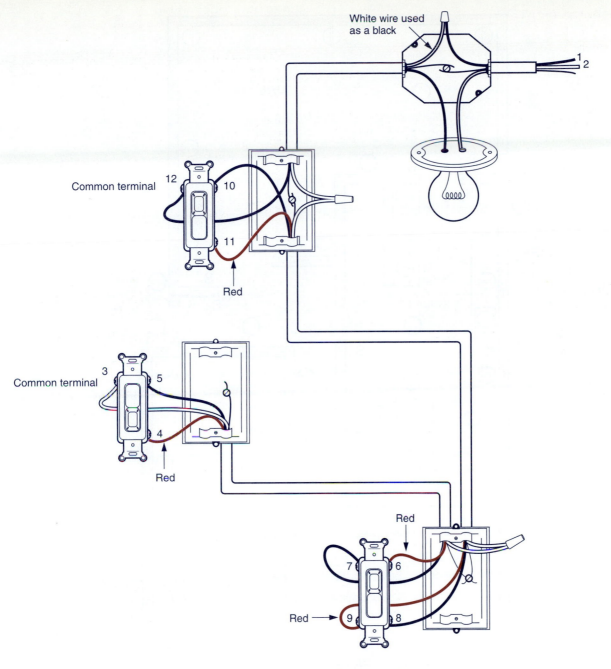

White wire used as a black

Common terminal 12   10

11

Red

Common terminal 3   5

4

Red

Red

7   6

Red → 9   8

Four-way switch

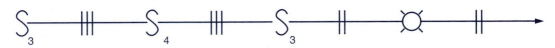

B. Wiring diagram

Figure 24-11   Continued.

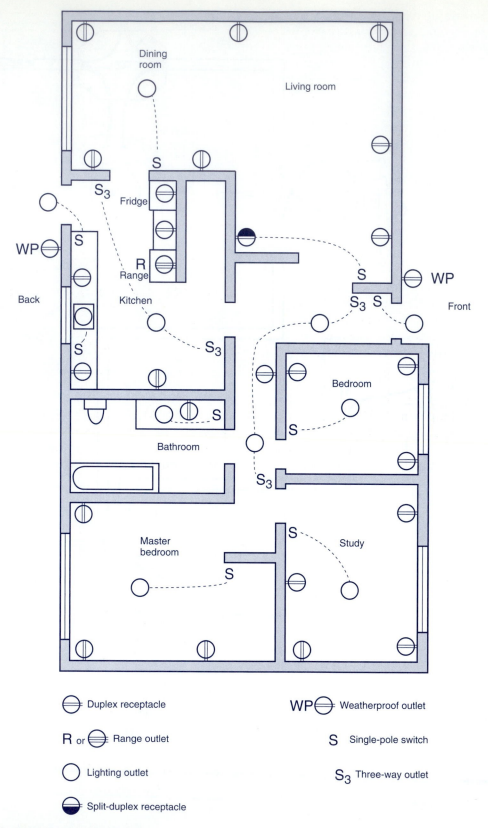

Dining room

Living room

Fridge

S

S₃

WP

S

Back

R
Range

Kitchen

S

S₃

Front

S

WP

S₃ S

S

S₃

Bedroom

S

Bathroom

S₃

Master
bedroom

S

S

Study

S

⊖ Duplex receptacle

WP⊖ Weatherproof outlet

R or ⊜ Range outlet

S Single-pole switch

◯ Lighting outlet

S₃ Three-way outlet

⬤ Split-duplex receptacle

Figure 24-12 Typical circuit layout plan.

assist in determining basic requirements such as the number of circuits and quantity of wire and materials needed. A dotted line running from a switch to a ceiling outlet, for example, only indicates that this switch controls this light. It does *not* imply that this is where the cable is supposed to be installed.

A cable-wiring diagram, of the type shown in Figure 24-13, shows the *sequence* in which the devices are to be connected and the *number* of wires required in the cable between the boxes. Graphical symbols are used to simplify the drawing of electrical devices in a circuit. Cables are represented by a single, solid line, with the number of insulated wires in the cable shown by short dashes across the cable line. For example, a two-wire cable is shown as ⎯++⎯, and a three-wire cable as ⎯+++⎯.

Boxes for switches should be installed so that the bottom of the box is from 44–48 in. above the floor. For receptacles, the usual height is from 12–18 in. above the floor as illustrated in Figure 24-14.

The NEC specifies the *maximum* number of wires permitted in a box (see Table 24-2). Deduct one from the numbers in the table if a box contains a fixture stud, a hickey, or one or more cable clamps. Deduct two from the numbers in the table for each switch, receptacle, or

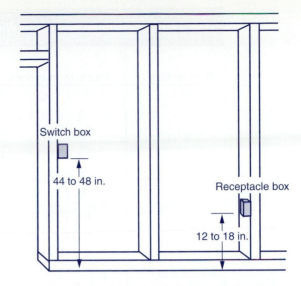

Figure 24-14   Location of switches and receptacles.

similar device to make room for each of the extra items. All grounding wires count as only one wire. Extra deep boxes are available if necessary. The point is not to crowd too many wires into too small a box.

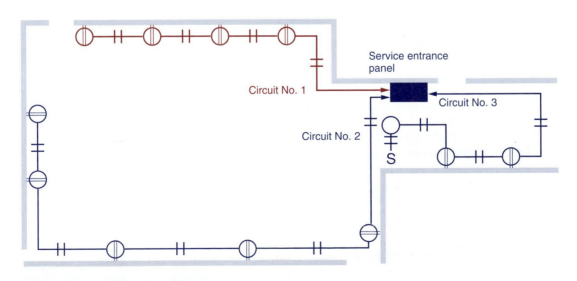

Figure 24-13   Typical cable wiring diagram.

Type of Box	Box Size in Inches		Maximum Number of Wires	
			No. 14	No. 12
Outlet box	$4 \times 1\frac{1}{4}$	Round	6	5
	$4 \times 1\frac{1}{2}$	or	7	6
	$4 \times 2\frac{1}{8}$	Octagonal	10	9
	$4 \times 1\frac{1}{4}$	Square	9	8
	$4 \times 1\frac{1}{2}$	Square	10	9
	$4 \times 2\frac{1}{4}$	Square	15	13
	$4\frac{11}{16} \times 1\frac{1}{4}$	Square	12	11
	$4\frac{11}{16} \times 1\frac{1}{2}$	Square	14	13
	$4\frac{11}{16} \times 2\frac{1}{8}$	Square	21	18
Switch box	$3 \times 2 \times 1\frac{1}{2}$		3	3
	$3 \times 2 \times 2$		5	4
	$3 \times 2 \times 2\frac{1}{4}$		5	4
	$3 \times 2 \times 2\frac{1}{2}$		6	5
	$3 \times 2 \times 2\frac{3}{4}$		7	6
	$3 \times 2 \times 3\frac{1}{2}$		9	8

TABLE 24-2 Maximum Number of Wires Permitted in a Box

HELP WANTED

Journeyman Electrician
Job Description
- Perform skilled electrical tasks in the installation and maintenance of electrical systems.
- Install and repair wiring, lighting systems, fixtures, switches, and outlets.
- Physical strength and agility to withstand the strain of manual work.
- Successful completion of a formalized electrician apprenticeship program.
- Current journeyman electrician license.

Review and Applications

1. Define the term "branch circuit" as it is used in the National Electrical Code (NEC).
2. What AWG conductor size and protection rating is most often used to wire branch circuits that supply voltage to lights and small appliances?
3. (a) What is the maximum number of outlets that may be wired in a single branch circuit?
   (b) Under what condition must this number be reduced?
4. What is the NEC requirement for a duplex receptacle installed in the laundry or utility area?
5. When a duplex receptacle is located outdoors or in a bathroom, what special protection device is required?
6. In a bedroom or family room, what is the maximum horizontal distance allowed between receptacles? Why?
7. What is the NEC small appliance branch circuit requirement for the food-preparation area?
8. In what two ways is an approved range receptacle different from an approved dryer receptacle?
9. Electric clothes dryers are supplied with both 240-V and 120-V circuits. Why?
10. What type of circuit breaker is required for a 240-V branch circuit? Why?
11. State the established NEC insulation color code for each of the following identified conductors:
    (a) neutral wire
    (b) live or hot wire
    (c) insulated ground wire
12. In a house branch circuit, when two or more lamps are to be operated by a single switch, what type of connection of lamps is used?
13. When it is desirable to have a switch control one half of a duplex receptacle, what change must be made to the receptacle?
14. You are required to operate a lamp from four locations. What switches are required?
15. What is the recommended mounting height from the floor to the bottom of a:
    (a) switch box?
    (b) receptacle box?

Problems

1. Refer to Figure 24-1. Why are the switches not counted as outlets?
2. Refer to Figure 24-3. Why is a double-pole circuit breaker required?

Critical Thinking

1. A switch must be connected into the live wire of a branch circuit. Why?
2. Refer to Figure 24-4. Assume the lamp burns open. What effect, if any, would this have on the receptacle? Why?
3. Refer to Figure 24-5C. Assume the black and white wires connected to the switch are reversed so that the black wire connects to the top terminal and the white to the bottom. What effect, if any, would this have on the operation of the circuit? Why?
4. Refer to Figure 24-6B. Assume a third light fixture is to be connected into the circuit and controlled by the switch. If a two-conductor cable run from the new fixture to the switch box for this addition, to what two wires in the switch box would the black and white wires of this cable be connected?
5. Refer to Figure 24-7C. Assume a second receptacle is to be connected into the circuit using a two-conductor cable run to the switch box. Why will this not work?
6. Refer to Figure 24-10B. Assume wires No. 3 and No. 4 of the 3-way switch had their terminal connections reversed. What effect, if any, would this have on the operation of the circuit?

Portfolio Project

Obtain the floor plan of a single-family home or cottage and design the electrical installation for it. Your design should include:
- Location of all outlets, lights, switches, and special-purpose circuits.
- Grouping of receptacles, lights, and switches to form designated branch circuits.
- Single-line wiring diagram showing the cable runs.

- Directory detailing all branch circuits to be installed.
- List of electrical supplies needed.

Circuit Challenge

Using a Simulator

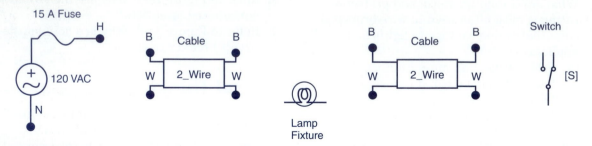

Procedure

The branch circuit shown is that of a lamp controlled by one switch. Using whatever simulation software package is available to you, complete the wiring of the circuit using the cables run as shown.

Appliance Cords and Connections

Electrical and electronic devices are connected to a receptacle by means of flexible line cords. These cords have two or three copper conductors that are stranded to provide maximum flexibility. The classes of cords covered in this chapter include lamp cords, appliance cords, lamp extension cords, and power extension cords. Also discussed are proper grounding techniques and the operation of a ground-fault circuit interrupter (GFCI).

Objectives

After studying this chapter, you should be able to:
1. Identify and select the proper cord for common applications.
2. Properly connect a two-prong plug cap.
3. Properly connect a three-prong plug cap.
4. Explain the purpose and method of grounding appliances.
5. Distinguish between the live, neutral, and grounding conductors of an appliance circuit.
6. Explain the operation of ground-fault circuit interrupter (GFCI).

Key Terms

- lamp cord
- heater cord
- two-prong plug cap
- polarized plug cap
- three-prong plug cap
- dead front plug cap
- weatherproof fixture
- cord strain relief
- dryer cord set
- range cord set
- grounded appliances
- grounding wire
- double-insulated appliances
- GFCI receptacle
- GFCI circuit breaker
- vacuum cleaner cord
- hard service cord

- Panel Components
- Cords Canada
- Crouse-Hinds

25-1 Types of Electric Cords

Any portable electric device that receives its electric energy from a receptacle must be equipped with an approved cord set. Too often an electric cord is regarded as simply "any old piece of wire." As a result, electric cords rank high as common causes of electrical fires and accidental electric shocks. Hazards may be avoided if cord sets are properly chosen, installed, and maintained.

Electric cords usually consist of two or three separately insulated copper conductors. Flexibility is achieved through the use of many strands of wire for each conductor. They usually come with a molded plug that is part of the cord. Cords are available in different *AWG* conductor sizes for different current ratings.

The selection of a type of appliance cord depends upon the type of appliance to be used and the service conditions under which the cord is to be used. Heating appliances, such as electric clothes irons, must be equipped with a special *insulated* cord which will resist the heat conducted from the appliance. Under no circumstances should any type of electric cord be used as a substitute for fixed wiring in a home. Article 400 of the *National Electrical Code* outlines the specific requirements regarding the use of cords. Tables 400-4 and 400-5 of the Code deal with the specific ratings of cords. Figures 25-1 through 25-5 show a few of the many cord types that are available.

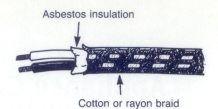

Asbestos insulation

Cotton or rayon braid

INSULATION:	Twisted thermoset conductors, asbestos fill, and cotton or rayon braid cover.
APPLICATIONS:	Portable heaters.
USAGE:	Not for hard usage. May be used in dry locations only.

Figure 25-2 Type HPD heater cord.

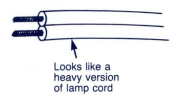

Looks like a heavy version of lamp cord

INSULATION:	Thermosetting
APPLICATIONS:	Heater type appliances such as toasters and frying pans.
USAGE:	Not for hard usage. May be used in dry or damp locations.

Figure 25-3 Type HPN parallel heater cord.

Never yank on an extension cord to pull it out of its receptacle. Instead, grasp the plug itself and pull it out of the receptacle. This method avoids putting strain on the fine wires inside the cord. Don't repair a worn extension cord—replace it! An extension cord with worn or damaged insulation is a potential fire and shock hazard.

Two-wire *lamp cord* usually comes with a molded plug that is part of the cord. This cord is commonly available in wire sizes of No. 16 and No. 18 AWG. Three-wire cords have stranded conductors with black, white, and

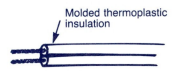

Molded thermoplastic insulation

INSULATION:	Thermoplastic
APPLICATIONS:	Lamps, radios, and small (non-heater type) appliances.
USAGE:	Not for hard usage. May be used in dry or damp locations.

Figure 25-1 Type SPT parallel lamp cord.

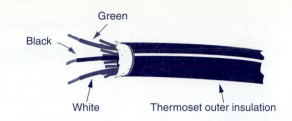

Green

Black

White

Thermoset outer insulation

INSULATION:	Twisted thermoset insulated conductors, cotton fill, rubber jacket.
APPLICATIONS:	Portable tools such as drills, saws, and sanders. Also non-heating appliances.
USAGE:	Approved for hard usage. May be used in dry or damp locations.

Figure 25-4 Type SJ junior hard service cord.

Thermoplastic outer jacket

INSULATION:	Thermoset or thermoplastic with thermoplastic outer covering.
APPLICATIONS:	Non-heating appliances such as vacuum cleaners and floor polishers.
USAGE:	Not for hard usage. May be used in dry or damp locations.

Figure 25-5 Type SVT vacuum cleaner cord.

green insulation to identify the live, neutral, and ground wire. Common sizes are Nos. 14, 16, and 18 AWG, but heavier cables are available for industrial use. Four-wire heavy duty cables for dryers and ranges differ in the size of their conductors and in the design of their plug caps. The dryer cable normally has three No. 10 AWG stranded conductors plus ground, and the range cable has three No. 8 AWG wires plus ground.

25-2 **Cord Connectors**

Two-Prong Plug Cap   The plug cap of an appliance cord is usually subjected to a lot of rough treatment. Many cord troubles can be traced to the plug cap itself; therefore, it is often necessary to replace the original molded plug cap supplied with the cord.

Polarized ***two-prong plug caps*** (Figure 25-6) are used on lamps, radios, televisions, and other appliances that are required to be properly grounded. This cap can be made of Bakelite, rubber, or plastic and has parallel brass prongs that receive the current. One prong is wider than the other so that it can be plugged into an outlet in only one way. The wide prong is wired to the side of the appliance that must be grounded.

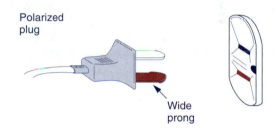

Polarized plug

Wide prong

Figure 25-6 Polarized molded two-prong plug cap.

The safest kinds of replacement plug caps are the ***dead-front types*** (Figure 25-7). They are constructed with the prongs and cap assembled as a removable unit. The cord is inserted through the cap and connected to the prong

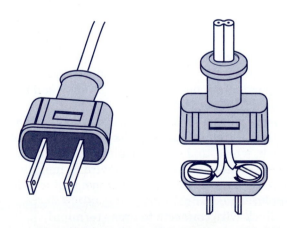

Figure 25-7 Nonpolarized replacement two-prong plug cap.

terminals. When assembled, the front is completely insulated with no terminals or conductors exposed. ***Nonpolarized plug caps*** have two prongs of the same (narrow) width. These plug caps can be used on appliances for which grounding is not required.

Three-Wire Plug Cap
Power tools and heavy appliances that must be grounded come equipped with a three-wire cord. A three-wire plug cap must be connected to this type of cord (Figure 25-8).

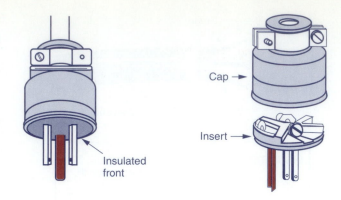

Figure 25-9   Dead-front ground plug cap (three-wire).

Three-Wire Extension Cord
Extension cords are used to reach receptacles that are not close enough to plug appliances directly into them. Appliances and power tools that have ground plug caps should be used only with a three-wire grounding type of extension cord. A three-wire extension cord is constructed using a three-wire cord, ground plug cap, and ground connector. The ground connector is wired using the same color code as was used for the plug (Figure 25-10).

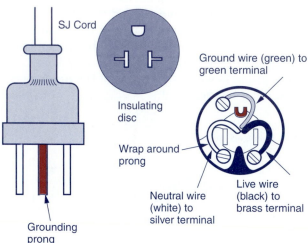

Figure 25-8   Standard three-wire plug cap.

The standard ***three-wire plug cap*** has three prongs. Two of the prongs are parallel, one colored silver and the other brass. The third prong is brass in color and sometimes has a U-shape. The *terminal* that is connected to this prong is colored green and is called the grounding terminal. This type of plug is also called a ***polarized plug*** because the prongs are always inserted into the receptacle in the same position. When connecting the hot and neutral wires they must be connected to the brass and silver terminals, respectively. A dead-front ground plug cap is shown in Figure 25-9.

Type SJ three-wire cable is commonly used with the ground plug cap. Its conductors have black, white, and green insulation to identify the live, neutral, and ground wire. When connecting a ground plug the following color codes apply:
- Black wire connects to brass terminal;
- White wire connects to silver terminal; and
- Green wire connects to green terminal.

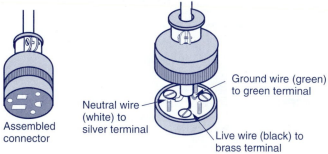

Figure 25-10   Attaching a three-wire ground connector.

Appliance Plug
Some heating appliances, such as frying pans, come equipped with a removable cord set. This set consists of an approved ***heater-type cord***, a two-prong plug cap at one end, and an appliance plug at the other end (Figure 25-11). The appliance plug is constructed using a two-piece bakelite cover fastened with nuts and bolts. Two nonpolarized prongs connect to the cord wires. A spring attached to the plug provides strain relief for the cord.

Protecting the Cord from Stress
A ***strain relief*** must be provided where the cord enters

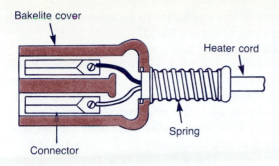

Figure 25-11　Attaching an appliance plug.

the appliance. This is to prevent the cord from being pulled out of the appliance. One method used is to pass the cord through a protective sleeve. The sleeve is then fastened to the appliance and holds the cord in place (Figure 25-12).

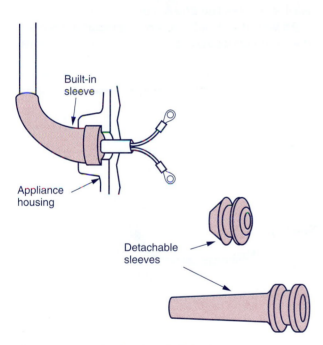

Figure 25-12　Cord strain-relief sleeve.

Dryer and Range Cables　Special heavy-duty cord sets are approved for use with electric dryers and ranges (Figure 25-13). Use of cord sets with these large appliances allows the units to be quickly disconnected for servicing. These units require a 120/240-V source for their operation. The approved *dryer cord set* is rated for 30 A. The approved *range cord set* is rated for 50 A. Each uses different plug caps and receptacles to prevent wrong usage.

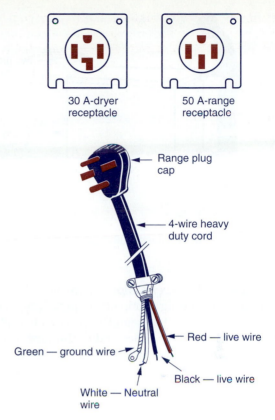

Figure 25-13　Range and dryer receptacles.

25-3 Grounding Appliances

The exposed metal case of an *ungrounded* appliance can become live if an electrical fault develops within the device. Suppose, for example, that the insulation on the live wire wears through and makes contact with the steel case. Immediately, the entire case of the appliance becomes live and a potential source of shock, death, or fire. When this happens, a person touching the case of the appliance can provide an alternate current path through his or her body to the ground (Figure 25-14). In the process, the person may receive a severe, if not fatal, shock! Moisture of any sort tends to reduce the contact resistance and increases the severity of the shock.

Grounded appliances come equipped with a polarized three-prong ground plug cap. As discussed before, the metal hardware of the entire electrical system is grounded. When a ground appliance cord is plugged into this circuit, it continues this grounding system through to all the metal parts of the appliance. The green

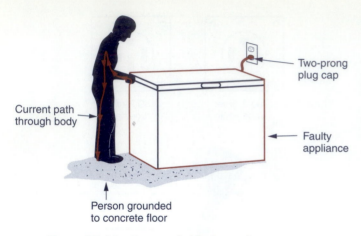

Current path through body

Two-prong plug cap

Faulty appliance

Person grounded to concrete floor

Figure 25-14 Ungrounded faulty appliance.

ground wire of the cord, along with the U-shaped grounding prong of the plug cap, form the grounding link between the receptacle and the appliance. At the other end of the cord, the green ground wire is connected to the metal

frame of the appliance to complete the grounding circuit (Figure 25-15).

The National Electrical Code distinguishes between a grounded wire and *grounding wire*. The wire which carries current during the normal operation of the circuit and is grounded is called just that; the **grounded wire** (Figure 25-16). This wire is also known as the neutral wire and in a three-wire cord is always the white wire. The wire that is in the circuit only as a safety measure and does not carry current during normal operation of the circuit is called the *grounding wire*. This wire is also known as the *ground* wire and in a three-wire cord is always the *green* wire. The ungrounded wire that carries current during the normal operation of the circuit is called the *live* (or *hot*) wire, because if you touch the wire itself you may get a shock. The *live* (or *hot*) *wire* in a three-wire cord is always the *black* wire.

As you have read, the green grounding wire does not carry current during the normal opera-

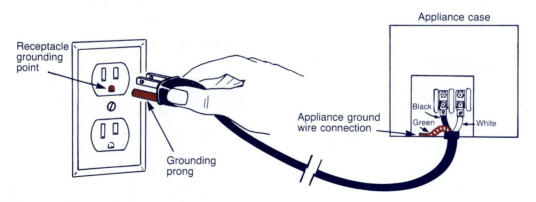

Receptacle grounding point

Appliance case

Appliance ground wire connection

Black
Green
White

Grounding prong

Figure 25-15 Appliance grounding connection.

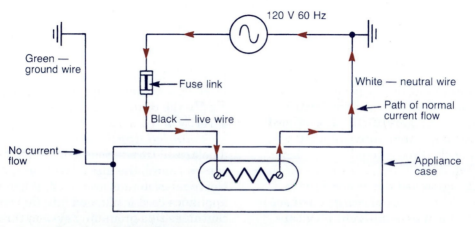

120 V 60 Hz

Green — ground wire

Fuse link

Black — live wire

White — neutral wire

Path of normal current flow

No current flow

Appliance case

Figure 25-16 Current flow through the grounding circuit.

tion of the appliance. Normally current flows through the (black) live and (white) neutral wires only. If the live wire makes contact with the grounded wire, a new circuit that leads from the live wire to the grounding wire is created. This new circuit is a *short circuit.* It contains no load to limit the current flow (Figure 25-17). A large surge of current will pass through the live and grounding wire, operating an automatic safety device in the main panelboard. A fuse will blow or a *circuit breaker* will trip. What is more, a replacement fuse also will blow and a circuit breaker will continue to trip until the fault in the appliance is found and corrected. The Code requires all noncurrent-carrying metal parts of tools and appliances to be grounded in order to protect against electric shock and fire.

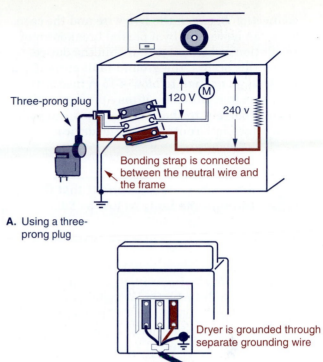

A. Using a three-prong plug

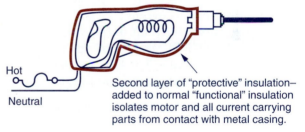

B. Using a four-prong plug

Figure 25-18 Grounding a dryer.

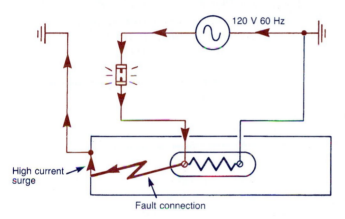

Figure 25-17 Ground fault current.

In the case of electric ranges and dryers, their frames may be grounded to the neutral circuit conductor, provided it is No. 10 AWG or heavier. When this is the case, a bonding strap is connected between the neutral wire and the frame (Figure 25-18). If local codes do not permit grounding to the neutral, the bonding strap from the frame to the neutral must be removed and a four-wire cord and plug must be used.

Double-Insulated Appliances Some metal-clad tools and small appliances are made using an approved extra layer of insulation and plastic. This secondary insulation is such that if the unit becomes live there is no risk of electric shock (Figure 25-19). These tools and appliances are called **double-insulated**. Under approved conditions the Code will allow double-

Figure 25-19 Double-insulated electric drill.

insulated, metal-clad tools and appliances to be used without being grounded.

25-4 Ground-Fault Circuit Interrupter (GFCI)

The use of a grounding wire in a three-wire cord with a three-prong plug cap and a grounding receptacle reduces the danger of shock, but does not eliminate the danger 100 percent. Sometimes a tool or appliance develops a ground fault that is not a solid or direct connection between the live wire and the case. This can be caused by a partial breakdown of insulation or by moisture within the device. When this occurs, the ground-fault current may not be high enough to blow a 15-A fuse or trip a 15-A breaker. However, it may produce a high enough current to shock or electrocute anyone who comes into contact with the device.

> ✋ **Remember, electric current that flows through the body as low as 50 mA (0.05 A) can be fatal!**

A *ground-fault circuit interrupter (GFCI)* is designed to minimize the probability of electrocution under the conditions just described.

Under normal conditions, the current in the hot wire and the neutral wire are absolutely identical. However, if wiring, a tool, or an appliance is defective and allows some current to leak to ground, then a GFCI will sense the difference in the two wires. The GFCI is fast-acting: The unit will shut off the current or interrupt the circuit within 1/40 sec after its sensor detects a leakage as small as 5 mA.

A GFCI is *not* to be considered as a substitute for grounding, but as supplementary protection that senses leakage currents too small to operate ordinary branch circuit fuses or circuit breakers. Both GFCI receptacles and circuit breakers are available. The **GFCI receptacle** provides ground-fault protection to users of any electric equipment plugged into the receptacle. Wire GFCIs in the same way as a standard receptacle, with the black hot wire going to the brass terminal, the white neutral wire to the silver-colored terminal, and the ground wire to the green terminal. Some GFCIs come with factory wiring but connect the same way— black-to-black, white-to-white, and ground-to-green (Figure 25-20). This receptacle is equipped with test and reset buttons. Pressing the test button simulates a ground fault and should cause the relay within the receptacle to react and open the circuit. Pressing the reset button back into place reactivates the circuit. The unit should be tested at least once a month for proper operation.

The GFCI receptacle will only provide ground-fault protection; it will not protect against short

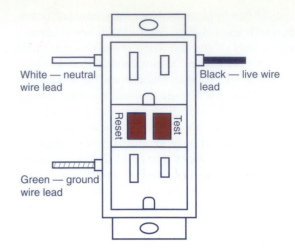

Figure 25-20   GFCI receptacle.

circuits or overloads—the branch circuit breaker or fuse must provide that protection. The **GFCI circuit breaker** provides both overcurrent and ground-fault protection to the entire branch circuit it protects. Externally it looks like a circuit breaker with a test button added to manually test the ground-detection circuit (Figure 25-21). Basically it works on the same principle as the GFCI receptacle by monitoring any difference in current flow between the live and neutral conductor. They are available in different current ratings to suit most circuits. Each GFCI circuit breaker has a white pigtail, which you must connect to the neutral busbar.

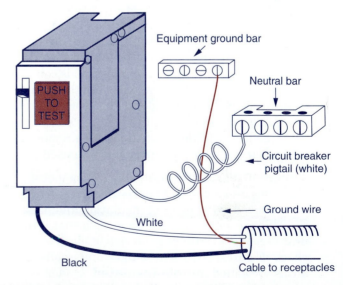

Figure 25-21   Typical GFCI circuit breaker connection.

In addition, you must connect the white neutral wire of the circuit to a terminal provided for it on the breaker. You can replace an ordinary breaker in your service panel with a GFCI type.

The GFCI duplex receptacle terminals do not include provisions for separation of its receptacles for use in split-circuit applications. However, most have provisions to connect to other regular receptacles. This is an option that will provide GFCI protection to any receptacle installed downstream in the circuit (Figure 25-22). The feed cable from the service panel must be joined to the leads marked **LINE,** while the outgoing cable leading to the rest of the circuit must be hooked to the leads marked **LOAD.** With factory-wired units, if the GFCI receptacle is the only fixture in the circuit, connect the **LINE** leads in the ordinary way, but cover each of the **LOAD** leads with a wire cap and fold them into the box.

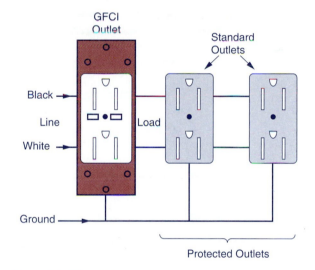

Figure 25-22 GFCI receptacle connected to provide protection to other conventional receptacles.

Standing on a concrete floor, standing outdoors on grass or earth, or coming in contact with metal plumbing places a person in direct contact with the earth or ground. For this reason, the Code in section 210-8 requires GFCI protection in residential bathrooms, garages, outdoor locations, basements, kitchens, and swimming pool installations.

25-5 Weatherproof Fixtures

As already mentioned, water mixed with electricity can lead to painful and even fatal shocks. All outdoor receptacles must be GFCI protected. In addition, all outdoor receptacles, lamp fixtures, and switches must be **weatherproof**. Outdoor parts are either built to be impervious to rain and snow, or designed for assembly in a weatherproof container.

About   Electronics

Propellers on arctic ice-breaking ships are outfitted with fiber optics that transmit data about the strain on the ice from 54 points along the propeller. This device provides crucial information to help the ship's captain decide whether it is safe to continue.

Figure 25-23 shows a weatherproof receptacle. The outlet box is made of a heavy metal casting. Instead of knockouts for incoming wires, the box has threaded holes into which metal plugs or the ends of conduit are screwed. A special cover makes it possible to install an ordinary indoor receptacle in the box. Two gasketed doors seal the receptacle sections when they are not in use. A larger gasket seals the gap between the cover plate and the rim of the box.

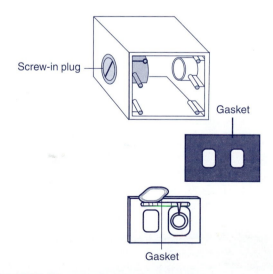

Figure 25-23 Weatherproof receptacle.

HELP WANTED

Electrical Sales

Job Description

- Required by a manufacturer of electrical controls.
- Candidates should have basic knowledge of electrical/electronic controls.
- Growth position with opportunities for advancement.
- Extensive travel is required.

Review and Applications

Review Questions

1. Name one typical approved application for each of the following cord types:
   (a) type SPT thermoplastic cord
   (b) type HPN neoprene cord
   (c) type SJ rubber cord
2. How is flexibility of an appliance cord achieved?
3. State two important factors that must be considered when deciding on the type of cord to be used for a particular appliance.
4. Outline the terminal color coding that is followed when connecting a three-wire grounding plug cap.
5. In what way is the current rating of a heavy-duty dryer cord set different from that of the range cord set?
6. What is the purpose of grounding an appliance?
7. How does the grounding circuit act to protect a person in case of a direct ground fault?
8. How does the National Electrical Code define each of the following:
   (a) grounded wire
   (b) grounding wire
9. (a) What type of metal-clad tool or appliance can be used without being grounded?
   (b) Why is this so?
10. What is the approximate speed of response and current-leakage sensitivity of a GFCI?
11. Explain the function of the "test" and "reset" buttons on a GFCI receptacle.

Problems

1. Cord sets are rated in volts and amperes. The rating of the unit must be matched to the requirements of the circuit. Why?
2. Assume the hot live wire in a switch box comes loose from its terminal and is touching the metal box.
   (a) Explain the potential hazard that would result if the metal box were not grounded.
   (b) Explain how grounding of the box eliminates this hazard.

Critical Thinking

1. How is the wire connection to a plug cap terminal made to prevent loose strands of the flexible wire from slipping out of the terminal connection?
2. The test button on a GFCI receptacle is pressed but the unit does not respond and, therefore, does not open the circuit. What adjustments, if any, should be made to the unit?

Portfolio Project

Search for examples of different types of line cords and plug caps. Report on the different applications of each and methods of grounding. Include any potentially hazardous situations, involving the use of line cords, you may have discovered in your search.

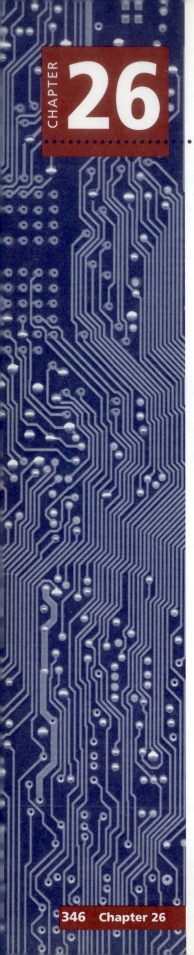

Lighting Equipment

The number of different types of lighting fixtures available staggers the imagination. In this chapter we will investigate the construction and operation of lighting equipment. We will focus our study on the following devices: incandescent lamps, gas-filled lamps, and lighting fixtures.

Objectives

After studying this chapter, you should be able to:
1. Describe the operating principle of an incandescent lamp.
2. Repair a table lamp.
3. Install a light fixture.
4. Describe the operating principle of a fluorescent light fixture.
5. Describe the operating principles of high-intensity discharge lamps.
6. Install security lighting devices.

Key Terms

- **incandescent lamp**
- **halogen lamp**
- **lumens**
- **bayonet base**
- **miniature base**
- **candelabra base**
- **medium base**
- **mogul base**
- **trilight lamp**
- **general-service filament**
- **vibration-service filament**

- **rough-service filament**
- **tab-mounted fixture**
- **strap-mounted fixture**
- **center-mounted fixture**
- **recessed fixture**
- **track lighting**
- **reflector bulb**
- **fluorescent lamp**
- **gas-discharge lamp**
- **ballast**

- **preheat fluorescent fixture**
- **rapid-start fluorescent fixture**
- **instant-start fluorescent fixture**
- **compact fluorescent bulb**
- **high-discharge lamp**
- **programmable timer**
- **motion-sensor light**
- **wireless remote control system**

RESEARCH
INTERNET
SITES

- **Noma**
- **Sylvania**
- **Union Lighting**

26-1 Incandescent Lamps

The *incandescent lamp* or light bulb was one of the first types of light sources developed, and it is still the most widely used today. An incandescent lamp produces light by using an electric current to heat a metallic *filament* to a high temperature. A **tungsten filament** is used because of its high melting point and low rate of evaporation at high temperature. The filament's resistance to the current causes it to become incandescent or white hot. The light which comes from an incandescent lamp is the glow of this white-hot filament.

During the manufacture of an incandescent lamp, the air is removed from the glass bulb before it is sealed. This is done to prevent the oxidation of the filament. The filament would burn out quickly if oxygen were present within the bulb. The glass bulb is filled with inert gas (e.g., nitrogen/argon mixture) at low pressure. The inert gas permits operation at higher temperatures, compared to a vacuum, resulting in a smaller evaporation rate and longer filament life. Often, the bulbs are frosted on the inside to provide a diffused light, instead of the glaring brightness of the unconcealed filament.

The ends of the filament are brought out to a screw-type base which makes replacement simple. In the base, the center contact is insulated from the outer metal part of the base. This produces two terminals, one for each of the two wires leading up to the filament (Figure 26-1).

Halogen lamps are a special type of incandescent lamp. Halogen bulbs are filled with halogen gas, which protects the tungsten filament and allows it to operate at a higher temperature. As a result, compared with conventional tungsten, halogen bulbs increase light output by about 25 percent for the same amount of current flow. The bulb of the halogen lamp is normally made of quartz glass, to withstand the lamp's high-temperature operating conditions (Figure 26-2). Common usage is automobile headlights.

A halogen bulb gets rid of heat by transferring heat and light out by way of the gas inside the bulb and the glass itself. Anything that interferes with this will cause the bulb to overheat, shortening its life. Manufacturers recommend that you do not touch the glass when you install the bulb, because the oil from your skin can cause the light to be reflected back into the bulb, raising its temperature.

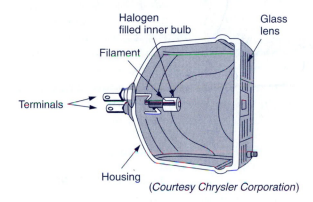

Figure 26-2 Sealed halogen headlight.

(*Courtesy Chrysler Corporation*)

Lamp Ratings

Light bulbs are rated for voltage, power, light output, light efficiency, and life expectancy (Figure 26-3).

The **voltage** rating indicates the maximum voltage that should be applied to the bulb. The most common voltage rating for general house lighting is 120 V. Operating an incandescent bulb at just a few volts higher than its rated voltage will reduce its life expectancy. The reverse is also true. For example, a lamp designed for 135 V, but operated at 120 V, will last about four times longer.

The *wattage* rating of a light bulb is directly proportional to the amount of light and heat it will produce. Rated power is obtained from an incandescent lamp only when the bulb is operated at its rated voltage. A fixture with a single incandescent lamp is more economical to operate and will give more useful life than one with several bulbs that have the same total wattage.

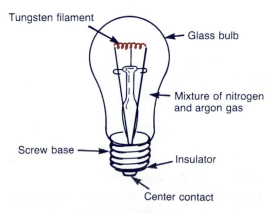

Figure 26-1 Incandescent lamp.

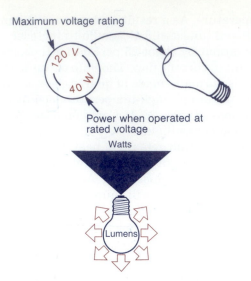

Maximum voltage rating

120 V

40 W

Power when operated at rated voltage

Watts

Lumens

Figure 26-3   Lamp ratings.

For example, *one 100-W incandescent bulb will give off 1 ¹/₂ times as much light as four 25-W bulbs.*

The amount of light output emitted by a lamp is rated in **lumens** (lm). The efficiency of a light source is the ratio of the light output (lumens) to the energy input (watts). The efficiency of different light sources varies dramatically, from less than 10 lm per watt to more than 200 lm per watt. Incandescent lamps have the lowest efficiency of all light sources. There are energy-saving incandescent lamps on the market with slightly more efficient designs and reduced wattages (34-, 52-, and 90-W bulbs that replace 40-, 60-, and 100-W bulbs). For example, replacing a 100-W lamp with a 90-W lamp will certainly save energy and you probably will not notice the slight difference in light output.

Lamp life ratings are based on an average. Many bulbs actually fail before they reach the rated life, while others burn for a longer length of time. Incandescent lamps have the shortest life of all light sources. There are incandescent bulbs available known as **long-life bulbs.** Although they last from 1500 to 10,000 hours as compared to 750 to 1000 hours for a normal bulb, they put out 30 percent less light while using the same amount of energy.

Lamp Bases   Lamps are manufactured in a variety of base sizes for different applications. Small lamps, such as those used in flashlights, are usually fitted with **miniature** or **bayonet-type bases.** To put a bayonet lamp in a socket,

line up the pins with the grooves in the socket, and then press the lamp in and turn it clockwise until it locks in place. **Candelabra-base** bulbs are often used for decorative lighting. The standard, or **medium-base** bulb is used with general-purpose lamps rated for 300 W or less. Lamps larger than 300 W are fitted with a **mogul base** (Figure 26-4).

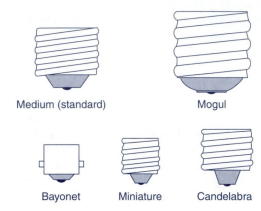

Medium (standard)   Mogul

Bayonet   Miniature   Candelabra

Figure 26-4   Common incandescent lamp bases.

Trilight Lamp   A **trilight lamp** makes it possible to obtain three different levels of light from a single lamp. Such a lamp contains two separate filaments within the same bulb. In the 40-60-100 W lamp, for example, one of these is a 40-W filament and the other is a 60-W filament (Figure 26-5). A separate rotary three-way switch or a combination socket and switch is used to operate the circuit. The switching sequence is as follows:

1. Off: no filaments connected
2. Low: 40-W filament connected

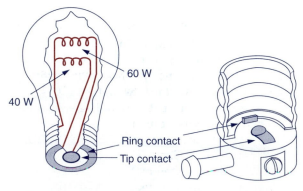

60 W

40 W

Ring contact

Tip contact

A. Three-way lamp   **B.** Combination socket and switch

Figure 26-5   Trilight lamp and socket.

3. Medium: 60-W filament connected
4. High: 60- and 40-W filaments connected in parallel

Lamp Filaments

Incandescent lamp filaments are coiled to shorten the overall length and to reduce heat losses. Filaments are rated according to the amount of filament support used in their construction (Figure 26-6). Shaking any bulb to test it can permanently damage the filament. Each filament support conducts heat away from the filament making it less efficient, but more rigid. ***General-service filaments*** have the minimum number of supports and are used for general household lighting. ***Vibration-service filaments*** offer more support and are used for lighting on motorized machines such as sewing machines or drill presses. ***Rough-service filaments*** offer maximum filament support and should be used for such applications as extension cord lights.

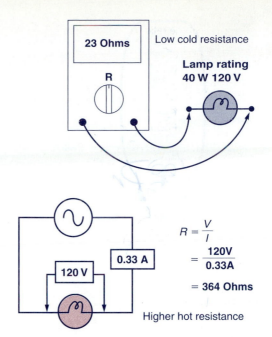

Figure 26-7 Filament resistance.

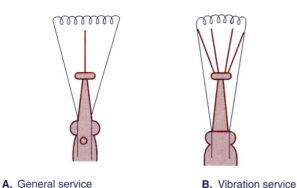

A. General service   **B.** Vibration service

Figure 26-6 Filament supports.

The resistance of a light-bulb filament varies directly with the filament temperature. As a result, the cold resistance of the filament is much lower than its normal operating resistance. The cold resistance can be measured directly with an ohmmeter, while the hot resistance can be calculated using Ohm's law and measured operating values of voltage and current (Figure 26-7). The power rating of the bulb is also determined by filament resistance. For a given voltage rating, the higher the power rating, the lower the bulb filament resistance.

26-2 Repairing a Table Lamp

Table lamps often stop working completely, work intermittently, work only when the cord is held in a certain position, or flicker ON and OFF. Regardless of the type or model of lamp, they all have a socket, switch, cord, and plug cap (Figure 26-8). These are the four things that may wear out, preventing the lamp from working. Replacing the defective part is usually all that is necessary to fix the lamp.

Troubleshooting a Table Lamp

1. If a lamp stops working, the first thing to check is the bulb. Make sure the light bulb is a good one and check that it is screwed in as far as it will go. Replace the bulb if necessary.
2. Next, unplug the lamp before doing any further checking. Examine the cord and plug cap for breaks or frayed areas. A cord with badly frayed or brittle insulation is a hazard. It should be thrown away and replaced, not repaired. A plug cap with a broken or wobbling prong also should be replaced. It is possible, through constant flexing of the cord, to develop a break in the conductor under the insulation covering. This type of break cannot be seen and

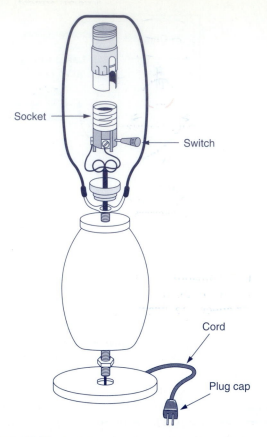

Socket

Switch

Cord

Plug cap

Figure 26-8 Basic parts and assembly of a table lamp.

therefore must be checked by making a continuity test of the cord.

3. Finally, check the socket and switch. The socket and switch are usually contained in one unit. Sometimes bulbs of a higher than approved power rating are used in the lamp. This produces extra heat that can damage socket contacts. Also, extra current can damage switch contacts. A visual check of the socket can tell you if it is suffering from overheating. Sometimes cleaning the center contact tip and adjusting its spring

tension will cure the problem. The switch contacts can be checked by making a continuity test of the switch.

Replacing a Lamp Socket   Unplug the lamp before doing any work on it. Remove the old socket, leaving the fixture knot and socket cap in place (Figure 26-9). To do this, first look for the word "press" on the socket shell. Remove the socket shell by squeezing and lifting near the switch. Remove the insulating sleeve, then loosen the terminal screws and disconnect the cord.

The new socket will include a socket cap but there is often no need to replace the existing one. Loop the wires clockwise around the terminal screws of the new socket and tighten. Place the insulating sleeve over the socket, and then put on the socket shell. Make sure that the corrugated edges of the shell fit inside the rim of the cap; then push them together until you hear the shell click into place.

Replacing a Lamp Cord   Make sure the lamp is unplugged. First, remove the socket shell and insulating sleeve. Next, remove the wires from the terminal screws and untie the fixture knot.

Attach a string to the old cord end. Pulling carefully on the old cord, thread the string through the lamp. Detach the old cord. The string should now be hanging out both ends of the lamp.

Tie the new cord to the end of the string and carefully pull it through the lamp. Tie a fixture knot in the cord and connect the ends to the socket terminals (Figure 26-10). (Soldering the ends of the wire will prevent the wires from flaring out from under the screw terminal.) After connecting the wires to the socket terminals, pull the socket back into the lamp base. Press the shell into the cap until it snaps into place. Connect a two-prong plug cap to the other end of the cord.

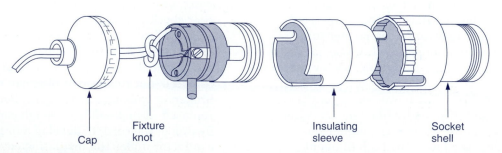

Cap

Fixture knot

Insulating sleeve

Socket shell

Figure 26-9 Lamp socket assembly.

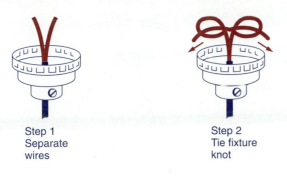

Step 1
Separate
wires

Step 2
Tie fixture
knot

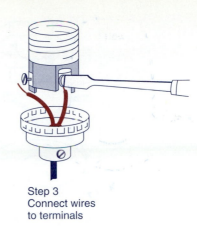

Step 3
Connect wires
to terminals

Figure 26-10   Connecting a new cord to the lamp socket.

26-3 Installing Light Fixtures

Electric Connections   Although light fixtures do not require repair very often, they sometimes become outmoded and a replacement is required. The replacement job consists of three operations: removing the old fixture, wiring the new one, and then mounting it to the box. Wiring is essentially the same for all fixtures, but the hardware needed for mounting varies.

About ◀▮▶ Electronics

New "electrodeless" lamps work on inductively coupled plasma. Compared with traditional light sources, these lights use less energy, last longer, and are easier to maintain—especially in places where maintenance is difficult such as on bridges.

All light fixtures must have the power rating marked on the inside of the structure. This is the approved rating and must not be exceeded. If this limit is exceeded, the conductors in the outlet behind the fixture will overheat and become brittle over a prolonged period of time.

Most single-bulb fixtures come equipped with one black and one white wire already connected to the bulb socket. All that is required is to correctly connect these two wires to the house electrical system via the electric-outlet box. Before starting the job, turn OFF the power to the fixture circuit at the service panel. Merely turning the wall switch off is *not* sufficient, since, in

certain situations, the switch may not turn OFF power to all wires in the box. Remember to always check the circuit with a voltmeter before starting to work on it. Connections are made by means of solderless mechanical connectors. Connect the black wire of the fixture to the black (live) wire of the house's electrical system (Figure 26-11A). Connect the white wire of the fixture to the white (neutral) wire of the house's electrical system. The white wire on the fixture connects to the screw part of the socket. This makes the screw shell neutral at all times, decreasing the possibility of receiving an electric shock when changing the bulb (Figure 26-11B).

Some pendant or hanging types of light fixtures are equipped with a grounding wire in addition to the white and black wires. The grounding wire can be a bare conductor or a green-colored (or green-yellow combination) conductor run with the circuit conductors. This grounding wire is connected to the ground screw of the electric-outlet box. Its purpose is to effectively ground the noncurrent-carrying metal parts of the hanging fixture.

Mounting Fixtures   Often, the most difficult part in installing a new fixture is to secure the fixture to the electric-outlet box. The hardware for holding a fixture in place is determined by the size and weight of the fixture. The smallest are *tab-mounted fixtures* (Figure 26-12). This fixture is held in place by two mounting screws that are fastened directly to the tabs of the outlet box.

A *strap-mounted fixture* is used for mounting fixtures with holes too far apart to fit the outlet box tabs (Figure 26-13). To install this

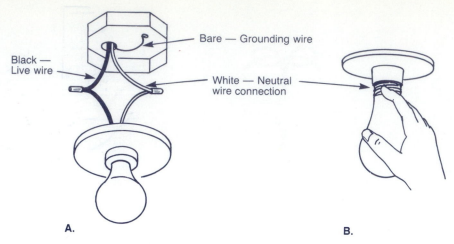

Figure 26-11   Fixture wire connections.

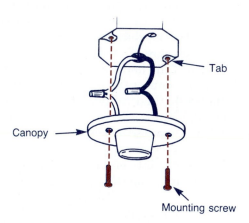

Figure 26-12   Tab-mounted fixture.

kind of fixture, first screw the fixture strap to the box tabs. Make sure the strap extends equally on both sides of the box. Connect the wires and then screw the canopy to the strap.

A **center-mounted fixture** uses a threaded nipple and cap nut to hold the fixture in place (Figure 26-14). To install this type of fixture, first fasten the strap to the box by screwing it to the box tab. Next, screw the nipple into the strap far enough so that when the cap nut is in place it will hold the canopy of the fixture snugly against the ceiling. Secure the nipple in this position with the locknut. Connect the wires. Place the fixture over the box so that the nipple protrudes through the center hole of the fixture. Fasten the fixture to the nipple with the cap nut.

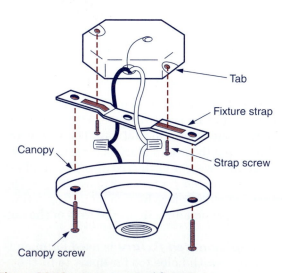

Figure 26-13   Strap-mounted fixture.

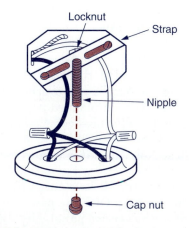

Figure 26-14   Center-mounted fixture.

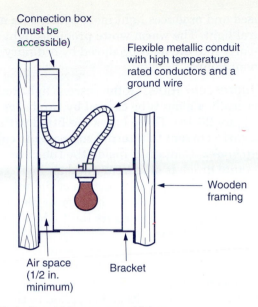

Connection box (must be accessible)

Flexible metallic conduit with high temperature rated conductors and a ground wire

Wooden framing

Air space (1/2 in. minimum)

Bracket

A. Typical mount for a recessed fixture

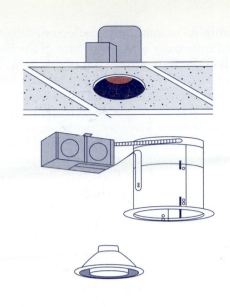

B. Pot light fixture

Figure 26-15 Recessed fixture.

Recessed Fixtures   *Recessed pot lights and fixtures* installed between joists can get quite hot. They must be properly enclosed and must be away from any insulation that might be flammable (Figure 26-15). The ventilation of a recessed fixture is very poor in comparison to a surface-mounted fixture. The extra heat developed by this type of fixture must be prevented from reaching combustible materials and circuit conductors. The National Electrical Code and local codes require that special procedures be followed when installing this type of fixture. Refer to the appropriate code for these specific requirements.

  Caution: Improperly installed recessed fixtures are a fire hazard!

Track Lighting   *Track lighting* is *surface-mounted* with lights that are adjustable both in their location along the track and in the direction that each fixture can be pointed (Figure 26-16). Power to the track is supplied from a ceiling outlet. Both two- and three-conductor tracks are available. The hardware and installation of the tracks varies with different manufacturers.

Reflector bulbs (Figure 26-17) are used for most track and pot light installations. Available

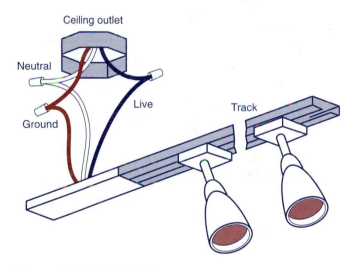

Ceiling outlet

Neutral

Ground

Live

Track

Figure 26-16   Track lighting.

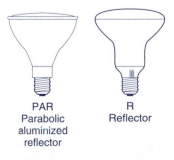

PAR
Parabolic aluminized reflector

R
Reflector

Figure 26-17   Reflector bulbs.

as floodlights or spotlights, reflector bulbs are silvered internally along the sides of the bulb so that all of the light shines out of the front. This makes them more efficient than regular household bulbs because they direct most of the light where you want it rather than trapping it in the pot or fixture. The spot lamps, in general, have a relatively narrow beam spread that is well suited to emphasizing specific areas, while the wider beams of the flood-type lamps are used more for general-purpose lighting.

26-4 Fluorescent Lighting

The *fluorescent lamp* belongs to a group of light sources known as *electric- or gas-discharge lamps*. Fluorescent lighting has many advantages over incandescent lighting. First of all, it produces far more light per watt of power used. Fluorescent lamps last much longer than ordinary incandescent lamps.

Fluorescent tubes produce less heat than incandescent lamps due to their higher light efficiency. If you touch a fluorescent tube after it has been ON for some time you will note that it is cool to the touch. Large incandescent bulbs will burn anyone trying to remove them after they have been in operation for some time.

Under ordinary operating conditions, fluorescent lamps last 5 to 15 times longer than standard incandescent lamps. However, the more often a fluorescent lamp is turned ON and OFF, the shorter its life span. The major disadvantage of fluorescent lighting is the higher initial cost of the fixture. This extra cost is due to the auxiliary hardware required to operate the fluorescent lamp circuit.

Fluorescent lamps are available in different kinds of "white," ranging from warm white to cool white. The cool white is most commonly used and produces light most nearly like natural light. The warm white produces light more nearly like the light produced by ordinary filament lamps.

Fluorescent Tube   A fluorescent light bulb is basically a glass tube capped by two bases (Figure 26-18). These bases are fitted with pins to carry current to internal components called *cathodes*. Contained inside the tube are minute droplets of mercury and an inert gas (e.g., argon). The inner surface of the tube is coated with a fluorescent powder or phosphorus. This phosphorus emits light when exposed to ultraviolet rays.

The fluorescent bulb produces light by a two-step process. First, invisible ultraviolet rays are produced. These rays are then converted to visible light. When a voltage is applied to the cathodes, they emit electrons, ionizing (electrically charging) the gas atoms inside the tube (Figure 26-19). The ionized gas is an electric conductor, and a current called an *arc* begins to flow between the two cathodes. The heat of this arc vaporizes the mercury droplets in the tube. Electrons in the arc strike the atoms of vaporized mercury, causing them to emit ultraviolet rays. When the ultraviolet rays strike the atoms of the tube's phosphorus coating, the atoms emit visible light and the whole phosphorus coating glows brightly and uniformly.

26-5 Fluorescent Light Fixture Circuits

Ballast   Every fluorescent fixture has a *ballast* housed inside it. The ballast provides the necessary starting and operating electric conditions. It steps up the supply voltage to strike an arc and start the tube conducting. Once the arc has been struck, the resistance of a gas discharge lamp becomes very low and the ballast

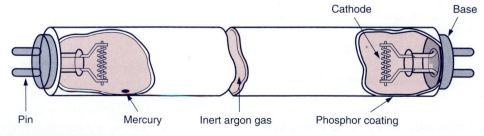

Figure 26-18   Fluorescent tube.

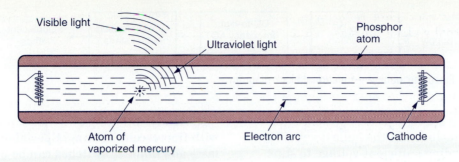

Figure 26-19   Producing light in a fluorescent tube.

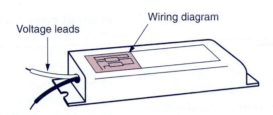

A. Construction

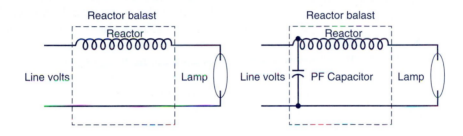

B. Schematic

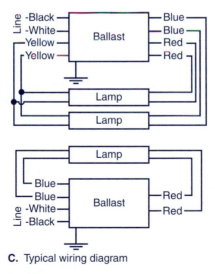

C. Typical wiring diagram

Figure 26-20   Standard fluorescent fixture ballast.

is used to limit the flow of current to the tube cathodes to keep them from burning out. The ballast is factory-wired according to the wiring diagram on the ballast label.

A simple standard ballast consists of an electric coil wound on a steel core (Figure 26-20). The core and coil assembly is impregnated with a nonconducting material to provide electrical

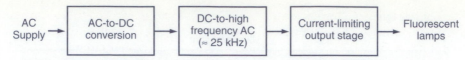

Figure 26-21 Electronic ballast—functional block diagram.

insulation and aid in heat dissipation. Capacitors may be included in the ballast circuit to assist in providing sufficient voltage to start the lamp and/or to correct power factor (PF). Each ballast is designed to be used with a specific type and size (wattage) of lamp. All core-coil ballasts produce a sound that is commonly described as *hum*. Manufacturers give their ballasts a sound rating from *A* to *F*. An *A* ballast produces the least hum and should be used in quiet areas (office and homes, for example).

The ballast, as part of the electric circuit, has energy losses. Energy-efficient ballasts (core-coil) have lower losses (better efficiency). Electronic ballasts (Figure 26-21) have the best light efficiency, producing more light per watt of power used. When a standard core-coil ballast is used, the lamp operates at 60 Hz (cycles per second) and the arc is struck 120 times per second—once on the positive peak and once on the negative peak. Conversely, electronic ballasts operate the lamp at high frequency, typically from 25-40 kHz. Because the lamp operates at a higher frequency, the arc is struck many more times per second to deliver equal light output and uses substantially less current.

In comparison to the standard ballasts, the electronic ballasts weigh less, operate at lower temperatures, operate at lower noise level, are more energy efficient, and cost more.

Normally, dimming switches *cannot* be used with fluorescent lighting. Special dimming ballasts and dimmer switches can be purchased for use with fluorescent lighting.

Preheat Fluorescent Light Circuit The simplest type of fluorescent fixture is the **preheat fluorescent light** with manual push button starter. The schematic circuit for this fixture is shown in Figure 26-22. When the line switch is first closed, no current flows through the tube because the resistance between the two cathodes is too high to permit an arc to start. Pressing the starter push button allows current to flow through the cathode filaments at the ends of the tube and through the ballast. This causes the heating of the end filaments and then warms the mercury vapor. Electrons are emitted from the hot filament, making the tube ready to conduct. When the starter push button is released, the collapsing magnetic field in the ballast coil induces a momentary high-voltage surge in the circuit. This produces a current flow across the gas vapors that are inside the tube and creates a new path for the current through the tube and ballast. During operation, the voltage across the tube may be from 50–100 V, depending on the size of the tube. The ballast regulates the tube's normal operating voltage and current.

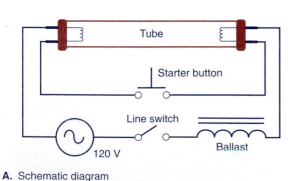

A. Schematic diagram

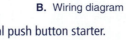

B. Wiring diagram

Figure 26-22 Preheat fluorescent fixture with manual push button starter.

Most preheat fluorescent fixtures have an automatic glow-switch starter, which replaces the manual push button starter (Figure 26-23). The glow switch is a neon-filled glass bulb, containing a bimetallic strip and a fixed, normally open contact. When the line switch is first closed, full line voltage is applied across the starter. An arc forms between the contacts. This arc (or glow) produces enough heat to cause the bimetallic strip to bend, closing the contact points. Closing of the starting circuit is now accomplished and the tube filaments are heated as before. The closing of the contact in the glow switch also stops the glow tube arc. Then the bimetallic strip cools, and the contacts open. At the instant of opening, the ballast delivers a high-voltage surge that starts current conduction through the fluorescent tube (Figure 26-24).

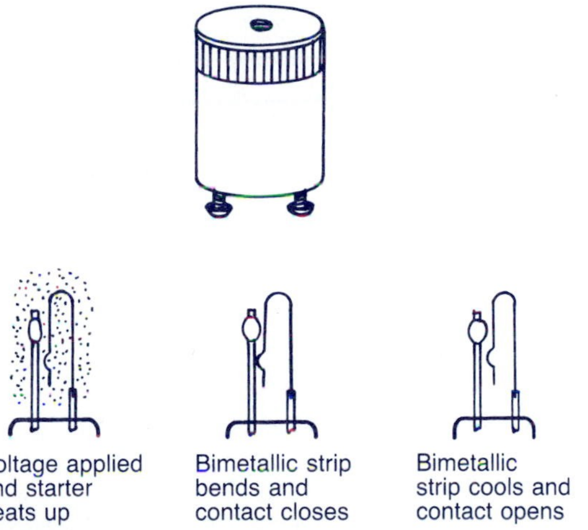

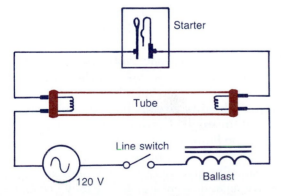

Figure 26-23   Glow-switch starter operation.

| Voltage applied and starter heats up | Bimetallic strip bends and contact closes | Bimetallic strip cools and contact opens |

Figure 26-24   Preheat fluorescent fixture with automatic starter.

Preheat fluorescent tubes have a bipin (double pin) base at each end. They operate normally in a preheat circuit and can be used in rapid-start circuits as well. Preheat fixtures are not widely used today.

Rapid-Start Fluorescent Light Circuit   A *rapid-start fluorescent fixture* does not use a manual push button or glow-switch starter. Instead, the ballast quickly heats the cathodes, causing sufficient ionization in the lamp for the arc to strike (Figure 26-25). The cathodes may or may not be continuously heated after lamp starting. This depends on ballast design. Rapid-start lamps start almost instantly (in one or two seconds).

Rapid-start fixtures make use of the voltage that exists between the cathodes and the metal frame of the fixture to aid in reliable starting. As a result, the fixture must be properly grounded. If not properly grounded, they will often fail to start in cold weather.

Rapid-start fluorescent tubes have a bipin base at each end and normally operate only in rapid-start circuits. The combination of quick-starting, bright light and compact ballast has made the rapid-start fixture today's most popular fluorescent light.

> **Caution:** Never dispose of a worn-out fluorescent tube by breaking it up. The mercury it contains is poisonous! Also, the exploding glass can cause dangerous cuts! For the same reasons, discarded tubes should never be thrown into incinerators or other fires.

Instant-Start Fluorescent Light Circuit   An *instant-start fluorescent fixture* uses a fluorescent tube with only a single pin at each end and no starter circuit to preheat the tube's cathodes (Figure 26-26). It comes equipped with a special ballast that initially produces a very high voltage across the cathodes, which immediately ionizes the gas in the tube. As the cathodes heat up and emit more electrons, the arc needs less voltage to sustain it. At this point, the ballast reduces its output voltage to the normal operating level.

The high voltage used to strike the arc in an instant-start bulb can be dangerous. As a

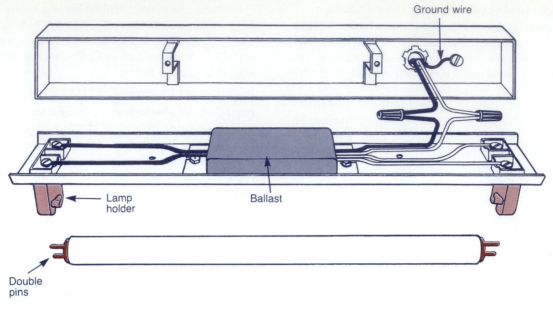

A. Wiring diagram

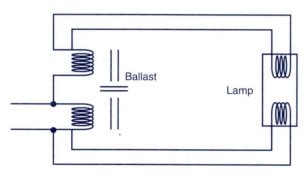

B. Schematic diagram

Figure 26-25 Rapid-start fluorescent fixture.

result, most lampholder designs incorporate a switch at the tube socket that cuts off all voltage to the ballast as the tube is removed. This reduces the hazard of shock to anyone who is changing a tube.

Instant-start circuits create their high voltage by using bulkier, more expensive ballasts than the rapid-start fixtures do. Because of these factors, they are not as popular as the rapid-start type.

Troubleshooting Fluorescent Lighting Systems
Table 26-1 on page 360 lists some troubles common to fluorescent lights and lists possible causes and solutions.

Compact Fluorescent Bulbs
In recent years there have been great advancements in fluorescent lighting technology. The long, standard fluorescent tubes that we are familiar with have been made more compact so that they fit into standard light fixtures. Some of these new bulbs come with separate adaptors, while others have adaptors already built in. These adaptors contain small electronic ballasts used to operate the bulb. They are available in a variety of shapes, sizes, and wattages.

These bulbs give you the same light as higher wattage incandescent bulbs, but they use just a fraction of the energy. Now it only takes a 13-W *compact fluorescent bulb* to provide the light

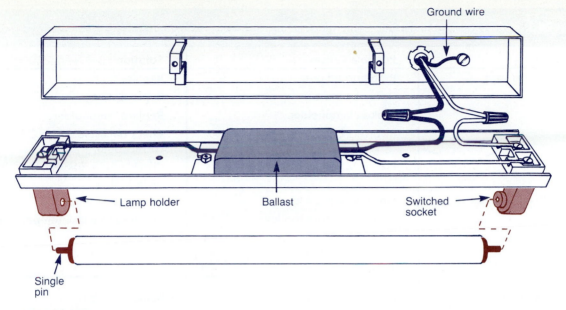

Ground wire

Lamp holder   Ballast   Switched
socket

Single
pin

A. Wiring diagram

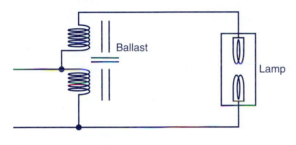

Ballast

Lamp

B. Schematic diagram

Figure 26-26   Instant-start fluorescent fixture.

of a 60-W incandescent bulb (using only a quarter of the energy). That is a considerable energy savings (Figure 26-27).

Compact fluorescents last up to 10 times longer than regular incandescents, and they offer convenience and savings! Not only does one compact fluorescent bulb save the cost of nine regular replacement bulbs, but in hard-to-reach places, bulbs do not need changing nearly so often.

26-6 High-Intensity Discharge (HID) Lamps

High-intensity discharge (HID) lamps, like fluorescent lamps, produce light by means of a gaseous-discharge arc tube. Unlike fluorescent lamps, these lamps operate at a much higher

current, which flows through a much shorter *arc tube* that is enclosed in an *outer bulb* (a bulb within a bulb). The gas pressure in the tube is much higher than in a fluorescent lamp. There are three different types in common use: the mercury vapor lamp, the metal halide lamp, and the high-pressure sodium lamp (Figure 26-28).

None of these lamps can be directly connected to an electric circuit; they must be operated with a matching ballast. When first turned ON, the lamp starts at low brightness and gradually increases in intensity. If turned OFF, it must cool off before it will relight.

The main advantages of high-intensity discharge lamps are: high light efficiency, very long life, and high watt output from single fixtures. They are generally used in locations such as factories and parking lots where large areas

TABLE 26-1	FLUORESCENT LIGHT TROUBLESHOOTING CHART	
Problem	**Possible Causes**	**Solution**
Light fails to start.	Bulb burned out.	Replace bulb.
	Defective starter.	Replace starter.
	Broken lampholder.	Replace lampholder.
	Incorrect bulb for ballast.	Check ballast label for correct bulb.
	Fixture wired incorrectly.	Check wiring diagram on ballast label.
	Line voltage too low.	Call power utility.
	Air temperature too low.	Install special low-temperature ballast.
	Defective ballast.	Replace ballast.
Ends of bulb glow, but center does not.	Defective starter.	Replace starter.
	Fixture wired incorrectly.	Rewire according to ballast wiring diagram.
	Fixture not adequately grounded.	Check ground connection on fixture.
Ends of bulb are black.	Bulb nearly burned out.	Replace bulb.
Bulb flickers or blinks.	Tube pins making poor contact.	Clean prongs and tighten tube in lampholder.
	Bulb nearly burned out.	Replace bulb.
	Defective or incorrect starter.	Replace starter.
	Air temperature too low.	Warm room; if necessary install special low temperature ballast.
Cycling ballast, which turns power OFF and ON.	Ballast is operating above maximum temperature, causing the thermal protective switch in the ballast to open and close.	Replace ballast.

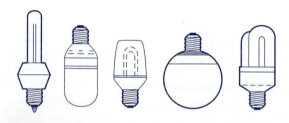

A. Typical shapes

13 W = 60 W

B. Energy savings

Figure 26-27 Compact fluorescent bulbs.

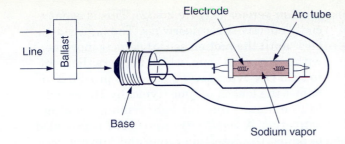

Figure 26-28   Sodium lamp construction.

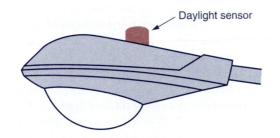

Figure 26-29   High-intensity highway lamp.

cuit can be wired on a solderless breadboard to simulate the operating principle of an automatic ON/OFF lighting system. A cadmium sulphide (CdS) cell is wired in series with a dc voltage source and a sensitive dc relay *coil*. The resistance of the CdS cell changes with the amount of light that falls on its face. When light shines on the face of the CdS cell, its resistance is low and maximum current flows through the relay coil energizing it. The normally closed contacts of the relay open and the light is automatically switched OFF. When light to the CdS cell is removed its resistance increases, reducing the current in the circuit and de-energizing the coil. The normally closed contacts of the relay close and the light is automatically switched ON.

About ⬛ Electronics

HID lights generate light by a gas arc between two electrodes. Compared to halogen lights, HIDs make three times more light, cover twice as much distance, use two-thirds less energy, and last three times longer. Also, when used as headlights HIDs make embedded lane reflectors more visible.

must be lighted. Used along with a built-in photoelectric-control unit, they are used for automatic street and highway lighting (Figure 26-29).

The schematic for a simple photoelectric-control unit is shown in Figure 26-30. This cir-

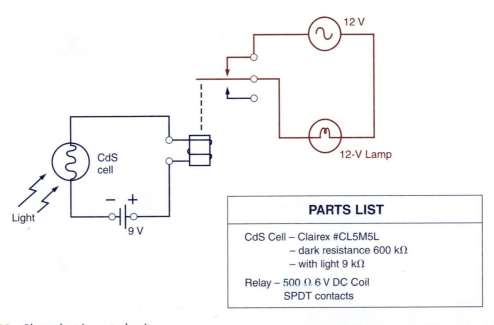

PARTS LIST
CdS Cell – Clairex #CL5M5L
– dark resistance 600 kΩ
– with light 9 kΩ
Relay – 500 Ω 6 V DC Coil
SPDT contacts

Figure 26-30   Photoelectric control unit.

26-7 Security Lighting

There are many ways that lights can be used as a security measure both inside and outside the house. To make your house look occupied when you are out, without burning up your energy budget, install **programmable timers** to turn selected lights ON and OFF at specified times. Portable timers plug into the wall, while permanent types replace light switches altogether.

A light timer is basically a switch that is operated by a 24-hour clock (Figure 26-31). Mechanical, motor driven timers require a neutral wire in addition to the switched hot wires so that the clock motor is provided with power at all times. Certain electronic timers, designed as the direct replacement for switches, are wired in exactly the same manner as the switch they replace and require no neutral wire connection.

If you want outdoor lights to go ON and stay ON all night, a **photoelectric cell** control module will automatically switch the lights ON at dusk and OFF at dawn.

Another effective outdoor security device is a fixture that incorporates a passive infrared **motion sensor** (Figure 26-32). This is especially effective over doorways. The light stays OFF until the motion sensor detects movement within a certain zone, which triggers the light to turn ON. A *delay adjustment* is provided to adjust the length of time (typically 10 seconds to 15 minutes) that the light remains *on* after activation. A *sensitivity adjustment* is provided to adjust the detection range and prevent accidental triggering by family pets or other interference. A photocell is integrated within the system to prevent lights from coming on during daylight hours. Most security motion-sensor light units work only with incandescent lights and can replace existing outdoor lighting that already uses a wall switch.

You can control lights and appliances in and around your home from one convenient location with a **wireless remote-control lighting system** (Figure 26-33). A *program controller* sends signals through existing house wiring to remote module receptacles and switches to dim/brighten lights and turn appliances ON/OFF. Programming selected lights and appliances to go ON and OFF gives the residence an occupied appearance when empty. No wiring is required,

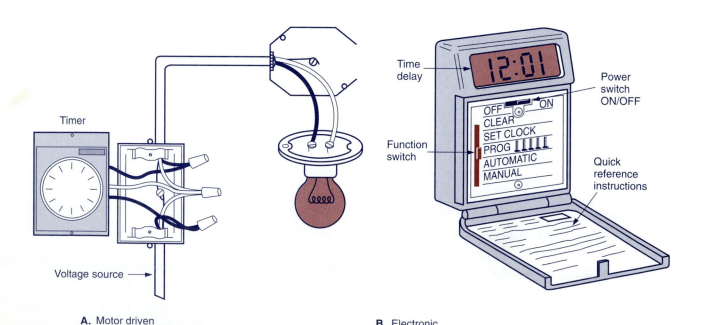

A. Motor driven

B. Electronic

Figure 26-31 Automatic light switch timer.

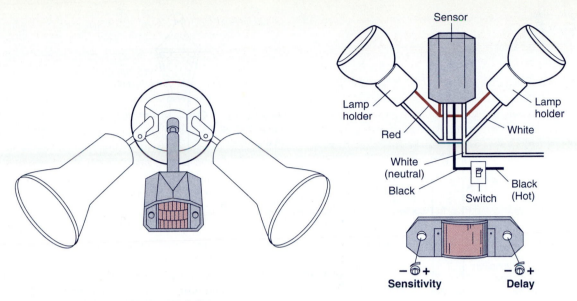

Figure 26-32   Motion-sensor security light.

Outdoor Lighting   Examples of outdoor lighting include decorative landscape lighting and post lights. Often this type of lighting involves running wires underground. Underground wiring is done by using conduit or special cables such as Type UF (underground feeder) designed for direct burial. Article 339 of the Code covers wiring with UF cable and Table 300-5 lists the minimum depths required for various types of wiring methods. Always check local Codes for rules that apply to outdoor wiring in your community. Figure 26-34 shows a typical post light installation.

as the controller and modules use existing ac outlets.

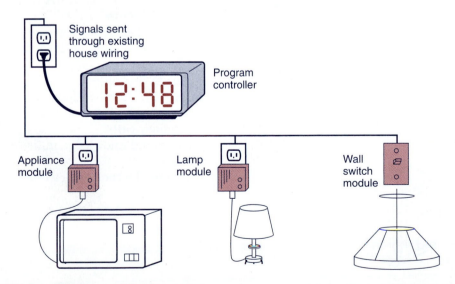

Figure 26-33   Wireless remote-control lighting system.

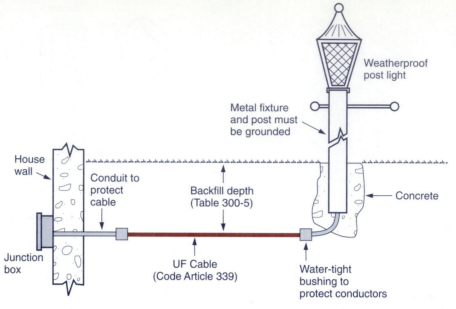

Weatherproof
post light

Metal fixture
and post must
be grounded

Concrete

House
wall

Conduit to
protect
cable

Backfill depth
(Table 300-5)

Junction
box

UF Cable
(Code Article 339)

Water-tight
bushing to
protect conductors

Figure 26-34 Typical post light installation.

A typical low-voltage decorative lighting system is shown in Figure 26-35. The light fixtures operate in the 6-14 V range and are safer and easier to install than traditional 120 V systems. Outdoor power to the system is supplied via a weatherproof receptacle that is GFCI protected.

A step-down transformer is used to lower the voltage to the parallel connected light fixture bulbs. The transformer unit normally has integrated within it a circuit breaker for overcurrent protection and a daylight sensor for automatic control of the lights.

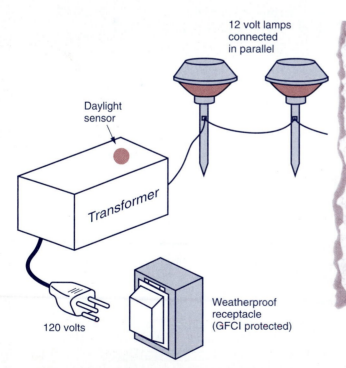

12 volt lamps
connected
in parallel

Daylight
sensor

Transformer

120 volts

Weatherproof
receptacle
(GFCI protected)

Figure 26-35 Low-voltage decorative lighting.

Review and Applications

Review Questions

1. How is light produced in an incandescent lamp?
2. (a) In what way are halogen bulbs constructed and operated differently from conventional tungsten types?
   (b) What advantage does this result in?
3. Name five ways in which lamps are rated.
4. A typical dual-filament incandescent lamp has rated filaments of 40 W and 60 W. Explain the order in which the filaments are operated to produce three different light levels.
5. Explain the difference in construction and application between a general-service bulb and vibration-service incandescent bulb.
6. How does the cold resistance of an incandescent bulb filament compare with its hot resistance?
7. Why are special NEC requirements specified for recessed-type light fixtures?
8. What type of incandescent bulbs are best to use with pot light fixtures? Why?
9. List three advantages of using fluorescent lighting rather than standard incandescent lighting.
10. Explain how light is produced in a fluorescent tube.
11. What potential hazard is involved in the disposal of fluorescent tubes?
12. State two basic functions performed by the ballast of any type of fluorescent fixture.
13. Compare core-coil and electronic ballasts with regard to light efficiency, weight, noise level, operating temperature, and cost.
14. (a) Name the three types of fluorescent light circuits.
    (b) Which is the most widely used?
15. State two ways in which the rapid-start fluorescent fixture circuit is different from that of the preheat fixture.
16. Why do instant-start fluorescent fixtures potentially pose a greater electrical hazard than other types do?

17. (a) List two indications of a possible defective glow-switch starter.
    (b) What does a blackened fluorescent tube end indicate?
    (c) How does air temperature affect the operation of fluorescent fixtures?
    (d) What type of fluorescent fixture requires a properly grounded chassis for reliable starting?
    (e) Where can you often find the wiring diagram for a fluorescent fixture that is under repair?
    (f) What is the possible cause of a ballast that is cycling power OFF and ON?
18. Compare standard fluorescent lamps and high-intensity discharge lamps with regard to:
    (a) current flow
    (b) length of the arc tube
    (c) light efficiency
    (d) life expectancy
19. (a) What does a light timer consist of?
    (b) What device is used to automatically switch lights ON at dusk and OFF at dawn?
    (c) Explain the function of the delay- and sensitivity-adjustment screws used on motion-sensor units.
    (d) Explain how a wireless remote-control lighting system operates.

Problems

1. A 40 W 130 V standard incandescent lamp is replaced by a 25 W 130 V halogen lamp. How does this change:
   (a) the amount of light output?
   (b) the amount of current flow?
2. A motion-sensor light fixture keeps shining for too long of a period after motion stops. The sensitivity control is adjusted for less sensitivity. Will this solve the problem? Why?

Critical Thinking

1. A table lamp works only when the cord attached to the plug cap is held in a certain position.
   (a) What is the most likely cause of this problem?
   (b) What should be done to correct the problem?
2. The circuit breaker located on the transformer of a newly installed low-voltage decorative lighting system trips after the system has been operational for several minutes. The wiring is checked and no problems are found. Assuming the transformer and circuit breaker are OK, what else could be the problem?

Portfolio Projects

- Construct a display of the various types of lights discussed in this chapter.
- Completely rewire a discarded table lamp.
- Make a plan of a wireless remote-control lighting system that could be used in a home. Include a bill of material and directory of all modules used.

Electric Motors

An electric motor converts electric energy to mechanical energy by using interacting magnetic fields. Electric motors are used for a wide variety of residential, commercial, and industrial operations. This chapter deals with the operating principles of dc, universal, and ac electric motors.

Objectives

After studying this chapter, you should be able to:
1. Explain the basic motor operating principle.
2. Describe the construction, connection, and operating characteristics of dc motors.
3. Describe the construction, connection, and operating characteristics of the universal motor.
4. Describe the construction, connection, and operating characteristics of ac single-phase and polyphase motors.
5. Explain the relationship between motor power, speed, and torque.
6. Diagnose common motor problems.

Key Terms

- right-hand motor rule
- torque
- permanent magnet dc motor
- series-type dc motor
- shunt-type dc motor
- compound-type dc motor
- armature
- rotor
- commutator
- stator

- pole
- universal motor
- single-phase
- three-phase
- polyphase
- rotating magnetic field
- synchronous speed
- induced voltage
- induction motor
- slip

- starting winding
- centrifugal switch
- shaded-pole motor
- squirrel-cage motor
- wound rotor motor
- split-phase motor
- capacitor-start motor
- permanent-capacitor motor
- synchronous motor
- energy-efficient motor

RESEARCH INTERNET SITES

- **Western Electric**
- **Reliance Electric**
- **U.S. Motors**

27-1 Motor Principle

Electric motors are used to convert *electric energy into mechanical energy*. The motor represents one of the most useful and labor-saving inventions in electrical technology. Over 50 percent of the electricity produced in the United States is used to power motors. Motors are used to turn the wheels of industry. In the home, more than half of all the appliances are operated by electric motors.

The operation of a motor depends upon the interaction of two *magnetic fields*. Simply stated, the motor operates on the principle that *two magnetic fields can be made to interact to produce motion*. The magnetic poles of a moveable armature are made to repel and attract the magnetic poles of a fixed stator. Attraction and repulsion of these magnetic poles produce the desired rotary movement (Figure 27-1).

A current-carrying conductor, placed in, and at right angles to a magnetic field tends to move at right angles to the field. The amount of force exerted to move it varies directly with the strength of the magnetic field, the amount of current flowing through the conductor, and the length of the conductor.

To determine the direction of movement of a conductor carrying current in a magnetic field, the ***right-hand motor rule*** is used. The thumb and first two fingers of the right hand are arranged to be at right angles to each other with the forefinger pointing in the direction of the magnetic lines of force of the field, the middle finger pointing in the direction of current flow (− to +) in the conductor. The thumb will be pointing in the direction of movement of the conductor as shown in Figure 27-2.

Figure 27-3 illustrates how motor *torque* is produced by a current-carrying coil or loop of wire placed in a magnetic field. The interaction

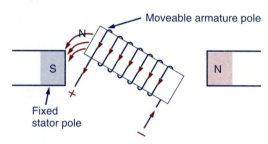

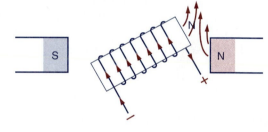

A. Attraction of poles

B. Repulsion of poles

Figure 27-1   Motor principle.

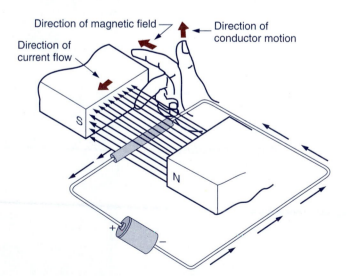

Figure 27-2   Right-hand motor rule.

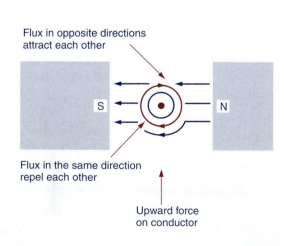

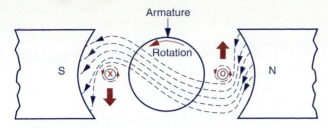

Figure 27-3 Developing motor torque.

between the two magnetic fields causes a bending of the lines of force. When the lines tend to straighten out, they cause the loop to undergo a rotating motion. The left conductor is forced downward, and the right conductor is forced upward, causing a counterclockwise rotation of the armature.

In general, motors are classified according to the type of power used (ac or dc) and the motor's principle of operation. There are several major classifications of the motors in common use; each will specify characteristics that suit it

to particular applications. Figure 27-4 shows a family tree of common types of motors.

27-2 Direct Current (DC) Motors

Permanent-Magnet Motor
A *permanent magnet dc motor* is constructed the same as a permanent-magnet *dc* generator. It has a pair of *permanent-magnet* poles, a wire-wound *armature*, and a commutator with a set of brushes. Any dc generator may be run as a motor and vice versa. Figure 27-5 illustrates the operation of the permanent-magnet motor.

The direction of rotation of a permanent-magnet dc motor is determined by the direction of the current flow through the armature coil. Reversing the leads to the armature will reverse the direction of rotation. A reversing switch can be wired to control the direction of rotation of the motor using a double-pole, double-throw (DPDT) switch. A diagram of the circuit connection is shown in Figure 27-6.

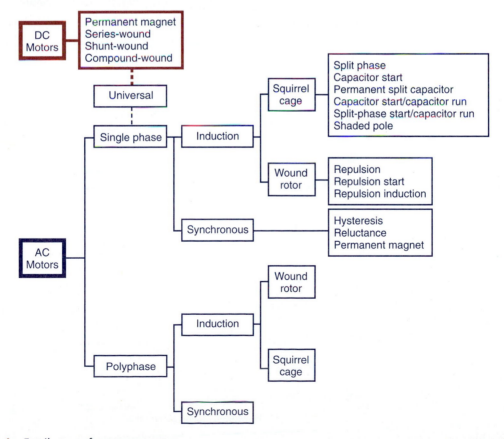

Figure 27-4 Family tree of common motors.

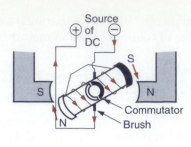

A.

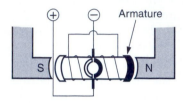

B.

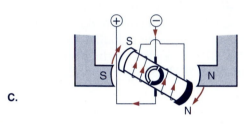

C.

Figure 27-5 Permanent-magnet dc motor operation.

A•Current flows through amature coil from dc voltage source. •North and south poles are set up in armature coil. •Magnetic attraction and repulsion between poles of the armature and poles of the main field cause armature to rotate in clockwise direction.

B•When armature poles are in line with field poles, brushes are at gap in commutator and no current flows in the armature. •Thus, forces of magnetic attraction and repulsion stop.

C•Then inertia carries the armature past this neutral point. •Current flows through armature coil in reverse direction due to commutator's reversing action. •A north and south pole each set up in armature coil. •Then forces of attraction and repulsion keep the armature rotating in same clockwise direction.

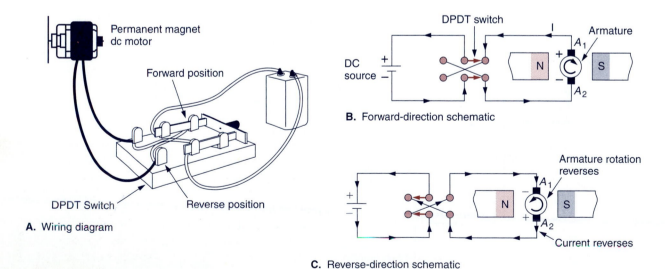

A. Wiring diagram

B. Forward-direction schematic

C. Reverse-direction schematic

Figure 27-6 Reversing the direction of rotation on a permanent-magnet motor.

One of the features of a dc motor is that its speed can be easily controlled. The speed of a permanent-magnet motor is directly proportional to the value of the voltage applied to the armature. The greater the armature voltage, the faster the speed of the motor.

Figure 27-7 shows the schematic for a three-speed dc automobile blower-motor circuit. The resistors and switch control motor speed by varying the amount of voltage that is applied to the motor armature. With the switch in the *Lo* position, current flows through both resistors, reducing both the voltage at the armature and the motor speed. As the switch is moved from *Lo*, to *Med*, to *Hi*, the voltage across the motor is increased and the blower speed increases.

Electromagnetic DC Motors

Small permanent-magnet dc motors are used for light-duty applications such as power windows on an automobile, electronic-control devices, and various toys and battery-operated equipment. The magnetism needed to operate larger dc motors is produced by sets of coils or windings. This design adds greater operating flexibility, since the strength of the magnetic field can be varied by controlling the amount of current supplied to the field windings. There are three types of electromagnetic dc motors: *series*, *shunt*, and *compound*. This refers to the method in which the field coils are connected with respect to the armature.

In an electromagnetic dc motor, the *stator* (field) winding is supplied by dc and the **rotor** (armature) winding is connected to dc by brushes and a *commutator*. Interaction of the armature current and the stator field produces a torque which rotates the armature. The commutator can be regarded as a switch that maintains the proper direction of current in the rotor conductors to give a constant unidirectional torque (Figure 27-8).

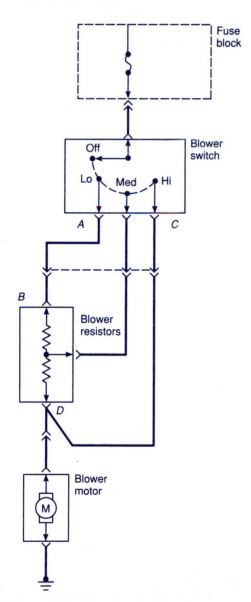

Figure 27-7 Blower-motor schematic.

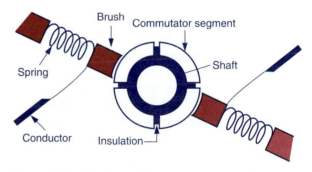

Figure 27-8 Brushes and commutator.

In the **series-type dc motor**, the field coils are made of a few turns of very large wire and, therefore, have *low resistance*. The field winding is connected in series with the armature (Figure 27-9A). The most important feature of this motor is its ability to start very heavy loads. As the load to the motor increases, the torque (or turning effort) increases, but the speed decreases. This means that a series motor runs fast with a light load and runs slower as the load is increased. The automobile starting motor is a series type. Such motors are also used in cranes, hoists, and elevators.

In the **shunt-type dc motor**, the field coils are made of many turns of small wire and, therefore, have *high resistance*. This field winding is connected across the voltage source and receives full voltage at all times (Figure 27-9B). A shunt motor has good speed regulation (speed does not vary much under different load conditions). The shunt motor is used where constant speed is desirable in the driven machine. By adding a rheostat in series with the shunt field circuit, the speed of the motor can be controlled above base speed. The speed of the motor will be inversely proportional to the field current. This means that a shunt motor runs fast with a low field current and runs slower as the current is increased. Shunt motors can race to dangerously high speeds if current to the field coil is lost.

A **compound-type dc motor** uses both series and shunt field windings (Figure 27-9C), which are connected so their fields add cumulatively. This two-winding connection produces characteristics intermediate to the shunt field and series field motors. The speed of these motors varies a little more than that of shunt motors, but not as much as series motors. Also, they have a fairly large starting torque, much more than shunt motors do, but less than series motors do. These combined features give the compound motor a wide variety of uses.

DC motors are often used where a wide range of torque and speed control is required to match the needs of the application. The brush-commutator arrangement, however, presents brush maintenance and electrical arcing problems.

27-3 Universal Motors

The **universal motor** is constructed in much the same way as a series-type dc motor. Like the dc-series motor, its armature and field coils are connected in series. Universal motors are motors which can be operated with either direct current or alternating current. The reason for this is that a dc motor will continue to turn in the same direction if the current through the armature and field are reversed at the same time. This is exactly what happens when the motor is connected to an ac source. Universal motors are motors that are easily recognized

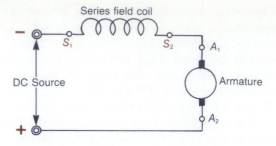

A. Series motor connection

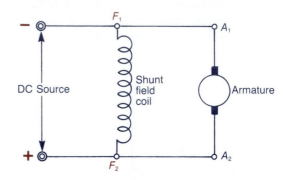

B. Shunt motor connection

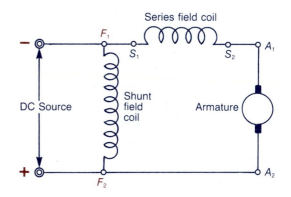

C. Compound motor connection

■ To reverse direction of rotation, reverse leads to the series field coil or armature, but *not* both.

■ To reverse direction of rotation, reverse leads to shunt field coil or armature, but *not* both.

■ Series and shunt field windings are connected so their fields add cumulatively.

■ To reverse direction of rotation, reverse leads to the armature or to *both* the series and shunt fields.

Figure 27-9 DC series, shunt, and compound motor connections.

because they use a commutator and brushes (Figure 27-10).

Like the dc-series motor, the speed of a universal motor drops with increasing load. The

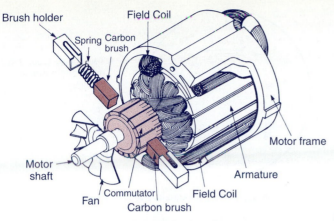

A. Construction

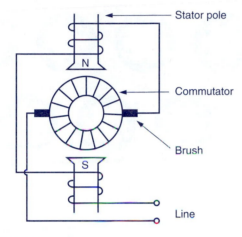

B. Internal connection

Figure 27-10 Universal motor.

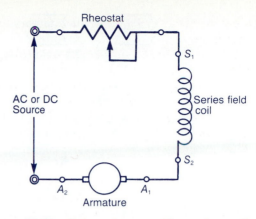

Figure 27-11 Universal motor with speed control.

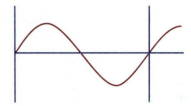

A. Single-phase

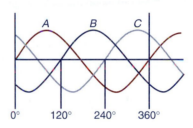

B. Three-phase

Figure 27-12 AC power sources.

speed can be controlled by varying the voltage that is applied to the motor. This can take the form of a solid-state voltage controller or rheostat connected in series with one of the line leads (Figure 27-11). The direction of rotation can be changed by reversing the connections to the armature brushes. This is usually done with a reversing switch. This type of motor is used in sewing machines, vacuum cleaners, drills, saws, and small appliances.

27-4 Alternating Current (AC) Motors

Alternating current (ac) motors are classified according to the type of ac power used as being *single-phase* or *three-phase* (polyphase). As the names indicate, **single-phase** motors operate with single-phase ac, while **polyphase** (*three-*

phase) motors operate with three-phase ac (Figure 27-12).

A common feature of all ac motors is a *rotating magnetic field* that is set up by the stator windings. Figure 27-13 illustrates the concept of a **rotating magnetic field**. A conductor is placed in the magnetic field at point 1 (Figure 27-13A). If a current is sent through the conductor, the conductor will move to point 2, and will stop there. However, if the field is moved to the position shown in Figure 27-13B, the conductor will again move, but when it has reached point 3, it will again stop. If the field is moved another time to the position shown in Figure 27-13C, the conductor will move again until it gets to point 4. From this, we can see that if the field continues to be turned, the conductor will follow along.

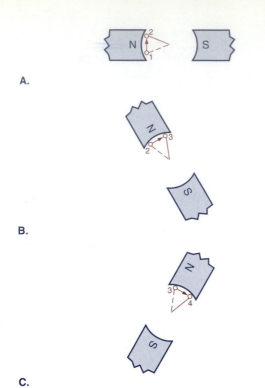

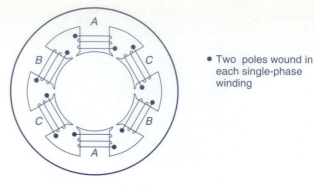

A. Placement of electromagnetic stator winding

- Two poles wound in each single-phase winding

Resultant magnetic field

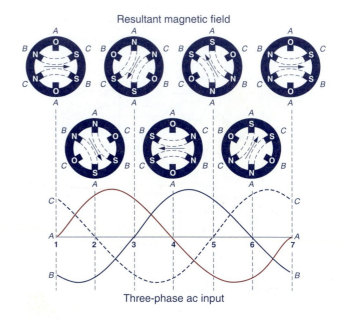

Three-phase ac input

$$S = \frac{120\ F}{P}$$

$$= \frac{120 \times 60}{2}$$

$S = 3600$ rpm, synchronous speed of field

- Direction of rotation of the magnetic field is reversed by interchanging any two of the three-phase input leads to the stator.

B. Resultant rotating magnetic field pattern

Figure 27-14 Development of a rotating magnetic field.

A.

B.

C.

Figure 27-13 Rotating magnetic field.

With the *pole* pieces stationary, as they actually are in a motor, the field can be made to rotate by electric means, rather than by mechanical means. This concept can be illustrated for three-phase motors by considering three coils placed 120° apart. Each coil is connected to one phase of a three-phase power supply (Figure 27-14). When a three-phase current passes through these windings, a rotating magnetic field effect is set up, which travels around the inside of the stator core. The speed of this rotating magnetic field depends on the number of stator poles and the frequency of the power source. This speed, called the *synchronous speed*, is determined by the formula:

$$S = \frac{120\ F}{P}$$

where:

S = SYNCHRONOUS SPEED IN RPM
F = FREQUENCY IN Hz OF THE
  POWER SUPPLY
P = NUMBER OF POLES WOUND IN
  EACH OF THE SINGLE-PHASE
  WINDINGS

AC motors are classified by operating principle as being either induction or synchronous motors. The ac *induction motor* is by far the most commonly used type of motor because it is relatively simple and can be built at less cost than other types. Induction motors are made in both three-phase and single-phase types. The induction motor is so named because *no external voltage is applied* to its rotor. Instead, the ac current in the stator *induces a voltage* across an air gap and into the rotor winding to produce rotor current and magnetic field. The stator and rotor magnetic fields then interact and cause the rotor to turn (Figure 27-15). Most ac induction motors are normally designed to operate at one fixed speed. Figure 27-16 describes the operation of a fixed speed three-phase squirrel-cage induction motor. Figure 27-17 describes the operation of a variable speed three-phase wound-rotor induction motor.

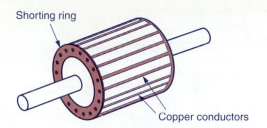

Shorting ring

Copper conductors

A.

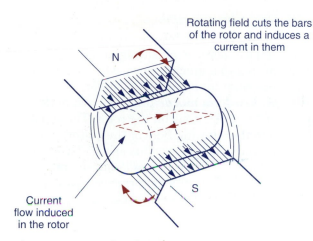

Rotating field cuts the bars of the rotor and induces a current in them

N

S

Current flow induced in the rotor

Figure 27-15 Induced rotor current.

The rotor of an induction motor does not rotate at synchronous speed, but lags it slightly. For example, an induction motor having a synchronous speed of 1800 rpm will often have a rated speed of 1750 rpm at rated horsepower. The lag is usually expressed as a percentage of the synchronous speed called the *slip*.

$$\% \text{ Slip} = \frac{\text{Synchronous speed} - \text{Running speed}}{\text{Synchronous speed}} \times 100$$

$$= \frac{1800 - 1750}{1800} \times 100$$

$$= 2.78 \text{ percent}$$

If the rotor turns at the same speed at which the field rotates, there will be no relative motion between the rotor and the field and no voltage induced. Because the rotor *slips* with respect to the rotating magnetic field of the stator, voltage and current are induced in the rotor. As load increases, slip and torque also increase.

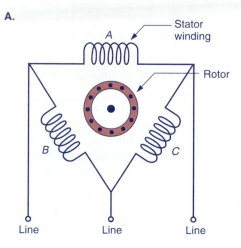

Stator winding

A

Rotor

B

C

Line   Line   Line

B.

- The rotor consists of copper conductors embedded in a solid core with the ends shorted to resemble a squirrel cage.
- Speed is essentially constant.
- Large starting currents required by this motor can cause voltage fluctuations.
- Direction of rotation can be reversed by merely interchanging any two of the three main power lines to the motor.
- Power factor tends to be poor for reduced loads.
- Simplest and most reliable motor.
- When voltage is applied to the stator winding, a rotating magnetic field is produced, which induces a voltage in the rotor. This voltage, in turn, creates a large current flow in the rotor. The high current in the rotor creates a magnetic field of its own. The rotor field and the stator field tend to attract each other. This situation creates a torque which spins the rotor in the same direction as the rotation of the magnetic field produced by the stator.
- Once started, the motor will continue to run with a phase loss.

Figure 27-16 Three-phase *squirrel-cage induction motor.*

While a three-phase induction motor sets up a rotating field that can start the motor, *a single-phase motor needs an auxiliary means of starting.* Once a single-phase induction motor is running, it develops a rotating magnetic field. However, before the rotor begins to turn, the

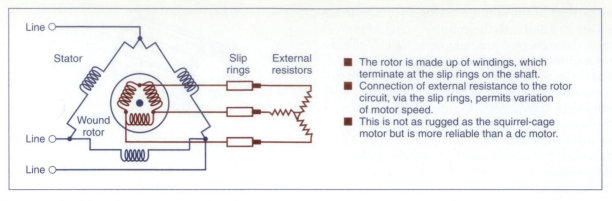

Figure 27-17 Three-phase *wound-rotor induction motor.*

stator produces only a pulsating, stationary field.

To produce a rotating field, and thus a starting torque, an auxiliary **starting winding** is placed at right angles to the main stator winding so that the currents through them are out of phase by 90°. Once the motor has started, the auxiliary winding is often removed from the circuit by a **centrifugal switch**.

Single-phase induction motors are used in applications where three-phase power is not available and are generally in the fractional to 10-hp range. See Figure 27-18 to 27-22.

When running, the torque produced by a single-phase motor is pulsating and irregular, contributing to a much lower power factor and efficiency than for a polyphase motor.

A **synchronous motor** produces magnetic poles at fixed positions on the rotor. These poles lock on to the rotating field of the stator and turn the rotor at synchronous speed. There are several different types of single-phase and polyphase synchronous motors. See Figures 27-23 and 27-24.

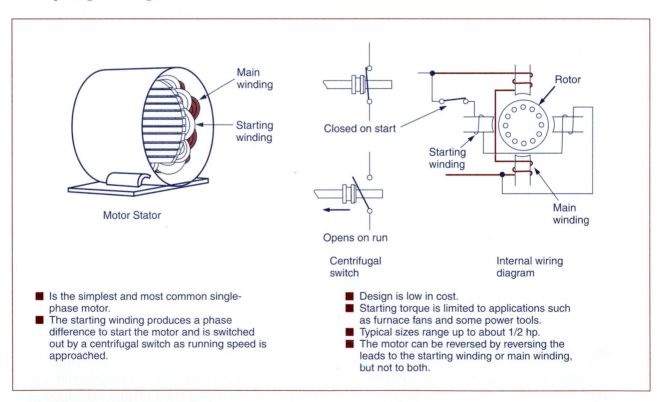

Figure 27-18 *Split-phase motor.*

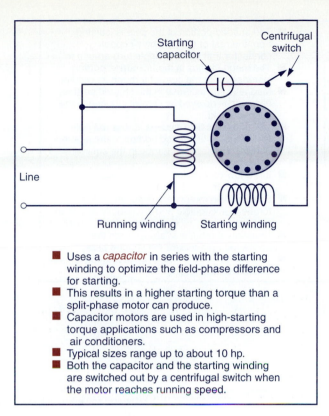

- Uses a *capacitor* in series with the starting winding to optimize the field-phase difference for starting.
- This results in a higher starting torque than a split-phase motor can produce.
- Capacitor motors are used in high-starting torque applications such as compressors and air conditioners.
- Typical sizes range up to about 10 hp.
- Both the capacitor and the starting winding are switched out by a centrifugal switch when the motor reaches running speed.

Figure 27-19 *Capacitor-start motor.*

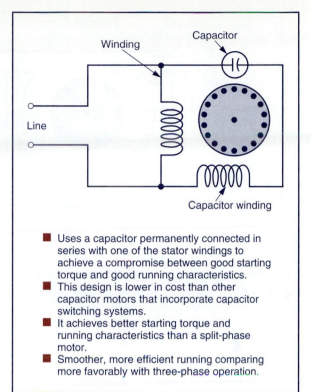

- Uses a capacitor permanently connected in series with one of the stator windings to achieve a compromise between good starting torque and good running characteristics.
- This design is lower in cost than other capacitor motors that incorporate capacitor switching systems.
- It achieves better starting torque and running characteristics than a split-phase motor.
- Smoother, more efficient running comparing more favorably with three-phase operation.

Figure 27-20 *Permanent-capacitor motor.*

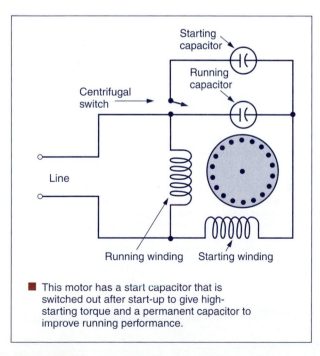

- This motor has a start capacitor that is switched out after start-up to give high-starting torque and a permanent capacitor to improve running performance.

Figure 27-21 Capacitor-start capacitor-run motor.

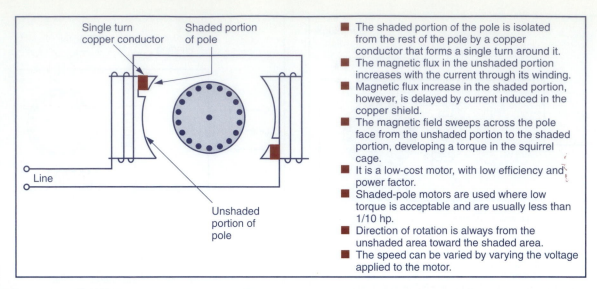

■ The shaded portion of the pole is isolated from the rest of the pole by a copper conductor that forms a single turn around it.

■ The magnetic flux in the unshaded portion increases with the current through its winding.

■ Magnetic flux increase in the shaded portion, however, is delayed by current induced in the copper shield.

■ The magnetic field sweeps across the pole face from the unshaded portion to the shaded portion, developing a torque in the squirrel cage.

■ It is a low-cost motor, with low efficiency and power factor.

■ Shaded-pole motors are used where low torque is acceptable and are usually less than 1/10 hp.

■ Direction of rotation is always from the unshaded area toward the shaded area.

■ The speed can be varied by varying the voltage applied to the motor.

Figure 27-22 *Shaded-pole motor.*

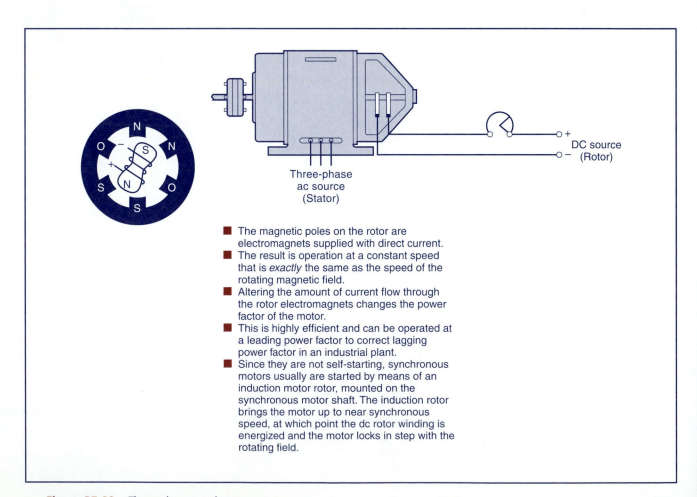

■ The magnetic poles on the rotor are electromagnets supplied with direct current.

■ The result is operation at a constant speed that is *exactly* the same as the speed of the rotating magnetic field.

■ Altering the amount of current flow through the rotor electromagnets changes the power factor of the motor.

■ This is highly efficient and can be operated at a leading power factor to correct lagging power factor in an industrial plant.

■ Since they are not self-starting, synchronous motors usually are started by means of an induction motor rotor, mounted on the synchronous motor shaft. The induction rotor brings the motor up to near synchronous speed, at which point the dc rotor winding is energized and the motor locks in step with the rotating field.

Figure 27-23 Three-phase synchronous motor.

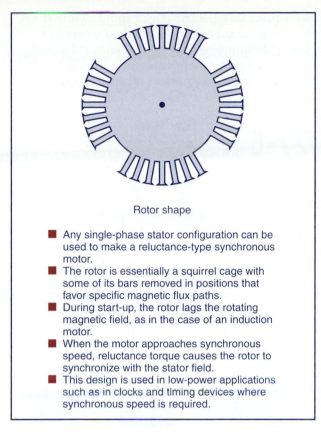

Rotor shape

- Any single-phase stator configuration can be used to make a reluctance-type synchronous motor.
- The rotor is essentially a squirrel cage with some of its bars removed in positions that favor specific magnetic flux paths.
- During start-up, the rotor lags the rotating magnetic field, as in the case of an induction motor.
- When the motor approaches synchronous speed, reluctance torque causes the rotor to synchronize with the stator field.
- This design is used in low-power applications such as in clocks and timing devices where synchronous speed is required.

Figure 27-24 Single-phase reluctance synchronous.

27-5 Motor Power and Torque

The mechanical power rating of motors is expressed in either horsepower (hp) or watts (W): 1 hp = 746 W. Two important factors that determine mechanical power output are torque and speed. *Torque* is the amount of twist or turning power and is often stated in foot/pound (ft/lb). Motor *speed* is commonly stated in *revolutions per minute* (rpm).

$$\text{Horsepower} = \frac{\text{Speed (in rpm)} \times \text{Torque (in ft/lb)}}{5252}$$

Thus for any given motor, the horsepower depends upon the speed. The slower the motor operates, the more torque it must produce to deliver the same amount of power. To withstand the greater torque, slow motors need stronger components than those of higher speed motors of the same power rating. Slower motors are generally larger, heavier, and more expensive than faster motors of equivalent power rating.

The amount of torque produced by a motor generally varies with speed and depends on the type and design of the motor. Figure 27-25 shows a typical motor torque-speed graph.

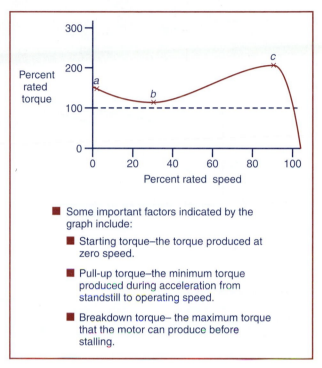

- Some important factors indicated by the graph include:
  - Starting torque–the torque produced at zero speed.
  - Pull-up torque–the minimum torque produced during acceleration from standstill to operating speed.
  - Breakdown torque– the maximum torque that the motor can produce before stalling.

Figure 27-25 Typical motor torque-speed graph.

The power efficiency (Figure 27-26) of an electric motor is defined as:

$$\% \text{ Efficiency} = \frac{\text{Output}}{\text{Input}} \times 100$$

The power input to the motor is either transferred to the shaft as power output or is lost as heat through the body of the motor. **Energy-efficient motors** have reduced heat losses and

About   Electronics

Quiet and odorless, electricity is being used to power more and more things. With electric boats, you don't have to start the engine, the ac electric motor fed by batteries eliminates the need for gearboxes, and you can change from forward to reverse instantly to maneuver in small areas. It is also easy to convert pontoon and deck boats to electric.

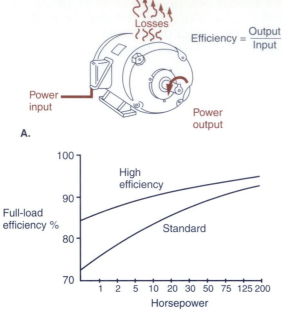

Efficiency = $\dfrac{\text{Output}}{\text{Input}}$

Power input

Power output

A.

Full-load efficiency %

High efficiency

Standard

Horsepower

B. Motor efficiency comparison

Figure 27-26 Motor efficiency.

so require less electric power input, while they still provide the same mechanical power output. This improvement in efficiency is basically accomplished by using more material in the motor, using better material, and implementing design changes.

27-6 Troubleshooting Motors

Insulation failure, worn brushes, and bearing failure are the two most common types of motor failures. Some common motor problems, as well as their possible causes and solutions, are given in Table 27-1.

TABLE 27-1	TROUBLESHOOTING SMALL MOTORS	
Problem	**Possible Causes**	**Solution**
Motor does not run.	Voltage not reaching motor.	Check fuse or circuit breaker; receptacle; cord set.
	Faulty ON/OFF switch.	Replace switch.
	Worn brushes.	Replace brushes.
	Weakened brush springs.	Replace springs if they no longer hold brushes firmly against commutator.
Motor runs continuously.	Shorted ON/OFF switch.	Replace switch.
Motor overheats.	Dirty motor or clogged air intake.	Clean with vacuum cleaner.
	Tight or dry bearings.	Lubricate bearings.
Motor is noisy.	Worn bearings.	Replace bearings.
Split-phase motor hums, but will not turn.	Faulty centrifugal switch.	Clean or replace centrifugal switch contacts.
Universal motor sparks.	New brushes not properly seated.	Seat brushes with fine sandpaper to fit contour of commutator.
	Worn or sticky brushes.	Replace or clean brush holder.
	Dirty commutator.	Clean with fine sandpaper.
	Open or shorted armature coils.	Replace armature.
Motor blows fuses.	Shorted cord set.	Replace cord.
	Rotor jammed in nonturning bearings.	Free shaft and lubricate bearings.
	Shorted or open windings.	Replace motor.
Motor smokes.	Burned or partially short-circuited windings.	Replace motor.

HELP WANTED

Electric Motor Repair Person
Job Description
- Troubleshoot and complete motor repairs to large ac and dc motors.
- Rewinding of motors.
- Motor dipping and baking.
- Dynamic balancing of motors.
- Minimum 2 years experience in service and repair of electric motors.

Review and Applications

Related Formulas

$$\text{Synchronous speed} = \frac{120\,F}{P}$$

$$\text{Horsepower} = \frac{\text{Speed (in rpm} \times \text{Torque (in ft/lb)}}{5252}$$

$$\%\text{ Slip} = \frac{\text{Synchronous speed} - \text{Running speed}}{\text{Synchronous speed}} \times 100$$

$$1 \text{ hp} = 746\ W$$

$$\%\text{ Efficiency} = \frac{\text{Output}}{\text{Input}} \times 100$$

Review Questions

1. Explain the basic principle of operation of a motor.
2. When a current-carrying conductor is placed in a magnetic field, what three factors determine the amount of force exerted on it to move?
3. In general, in what two ways are motors classified?
4. (a) How is the direction of rotation of a permanent-magnet dc motor changed?
   (b) How is the speed of a permanent-magnet motor controlled?
5. Name the three types of electromagnetic dc motors.
6. Explain the function of the commutator in the operation of a dc motor.
7. Compare the dc series and shunt motors with regard to:
   (a) starting torque
   (b) speed regulation
8. In a dc shunt motor, how does a decrease in shunt field current affect the speed of the motor?
9. (a) What type of dc motor is constructed basically the same as the universal motor?
   (b) Why can the universal motor operate from either an ac or dc source?
   (c) How can the speed of a universal motor be controlled?
   (d) How can the direction of rotation of a universal motor be changed?
   (e) Name two common applications of the universal motor.
10. In what way are ac motors classified according to the type of ac power used?
11. Explain how the rotating magnetic field is produced in a three-phase ac motor.
12. Why is the induction motor so named?
13. Outline the principle of operation of a three-phase, squirrel-cage induction motor.
14. (a) How is the speed of a three-phase wound rotor induction varied?
    (b) How is its direction of rotation reversed?
15. What is the major difference between the starting requirement for a three-phase and single-phase induction motor?
16. (a) Outline the starting sequence for a split-phase induction motor.
    (b) How can the direction of rotation of this motor be changed?
17. What is the main advantage of connecting a capacitor in series with the starting winding of a single-phase induction motor?
18. (a) In what way are the magnetic poles on a three-phase synchronous motor different from those of an induction motor?
    (b) State two advantages of using synchronous motor drives in an industrial plant.
19. What two factors determine mechanical power output of a motor?
20. In a motor, not all of the electric power input is transferred to the shaft as mechanical power output. What accounts for most of the power lost?
21. What are the two most common types of motor failures?

Problems

1. Calculate the synchronous speed of the rotating magnetic field of a four-pole, three-phase ac motor feed from a standard voltage source.
2. Calculate the percentage of slip of a 60-Hz, eight-pole motor operating at 840 rpm.

Critical Thinking

1. (a) List three possible causes of excessive sparking between the commutator and brushes of a universal motor.
   (b) A split-phase induction motor hums, but will not start. What is one possible cause of the problem?
   (c) A motor runs very hot when normal load is applied to it. Suggest two possible causes of this problem.

2. Assume a motor that produces an output of 4.75 HP uses 3,750 watts of power.
   (a) Determine the output of the motor in watts.
   (b) Determine the efficiency of the motor.

Portfolio Projects

• Obtain a small permanent magnet dc motor from a discarded toy or electronics store. Carefully disassemble it and report on its construction.
• Use a small permanent magnet dc motor to demonstrate motor speed and direction control.
• Build a small permanent magnet dc motor from scratch or using a kit.

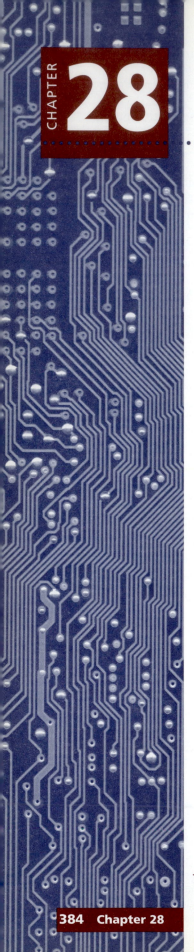

Relays

A *relay is a device that is used to perform switching functions. The relay performs the same function as a switch, except that it is electrically operated instead of manually operated. Since relays are electrically operated, unlike traditional switches, they can be opened or closed from a remote location. In this chapter, you will learn about the different types of relays and their operating characteristics.*

Objectives

After studying this chapter, you should be able to:
1. Compare electromagnetic and solid-state relays.
2. Identify relay symbols used on schematic diagrams.
3. Describe the different ways in which relays are used.
4. Explain how relays are rated.
5. Describe the operation of ON-delay and OFF-delay timer relays.

Key Terms

- electromechanical relay (EMR)
- relay coil
- relay contact
- normally open contact
- normally closed contact
- remote-control lighting
- controlling high voltages
- controlling high currents
- multiple-switching
- magnetic-reed relay
- solid-state relay (SSR)
- timing relay
- pneumatic timing relay
- solid-state timing relay
- on-delay timer
- off-delay timer

RESEARCH INTERNET SITES

- **Aromat**
- **Grayhill**
- **Guardian Electric**

28-1 Electromechanical Relay

An *electromechanical relay* (*EMR*) is essentially a remote controlled switch. It turns a load circuit ON or OFF by energizing an electromagnet, which opens or closes contacts in the circuit. The electromagnetic relay has a large variety of applications in both electric and electronic circuits. In switching and control applications requiring circuit isolation, relays cannot be surpassed for their ruggedness and performance.

A *relay* will usually have only one coil, but may have any number of different contacts. Figure 28-1 illustrates the operation of a typical electromechanical relay. With no current flow through the coil (de-energized), the *armature* is held away from the core of the **coil** by spring tension. When the coil is energized, it produces an electromagnetic field. Action of this field, in turn, causes the physical movement of the armature. Movement of the armature causes the *contact points of the relay* to open and close alternately. The coil and contacts are insulated from each other; therefore, under normal conditions, no electric circuit will exist between them.

A typical symbol used to represent an electromechanical relay is shown in Figure 28-2. The contacts are represented by a pair of short parallel lines and are identified with the coil by means of the same number and letters (CR). Both a N.O. and a N.C. contact are shown. **Normally open (N.O.) contacts** are defined as those contacts that are *open* when no current flows through the coil but *close* as soon as the coil conducts a current or is energized. **Normally closed (N.C.) contacts** are those contacts that are *closed* when the coil is de-energized and *open* when the coil is energized. Each contact is usually drawn as it would appear with the coil *de-energized*.

Controlling High Voltages

The electromechanical relay can be used to **control a high-voltage** *load circuit with a low-voltage control circuit.* This is possible because the coil and contacts of the relay are electrically insulated from each other. From a safety point of view this circuit provides extra protection for the operator. For example, assume you wanted to use a relay to control a 120-V lamp circuit with a 12-V control circuit. The lamp would be

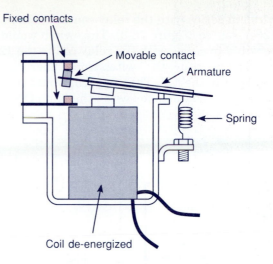

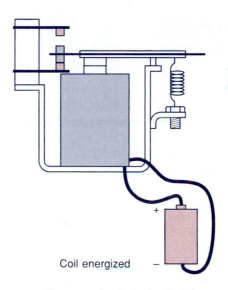

Figure 28-1 Electromechanical relay (EMR) operation.

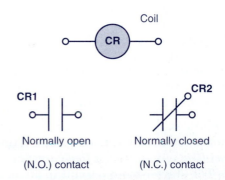

Figure 28-2 Electromechanical relay (EMR) symbol.

wired in series with the relay contact to the 120-V source (Figure 28-3). The switch would be wired in series with the relay coil to the 12-V source. Operating the switch would energize or de-energize the coil. This, in turn, would close or open the contacts to switch the lamp ON or OFF.

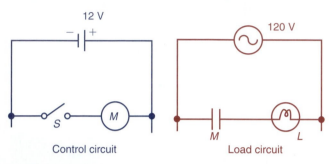

A. Schematic diagram

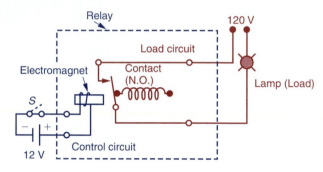

B. Wiring diagram

Figure 28-3   Using a relay to control a high-voltage load circuit with a low-voltage control circuit.

Remote-control lighting systems are operated by low-voltage relay circuits. This type of lighting control is used in some homes and commercial buildings (Figure 28-4). The system consists of a low-voltage 24-V relay, which is energized by closing the switch. This permits 120 V to pass through the relay contacts. A transformer is used to step down the 120 V to 24 V for operating the low-voltage control relay. In addition to the added safety feature, this system of wiring makes it simpler to control lights from a number of locations.

Controlling High Currents   Another basic application for a relay is to **control a high-current** *load circuit with a low-current control circuit.* The starter-motor circuit in an automobile uses a relay for this purpose (Figure 28-5).

The starting motor of a vehicle draws *hundreds of amperes* of current upon starting. A *solenoid*, or *relay*, is used to avoid having to run heavy cables to an equally heavy-duty ignition switch. Instead the ignition switch is wired in such a way so as to control the low current to the solenoid or relay coil. This is known as the **control circuit**, and it is wired using light-gauge wire. The *motor feed*, or *power circuit* is wired using a heavy-gauge cable connecting the battery, contacts, and starter motor in series. The solenoid or relay coil is energized to close the contacts and start the motor. Thus, the motor can be controlled from a remote location with a relatively small amount of control current.

Relay coils are also capable of being controlled by low-circuit signals from integrated circuits and transistors. This is illustrated in Figure 28-6. In this circuit, the electronic-control signal switches the transistor ON or OFF, which in turn, causes the relay coil to energize or de-energize. The current in the control circuit, which consists of the transistor and relay coil, is quite small. The current in the power circuit, which consists of the contacts and motor, is much larger in comparison.

Multiple-Switching Operation   Many electromechanical relays contain several sets of contacts operated by a single coil. Such relays are used to control several switching operations by a single, separate current. This type of relay is often used in industrial-control systems to automatically control machine operations.

A typical control relay used to control two pilot lights is shown in Figure 28-7. With the switch *open*, coil CR is de-energized. The circuit to the green pilot light is completed through N.C. contact CR2, so this light will be ON. At the same time, the circuit to the red pilot light is opened through N.O. contact CR1, so this light will be OFF. With the switch closed the coil is energized. The N.O. contact CR1 closes to switch the red pilot light ON. At the same time, the NC contact CR2 opens to switch the green pilot light OFF.

Relay Ratings   Electromechanical relays are made in a variety of types for different applications. Relay coils and contacts have separate ratings. Relay coils are usually rated for type of operating current (dc or ac), normal operating voltage or current, resistance, and power. Very sensitive relay coils, rated in the low

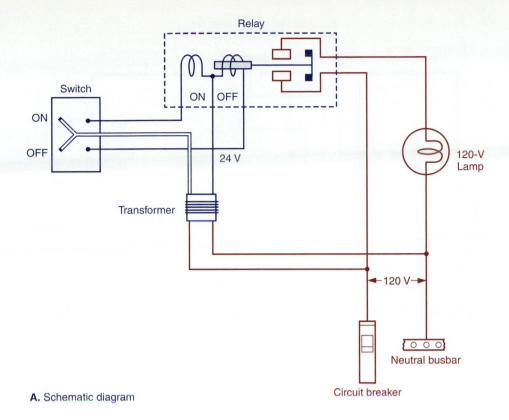

Relay

Switch

ON

OFF

ON   OFF

24 V

Transformer

120-V
Lamp

120 V

Circuit breaker

Neutral busbar

A. Schematic diagram

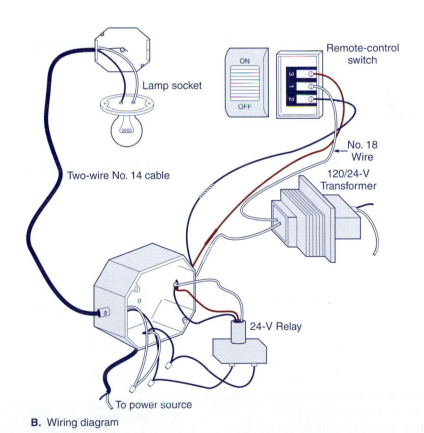

Lamp socket

ON

OFF

Remote-control
switch

3 1 2

No. 18
Wire

120/24-V
Transformer

Two-wire No. 14 cable

24-V Relay

To power source

B. Wiring diagram

Figure 28-4   Remote-control lighting system.

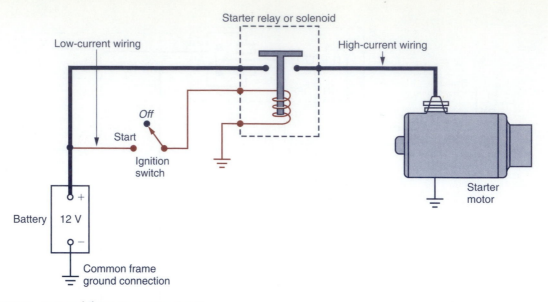

Figure 28-5 Automobile starter motor circuit.

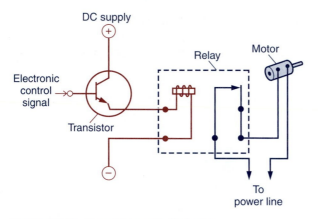

Figure 28-6 Transistor-controlled relay.

milliampere range, are often operated from transistor or integrated circuits.

The most important relay contact specification is its current rating. This indicates the maximum amount of current the contacts are capable of handling. Contacts are also rated for the maximum ac- or dc-voltage level at which they can operate. The number and arrangement of switching contacts required must also be specified (Figure 28-8).

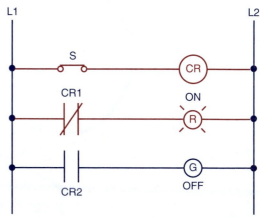

A. Switch open—coil de-energized

B. Switch closed—coil energized

Figure 28-7 Relay multiple-switching circuit.

Single-pole, double-throw (SPDT)   Double-pole, single-throw (DPST)   Double-pole, double-throw (DPDT)

Figure 28-8 Typical relay contact arrangements.

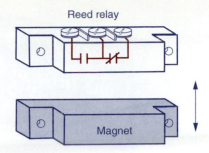

Reed relay

Magnet

A. Proximity Motion: Movement of the relay or magnet will activate the relay.

28-2 Magnetic-Reed Relay

In place of an armature, the ***magnetic-reed relay*** uses magnetically sensitive metal-reed contacts sealed in a glass tube (Figure 28-9). These contacts will make and break when they come under the influence of a magnetic field. If a permanent magnet is brought near the glass tube, the armature reed breaks (opens) the normally closed contact and makes (closes) the normally open contact. The reed relay can also be actuated by a dc electromagnet.

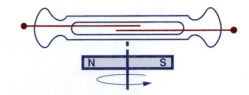

B. Rotary Motion: Relay is activated twice for each complete revolution.

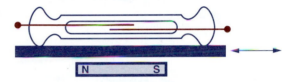

C. Shielding: The ferromagnetic (iron based) shield short circuits the magnetic field holding the contacts. Relay is activated by removal of the shield.

Figure 28-10 Reed relay activation.

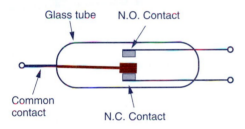

Glass tube   N.O. Contact

Common contact   N.C. Contact

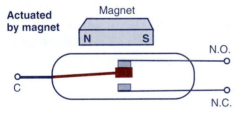

Actuated by magnet   Magnet

N   S

N.O.

C

N.C.

Figure 28-9 Magnetic-reed relay.

A permanent magnet is the most common actuator for a reed relay. Permanent magnet actuation can be arranged in several ways dependent upon the switching requirement. Typically, the most often used arrangements are proximity motion, rotation, and the shielding method (Figure 28-10).

Reed relays are faster, more reliable, and produce less arcing than conventional electro-

mechanical relays do. However, the current-handling capabilities of the reed relay are limited.

28-3 Solid-State Relays

After performing switching tasks for several decades, the electromagnetic relay is being replaced in some applications by a new type of relay, the ***solid-state relay (SSR)*** (Figure 28-11). Although electromagnetic relays and *solid-state* relays are designed to perform similar functions, each accomplishes the final results in different ways. Unlike electromagnetic relays, SSRs do not have actual coils and contacts. Instead, they use semiconductor switching devices such as *transistors*, silicon-controlled rectifiers (SCRs) or *triacs*. An SSR has no moving parts!

A. Printed circuit board mounting

B. Bulkhead mounting

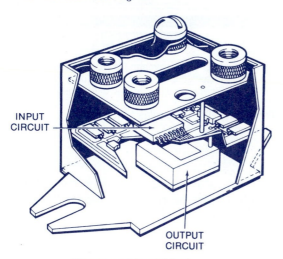

INPUT
CIRCUIT

OUTPUT
CIRCUIT

(Courtesy of Grayhill Inc.)

C. Internal construction

Figure 28-11   Typical solid-state relays.

Just like EMRs, SSRs find application in isolating a low-voltage control circuit from a high-power load circuit. A block diagram of an *optically coupled* solid-state relay is illustrated in Figure 28-12. A light-emitting diode (LED) incorporated in the input circuit glows when the conditions in that circuit are correct to actuate the relay. The light-emitting diode shines on a phototransistor which then conducts, causing the trigger current to be applied to the triac. Thus, the output is isolated from the input by the simple LED and phototransistor arrangement, just as the electromagnet isolated the input from the switching contacts in the conventional electromechanical relay. Also available are hybrid solid-state relays that incorporate a small reed relay or transformer to serve as the actuating device.

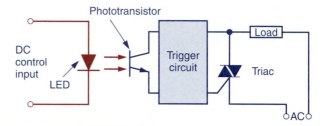

Figure 28-12   Optically-coupled solid-state relay.

Most often, the *black-box* approach is used to symbolize an SSR. That is, a square or rectangle will be used on the schematic to represent the relay. The internal circuitry will not be shown, and only the input and output connections to the box will be given (Figure 28-13).

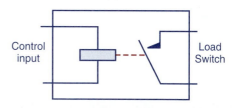

Figure 28-13   Typical SSR symbol.

The SSR has several advantages over the electromechanical relay. The SSR is more reliable and has a longer life because it has no moving parts. It is compatible with transistor and *integrated-circuit* circuitry and does not generate as much electromagnetic interference. The SSR is more resistant to shock and vibration, has a much faster response time, and does not exhibit contact bounce.

As in every device, SSRs do have some disadvantages. The SSR contains semiconductors that are susceptible to damage from voltage and current spikes. Unlike the EMR contacts, the SSR switching semiconductor has a significant on-state resistance and off-state leakage current.

28-4 Timing Relays

Timing relays are conventional relays that are equipped with an additional hardware mechanism or circuitry to delay the opening or closing of load contacts.

A ***pneumatic- (air) timing relay*** uses mechanical linkage and an air bellows system to achieve its timing cycle (Figure 28-14). The bellows design allows air to enter through a needle valve at a predetermined rate to provide the different time-delay increments and to switch a contact output.

(Courtesy of Allen-Bradley Canada)

Figure 28-14   Pneumatic- (air) timing relay.

A ***solid-state timing relay*** (Figure 28-15) uses electronic circuitry to achieve its timing cycle. A resistor/capacitor (*RC*) oscillator network generates a highly stable and accurate pulse that is used to provide the different time-delay increments and switch a contact output.

(Courtesy of Allen-Bradley Canada)

Figure 28-15   Solid-state timing relay.

Typical applications for solid-state timing relays are in industrial controls, appliances, and machines in which the start of an event must be delayed until another event has occurred. For example, a mixing machine may be delayed until the liquid has been heated, or a fan may remain OFF until a heating coil has heated the surrounding air to a specified temperature.

As its name implies, a time-delay removes power from, or applies power to, a component or circuit after a certain preset time interval has elapsed. Time-delay relays can be classified into two basic groups: ***ON-delay*** and ***OFF-delay***. Figure 28-16 illustrates the standard relay diagram symbols used for timed contacts. The circuits shown in Figures 28-17, 28-18, 28-19, and 28-20, are designed to illustrate the basic timed-contact functions. In each circuit the time-delay setting of the timing relay is assumed to be 10 seconds.

Electrical-Mechanical Assembly Person
Job Description
- Electrical/electronic and mechanical assembly of products.
- Repair product after inspection or test.
- Must have a working knowledge of electrical/electronic assembly from components to final assembly.
- Must be able to work from drawings, written, or oral instructions.

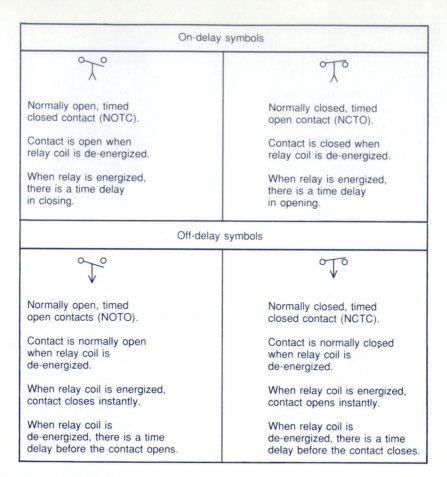

On-delay symbols	
Normally open, timed closed contact (NOTC).	Normally closed, timed open contact (NCTO).
Contact is open when relay coil is de-energized.	Contact is closed when relay coil is de-energized.
When relay is energized, there is a time delay in closing.	When relay is energized, there is a time delay in opening.
Off-delay symbols	
Normally open, timed open contacts (NOTO).	Normally closed, timed closed contact (NCTC).
Contact is normally open when relay coil is de-energized.	Contact is normally closed when relay coil is de-energized.
When relay coil is energized, contact closes instantly.	When relay coil is energized, contact opens instantly.
When relay coil is de-energized, there is a time delay before the contact opens.	When relay coil is de-energized, there is a time delay before the contact closes.

Figure 28-16 Timed-contact symbols.

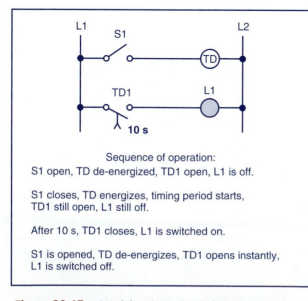

Sequence of operation:

S1 open, TD de-energized, TD1 open, L1 is off.

S1 closes, TD energizes, timing period starts, TD1 still open, L1 still off.

After 10 s, TD1 closes, L1 is switched on.

S1 is opened, TD de-energizes, TD1 opens instantly, L1 is switched off.

Figure 28-17 ON-delay timer circuit (NOTC contact).

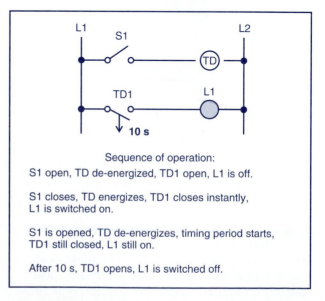

Sequence of operation:

S1 open, TD de-energized, TD1 open, L1 is off.

S1 closes, TD energizes, TD1 closes instantly, L1 is switched on.

S1 is opened, TD de-energizes, timing period starts, TD1 still closed, L1 still on.

After 10 s, TD1 opens, L1 is switched off.

Figure 28-18 OFF-delay timer circuit (NOTO contact).

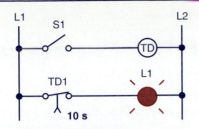

Sequence of operation:

S1 open, TD de-energized, TD1 closed, L1 is on.

S1 closes, TD energizes, timing period starts,
TD1 still closed, L1 still on.

After 10 s, TD1 opens, L1 is switched off.

S1 is opened, TD de-energizes, TD1 closes instantly,
L1 is switched on.

Figure 28-19 ON-delay timer circuit (NCTO contact).

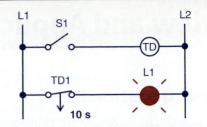

Sequence of operation:

S1 open, TD de-energized, TD1 closed, L1 is on.

S1 closes, TD energizes, TD1 opens instantly,
L1 is switched off.

S1 is opened, TD de-energizes, timing period starts,
TD1 still open, L1 still off.

After 10 s, TD1 closes, L1 is switched on.

Figure 28-20 OFF-delay timer circuit (NCTC contact).

Review and Applications

Review Questions

1. Explain the basic operating principle of an electromechanical relay.
2. Define the terms normally-open and normally-closed contact as they apply to a relay.
3. List three ways in which relays are put to use in electric and electronic circuits.
4. A certain electromechanical relay (EMR) coil is rated for 250 mA, while the contacts are rated for 10 A. Explain what these ratings mean.
5. (a) Explain the basic operating principle of a reed relay.
   (b) List three methods used for permanent magnet actuation of reed relays.
   (c) List one advantage and one limitation of reed relays.
6. (a) How is switching of the load circuit accomplished in a solid-state relay?
   (b) List two advantages and two limitations of solid-state relays.
7. (a) Explain the function of a time-delay relay.
   (b) Into what two basic groups can time-delay relays be classified?

Problems

1. Refer to Figure 28-1. The contacts of the relay shown are suspected of being bad and are to be checked using an ohmmeter. Outline the procedure to be followed.
2. Refer to Figure 28-4. The lamp fails to operate when the remote switch is momentarily pressed to the "on" position. Assuming the wiring is OK, list the circuit checks you would make (in the most likely order of priority) to determine the problem.

3. Refer to Figure 28-17. Power is first applied, switch S is closed for 1 minute then opened. Describe the timing sequence of light L_1 that takes place.
4. Refer to Figure 28-20. Power is first applied, switch S is closed for 1 minute then opened. Describe the timing sequence of light L_1 that takes place.

Critical Thinking

1. Refer to Figure 28-3. Assume that the two voltage sources were incorrectly connected so that the 12 volts dc was applied to the control circuit and the 120 volts ac was applied to the load circuit.
   (a) How would this affect the operation of the load circuit?
   (b) How would this affect the operation of the control circuit?
2. Refer to Figure 28-7. A third yellow light is to be connected to the circuit to come "on" whenever the relay coil is energized. How would this light be connected? Redraw the circuit schematic with this light properly connected.

Portfolio Projects

- Obtain an old automobile relay from a used auto parts dealer. Carefully dismantle it. Examine it to see how the coil actually operates the contacts. Report on your findings.
- Design a circuit that uses a relay to automatically switch on a battery operated light in case of a power failure. Include a detailed bill of material of all parts required.

Circuit Challenge

Using a Simulator

Procedure

Using whatever simulation software package is available to you, wire the circuit using the relay contacts to operate the lights as follows:

- With switch S open (relay coil de-energized) light L_1 is "on" and light L_2 is "off"
- With switch S closed (relay coil energized) light L_1 is "off" and light L_2 is "on"

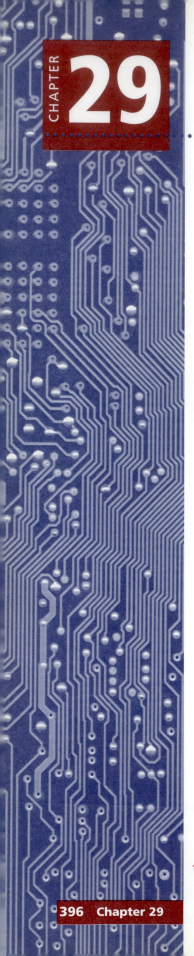

Motor Controls

Without properly designed, installed, and maintained motor control systems, industries would cease to function. Motor control is that portion of the electrical system which starts, stops, and reverses motors that drive various pieces of equipment. In addition, motor control equipment is designed to vary speed, limit starting current, and control starting torque. This chapter provides an overview of the more common motor control systems.

Objectives

After studying this chapter, you should be able to:
1. Explain the principles and connections of motor protection circuits.
2. List and describe the methods by which a motor may be started.
3. Describe the operation of reversing and jogging motor-control circuits.
4. List the methods of stopping a motor.
5. Explain the operating principles of variable-speed motor drives.
6. Describe the operation of motor pilot-devices.

Key Terms

- disconnect switch
- overcurrent protection
- overload protection
- internal-overload protection
- external-overload protection
- magnetic-overload relay

- thermal-overload relay
- electronic-overload relay
- time-delay fuses
- low-voltage protection
- low-voltage release protection

- phase-failure protection
- phase-reversal protection
- ground-fault protection
- counter-emf (cemf)
- locked-rotor current
- contactor

(continued)

RESEARCH
INTERNET
SITES

- Square D
- Westinghouse
- Omron

Motor Controls

Key Terms (*continued*)

- across-the-line starter
- manual starter
- reduced-voltage starter
- primary-resistance starter
- autotransformer starter
- wye-delta starter
- solid-state starter
- soft start
- interlocking

- jogging
- plugging
- dynamic braking
- electric braking
- friction brake
- eddy-current brake
- open loop
- closed loop
- variable-frequency drive
- pulse width modulation

- wound-rotor motor
- regenerative dc drives
- pilot devices
- two-wire control
- three-wire control
- thermostat
- pressure switch
- level switch
- proximity switch
- drum switch

29-1 Motor Protection

Motor protection safeguards the motor, the supply system, and personnel from various upset conditions of the driven load, the supply system, or the motor itself. A suitable *disconnect device* of sufficient capacity is required, as part of a motor branch circuit, within sight of the motor. Figure 29-1, shows a typical motor branch circuit. The purpose of the disconnect device is to open the supply conductors to the motor, allowing personnel to work safely on the installation.

A motor *disconnect switch* is not designed to open the circuit to the motor while it is operating. This is because the motor contains considerable *inductance*, and when the switch is opened under load, a spark will jump across the switch as it opens. This spark can burn the switch blades, or can even fuse them to their contacts. For this reason, an ordinary disconnect switch is not to be used for normal motor starting and stopping.

Motor *overcurrent protection* is *not* designed to protect the motor, but rather to protect the electrical *supply system* if the motor or circuit shorts out. Circuit breakers and fuses

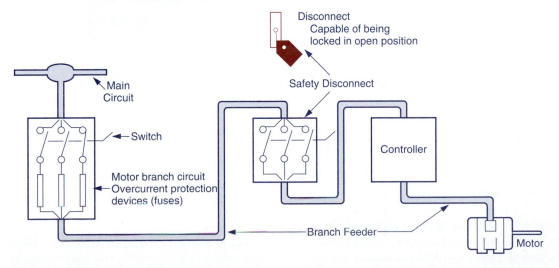

Figure 29-1 Typical motor branch circuit.

are used for overcurrent protection against the devastating effects of a short. They trip or blow to limit further damage that can occur from high short-circuit currents. Such currrents are not related to the motor current. Magnitudes of 10,000–50,000 A are not uncommon for short-circuit currents in a modern distribution system. Figure 29-2 illustrates how motor overcurrent protection works.

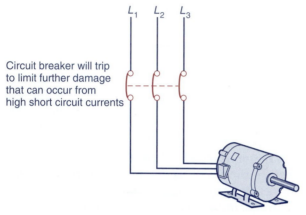

Circuit breaker will trip to limit further damage that can occur from high short circuit currents

A. Overcurrent-protection circuit

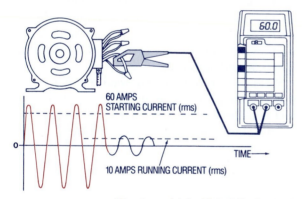

60 AMPS STARTING CURRENT (rms)

TIME→

10 AMPS RUNNING CURRENT (rms)

(Courtesy of John Fluke Mfg. Co. Inc.)

B. Starting current is much higher than running current

Figure 29-2   Motor overcurrent protection.

Overcurrent protection provided by circuit breakers and fuses provides short-circuit protection for the motors they serve, but they have *too large* a current-carrying capacity to protect the motor from the excessive heat that may develop when the motor is running.

Overload protection devices safeguard the motor from mechanical overload conditions. A motor can be protected from overload in two ways—by internal protection that is mounted in the motor; and by external protection that is mounted in, or near, the motor starter.

Internal overload protection is usually a temperature detector in the motor windings. There are four common devices:

- *Thermostatic devices:* These are usually embedded in the windings and directly connected to the control circuitry.
- *RTD:* A Resistance Temperature Detector (RTD) is used to indicate accurately temperature in the windings.
- *Thermocouple*: This is a pair of dissimilar conductors, joined at the end.
- *Thermistor*: This is a semiconductor that changes resistance significantly with changes in temperature.

External-overload protection devices, mounted in the starter, attempt to simulate the heating and cooling of a motor by sensing the current flowing to it. Four common forms are: magnetic-overload relays, thermal-overload relays, electronic-overload relays, and fuses.

Magnetic-overload relays (Figure 29-3) operate on the magnetic action of the load current that is flowing through a coil. When the load current becomes too high, a plunger is pulled up into the coil, interrupting the circuit. The tripping current is adjusted by altering the initial position of the plunger with respect to the coil.

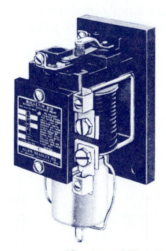

(Courtesy of Allen-Bradley Canada.)

Figure 29-3   Magnetic-overload relay.

A *thermal-overload relay* uses a heater connected in series with the motor supply (Figure 29-4). The amount of heat produced

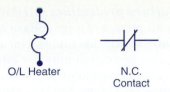

O/L Heater N.C. Contact

Figure 29-4 Thermal-overload relay symbol.

increases with supply current. If an overload occurs, the heat produced causes a set of contacts to open, interrupting the circuit. The tripping current is changed by installing a different heater for the required trip point. This type of

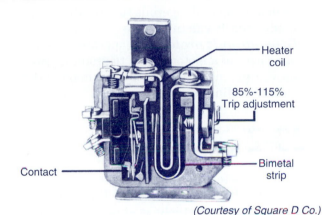

Heater coil

85%-115% Trip adjustment

Contact

Bimetal strip

(Courtesy of Square D Co.)

Figure 29-5 Bimetallic thermal-overload relays.

protection is very effective because the heater closely approximates the actual heating within the windings of the motor and has a thermal "memory" to prevent immediate reset and restarting. There are two common types: (1) the bimetallic type, which utilizes a bimetallic strip (Figure 29-5), and (2) the melting-alloy type, which utilizes the principle of heating solder to its melting point (Figure 29-6).

An ***electronic overload relay*** uses a current transformer that senses the current flowing to the motor and electronic circuitry that senses the changes in the motor current, then trips when the current reaches the full load (Figure 29-7). If an overload condition exists, the sensing circuit interrupts the power circuit. The tripping current can be adjusted to suit the particular application. Electronic overloads often perform additional protective functions such as ground-fault and phase-loss protection.

Dual-element or ***time-delay fuses*** also can be used to provide overload protection. They have the disadvantage of being nonrenewable and must be replaced each time they operate. An overload relay in the motor starter is only sensing the motor current in an attempt to determine the heating in the motor windings. This method of protection may not be as effective if the motor is covered in dirt.

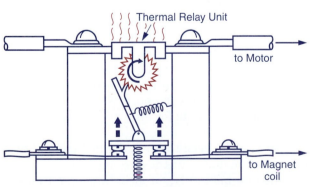

Thermal Relay Unit

to Motor

to Magnet coil

Drawing shows operation of melting alloy overload relay. As heat melts alloy, ratchet wheel is free to turn. The spring then pushes contacts open.

Figure 29-6 Melting-alloy thermal-overload relay.

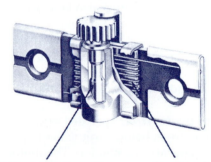

Solder pot (heat sensitive element) is an integral part of the thermal unit. It provides accurate response to overload current, yet prevents nuisance tripping.

Heating winding (heat producing element) is permanently joined to the solder pot, so proper heat transfer is always ensured. No chance of misalignment in the field.

(Courtesy of Square D Co.)

Figure 29-7 Electronic overload relay mounted within a starter. (*Courtesy Furnas Electric Company*)

External overload protection also is not sensitive to changes in ambient temperature around the motor. This means a motor can be overheated and the external overload protection will not trip. Internal protectors are recommended for applications where the ambient (surrounding) temperature can change drastically, or where the cleanliness of the motor cannot be guaranteed.

External-overload relays are also less effective if the motor is started and stopped frequently, or if it is restarted too quickly after an overload trip. Internal-thermal protection is the best for frequent starting applications.

Other types of motor protection circuits include:

- **Low-voltage protection:** operates when the supply voltage drops below a set value to provide machine-operator protection. The motor must be *restarted* upon resumption of normal supply voltage.
- **Low-voltage release protection:** interrupts the circuit when the supply voltage drops below a set value, and re-establishes the circuit when the supply voltage returns to normal.

- **Phase-failure protection:** interrupts the power in all phases of a polyphase circuit upon failure of any one phase. Normal fusing and overload protection may not adequately protect a polyphase motor from damaging single-phase operation. Without this protection, the motor will continue to operate if one phase is lost. Large negative-sequence currents are developed in the rotor circuit, causing excessive current and heating in the stator circuit, which will eventually burn out.
- **Phase reversal protection:** operates upon detection of a phase reversal in a polyphase circuit. This type of protection is used in applications such as elevators where it would be damaging or dangerous to have the motor inadvertently run in reverse.
- **Ground-fault protection:** operates when one phase of a motor shorts to ground, thus preventing high currents from damaging the stator windings and the iron core.

29-2 Motor Starting

When a motor is running, a *counter-emf (cemf)* or *voltage* is generated by the rotor cutting through magnetic lines of force and this reduces the current supplied to the motor. However, upon being started and before the motor has begun to turn, there is no counter-emf to limit the current, and initially there is a high inrush of current. The magnitude of the current that flows before a motor begins to turn is called its **locked-rotor current**. Since this current occurs before the rotor has moved, the locked-rotor current will be the same whether the motor is being started at no load or at full load. The term *locked-rotor current* is derived from the fact that it is determined by locking the motor shaft so it cannot turn, then applying rated voltage to the motor, and measuring the current. Although the locked-rotor current may be perhaps five times the normal running current, it normally lasts for only a fraction of a second.

Relays can be used to carry the line current of very small motors, but they are basically intended for pilot duty, for switching operations in control circuits. This is because the contacts are intended to interrupt only small currents. When a relay is made so it is capable of opening and closing an ac circuit supplying a load such

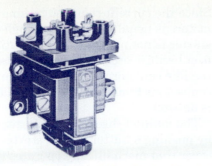

(Courtesy of Allen-Bradley Canada.)

A. Typical contactor

as a motor, it is called a ***contactor*** (Figure 29-8). A contactor will have the same general appearance as a relay, except that the upper portion containing the contacts is enclosed. As previously discussed, an ordinary disconnect switch is not suitable for interrupting the load current of a motor because of the arc that occurs. The section of the contactor that contains the contacts is specially constructed with passages and baffles designed to snuff out the arc. This enables the contactor to interrupt a motor current without being damaged. DC arcs

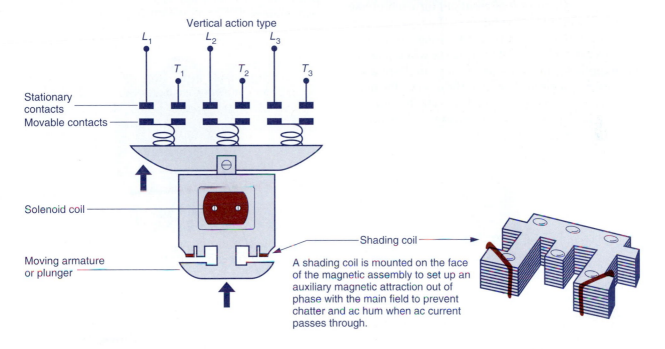

Vertical action type

Stationary contacts

Movable contacts

Solenoid coil

Moving armature or plunger

Shading coil

A shading coil is mounted on the face of the magnetic assembly to set up an auxiliary magnetic attraction out of phase with the main field to prevent chatter and ac hum when ac current passes through.

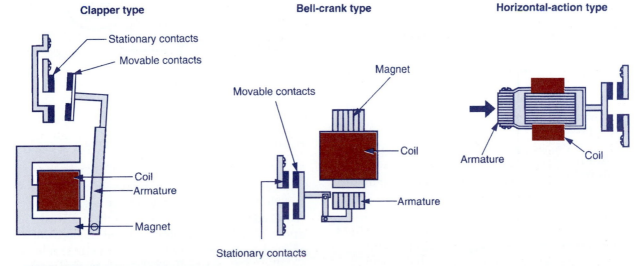

Clapper type

Stationary contacts

Movable contacts

Coil
Armature

Magnet

Bell-crank type

Movable contacts

Magnet

Coil

Armature

Stationary contacts

Horizontal-action type

Armature

Coil

B. Solenoid action

Figure 29-8 Motor contactor.

are more difficult to extinguish than ac arcs because the continuous dc supply causes current to flow constantly.

Across-the-Line Starting of AC Induction Motors

A *motor starter* is a device that combines a contactor and thermal-overload relays. An *across-the-line starter* is designed to apply full line voltage to the motor upon starting. Across-the-line starters may be of either manual or magnetic type.

Manual starters are often used for smaller motors, those up to about 10 hp. They consist of a switch with one set of contacts for each phase and a thermal-overload device. The starter contacts remain closed if power is removed from the circuit and the motor restarts when power is reapplied. Manual starters for single-phase motors are popular because they serve for both motor protectors and branch feeder disconnects. Figure 29-9 shows a single-phase starter having one overload relay and a toggle switch.

(Courtesy of Allen-Bradley Canada.)

A. Switch

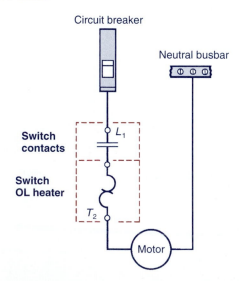

B. Schematic

Figure 29-9 Single-phase manual starting switch.

Since the current required by single-phase motors is usually small, the single-phase starter is not needed as an operational disconnect because ordinary thermostats, relays, and the like can normally serve to connect and disconnect such motors when automatic operation is desired.

The manual three-phase starter is operated by pushing a button on the starter enclosure cover that mechanically operates the starter (Figure 29-10). These types of starters are used in applications that do not require undervoltage protection or operation by remotely-located control devices.

(Courtesy of Allen-Bradley Canada.)

A. Manual starter

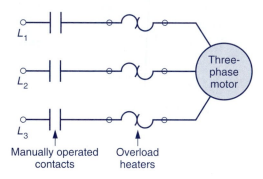

B. Schematic

Figure 29-10 Three-phase manual starter.

Larger motors require the use of contactors to handle their load current, and it is generally required that starters operate automatically under control of remotely-located devices such as push buttons, switches, thermostats, relays, or solid-state controllers. Therefore, the most common type of motor starter is the *magnetic across-the-line starter*, which is operated by an electromagnet or a solenoid. Starters are rated for various operating voltages and are sized according to motor hp and the type of duty expected.

Figure 29-11, shows the schematic diagram for a typical *three-phase*, magnetically operated, across-the-line ac starter. When the START button is pressed, coil *M* is energized. When coil *M* is energized, it closes *all M* contacts. The *M* contacts in series with the motor close to complete the current path to the motor. These contacts are part of the *power circuit* and must be designed to handle the full load current of the motor. Control contact *M* (across START button) closes to *seal in* the coil circuit when the START button is released. This contact is part of the *control circuit* and, as such, is required to

handle the small amount of current needed to energize the coil. The starter has three overload heaters, one in each phase. The normally closed (N.C.) relay contact OL opens automatically when an overload current is sensed on any phase to de-energize the *M* coil and stop the motor.

Reduced-Voltage Starters   The large starting inrush current of a big motor could cause an excessive voltage drop through an electric utility's transformer and distribution system. If the driven load or the power distribution system cannot accept a full voltage start, some type of reduced voltage or soft starting scheme must be used. Typical *reduced voltage starters* are: primary-resistance starters, autotransformers,

(Courtesy of Allen-Bradley Canada.)

A. Typical ac across-the-line starter

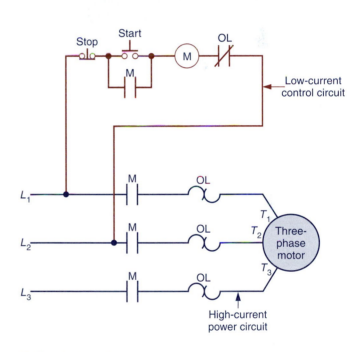

B. Circuit schematic

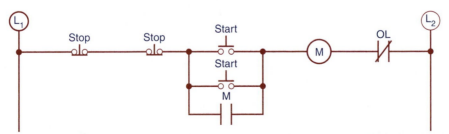

C. The motor can be started or stopped from a number of locations by connecting additional start buttons in parallel with one another and stop buttons in series with another.

Figure 29-11   Three-phase magnetic starter.

part-winding starters, wye-delta, and solid-state starters. These devices can be used only where low starting torque is acceptable.

Figure 29-12 shows the general arrangement of a **primary-resistance starter**. The ac motor is often compared to a transformer in which the stator winding is the primary. The primary-resistance starter adds resistance to the stator circuit during the starting period, thus reducing the current drawn from the line. Closing the contacts at A connects the motor to the supply via resistors which provide a voltage drop to reduce the starting voltage available to the motor. The resistors' value is chosen to provide adequate starting torque while minimizing starting current. Motor inrush current declines during acceleration, reducing the voltage drop across the resistors and providing more motor torque. This results in smooth acceleration. After a set period of time, contacts A open and the resistors are shorted out by the closing of contacts B, applying full voltage to the motor.

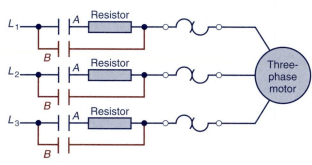

Figure 29-12   Primary resistance starter.

One type of **autotransformer starter** is shown in Figure 29-13. This type of starter uses autotransformers to reduce the voltage at start-up, and when the motor approaches full speed, the autotransformers are bypassed. An autotransformer is a single-winding transformer on a laminated core with taps at various points on the winding. The taps are usually expressed as a percentage of the total number of turns and thus percentage of applied voltage output. The three autotransformers are connected in a wye configuration, with taps selected to provide adequate starting current.

The motor is first energized at a reduced voltage by closing contacts A. After a short time, the autotransformers are switched out of the circuit by opening contacts A and closing con-

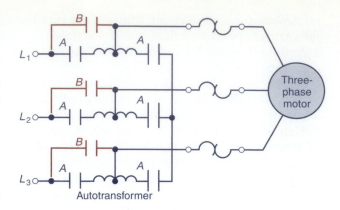

Figure 29-13   Autotransformer starter.

tacts B, thus applying full voltage to the motor. The autotransformers need not have high capacity as they are only used for a very short period of time.

Wye-delta starters (Figure 29-14) can be used with motors where all six leads of the stator windings are available (on some motors only three leads are accessible). The major advantage is the absence of starting resistors or transformers. By first closing contacts A and B, the windings are connected in a *wye configuration*, which presents only 58 percent of rated voltage to the motor.

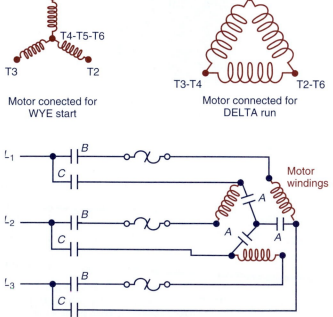

Figure 29-14   Wye-delta starter.

Full voltage is then applied by reconnecting the motor in a *delta configuration* by closing contacts *C* and opening those at *A*.

The starting current and torque are 33 percent of their full voltage ratings, limiting applications to loads that require very low starting torque.

Part-winding starters are sometimes used on motors wound for dual voltage operation such as a 230/460-V motor. These motors have two sets of windings connected in parallel for low voltage and connected in series for high-voltage operation.

When used on the lower voltage, they can be started by first energizing only one winding, limiting starting current and torque to approximately one half of the full voltage values.

The second winding is then connected normally once the motor nears operating speed.

Solid-state reduced-voltage *starters* (Figure 29-15) typically use high-power semiconductors such as triacs or *silicon-controlled rectifiers (SCRs)* to control the voltage applied to the motor. Known in industry as **soft start**, the voltage is increased gradually as the motor starts, providing smooth, stepless, torque-controlled acceleration. Solid-state starters use microcomputer-based circuitry and can provide useful additional features such as:

- Fault indicators for stalled motor, phase loss, high temperature, and the like.
- Overload protection.
- Current limiting.
- Energy-saving feature for lightly loaded motors.
- Controlled stopping (braking or extended stop time).

(*Courtesy of Allen-Bradley Canada.*)

A. Starter module

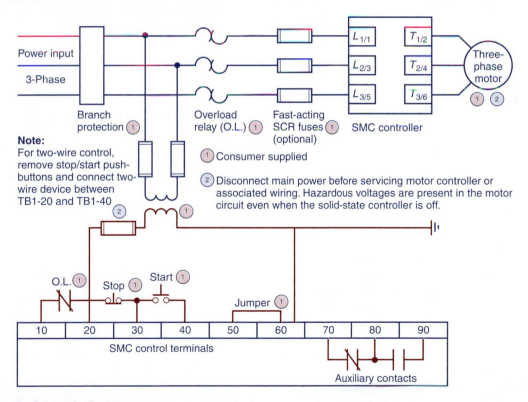

B. Schematic diagram

Figure 29-15 Typical solid-state starter.

29-3 Motor Reversing and Jogging

Interchanging any two leads to a three-phase induction motor will cause it to run in the *reverse* direction. A three-phase reversing starter (Figure 29-16) consists of *two contactors* enclosed in the same cabinet. As seen in the power circuit, the contacts (F) of the forward contactor, when closed, connect L_1, L_2, and L_3 to motor terminals T_1 T_2, and T_3, respectively. The contacts (R) of the reverse contactor, when closed, connect L_1 to motor terminal T_3 and connect L_3 to motor terminal T_1, causing the motor to run in the opposite direction. Whether operating through either the forward or reverse contactor, the power connections are run through the same set of overload relays.

Mechanical and electrical **interlocking** are used to prevent the forward and reverse contactors from being activated at the same time, which would cause a short circuit (Figure 29-17). With the *mechanical interlock,* the first

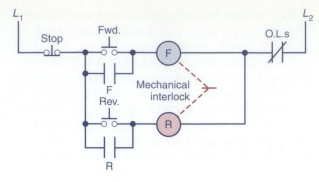

A. Mechanical interlock

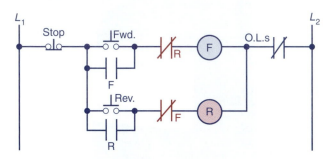

B. Electrical auxiliary contact interlocks

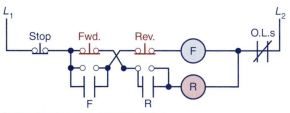

C. Electrical push-button interlocks

Figure 29-17   Reversing-starter mechanical and electrical interlocks.

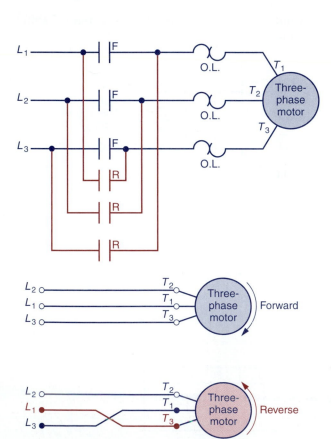

Figure 29-16   Three-phase reversing starter-power circuit.

coil to close moves a lever to a position that prevents the other coil from closing its contacts when it is energized. *Electrical auxiliary-contact interlocks* are simply N.C. (normally-closed) auxiliary contacts from one contactor that are wired in series with the control circuit of the other contactor. *Electric push-button interlocks* use double-contact (N.O. and N.C.) pushbuttons. When the forward pushbutton is pressed, the N.C. (normally closed) contacts open the reverse-coil circuit. There is no need to press the stop button before changing the direction of rotation. If the forward button is pressed while the motor is running in the reverse direction, the reverse control circuit is de-energized and the forward contactor is energized and held closed.

Jogging (sometimes called *inching*) is the momentary operation of a motor for the purpose of accomplishing small movement of the driven machine. Jogging is used where motors must be operated momentarily, in cases such as for machine tool set up. Figure 29-18 illustrates one method of wiring a jogging circuit. When the JOG button is pressed, the seal in circuit to the starter coil is opened by the N.C. contacts of the jog push-button. As a result, the starter coil will not lock in but instead can only stay energized as long as the jog button is fully depressed.

(Courtesy of Allen-Bradley Canada.)

A. Zero-speed, or plugging, switch

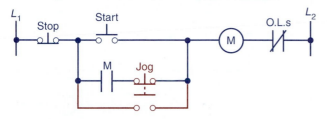

Figure 29-18   Jogging-control circuit.

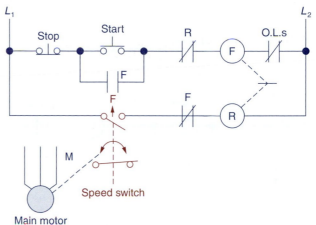

B. Schematic diagram

Figure 29-19   Plugging a motor to a stop.

29-4 Motor Stopping

The most common method of stopping a motor is to remove the supply voltage and allow the motor and load to coast to a stop. In some applications, however, the motor must be stopped more quickly or held in position by some sort of braking device. Electric braking uses the windings of the motor to produce a retarding torque. The kinetic energy of the rotor and the load is dissipated as heat in the rotor bars of the motor. Two means of electric braking are plugging and dynamic braking.

Plugging brings an induction motor to a very quick stop by connecting the motor for reverse rotation while it is still running in the forward direction. To prevent the motor from reversing after it has come to a stop, the power is removed by means of a **zero-speed switch** (also known as a plugging switch).

The control schematic of Figure 29-19 shows one method of plugging a motor to stop from one direction only. Pushing the START button closes the forward contactor and the motor runs forward. The normally-closed contact *F* opens the circuit to the reverse contactor. The forward contact on the speed switch closes. Pushing the STOP button drops out the forward contactor. The reverse contactor is energized and the

motor is plugged. The motor speed decreases to the setting of the speed switch, at which point its contact opens and drops out the reverse contactor. This contactor is used only to stop the motor using the plugging operation; it is not used to run the motor in reverse.

Dynamic braking can be achieved if a motor that is running is reconnected to act as a *generator* immediately after it is turned OFF. Connecting the motor in this way makes it act as a loaded generator that develops a retarding torque, which rapidly stops the motor. The generator action converts the mechanical energy of rotation to electric energy that can be dissipated as heat in a resistor.

The circuit shown in Figure 29-20 illustrates how dynamic braking is applied to a dc motor. When the stop button is depressed, normally closed contact *M* completes the braking circuit through the braking resistor, which acts as a load. The shunt field windings of the dc motor are left connected to the power supply. The armature generates a voltage referred to as counter-electromotive force (cemf). This cemf

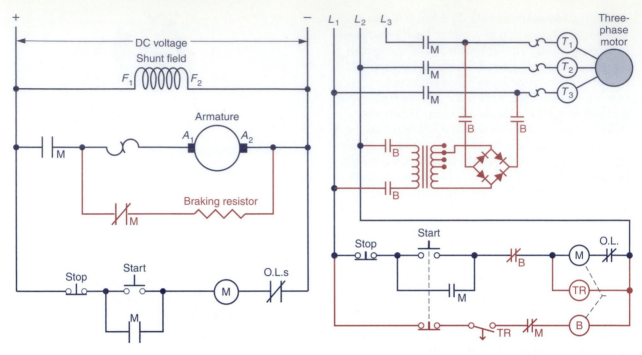

Figure 29-20 Dynamic braking applied to a dc motor.

Figure 29-21 Electric braking applied to a three-phase motor.

causes current to flow through the resistor and armature. The smaller the ohmic value of the braking resistor the greater the rate of energy dissipation and the faster the motor comes to rest. Neither plugging nor dynamic braking can hold the motor stationary after it has stopped.

Electric braking can be achieved with a three-phase induction motor by removing the ac power supply from the motor and applying direct current to one of the stator phases.

Figure 29-21 illustrates one method used to apply dc to the motor after the ac is removed. This circuit uses a bridge rectifier circuit to change the ac into dc. An off-delay timer is connected in parallel with the motor-starter coil. This off-delay timer controls a normally open (N.O.) contact that is used to apply power to the braking contactor for a short period of time after the stop push button is pressed. This timing contact is adjusted to remain closed until the motor comes to a complete stop. A transformer with tapped windings is used in this circuit to adjust the amount of braking torque applied to the motor. The motor starter (*M*) and braking contactor (*B*) are mechanically and electrically interlocked so that the ac and dc

supplies are not connected to the motor at the same time.

The term *electromechanical* **friction brake** refers to a device external to the motor that provides retarding torque. Most rely on friction in a drum or a disc brake arrangement, and are set with a spring and released by a solenoid (Figure 29-22). This device has the ability to hold a motor stationary and is used in applications that require the load to be held.

An *eddy-current brake* is an electromechanical device that provides a retarding torque by inducing eddy currents in a drum via an electromagnetic rotor attached to the motor shaft. The amount of braking force can be controlled by altering the rotor current. Eddy-current brakes cannot hold the motor stationary.

29-5 Motor Drives

A **motor drive** is a mechanical or electric/electronic device used to control the speed, torque, horsepower, and the direction of rotation of a motor. Drive systems, ac or dc, are used in any application where simple starter

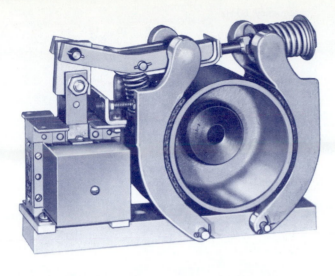

(Courtesy Eaton Corporation, Cutler-Hammer Products.)

Figure 29-22 Solenoid-operated brake used on machine tools, conveyors, and small hoists.

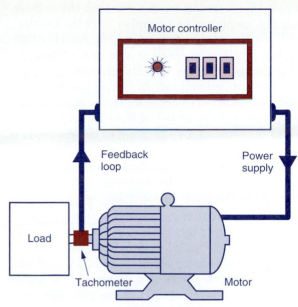

Figure 29-23 Closed-loop motor drive.

control of a motor is inadequate. This could involve, as an example, operator-controlled variance of speed, or the use of control-feedback systems to maintain steady motor speed in spite of load fluctuations, or other disturbances. Speed control can be **open loop** where no feedback of actual motor speed is used, or **closed loop** where feedback is used for more accurate speed regulation (Figure 29-23). In closed-loop control, a change in demand is compensated by a change in the power supplied to the motor, and thus a change in motor speed (within regulation capability).

Selecting the proper drive system is dependent on the application at hand. ac and dc motors each have inherent characteristics that govern their uses. AC motor characteristics include:

- Lower cost for use
- Less maintenance required
- Various enclosures readily available for different operating environments
- Can withstand harsh operating environments
- Physically smaller than dc motors of the same horsepower
- Repairs less costly
- Can be run at speeds above the nameplate rating

DC motor characteristics include:

- High torque at low speed
- Good speed control over full range (no low-end cogging)
- Better overload capability
- More expensive than ac motors
- Physically larger than ac motors of the same horsepower
- Require more routine maintenance and repairs

Over 90 percent of all motors used run on alternating current (ac). AC induction motors are generally considered constant-speed devices operating at neat *synchronous speed*. The synchronous speed of an induction motor is expressed by the following formula:

$$\text{Synchronous speed (rpm)} = \frac{120 \times F}{P}$$

F = Supply frequency in Hz

P = Number of poles in motor winding

From this formula, we can see that the voltage-supply frequency and number of poles are the only variables that determine the speed of the motor. Unlike the speed of a dc motor, the speed of an ac motor is not usually changed by varying the applied voltage. In fact, if the supply voltage is varied to an ac motor by more than 10 percent above or below the rated nameplate voltage, damage may be done to the motor. It is, however, possible in some special applications when the motor is connected to a load, to control the speed of the load by varying the applied voltage to the motor.

Since the frequency or number of poles must be changed to change the speed of an ac motor, two methods of speed control are available. These are: (1) changing the frequency applied to the motor, or (2) using a motor with windings that may be reconnected to form different numbers of poles.

Multispeed ac motors, designed to be operated at constant frequency, are provided with stator windings that can be reconnected to provide a change in *the number of poles* and, thus, a change in the speed. These multispeed motors are available in two or more fixed speeds, which are determined by the connections made to the motor. Two-speed motors usually have one winding that may be connected to provide two speeds, one of which is half the other (Figure 29-24). They are often found in applications such as ventilating fans and pumps.

Speed control of ac squirrel-cage motors can be accomplished if the *frequency* of the voltage applied to the stator is varied to change the synchronous speed.

High-power, solid-state electronics have made efficient and accurate **variable frequency drives** possible. The voltage and frequency both are controlled by the motor drive to suit the motor's speed and load conditions.

To keep variable frequency motors running efficiently and to prevent them from overheating, the ratio of voltage to frequency (V/Hz) must be maintained. As the frequency is reduced, the voltage must also be reduced to produce adequate torque. This type of speed control, when used with induction motors, is becoming the most cost effective and popular system.

Figure 29-25 shows the general arrangement of a typical variable-frequency, adjustable-speed ac drive. The circuit has two states of

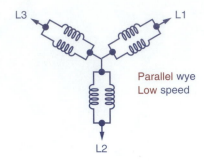

Frequency: 60 Hz

Poles:	2	4	6	8	10	12	14	16
RPM:	3600	1800	1200	900	720	600	514	450

Parallel wye
Low speed

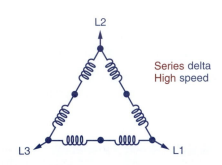

Series delta
High speed

Figure 29-24   Multispeed ac motor-winding connections.

power conversion: a *rectifier* and an *inverter*. The rectifier changes the incoming three-phase ac power to dc power and delivers this power to the inverter circuit. The inverter circuit changes the dc power back to an adjustable frequency ac output that controls the speed of the motor. The inverter is composed of electronic switches (thyristors or transistors) that switch the dc power ON and OFF to produce a controllable ac power output at the desired frequency and voltage. A **regulator** modifies the inverter switching characteristics so that the output frequency can be controlled. Its inputs include sensors to measure the control variables.

The voltage and current waveforms produced by inverter systems approximate, to varying degrees, the pure sine wave. The **pulse width modulated (PWM) inverter** is one of the most commonly used types of inverters (Figure 29-26). Diode rectifiers provide constant dc voltage. Since the inverter receives a fixed voltage, amplitude of output waveform is fixed. The inverter adjusts the width of output voltage pulses as well as frequency, so that voltage is

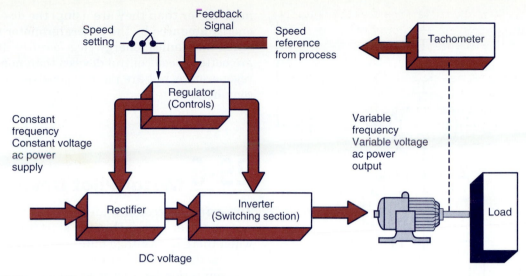

Figure 29-25 Variable-frequency adjustable-speed ac drive.

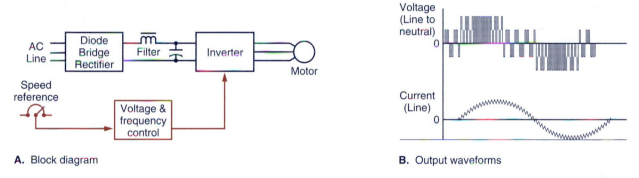

A. Block diagram

B. Output waveforms

Figure 29-26 Pulse width modulated (PWM) inverter.

approximately sinusoidal. PWM drives have near unity *power factor* throughout the speed range.

Wound-rotor motor controllers are used to control the speed of wound-rotor induction motors. By changing the amount of external resistance connected to the rotor circuit through the slip rings, the motor speed can be varied (the lower the resistance, the higher the speed). Figure 29-27 shows the power circuit for a typical wound-rotor motor controller. It consists of a magnetic starter (M), which connects the primary circuit to the line and two secondary accelerating contactors (S and H), which control the speed. When operating at low speed, contactors S and H are both open and full resistance is inserted in the rotor secondary circuit. When contactor S closes, it shunts out part of

the total resistance in the rotor circuit, and as a result, the speed increases. When contactor H closes, all resistance in the secondary circuit of the motor is bypassed, and, as a result the motor runs at maximum speed.

The speed of a direct current (dc) motor is the simplest to control. Its speed can be varied over a very wide range. The speed of a dc motor is controlled by varying the applied voltage across the armature, the field, or both the armature and the field. Figure 29-28 illustrates one simple method of electronic speed control of a dc motor. Motor speed is directly proportional to the voltage applied to the armature. The silicon-controlled rectifier (SCR) is the major power control element of the circuit. The conduction of the SCR is controlled by the setting of the speed-reference potentiometer, which

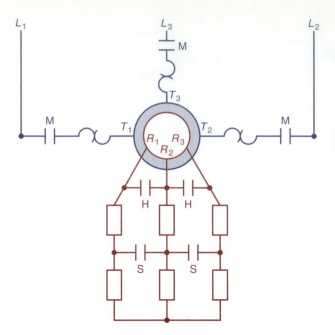

Figure 29-27   Wound-rotor motor controller.

varies the ON time of the SCR per each positive half-cycle and so varies the voltage applied to the armature. The ac input is applied directly to the SCR, since it will rectify (change to dc), as well as control the voltage. A bridge rectifier is used to convert the ac to dc that is required for the field circuit to operate.

Armature voltage-controlled dc drives are constant-torque drives, capable of rated motor torque at any speed up to rated motor base speed. Field voltage-controlled dc drives provide constant hp, and variable torque. Because all dc motors are dc generators as well, *regenerative dc drives* can *invert* (return

more power than they are using) the dc electric energy produced by the generator/motor rotational mechanical energy. *Regenerative* drives are better speed control devices than non-regenerative, but are more expensive and complicated.

29-6 Motor Pilot Devices

Two-wire control or *undervoltage release* of a motor starter means that the starter drops out when there is a voltage failure and picks up as soon as the voltage returns. Figure 29-29 shows the connection for a basic two-wire control circuit. Typically, these circuits involve automatic *pilot devices* such as thermostats, float switches, and pressure switches.

Three-wire control or *undervoltage protection* of a motor starter means that the starter will drop out when there is a voltage failure, but will *not* pick up automatically when voltage returns. Figure 29-30 shows the connection for a basic three-wire control circuit. The control circuit is completed through the stop button and also through a holding contact (2-3) on the starter. When the starter drops out, the holding contact opens and breaks the control circuit until the start button is pressed to restart the motor. In the event of power failure, these circuits are designed to protect against automatic restarting when the power returns. This type of protection should be used where accidents or damage might result from unexpected starts.

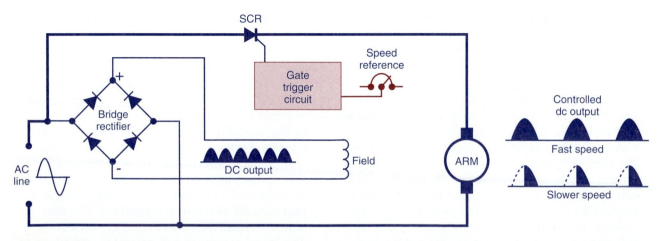

Figure 29-28   Electronic speed control of a dc motor.

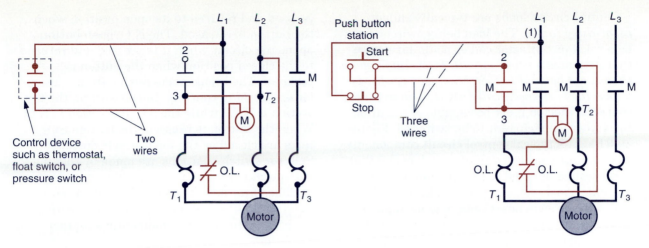

Figure 29-29   Two-wire control.

Figure 29-30   Three-wire control.

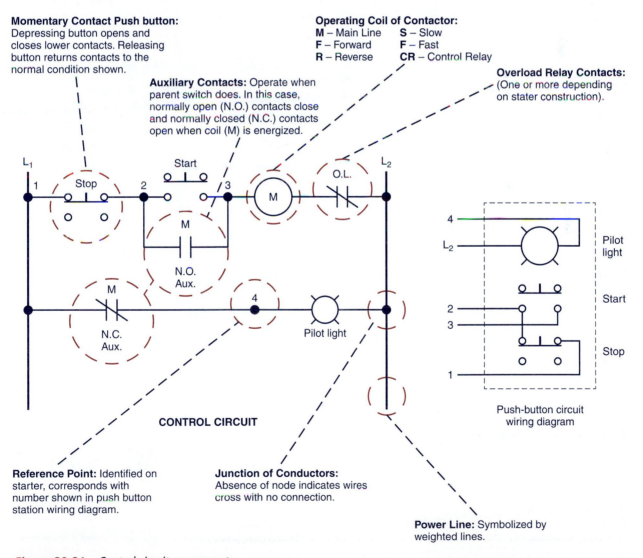

Momentary Contact Push button: Depressing button opens and closes lower contacts. Releasing button returns contacts to the normal condition shown.

Auxiliary Contacts: Operate when parent switch does. In this case, normally open (N.O.) contacts close and normally closed (N.C.) contacts open when coil (M) is energized.

Operating Coil of Contactor:
M – Main Line   **S** – Slow
F – Forward   **F** – Fast
R – Reverse   **CR** – Control Relay

Overload Relay Contacts: (One or more depending on stater construction).

CONTROL CIRCUIT

Push-button circuit wiring diagram

Reference Point: Identified on starter, corresponds with number shown in push button station wiring diagram.

Junction of Conductors: Absence of node indicates wires cross with no connection.

Power Line: Symbolized by weighted lines.

Figure 29-31   Control circuit components.

Control circuit loads are typically called *pilot-duty loads*. The load being switched can be a relay or contactor coil or a similar device that activates a *power circuit. Pilot-duty devices should not be used to switch horsepower loads unless they are specifically rated to do so.* The contacts selected must be capable of handling the voltage and current to be switched. Figure 29-31 shows typical control circuit components.

A *manually-operated pilot switch* is one that is controlled by hand. These include toggle switches, push-button switches, knife switches, and selector switches. Commonly used push button and selector switches are illustrated by their symbols in Figure 29-32. Most push buttons are of the momentary contact type. The N.O. push button makes a circuit when it is pressed and returns to its open position when the button is released. The N.C. push button opens the circuit when it is pressed and returns to the closed position when the button is released. The break-make push button is used for *interlocking* controls. In this switch, the top section is N.C., while the bottom section is N.O. When the button is pressed, the bottom contacts are closed as the top contacts open. Selector switch positions are made by turning the operator knob—not by pushing it. Selector switches may have two or more selector positions with either maintained contact position or spring return to give momentary contact operation.

A *mechanically operated pilot switch* is one that is controlled automatically by such fac-

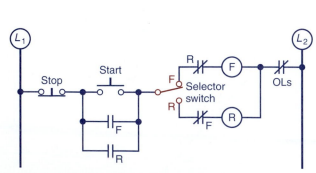

Normally open (N.O.) push button

Normally closed (N.C.) push button

Break-make push button

Note: The abbreviations N.O. and N.C. represent the electrical state of the switch contacts when the switch is not actuated.

A. Symbols

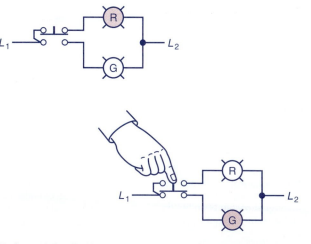

(Courtesy of Allen-Bradley Canada.)

B. Typical push-button-control station

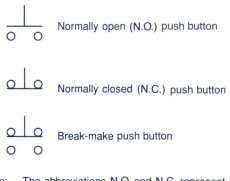

C. Control circuit for selector switch used for reversing the direction of rotation of a motor

D. Control circuit using a combination break-make push-button

Figure 29-32 Push-button and selector switches.

Symbols

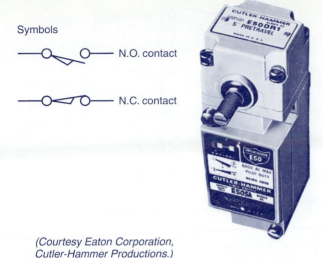

—o◦—o— N.O. contact

—o◦—o— N.C. contact

(Courtesy Eaton Corporation,
Cutler-Hammer Productions.)

A. Switch and symbols

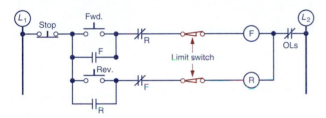

B. Control circuit for starting and stopping a motor in forward and reverse with limit switches providing overtravel protection

Figure 29-33 Limit switch.

tors as pressure, position, or temperature. The *limit switch*, shown in Figure 29-33, is a very common industrial-control device. Limit switches are designed to operate only when a predetermined limit is reached, and they are usually actuated by contact with an object such as a cam. These devices take the place of a human operator. They are often used in the control circuits of machine processes to govern the starting, stopping, or reversal of motors.

The *temperature switch*, or *thermostat*, shown in Figure 29-34, is used to sense temperature changes. Although there are many types available, they are all actuated by some specific environmental temperature change. Temperature switches open or close when a designated temperature is reached. Industrial applications for these devices include maintaining the desired temperature range of air, gases, liquids, or solids.

The *pressure switch* shown in Figure 29-35 is used to control the pressure of liquids and

gases. Again, although many types are available, they are all basically designed to actuate (open or close) their contacts when the specified pressure is reached.

Level switches, such as the one illustrated in Figure 29-36, are used to sense the height of a liquid. The raising or lowering of a float, which is mechanically attached to the level switch, trips the level switch; the level switch, itself, is used to control motor-driven pumps that empty or fill tanks. Level switches are also used to open or close piping solenoid valves to control fluids.

A newer type of sensor switch that is becoming increasingly popular is the *proximity switch*. Proximity switches are part of a series of solid-state sensors. They sense the presence or absence of a target *without physical contact*. The six basic types are magnetic, capacitive, ultrasonic, inductive, air jet stream, and photoelectric. Figure 29-37 illustrates typical examples of industrial processes that use proximity-

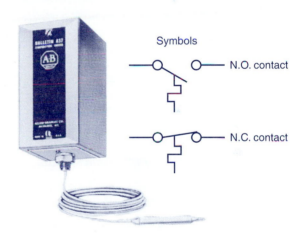

Symbols

—o◦—o— N.O. contact

—o◦—o— N.C. contact

A. Switch and symbols

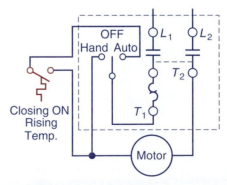

B. Temperature switch used to automatically control a motor

Figure 29-34 Temperature switch.

Symbols

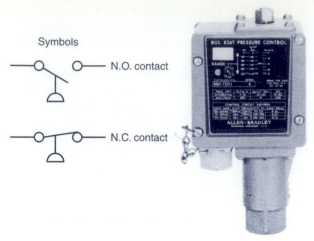

— N.O. contact

— N.C. contact

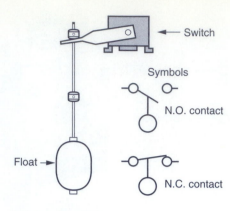

← Switch

Symbols

N.O. contact

Float →

N.C. contact

A. Switch and symbols

A. Switch and symbols

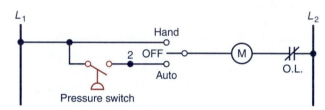

L_1

Hand

2 OFF

Auto

M

L_2

O.L.

Pressure switch

B. Starter operated by pressure switch

Figure 29-35 Pressure switch.

L_1

Float
switch

M

L_2

O.L.

B. Two-wire level switch control starter

Figure 29-36 Level switch.

(Courtesy of Allen-Bradley Canada.)

Typical proximity limit switch designed for industrial environments in applications where it is required to sense the presence of metal objects without touching them.

A. Switch

Figure 29-37 Proximity-switch applications.

(Courtesy of Rechmer Electronics Industries, Inc.)

B. Applications

(Courtesy of Furnas Electric Company.)

A. Switch

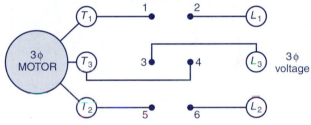

B. Internal switching arrangement

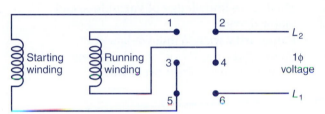

C. Single-phase motor reversing—Wiring diagram

D. Three–phase motor reversing-wiring diagram

Figure 29-38   Reversing drum switch.

switch sensors. The symbols for these switches are usually the same as those used for limit switches.

A *drum switch* consists of a set of moving contacts mounted on and insulated from a rotating shaft. The switch also has a set of stationary contacts that make and break contact with the moving contacts as the shaft is rotated. Drum switches (Figure 29-38) are used for starting and reversing squirrel-cage motors, single-phase motors that are designed for reversing service, and dc shunt and compound-wound motors.

In some cases, the starter coil is operated at a voltage lower than line voltage. This is usually done for safety reasons. This requires the use of a *stepdown transformer* in the pilot circuit (Figure 29-39).

Sequence control is the method by which starters are connected so that one cannot be started until the other is energized. One application is a power distribution system that does not have sufficient capacity to start several motors simultaneously. If several motors are to

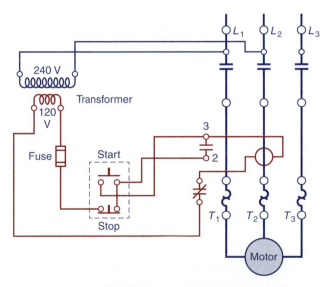

Figure 29-39   Step-down transformer in the control circuit.

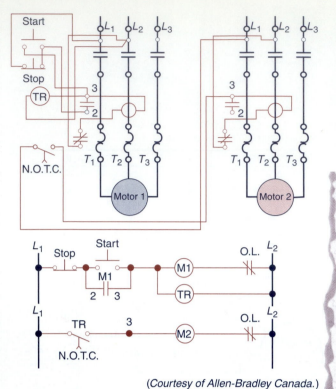

be started from the same push button station under these conditions a time delay can be provided between the operation of the motor starters. The sequence control circuit of Figure 29-40 illustrates such a case. When the START button is pushed, the first starter is energized along with a timing relay. When the timing relay times out, it operates a contact that closes the control circuit of the second starter.

(Courtesy of Allen-Bradley Canada.)

Figure 29-40 Sequence control of two motors.

Review and Applications

Review Questions

1. State the function of each of the following devices as part of the protection for a motor branch circuit:
   (a) disconnect device
   (b) overcurrent-protection device
   (c) overload-protection device
2. Name four commonly used internal overload-protection devices.
3. (a) How do external overload-protection devices determine the heating of a motor?
   (b) Name four commonly used external overload-protection devices.
4. Explain the type of protection provided by each of the following motor protection circuits:
   (a) low-voltage protection
   (b) low-voltage release protection
   (c) phase-failure protection
   (d) phase-reversal protection
   (e) ground-fault protection
5. Why is the starting current of a motor much greater than the normal running current?
6. In what way is a contactor different from a relay?
7. A motor starter is a combination of what two devices?
8. Define *across-the-line* starting.
9. What type of motor protection is not normally provided by manual starters?
10. What are the two circuits normally associated with a typical magnetically operated starter?
11. List five types of reduced-voltage starters.
12. (a) How is the direction of rotation of a three-phase induction motor reversed?
    (b) What is the purpose of the mechanical and electrical interlocks used with three-phase reversing starters?
13. Define jogging as it applies to motor control.
14. Explain how a motor is brought to a stop in each of the following instances:
    (a) plugging of a three-phase motor
    (b) dynamic braking of a dc motor

(c) electric braking of a three-phase motor
   (d) electromechanical braking
   (e) eddy-current braking
15. Define *motor drive*.
16. Compare open-loop and closed-loop speed control.
17. What are the two variables that determine the speed of ac induction motors?
18. Explain the two stages of power conversion that take place in a variable frequency ac drive.
19. How is motor speed controlled using a wound-rotor motor controller?
20. How is the speed of a dc motor controlled?
21. Compare the operation of a motor starter with two-wire and three-wire control when voltage is lost to a running motor and then restored.
22. Why should pilot-duty devices not normally be used to switch horsepower loads?
23. Define *manually-operated switch*.
24. Define *mechanically-operated switch*.
25. What are drum switches used for?
26. Define *sequence control*.

Problems

1. Refer to Figure 29-11B. Coil *M* fails to energize when the start button is pressed. A voltmeter connected across the coil indicates that line voltage is present when the start button is pressed. What is the most likely problem? Why?
2. Refer to Figure 29-11C. Suppose one of the two stop push buttons is defective. The normally-closed contact of the defective button is assumed to be open. The voltage is measured across the terminals of each stop button.
   (a) What would the value of the voltage be across the good closed push button?
   (b) What would the value of the voltage be across the defective open push button?
3. Refer to Figure 29-16. Assume that motor terminals T_1 and T_2 are reversed. What will happen when forward contacts "*F*" close?

4. Refer to Figure 29-18. Assume that while the motor is operating (*M* coil energized), the jog button is momentarily actuated. What happens? Why?

5. Refer to Figure 29-20. Assume no dynamic braking occurs when the motor is stopped. State the two most likely faults to look for.

6. Refer to Figure 29-28. Assume that one of the diodes of the bridge rectifier circuit becomes open circuited resulting in only half the normal dc voltage being applied to the field. How will this affect the normal speed of the motor? Why?

7. Refer to Figure 29-30. Assume that the wire connected between terminal No. 2 of the starter and the start terminal of the push-button becomes open. Explain how this would affect the operation of the circuit.

Critical Thinking

1. Refer to Figure 29-11B. Assume that normally-open contact *M* connected across the start push button is bad and fails to close when the coil is energized. Describe how the circuit would operate when attempting to start the motor by closing the start push button.

2. Refer to Figure 29-12. Assume one of the starting resistors is open. Explain what would happen when the contacts *A* and *B* are closed in sequence.

3. Refer to Figure 29-17B. Assume the motor is running in the forward direction and the reverse push button is pressed. What happens? Why?

4. Refer to Figure 29-19B. Assume an open is keeping the speed switch contact *F* closed at all times. Explain how the operation of the circuit would be affected?

5. Refer to Figure 29-21. The amount of braking torque applied to the motor is to be reduced. What adjustment of the transformer needs to be made?

6. Refer to Figure 29-29. A single-pole double-throw auto/manual switch is to be wired into the circuit for manual or automatic control of the motor. Redraw the circuit showing this switch properly connected to the existing control device.

7. Refer to Figure 29-32C. Assume that the motor is running in the forward direction and the selector switch is moved from *F* to *R*. Would the motor immediately start turning in reverse? Why?

Portfolio Project

- Two three-phase magnetic starters, each controlling a three-phase motor, are to be controlled by a *common* START-STOP push-button station. Pressing the start button starts both motors and pressing the stop button stops both motors. If an overload occurs on either one of the two motors, both starters will be automatically disconnected. Design a motor control circuit that will accomplish this task.

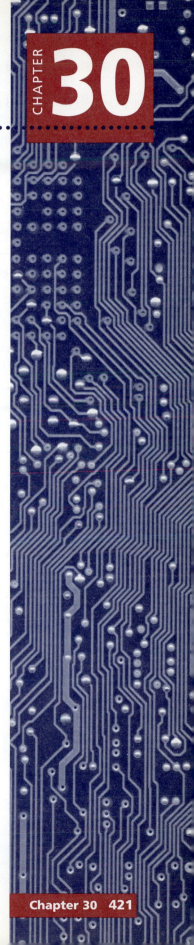

Inductance and Capacitance

Resistance, inductance, and capacitance are the three basic circuit properties that we use to control voltages and currents in electric circuits. Each behaves in a different way. Resistance opposes current, inductance opposes any change in current, while capacitance opposes any change in voltage. Also, resistance dissipates energy, while inductance and capacitance both store energy. In this chapter, we will study the unique electrical properties of inductance and capacitance.

Objectives

After studying this chapter, you should be able to:

1. List the physical factors that affect inductance and capacitance.
2. Calculate the inductive reactance of a coil and capacitive reactance of a capacitor.
3. Compare phase shift in an inductor and capacitor.
4. Define and calculate RC-time constants.
5. Explain the factors that determine the impedance of an ac circuit.
6. Define resonance.
7. Compare the real, reactive, and apparent power of an ac circuit.

Key Terms

- inductor
- air-core
- iron-core
- ferrite-core
- inductance (*L*)
- self-inductance

- mutual-inductance
- Lenz's law
- henry (H)
- inductive reactance (X_L)
- in phase

- out of phase
- capacitance (*C*)
- dielectric
- farad (F)
- microfarad (μF)
- variable capacitor

(*continued*)

RESEARCH INTERNET SITES

- ISI Inductor Supply Inc.
- North America Capacitor Co. (NACC)
- CK Corporation

- **fixed nonpolarized capacitor**
- **fixed polarized capacitor**
- **RC-time constant**
- **capacitive reactance (X_C)**

- **leading**
- **lagging**
- **impedance (*Z*)**
- **net reactance**
- **resonance**
- **resonant frequency**

- **real power (Watts)**
- **reactive power (VAR)**
- **apparent power (VA)**
- **power factor (PF)**

30-1 Types of Inductors

An *inductor* is a coil of wire sometimes referred to by such names as choke, impedance coil, or inductive reactor. There are three popular types of inductors: air-coil, iron-core, and ferrite-core inductors. The **air-core** type consists of a coil wrapped around a form with nothing but air in the middle (Figure (30-1). Air-core inductors are used in high-frequency radio circuits.

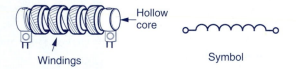

Figure 30-1   Air-core inductor.

Iron-core inductors are constructed with the coil wound around a laminated shell-type core (Figure 30-2). They are used in circuits that operate at relatively low ac frequencies which are called *audio frequencies* (from about 20 Hz to 20 kHz).

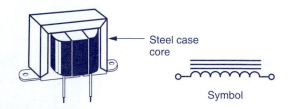

Figure 30-2   Iron-core inductor.

Ferrite-core inductors use cores made up of ceramic materials called ferrites (Figure 30-3). They are designed primarily for use in

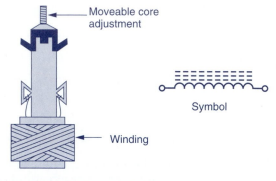

Figure 30-3   Ferrite-core inductor.

high-frequency radio circuits. This type of inductor is available with a movable ferrite slug that can be adjusted for different circuit tuning applications.

30-2 Inductance

Inductance (*L*) is the ability of an electric circuit or component to oppose any *change* in current flow. The ability of the circuit or component to oppose changes in current is due to its ability to store and release energy that it has stored in a *magnetic field*. When direct current is initially applied to a coil or inductor, a magnetic field builds up when the circuit is energized. Expansion of the magnetic field cutting across the coil windings causes a countervoltage to be induced in the coil. This countervoltage tends to oppose the original applied voltage (Figure 30-4). Once the current reaches its Ohm's law value, it remains constant and the inductive effect stops.

If the dc voltage applied to the coil is switched OFF, inductance occurs again. When the switch is opened, there will be a noticeable

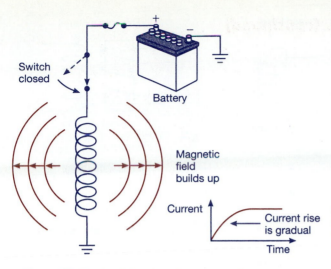

Figure 30-4 Applying dc voltage to a coil.

arc at the switch contacts (Figure 30-5). As the current falls to a zero value, the collapsing magnetic field causes a high voltage to be induced in the coil. The arc produced at the contacts is a result of the induced voltage attempting to maintain the current in the circuit.

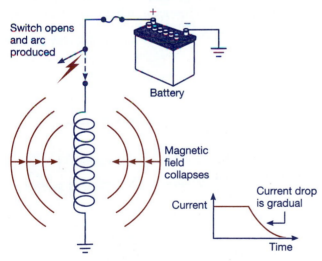

Figure 30-5 Removing dc voltage from a coil.

Inductance is present only when the *current changes*. In a dc circuit, this occurs each time the circuit is turned ON or OFF. The inductive effect just described takes place in the coil itself and is called **self-inductance**. If a second coil is magnetically linked to the first, the inductive effect will also be felt in the second coil. **Mutual-inductance** is the term used when the

effect of induction is such that current change in one circuit produces an induced voltage in another circuit (Figure 30-6).

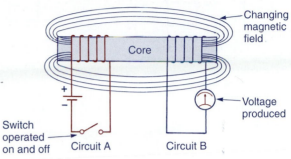

Figure 30-6 Mutual-inductance.

In any type of inductive circuit there is an important relationship between the direction of the current change and the induced voltage. This relationship is summarized by *Lenz's law* and stated as follows:

THE INDUCED VOLTAGE ALWAYS ACTS IN A DIRECTION TO OPPOSE THE CURRENT CHANGE THAT PRODUCED IT.

The inductance rating of an inductor refers to its ability to generate a countervoltage that opposes a change in the current flow. The amount of inductance of a coil is measured by a unit called the *henry* (H). One henry (1 H) represents the inductance of a coil in which a current change of one ampere per second (1 A/s) will produce a countervoltage of one volt (1 V). Practical values of inductors used in electronic circuits vary from several henries (H) down to a few microhenries (μH). In addition to being rated for inductance, inductors carry a maximum current rating. This current rating indicates the maximum current flow that the inductor can handle without overheating the coil.

An easy way to determine if an inductor is open or shorted is to make a dc resistance measurement with an ohmmeter. Normal resistance values depend on the wire size (diameter) and number of turns (length) of the wire that makes up the inductor. Some coils with fine wire and a large number of turns will have hundreds of

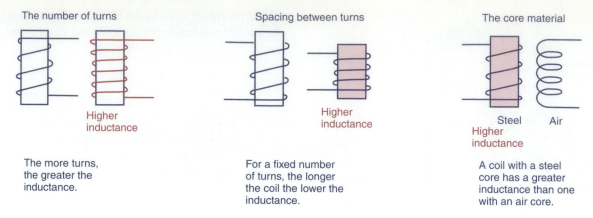

The number of turns

Higher
inductance

The more turns,
the greater the
inductance.

Spacing between turns

Higher
inductance

For a fixed number
of turns, the longer
the coil the lower the
inductance.

The core material

Steel   Air
Higher
inductance

A coil with a steel
core has a greater
inductance than one
with an air core.

Figure 30-7 Factors that determine inductance.

ohms of resistance; large coils with large wire and a small number of turns will have tens of ohms of resistance. If no resistance is measured at all, the inductance is open.

Inductance itself can only be measured with special meters and depends entirely on the physical construction of both the core and the windings around the core. Figure 30-7, illustrates these factors. The more turns, the better the magnetic-core material, the larger core cross-section area, and the shorter the coil length. All these factors increase the inductance. Many differently constructed coils could have an inductance of one henry, and each would have the same effect in the circuit. Certain semiconductor circuits permit the simulation of inductance electronically without the need for bulky inductors.

30-3 Inductive Reactance

In a dc circuit, the only changes in current occur when the circuit is closed to start current, and when it is opened to stop current. However, in an ac circuit, the current is continually changing each time the voltage alternates. Since inductance in a circuit opposes a change in current and ac is continually changing, there is an opposition offered by the inductor to the ac current that is called *inductive reactance*.

Inductance reactance is measured in *ohms* and is represented by the symbol X_L. Current flow through a coil connected to a dc source is limited by the wire resistance of the coil only. Current flow through the same coil connected

to an ac source is limited by both the wire resistance and the inductive reactance.

Example 30-1

An inductor and lamp are connected in series to a dc and ac circuit (Figure 30-8). The steady dc voltage produces no varying magnetic field on the inductor, which generates no opposing countervoltage. The result is that compared to the ac circuit, the current flow and the brightness of the lamp are greater for the dc circuit.

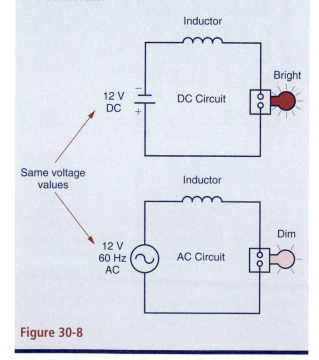

Inductor

12 V
DC

DC Circuit

Bright

Same voltage
values

Inductor

12 V
60 Hz
AC

AC Circuit

Dim

Figure 30-8

The inductive reactance of a coil is directly proportional to the inductance rating of the coil. That is, the higher the inductance (L), the greater the inductance reactance (X_L).

Example 30-2

A 5-H and 10-H inductance are connected in series with a lamp to an ac circuit (Figure 30-9). The larger 10-H inductance will produce more inductive reactance than the 5-H inductance. The result is that compared to the 10-H circuit, the current flow and the brightness of the lamp is greater for the 5-H circuit.

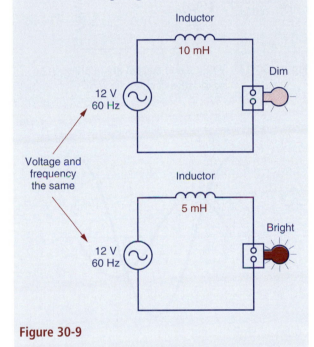

Figure 30-9

The inductive reactance of a coil is also directly proportional to the frequency of the ac supply source. For any given coil, increasing the frequency of the voltage source increases the rate of change of current through the coil. This results in the lines of force created by the current cutting across the coil windings at an increased rate. This, in turn, increases the countervoltage produced.

The inductive reactance (X_L) of a coil can be calculated using the formula:

$$X_L = 2\pi fL$$

Example 30-3

An inductance and lamp connected in series are operated from a 60-Hz and 300-Hz ac source (Figure 30-10). The higher frequency generates more inductive reactance with more ohms of opposition. The result is when the circuit is operated at 60 Hz, the current flow and brightness of the lamp are greater than when operated at 300 Hz.

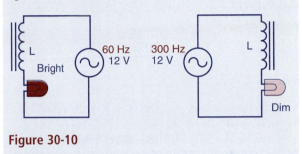

Figure 30-10

where

X_L = the inductive reactance in ohms
f = the frequency of the AC in hertz
L = the inductance in henries

(The quantity of $2\pi f$ represents the rate of change of current in radians per second.)

Example 30-4

Problem At a frequency of 1000 Hz, what is the inductive reactance of a 200 mH inductor?

Solution Use the formula

$$X_L = 2\pi fL$$

Convert 200 mH to henries: 0.2 H
Enter known values

$$X_L = (2)(3.14)(1000\,\text{Hz})(0.2\,\text{H})$$

Therefore, $X_L = 1256\Omega$

In ac circuits that contain only pure inductance, the inductive reactance is the only thing that limits the current. The current is determined by Ohm's law with X_L replacing R, as follows:

$$I = \frac{V}{X_L}$$

If an ac circuit contains both resistance and inductance (reactance), then the total opposition to current flow is termed *impedance* and is designated by the letter Z. When a voltage V is applied to a circuit that has an impedance Z, the current I is:

$$I = \frac{V}{Z}$$

where

$$Z = \sqrt{R^2 + X_L^2}$$

Finding the total inductance of a series circuit, composed totally of inductors, is the same as finding the total resistance of a series resistor circuit. You simply add all of the individual inductances. This is assuming there is no magnetic interaction between the inductors:

$$L_T(\text{series}) = L_1 + L_2 + L_3 \ldots$$

Inductors placed in parallel are also treated like resistors in parallel. You have your choice of three formulas for solving parallel inductance: the *same-values formula*, the *product-over-sum formula*, and the *reciprocal formula*.
The product-over-the-sum formula

$$L_T(\text{parallel}) = \frac{L_1 \times L_2}{L_1 + L_2}$$

30-4 Phase Shift in Inductance

In an ac circuit containing only resistance, the voltage sine wave and the current sine wave are in phase. That is, when the voltage reaches its peak value, the current also reaches its peak value (Figure 30-11). When the voltage is zero, the current is zero, and so on. The voltage and current are ***in phase*** in an ac circuit containing resistance only.

In an ac inductive circuit, the voltage and current are ***out of phase***. In relation to time, the current reaches its peak slightly later than the voltage does (Figure 30-12). This is due to

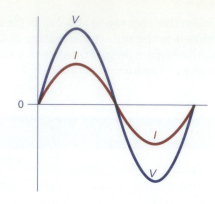

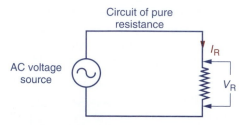

Figure 30-11 Voltage and current are in phase in an ac resistive circuit.

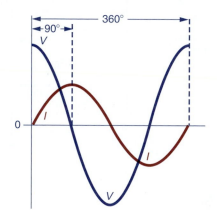

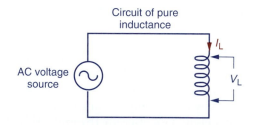

Figure 30-12 Current lags behind voltage in an ac inductive circuit.

the fact that the alternating current produces a self-induced voltage across the inductor which is said to be 90° out of phase with the current through the inductor (a complete cycle of an ac wave is considered to be 360°). In an ideal inductor (one with no resistance), the *voltage wave leads the current by 90°,* or, in opposite terms, *the current wave lags behind the voltage by 90°.* In a circuit containing both inductance and resistance, the ac current wave will lag the voltage wave by an amount between zero degrees and 90°.

30-5 Capacitance

Capacitance (C) is the ability of an electric circuit or component *to store electric energy by means of an electrostatic field.* The **capacitor** is an electrical device specially designed for this purpose. The capacitor has the ability to store electrons and release them at a later time. Basically, a capacitor consists of two metal plates (conductors) placed near each other and separated by an insulating material called the *dielectric* (Figure 30-13). The dielectric can be air or any nonconducting material such as paper, mica, or ceramic.

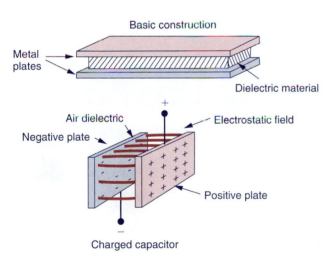

Figure 30-13   Capacitor.

The operation of a capacitor depends on the *electrostatic field* that is built up between the two oppositely charged parallel plates. A capacitor with a field built up is said to be charged, while one with no field is discharged. A capacitor is *charged* by connecting its two plate leads to a *dc voltage source.* The positive terminal of the voltage source attracts electrons from the plate connected to it. At the same time, the

negative terminal of the voltage source repels an equal number of electrons into the other plate. This charging or flow of electrons continues until the voltage across the charged plates is equal to the applied voltage. *Note that there is no flow of electrons through the dielectric.* Electrons are simply removed from one plate and deposited on the other plate through the circuit that connects them (Figure 30-14).

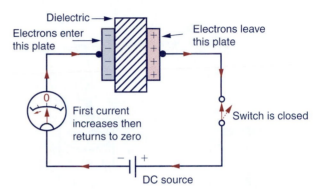

Figure 30-14   Capacitor charging circuit.

Once charged, the capacitor can be disconnected from the voltage source, and the energy will remain stored in the electrostatic field between its plates. To discharge the capacitor, the two charged plate leads are connected together (Figure 30-15). Electrons now flow in the opposite direction, from the negatively charged plate to the positively charged plate. Both plates then become neutralized and the capacitor is said to be discharged.

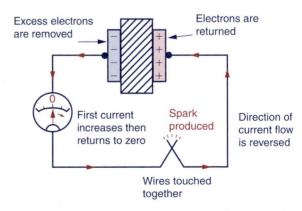

Figure 30-15   Capacitor discharging circuit.

Capacitors connected in live dc circuits can store a charge for a considerable time after the voltage to the circuits has been switched OFF. *High voltage capacitors like those used in TV sets can store a lethal charge! Never touch leads of such a capacitor before it is carefully discharged.* An insulated jumper probe with a built-in resistor can be used to safely discharge a capacitor. The resistor slows the discharge flow of electrons to avoid a damaging current surge (Figure 30-16).

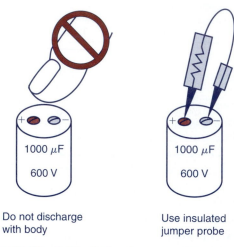

Do not discharge with body

Use insulated jumper probe

Figure 30-16   Safely discharging a capacitor.

About ◀▬▶ Electronics

The unit of measure for capacitance, the farad, was named for Michael Faraday, an English chemist and physicist who discovered the principle of induction (1 F is the unit of capacitance that will store 1 C of charge when 1 V is applied).

Capacitance exists between any two conductors separated by an insulating dielectric. The conductors do not have to be plates. They may be wires, or conductors, of any shape. The dielectric may be air or any other material that is an insulator. In any two-wire cable, there will be capacitance between the two wires. Capacitance will also exist between circuit wiring and a metal chassis, between conductors on a printed circuit board, or between the leads of component parts. Capacitance resulting from these and other unwanted sources is referred to as **stray capacitance**.

30-6 Capacitor Ratings

The number of electrons that a capacitor can store for a given applied voltage is a measure of its *capacitance*. The unit used to measure capacitance is the *farad (F)*. A capacitor has a capacitance of one farad when it stores a charge of one coulomb (1 C) when a voltage of one volt (1 V) is applied to it. The farad is a very large unit of capacitance and is not often used for practical applications, therefore, smaller values are used. A **microfarad** is 10^{-6} farads; a nanofarad is 10^{-9} farads; and a picofarad is 10^{-12} farads. Microfarads and picofarads are very common in electronic circuits.

If the value of the applied voltage and capacitance of the capacitor are known, the amount of charge stored in the capacitor can be calculated. The equation used is

$$Q = CV$$

where

Q = charge in coulombs
C = capacitance in farads
V = voltage in volts

Example 30-5

Suppose a 500-μF capacitor was charged from a 100-V dc source. The amount of charge stored in the capacitor would be:

$$Q = CV$$
$$Q = (500\,\mu\text{F})(10^{-6})(100\,\text{V})$$
$$Q = 0.05\,\text{C (coulombs)}$$

Capacitors are rated for both capacitance and voltage. The capacitance value of a capacitor depends on (Figure 30-17):
• The area of the plates (the greater the plate area, the higher the capacitance value).
• Type of dielectric used (the better the dielectric material, the higher the capacitance value).
• The spacing between plates (the closer the plates, the higher the capacitance value).

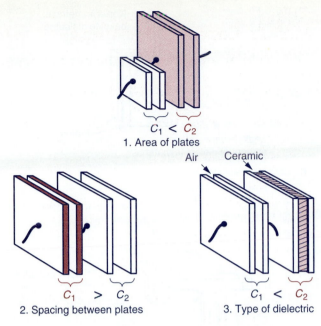

1. Area of plates $C_1 < C_2$

2. Spacing between plates $C_1 > C_2$

3. Type of dielectric $C_1 < C_2$

Air   Ceramic

Figure 30-17   Factors that determine capacitance of a capacitor.

The voltage rating of a capacitor indicates the maximum voltage that can be safely applied to its plates (Figure 30-18). Voltages in excess of this value may break down the insulating dielectric material and permanently damage the capacitor. Always use a capacitor that has a voltage rating higher than the voltage of the circuit. A 50-V capacitor can be used on a 10-V or 25-V circuit, but a capacitor rated for only 25 V cannot be used on a 50-V circuit.

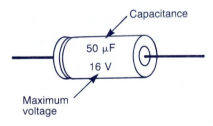

Figure 30-18   Capacitor ratings.

Capacitors are connected in parallel to obtain a greater total capacitance than is available in one unit (Figure 30-19). When a number of capacitors are connected in parallel, the total capacitance of the group is equal to the *sum* of the individual capacitances. The largest voltage that can be applied safely to a group of capacitors in parallel can be determined easily. It is

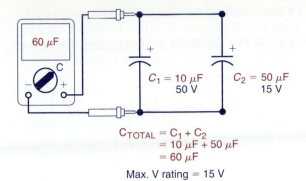

$$C_{TOTAL} = C_1 + C_2$$
$$= 10\ \mu F + 50\ \mu F$$
$$= 60\ \mu F$$

Max. V rating = 15 V

Figure 30-19   Connecting capacitors in parallel.

the voltage that can be applied safely to the capacitor having the lowest voltage rating.

Capacitors are connected in series to enable the group to withstand a higher voltage than any individual capacitor is rated for (Figure 30-20). However, this is accomplished at the expense of decreased total capacitance. The formula for calculating the total capacitance of capacitors in series is similar to that used to calculate the total resistance of resistors connected in parallel.

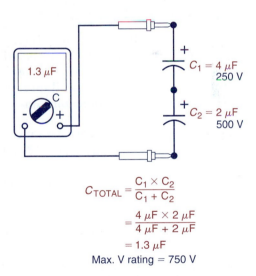

$$C_{TOTAL} = \frac{C_1 \times C_2}{C_1 + C_2}$$
$$= \frac{4\ \mu F \times 2\ \mu F}{4\ \mu F + 2\ \mu F}$$
$$= 1.3\ \mu F$$

Max. V rating = 750 V

Figure 30-20   Connecting capacitors in series.

Capacitance of a capacitor can be measured directly using a digital multimeter with a capacitance function. Two capacitors can be compared as to their relative capacitance by using an analog ohmmeter. The amount of needle deflection of the ohmmeter can be used to indicate a relative amount of capacitance. By connecting the ohmmeter to the capacitor as shown in Figure 30-21, the ohmmeter battery charges the capacitor to its voltage. The meter will deflect initially and then fall back to infinity as the capacitor charges.

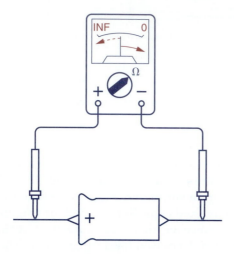

Figure 30-21   Using an analog ohmmeter to measure relative amount of capacitance.

30-7 Types of Capacitors

Capacitors are generally classified by their dielectric. The names mica capacitors, paper capacitors, air capacitors, and ceramic capacitors all refer to the dielectric used. Tantalum and aluminum electrolytic capacitors, on the other hand, refer to the plate material. Capacitors come in three general classifications: variable, fixed nonpolarized, and fixed polarized.

Variable capacitors have some means of rotating one set of plates, attached to a rotating shaft, past another set of plates. A dielectric, air or mylar, is between each pair of plates. As the movable set slides in between the fixed set, the distance between plates is reduced and the area of the plates at the reduced distance increases, so the capacitance increases (Figure 30-22).

Ceramic, mica, and mylar capacitors, among others, are *fixed nonpolarized* types. This means they can be inserted in the circuit with either lead toward the positive voltage. They generally exhibit less capacitance per unit volume than do polarized types. They are generally used, therefore, in situations where small capacitance values are required (Figure 30-23).

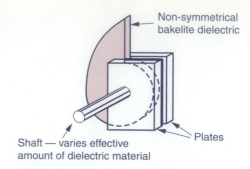

A. Movable dielectric type

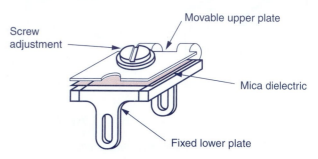

B. Adjustable pressure type

C. Symbols

Figure 30-22   Variable capacitors.

Electrolytic and some tantalum capacitors are *polarized capacitors*. They must be installed with their positive terminal connected to the positive voltage in the circuit or else they will self destruct. Their capacitance per unit volume is relatively large, so they are used in applications, such as power-supply filtering, that require large capacitance (Figure 30-24).

30-8 *RC*-Time Constant

When dc is applied directly to a capacitor, it charges almost instantly to the value of the source voltage (Figure 30-25). The time required to charge a capacitor can be controlled by connecting a resistor in series with the capacitor. A high-value resistor will give us a long charging time. A low-value resistor used with the same capacitor will give a shorter charging time.

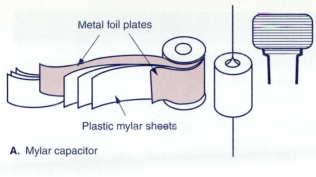

Metal foil plates

Plastic mylar sheets

A. Mylar capacitor

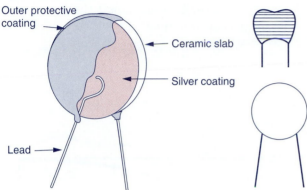

Outer protective coating

Ceramic slab

Silver coating

Lead

B. Ceramic capacitor

Figure 30-23 Fixed nonpolarized capacitors.

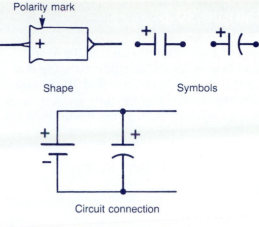

Polarity mark

Shape   Symbols

Circuit connection

Figure 30-24 Electrolytic capacitor.

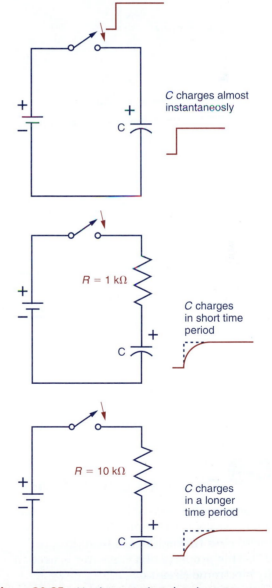

C charges almost instantaneosly

$R = 1\ k\Omega$

C charges in short time period

$R = 10\ k\Omega$

C charges in a longer time period

Figure 30-25 Varying capacitor charging rate.

The charging rate of a resistor and capacitor connected in series is called the **RC-time constant**. When the resistance in megohms (MΩ) is multiplied by the capacitance in microfarads (μF), the product is called the *RC-time constant in seconds*.

$$t = RC$$

where

t = time constant in seconds

R = resistance in MΩ

C = capacitance in μF

The *RC*-time constant represents the time, in seconds, it takes to charge a capacitor to 63.2 percent of its applied-voltage value. It requires approximately five *RC*-time constants for the capacitor to reach the applied voltage value.

The discharge rate of a charged capacitor connected in series with a resistor is also predictable. The same *RC*-time constant formula is used and the only difference is that the

Example 30-6

Assume a dc voltage source of 100 V is applied to a series RC circuit consisting of a 500,000-Ω (0.5-MΩ) resistor and 10-μF capacitor (Figure 30-26). The RC-time constant and charging and discharging rate would then be:

$$t = RC$$
$$t = (0.5\,\text{M}\Omega)(10\,\mu\text{F})$$
$$t = 5\,\text{s}$$

If the switch is initially moved from the OFF to the CHARGE position, after 5 seconds (or

1 time-constant period), the voltage across the capacitor would be:

$$V_C = 100\,\text{V} \times 63.2\%$$
$$V_C = 63.2\,\text{V}$$

Once the capacitor is fully charged, if switched to DISCHARGE, the voltage across the capacitor after 5 seconds (or 1 time-constant period) would be:

$$V_C = 100\,\text{V} - 63.2\,\text{V}$$
$$V_C = 36.8\,\text{V}$$

$$V_C = 100\,\text{V} - 63.2\,\text{V}$$
$$V_C = 36.8\,\text{V}$$

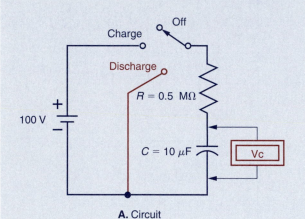

A. Circuit

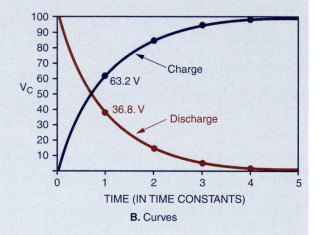

B. Curves

Figure 30-26

capacitor loses 63.2 percent of its charge in each time constant. This characteristic makes a capacitor very useful in timing circuits.

30-9 Capacitive Reactance

A capacitor will charge to block the flow of dc in a circuit, while at the same time, it will allow the ac to pass through to the load (Figure 30-27). This property of a capacitor is used in many electronic circuits.

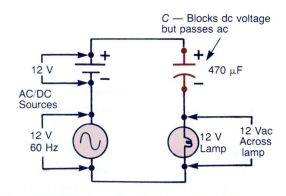

Figure 30-27   Capacitor blocks the flow of dc.

When a capacitor is connected to an ac circuit the charges upon the plates reverse with each change of the applied-voltage polarity. The plates then are alternately charged and discharged. This results in a constant ac current flow (Figure 30-28). Again, electrons flow in and out of the plates through the external circuit. The circuit current does not pass through the dielectric but only *appears* to. This is why we say *capacitors pass ac*.

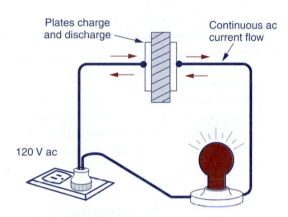

Figure 30-28 Capacitor action in an ac circuit.

Different sized capacitors offer a different amount of opposition to the flow of current in an ac circuit. The opposition to the flow of ac current offered by a capacitor is called *capacitive reactance*. Capacitive reactance is measured in *ohms* and is represented by the symbol X_C. The capacitive reactance of a capacitor is inversely proportional to the capacitance of the capacitor. That is to say, the larger the capacitance of the capacitor, the lower is its capacitive reactance or opposition to ac current flow.

The capacitive reactance of a capacitor is inversely proportional to the frequency of the ac supply source. For any given capacitor, increasing the frequency of the voltage source increases the rate at which the capacitor charges and discharges. This results in less capacitive reactance and a greater ac current flow.

The capacitive reactance (X_C) of a capacitor can be calculated using the formula:

$$X_C = \frac{1}{2\pi f C}$$

where

$$X_C = \text{capacitive reactance in ohms } (\Omega)$$
$$f = \text{frequency in hertz (Hz)}$$
$$C = \text{capacitance in farads (F)}$$

In ac circuits that contain only capacitance, the capacitive reactance is the only thing that limits the current. The current is determined by Ohm's law with X_C replacing R as follows:

$$I = \frac{V}{R} = \frac{V}{X_C}$$

Example 30-7

A 10-μF and 50-μF capacitor are connected in series with a lamp to an ac circuit. The larger capacitance of the 50-μF capacitor in the circuit can store more charge and discharge more current. The result is that compared to the 10-μF circuit, the current flow and brightness of the lamp are greater for the 50-μF circuit (Figure 30-29).

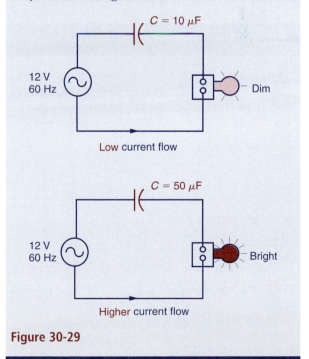

Figure 30-29

Example 30-8

A capacitor and lamp are connected in series and operated from a 60-Hz and 60-kHz ac source. When the generator frequency is increased from 60 Hz to 60 kHz, with the same capacitance in the circuit, the current flow and brightness of the lamp increase (Figure 30-30).

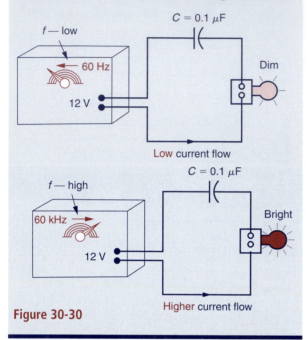

Figure 30-30

Example 30-9

Suppose a 10-μF capacitor is connected to a 120-V, 60-Hz source. The capacitive reactance and ac flow through the capacitor would be:

$$X_C = \frac{1}{2\pi f C}$$

$$X_C = \frac{1}{(2)(3.14)(60\,\text{Hz})(10\,\mu\text{F})(10^{-6})}$$

$$X_C = \frac{10^6}{(2)(3.14)(60\,\text{Hz})(10\,\mu\text{F})}$$

$$X_C = 265\,\Omega$$

$$I = \frac{V}{X_C}$$

$$I = \frac{120\,\text{V}}{265\,\Omega}$$

$$I = 0.453\,\text{A}$$

30-10 Phase Shift in Capacitance

Like inductive reactance, capacitive reactance causes the voltage and current to be out of phase. However, in the case of capacitive reactance, the current **leads** the voltage. This is due to the fact that when a dc voltage is first applied to a capacitor, the current flow is maximum and then tapers off as the charge voltage across the capacitor increases. In other words, current is leading voltage.

Figure 30-31 illustrates the phase relationship between current flow in the external circuit and the voltage across the capacitor when an ac source is applied. *The current leads the voltage in the ideal capacitor by exactly 90°.* This is so because the maximum value of current corresponds to the time when the capacitor is fully discharged (has zero volts across it). When the capacitor is fully charged, current flow stops or is at zero. Since an ac source is used, the voltage applied to the capacitor is constantly causing the capacitor to charge in one direction, then discharge in the other.

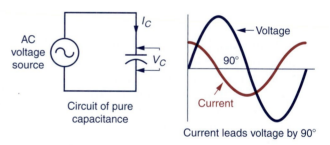

Figure 30-31   Current leads voltage in an ac capacitive circuit.

30-11 Impedance

In a dc circuit, current flow is opposed only by the resistance (R) of the circuit. In an alternating-current circuit, not only resistance but also inductive reactance (X_L) and capacitive reactance (X_C) oppose current flow. The total opposition to current flow in an ac circuit is called *impedance* (Z) and is measured in ohms.

Since inductive reactance causes a *lagging* current and capacitive reactance causes a leading current, they *cancel each other out*. In series, the ohms of X_C and X_L cancel. In parallel, the capacitive and inductive branch currents I_C and I_L cancel. Whichever of the two is larger determines whether the current leads or lags the voltage.

In an ac circuit, current (I), voltage (V), and impedance (Z), obey Ohm's law in the following relationships:

$$Z = \frac{V}{I}$$

$$V = IZ$$

$$I = \frac{V}{Z}$$

Because of the different phase relationships we cannot simply add reactances to resistance to obtain the total impedance. However, capacitive and inductive reactances can be combined

Example 30-10

Problem   Find the **net reactance**, impedance, current, and voltage drops in a series circuit containing a coil with a 10-Ω inductive reactance, a capacitor with a 7-Ω capacitive reactance, and a resistor with a 4-Ω resistance (Figure 30-32). The applied voltage is 24-V ac.

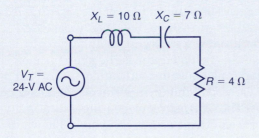

$X_L = 10\ \Omega$ $X_C = 7\ \Omega$

$V_T = $ 24-V AC

$R = 4\ \Omega$

Figure 30-32

Solution   The net reactance in this circuit is:

$$X = X_L - X_C$$
$$X = 10\,\Omega - 7\,\Omega$$
$$X = 3\text{-}\Omega \text{ inductive reactance}$$

The impedance is:

$$Z = \sqrt{R^2 + X^2}$$
$$= \sqrt{4^2 + 3^2} = \sqrt{16 + 9}$$
$$= \sqrt{25} = 5\,\Omega$$

The current is:

$$I = \frac{V}{Z} = \frac{24\,\text{V}}{5\,\Omega}$$
$$= 4.8\,\text{A}$$

The voltage drops are:

$$V_R = IR = 4.8\,\text{A} \times 4\,\Omega = 19.2\,\text{V}$$
$$V_C = IX_C = 4.8\,\text{A} \times 7\,\Omega = 33.6\,\text{V}$$
$$V_L = IX_L = 4.8\,\text{A} \times 10\,\Omega = 48\,\text{V}$$

(Note that the voltages across the capacitor and coil *exceed* the value of the applied voltage.)

Since X_L is greater than X_C the line current lags the line voltage. The current through the resistor, however, is in phase with the voltage across the resistor. The same *value* of current passing through the capacitor shifts and leads the voltage across the capacitor by 90°. Finally, the same value of current will shift again as it passes through the coil. This time it will lag the voltage by 90° (Figure 30-33).

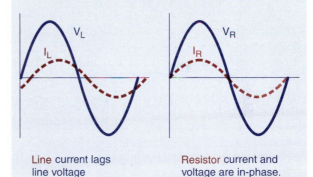

V_L

I_L

Line current lags line voltage

V_R

I_R

Resistor current and voltage are in-phase.

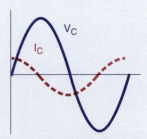

V_L

I_L

Inductor current lags inductor voltage by 90 degrees.

V_C

I_C

Capacitor current leads capacitor voltage by 90 degrees.

Figure 30-33

merely by finding the difference between their values. For example, the total reactance in a circuit that has 20-Ω inductive reactance in series with 15-Ω capacitive reactance is $20 - 15 = 5\Omega$. Since the inductive reactance was greater than the capacitive reactance, the net effect is 5-Ω inductive reactance. If there was also series resistance in the circuit, it would have to be combined with the reactance to find the total impedance of the circuit. The formula for finding impedance in a series circuit is:

$$Z = \sqrt{R^2 + X^2}$$

where

Z = impedance

R = resistance

X = difference between X_L and X_C

For many ac circuits it is very difficult to determine the total impedance by working through calculations. Special *impedance meters* are available, which make impedance measurement quick and accurate. The impedance meters have a broad range frequency generator built into them. This frequency is applied to the circuit under test through the test probes. Thus, the impedance of any circuit can be measured at any frequency within the range of the signal generator.

As previously noted, inductive reactance increases as the frequency is increased, but capacitive reactance decreases with higher frequencies. Because of these opposite characteristics, for any LC combination there is a frequency at which the X_L equals the X_C. This case of equal and opposite reactance is called *resonance,* and the ac circuit is then a *resonant circuit.*

The frequency at which the opposite reactances are equal is the **resonant frequency**. This frequency can be calculated as

$$f_r = \frac{1}{2\pi\sqrt{LC}}$$

where L is the inductance in henries, C is the capacitance in farads, and f_r is the resonant frequency in hertz that makes $X_L = X_C$.

The resonance effect occurs when the inductive and capacitive reactances are equal. The main application of resonance is for *tuning* in

radio and television receivers. In this use, the LC circuit provides maximum voltage output at the resonant frequency, compared with the amount of output at any other frequency either below or above resonance (Figure 30-34).

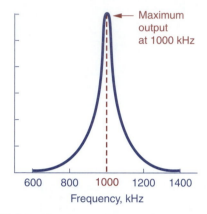

Figure 30-34   Circuit tuned for resonance.

30-12 Power in AC Circuits

Real power (watts), sometimes called *true power* or *average power,* in an ac circuit is the electric power that is actually converted into heat. In the dc circuit, we learn that power in watts is equal to the voltage multiplied by the current. The same is true of the *resistive ac circuit.* In Figure 30-35, however, we see that V and I are always varying in an ac circuit. The power in such a circuit is the average of all the instantaneous values of voltage multiplied by the corresponding instantaneous values of current. The *true power,* or heating effect, is the average of these areas, and can be shown to be equal to the product of the voltage (V) multiplied by the current (I). The unit used to measure real power is the watt. Real power is measured with a wattmeter.

Reactive power (VAR) is the power that a capacitor or inductor *seems* to be using. However, in a sense, capacitors and inductors *do not use power.* Inductive reactance takes power from the source to build its magnetic field, but then when the magnetic field is collapsing, it returns the power to the source. The effect is that no net power from the source is consumed. Similarly capacitance reactance takes power from the source to charge its plates. During discharge, the power is returned

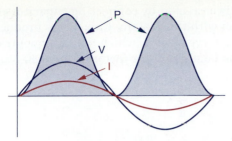

A. AC power waveform for circuit with resistance

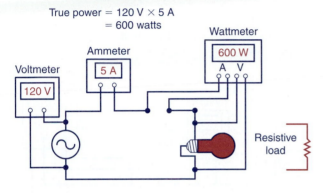

True power = 120 V × 5 A
= 600 watts

B. Wattmeter measures real power

Figure 30-35 Real power (watts).

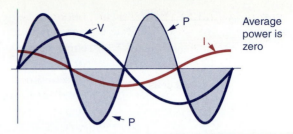

Average power is zero

A. AC power waveform for circuit with pure capacitance

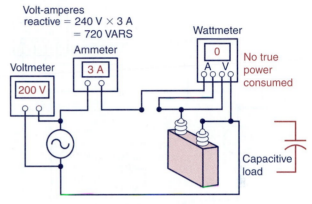

Volt-amperes reactive = 240 V × 3 A
= 720 VARS

No true power consumed

B. No reactive power measured by wattmeter

Figure 30-36 Reactive power (VAR).

to the source. Again the effect is that no net power is consumed from the source. On the other hand, resistance converts power to heat and the heat is lost. Thus, the power used by the resistance is not returned to the source, and there is a net consumption of power. In the capacitor circuit of Figure 30-36, we see that although the voltmeter and ammeter both register readings, the reading on the wattmeter is *zero*, indicating zero power consumption. To distinguish reactive power from real power, the watt power unit is not used. Instead, *volt-amperes reactive* (VAR) is used. The product of the capacitor voltage and current is used to calculate reactive power with VAR used as the unit of measurement.

Apparent power (VA) is the total power that the entire ac circuit is apparently using. It is the total power supplied to the circuit from the source including real and reactive power. Apparent power and real power for ac circuits are equal only when the circuits consist entirely of pure resistance. In the motor circuit of Figure 30-37, we see that of the total apparent power applied to the motor, some of it is being

converted to heat and work, while some is stored and returned to the circuit. Apparent power is measured in *volt-ampere* (VA) units to distinguish it from real power.

Power factor (PF) is the ratio of real power to apparent power:

$$PF = \frac{Watts}{Volt\text{-}ampere} = \frac{Real\ power}{Apparent\ power}$$

The power factor of a circuit is an indication of the portion of volt-amperes that are actually real power. In practice, the power factor of a circuit can be found by dividing the reading of a wattmeter by the product of a voltmeter reading that has been multiplied by the ammeter reading (Figure 30-37).

Power factor is, therefore, real power divided by apparent power. It is commonly expressed as a percentage. When real power is equal to apparent power—such as in a circuit having only resistance or one in which the reactances exactly cancel one another—the power factor is equal to 1.00 or 100 percent.

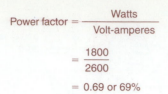

$$\text{Power factor} = \frac{\text{Watts}}{\text{Volt-amperes}}$$

$$= \frac{1800}{2600}$$

$$= 0.69 \text{ or } 69\%$$

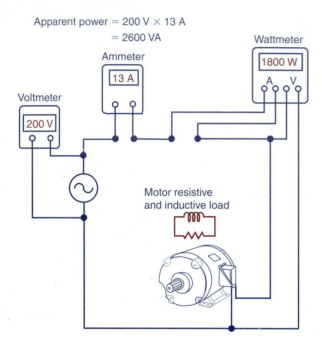

Apparent power = 200 V × 13 A
= 2600 VA

Voltmeter
200 V

Ammeter
13 A

Wattmeter
1800 W
A   V

Motor resistive
and inductive load

Figure 30-37 Apparent power (VA) and power factor (PF).

A power factor of less than 100 percent indicates that not all the current is doing useful work. Yet the conductors must be sized to carry the total current, and the capacity of the equipment in the system such as generators and transformers must be capable of handling this total current. The power factor is leading for a capacitance load, lagging for an inductive load, or in phase for a resistive load.

Most electrical systems have a lagging-power factor owing to motors, coils, fluorescent-lighting ballasts, and the like. When the power factor of an entire system drops below 80 percent, measures must be taken to increase the power factor. Since a low power factor makes increased demands on the power company without consuming power (on which the utility bases its bill), power companies often charge penalty fees for customers having excessively low power factors (below 70 percent lagging, for example). In most cases, a low lagging-power

factor is corrected by connecting capacitors in parallel with the system. Capacitors produce a leading-power factor, which partially neutralizes the lagging-power factor (Figure 30-38).

$$\begin{array}{l}\text{Power factor} =\\ \text{(improved)}\end{array} \frac{\text{Watts}}{\text{Volt-amperes}}$$

$$= \frac{1800}{2600}$$

$$= 0.69 \text{ or } 90\%$$

$$\begin{array}{l}\text{Apparent power} = 200 \text{ V} \times 10 \text{ A}\\ \text{(reduced)} \qquad = 2000 \text{ VA}\end{array}$$

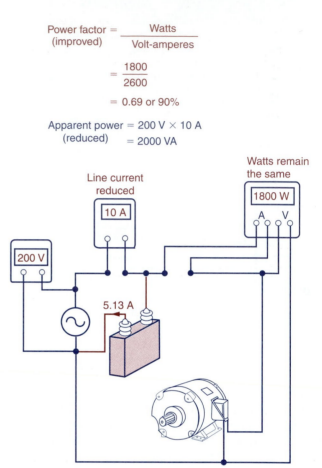

Line current
reduced
10 A

200 V

Watts remain
the same
1800 W
A   V

5.13 A

Figure 30-38 Power factor correction (capacitive current cancels inductive current).

30-13 Troubleshooting Inductors and Capacitors

Inductors almost always fail due to open-circuit or short-circuit conditions. Therefore, the ohmmeter can be used to test these devices. When testing the resistance of an inductor, the ohmic value should be low, but never zero ohms. The ohmic value of an inductor depends on the length of the coil and the wire size. Although inductors can be manufactured with resistances of several hundred ohms, in electronic circuits the resistance of an inductor is typically under 50 ohms.

The test procedure for capacitors is significantly different from the procedure used to test inductors. A very important consideration when troubleshooting capacitors is the dielectric *leakage current*. Since a perfect insulator does not exist, a certain amount of current trickles through the dielectric even in the best of capacitors. Normally leakage current is so small that it is hardly measurable and has no effect on the operation of the circuit. However, as a capacitor ages, its dielectric resistance may decrease resulting in high values of leakage current. High leakage current can alter the normal operation of a circuit because the faulty capacitor is now providing a dc current path instead of having infinite resistance. A quick way to determine if a capacitor is shorted or if the dielectric is bad, is to take a dc resistance measurement of it with an ohmmeter. If the capacitor is shorted or there is leakage in the dielectric, the ohmmeter will read near zero ohms. If checking by measuring capacitance with a capacitance meter, the capacitance will also change as the dielectric deteriorates.

Before checking a capacitor with an ohmmeter, the capacitor should be discharged by connecting a jumper across it. Because some capacitors are capable of storing a charge for long periods of time, it should never be assumed that a capacitor is discharged. The leakage current is the reason why it is impossible for a capacitor to maintain a charge indefinitely. However, some capacitors are capable of retaining their charge for days or weeks after power has been removed.

When testing electrolytic capacitors with an ohmmeter, ensure that proper polarity is observed when connecting the terminals of the ohmmeter to the capacitor. The negative terminal of the meter must be connected to the negative terminal of the capacitor. If the terminals are reversed, the resistance readings will be false and, in the case of low-voltage capacitors, it is possible to destroy the capacitor by applying reverse-polarity voltage.

HELP WANTED

Electronic Design Technologist
Job Description
- Required by a rapid transit authority.
- Prepare drawings, specifications, and estimates for new installations.
- Provide technical support for existing systems.
- Sound knowledge of electrical and mechanical engineering principles.
- College diploma in Electrical Engineering or equivalent education and relevant experience.

Review and Applications

Related Formulas

$X_L = 2\pi f L$

$I = \dfrac{V}{X_L}$

$I = \dfrac{V}{Z}$

$Z(\text{series}) = \sqrt{R^2 + X_L^2}$

$L_T(\text{series}) = L_1 + L_2 + L_3 \ldots$

$\dfrac{1}{L_T}(\text{parallel}) = \dfrac{1}{L_1} + \dfrac{1}{L_2} + \dfrac{1}{L_3} + \ldots$

$L_T(\text{parallel}) = \dfrac{L_1 \times L_2}{L_1 + L_2}(\text{special case})$

$Q = CV$

$C_T(\text{parallel}) = C_1 + C_2 + C_3 \ldots$

$\dfrac{1}{C_T}(\text{series}) = \dfrac{1}{C_1} + \dfrac{1}{C_2} + \dfrac{1}{C_3} + \ldots$

$C_T(\text{series}) = \dfrac{C_1 \times C_2}{C_1 + C_2}(\text{special case})$

$t = RC$

$X_C = \dfrac{1}{2\pi f C}$

$Z = \sqrt{R^2 + X_C^2}$

$Z = \dfrac{V}{I}$

$V = IZ$

$I = \dfrac{V}{Z}$

$f_r = \dfrac{1}{2\pi \sqrt{LC}}$

$\text{PF} = \dfrac{\text{Watts}}{\text{Volt-ampere}} = \dfrac{\text{Real power}}{\text{Apparent power}}$

$Z = \sqrt{R^2 + (X_L - X_C)^2}$

Review Questions

1. Name three basic types of inductors.
2. How is energy stored and released in an inductor?
3. Define the term *inductance*.
4. When an inductor is connected into a dc circuit, under what conditions is the inductive effect present?
5. Why is the inductive effect present at all times when the inductor is connected into an ac circuit?
6. Explain the difference between self inductance and mutual inductance.
7. State Lenz's law as it applies to the direction of the current change and the induced voltage in a coil.
8. (a) Define the term inductive reactance.
   (b) What is the unit used to measure inductive reactance?
9. What is the unit used to measure inductance?
10. List four factors that determine the inductance of a coil.
11. State whether the inductive reactance increases or decreases with each of the following changes:
    (a) increase in the frequency of the ac supply source
    (b) decrease in the inductance of the coil
12. What is the phase relationship between the voltage across and the current flow through an ideal inductor?
13. What is capacitance?
14. How is energy stored and released in a capacitor?
15. Describe the potential hazard involved in the use of high-voltage capacitors in dc circuits.
16. What is the unit used to measure capacitance?
17. List three factors that determine the capacitance of a capacitor.

18. What are capacitors rated for in addition to capacitance?
19. How would two $50\,\mu\text{F}$ capacitors be connected to provide a total capacitance of $100\,\mu\text{F}$?
20. What are the three general classifications of capacitors?
21. (a) What is the main advantage of polarized capacitors?
    (b) When connecting polarized capacitors into dc circuits, what polarity rule is followed?
22. In general, explain how a capacitor affects the flow of ac and dc currents in a circuit.
23. (a) What is capacitive reactance?
    (b) What is the unit used to measure capacitive reactance?
    (c) In what way is capacitive reactance affected by capacitance?
    (d) In what way is capacitive reactance affected by the frequency of the voltage source?
24. What is the phase relationship between the voltage and current of an ideal capacitor circuit?
25. (a) Define impedance of an ac circuit.
    (b) What unit is used to measure impedance?
    (c) What three opposing factors determine the impedance of an ac circuit?
26. (a) What is a resonant circuit?
    (b) Why is the net reactance of a resonant circuit zero?
27. The apparent power of an ac circuit is a combination of what other two types of power?
28. (a) What is *power factor*?
    (b) What is the PF of a purely resistive load?
    (c) How can lagging PF in a system be improved?

Problems

1. Calculate the inductive reactance (X_L) of a 2.5-mH inductor when the frequency is 100 kHz.
2. A 6-H inductor is connected to a 12 V dc-voltage source. What is its inductive reactance?
3. (a) A 6-H and 4-H coil are connected in series. What is the value of the total inductance?

(b) What is the value of the total inductance when the same two coils are connected in parallel?
4. Calculate the amount of charge stored in coulombs, when a 300-μF capacitor is charged from a 450 V-dc source.
5. (a) Define the term RC-time constant.
   (b) Calculate the RC-time constant for a 25-kΩ resistor connected in series with a 1000-μF capacitor operated from a 100-V dc source.
6. Calculate the capacitive reactance (X_C) of a 0.01-μF capacitor at 400 Hz.

Critical Thinking

1. An ac voltage of 120 volts with a frequency of 60 Hz is applied to a 3-H inductor. Assuming that the inductor was pure or ideal (no resistance; inductive reactance only), how much current would flow through it?
2. An ac voltage source of 10 volts is applied to a series circuit consisting of a resistor with a resistance (R) of 3 Ω and a inductor with an inductive reactance (X_L) of 4 Ω. If the resultant current flow is 2 A, what is the impedance of the circuit?
3. Determine the total capacitance and maximum voltage rating for two 220$\,\mu\text{F}$, 300 V capacitors connected in series.
4. An ac voltage of 12 volts with a frequency of 400 Hz is applied to a 100$\,\mu\text{F}$ capacitor. Determine the amount of ac current flow in the circuit.
5. Refer to Figure 30-32. Assume the 24 V ac voltage supply is changed to 120 V ac voltage supply of the same frequency. Determine the value of:
   (a) net reactance
   (b) impedance
   (c) current
   (d) V_R
   (e) V_C
   (f) V_L.
6. Determine the resonant frequency of a 10-mH inductor and 0.005-μF capacitor.
7. Refer to Figure 30-38. To what minimum value can the line current be reduced by adding more capacitive current?

Portfolio Projects

- Examine the circuit board of a discarded electronic device and identify all inductors and capacitors.
- Obtain a variety of junk-box inductors and capacitors. Measure the resistance of the inductor windings with an ohmmeter. Use the ohmmeter to check for shorted or open capacitors. Use a capacitance meter to find out the tolerance capacitors.

Circuit Challenge

Using a Simulator

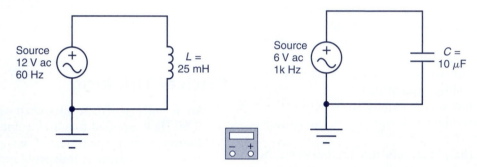

Procedure

(a) Calculate the inductive reactance and current flow of the inductor assuming it has negligible resistance.

(b) Calculate the capacitive reactance and current flow of the capacitor.

(c) Using whatever simulation software package is available to you, wire the circuits and record the actual amount of current flow in the inductor and capacitor circuits.

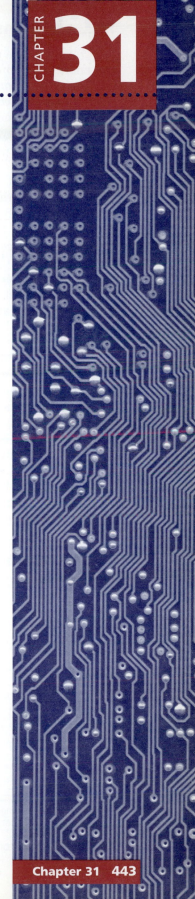

Transformers

The transformer is a device which changes alternating current to different voltage and current levels. Some of the many applications for transformers include their use in long-distance power distribution systems, power supplies, and automotive ignition systems. In this chapter, we will study transformer action, voltage and current ratios, and common applications.

Objectives

After studying this chapter, you should be able to:

1. Explain the principle of operation of transformers.
2. Identify different types of transformers and their specific applications.
3. Use the basic transformer equations to calculate turns ratio, voltages, and currents.
4. Troubleshoot a transformer using an ohmmeter and a voltmeter.

Key Terms

- transformer
- mutual induction
- primary winding
- secondary winding
- laminations
- eddy currents
- hysteresis
- turns ratio
- voltage ratio
- current ratio

- step-up transformer
- step-down transformer
- isolation transformer
- terminal markings
- ignition coil
- power transformer
- autotransformer
- audio transformer

- energy-limiting transformer
- toroidal transformer
- RF transformer
- current transformer
- delta
- wye
- line voltage
- phase voltage

RESEARCH
INTERNET
SITES

- Hammond Manufacturing
- Toroid Transformer
- Electric Net

31-1 Transformer Action

A *transformer* is a devise used to transfer electric energy from one circuit to another by **mutual induction**. The basic transformer consists of two fixed coils that are wound on a common iron core. Since transformers have no moving parts, they operate at a very high efficiency level and require a minimum of maintenance.

The experiment illustrated in Figure 31-1 demonstrates a transformer action. A battery is connected to the first coil through a switch. This coil is called the *primary coil*. A second coil, called the *secondary coil*, is placed next to the primary coil. Connected across the secondary coil is a very sensitive *galvanometer*.

When the switch is first closed, current flows through the primary coil producing an expanding magnetic field around this coil. Lines of force from this magnetic field cut across the wires of the secondary coil. The electrons of the atoms of this coil gain energy from the magnetic field, become excited, and leave their valence shells. The movement of these electrons can be detected by the needle deflection of the galvanometer. As a result of the energy transfer by the magnetic field to the electrons, a voltage has been induced.

If the switch is held down, the induced current in the secondary coil soon stops flowing.

When the switch is released and the circuit is opened, the magnetic lines of force collapse, thus cutting the wires of the secondary coil in the opposite direction. A current is again induced in the secondary coil but in the opposite direction. Each time the switch is opened or closed, electric energy is transferred from the primary coil to the secondary coil.

Practical transformers are operated from *alternating current* or pulsating *direct current* sources (pure, unvarying dc will not work). See Figure 31-2. The simplest transformer is made of two separate insulated coils wound on the same core. The input coil is the primary, and the output coil is the secondary. When an ac or pulsating dc voltage is applied to the primary coil the resultant current flow sets up a magnetic field that is constantly changing. As this field expands and collapses, it causes an ac voltage to be induced in the secondary winding. There is no physical connection between the primary winding and the secondary winding; energy is transferred by movement of the magnetic field. The value of the induced voltage depends on both the strength of the applied voltage and the ratio of the secondary turns to primary turns. The frequency at the output, or secondary winding is equal to the frequency at the input or primary winding.

Transformers are wound on cores made from stacks or **laminations**, of sheet steel. The core ensures a good magnetic linkage between the

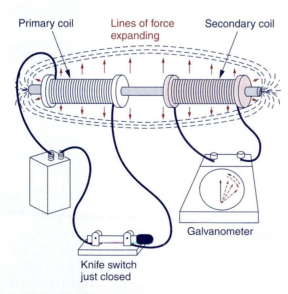
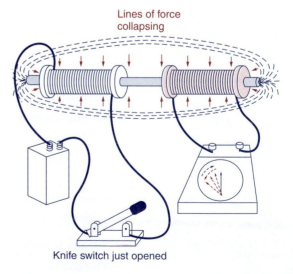

Primary coil   Lines of force expanding   Secondary coil

Lines of force collapsing

Galvanometer

Knife switch just closed

Knife switch just opened

Figure 31-1   Transformer action.

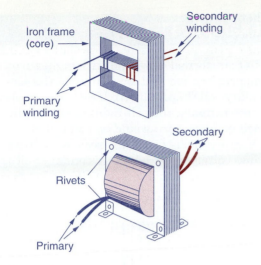

Iron frame (core)

Secondary winding

Primary winding

Secondary

Rivets

Primary

A. Construction

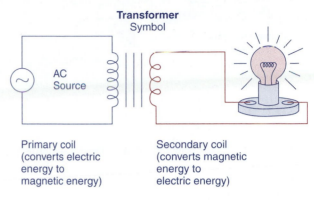

Transformer
Symbol

AC Source

Primary coil (converts electric energy to magnetic energy)

Secondary coil (converts magnetic energy to electric energy)

B. Symbol

Figure 31-2 Practical transformer.

primary and secondary windings. *Eddy currents* are caused by the alternating current inducing a voltage in the core of the transformer itself. Since the iron core is a conductor, it produces a current by the induced voltage. By laminating the core, the paths of the eddy currents are reduced considerably as seen in Figure 31-3, thereby reducing heat and power loss. Eddy currents are prevented from flowing from lamination to lamination by a thin coating of insulating material on the flat surfaces of the lamination. The eddy currents that do exist are very small and represent wasted power dissipated as heat in the core.

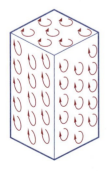

A. Eddy currents in a solid piece of metal

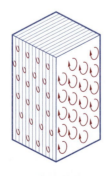

B. Laminations reduce the path of circulating eddy currents

Figure 31-3 Eddy currents in steel core.

31-2 Transformer Voltages, Current, and Power Relationships

Transformers increase the voltage, decrease the voltage, or allow the voltage to remain the same between the primary and secondary windings without significant power loss. Transformer power output equals transformer power input minus the internal losses and is the product of voltage and current.

There is no gain or loss in energy in an ideal transformer. Energy is transferred from the primary circuit to the secondary circuit. This means that the volts multiplied by the amperes of the primary circuit equal the volts multiplied by the amperes of the secondary circuit. In other words, in the ideal transformer, the power output must equal the power input (Figure 31-4). There is, in fact, some power loss in a practical transformer, but the average efficiency of a transformer is over 90 percent. Efficiency losses result from: the ohmic resistance (copper losses) of the windings; core losses, which are caused by the induction of eddy currents in the core material; and *hysteresis*, or molecular friction, which is caused by changes in the polarity of the applied current.

The maximum-power rating of a transformer is often included on the transformer's specification plate. Transformer power is rated in

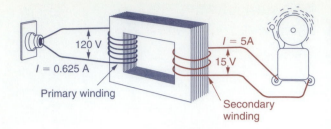

$$\text{Power in} = \text{Power out}$$
$$V \times I \text{ primary} = V \times I \text{ secondary}$$
$$(120 \text{ V})(0.625 \text{ A}) = (15 \text{ V})(5 \text{ A})$$
$$75 \text{ VA} = 75 \text{ VA}$$

Figure 31-4 Power transformer in an ideal transformer.

volt-amperes (VA) and *not* in watts. This is because transformer power is **apparent power**. It does not convert the power to heat, but merely transfers the *power* from a source to a load.

The ratio of the number of turns in the primary to the number in the secondary is the **turns ratio** of a transformer:

$$\text{Turns ratio} = \frac{N_P}{N_S}$$

where

N_P = number of turns in primary

N_S = number of turns in secondary

In the ideal transformer, the voltage induced in each turn of the secondary is the same as the self-induced voltage of each turn in the primary. The voltage that is self-induced in each turn of the primary equals the voltage applied to the primary, divided by the number of turns in the primary.

Therefore, it is seen that the **voltage ratio** of a transformer is equal to its turns ratio. This can be written as a formula:

$$\text{Turns ratio} = \text{voltage ratio} = \frac{N_P}{N_S} = \frac{V_P}{V_S}$$

where

N_P = number of turns in primary

N_S = number of turns in secondary

V_P = primary voltage

V_S = secondary voltage

Example 31-1

A transformer is wound with eight turns on the primary coil and four turns on the secondary coil (Figure 31-5). If 8-V ac is applied to the primary, the self-induced voltage in each primary turn would be equal to 8 ÷ 8, or 1 V. Since each turn of the secondary has the same voltage induced, the secondary voltage would be equal to 4 × 1 or 4 V.

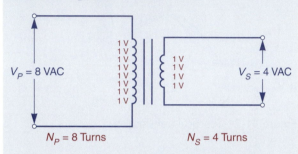

Figure 31-5 Self-induced voltage in primary turns.

The voltage either steps up or steps down across the transformer in proportion to the turns ratio. For example, if the number of secondary turns is twice the number of primary turns, the second voltage will be twice the primary voltage. However, if the number of primary turns is twice the number of secondary turns, the secondary voltage will be half the primary voltage.

Transformers are classified as step-up or step-down in relation to their effect on voltage. A *step-up transformer* is one in which the secondary-coil output voltage is greater than the primary-coil output voltage. This type of transformer has *more* turns in the secondary coil than in the primary coil. The ratio of the primary to secondary turns determines the input-to-output voltage ratio of the transformer.

A *step-down transformer* is one in which the secondary-coil output voltage is less than the primary-coil input voltage. This type of transformer has *less* turns in the secondary coil than in the primary coil. Again, the ratio of the primary to secondary determines the input-to-output voltage ratio of the transformer.

A transformer automatically adjusts its input current to meet the requirements of its output or load current. Thus, when no current is being used from the *secondary winding*, no current flows in the primary except excitation current.

Example 31-2

Problem   A step-up transformer has 50 turns for the primary winding and 100 turns for the secondary winding (Figure 31-6). What is the turns ratio? How much is the secondary voltage?

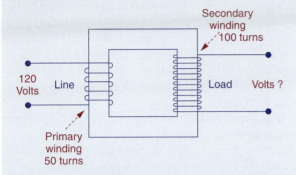

Figure 31-6   Step-up transformer.

Solution

$$\text{Turns ratio} = \frac{N_P}{N_S} = \frac{50}{100} = \frac{1}{2} \text{ or } 1{:}2$$

Therefore, V_P is stepped up by a factor of 2, making V_S equal to 2×120 *or* 240 V.

Example 31-3

Problem   A step-down transformer has 100 turns for the primary winding and 5 turns for the secondary winding (Figure 31-7). What is the turns ratio? How much is the secondary voltage when there is primary voltage of 240 V?

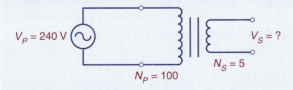

Figure 31-7   Step-down transformer.

Solution

$$\text{Turns ratio} = \frac{N_P}{N_S} = \frac{100}{5} = \frac{20}{1} \text{ or } 20{:}1$$

Therefore, V_P is stepped down by a factor of 20, making V_S equal to $240 \div 20$ *or* 12 V.

Example 31-4

Problem   A step-up transformer with a 1:5 turns ratio has a secondary voltage of 60 V across a 25-Ω resistive load (Figure 31-8). (a) How much is I_S? (b) Calculate the value of I_P.

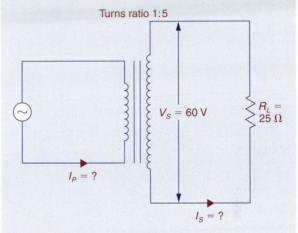

Figure 31-8   Step-up transformer under load.

Solution

(a) $I_S = \dfrac{V_S}{R_L} = \dfrac{60 \text{ V}}{25 \text{ }\Omega} = 2.4 \text{ A}$

(b) With a turns ratio of 1:5, the ***current ratio*** is 5:1. Therefore:

$$I_P = 5 \times I_S = 5 \times 2.4 \text{ A} = 12 \text{ A}$$

Example 31-5

Problem   A step-down soldering gun transformer has a turns ratio of 200:1 and a secondary heating current of 400 A (Figure 31-9). (a) How much is the primary current? (b) If the primary is operated from a 120-V source, calculate the secondary voltage.

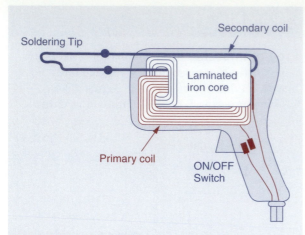

Figure 31-9   Transformer-type soldering gun.

Solution

(a) $I_P = \dfrac{I_S}{200} = \dfrac{400\,\text{A}}{200} = 2\,\text{A}$

(b) $V_S = \dfrac{V_P}{200} = \dfrac{120\,\text{V}}{200} = 0.6\,\text{V}$

Excitation current is the very small amount of current that is necessary to maintain the magnetic circuit. For self-regulation, a transformer depends on counter-electromotive force generated in its primary winding by its own magnetism and an opposing magnetism produced by the current drawn by the load on the secondary winding.

When viewed from the primary side, a resistive load connected across the secondary of a transformer appears to have a value that depends on the turns ratio. The load that is *reflected* to the primary side is what the source effectively sees, and it determines the amount of primary current.

If the secondary circuit of the transformer becomes shorted, the high current that results generates a great opposition to the primary winding flux. As a result, the cemf of the primary is made to drop very low and the primary current increases dramatically also. It is for this reason that a fuse placed in series with the *primary winding* protects both the primary and secondary circuits from excessive current.

By Ohm's law, the amount of secondary current equals the secondary voltage divided by the resistance in the secondary circuit (a negligible coil resistance assumed).

When a transformer steps up the voltage, the secondary-coil current is correspondingly less than the primary-coil current so that the power (voltage multiplied by current) is the same in both windings. The ratio of primary to secondary current is *inversely proportional to the voltage or turns ratio.*

$$V_S \times I_S = V_P \times I_P$$

or

$$\frac{I_S}{I_P} = \frac{V_P}{V_S}$$

Isolation transformers are specified to eliminate direct electric connections between the primary and secondary circuits without changing the voltage and current ratings. Figure 31-10 illustrates this application. The transformer's turns ratio is 1:1. Using the isolation transformer, the load is isolated from the voltage source so there is no chance that the chassis might accidentally become hot due to improper plug placement.

Transformers can reverse the *phase* of the input voltage by winding the secondary coil in

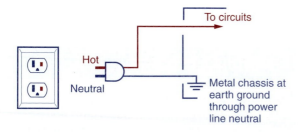

A. With a direct connection improper plug placement may cause the chassis to become hot.

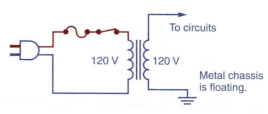

B. When an isolation transformer is used, the chassis is isolated from the input ac line.

Figure 31-10   Isolation transformer.

the opposite direction of the primary coil. Figure 31-11 illustrates the method used to identify the relative phase relationship of transformer inputs and outputs on schematic diagrams. Notice that when the schematic does not contain "dots" no phase inversion takes place.

A. No phase inversion

B. No phase inversion

C. Phase inversion

Figure 31-11   Transformer phase inversion.

The high-voltage winding leads of transformers are marked H_1 and H_2 and the low-voltage winding leads are marked X_1 and X_2 (Figure 31-12). When H_1 is instantaneously positive, X_1 also is instantaneously positive. These **terminal markings** are used in establishing the proper terminal connections when connecting single-phase transformers in parallel, series, or three-phase configurations.

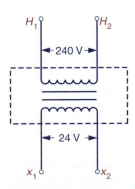

Figure 31-12   Transformer terminal markings.

31-3 Ignition Coil

An automobile ***ignition coil*** is used to transform the 12 V from the battery to the 40,000 V or more required to jump the gap in the spark plugs. An ignition coil is basically a step-up transformer designed to operate with pulsating *direct current.* Figure 31-13 illustrates the operation of the ignition coil from a dc battery source.

– No current flows through the primary coil.
– No magnetic field is produced.
– Zero voltage is induced in the secondary coil.

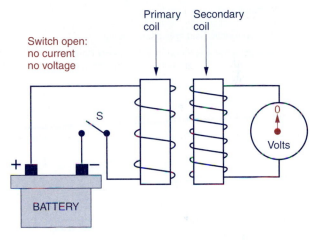

A. Switch open

— Current flows through primary coil.
— The magnetic field expands or builds up *slowly*.
— A *small* voltage is induced in the secondary coil (this low voltage cannot overcome the resistance of the spark plug gap, so no spark is produced as the magnetic field builds up).

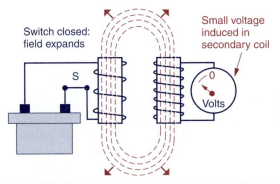

B. Switch closed

Figure 31-13   Operation of an ignition coil from a dc battery source.

— After a short period of time, the magnetic field reaches maximum strength and stabilizes.

— Zero voltage is induced in the secondary coil.

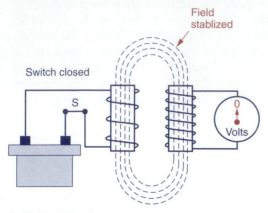

C. Field stabilized

— The magnetic field collapses *quickly*.

— A *high* voltage is induced in the secondary coil (this high secondary voltage jumps the gap of the spark plug).

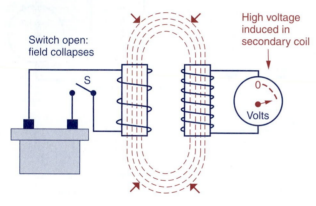

D. Switch opened

Figure 31-13   (Continued).

Typically, the *primary winding* of an ignition coil consists of 100–180 turns of heavy No. 20 copper wire, whereas the *secondary winding* consists of about 18,000 turns of fine No. 38 copper wire. Since the turns ratio is the inverse of the current ratio, the secondary coil operates at a high voltage, but a *low current level* of short duration. A high-power *transistor* contained within the electronic-control module is used to switch the primary current ON and OFF (Figure 31-14).

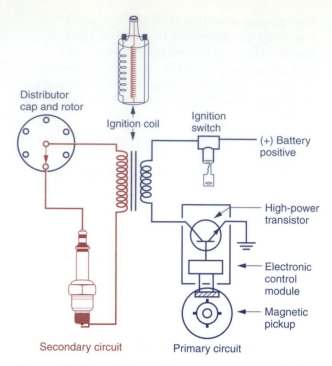

Figure 31-14   Ignition module transistor used to switch primary current ON and OFF.

31-4 Types of Transformers

Many different varieties of transformers are used in electric and electronic circuits. In most cases, these transformers are classified by application. Some are designed to transfer power, while others are built to transfer only signal voltages (Figure 31-15). Signal transformers are usually rated by their current, power, input/output impedances, and frequency range. Power transformers are rated by their primary-secondary voltages, current, or VA capacity.

Power Transformers (A)   change voltage and current levels to meet circuit power requirements. In most cases, *power transformers* are designed for operation at alternating current line frequencies of 50/60 Hz.

Autotransformers (B)   are normally connected for step-down or step-up applications in 50/60 Hz line circuits. When it is not necessary to have electrical isolation between the primary and secondary circuits, and when direct connection between them is permissible,

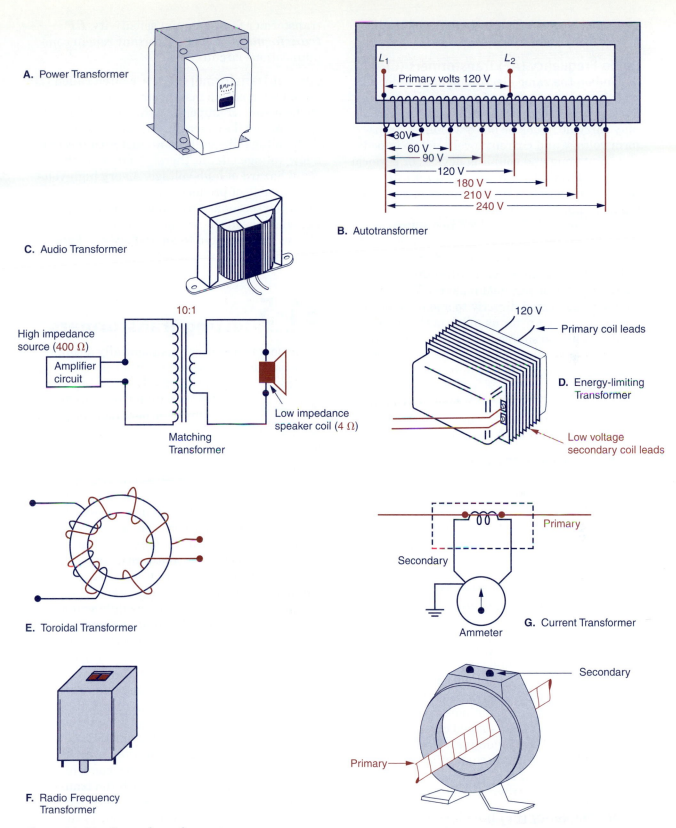

A. Power Transformer

B. Autotransformer

Primary volts 120 V

30V
60 V
90 V
120 V
180 V
210 V
240 V

C. Audio Transformer

10:1

High impedance
source (400 Ω)

Amplifier
circuit

Low impedance
speaker coil (4 Ω)

Matching
Transformer

120 V — Primary coil leads

D. Energy-limiting
Transformer

Low voltage
secondary coil leads

E. Toroidal Transformer

Primary

Secondary

Ammeter

G. Current Transformer

Secondary

Primary

F. Radio Frequency
Transformer

Figure 31-15 Types of transformers.

autotransformers offer the benefits of smaller size, lower weight, and lower cost.

Audio Frequency (AF) Transformers (C)

depend on the same electromagnetic induction principle as power transformers but generally operate over a wider frequency range. They may carry dc in one or more windings, transform voltage and current levels, act as impedance matching and coupling devices or perform filtering. In addition, *audio frequency transformers* may pass a limited range of voice frequencies (20 to 20,000 Hz).

Impedance matching a load to a source is done in order to achieve maximum transfer of power from the source to the load. Some internal resistance is inherent in all sources due to their circuitry or physical makeup. When the source is connected directly to a load, often the objective is to transfer as much of the power produced by the source to the load as possible. However, a certain amount of the power produced by the source is lost in its internal resistance and the remaining power goes to the load. With a source connected to a load, maximum power is delivered to the load when the *load resistance is equal to the source resistance.*

Energy-Limiting Transformers (D)

are widely used in consumer and commercial electronic products in which exceeding the rated voltage and current could present a safety hazard. Class 2 *energy-limiting transformers* include a fuse or other heat-sensitive element; the element opens when the specified ratings are exceeded. Others contain sufficient impedance to limit the secondary current to the extent that a short circuit on the secondary will not burn them out or create a fire hazard.

Toroidal Transformers (E)

are wound on a closed, ring-shaped solid core with both primary and secondary windings on the same core. They may be used in the same applications as conventionally wound, laminated core transformers, but their ring-shaped cores eliminate certain losses found in laminated core transformers, thus increasing their efficiencies. *Toroidally wound transformers* offer higher power output in smaller, lighter packages.

Radio Frequency (RF) Transformers (F)

are used to select and couple signal circuitry of radio and television receivers. They have a threaded ferrite core that is used to adjust the transformer for maximum sensitivity. *RF transformers* play an important role in communications circuitry.

Current Transformers (G)

are often called CTs and make possible measurement of current up to thousands of amperes with a low-range ammeter, or they measure the current in a high-voltage line with an ammeter connected to a low-voltage circuit, thereby removing the personal hazard of high voltage. A very high voltage, capable of producing a fatal shock, builds up in the secondary winding when it is open. For this reason the secondary lead should always be connected to an ammeter or kept short-circuited if the meter is removed.

31-5 Testing Transformers

Because a transformer is essentially a coil, testing procedures are similar to that used for inductors. Faults are typically caused by either open-circuits or short-circuits. Open-circuits can occur at any point in either the primary or secondary winding. Open-circuits that develop at the point where the winding is connected to the transformer can often be detected by visual inspection and repaired by resoldering the terminal.

A transformer is not normally repairable if it is defective. Transformers can be checked while out of the circuit with an *ohmmeter*. Each winding can be checked for continuity. Since most windings are rather small, the dc resistance will usually be low—even less than one ohm in many cases. As a result, one usually will have to check for shorts with a milliohmmeter.

Short-circuit conditions inside the transformer are often caused by overloading of the transformer. Overload current flows through the windings causes the operating temperature of the transformer to rise. As a result, the insulation of the windings melts and the turns of wire are shorted together. The insulation will also produce a distinct smell when burned.

Power transformers can be checked while in the circuit with a *voltmeter*. With the transformer primary coil connected to its rated source of voltage, the output voltage of the secondary coil is measured with a voltmeter. The output voltage will often be higher than specified if the transformer is *not* connected to a load. The output will drop slightly when the load is connected; this is due to the internal resistance of the transformer.

Because transformers operate on the principle of magnetic induction, an open-circuit condition in the primary winding will result in zero voltage across the secondary winding.

31-6 Three-Phase Transformer Systems

Transformers become parts of three-phase circuits, and from the point of view of the generator, a transformer is a load; and from the point of view of the load, a transformer is a source. In either case, there must be a winding for each phase (Figure 31-16).

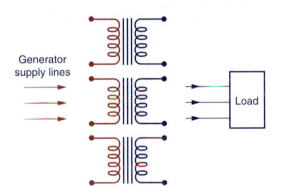

Figure 31-16 Transfromers as part of a three-phase system.

Large amounts of power are generated using a three-phase system. This generated voltage will be stepped up and down many times before it reaches the loads in your home or plant. This transformation can be accomplished by using *wye-* or *delta-connection transformers* or a combination of both wye- and delta-transformers along with differing voltage ratio transformers.

Figure 31-17 is a schematic diagram of the three-wire delta-delta, three-phase, step-down transformer bank. The three single-phase transformers have both primary and secondary windings connected in delta. If the primary has 2400 V and the transformer ratio is 10:1, the secondary voltage in relation to the primary voltage is determined solely by the ratio, since delta **line** and **phase voltages** are the same.

Figure 31-18 is a schematic diagram of the four-wire, wye-wye, three-phase connection

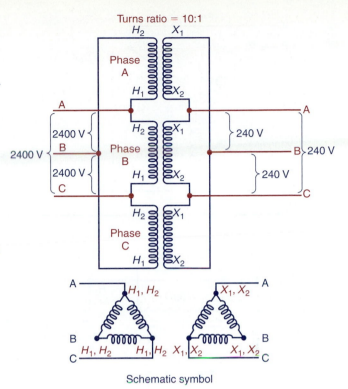

Figure 31-17 Three-wire, delta-delta, three-phase transformer bank.

that is commonly used for supplying three-phase and single-phase service. Single-phase transformers with 2400-V primaries, when connected wye in a three-phase bank, require 4160 line volts to give 2400 phase volts.

$$V_{ph} = \frac{V_l}{1.73}$$
$$= \frac{4160}{1.73}$$
$$= 2400 \text{ V}$$

If the primary has 2400 V and the transformer ratio is 20:1, the secondary phase voltage will be 120 V and the line voltage will be 208 V (120 V × 1.73). Three-phase loads are supplied at 208 V. The voltage for single-phase loads is 208 V or 120 V.

The capacity of large transformers is rated in kilovolt-amperes (kVA). The total kVA capacity of a transformer bank is found by adding the individual kVA rating of each transformer in the bank.

Losses occurring in transformers appear as *heat*. The wire of which the windings are made

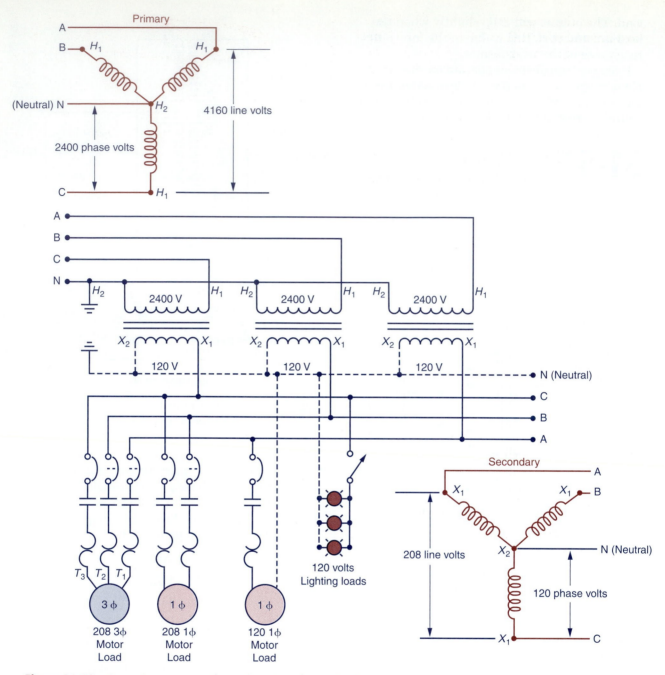

Figure 31-18 Four-wire, wye-wye, three-phase transformer bank.

must, of course, be large enough to carry the intended current, and its insulation must be able to withstand the intended voltage. But in addition, a transformer's rating must include its ability to dissipate the heat caused by the losses. Transformers can be cooled by forced air, natural circulation of air, water, or oil. Smaller

transformers are usually air cooled by natural convection currents.

Large power system transformers are usually oil cooled. The windings are immersed in oil, which acts as an insulator as well as a cooling agent. Vertical pipes, around the outside of a steel housing and enclosing the transformer, connect to the upper and lower parts of the housing and permit the oil to be cooled by natural convection and gravity circulation of the oil. Very large oil-cooled transformers may have forced circulation, using a pump to circulate the oil.

Power can be transmitted using only three wires if a delta configuration is used. Thus the power utility can save the cost of the fourth wire. When the voltage is transformed down the wye, configuration is used for the secondary. This provides the fourth wire used for single phase voltages.

HELP WANTED

Field Test Technician
Job Description
- Field testing of transformers, protective relays, circuit breakers, and switchgear.
- Journeyman electrician license needed.
- Minimum of five years experience in the power systems industry.
- Candidates will adapt readily to change and enjoy the challenge of a fast-paced workplace.

Review and Applications

Related Formulas

Turns ratio = voltage ratio

$$\frac{N_P}{N_S} = \frac{V_P}{V_S}$$

$$\text{Turns ratio} = \frac{1}{\text{current ratio}}$$

$$V_S \times I_S = V_P \times I_P$$

$$\frac{I_S}{I_P} = \frac{V_P}{V_S}$$

$$\text{V phase (wye)} = \frac{\text{V line}}{1.73}$$

$$\text{V line (wye)} = \text{V phase} \times 1.73$$

Review Questions

1. What is the purpose of a transformer?
2. Why must a transformer be operated from an ac or pulsating dc voltage source?
3. (a) What are eddy currents?
   (b) In what way is the core of a transformer constructed to reduce eddy currents?
4. In an ideal transformer:
   (a) what is the relationship between the turns ratio and voltage ratio?
   (b) what is the relationship between the voltage ratio and current ratio?
   (c) what is the relationship between the primary power and secondary power?
5. What is the difference between a step-up and step-down transformer?
6. (a) What is the main purpose of an isolation transformer?
   (b) What is the turns ratio for an isolation transformer?
7. (a) What type of transformer is an ignition coil?
   (b) What type of source is it designed to operate from?
   (c) At what point in its operating cycle is a high voltage induced in the secondary coil?
   (d) Although the secondary voltage is in the 40,000 or more voltage range, why is the shock produced by this voltage not normally fatal?
8. Compare the operating frequency of power and signal transformers.
9. (a) In what way is an autotransformer constructed differently from a conventional power transformer?
   (b) What benefits do autotransformers offer?
10. What built-in safety feature do class-two energy-limiting transformers have?
11. (a) In what way is a toroidal transformer constructed differently from a conventional transformer?
    (b) What benefits do toroidal transformers offer?
12. What are RF transformers used for?
13. What are the benefits to metering of current using a current transformer?
14. What check can be made on a power transformer with:
    (a) an ohmmeter?
    (b) a voltmeter?
15. (a) Name the two types of transformer connections commonly used when connecting three single-phase transformers to a three-phase system.
    (b) Which of the two is commonly used for supplying three-phase and single-phase service?
16. (a) What do losses that occur in transformers appear as?
    (b) List four ways in which transformers can be cooled.

Problems

1. A step-down transformer with a turns ratio of 10 to 1 has 120 V applied to its primary coil. A 3-Ω load resistor is connected across the secondary coil. Calculate the following:
   (a) secondary coil voltage
   (b) secondary coil current
   (c) primary coil current

2. A step-up transformer has a primary coil current of 32 A and an applied voltage of 240 V. The secondary coil current is 2 A. Calculate the following:
   (a) power input of primary coil
   (b) power output of secondary coil
   (c) secondary coil voltage
   (d) turns ratio
3. A certain single-phase transformer is rated at 6 kVA. If it has a secondary coil voltage rating of 240 V, what is the rated secondary coil current capacity?

Critical Thinking

1. Refer to Figure 31-1. Assume the knife has been closed for 1 minute. What would the galvanometer reading be? Why?
2. Refer to Figure 31-5. Assume the primary is wound with 16 turns and the secondary with 8 turns. Would the secondary voltage increase, decrease, or stay the same? Why?
3. Refer to Figure 31-6. Assume the number turns on the secondary winding are increased to 150. Determine the value of the secondary voltage.
4. Refer to Figure 31-7. Assume the number turns on the primary winding are decreased

to 20. Determine the value of the secondary voltage.
5. Refer to Figure 31-8. Assume the load resistance is changed to 30 Ω. Determine the value of the secondary and primary currents.
6. Refer to Figure 31-17. Assume the 3-phase primary line voltage is changed to 1200 V. Determine the value of the 3-phase secondary line voltage.
7. Refer to Figure 31-18. Assume the 3-phase primary line voltage is changed to 2820 V. Determine: (a) the primary phase voltage (b) the secondary phase voltage (c) the secondary line voltage.

Portfolio Project

Obtain a small power transformer and conduct the following tests on it:
• measure the dc resistance of the windings and record the difference in resistance between the high-voltage and low-voltage winding
• (under supervision of the instructor) apply rated ac voltage to the primary and connect different resistive loads to the secondary to verify voltage and current ratios

Circuit Challenge

Using a Simulator

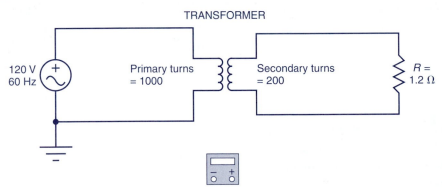

TRANSFORMER

120 V
60 Hz
Primary turns = 1000
Secondary turns = 200
R = 1.2 Ω

Procedure

(a) For the transformer circuit shown calculate:
   (i) the secondary voltage
   (ii) the secondary current
   (iii) the primary current

(b) Wire the transformer circuit shown, using whatever simulation software package is available to you. Connect the multimeter to verify your calculated values.

Signal Sources

*S*ignal sources are used to communicate information to a circuit. They range in complexity from a switch, which may provide one bit of information, to a computer disk, which may provide millions of bits of information. In this chapter, we will study the operation and application of a wide variety of signal sources.

Objectives

After studying this chapter, you should be able to:
1. Define what a transducer is.
2. Define what a sensor is.
3. Explain how energy is converted in loudspeakers, microphones, and phonograph cartridges.
4. Describe the process of storing and retrieving information used in a tape recorder, computer disk drive, and compact disc player.
5. Explain the principles of operation for light sensors, thermistors, Hall-effect sensors, capacitive sensors, ultrasonic sensors, and bar codes.

Key Terms

- transducer
- loudspeaker
- headphone
- carbon microphone
- dynamic microphone
- crystal microphone
- capacitor microphone
- crystal phonograph cartridge
- magnetic phonograph cartridge

- record/play tape head
- disk drive
- stepper motor
- floppy disk
- formatting
- ZIP disk
- hard drive
- CD-ROM
- CD-R
- CD-RW

- CD drives
- DVD
- solar cell
- photoconductive cell
- thermistor
- Hall-effect sensor
- proximity sensor
- ultrasonic sensor
- bar-code

RESEARCH
INTERNET
SITES

- Keyence
- Maxwell
- SNX

Signal Sources

32-1 Transducers

A *transducer* is any device that converts energy from one form to another. Transducers may be divided into two classes: ***input transducers*** and ***output transducers*** (Figure 32-1). Electric-input transducers convert nonelectric energy such as sound or light to electric energy. Electric-output transducers work in the opposite way. They convert electric energy to forms of nonelectric energy. In this chapter, we will describe different types of electric transducers and the principles by which they operate.

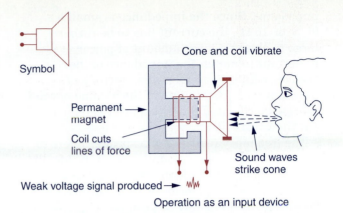

Figure 32-2 Permanent-magnet speaker.

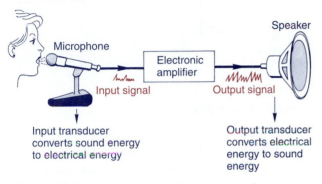

Figure 32-1 Electric-input and -output transducers.

32-2 Loudspeakers

A ***loudspeaker*** is a transducer that may serve as either an input or output device. The permanent-magnet speaker is the most commonly used type. It is made using one permanent magnet and one electromagnet or coil. The electromagnetic voice coil is suspended in the magnetic field of the permanent magnet and is attached to a flexible paper speaker cone. Now, operating as an input device, a mechanical force such as a sound wave causes the coil to vibrate and to cut the lines of force of the permanent magnet, thus inducing a voltage in the coil (Figure 32-2). In this way, mechanical energy is converted to electric energy and the loudspeaker can be used as a microphone.

When a varying voltage is applied to the terminals of the same voice coil, current flows through the coil to establish a second magnetic field (Figure 32-3). The two magnetic fields react with each other causing the coil to vibrate, or move, mechanically. This vibration depends on the frequency and amplitude of the

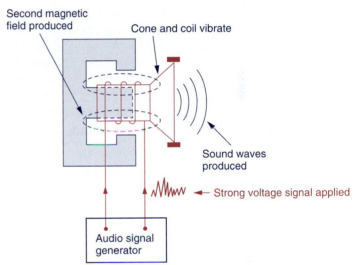

Figure 32-3 Speaker opeation as an output device.

applied voltage. Thus, electric energy is converted to mechanical energy. Speakers are often used for both the input and output of intercom amplifier systems. In a circuit such as this, the two speakers are switched from input to output to provide two-way communication.

Permanent-magnet (PM) loudspeakers are perhaps the most common of all electromagnetic transducers. The permanent magnets used for these speakers are made from various combinations of ferromagnetic materials. Loudspeakers can be designed to handle power ranging from a fraction of a watt to hundreds of watts. The size of wire used in the voice coil determines the power rating of the speaker, since the size of the wire is directly related to the amount of current it can handle. Power in watts is equal to the current squared times impedance. Since the impedance is small e.g. (4, 8, or 16 Ω), the current has to be rather large to obtain a large amount of power. It is important to match the impedance of the speaker with the receiver. Mismatching of impedances or turning up of the volume beyond the speaker's rated power capacity can burn out the voice coil.

The frequency response of a speaker refers to its ability to reproduce sound accurately. Ideally speakers should be able to respond to the whole audio range of frequencies of 16 to 16,000 Hertz. However, speakers may be designed to handle a certain range of frequencies within this audio range. Typically, the 12-inch speaker is called a *woofer* and provides a higher quality of reproduction to low frequencies of about 30 to 300 Hertz. The *midrange* speaker is normally an 8-inch speaker and handles frequencies in the 300 to 3000 Hertz range. The tweeter, or horn, is a 4-inch speaker and handles high frequencies in the 3000 to 20,000 Hertz range. A crossover circuit board consisting of a filter network with a combination of capacitance, resistance, and inductance can be used to direct the proper frequencies to the proper speakers.

Headphones are used as output transducers, like miniature speakers, to convert electric energy into sound energy. They require much less electric energy to operate than a speaker. In addition to their use for private listening, they can be used to detect weak ac voltages in the audio frequency range.

The principle of the operation of headphones is similar to that of the permanent-magnet speaker when it is used as an output device. Inside each headphone is a small, permanent horseshoe magnet with two coils wound on its pole pieces (Figure 32-4). A flexible, soft-iron diaphragm is suspended in front of the poles of the horseshoe magnet. The incoming ac audio signal is applied to the coil and produces a varying magnetic field that reacts with the magnetic field of the permanent magnet, causing the metal diaphragm to vibrate. The moving diaphragm produces sound energy of the same frequency and amplitude as the incoming ac audio signal.

32-3 Microphones

Sound energy can be converted to electric energy by use of a microphone. When you speak

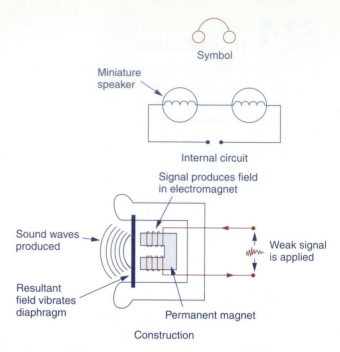

Figure 32-4   Headphones.

into a microphone, the sound waves cause a diaphragm to move. The diaphragm is attached to a device that causes current to flow in proportion to the sound waveform.

Microphones may be rated according to their frequency response, impedance, and sensitivity. For good quality, the electric output from the microphone must correspond closely to the magnitude and frequency of the sound waves that cause them. The impedance of the microphone must match that of the load impedance into which the microphone will operate for maximum transfer of power.

Carbon Microphone   A microphone is an input device that converts sound wave variations into corresponding variations in an electric current. The sound wave variations are first converted to the back-and-forth movements of a foil called a *diaphragm*. The **carbon microphone** consists of a cup-shaped housing filled with carbon granules and fitted with a diaphragm (Figure 32-5). When sound waves strike the diaphragm of the microphone, the diaphragm vibrates at the frequency and amplitude of the sound waves. This causes corresponding changes in the pressure and resistance of the microphone. Thus, the sound waves applied to the diaphragm are converted to changes of electrical resistance. The carbon

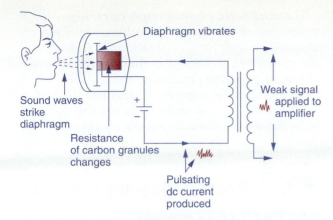

Figure 32-5 Carbon microphone.

microphone has low impedance and requires a dc source for its operation.

Dynamic Microphone

The *dynamic microphone* functions on the same principle as the permanent-magnet loudspeaker (Figure 32-6). Sound waves striking the diaphragm of the microphone cause the coil attached to it to move back and forth within the field of a permanent magnet. As the coil moves in the magnetic field, a voltage is induced in the coil. This voltage depends on the frequency and amplitude of the sound waves that move the coil. Although a dynamic microphone actually generates a voltage, this voltage is not sufficient to operate the output transducer directly. Therefore, it is necessary to amplify this voltage by means of an amplifier.

The dynamic microphone has a low impedance of between 50 and 100 ohms. The name *dynamic* comes from the fact that the microphone has moving parts, namely the voice coil

and diaphragm. It does not require an external source of dc for its operation.

Crystal Microphone

The *crystal microphone* operates on the principle that certain crystal substances will produce a voltage when stress or pressure is applied to them. In the case of the crystal microphone, sound waves strike a diaphragm, which in turn applies varying pressure to a *crystal* (Figure 32-7). This causes the crystal to produce an ac voltage that varies with the frequency and amplitude of the sound waves striking the diaphragm. The signal from the crystal is weak and must be amplified to operate an output transducer.

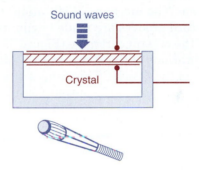

Figure 32-7 Crystal microphone.

Capacitor or Condenser Microphone

The *capacitor or condenser microphone* operates on the principle that changing the distance between the two plates of a capacitor will vary the capacitance. This type of microphone is made using a flexible diaphragm and a fixed metal plate for the two capacitor plates, and air is used for the dielectric (Figure 32-8). When sound waves strike the diaphragm, it vibrates.

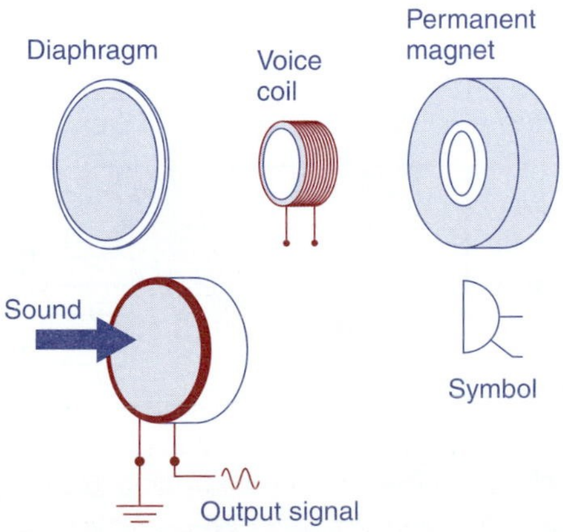

Figure 32-6 Dynamic microphone.

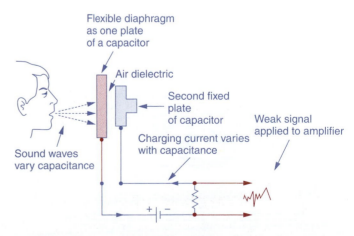

Figure 32-8 Capacitor microphone.

The vibrating diaphragm causes a change in distance between the plates, which varies the capacitance of the microphone. In this way, the sound waves applied to the diaphragm are converted to changes of electrical capacitance.

32-4 Phonograph Cartridges

The phonograph cartridge or pickup is an input transducer that converts the mechanical vibrations caused by the grooves of a record into electric energy. One common type is the *crystal phonograph cartridge*. The crystal cartridge operates on the same principle as the crystal microphone. The mechanical vibrations caused by the record grooves are made to vibrate the stylus of the crystal cartridge (Figure 32-9). These vibrations of the stylus exert a varying pressure on the crystal. This causes a corresponding variation in the output voltage of the crystal, which is then applied to an amplifier.

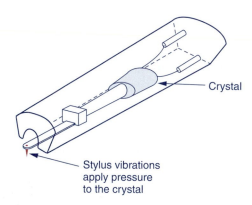

Figure 32-9 Crystal cartridge.

The *magnetic phonograph cartridge* makes use of changes in the reluctance of a magnetic circuit for its operation. In this unit, the movement of the stylus changes the length of the air gap in the magnetic circuit. This, in turn, changes the number of magnetic lines of force cutting the coil and causes a small voltage to be induced in the coil (Figure 32-10).

32-5 Record/Play Tape Head

A *record/play tape head* is an *electromagnet* that converts electric energy to magnetic energy and vice versa. In the *record mode* (Figure 32-11A) the tape head uses the amplified signal from a microphone to magnetize very fine magnetic particles on a thin plastic tape. The tape becomes magnetized by induction, as it passes by the head, from the magnetic field produced by the current flow in the tape head winding.

An *analog tape* stores the sound signals as a continuous stream of magnetism. The magnetism may have any value within a limited range, varying by the same amount as the sound signal voltage. In a *digital tape,* the sound signal is stored as a precise sequence of areas with high and low magnetism. These represent the ones and zeros of the binary code. Two reels enclosed in a plastic container called a *cassette* are often used for tape storage. Both recording audio cassettes and video cassettes are in widespread use.

On *playback* the process is reversed. The magnetic field of the moving magnetized tape induces a small voltage in the coil. This small voltage (low millivolts) is then amplified and

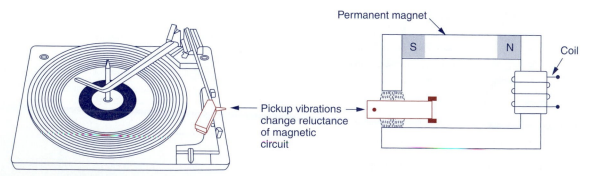

Figure 32-10 Magnetic cartridge.

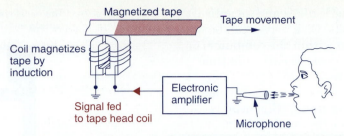

A. Magnetic tape head—record mode

Magnetized tape

Coil magnetizes tape by induction

Tape movement

Signal fed to tape head coil

Electronic amplifier

Microphone

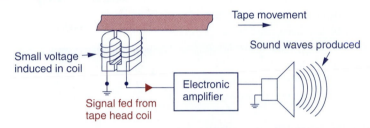

B. Magnetic tape head—playback mode

Tape movement

Sound waves produced

Small voltage induced in coil

Signal fed from tape head coil

Electronic amplifier

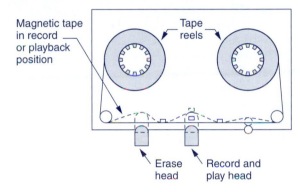

C. Cassette recorder

Magnetic tape in record or playback position

Tape reels

Erase head

Record and play head

Figure 32-11   Tape recorder.

sent to a loudspeaker (Figure 32-11B), which converts the voltage back to sound waves.

Pressing the PLAY button brings the playback head and drive mechanism into contact with the tape. The tape moves and the head records or plays the sounds (Figure 32-11C). A high-frequency electric signal is fed to the *erase* head when recording. This produces a magnetic field that alternates rapidly, disorienting the magnetic particles to demagnetize the tape and erase any previous recording.

32-6 Magnetic Disk Drives

A *disk drive* (Figure 32-12) is a machine that reads data from and writes data onto a *disk*. The disk drive rotates the disk very fast and

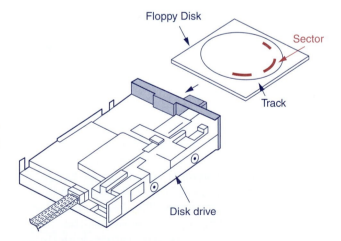

Floppy Disk

Sector

Track

Disk drive

Figure 32-12   Floppy disk drive.

has one or more heads that read and write data. There are different types of disk drives for different types of disks. For example, a magnetic disk drive reads magnetic disks and an optical drive reads optical disks. Disk drives can be either internal (housed within the computer) or external (housed in a separate enclosure that connects to the computer).

Computer magnetic disk drives use an electromagnetic *read/write head*. This head is mounted on an actuator that resembles a record needle pickup. It operates similar to the read/write head used in tape recorders. The read/write head "writes" data by converting digital signals from the computer into magnetic codes on the surface of the disk. The drive then reverses this process to "read" the disk. A disk drive contains two actuators—one to rotate the disk at high speed and one to move the head across the disk.

The 3.5-inch *floppy disk* (*diskette*) is a small, portable plastic disk coated in a magnetisable substance which can be magnetized. The physical size of the floppy disk has decreased from the early $5\frac{1}{4}$ inch to $3\frac{1}{2}$ inch while the data capacity has risen. These disks are known as "floppy" disks because of the soft flexible plastic disk found within the hard plastic shell. The floppy disk is still the most universal means of transferring data from one computer to another. The hole through the case is for write enable/protection (Figure 32-13). When this hole is *open,* nothing can be written on the disk. The floppy disk is used for storing and backing up small amounts of computer data. Floppy disks contain *tracks,* which are concentric circles radiating out from the center. The information is stored on these tracks. The standard floppy disk in current use is the 3.5 inch HD (high density) disk, which stores 1.44 MB of data on two sides, with 80 tracks per side and 18 sectors per track. *Formatting* refers to the magnetic track pattern that specifies the locations of the tracks and sectors. This information must exist on a disk before it can store any user data. Formatting erases any previously stored data.

Currently, all conventional floppy disks are double-sided requiring two read/write heads, one per side, on the drive. The *head actuator* is the device that physically positions the read/write heads over the correct track on the surface of the disk and is driven by a stepper motor. The *spindle motor* is what spins the floppy disk when it is in the drive. A very slow spindle speed of 300 rpm is used and this

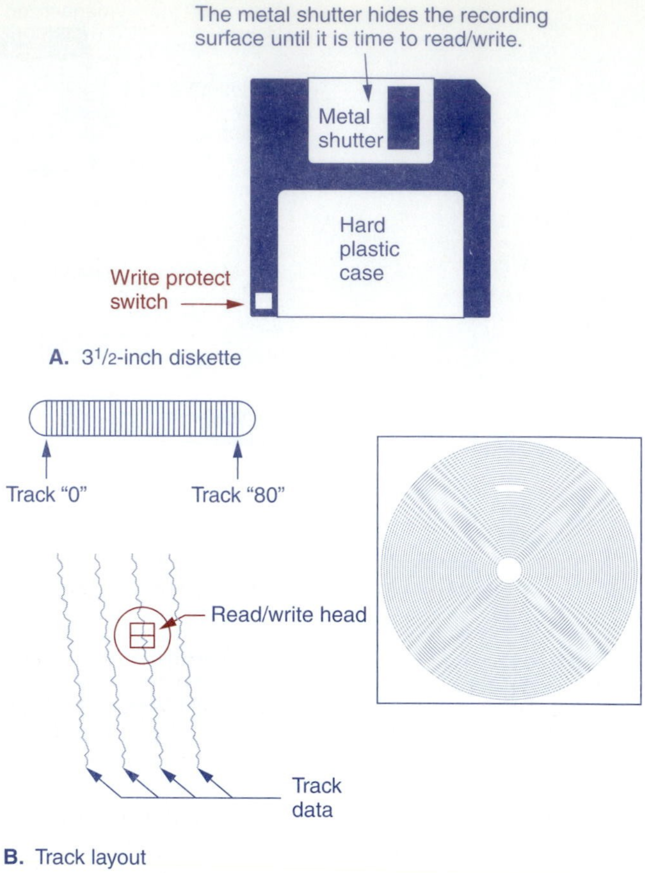

The metal shutter hides the recording surface until it is time to read/write.

Metal shutter

Hard plastic case

Write protect switch

A. 3$\frac{1}{2}$-inch diskette

Track "0"   Track "80"

Read/write head

Track data

B. Track layout

Figure 32-13   Floppy disks.

accounts for their slower transfer times in comparison to other storage media. It is this slow speed however that allows the heads to ride contacting the surface of the media without causing the floppy disk's magnetic coating to wear off.

Higher capacity floppy disk drives are also available. One such drive is the ZIP drive. *ZIP disks* are slightly larger than conventional floppy disks and about twice as thick. They can hold 100 MB of data and have become a popular media for backing up hard disks and for transporting large files.

Hard disks store much more information than floppy disks do. Also, because of a higher rotational speed, the data-transfer rates are much faster. The hard disk retains data as magnetic patterns on a rigid disk, usually made of a magnetic thin film deposited on an aluminum or glass platter. Typically a *hard disk drive* consists of several platters stacked on top of each other in a *sealed-drive package* (Figure 32-14). The read/write head floats a fraction of

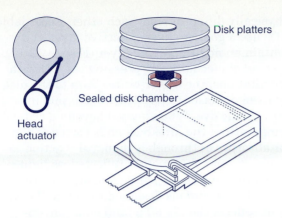

Disk platters

Sealed disk chamber

Head
actuator

Figure 32-14 Hard disk drive unit.

a millimeter above the disk surface on a layer of air, landing gently on it when the drive is powered down. Because of the extremely small space between the disk and the head, a very clean environment is essential to reliable operation. A small particle of dirt could cause a head to 'crash', touching the disk and scraping off the magnetic coating. This is the reason for using a sealed enclosure.

Each platter of the hard disk is double-sided so there is a read/write head for each side of each platter, mounted on arms that control their movement. The arms are moved by a head actuator which contains a *voice coil*—an electromagnetic coil that can move a magnet very rapidly. Loudspeaker cones are vibrated using a similar mechanism. Voice coil actuators are more durable than their stepper counterparts, since fewer parts are subject to daily wear and tear thereby providing higher performance. Hard disks can store anywhere from 20 MB to more than 10 GB of data and are from 10 to 100 times faster than floppy disks.

32-7 Optical Disk Drives

An *optical disk* is a storage medium which uses lasers for reading and writing data. A single optical disk can store up to 6 GB of data which is much more than floppies. There are three basic types of optical disks:

CD-ROM: *CD-ROM* (Compact Disk — Read Only Memory) has data already encoded onto them. The data is permanent and can be read any

number of times, but CD-ROMs cannot be modified.

CD-R: *CD-R* (Compact Disk — Recordable) allows only one shot at making a disk. Once data is written onto the disk, it can't be altered. After that, the CD-R disk is just like a CD-ROM.

CD-RW: *CD-RW* (Compact Disk — Recordable Writeable) disks can be erased and loaded with new data, just like magnetic disks. Each type of these optical disks may require a different type of disk drive and disk. Even within one category, there are many competing formats, although CD-ROMs are relatively standardized.

The compact disk has been one of the major successes of modern times. CDs can hold both music and computer data. There are many games that now come on CD-ROM, and they use graphics and sounds as they never could before. The same is true for educational programs. Clearly the CD is the storage media of choice for the future.

The standard 12 cm-diameter CD-ROM is made up of a 1.2 mm sandwich of three coatings: a black layer of clear polycarbonate plastic; a thin sheet of aluminum which gives the disk its distinctive silver coloring and increases its reflectivity; and finally a lacquer coating to protect the disk from external scratches and dust (Figure 32-15).

Commercial CDs are made using an injection molding technique and a stamper. The layer of aluminum is stamped with millions of tiny indentations called *pits*, molded in a spiral from the center of the disk outward. The encoded data is recorded onto the disk as a series of these small pits of varying lengths.

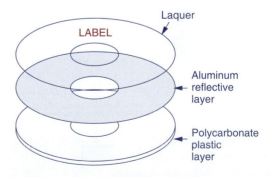

Laquer

LABEL

Aluminum
reflective
layer

Polycarbonate
plastic
layer

Figure 32-15 Compact disk (CD-ROM).

The area around the pits called *lands*, also play an important part in the process. Unravelled and laid in a straight line the data would stretch four miles.

The operation of a **CD-ROM drive** is basically the same as that of a CD audio player. CD players use a laser beam spot of about 1.6 micrometer to read the pits on the disk. By rotating the disk and shining the laser beam on the series of pits and lands, an optosensor can be used to detect the presence or absence of pits within a fixed period of time. If the laser beam is reflected from its path unchanged, there is no pit. Therefore, a binary 0 is represented at the particular spot on the disk. If there is a change in the reflected beam, caused by a pit, a binary 1 is represented at that particular spot on the disk (Figure 32-16).

Recordable CDs use a type of technology that can be written on to only once. The drive that writes the CD-R disk is often called a one-off machine and can also be used as a regular CD-ROM reader. CD-ROM and CD-R disks are physically different from each other. Recordable CDs use a gold or silver reflective layer and contain an additional **dye** layer (Figure 32-17). Instead of mechanically pressing the CD with indentations, the driver writes data to the disk by using its laser to physically *burn* pits into the organic dye. When heated beyond a critical temperature, the area burned becomes opaque (or absorptive) through a chemical reaction to the heat and subsequently reflects less light than areas that have not been heated by the laser. This system is designed to mimic the way light reflects cleanly off a land on a conventional CD, but is scattered by a pit. Consequently, a CD-R disk can generally be used in a conventional CD player.

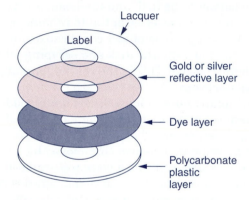

Figure 32-17 Recordable compact disk (CD-R).

Although CD-R drives look just like CD-ROM drives, the disks are easily identifiable. The underside of a CD-R disk typically has a greenish-gold or silver-blue color, whereas CD-ROMs have a distinctive silver color. CD-R disks are used for beta versions and original masters of CD-ROM material as well as a means to distribute large amounts of data to a small number of recipients. A major advantage over other media is that they can read in any CD-ROM reader.

CD-RW (Rewriteable) disks use a much more advanced technology in order to accomplish the goal of making a compact disk both writeable and rewriteable. There are, therefore, many issues related to the operation of these drives, and also compatibility, that aren't applicable to

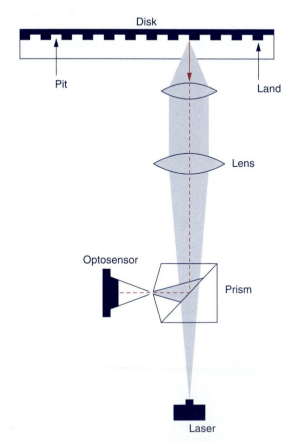

Figure 32-16 Compact disk optical pick-up.

most other CD formats. Still, using CD-RW drives and disks, you can treat the optical disk just like a floppy or hard disk, writing data onto it multiple times.

The CD-RW uses a disk with a polycrystalline layer, which can be altered between *amorphous* and *ordered* states by heating with the drive's laser beam. Any data on the disk can be replaced by simply overwriting, but there is a limit as to how many times this can be done i.e. (most CD-RW disks guarantee at least 1,000 rewrites). Special writing strategies are needed to make sure that portions of the disk are not overused with respect to others, causing them to deteriorate faster.

The **DVD** (short for digital versatile disk or digital video disk) is a new type of CD-ROM that holds a minimum of 4.7 GB, enough for a full-length movie. A DVD looks like today's CD: it is a silvery platter, 4.75 inches in diameter, with a hole in the center. Like a CD, data is recorded on the disk in a spiral trail of tiny pits, and the disks are read using a laser beam. The DVD's larger capacity is achieved by making the pits smaller and the spiral tighter, and by recording the data in as many as four layers, two on each side of the disk.

A DVD has 540 horizontal lines of resolution making for much sharper TV images than the standard VCR format, which has a resolution of 210 lines. DVD sounds better, too; digital sound can be separated into discrete channels, making surround sound possible. Many experts believe that DVD disks will eventually replace CD-ROMs as well as VHS video cassettes. There are different types of DVD platforms, some of which are as follows:

DVD-Video: Disks designed to be used in DVD-video players, mainly as a medium for home entertainment.

DVD-ROM: Disks designed for computer applications, incorporating high-quality video, audio, and multimedia applications.

DVD-R, DVD-RW, DVD-RAM: Recordable DVD formats designed to work with DVD-ROM capable computers.

DVD-Audio: A high-quality successor to the music CD.

DIVX: A modified DVD-video player with the DIVX "feature" that allows for "pay per view" viewing.

32-8 Sensors and Detectors

Sensors and detectors are devices that are used to detect, and often to measure, the presence of something. They are a type of transducer used to convert mechanical, magnetic, thermal, optical, and chemical variations into electric voltages and currents. Sensors and detectors are usually categorized by what they measure and play an important role in modern manufacturing process control (Figure 32-18).

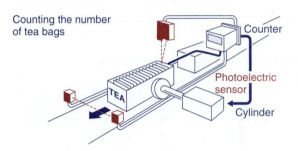

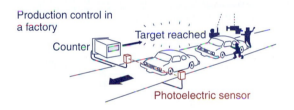

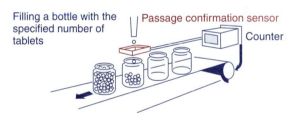

(Courtesy of Keyence Corp. of America.)

Figure 32-18   Sensors used in manufacturing process control.

Light Sensors   The **photovoltaic** or **solar cell** is a common light-sensor device that converts light energy directly into electric energy (Figure 32-19). Modern silicon solar cells are basically PN junctions with a transparent P-layer. Shining light on the transparent P-layer causes a movement of electrons between P- and N-sections producing a small dc *voltage*. Typical output voltage is about 0.5 V per cell in full sunlight.

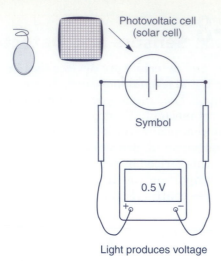

Photovoltaic cell
(solar cell)

Symbol

0.5 V

+   −

Light produces voltage

Figure 32-19   Photovoltaic or solar cell.

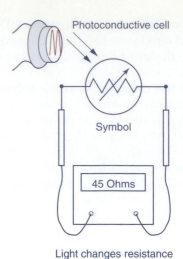

Photoconductive cell

Symbol

45 Ohms

Light changes resistance

Figure 32-20   Photoconductive cell.

The **photoconductive cell** (also called photoresistive cell) is another popular type of light transducer (Figure 32-20). Light energy falling on a photoconductive cell will cause a change in the *resistance* of the cell. One of the more popular types is the *cadmium sulphide photocell.*

When the surface of this device is dark, the resistance of the device is high. When brightly lit, its resistance drops to a very low value.

There are two main types of light-sensor switches that use the photoelectric technique to sense position (Figure 32-21). One must have

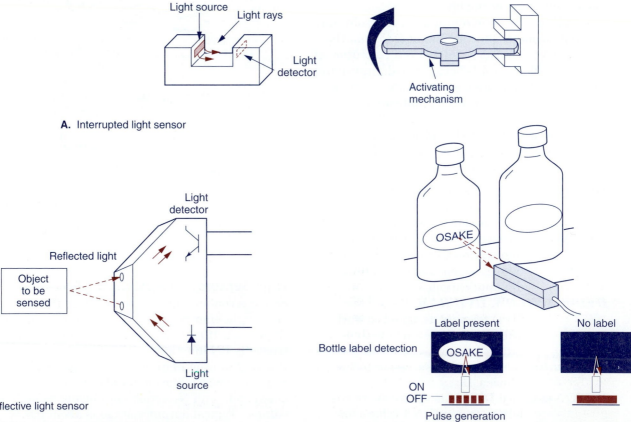

Light source
Light rays

Light detector

A. Interrupted light sensor

Activating mechanism

Light detector

Reflected light

Object to be sensed

Light source

B. Reflective light sensor

Figure 32-21   Light sensor switches.

OSAKE

Bottle label detection

Label present

OSAKE

No label

ON
OFF

Pulse generation

its light beam directly interrupted, while the other uses a reflection of its light beam. An infrared light emitter, usually a light-emitting diode (LED), emits light. A detector senses the light, either directly or reflected, and produces an output.

Thermistors A *thermistor* is a semiconductor heat sensor whose resistance varies with temperature. The resistance of negative-temperature-coefficient (NTC) thermistors varies inversely with temperature. In other words, the *higher* the temperature of the thermistor, the *lower* its resistance. Positive-temperature-coefficient (PTC) thermistors are used also. With this type of thermistor, the resistance varies directly with temperature. That is to say, the *higher* the temperature of the thermistor, the *higher* its resistance. Figure 32-22 shows a typical automobile coolant temperature-sending unit used with a coolant temperature gauge. The heart of the sending unit is a thermistor, which changes resistance with temperature.

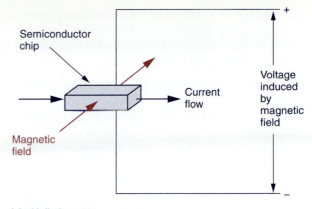

(a) Hall element

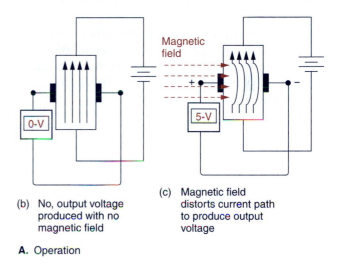

(b) No, output voltage produced with no magnetic field

(c) Magnetic field distorts current path to produce output voltage

A. Operation

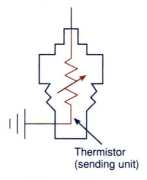

Thermistor (sending unit)

Figure 32-22 Typical automotive coolant temperature-sensing unit.

Hall-Effect Sensors The ***Hall-effect sensor*** is designed to sense the presence of a magnetic object, usually a permanent magnet. It is used to signal the *position* of a component. Due to its *accuracy* in sensing position, the *Hall-effect switch* is a popular type of sensing device. Figure 32-23 illustrates how a Hall element operates. The Hall element is a small, thin, flat slab of semiconductor material. When a current is passed through this slab and *no* magnetic field is present, zero output voltage is produced. When a magnet is brought close to the semiconductor material, the current path is distorted. This distortion causes electrons to be forced to

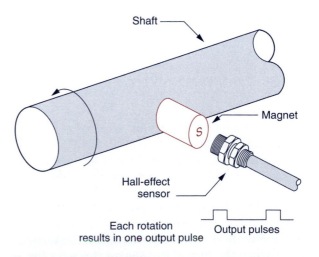

B. Used as a speed sensor

Figure 32-23 Hall-effect sensor.

the right side of the material, which produces a voltage across the sides of the device. Hall-effect devices use two terminals for excitation and two for output voltage.

Proximity Sensors
Proximity sensors are used to detect the presence of an object without making physical contact. Normally the output of the proximity sensor switches on or off rather than producing a linear output proportional to the distance of the object to the sensor. The two major types of proximity sensors are inductive and capacitive. The inductive proximity switch detects the presence of metallic materials. The capacitive proximity switch detects the presence of conductive and non-conductive targets.

In principle, an *inductive sensor* consists of a coil, an oscillator and detector circuit, and a solid-state output (Figure 32-24). When energy is supplied, the oscillator operates to generate a high-frequency field. At this moment, there must not be any conductive material in the high-frequency field. When a metal object enters the high-frequency field, eddy currents are induced in the surface of the target. This results in a loss of energy in the oscillator circuit and, consequently, this causes a small amplitude of oscillation. The detector circuit recognizes a specific change in amplitude and generates a signal that will turn the solid-state output ON or OFF. When the metal object leaves the sensing area, the oscillator regenerates, allowing the sensor to return to its normal state.

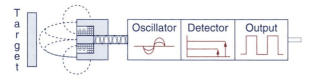

Figure 32-24   Inductive sensor operation.

The operation of *capacitive sensors* is also based on the principle of an oscillator. However, instead of a coil, the active face of a capacitive sensor is formed by two metallic electrodes—rather like an "opened" capacitor (Figure 32-25). The electrodes are placed in the feedback loop of a high-frequency oscillator that is inactive with "*no* target present." As the target approaches the face of the sensor it enters the electrostatic field that is formed by the electrodes. This causes an increase in the coupling

Figure 32-25   Capacitive sensor.

capacitance and the circuit begins to oscillate. The amplitude of these oscillations is measured by an evaluating circuit that generates a signal to turn the solid-state output ON or OFF.

In order the actuate inductive sensors, we need a conductive material. Capacitive sensors may be actuated by both conductive and non-conductive materials such as wood, plastics, liquids, sugar, flour, wheat, and the like. However, this advantage of the capacitive sensor compared to the inductive sensor brings some disadvantages with it. Whereas, inductive proximity switches may be actuated only by a metal and are insensitive to humidity, dust, dirt, and the like, the capacitive proximity switches also are actuated by any dirt in their environment. For general applications, the capacitive proximity switches are not really an alternative, but a *supplement to* the inductive proximity switches. They are a supplement when there is no metal available for the actuation, for example, for woodworking machines and for determination of grain levels in silos.

Ultrasonic Sensors
An *ultrasonic sensor* operates by sending sound waves toward the target and measuring the time it takes for the pulses to bounce back. The time taken for this echo to return to the sensor is directly proportional to the distance or height of the object because sound has a constant velocity (Figure 32-26).

Bar Codes
Bar-code technology is widely implemented in the retail marketplace, and is rapidly gaining in a broad range of applications. It is easy to use, can be used to enter data much more quickly than manual methods, and is very accurate.

A bar-code system consists of three basic elements: the bar-code symbol, a scanner, and a decoder.

The *bar-code symbol* contains up to 30 characters, which are encoded in a machine-readable form. The characters are usually printed above or below the bar code, so data can be entered manually if a symbol cannot be read

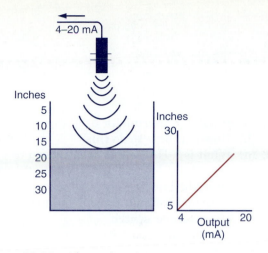

Figure 32-26 Ultrasonic sensor.

by the machine. The blank space on either side of the bar code symbol, called the quiet zone, along with the start and stop characters, lets the scanner know where data begins and ends (Figure 32-27).

Figure 32-27 Bar-code symbol.

There are several different kinds of bar codes. In each one, a number, letter, or other character is formed by a certain number of bars and spaces. The Universal Product Code (UPC) is the standard bar-code symbol for retail food packaging in the United States. The UPC symbol (Figure 32-28) contains all of the encoded information in one symbol. It is strictly a numeric code that contains the:
- UPC type (1 character)
- UPC manufacturer, or vendor, ID number (5 characters)
- UPC item number (5 characters)
- Check digit (1 character) used to mathematically check the accuracy of the read
    Bar-code scanners are the eyes of the data collection system. A light source within the scanner illuminates the bar-code symbol; the bars absorb

Figure 32-28 UPC bar-code symbol.

light, and spaces reflect the light. A photo detector collects this light in the form of an electronic-signal pattern representing the printed symbol. The *decoder* receives the signal from the scanner and converts this data into the character data representation of the symbol's code.

Although the scanner and decoder operate as a team, they can be integrated or separate, depending on the application (Figure 32-29).

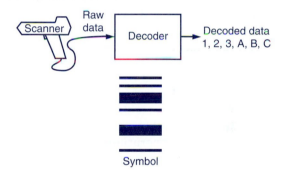

Figure 32-29 Bar-code scanner and decoder.

Aerospace Electronics Technician
Job Description
- Design, test, and construct, specialized electronic devices.
- Provide technical support for aircraft control systems.
- Working knowledge of metal fabrication and other manufacturing processes.
- Graduate of a college or university program in Electrical/Electronic Technology.

CHAPTER 32
Review and Applications

Review Questions

1. What are the two basic functions of electric transducers?
2. (a) How is electric energy converted into mechanical energy in a permanent-magnet speaker?
   (b) How is mechanical energy converted into electric energy in a permanent-magnet speaker?
3. State the type of electric-output signal produced by each of the following microphones:
   (a) carbon microphone
   (b) dynamic microphone
   (c) crystal microphone
   (d) capacitor or condenser microphone
4. Why would a headphone, rather than a speaker, be used in testing the weak ac signal voltage produced by a dynamic microphone?
5. How is energy converted in a crystal phonograph cartridge?
6. (a) What type of varying electric-output signal is produced by the magnetic phonograph cartridge?
   (b) Explain how this signal is produced.
7. (a) Explain the operation of a tape recorder record-and-play head in the record mode.
   (b) Explain the operation of a tape recorder record-and-play head in the playback mode.
   (c) Explain how the erase head of a tape recorder operates to erase a recording.
8. Explain how a computer disk drive writes and reads data.
9. (a) How is a digital data recorded on the surface of an optical disk?
   (b) How is this information retrieved by the compact disk player?
10. List five types of nonelectric variations that sensors are required to recognize.
11. (a) Compare the types of varying electric-output voltages produced by the photovoltaic and photoconductive cell.
    (b) State what effect an increase in light intensity would have on each device.

12. (a) What is a thermistor used to sense?
    (b) How is it able to do this?
13. (a) What is a Hall-effect sensor used to sense?
    (b) Describe how the output voltage is switched from OFF to ON in a Hall-effect sensor switch.
14. (a) What is an inductive sensor used to sense?
    (b) Describe how the output circuit is switched in an inductive sensor.
15. (a) What is a capacitive sensor used to sense?
    (b) Describe how the output circuit is switched in a capacitive sensor.
16. (a) What is an ultrasonic sensor used to sense?
    (b) Describe how sensing is accomplished in an ultrasonic sensor.
17. (a) List three advantages of using bar-code technology.
    (b) What are the three basic elements of a bar-code system?

Problems

1. A capacitor microphone is suspected of being defective. The dc resistance between the output leads of the microphone is measured and found to be of infinite value. What fault, if any, does this indicate? Why?
2. Refer to Figure 32-10. Should any voltage be measured across the coil when the pickup is in the holder? Why?
3. Refer to Figure 32-19. How many solar cells of the type shown would be required to produce an output voltage of 12 volts? How would they be connected?
4. Refer to Figure 32-23. Assume a scope connected to the output of the Hall-effect sensor indicates a wave pattern frequency of 50 Hz. What would the rpm of the shaft be?

Critical Thinking

1. The plug jack or coil of a dynamic microphone is suspected of being bad. How could this be verified using an ohmmeter?
2. Refer to Figure 32-13. How many 3.5 inch high-density disks would be required to store 19 MB of data?
3. Can a CD-ROM disk be used for recording purposes? Why?
4. A thermistor sensor, with a negative-temperature-coefficient, is connected as a signal input to the engine computer to monitor engine temperature. Assume a bad, high-resistance connection develops at the output terminal of the thermistor. How might the computer interpret this fault. Why?

Portfolio Project

Obtain a discarded floppy disk drive from an old computer. Carefully disassemble it and suitably display the major components. Write a report on how the unit operates.

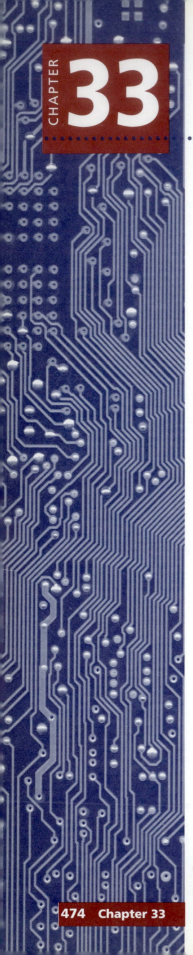

Printed Circuits

The days of metal chassis are long behind us, having given way to the printed circuit board. Miniaturization of electronic equipment has led to the extensive use of printed circuits in all kinds of electrical and electronic products. Printed-circuit board (PCB), consists of a thin plate on which components are placed. In this chapter, you will study the basic process used in the manufacture of printed circuits and how to make repairs to PCBs.

Objectives

After studying this chapter, you should be able to:
1. State the advantages of using printed circuit boards.
2. Describe the different ways on which a PCB is constructed.
3. Explain the photographic and print-and-etch processes.
4. Properly mount and solder components to a PCB.
5. Properly desolder and replace components on a PCB.

Key Terms

- **printed-circuit board**
- **hand-wired**
- **laminate**
- **cladding**
- **plated through hole**
- **surface-mount technology**
- **parts-layout**
- **foil-pattern layout**
- **computer-aided design (CAD)**

- **resist**
- **etch**
- **photographic printing**
- **print-and-etch**
- **soldering**
- **solder**
- **flux**
- **soldering iron**
- **soldering tip**

- **cold-solder joint**
- **bridging**
- **desoldering tools**
- **solder wick**
- **heat sink**
- **heat sink grease**
- **hook-type probe**

RESEARCH
INTERNET
SITES

- **Omni Graphics**
- **Printed Circuit Board Fabricators (PCBFAB)**
- **International Circuits and Components Incorporated (ICCI)**

33-1 PC Board Construction

Printed-circuit (*PC*) *boards* are used extensively in all types of electronic equipment. Prior to printed circuits, all electronic equipment was assembled using a ***hand-wired*** type of metal box or chassis (Figure 33-1). Electronic components were soldered to sockets or terminal strips and interconnected with soldered wire connections. Today, most of the circuitry is contained on printed-circuit boards. Only large components such as transformers that require considerable mounting space are *hand-wired*.

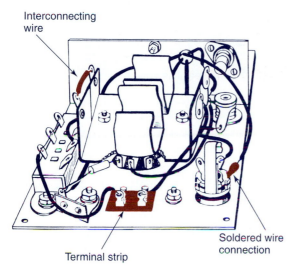

Figure 33-1   Hand-wired chassis.

Printed-circuit boards have helped in producing smaller and more reliable electronic equipment. A PC board serves two basic functions. It provides both the mounting space for the components and connects them together into a working circuit. These boards are designed so that many components can be mounted in a small area. The interconnecting printed circuit is often more reliable than the same hand-wired circuit. Also, many identical boards can be made from one set of diagrams. This makes the mass production of the electronic device much less expensive than if it were assembled entirely by hand.

The basic unprocessed PC board consists of a thick layer of insulating material and a very thin layer of conducting material. The insulating material is called the ***laminate*** and the conductive material is called the ***cladding*** (Figure 33-2). There are various types of laminates and cladding. The laminate material used most often is phenolic paper or fiberglass. The cladding used is generally copper.

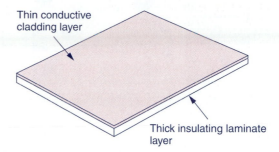

Figure 33-2   Unprocessed copper-clad PC board.

The simplest type of PC board is the *single-sided board* (Figure 33-3). It consists of a thin conductive foil cladding on one side of a rigid nonconductive laminate. Components are positioned on the insulator side of the board. Lead access holes are drilled through both the laminate material and conductive foil for each component lead. These leads are passed through the holes from the insulator side to the foil side. Then they are bent, soldered to the foil, and the excess lead is trimmed. The foil cladding is previously processed into conductive paths to provide the required wiring for the working circuit.

Commercially mass-produced PC boards can be populated using either through-hole or surface-mount components. The ***plated through hole*** (***PTH***) type is basically a single sided board with one side carrying the components and the other used for soldering. Holes drilled through the board are plated with tin-lead. As components are mounted, the leads extend through the holes to protrude on the solder side. Components are mounted by hand or by pick and place machines. When soldered, the solder flows through the plated hole to provide a secure and reliable connection. Typically, soldering is done by *wave* soldering where the board passes over a wave of molten solder.

With ***surface-mount technology*** (***SMT***) components sit on pads (areas of exposed tin-lead plated copper) on the board, so that the component and solder sides of the board are actually the same. Instead of wave soldering, solder paste is applied to the pads. The components are placed in the desired location, with

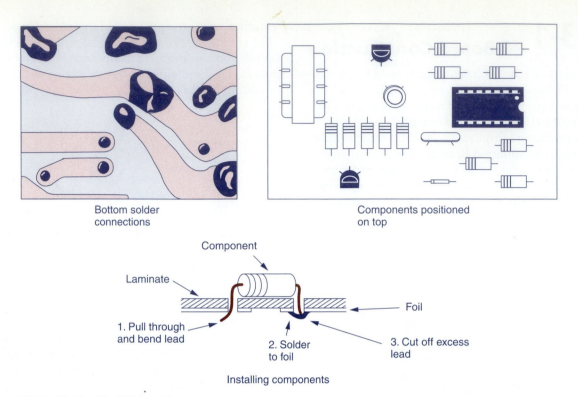

Bottom solder connections

Components positioned on top

Component

Laminate

Foil

1. Pull through and bend lead

2. Solder to foil

3. Cut off excess lead

Installing components

Figure 33-3   Single-sided PC board.

the paste holding them in place. The solder flows by heating the board in a hot gas or by some other technique.

SMT components must be much smaller than through-hole components, which results in high density. Both sides of the SMT board are used for components, and weight is lower. Automated assembly is much easier using SMT. Because component choices are limited with surface-mount devices, many circuit boards are hybrids. These contain a mixture of surface-mounted and through-hole components. For example, transformers, electrolytic capacitors, and relays may be PTH devices while integrated circuits and resistors may be surface mounted on the same board.

33-2 Planning and Layout

There is no standard printed wiring board configuration; the conductive line patterns are determined by the specific electronic requirements of each board design, and each design is considered unique.

The term *printed* refers to the techniques by which conductor lines and traces are reproduced on the cladding. Much of the technology

was adapted from methods such as silk screening and photolithography, both of which originated in the printing industry.

Different methods are used for producing the conductive foil pattern on a PC board. Basic to all the methods is the preparation of the artwork. Two pictorial diagrams are usually required. The first is a ***parts-layout*** diagram that shows all components fitted onto the insulator side of the board. The other is a ***foil-pattern layout*** diagram that shows all necessary conductor routing on the foil side of the board. Both of these diagrams are normally drawn to exact size.

To start laying out the circuit, you need a schematic of the circuit, as well as a physical size list of the components and PC board to be used (Figure 33-4). Study the schematic and plan the physical layout of the circuit by sketching the shape of each component on paper.

REARRANGEMENT OF THE COMPONENTS AND REDRAFTING OF SKETCHES ARE OFTEN REQUIRED FOR THE BEST LAYOUT OF PARTS.

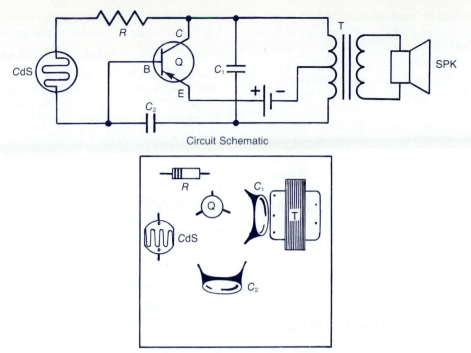

Circuit Schematic

Parts layout drawn to actual size

Figure 33-4 Typical schematic and parts layout diagram.

A foil-pattern layout is produced using the component layout diagram as a guide. The foil pattern represents the conductor paths and terminal pad areas that will be processed into a conducting foil pattern on the PC board (Figure 33-5). Following the schematic, lines are drawn between each of the components that are connected. Lines that cross, but do not connect, on the schematic cannot cross on the foil pattern. Because the conductors on the circuit board cannot cross in this manner, considerable planning is necessary to avoid the problem of conductor crossovers. The connecting lines on the foil pattern can, however, be drawn through the

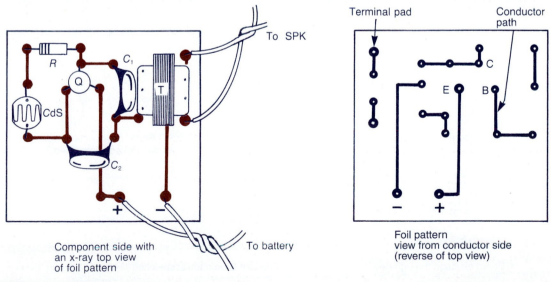

Component side with an x-ray top view of foil pattern

To battery

Foil pattern view from conductor side (reverse of top view)

Figure 33-5 Foil-pattern layout.

outline of the parts, since the parts will be on the top of the board and the conductors will be on the bottom.

Today, most printed wiring-board master artwork is produced with **computer-aided design (CAD)** equipment to reduce design and draft time.

A graphics display terminal operates in concert with a computer to permit the rapid generation of printed wiring-board artwork. Design rules and other parameters are entered into the system, where a library of standard design aids may reside to help expedite the process. The tape, disk, or other type of output from the CAD system is fed into a plotter to generate the artwork.

33-3 Printing the PC Board

Printing of a PC board refers to transferring the foil pattern onto the copper cladding. This process involves the removal of the copper cladding from the places where it is not wanted. This is done by first transferring the pattern to the copper cladding with a material called resist. **Resist** is a material that is resistant to the chemical reaction used to remove the unwanted copper. Once the resist has been applied to the required conductive paths, the board is **etched**. This is done by dipping the PC board into a specially prepared chemical known as the *etchant*. This chemical etches (eats away) the unwanted nonprotected copper. Only the

metal foil under the resist remains, thus forming the conductive paths of the circuit.

Photographic Printing
The **photographic process of printing** the resist onto the PC board is the one most often used in industry (Figure 33-6). First, a full-size negative of the foil pattern is produced. Next, the copper cladding is cleaned and sensitized with a thin layer of light-sensitive plastic resist. Then the negative is placed over the sensitized board and exposed to ultraviolet light. The board then is developed just like a photograph and etched.

Print-and-Etch Technique
The **print-and-etch** technique refers to all methods of producing printed circuits other than the photo-etch method. Also known as direct printing, it involves putting the resist material directly onto the copper cladding. The direct method is an easier and less expensive process. It is convenient to use when a one-of-a-kind project is to be built. Either resist tape or resist liquid can be used with this process (Figure 33-7). **Resist tape** comes in various widths and shapes that are stuck to the copper cladding. Tape pads or circular enlargements of the conductor are used at each point where a component lead is to be soldered. The size of the pads and width of the conductive path used depends on the current requirements of the circuit to be made. If a **liquid resist** is used, it is applied with an artist's fine brush or a special felt pen that contains an acid-resist ink.

Once the resist pattern has been applied, the board is etched to form the conductive paths

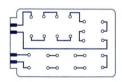
1. Foil pattern laid out on paper

2. Negative of foil pattern made

3. PC Board copper-cladding is cleaned and sensitized

4. Board exposed through negative

5. Board developed and etched

Figure 33-6 Photographic printing.

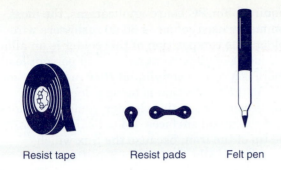

Resist tape   Resist pads   Felt pen

Figure 33-7   Resist materials.

(Figure 33-8). There are numerous types of etching solutions used to remove exposed copper foil: *All are dangerous chemicals that need to be handled with extreme care.* The simplest method of etching is accomplished by submerging the PC board into an etching tray with the copper side up. Always wear eye protection and plastic gloves during the etching process! Gently rocking the etching tray will allow fresh etchant to wash over the copper surface and will aid in decreasing etching time. Etching is completed when all of the undesired copper foil is removed from the board. It is then removed from the tank, flushed with cold water for several minutes, and dried with a towel.

Once the board has been etched, the resist protecting the conductor pattern on the copper foil has served its purpose. It is now necessary to strip completely the resist from the underlying copper pattern. The board is first placed in a solvent-type stripping solution for a few minutes. The board is then removed and scrubbed with a nylon bristle brush under cold tap water. Finally, it is dried with a towel to complete the fabrication.

33-4 Component Assembly and PC Board Soldering

Once the PC board has been fabricated it is ready for the assembly of the components and the ***soldering*** of these components to the board. Often, electronic kit projects are sold in an unassembled form with the printed circuit board and the components supplied. The same basic rules of assembly apply.

Drilling   Lead access holes must be drilled through the centers of each copper terminal pad where a component lead will pass through the PC board. Normally, holes drilled in PC boards are of a very small diameter. Only moderate pressure should be applied when drilling to avoid breaking the bit during the drilling process. The size of drill used is determined by the diameter of the lead component. Generally, a No. 60 drill is used for drilling lead-access holes for capacitors, diodes, transistors, and resistors.

Copper Cleaning   In general, all components are mounted before the soldering operation is started. Precleaning is required just prior to component assembly and soldering. The simplest method is to polish the copper with fine steel wool (Figure 33-9). *Cleaning the copper foil just prior to soldering is essential; this cannot be emphasized enough.* Poor soldering will result if the copper foil is contaminated with any type of film or copper oxide. In fact, the circuit may not even function at all due to poor solder connections. Poor solder connections can act as open circuits.

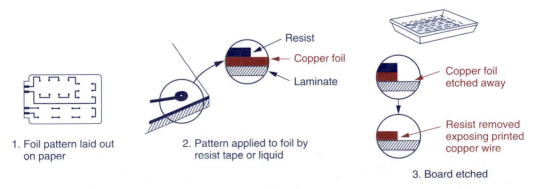

1. Foil pattern laid out on paper

2. Pattern applied to foil by resist tape or liquid

Resist
Copper foil
Laminate

Copper foil etched away

Resist removed exposing printed copper wire

3. Board etched

Figure 33-8   Print-and-etch process.

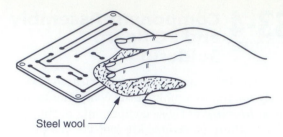

Steel wool

Figure 33-9 Steel wool should be used for cleaning prior to soldering.

Component Mounting

In general, all component bodies should be mounted flush to the insulated side of the PC board. The leads of axial lead components are bent at a 90° angle to the body at a distance of about 1/8 in. from the end of the component body (Figure 33-10). This distance should be in agreement with the previously drawn component-layout diagram. Fine needle-nose pliers can be used for forming the bends. Some vertical-mounted components, such as disc capacitors and transistors, do not require their leads to be bent prior to insertion into access holes. Their leads are simply fed straight through into the holes. Once the component is properly positioned, its leads are bent to hold them in place before soldering.

In an alternate mounting technique called *surface-mount technology* (SMT) solder connections are made to pads on the top side (component side) of the board. Instead of components having leads that are bent and stuck through mounting holes in a PC board, the components are leadless and mounted directly to the top surface of the PC board (Figure 33-11).

PC Board Soldering

In order to form a good electrical connection between a component lead and its terminal pad, proper soldering is required. For PC board applications, the most commonly used *solder* is 60/40 resin-core wire solder. The composition of this solder is an alloy consisting of 60 percent tin and 40 percent lead. There is a core of semiliquid **flux** inside the solder. As the connection is being soldered, the flux becomes an active cleaner, removing oxides from the metal surfaces. Do not carry solder on the tip of the iron, because the flux which should normally clean the connection will evaporate.

Once the soldered connection has cooled, the resin solidifies and returns to an inactive state, and has no adverse effect on the connection. However, since there may be some conductive particles trapped in the resin residue, it should be removed with a flux cleaner. Acid-core solder should *never* be used because the acid flux is highly corrosive. The diameter of the wire solder to be used will depend upon the size of the component leads and terminal pads to be soldered. Use of a large diameter solder requires excessive heat for melting. A diameter of about 1/32 in. is suitable for most PC board soldering.

The proper selection of the **soldering iron** and **tip** is important for obtaining good quality soldered connections. A soldering gun or high-wattage iron should never be used for PC board soldering because their high heat level may damage components and foil. For general-purpose PC board soldering a soldering iron in the 20–35 W range with a 1/16–1/8 in. chisel tip is recommended.

In order to minimize damage due to overheating, a *temperature-controlled soldering iron* is recommended for PC board soldering (Figure 33-12). Never solder on a PC board that has power still applied to it. Soldering on a "hot" board could cause extensive damage due to accidental grounding of the circuit.

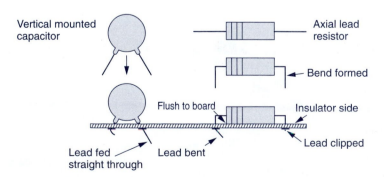

Vertical mounted capacitor

Axial lead resistor

Bend formed

Flush to board

Insulator side

Lead fed straight through

Lead bent

Lead clipped

Figure 33-10 Component mounting.

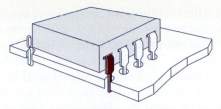

A. Conventional plated through hole (PTH) printed circuit board

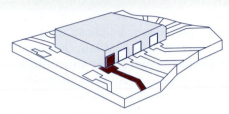

B. Surface-mount technology (SMT) printed circuit board

Figure 33-11   Plated through-hole and surface-mount PC boards.

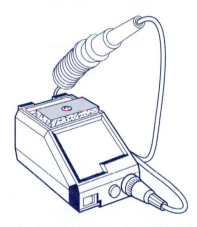

Figure 33-12   Temperature-controlled soldering system.

To solder effectively, the tip of the iron must be kept *clean* at all times. Before starting to solder, the tip is heated and wiped across a damp sponge. Next, a small amount of solder is applied to the flat faces of the tip and wiped clean. This **tinning** process is used to coat the tip with a layer of solder and will help to maximize the transfer of heat to the connection. As you are soldering, the tip gets dirty quite rapidly because the wax, grease, dirt, and burned flux on the connections all get picked up by the tip. Therefore, you should quickly clean the tip by wiping it across a small wet sponge each time you pick up the iron. Most good-quality soldering pencils have *coated tips*, which have been plated with a protective

iron-clad coating that resists pitting. These tips should not be filed clean. A badly worn and pitted plated tip should be replaced (Figure 33-13).

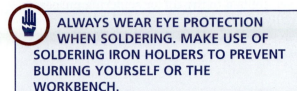

✋ **ALWAYS WEAR EYE PROTECTION WHEN SOLDERING. MAKE USE OF SOLDERING IRON HOLDERS TO PREVENT BURNING YOURSELF OR THE WORKBENCH.**

To properly solder a connection, the tip is first wiped across a damp sponge. Next, a small amount of solder is applied to one of the flat faces of the tip. This wetted tip surface is then placed simultaneously in contact with the terminal pad area of the PC board and one side of the lead to be soldered. The small amount of solder that has been applied to the tip aids in improving heat transfer from the tip to the metals being joined. Contact between the soldering iron tip, the lead, and pad connection is maintained for 2–3 seconds to bring the connection

A. Wipe with damp sponge

B. Never clean coated tips with a file

C. Make use of iron holder

Figure 33-13   Cleaning the tip of the iron.

up to proper temperature for soldering (Figure 33-14A). With the soldering iron in place, a small amount of solder is then applied to the opposite side of the lead and terminal pad (Figure 33-14B). Solder is always applied to the junction of the lead and terminal pad area. It is never applied to the tip for soldering purposes. Finally, the solder is removed and the iron is then withdrawn from the connection (Figure 33-14C). Do not disturb the joint until it cools.

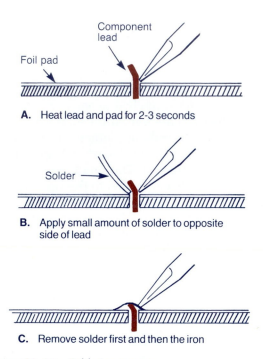

A. Heat lead and pad for 2-3 seconds

B. Apply small amount of solder to opposite side of lead

C. Remove solder first and then the iron

Figure 33-14 Soldering process.

The *soldering-melting temperature* is reached in a matter of seconds. The bond between the copper foil and the laminate will be destroyed if heat is applied for an *excessive period of time*. Overheating causes the foil to separate from the laminate, rendering the PC board useless.

The solder of a properly soldered connection will appear shiny and will flow evenly about the entire lead and pad. There should be no pin holes or copper showing. If a solder connection has a dull gray crystalline appearance, it is an indication of a ***cold-solder joint***, which exhibits a high resistance and could act as an open circuit. This is easily corrected with the reapplication of heat.

Another major problem created when soldering is bridging. The ***bridging*** problem is caused when solder joins, or bridges, two conducting paths that normally would be separated. To avoid bridging, heat from the soldering pencil should be applied away from other conductors whenever possible. If bridging does occur, the solder may be drawn away by heat in some cases. In other cases, the solder must be removed, the surface scraped clean between the conductors, and the joint resoldered.

Once the component has been properly soldered in place, excess lead lengths should be clipped. Fine diagonal pliers should be used to clip leads. Care must be taken to avoid nicking the foil pattern.

33-5 Servicing Printed Circuit Boards

At times it is necessary to desolder a component lead from the PC board. This could occur if you make a mistake in wiring or have to replace a defective part. When removing a component, the PC board is generally positioned so that the component leads that are to be unsoldered are facing downward. Components with leads must be removed by taking away as much of the solder as possible before wiggling the leads (one at a time) loose. It is often difficult to unsolder leads without damaging the component or terminal pad. Several tools and techniques are available to aid in the ***desoldering*** operation. One of the most commonly used methods is vacuum desoldering using a special ***vacuum desoldering bulb*** (Figure 33-15). First, apply heat to the solder with the iron while touching the tip of the bulb to the connection. The bulb should be squeezed before the solder melts, otherwise the air being forced out will splatter the molten solder. When the solder melts, squeeze the bulb and then release it. The

Desoldering bulb

Figure 33-15 Vacuum desoldering bulb.

bulb then removes the hot molten solder from the connection.

Another effective way to remove solder from a connection is to use **solder wick**. Solder wick consists of a braided mesh of fine copper wire strands. It is coated with a resin flux and absorbs solder. To use it, place some of it on the connection to be unsoldered and apply the soldering iron to the connection (Figure 33-16). When the solder melts, the heated flux cleans the braid thoroughly and the solder is drawn up into the wick by capillary action.

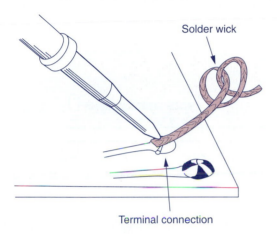

Figure 33-16 Desoldering using solder wick.

Straight-line, multiterminal connections can be removed by using a **bar-type desoldering tip** across all the connections at once, and the component will be removed in a single action (Figure 33-17). In all cases, be careful not to pull on the tips remaining in the board until the solder has melted completely through the board. It may be very easy to pull a trace right off the board if too much pull is applied.

In emergency situations or when access to the bottom of a printed circuit board is restricted, it is expedient to cut-and-crush a bad component and fasten the new component to the old leads (Figure 33-18).

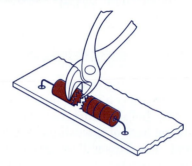

A. Cut in half

B. Crush away excess material

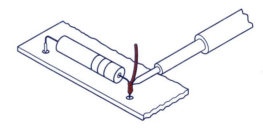

C. Solder new component to old leads

Figure 33-18 Replacing a component when the underside of the board is not accessible.

A *heat-sink tool* is used to help conduct the heat of soldering away from components and protect them from heat damage (Figure 33-19). Grip the component lead tightly with the heat sink tool and leave it connected until several seconds after the soldering operation is complete.

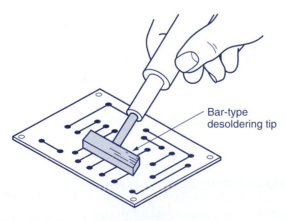

Figure 33-17 Bar-type desoldering tip.

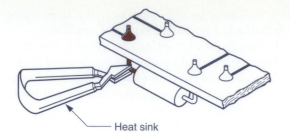

Figure 33-19 Heat-sink tool.

board circuit. If a PC board is suspected of having an open foil or trace (open circuit), an ohmmeter can also be used to check it (Figure 33-20). Lifted traces and open circuits may be repaired with a short piece of bare jumper wire soldered across the open foil to bridge the gap (Figure 33-21). Once the repair is complete a protective coating (varnish or approved equivalent) should be applied to the PC board.

When replacing power transistors and diodes, be sure to apply a small amount of **heat-sink grease** to help conduct away the normal operating heat of the component to the chassis or heat sink. This will greatly increase the life of the component, since it will not get as hot as it would without the compound.

A magnifying lens or light behind the PCB can be used to find cracks or breaks in a PC

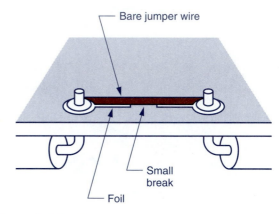

Figure 33-21 Repairing an open in a circuit board foil.

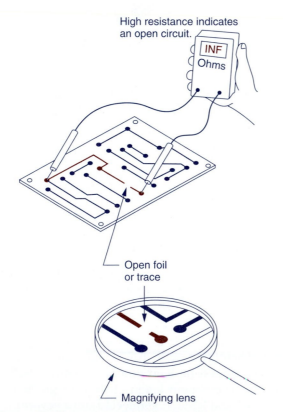

Figure 33-20 Locating a break in a PC board.

Most PC boards are covered with a nonconductive protective coating. For this reason, it is important when taking electrical measurements to use **needle-point** meter **probes** that penetrate the protective coating to give the correct reading. When taking measurements on live circuitry, be careful not to short out components by accidently bridging across component leads. Special **hook-type probes** are designed for use in close quarters in order to minimize slippage. *Circuit extenders* allow access to PC board components in excessively crowded situations (Figure 33-22).

It is best to handle printed circuit boards by holding them by their edges only between the thumb and forefinger. Touching the board elsewhere may cause contamination due to dirty fingers or damage due to static electricity from your body. Avoid flexing a printed circuit board as this may cause cracked paths.

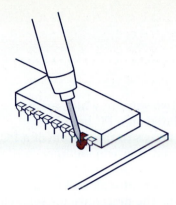

A. Use needle-point meter probes to penetrate nonconductive protective coating

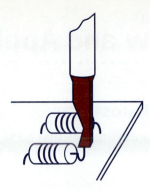

B. Use hood-type probes to minimize slippage

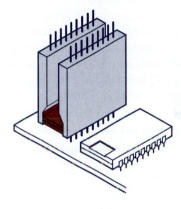

C. Use a circuit extender to access components

Figure 33-22 Taking electrical measurements on a PC board.

HELP WANTED

Tooling Engineer
Job Description
- Develop tooling required for the manufacturer of printed circuit boards.
- Knowledgeable in CAD/CAM operations.
- Technical degree required.
- Entry level, no experience required.

Review and Applications

Review Questions

1. List three advantages that a PC board has over the traditional hand-wired type of chassis.
2. (a) Describe the construction of a single-sided PC board.
   (b) On what side of the board are the components mounted?
   (c) How are the circuit connections made?
3. State the function of the *resist* and *etchant* as used in processing a PC board.
4. List the basic steps involved in the processing of a PC board using the photographic-printing method.
5. When is it a definite advantage to use the print-and-etch method rather than the photographic-printing process?
6. What must you be careful of when drilling lead-access holes in a PC board?
7. Why must the copper foil of a PC board be cleaned before the soldering is started?
8. Explain how axial-lead components are mounted to the PC board.
9. In what way is surface-mount technology different from conventional mounting methods?
10. (a) What is the purpose of the flux in the soldering process?
    (b) What type of flux should *not* be used for PC board soldering?
    (c) After soldering, why should flux residue be removed?
    (d) What type and diameter of solder is suitable for most PC board soldering?
11. What soldering iron power range is recommended for use with PC board soldering?
12. Why should a temperature-controlled soldering iron be used for PC board soldering?
13. When should the tip of the soldering iron be cleaned?
14. What damage can be done by leaving the soldering iron on the connection for too long a period of time?
15. Name two visual indications of a well-soldered connection.
16. Name two methods for removing solder from printed-circuit boards.
17. State two ways in which an open circuit can be isolated on a PC board.
18. When access to the bottom of a PC board is restricted, how can emergency repairs be made?
19. What is a heat-sink tool used for?
20. Describe the procedure for repairing a break in a PC foil.

Problems

1. After assembly of a printed circuit board you find the device does not operate properly. Upon examination of the soldered connections, you find that some appear to have a dull grainy texture. What, if anything, can be done to rectify this problem.
2. Will a resistor inserted backwards through the two holes provided for it in the PCB cause the circuit to work improperly? Why?

Critical Thinking

1. A newly assembled printed circuit board module blows a fuse when initially connected to its power source. An ohmmeter reading, taken across the two power input terminals of the module, indicates near zero resistance.
   (a) What circuit fault does this most likely indicate?
   (b) What soldering problem could create this type of fault?
2. Some printed circuit boards contain special test points connections that extended up from the surface of the board. When and how might you make use of these?

Portfolio Projects

• Obtain a junk-box printed circuit board and demonstrate your ability to replace components.
• Construct a working printed circuit board project from scratch or using a kit.

Electronic Test Instruments

When studying and troubleshooting electronic circuits, it is often necessary to use signal-testing and signal-generating test instruments. The oscilloscope is a signal-measuring instrument that provides a visual display of the signal. Oscilloscopes provide information regarding frequency, magnitude, and waveshape of a signal. Signal-generating test instruments are used to inject a signal of a specific shape at a specific frequency. This chapter gives a broad overview of the many different types of electronic test instruments available.

Objectives

After studying this chapter, you should be able to:
1. Operate an oscilloscope to observe and measure signals.
2. State the various functions of the basic front-panel controls of an oscilloscope.
3. Compare the output signals of audio, function, pulse, and radio frequency signal generators.
4. Define specifications related to lab power supplies.
5. Design and build a simple Wheatstone-bridge resistance measurement circuit.

Key Terms

- oscilloscope
- cathode-ray tube (CRT)
- horizontal deflection plates
- vertical deflection plates
- sawtooth waveform
- volts per division
- time per division
- intensity control
- focus control
- vertical position control
- horizontal position control
- trigger control
- dual-trace scope

(continued)

RESEARCH INTERNET SITES

- Tektronix
- Hewlett-Packard
- Leader Instruments

- digital-storage scope
- bandwidth
- rise time
- fall time
- pulse
- duty cycle
- phase angle

- sampling rate
- signal generator
- oscillator
- audio generator
- pulse generator
- RF generator

- amplitude modulation (AM)
- frequency modulation (FM)
- current limited supply
- bridge connection

34-1 The Oscilloscope

The *oscilloscope* is a versatile piece of test equipment primarily used to *measure and display voltages*. It can be used to measure ac and dc voltages like a voltmeter. In addition, it can provide information about the shape, time period, and *frequency* of voltage waveforms. All oscilloscopes function in accordance with the same set of fundamental rules. If you learn how one oscilloscope works and how it can be used, you can easily learn how to operate others.

The heart of an oscilloscope is the *cathode-ray tube* (CRT) which is a special type of electron tube (Figure 34-1). It is similar in shape to a TV picture tube and functions to convert a signal into an image. The CRT tubes consists of an electron gun for supplying a concentrated beam of electrons, vertical and horizontal deflection plates for changing the direction of the electron

beam, and a fluorescent screen that glows when struck by the electron beam. Electrons emitted by the heated cathode of the tube are focused and accelerated. They form a narrow beam of high velocity electrons. The direction this beam travels is then controlled, and it is allowed to strike a fluorescent screen. Light is emitted at the spot of impact with the screen and produces a visual indication of the beam position. A transparent ruled screen, called a graticule, is mounted in front of the fluorescent screen.

The main advantage of the oscilloscope is its ability to project a rapidly oscillating input signal on a screen to permit examining its waveform or shape. To accomplish this, a sawtooth voltage waveform, generated within the oscilloscope itself, is applied to the *horizontal-deflection plates* of the CRT (Figure 34-2). First, assume no voltage is applied to the vertical-deflection plates. As the horizontal sawtooth voltage increases linearly it moves the beam in

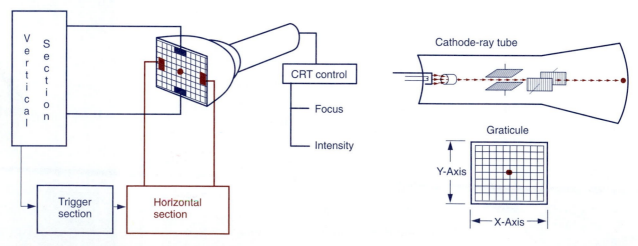

Figure 34-1 Cathode-ray tube and controls.

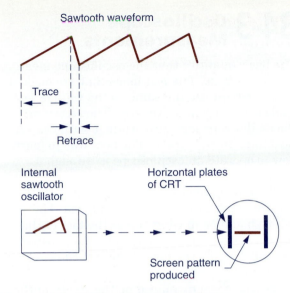

Sawtooth waveform

Trace

Retrace

Internal sawtooth oscillator

Horizontal plates of CRT

Screen pattern produced

Figure 34-2 Horizontal sawtooth voltage.

obtained on the screen, and so on (Figure 34-3B). In the same manner, other nonsinusoidal waves can be traced out on the screen. By adjustment of the sawtooth voltage frequency, it is possible to display more than one cycle, or just a small portion of one cycle.

Oscilloscopes are used primarily to display signals in the *time domain*. The sweep circuits in the oscilloscope deflect the electron beam in the CRT across the screen horizontally. This represents units of time. The input signal to be displayed is applied to deflect the beam vertically. Thus, electronic signals which are voltages occurring with respect to time are displayed on the oscilloscope screen. A slower sweep will allow you to see more of a signal on the screen, while a faster sweep will show you a smaller portion of a signal.

a straight line from left to right across the screen. At the end of the cycle, the beam is snapped back to the left of the screen by the retrace of the *sawtooth wave*. In this manner, a visible horizontal trace line is formed across the screen. The beam scans back and forth in response to the changing sawtooth voltage.

If, in addition to the sawtooth voltage applied to the horizontal plates, a sine wave of the same frequency is applied to the *vertical plates*, one complete cycle of the sine wave will be traced on the screen (Figure 34-3A). If the frequency of the applied signal is twice the sawtooth frequency, two complete cycles will be

34-2 Oscilloscope Front-Panel Controls

In every oscilloscope there are four main sections that provide the basic waveform display. These are:

- *Power supply:* provides power to the horizontal and vertical sections of the cathode-ray tube as well as to other operating circuits required for the proper display of the waveform.
- *Display section:* consists of the CRT display screen and the controls that affect the display.

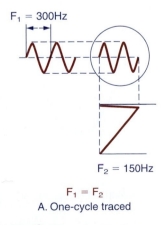

$F_1 = 300Hz$

$F_2 = 150Hz$

$F_1 = F_2$
A. One-cycle traced

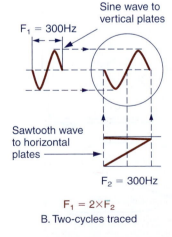

Sine wave to vertical plates

$F_1 = 300Hz$

Sawtooth wave to horizontal plates

$F_2 = 300Hz$

$F_1 = 2 \times F_2$
B. Two-cycles traced

Figure 34-3 Displaying a waveform.

- **Vertical section:** controls the amplitude of the applied waveform.
- **Horizontal section:** controls the sweep from left to right on the screen of the scope.

A number of control circuits are required in order for the oscilloscope to convert the voltage signal fed to it into a visual display on the CRT screen. The process involved is similar to that which takes place in a television set. Every oscilloscope has several front-panel controls used to adjust its circuits for proper operation (Figure 34-4). Names for these controls may be different for different makes and models. Extra controls are added to make the oscilloscope more flexible in its measurement of signals. These controls need not be adjusted for each measurement. In fact, most of them are set once and then left alone.

An understanding of the front-panel controls as shown in Table 34-1 is essential in order to use the oscilloscope properly.

34-3 Oscilloscope Measurements

The basic quantity that the oscilloscope measures is *voltage*. The test leads that are hooked up to a circuit are the same as the voltage test leads used with a multimeter. The screen on the oscilloscope measures voltage with the vertical lines on the screen. The best way to learn how to operate an oscilloscope is through a "hands-on" approach.

When using the oscilloscope, avoid operating it with a high-intensity spot displayed. The beam should be moving across the face of the screen. If it is not, you may burn the phosphorus and damage the face of the cathode-ray tube.

The *waveform* displayed on the screen of the oscilloscope represents *a plot of voltage versus time* (Figure 34-6). Therefore, it is possible to measure both voltage amplitude and the time period between voltage changes. The display screen on the oscilloscope is divided into grati-

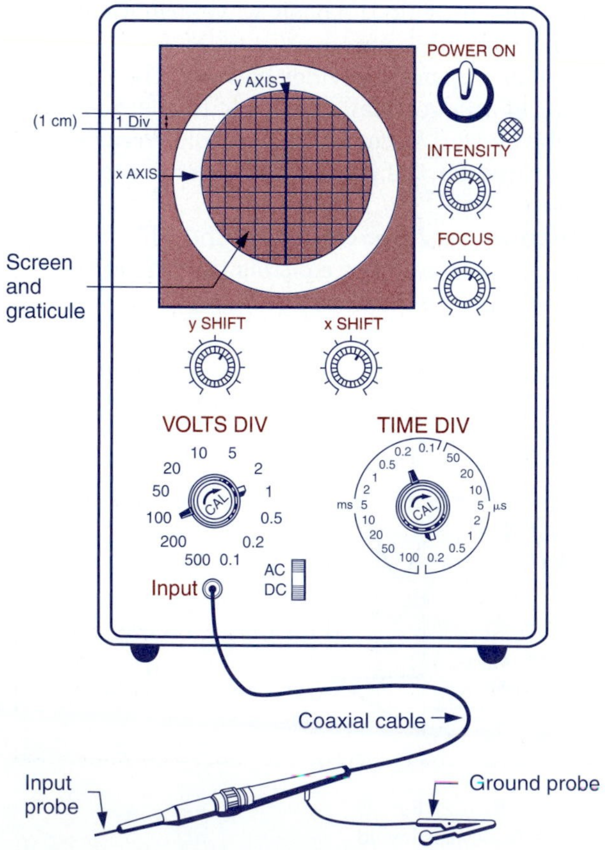

Figure 34-4 Standard front-panel controls.

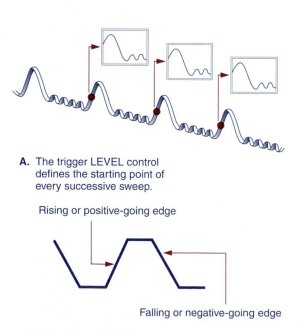

A. The trigger LEVEL control defines the starting point of every successive sweep.

Rising or positive-going edge

Falling or negative-going edge

B. The trigger SLOPE control permits the scope to trigger on the positive or negative half of the waveform.

Figure 34-5 Trigger control.

TABLE 34-1	Oscilloscope Controls and Their Functions
Control	**Function**
POWER ON	A switch for the instrument's main voltage supply ON/OFF control.
INTENSITY	A rotary control used to vary the brightness of the displayed waveform or trace.
FOCUS	A rotary control used to vary the sharpness or definition of the trace.
Y SHIFT (Vertical position)	A rotary control used to adjust the position of the displayed waveform vertically up or down.
X SHIFT (Horizontal position)	A rotary control used to adjust the position of the displayed waveform horizontally to the left or right.
VOLTS/DIV	A rotary switch used to select the height or amplitude of the display. It works in a way similar to the range switch of a voltmeter. The switch is calibrated in volts per division of the vertical axis. These calibration markings permit you to determine the magnitude of the voltage display. The small control knob at the center of the VOLTS/DIV knob adjusts the amplitude between the calibrated settings. When the VOLTS/DIV selection is to be used for measuring the amplitude of a waveform, the small center knob must be turned to its extreme clockwise (calibrate) position.
TIME/DIV	A rotary switch used to select the width of the display. The calibration markings on this control allow you to measure the elapsed time between any two horizontal points on the display. This time period measurement is used in calculating the frequency of the waveform. The small control knob at the center of the TIME/DIV knob adjusts the width of the display between the calibrated time settings. The calibrated time settings only apply if the center control knob is turned to the extreme clockwise (calibrate) position.
INPUT	The input signal to be displayed on the screen is fed into this jack. Normally a coaxial cable equipped with a connector and probes is used to make the electrical connection from the instrument to the circuit. The use of coaxial cable helps prevent unwanted, stray signals being picked up by the oscilloscope input.
AC/DC	A switch that permits the input signal to be directly connected (dc position) or connected via a capacitor (ac position). In the dc position, both the ac and dc components of the waveform would be displayed. In the ac position, the capacitor blocks the dc, so that only the ac component of the waveform would be displayed.
TRIGGER CONTROL (Figure 34-5)	Controls used to provide a stable (no-jitter) CRT display. Triggering defines the starting point of every successive sweep of the displayed trace. The four trigger controls are:
AUTO/NORMAL: SWITCH	Determines whether the time base will be triggered automatically or if it is to be operated in a free-running mode. If it is operated in the normal setting, the trigger signal is taken from the line to which the probe is connected. The scope is generally operated with the trigger set in the automatic position.
TRIGGER LEVEL: KNOB	Adjusts the instant in time at which the waveform display commences. The display may be made to commence exactly when the signal goes through its zero position or at some time shortly before or after this instant.
SLOPE CONTROL:	Permits the scope to trigger on the positive or negative half of the waveform.
INT-LINE-EXT: SWITCH	Selects the source of trigger signal to be used. The scope is generally operated in the internal (INT) mode in which the trigger signal is provided by the scope. In the line mode, the trigger signal is provided from a sample of the line. The external (EXT) mode permits an external trigger signal to be applied.

cules and divisions to provide for measurement in the vertical and horizontal axes. The general procedure to be followed is outlined below. It must be emphasized that, in making these measurements accurately, the variable controls of both the **VOLTS/DIV** and **TIME/DIV** must be set at their fully clockwise (calibrate) position.

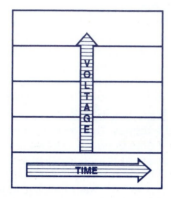

Figure 34-6 Oscilloscope measures voltage versus time.

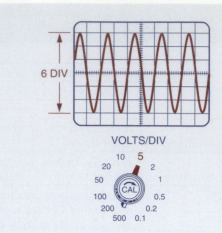

Figure 34-7 AC voltage measurement.

$$\therefore \text{ peak-to-peak voltage}$$
$$= 6 \times (5 \text{ V})$$
$$= 30 \text{ V}$$

$$\therefore \text{ rms voltage}$$
$$= \text{Peak V} \times 0.707$$
$$= (15 \text{ V}) \times 0.707$$
$$= 10.6 \text{ V}$$

Procedure

Measurements of AC Sine-Wave Voltages

1. Connect the ac waveform that is to be measured to the input.
2. Set the ac/dc switch to ac.
3. Set the VOLTS/DIV switch to display about 5 or 6 divisions of the waveform.
4. Set the TIME/DIV switch to display several cycles of the waveform.
5. Use the Y SHIFT control to set the lower edge of the waveform on one of the lower graticule lines so that the top edge of the waveform is in the graticule area.
6. Measure the vertical amplitude (divisions peak-to-peak) of the signal on the screen.
7. Multiply the amplitude in Step 6 by the VOLTS/DIV setting to find the peak-to-peak voltage value.
8. Multiply the peak value by 0.707 to find the rms voltage value.

   Using the information shown in Figure 34-7 assume a vertical deflection of 6 divisions and a VOLTS/DIV setting of 5 V.

Procedure

Measurement of DC Voltages

1. Set the ac/dc switch to dc.
2. Connect the input probe to the ground of the unit.
3. Adjust the Y SHIFT and TIME/DIV controls for a horizontal line in the center of the screen (this represents zero dc volts).
4. Remove the input probe from its ground connection and connect the dc signal to be measured to the input.
5. The horizontal line will move up or down depending on the polarity of the dc signal.
6. Set the VOLTS/DIV switch to display 3 or 4 divisions of the waveform.
7. Measure the vertical distance in divisions from the zero reference line.
8. Multiply the vertical distance (div) by the VOLTS/DIV setting to find the value of the dc voltage.

Using the information shown in Figure 34-8, assume a vertical deflection of 2.6 divisions and a VOLTS/DIV setting of 20 V.

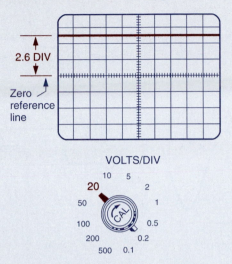

Figure 34-8   DC voltage measurement.

$$\therefore \text{dc voltage}$$
$$= 2.6 \times (20\ \text{V})$$
$$= 52\ \text{V}$$

Procedure

Measurement of Cycle Time Period and Frequency

1. Connect the waveform to be measured to the input.
2. Set the VOLTS/DIV switch to display a suitable vertical amplitude of the waveform.
3. Set the TIME/DIV switch to display approximately two cycles of the waveform to be measured.
4. Use the Y SHIFT control to move the trace so that the measurement points are on the horizontal center line.
5. Use the X SHIFT control to move the start of the measurement period to a convenient reference point.
6. Measure the distance (divisions) between the two points of the cycle.
7. Multiply the measured distance in Step 6 by the setting of the TIME/DIV switch to

find the value of the time period for one cycle.

8. To determine the frequency of the waveform, measure the time period for one complete cycle and apply the formula:

$$\text{Frequency} = \frac{1}{\text{Time period}}$$

where Frequency is expressed in hertz (Hz); Time period is expressed in seconds (s).

Using the information shown in Figure 34-9, assume one cycle occupies 5 divisions with the TIME/DIV setting of 0.2 ms. Then:

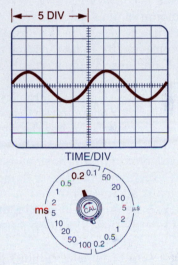

Figure 34-9   Period and frequency measurement.

$$\text{Time period} = 5 \times (0.2\ \text{ms})$$
$$= 1.0\ \text{ms}$$

$$\text{Frequency} = \frac{1}{\text{Time period}}$$
$$= \frac{1}{(1.0\ \text{ms}) \times 10^{-3}}$$
$$= 1\ \text{kHz}$$

Oscilloscopes allow you to actually "see" what you are measuring and their waveshapes can tell you a great deal about the signal. Anytime you see a change in the vertical dimension of a signal, you know that this amplitude change represents a change in voltage. If there's a flat horizontal line, there is no change in voltage for that length of time. Straight diagonal lines

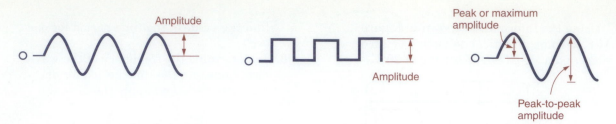

Figure 34-10 Amplitude of a waveform.

indicate a linear change, equal rise (or fall) of voltage for equal amounts of time. Sharp angles on a waveform mean a sudden change in voltage.

Amplitude is characteristic of all waveforms. It is the amount of displacement from a reference point without regard to the direction of the change (Figure 34-10). In oscilloscope measurement, amplitude usually means peak-to-peak amplitude.

Period is the time required for one cycle of a signal if the signal repeats itself (Figure 34-11). *Period* is a parameter whether the signal is symmetrically shaped (like a sine wave or square wave), or whether it has a more complex shape (like a rectangular wave or damped sine wave). Period is always expressed in units of time. One-time signals like a step or signals without a time relation, like noise, have no time period.

The parameters of a pulse and duty cycle are particularly important in digital circuitry. *Pulse* specifications (Figure 34-12) include: *rise time* (the transition time between 10% and 90% measured on the leading edge), *fall time* (the transition time between 90% and 10% on the trailing edge), *pulse width* (measured between the 50% points of the amplitude). The *duty cycle* (Figure 34-13) is the ratio of the pulse width to signal period expressed as a percentage. For square waves the duty cycle is always 50%. Duty factor is the same as duty cycle except it is expressed as a decimal, not a percentage. The pulse repetition rate is the same as frequency.

The *phase angle* of waveforms is used when describing the time relationship between two signals. A good analogy of this is to picture two clocks with their second hands sweeping the dial every 60 seconds. If the second hands both touch twelve at the same time, the clocks are *in phase;* if they don't, they are out of phase. Recall that a sine wave is based on the sine of all angles from 0° through 360°. The result is a

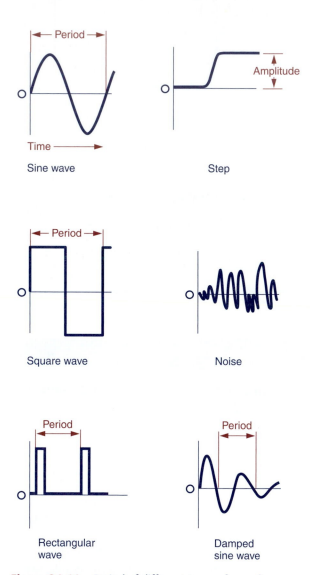

Figure 34-11 Period of different types of waveforms.

plot that changes from 0 at 0°, 1 at 90°, 0 again at 180°, minus 1 at 270°, and finally 0 again at 360° (Figure 34-14a). Consequently, it is useful to refer to the phase angle of a sine wave when you want to describe how much of the period has elapsed. Therefore you can use phase shift

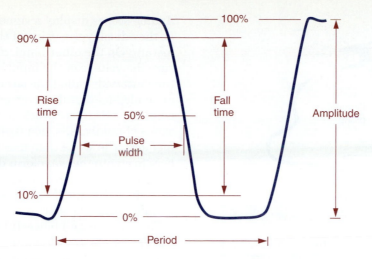

Figure 34-12 Parameters of a pulse.

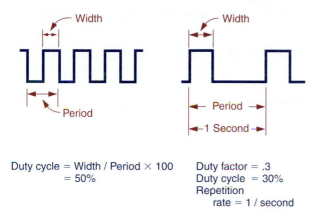

Duty cycle = Width / Period × 100
= 50%

Duty factor = .3
Duty cycle = 30%
Repetition
    rate = 1 / second

Figure 34-13 Duty cycle.

in degrees to express how far out of phase two waveforms are. For example, in a pure inductor, the current waveform lags the voltage waveform by 90° (Figure 34-14b).

Oscilloscopes are normally powered from a standard 120-V grounded receptacle. The test probe of the scope contains two leads. One is connected to the input circuit of the scope and the other *directly to ground*. When the scope is used to test a circuit that has one side grounded and the other side ungrounded, *care must be taken to ensure that the grounded conductor of the probe is connected to the grounded side of the circuit (Figure 34-15).* If the grounded conductor of the probe should be connected to the ungrounded side of the circuit, a direct *short* to ground through the probe lead and the case of the scope will be created. Should the ground become loose, all accessible conductive parts (including knobs that appear to be insulated) can give you a shock.

The **dual-trace oscilloscope** can display two different traces *at the same time*. It has two sets

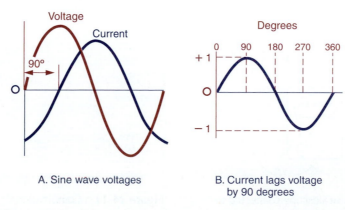

A. Sine wave voltages

B. Current lags voltage by 90 degrees

Figure 34-14 Phase angle.

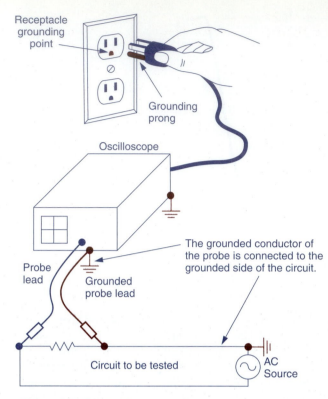

Figure 34-15 Oscilloscope probe connection.

Labels in figure: Receptacle grounding point / Grounding prong / Oscilloscope / Probe lead / Grounded probe lead / The grounded conductor of the probe is connected to the grounded side of the circuit. / Circuit to be tested / AC Source

of controls, one for each trace. The dual-trace oscilloscope is particularly useful in *comparing* two waveforms to show amplitude differences, distortion, or time (phase) differences. Figure 34-16 illustrates a dual-trace scope connected to compare waveforms in a digital logic circuit, where, along with signal levels, timing is of critical importance. In this example, the "external" triggering input is used as an extra input to the scope.

Generally speaking there are two types of oscilloscopes on the market today: analog and digital.

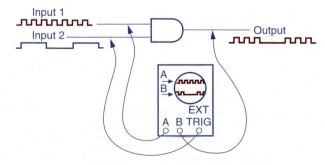

Figure 34-16 Dual-trace oscilloscope connected to compare waveforms.

Labels in figure: Input 1 / Input 2 / Output / A / B / EXT / A B TRIG

Analog scopes display a signal continuously. That is, they display the value of the signal at every instant. On the other hand, *digital* scopes measure the voltage of the input signal at discrete time intervals called the **sampling rate**. Using an analog-to-digital converter, the digital storage oscillscope (DSO) converts a waveform into a series of numbers, which it stores in a table in its memory. The scope then uses that table of numbers to create the waveform display. If the sampling rate is not high enough, the scope cannot reproduce the signal accurately and may display an inaccurate waveform.

The modern and powerful **digital-storage oscilloscopes (DSOs)** allow you to capture, store, and analyze a variety of repetitive or single-event signals (Figure 34-17). The benefits of DSOs over analog oscilloscopes are numerous and can include the following features:

Save/Recall   Overall digital and analog scope architectures are quite similar. One major difference is that analog scopes have an analog signal path to the display and digital scopes have a digital storage path (Figure 34-18). The digitized waveform of a digital scope is stored in digital memory, from which it can be recalled any time for processing, display, and measurement. This feature allows you to compare a waveform to a known good waveform for troubleshooting purposes.

AUTOSET

Puts a meaningful waveform on display at the touch of a single button. This feature auto-

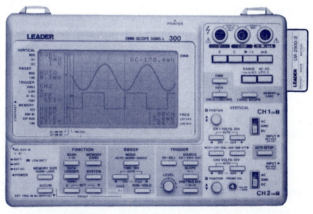

(Courtesy Leader Electronics Corp.)

Figure 34-17 Combination digital storage oscilloscope and digital multimeter.

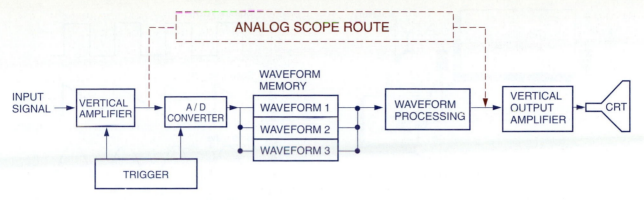

Figure 34-18 Digital and analog scope architectures.

matically sets voltage, time, triggering, and position.

STEP SEQUENCE
This allows different setups to be executed in a fixed sequence by pressing one button.

CURSORS
Direct measurement and readout of basic waveform parameters can be made without counting display divisions and scale-factor multiplication.

COUNTER/TIMER
This feature adds frequency, period, and event counting to the scope.

DMM
This feature adds digital multimeter measurements to the oscilloscope.

Waveform Analysis   Because DSOs contain a processor, and the waveform information is in digital form, they can analyze as well as measure a signal.

Glitch Triggering   Glitches are short pulses that occur as a result of noise or timing problems in a logic circuit. You can program a DSO to trigger on such short pulses. Because the scope captures signals both before and after a trigger, you can see events that may have caused the glitch as well as what happened to a circuit after the glitch.

Bandwidth, rise time, and sensitivity are the fundamental specifications for an oscilloscope; they determine what can be displayed on the screen. *Bandwidth* is the frequency range a scope can display. *Rise time* determines the ability of the instrument to deal with short events, such as rise times of glitches. *Sensitivity* is the amount

of signal amplitude (volts/division) the scope can display.

Connecting a signal to a scope is not simply a matter of running a wire from the circuit under test to the scope input. The right probe must be used to ensure correct reproduction of the signal. There are different probes designed for different applications and these include: high impedance probes, differential probes, current probes, and active probes. Attenuating, high-impedance probes are the most common, general-purpose probes. They are mechanically rugged and good for measurements up to 350 MHz.

34-4 Signal Generators

There are times when it is necessary to apply a standard signal to an electronic circuit for purposes of testing, troubleshooting, or calibrating. A **signal generator** is an instrument that serves as a source for such a signal. The heart of the signal generator is an *oscillator* of variable frequency and output. Signal generators may be classified according to their output-frequency range as being audio frequency (AF) generators or radio frequency (RF) generators.

Audio Frequency (AF) Generators   *Audio frequency generators* are capable of producing signals in the audio frequency range of about 20–20,000 Hz, and up to a maximum of about 100 kHz. Audio signal generators produce stable audio frequency signals for testing audio equipment. They are used to simulate the type of electrical signal that can normally be seen at the input or output of an audio amplifier. A simple audio signal generator application is illustrated in Figure 34-19. The AF generator is

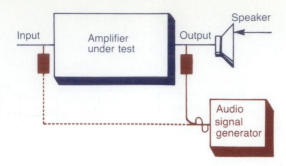

Figure 34-19 Audio signal generator used to test an amplifier.

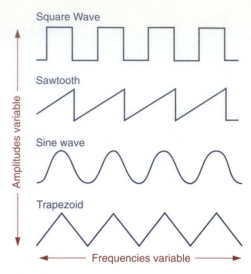

Figure 34-21 Waveforms available from a function generator.

connected first on the output and then on the input of the amplifier. An increase in the sound level is one indication that the amplifier is operating to increase the strength of the signal applied to it.

Audio frequency generators can produce either a sine-wave or a square-wave signal. The signal wave frequency is variable within several frequency ranges. A frequency range switch is used to select the desired frequency range. The frequency dial is used to select the frequency within the desired range. An amplitude control is used to adjust the voltage level of the output wave (Figure 34-20).

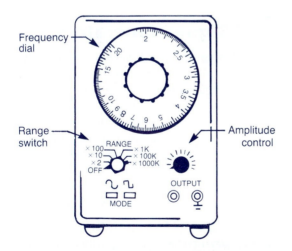

Figure 34-20 Typical audio signal generator controls.

Function Generators   *Function generators* are signal generators that are used to supply a wide variety of signals. These may include square wave, sawtooth, sine wave, trapezoid, ramp and pulse waveforms (Figure 34-21). Function generators are also capable of

supplying dc offset, which adds a positive or negative dc component to any of the waveforms produced by the generator. When signal tracing with a function generator, the frequency, waveshape, and amplitude of the generator are set to the actual values with which the circuit normally operates. This allows the signal to be easily identified at various points in the circuit. Consequently, any distortion or abnormal signal magnitudes can be detected. In audio circuits, the function generator can be used to replace the normal signal input, (e.g. a microphone) or at some intermediate point in the circuit.

Pulse Generators   A *pulse generator* is an oscillator that produces a voltage of short duration on a regular basis. Depending upon the individual equipment, pulse generators may be adjustable for setting the *pulse* amplitude, duration, and limit how often the pulses occur (Figure 34-22). The frequency of recurrence is called the *pulse-repetition rate*. Frequency is the equivalent term used mostly for sine waves.

Radio Frequency (RF) Generators   A *radio frequency (RF) signal generator* is used to simulate the type of high-frequency radio signal that is radiated from radio transmitters. It can develop frequencies from 100 kHz up to several hundred MHz. The most common application of an RF signal generator is its use in tuning, aligning, and troubleshooting radio and television receivers. In the typical RF signal

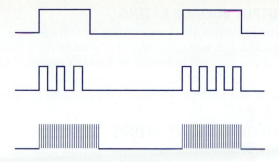

Figure 34-22 Typical pulse-generator waveforms.

generator, the audio signal from a 400-Hz oscillator circuit in the generator can be used to modulate the output RF signal. *Modulation* is the varying of the amplitude or the frequency of a carrier wave. In this case, the carrier wave is the sine-wave output voltage. When the amplitude is varied, it is called *amplitude modulation (AM)*. When the frequency is varied, it is called *frequency modulation (FM)*.

The design and output waveforms of an RF signal generator are substantially different from those of an RF signal generator. Radio frequency currents will transmit or radiate energy, while audio frequency currents will not. Some RF signal generators work like a miniature AM radio transmitter. These generators are capable of providing a high-frequency carrier wave output, a low-frequency audio sine-wave output, and an amplitude modulated (AM) radio signal output (Figure 34-23). The frequency range of the carrier wave output is matched to that of the radio receiver. For an AM radio, this would be in the 540-kHz to 1.6-MHz range.

The RF signal generator can be used to demonstrate the principles of AM radio transmission. Several loops of insulated wire, wound into a coil, and connected to the generator output will act as a transmitting antenna. If this coil is placed near an AM radio, the signal radiated by the coil will be picked up by the radio. By tuning the radio and signal generator to the same RF frequency, the audio signal being produced by the signal generator can be heard on the radio (Figure 34-24).

A. High-frequency carrier wave output
Variable within the broadcast band

B. Low-frequency audio output
Fixed at about 400 Hz

C. Amplitude modulated (AM) output

Figure 34-23 Radio frequency generator output waveforms.

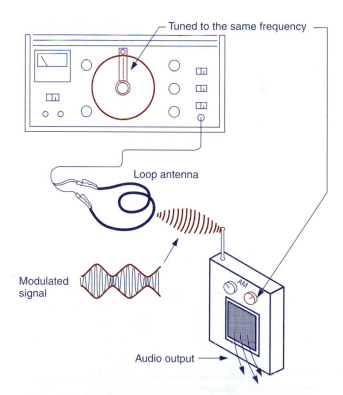

Figure 34-24 Transmitting a signal using an RF signal generator.

34-5 Frequency Counters

In addition to oscilloscopes and digital multimeters, frequency measurements can also be made with frequency counters. A digital frequency meter measures the unknown frequency by counting the number of cycles the frequency produces in a precisely controlled period of time. Digital electronic counters are used to measure the frequency of an unknown signal, the period of any signal, and the frequency ratio of two different signals. Frequency counters will typically provide a 5-digit readout of the frequency being measured, with a range of between 2 Hz and 300 kHz. These counters are often equipped with plug-in accessories that can extend this frequency range to beyond 12.5 GHz.

34-6 DC Voltage Supplies

The purpose of bench type dc voltage supplies is to convert the 120-V ac from an electric outlet to a dc voltage level required for the proper operation or testing of circuits. Electronic devices usually come equipped with their own built-in dc voltage supply. Typical voltage supply circuits are covered in a separate unit of this text.

Many different types of dc voltage supplies are used in electronic and electrical labs (Figure 34-25). Some of the more important terms that apply to the rating and features of dc voltage supplies are described below.

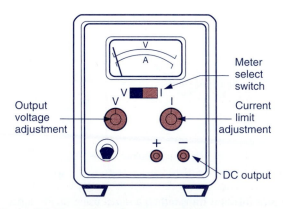

Output voltage adjustment

Meter select switch

Current limit adjustment

DC output

Figure 34-25 Typical dc voltage supply.

OUTPUT VOLTAGE RATING
This indicates the value of the dc voltage output. Output voltages may be variable from zero to some maximum value, or fixed at one or more specific value. A dual-output voltage supply is one that provides two separate dc output voltages.

OUTPUT CURRENT RATING
This indicates the maximum current that the voltage supply is capable of delivering at its rated voltage.

OUTPUT RIPPLE VOLTAGE RATING
This indicates the amount of ac voltage ripple present in the dc output wave.

UNREGULATED OR REGULATED
A *regulated voltage supply* is one that is equipped with a special circuit to keep the output voltage constant over wide variations in load current.

METERED OUTPUT
A single built-in meter is used to measure either output voltage or current. The voltage or current function is usually selected by a meter selection switch located on the front panel.

CURRENT LIMITED
A *current limited voltage supply* is one that is equipped with a special circuit which limits the output current to some preset or fixed value. It protects the voltage supply for all overloads including a direct short circuit placed across the output terminals.

34-7 Bridge Measuring Circuits

A *bridge circuit* consists of four sections connected to form two parallel circuits. The balanced-bridge method of making measurements is extremely accurate and very useful. It is often used in control circuits and test equipment for accurate measurement of resistance, capacitance, or inductance.

The resistance and voltage characteristics of a balanced-bridge circuit are as follows:

$$\frac{R_1}{R_2} = \frac{R_3}{R_4}$$
$$V_1 = V_3$$
$$V_2 = V_4$$

Wheatstone Bridge Circuit

The Wheatstone bridge circuit shown in Figure 34-26 can be used to measure resistance. It requires fixed, known resistance values for R_1 and R_2 and a variable resistor for R_3. Variable resistor R_3 is attached to a calibrated dial that shows its adjusted resistance value. R_4 represents the resistor whose resistance value is to be measured. To determine the unknown resistance of R_4, the variable resistance R_3 is adjusted until the galvanometer indicates zero current. Consequently, the galvanometer is usually termed a **null detector**. The zero galvanometer reading indicates that the bridge is balanced. When the galvanometer indicates a null condition, the voltage on each side of the galvanometer is equal. If R_1 and R_2 are the same value, then the calibrated dial reading of R_3 indicates the resistance value of R_4.

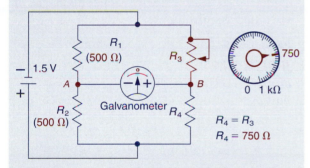

Figure 34-26   Balanced Wheatstone bridge circuit.

Assume for a balanced bridge (Figure 34-27) the following resistance values:

$$R_1 = 400 \ \Omega$$
$$R_2 = 800 \ \Omega$$
$$R_3 = 250 \ \Omega$$

Then the value of unknown resistor R_4 is:

$$R_4 = \frac{R_2}{R_1} \times R_3$$

$$R_4 = \frac{(800 \ \Omega)(250 \ \Omega)}{400 \ \Omega}$$

$$R_4 = 500 \ \Omega$$

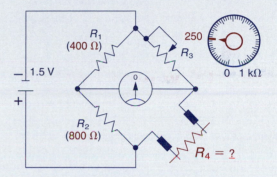

Figure 34-27   Resistance measurement with the Wheatstone bridge.

Therefore, if the values of R_1 and R_2 are different, the unknown resistance value for a balanced bridge can be calculated. The equation used is:

$$R_4 = \frac{R_2}{R_1} \times R_3$$

With a bridge circuit, more precise control can be obtained compared to a straight series or parallel connection. For example, a photoconductive cell light sensor is placed in one arm of the bridge, as shown in Figure 34-28, so that it can detect changes in light. The bridge current is then balanced by adjustment of potentiometer R_3. Any changes in light will change the

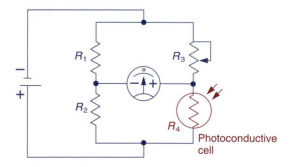

Figure 34-28   Light-sensor bridge circuit.

resistance of the photoconductive cell, and the bridge becomes unbalanced. With the bridge unbalanced, current flows through the galvanometer, which acts as a very sensitive light meter.

Figure 34-29 shows a temperature-sensor bridge circuit that contains two thermistor sensors. Here, two matched thermistors are used to form the bridge measuring circuit. In normal operation, one thermistor is placed where its temperature is kept at a fixed value, and the other thermistor is used as a monitor. The bridge is then balanced by the adjustment of R_3. As long as the two thermistors have equal temperature, their resistance values will be the same and the bridge circuit will remain balanced. However, should the monitoring thermistor detect an increase or decrease in temperature, its resistance will change and the bridge becomes unbalanced. With the bridge unbalanced, current flows through the galvanometer, which acts as a very sensitive temperature meter.

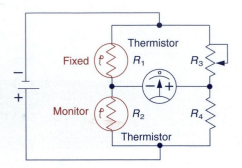

Figure 34-29 Heat-sensor bridge circuit.

Instrument Technical Support Specialist

Job Description

- Working knowledge of electronic test instruments such as oscilloscopes and signal generators.
- Ability to work independently with no immediate supervisor.
- Capable of performing hardware maintenance and repair.
- Associate degree or equivalent in a related field.
- Required to work a flexible schedule based on customer requirements.

Review and Applications

Related Formulas

RMS value = Peak value × 0.707

Peak-to-peak value = Peak value × 2

$$\text{Frequency} = \frac{1}{\text{Time period}}$$

(Balanced Bridge) $R_4 = \dfrac{R_2}{R_1} \times R_3$

$$\text{Percent duty cycle} = \frac{\text{Pulse width}}{\text{Pulse period}} \times 100$$

Review Questions

1. State three basic uses for an oscilloscope.
2. Explain the basic operation of a cathode-ray tube (CRT).
3. What shape of voltage waveform is applied to the horizontal deflection plates of the CRT?
4. Name the four main sections of an oscilloscope.
5. State the general function of each of the following oscilloscope front-panel controls:
   (a) INTENSITY
   (b) FOCUS
   (c) Y SHIFT
   (d) X SHIFT
   (e) VOLTS/DIV
   (f) TIME/DIV
   (g) AC/DC
   (h) TRIGGER CONTROLS
6. What precaution must be observed when connecting the test probe of an oscilloscope to test a circuit that has one side grounded?
7. For what applications are dual-trace oscilloscopes particularly useful?
8. Describe the main feature of a digital storage oscilloscope.
9. Explain each of the following oscilloscope specifications:
   (a) bandwidth
   (b) rise time
   (c) sensitivity
10. (a) State the purpose of a signal generator.
    (b) What type of circuit is the "heart" of all signal generators?
11. Compare the outputs of:
    (a) audio
    (b) function
    (c) pulse
    (d) radio frequency signal generators.
12. Explain what each of the following dc power supply specifications indicate:
    (a) output voltage
    (b) output current
    (c) ripple voltage
    (d) regulated supply
    (e) current limited supply
13. Answer each of the following with reference to a Wheatstone bridge circuit:
    (a) How many separate resistance sections are required for this circuit?
    (b) When a balanced-bridge null condition is indicated by the galvanometer, what are the resistance and voltage relationships of the circuit?

Problems

1. A sine-wave voltage is measured using an oscilloscope and found to be 24 V peak-to-peak. What is the rms value of this voltage?
2. The time period of a signal measured using an oscilloscope is found to be 14 ms for one complete cycle. What is the signal frequency?
3. An oscilloscope displays a pulse width of 2 ms and a period of 16 ms. What is the duty cycle?
4. A dual trace oscilloscope is used to compare the phase angle of two sine wave signals. If signal A is at maximum at the same time that signal B is at zero, what is the phase angle between the two waves?

Critical Thinking

1. The galvanometer in a balanced resistance bridge circuit reads zero when the applied voltage is 3 volts. What is the most likely change that will occur in the current flow if the applied voltage is increased to 6 volts? Why?

Portfolio Project

• Design and build a Wheatstone-bridge resistance measurement circuit. Use resistors from a discarded or obsolete electrical device. Make a calibrated dial for your variable-resistor that will read the value of the unknown resistor. Use a multimeter for the galvanometer.

Circuit Challenge

Using a Simulator

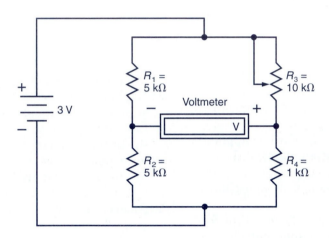

Procedure

(a) Construct the resistance bridge circuit shown, using whatever simulation software package is available to you.

(b) What setting of variable resistor R_3 will balance the bridge?

(c) With the polarity of the voltmeter as shown, if you read a positive voltage then which of the following is true:

$R_3 / R_4 = 1$

$R_3 / R_4 =$ greater than 1

$R_3 / R_4 =$ less than 1

(d) If R_2 is changed to 10 kΩ, what setting of variable resistor R_3 will balance the bridge?

Semiconductor Diodes

This chapter introduces the simplest semiconductor device, the diode. The diode is one of the basic building blocks of semiconductor devices. Transistors, integrated circuits (ICs), and microprocessors are all based on its theory and technology. Diode applications covered include the rectifier, clamper, voltage regulator, clipper, and light-emitting diode (LED).

Objectives

After studying this chapter, you should be able to:
1. Discuss the characteristics of semiconductor crystals.
2. Explain the semiconductor doping process.
3. Predict the conductivity of junction diodes under the conditions of forward and reverse bias.
4. Use a multimeter to test a diode.
5. Describe diode operating characteristics.
6. List several diode types and applications.

Key Terms

- semiconductor
- valence shell
- N-type material
- P-type material
- doping
- dopant
- carriers
- pentavalent
- holes
- trivalent
- PN-junction

- diode
- diffusion
- potential barrier
- depletion region
- forward bias
- reverse bias
- forward breakover voltage
- reverse breakdown voltage
- anode
- cathode

- thermal runaway
- rectifier diode
- circuit-control diode
- clamping diode
- varactor diode
- zener diode
- light-emitting diode (LED)
- laser diode
- photo diode
- fiber-optic cable

RESEARCH INTERNET SITES

- **Ledtronics**
- **Opto Power**
- **Electronix Express**

35-1 Semiconductors

Most electronic components are made from special crystals called *semiconductors*. The four most commonly used semiconductor devices are *diodes*, **transistors**, **thyristors**, and **integrated circuits**. The term *semiconductor* is used to describe any material having characteristics that fall between those of insulators and conductors. Depending on surrounding temperature conditions, a pure semiconductor can act like a conductor, or like an insulator.

There are many different semiconducting materials. The two most common are silicon (Si) and germanium (Ge). Of these two, *silicon* is by far the most commonly used. Silicon is cheap and plentiful. It is the main ingredient of sand. Silicon can be "grown" into large crystals, which are cut into wafers for making electronic components.

Pure semiconductors may be classified according to their atomic structure. Basically, they are all solids that have four electrons in their outer **valence shell** (Figure 35-1). Although each may have a different total number of electrons, they all have *four* electrons in their outer shell.

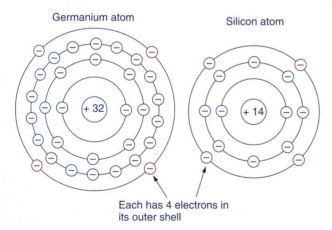

Figure 35-1 Atomic structure of semiconductors.

When atoms of semiconductors combine to form a solid, they arrange themselves in an orderly pattern called a crystal or lattice structure (Figure 35-2). If a pure silicon crystal is examined we find that the four electrons in the outer shell of each atom tend to share outer-shell electrons. This sharing is called a *covalent*

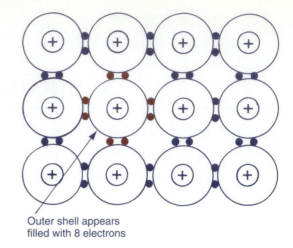

Outer shell appears
filled with 8 electrons

Figure 35-2 Lattice structure of silicon atoms.

bond. Each atom appears to have a full outer shell of eight electrons. The electrons have less tendency to drift or be pulled away from their orbits. *Thus, pure semiconductors, at room temperature, are poor conductors of electricity.*

35-2 N-Type and P-Type Material

N-Type Material In order to make the pure semiconductor crystal a better conductor, a small amount of some other element is added to it. The process of adding impurities to the crystal to change its structure is called *doping*. The impurity added is called the **dopant**.

To create **N-type material**, a crystal is doped with atoms from an element that has *more* electrons in its outer shell than the natural crystal. For example, an arsenic (As) atom with five valence electrons is added to pure silicon. Four of the valence electrons of the arsenic atoms join in covalent bonds with the valence electrons of the silicon atoms. Since the outer shell of the silicon is now complete, the extra electron from the arsenic atom is free to move about the crystal. This electron is called a *free electron*. The added impurity creates many of these free electrons which help move current and are called **carriers**. A crystal doped in this manner is an *N-type material* (Figure 35-3).

Some elements commonly used for creating N-type material are arsenic, bismuth, and antimony. The quantity of doping material used generally ranges from a few parts per billion to

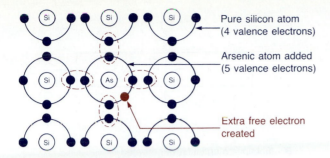

Figure 35-3   N-type material formation.

exists in the valence shell of the gallium atom. This valence shell can easily accept an electron into this hole (Figure 35-5). The valence shell is incomplete and acts like a positive ion although the atoms are neutral. It is because of this that the material is called a *P-type material*.

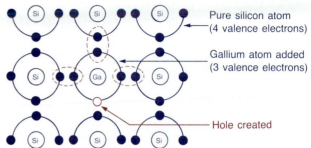

Figure 35-5   P-type material formation.

a few parts per million. By controlling even these small quantities of impurities in a crystal, the manufacturer can control the operating characteristics of the semiconductor.

The dopant atom used to form an N-type material is said to be **pentavalent**, having five valence electrons. It is called a *donor* impurity because the free electron in the dopant atom can easily be donated to the crystal. When an N-type material is connected into an electric circuit, it will conduct electricity by means of these free electrons much like a metal conductor (Figure 35-4). The more impurity atoms added, the better it will conduct.

Typical elements used for doping a crystal to create P-type material are gallium, boron, and indium. In P-type material, the *holes act as carriers*. When voltage is applied, the holes are filled with free electrons, since the free electrons move from negative potential to positive potential through the crystal (Figure 35-6). Movement of the electrons from one hole to the next makes the holes appear to move in the opposite direction. Hole flow is equal to and opposite of electron flow.

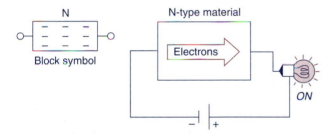

When a voltage is applied to N-type material, current flows from negative to positive using the free electrons called carriers.

Figure 35-4   Conduction through N-type material.

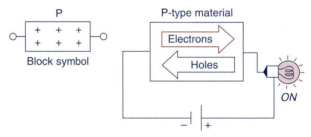

The direction of hole flow inside P-type material appears to move from positive to negative, while the current outside the P-type material flows from negative to positive.

Figure 35-6   Conduction through P-type material.

P-Type Material
To create **P-type material**, a crystal is doped with atoms from an element that has *fewer* electrons in its outer shell than the natural crystal. For example, a gallium (Ga) atom with three valence electrons is added to pure silicon. This time the three valence electrons of the gallium join in covalent bond with the valence electrons of the silicon atoms. But the silicon atoms are not complete; another electron is needed to complete the covalent bonding structure. As a result, a space or **hole**

The dopant atoms used to form a P-type crystal are said to be **trivalent** referring to their three valence electrons and are called an *acceptor* impurity. It is important to note that in the formation of both N- and P-type crystals, the total number of electrons always equals the total number of protons. The crystal neither gains nor loses electrons as a result of the doping process and so remains electrically *neutral*.

35-3 PN-Junction Diode

As we have seen, both P-type and N-type materials conduct electricity. They function like resistors and will conduct current in both directions. Their resistance is determined by the amount of dopant added to the pure crystal.

P-type and N-type materials are seldom used by themselves in producing electronic components. Placed together as a pair they form what is called a *PN-junction diode* (Figure 35-7). The P material is called the *anode* and the N material is called the *cathode*.

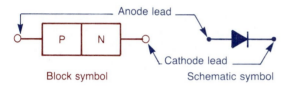

Figure 35-7   PN-junction diode.

Depletion Region   To understand the behavior of these crystals when they are joined together, it is helpful to look at the action that takes place at the junction point or boundary of the two materials. Electrons from the N material (mainly from the dopant atoms) are attracted across the boundary and fill holes near the boundary. This process is called **diffusion**.

The atoms in the P material that are gaining electrons become negative ions and the atoms in the N material that are losing electrons become positive ions (Figure 35-8A). This process does not continue for very long because the negative ions of the P material then repel additional electrons. Between the sides of the boundary there will now be a small voltage, usually in the mV range. This voltage can be represented by a small cell drawn with dotted connections across the junction (Figure 35-8B). The establishment of this voltage tends to offset any further diffusion of electrons as it acts like an *insulator* to oppose any further movement. This internal voltage that is set up is called the **potential barrier** and the area in which it is set up is the **depletion region** (Figure 35-8C).

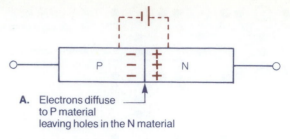

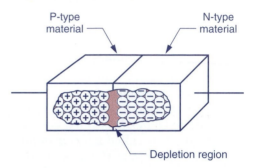

A. Electrons diffuse to P material leaving holes in the N material

B. Junction potential barrier created (about 0.7 V for silicon)

C. At the junction of the P-type and N-type material, the materials exchange carriers, and a thin area called a depletion region is formed.

Figure 35-8   PN-junction diode potential barrier and depletion region.

The barrier potential (voltage) depends on the extent of doping and is between 0.6 V and 0.7 V for silicon diodes between 0.2 V and 0.3 V for germanium diodes.

Forward and Reverse Bias   By controlling the size of the depletion region, we can control the conduction of a diode. Forward bias will eliminate the depletion region and cause a diode to pass current. Reverse bias will increase the size of the depletion region and, in turn, block current.

A diode is said to be *forward-biased* or *reverse-biased* depending on the *polarity* of the dc voltage applied across it. *Forward bias* exists when the voltage applied to the diode is such that the P-type material (anode) is positive with respect to the N-type material (cathode).

When a diode is connected in forward bias the positive polarity of the battery *repels* the holes in the P material and pushes them towards the depletion region (Figure 35-9). At the same time, the negative polarity of the battery *repels* the electrons in the N material and pushes them towards the depletion region. The carriers bridge the depletion region and cause it to close or collapse. In this condition, carriers are available

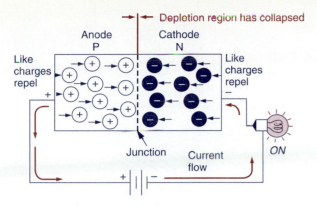

Figure 35-9 Forward-biased diode.

from one end of the diode to the other, allowing normal current flow with little resistance.

Reverse bias exists when the voltage applied to the diode is such that the P-type material (anode) is negative with respect to the N-type material (cathode).

When reverse bias is applied to a diode the positive battery potential will *attract* the electrons away from the junction area in the N-type material, and the negative battery potential will *attract* the holes away from the junction area in the P-type material. This action makes the depletion region wider and no current flow will result (Figure 35-10). In effect, the depletion region acts as an insulator and, thus, blocks current flow.

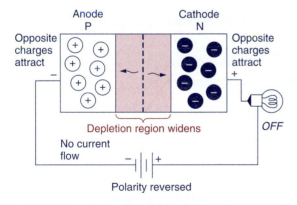

Figure 35-10 Reverse-biased diode.

35-4 Diode Operating Characteristics

The operation of a diode can be summarized by reference to its volt-ampere characteristic curve. As the forward-bias voltage is increased

from zero, current will not flow until the depletion region is closed. This voltage is often referred to as the *forward breakover voltage* (about 0.7 V for silicon and 0.2 V for germanium). Once conduction starts there is an approximate linear relationship between voltage and current (Figure 35-11).

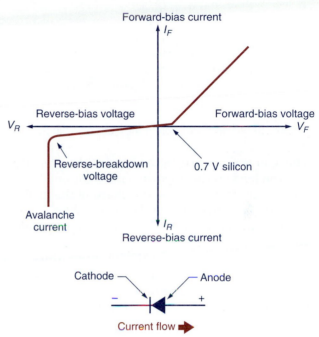

Figure 35-11 Diode characteristic curve.

When applying voltage in the reverse-bias direction, the voltage must be increased to a much higher value before the diode will finally break down and conduct. This reverse-bias voltage value is called the diode *reverse-breakdown voltage* (Figure 35-11). Reverse current flow is so heavy that it is called *avalanche current,* and it can destroy the diode. The reverse-breakdown voltage is often specified as the peak inverse voltage or PIV rating.

35-5 Diode Packages and Ratings

Diodes are available in many different case shapes and sizes. It is important to properly identify the anode and cathode leads of the diode for proper connection into circuits (Figure 35-12). Some diodes have the diode symbol printed on the case. Popular cylinder-shaped diodes use a dark band for lead identification.

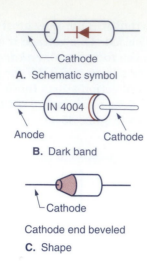

A. Schematic symbol

B. Dark band

Anode

Cathode

Cathode end beveled

C. Shape

Figure 35-12 Identification of the leads on a diode.

The lead end nearest to the band is the *cathode,* and the lead on the opposite end is the *anode.* Some manufacturers use the shape of the diode package to indicate the cathode end. When unsure, it is best to check the polarity of a diode with an ohmmeter.

Diode ratings represent the limiting values assigned by a manufacturer to various parameters of a diode. These parameters include temperature, voltage, current, and power. If any of these assigned values are exceeded, permanent damage may result to the diode.

Diodes are rated according to the **maximum current** they can safely conduct in the forward-bias direction. Exceeding this current rating will cause the diode to overheat and can permanently damage it. Generally, the larger the physical size of a diode the more current it can handle. **Power diodes** are sometimes mounted in *heat sinks* which helps to dissipate the heat generated and increases their current-carrying capacity (Figure 35-13).

Diodes are voltage rated for both *maximum forward-bias* and *reverse-bias voltages.* The forward-bias voltage rating indicates its maximum forward-bias operating voltage, usually at a definite forward-bias current. The maximum peak reverse-bias voltage indicates the maximum allowable reverse-bias voltage that can be applied across the diode without damaging it.

Exceeding the temperature rating of a diode can do permanent damage. In a typical diode, a 10°C increase in temperature will cause the reverse current to nearly double. The forward voltage drop and temperature of the diode are inversely proportional; as the temperature

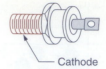

Cathode

A. Cathode end forms bolt for mounting into heat sink

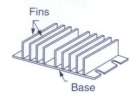

Fins

Base

B. Heat sink

Heat sink

C. Automobile alternator diodes mounted in a heat sink

Figure 35-13 Power diodes.

increases, the voltage drop decreases. All semiconductor devices have a negative temperature coefficient and are subject to the possibility of **thermal runaway**. As the temperature of a diode increases, its internal resistance decreases, allowing more current to flow through the device. An increase in current results in more heat being dissipated and an even further decrease in resistance. This cycle can continue until the semiconductor material overheats and the device fails.

35-6 Testing Diodes

A diode operates like a polarity-operated switch. When forward biased, it presents a relatively low resistance and acts like a closed switch (Figure 35-14). When reverse biased, it presents a high resistance and acts like an open switch.

About ⬭ Electronics

T rays are radiation waves from laser pulses. Semiconductor manufacturers use them with optical and electrical probes to check the integrity of high-speed electrical circuits.

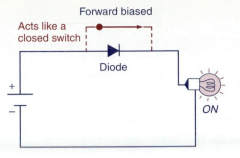

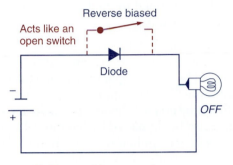

B. Reverse-bias connection

Figure 35-14 The diode as a polarity-operated switch.

The easiest way to test a diode is to check its resistance with an ohmmeter when not connected in circuit. A good diode will be indicated by a low resistance reading in one direction (forward bias) and a very high resistance reading in the other (reverse bias). The ratio of reverse-bias resistance to forward-bias resistance should be greater than 100:1. A defective shorted or leaky diode will show a low resistance reading in both directions. A defective open diode will show a very high resistance reading in both directions (Figure 35-15).

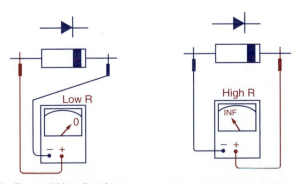

A. Forward-bias direction   **B.** Reverse-bias direction

Figure 35-15 Testing diodes with an analog ohmmeter.

Most digital multimeters have a special diode-function test feature. When set to the DIODE CHECK function, with leads properly connected, the forward voltage drop of the diode will be displayed. For silicon diodes, the instrument should display about 600 to 700 mV. If the digital display indicates an over-range reading with test leads in both positions, the diode is open. If a low reading (less than 1 V with test leads in both positions) occurs, the diode is shorted (Figure 35-16).

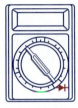

A. Set multimeter to the DIODE CHECK (▶▶—) function

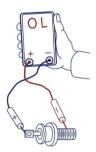

B. When connected in forward bias, the forward voltage drop of the diode is displayed

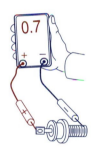

C. When connected in reverse bias, an over-range reading is displayed

Figure 35-16 Using the diode-check function of a digital multimeter to test a diode.

35-7 Practical Applications of Diodes

Rectifier Diode

Rectification is the process of changing alternating current (ac) to direct current (dc). Because of their ability to allow current to flow in only one direction, diodes are used as *rectifiers*.

Figure 35-17 shows the schematic for a single-phase half-wave rectifier circuit. During the positive half-cycle of the ac input wave, the anode side of the diode is positive. The diode is then forward biased, allowing it to conduct a current flow to the load. Since the diode acts as a closed switch during this time, the positive half-cycle is developed across the load. During the negative half-cycle of the ac input wave, the anode side of the diode is negative. The diode is now reverse biased; as a result, no current can flow through it. Since the diode acts as an open switch during this time, no voltage is produced across the load. Thus, applying a constant ac voltage produces a pulsating dc voltage across the load.

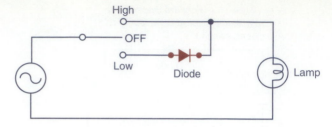

Figure 35-18   Diode lamp dimmer circuit.

voltage to be applied to the lamp, causing it to burn at half brightness.

Circuit-Control Diode

The diode can be used as a control device to direct current in a specific direction. The reverse-polarity protector circuit of Figure 35-19 is one application of this function. Some electronic devices are polarity sensitive and can be damaged by having voltage of a *reverse polarity* applied to them. By properly connecting a diode in series with the module, current flow is automatically blocked if the device is connected with reverse polarity.

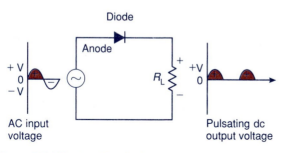

Figure 35-17   Rectifier diode.

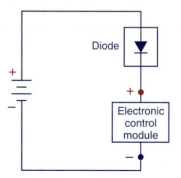

Figure 35-19   Diode used as a reverse-polarity protector.

Figure 35-18 shows a rectifier diode used as a dimmer control for a lamp. The switch is a single-pole, double-throw with a center-off position. When the switch is moved to the HIGH position, full voltage is connected to the lamp and it burns at full brightness. When the switch is moved to the LOW position, the diode is connected in series with the lamp. Since the diode permits current to flow through it in only one direction, half of the ac waveform is blocked during each cycle. This permits only half the

The diode can also be used to direct current in specific directions between circuits, as illustrated in the automotive headlight and ignition key circuit of Figure 35-20. The different modes of operation for the circuit are as follows:
- The warning *light* and *buzzer* come on as a reminder whenever the driver's door is opened while the headlight switch is ON.
- Only the buzzer comes on when the driver's door is opened while the key is in the ignition. The warning lamp can be grounded only through the headlamp switch. The position of the diode prevents the lamp's grounding

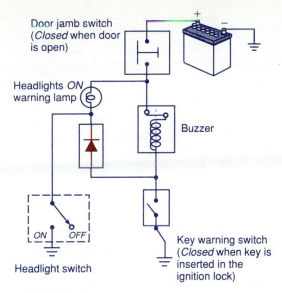

Figure 35-20 Diode used to direct current between circuits.

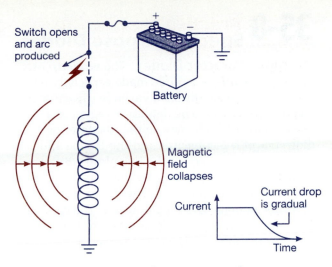

Figure 35-21 Opening the circuit to a coil.

through a closed key-warning switch. The position of the diode also allows the buzzer to be grounded through the headlight switch, even though the keywarning switch is open.
- If the key is removed from the lock and the headlights are OFF, all grounds are removed from the system, so neither light nor buzzer will operate.

Clamping or Despiking Diode
A *diode-clamping* or *despiking circuit* is a circuit that holds the voltage or current to a specific level. In certain electronic circuits, it is necessary to limit the amount of voltage present in the circuit. For example, if the dc voltage to a coil is switched OFF, there will be a noticeable arc at the switch contacts (Figure 35-21). As the current falls to a zero value, the collapsing magnetic field causes a high voltage to be induced in the coil. The arc produced at the contacts is a result of the induced voltage attempting to maintain the current in the circuit.

Of special concern are the coils controlled by electronic control modules that contain transistors and integrated circuits. If the high-voltage spike generated by switching OFF an inductive device is allowed to reach the electronic control module, it can destroy the module. The **clamping, or despiking, diode** (Figure 35-22) protects the circuit from surge currents developed when the current flowing through an electromagnetic device is abruptly interrupted. It is connected in *reverse bias* to block the battery's

positive voltage and, therefore, allow the current to pass through the coil. When the battery's positive voltage is switched OFF, the magnetic field of the coil collapses, inducing current flow in the coil. This current then passes through the *diode*. Thus, the induced-current flow is allowed to collapse upon itself. A spark is not produced, and the sensitive electronic circuit controlling the coil is protected against damage. It is important to note that the clamping diode *must* be connected in *reverse bias*. Operating the circuit with the diode connected in forward bias creates a high-current *short circuit* through the diode and solid-state control module, which can damage both the diode and module (Figure 35-22).

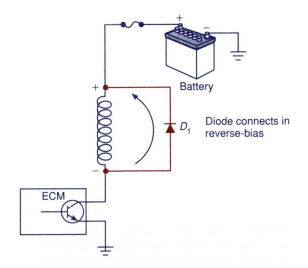

Figure 35-22 Clamping or despiking diode connection.

35-8 Special Purpose Diodes

Varicap or Varactor Diode   The *varicap* or *varactor diode* is a diode made especially for use as a replacement for a variable capacitor in electronic circuits. The depletion region of a *reverse-biased* diode acts as the insulator or dielectric of the capacitor and the P-type and N-type materials act as the two plates. By adjusting the reverse-bias voltage, the width of the depletion region is made wider or narrower and this changes the capacitance of the diode. A common application is electronic tuning for communications equipment (Figure 35-23).

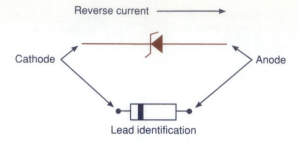

Figure 35-24   Zener diode symbol.

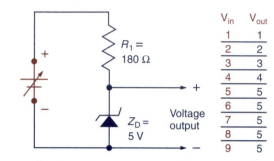

Figure 35-25   5-V zener diode voltage-regulator circuit.

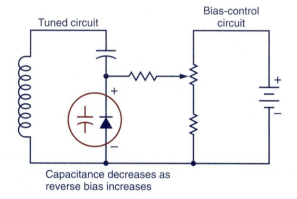

Figure 35-23   Tuning using a varicap diode.

Zener Diode   The *zener diode* is like an ordinary diode in that it allows current to flow in the forward direction. However it differs from an ordinary diode in that its reverse-direction break-down voltage is much lower than that of an ordinary rectifying diode. It is very heavily doped during manufacture. The large number of extra current-carriers allows the zener diode to *conduct current in the reverse direction*. This reverse-bias current would destroy a normal diode, but the zener is made to operate this way (Figure 35-24). The specified zener voltage rating of a zener diode indicates the voltage at which the diode begins to conduct when reverse biased.

Zener diodes are often used as part of voltage-regulator circuits. A simple zener-diode voltage-regulator circuit is shown in Figure 35-25. A 5-V zener diode is connected in series with a resistor R_1 to the variable dc input voltage. This input voltage is connected so that the

zener diode is *reverse biased*. The series (R_1) is required to drop all the input voltage not dropped across the zener diode. As the input voltage is increased from 0 V, the voltage across the zener diode increases until the 5-V zener voltage is reached. At this point the zener diode conducts and *maintains a constant* 5-V output as the input voltage varies from 5-9 V.

The zener diode can be used to *shape, or condition,* signals from sensors for use by a digital computer. For example, the waveform-clipper circuit of Figure 35-26 can be used to convert an incoming sine-wave signal to a near-square-wave signal that can be used by the computer. It clips both halves of the ac input wave equally when two zener diodes of the same voltage rating are used.

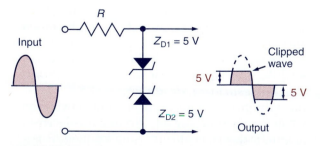

Figure 35-26   Zener diode waveshaping circuit.

Light-Emitting Diode (LED)

A *light-emitting diode (LED)* is a special semiconductor diode designed specifically to emit light when current flows through it. When forward biased, the energy of the electrons flowing through the resistance of the junction is converted directly to light energy. Because the LED is a diode, current will flow only when the LED is connected in forward bias. The diode itself is only a part of the LED package. It also requires leads and a plastic lens to diffuse the light (Figure 35-27).

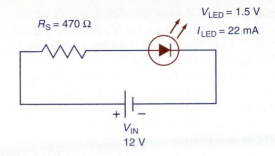

Figure 35-28 LED operating circuit.

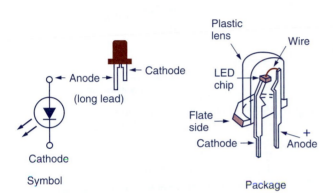

Figure 35-27 Light-emitting diode (LED).

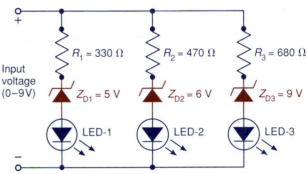

Figure 35-29 Voltage-indicator circuit.

Conventional silicon diodes convert energy to heat. *Gallium arsenide* diodes convert energy to heat and infrared light. This type of diode is called an *infrared-emitting diode (IRED)*. Infrared light is not visible to the human eye. By doping gallium arsenide with various materials, LEDs with visible outputs of red, green, or yellow light can be manufactured.

The LED must be operated within its specified voltage and current ratings to prevent irreversible damage. Most LEDs require approximately 1.5 to 2.2 V to forward bias them and can safely handle 20–30 mA.

An LED is usually connected in series with a resistor that limits the voltage and current to the desired value (Figure 35-28). Some LED packages contain built-in limiting or series-dropping resistors.

The main advantages of using an LED as a light source rather than an ordinary light bulb are a much lower power consumption and a much higher life expectancy.

Figure 35-29 shows a voltage-indicator circuit that uses different voltage-sensitive zener diodes to switch light-emitting diodes (LEDs)

Example 35-1

Referring to Figure 35-28, use this formula to determine the resistance of the series-dropping resistor R_S.

$$R_S = \frac{V_{IN} - V_{LED}}{I_{LED}}$$

$$R_S = \frac{12\ V - 1.5\ V}{22\ mA}$$

$$R_S = \frac{10.5\ V}{22\ mA}$$

$$R_S = 0.477\ k\Omega$$

$$R_S = 477\ \Omega$$

The closest standard resistance value is 470 Ω.

ON in sequence. As each zener diode reaches its specified voltage rating, it will conduct to complete the current path to the LED. As a result, the LEDs will glow in sequence as the input voltage rises from 0 to 9 V.

Laser Diode *Laser diodes* are specially made LEDs that are capable of operating as lasers. Laser stands for *light amplification by stimulated emission of radiation*. Unlike the LED, it has an *optical cavity*, which is required for laser production. The optical cavity is formed by coating opposite sides of a chip to create two highly reflective surfaces. Like the LED, it is a PN-junction diode, which at some current level will emit light. This emitted light is bounced back and forth internally between two reflecting surfaces. The bouncing back and forth of the light waves causes their intensity to reinforce and build up. The result is an incredibly high brilliance, single frequency light beam that is emitted from the junction. Laser diodes are used in applications such as fiber-optic communications and bar-code scanning systems. Never look directly into a laser diode.

Photodiode All PN-junction diodes are light sensitive. *Photodiodes* are PN-junction diodes specifically designed for light detection. The basic construction of a photodiode is shown in Figure 35-30. Light energy passes through the lens that exposes the junction.

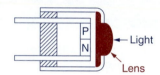

Figure 35-30 Photodiode.

The photodiode is designed to operate in the *reverse-bias mode.* In this device, the reverse-bias leakage current increases with an increase in the light level. Figure 35-31 shows the schematic symbol and the proper connection of

a photodiode. The inward arrows represent the incoming light. As the light becomes brighter, the reverse current increases. Typical current values are in the microamperes range.

The photodiode has a fast response time to variations in light levels. Photodiodes and light-emitting diodes are often used in conjunction with **fiber-optic cable** for the purpose of data transmission.

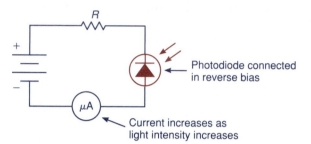

Figure 35-31 Photodiode connection.

A simple fiber-optic system consists of a transmitter with a light source, a length of fiber-optic cable, and a receiver with a light detector. Figure 35-32 illustrates transmission in a fiber-optic cable system. The *transmitter* consists of an LED, IRED laser diode light source. Within the transmitter, the input signal (normally digital pulses) is converted from electric to light energy by flashing a light source within the transmitter OFF and ON very rapidly. The light-beam pulses are directed into a length of *light-conducting fiber* where they reflect from wall to wall through the fiber core. At the other end of the fiber a *receiver* which contains a photodiode accepts the light-beam pulses and converts them back to their original electrical form.

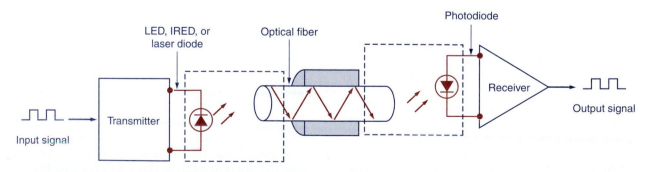

Figure 35-32 Transmission in a fiber-optic cable system.

Review and Applications

· ·

Related Formulas

Power = $V \times I$

$$R = \frac{V}{I}$$

Review Questions

1. In what way is the atomic structure of all pure semiconductor crystals similar?
2. What is the most common semiconducting material used in making electronic components?
3. (a) What is the sharing of outer-orbit electrons in semiconducting materials called?
   (b) What effect does this have on the resistance of the semiconductor?
4. Explain the process involved in forming an N-type semiconductor material.
5. Explain the process involved in forming a P-type semiconductor material.
6. In what way is current conducted through an N-type material?
7. In what way is current conducted through a P-type material?
8. Compare the way forward and reverse bias affect the depletion region and the conduction of a PN-junction diode.
9. (a) What determines whether a diode is forward or reverse biased?
   (b) Under what condition is a diode considered to be connected in forward bias?
10. Compare the amount of forward-bias voltage it takes to turn ON a germanium and silicon diode.
11. When will a silicon diode avalanche?
12. Which lead end of a cylinder-shaped diode is nearest to the identification band?
13. State three ways in which diodes are commonly rated.
14. A diode is tested using an ohmmeter. What readings would indicate that the diode is not defective?

15. What is the purpose of a rectifier diode?
16. What is the purpose of a circuit control diode?
17. What is the purpose of a clamping or despiking diode?
18. Explain the operating principle of a varicap or varactor diode.
19. In what way is the operation of a zener diode different from that of a conventional diode?
20. What is an LED specifically designed to do?
21. How is an infrared-emitting diode different from a conventional light-emitting diode?
22. What type of light beam is emitted from a laser diode?
23. Under what conditions is a photodiode designed to conduct current?
24. Name the three major components of a fiber-optic cable system.

Problems

1. A diode has a power rating of 5 W. During normal operation, the diode current and voltage are measured and found to be 2 amperes and 0.8 V. Will this amount of load placed on the diode cause it to overheat? Why?
2. An LED is rated for a voltage of 2 V and a current of 15 mA. Assuming it is to be operated from a 12 V battery, calculate the resistance of the series-dropping resistor required and the amount of power this resistor must dissipate.

Critical Thinking

1. (a) A silicon diode is connected in forward-bias to a 10 Vdc source and 1 kΩ series resistor. If a voltmeter is connected across the diode, approximately what voltage value should be indicated?
   (b) If the diode connection were changed to reverse-bias, approximately what voltage value should be indicated?

Portfolio Project

- Build a long lasting LED penlight. The circuit should consist of an LED, 3 V battery, ON/OFF switch, and resistor all connected in series. Assemble the circuit in an existing incandescent penlight or some other type of enclosure. Calculate the size of resistor required based on the voltage and current rating of the LED.

Circuit Challenge

Using a Simulator

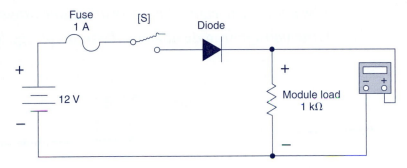

Procedure

(a) Construct the diode reverse-polarity protection circuit shown, using whatever simulation software package is available to you.
(b) Operate the circuit and record the value of the voltage across the load module when the switch is closed.
(c) Repeat, with the polarity of the dc voltage source reversed.

(d) Reconnect the diode in parallel to provide shunt protection of reverse-polarity.
(e) Operate the circuit and record the value of the voltage across the load module when the switch is closed.
(f) Reverse the polarity of the dc voltage source. Operate the circuit and make note of what happens when the switch is closed.

Power Supplies

Almost every piece of electronic equipment has a power supply circuit. The power supply circuit produces the dc voltage needed to operate electronic components. Of course, batteries can be and are used in portable equipment, but in larger systems, where considerable power is needed, batteries are an inconvenience and expensive. This chapter covers the job of the dc power supply, which is to convert readily available alternating-current (ac) line voltage into dc at some desired voltage level.

Objectives

After studying this chapter, you should be able to:
1. Identify the common rectifier circuits and explain how they work.
2. Wire and test power supply circuits.
3. Discuss the operation of a three-phase rectifier.
4. Measure and calculate power-supply voltage regulation and ripple.
5. Describe the action of voltage multipliers.
6. Explain how a switching power supply works.

Key Terms

- power supply
- step-down transformer
- step-up transformer
- center-tapped transformer
- autotransformer
- half-wave rectifier
- full-wave rectifier
- bridge rectifier
- three-phase rectifier
- filter
- ripple
- choke
- voltage regulation
- unregulated power supply
- regulated power supply
- dc-to-dc converter
- voltage multiplier
- switching power supply

RESEARCH **INTERNET** SITES

- Melcher-Power
- Power Integrations
- Vicor

36-1 Power Supplies

Electronic circuits normally require a different type and value of voltage than is available from a standard 120-V *ac* wall outlet. For example, transistors and integrated circuits require low *dc* voltages for their operation. For certain types of equipment, batteries are used to supply this needed voltage. For other equipment, however, an electronic circuit called a *power supply* provides the proper voltage that is necessary to operate the device (Figure 36-1).

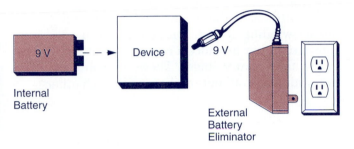

Figure 36-1 Common types of power supplies.

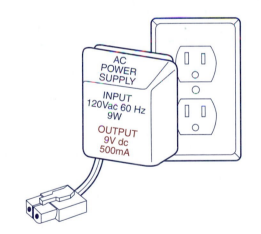

Figure 36-2 AC power supply.

The function of the electronic power supply is to take the input voltage and convert this to the proper type and value of voltage that is needed to operate circuits. There are many types of power supplies in use today. One method of classifying power supplies is to separate them according to the type of input voltage. If the input voltage is obtained by plugging the power supply into a wall outlet, then it is called an ac power supply (Figure 36-2).

An ac-to-dc power supply system can be divided into four different sections: transformer, rectifier, filter, and regulator. This chapter will explain in detail each of these sections (Figure 36-3).

36-2 Power Transformer Circuits

One of the major components in most ac power supply units is the **power transformer**. The power transformer performs one or more of the following functions for a power supply:

- Steps up voltages
- Steps down voltages
- Electrically isolates the primary circuit from the secondary circuit
- Supplies separate voltages that are out-of-phase with each other
- Supplies a variable ac voltage

Step-Down Transformer   When the transformer primary coil turns are greater than the secondary coil turns, the voltage is *stepped down* and the current is stepped up in proportion to the turns ratio. In the ideal transformer, (considered to be 100 percent efficient) the voltage multiplied by the current of the primary coil equals the voltage multiplied by the current of the secondary coil (Figure 36-4).

$$V_P \times I_P = V_S \times I_S$$

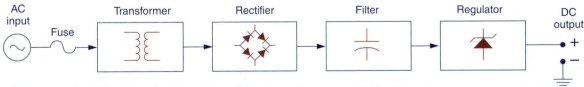

- Voltage transformation: Changing the ac line voltage into another more suitable voltage
- Rectification: Changing ac to dc
- Filtering: Smoothing the ripple of the rectified voltage
- Regulation: Controlling the output voltage to give a constant value with the line and load changes

Figure 36-3 AC-to-dc power supply block diagram.

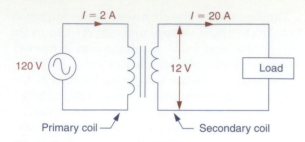

Figure 36-4 Step-down transformer.

Step-Up Transformer

When the transformer secondary coil turns are greater than the primary coil turns, the voltage is *stepped up* and the current is stepped down in proportion to the turns ratio (Figure 36-5). Again, voltage multiplied by amperes of the primary coil equals volts multiplied by amperes of the secondary coil.

$$V_P \times I_P = V_S \times I_S$$

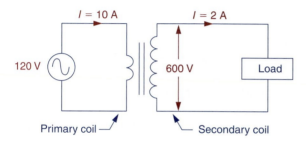

Figure 36-5 Step-up transformer.

Isolation

A standard transformer electrically isolates the primary and secondary circuits. *Isolation* gives electrical separation between ac neutral or ground input and the power supply output common.

In a transformer, the ac line voltage is connected to the primary coil to create a changing magnetic field that induces a voltage in the secondary coil. The primary and secondary coils are not connected physically, but simply are magnetically linked. If a high voltage is applied to the primary coil of a stepdown transformer, the low secondary voltage is completely isolated from the high primary voltage. If a person touched either wire of the secondary coil he/she would *not* be shocked by the high primary voltage (Figure 36-6).

Center-Tapped Transformer

A *center-tapped transformer* is built to provide two

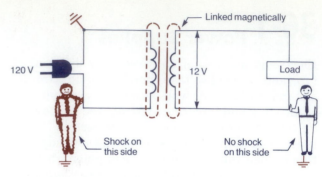

Figure 36-6 Transformer voltage isolation.

separate secondary voltages that are out-of-phase with each other (Figure 36-7). The secondary winding is made with a center-tap connection that provides a common point for two equal output voltages. These two voltages are said to be 180° out-of-phase with each other.

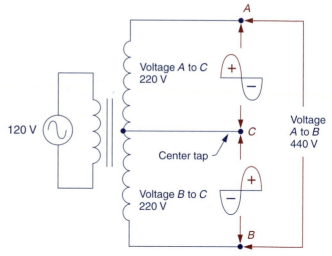

Figure 36-7 Center-tapped transformer.

Autotransformer

An *autotransformer* is made so that its primary and secondary coils are physically connected together. The theory of operation for the autotransformer is the same as for any other transformer except that there is no isolation between the primary and secondary coils. The entire winding is considered to be the primary coil, and part of the primary coil is also the secondary coil. If one end of the secondary winding is movable, the autotransformer can be used to provide a variable ac secondary coil voltage (Figure 36-8).

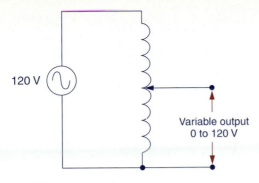

Figure 36-8 Autotransformer.

36-3 Rectifier Circuits

Half-Wave Rectifier Circuit Rectification is the process of changing ac to dc. *Diodes* are commonly used as rectifiers in power supplies. There are two major classifications of rectifier circuits: *half-wave* and *full-wave*.

The schematic for a simple ***half-wave rectifier*** circuit is shown in Figure 36-9. The diode acts like a *polarity*-operated switch, allowing current flow in one direction only. During the positive half-cycle of the ac input wave, the anode side of the diode is positive. The diode is then *forward-biased*, allowing it to conduct a current flow to the load. Since the diode acts as a closed switch during this time, the positive half-cycle is developed across the load.

During the negative half-cycle of the input sine wave, the anode side of the diode is negative. The diode is now *reverse-biased* and, as a result, no current can flow through it. Since the diode acts as an open switch during this time, no voltage is produced across the load. Thus, applying a constant ac voltage will produce a pulsating dc voltage across the load.

The ac supply that is available at ordinary wall outlets is 120 V and 60 hertz (Hz). Electronic circuits require lower voltages. Transformers are used to step-down the voltage to the level needed. Figure 36-10 shows a simple dc power supply using a step-down transformer and half-wave diode rectifier. Note that the diode polarity determines the load polarity.

The half-wave rectifier gets its name from the fact that it conducts during only half the input cycle. Its output is a series of pulses with a *frequency* that is the same as the input frequency. Thus, when operating from a 60-*Hz* line source,

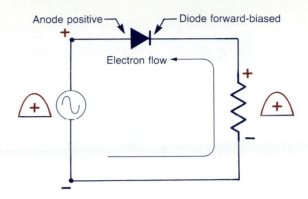

A. Conducting cycle

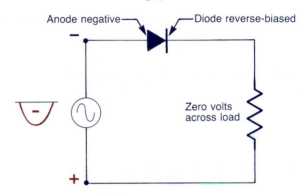

B. Nonconducting cycle

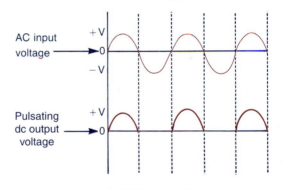

C. Voltage waveforms

Figure 36-9 Operation of a half-wave rectifier circuit.

the frequency of the pulses is 60 Hz. The frequency at which the pulses appear is called *ripple frequency.*

Full-Wave Rectifier Circuit Since the half-wave rectifier makes use of only one half of the ac input wave, its use is limited to low-power applications. A less pulsating and more powerful direct current can be produced by rectifying both half-cycles of the ac input wave. Such

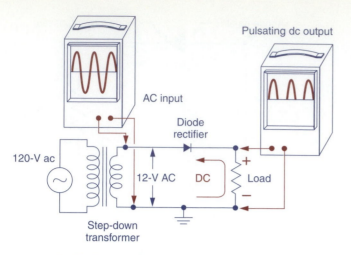

A. Oscilloscope connected to view input and output waveforms

B. Diode polarity determines load polarity

Figure 36-10   Simple half-wave dc power supply.

a rectifier circuit is known as a ***full-wave rectifier***.

The schematic for a full-wave rectifier circuit using a *center-tapped transformer* is shown in Figure 36-11. The center-tapped secondary winding of the transformer produces two separate ac voltages that are 180° out-of-phase with each other. The effect is that of two half-wave circuits. Each half of the transformer supplies opposite polarity to the anodes of the two diodes. During the positive half-cycle, the anode of D_1 is positive, while the anode of D_2 is negative. Therefore, only D_1 can conduct a current to the load. When D_1 conducts, it acts like a

closed switch so that the positive half-cycle appears across the load.

During the next half-cycle, the polarity of the voltage reverses. This causes the anode of D_1 to change to negative, while the anode of D_2 becomes positive. Therefore, only D_2 can conduct a current to the load. D_2 now acts like a closed switch so that a second pulse of current is delivered to the load. Notice that during this half-cycle the current still flows to the load in the same direction, producing a second positive pulse of current. Note that the full-wave rectifier is more efficient than the half-wave circuit. The dc output voltage is higher

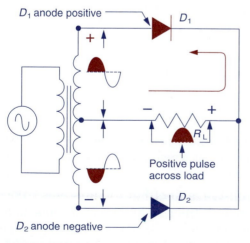

A. First half-cycle

B. Second half-cycle

Figure 36-11   Full-wave center-tapped rectifier.

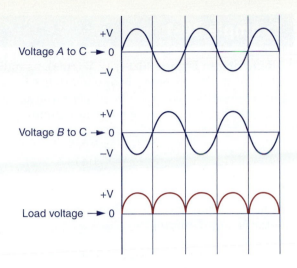

Voltage A to C

Voltage B to C

Load voltage

C. Voltage waveforms

Figure 36-11 Continued.

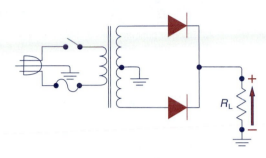

D. Circuit simplified using common ground connections.

and closer to pure dc than the half-wave output.

Another common type of full-wave rectifier is the **bridge rectifier** circuit. It can be used in transformer-type power supplies as well as line-operated power supplies. No center-tapped transformer is required. Instead, the circuit uses *four* diodes to obtain full-wave rectification.

The schematic for a full-wave, bridge rectifier circuit is shown in Figure 36-12. During the positive half-cycle, the anodes of D_1 and D_2 are positive (forward biased), while the anodes of D_3 and D_4 are negative (reverse biased). Electron flow is from the negative side of the line, through D_1, to the load, then through D_2, and back to the other side of the line.

During the next half-cycle, the polarity of the ac line voltage reverses. As a result, diodes D_3 and D_4 become forward biased. Electron flow is now from the negative side of the line through D_3, to the load, then through D_4, and back to the other side of the line. Note that during this half-cycle the current flows through the load in the same direction, producing a full-wave pulsating dc.

The diode bridge configuration can be wired using four individual diodes (Figure 36-13). When connecting these diodes, it is important to follow the schematic in order to avoid short circuiting the bridge. Bridge rectifiers are also available with the four diodes integrated in a single package.

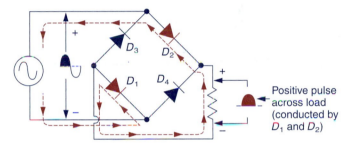

A. First half-cycle

Positive pulse across load (conducted by D_1 and D_2)

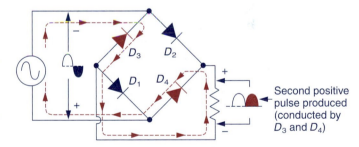

B. Second half-cycle

Second positive pulse produced (conducted by D_3 and D_4)

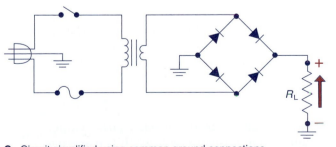

C. Circuit simplified using common ground connections.

Figure 36-12 Full-wave bridge rectifier.

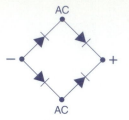

A. Bridge configuration

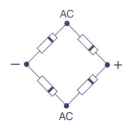

B. Wired using 4 diodes

C. Integrated in a single package

Figure 36-13 Bridge-rectifier packages.

The center-tapped transformer configuration uses only two diodes and has just one diode voltage drop along the path from the transformer to the load. These diodes, however, must withstand *twice* the reverse voltage compared to the diodes in the bridge configuration. The later circuit avoids the transformer center-tap, but has two diode voltage drops along the path from the transformer to the load. The full-wave rectifier produces two output pulses for each input cycle. Therefore, the output pulse frequency is *double* the line source frequency. Thus, when operating from a 60-Hz line source, the output ripple frequency is 120 Hz.

Computing Average DC Value
When ac voltage is rectified to dc, the output voltage of the rectifier is known as the *average value*. This can be measured by a dc voltmeter or *approximated* by multiplying the peak value (rms × 1.414) of the waveform by 0.637, or by multiplying the rms value of the applied alternating current by 0.9 or 90%. These values apply to a *full-wave rectifier*. If the average voltage is being calculated for a half-wave rectifier, multiply the rms voltage by 0.45 or 45%.

Example 36-1

Referring to Figure 36-14, 12 V (rms), as indicated by an ac voltmeter, is applied to the input of a bridge rectifier circuit. The ac peak voltage as indicated by an oscilloscope is approximately

$$V_{\text{PEAK}} = V_{\text{rms}} \times 1.414$$
$$V_{\text{PEAK}} = 12\text{V} \times 1.414$$
$$V_{\text{PEAK}} = 17\text{V}$$

The dc output voltage as indicated by a dc voltmeter is approximately

$$V_{\text{dc}} = V_{\text{PEAK}} \times 0.637$$
$$V_{\text{dc}} = 17\text{V} \times 0.637$$
$$V_{\text{dc}} = 10.8\text{V}$$

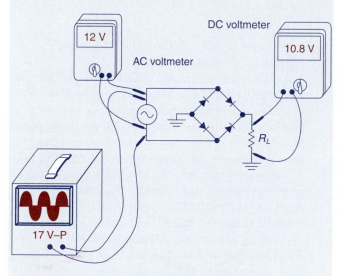

Figure 36-14 Circuit for Example 36-1.

36-4 Three-Phase Rectifier

For heavier load requirements, such as are required for industrial applications, it is necessary to change *three-phase* alternating current into direct current. The simplest configuration is the *three-phase, half-wave rectifier* shown in Figure 36-15. A diode is connected to each leg of the Y-secondary of the three-phase power transformer. The diodes are forward biased when the voltage of each line becomes positive and reverse biased when the voltage becomes nega-

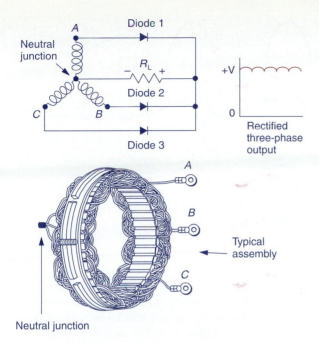

Figure 36-15   Three-phase half-wave rectifier.

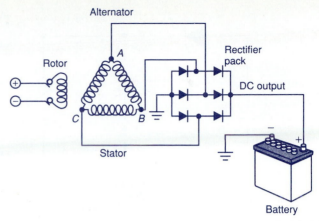

A. Delta-wound automobile alternator bridge rectifier circuit

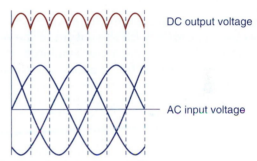

B. Input and output voltage waveforms

Figure 36-16   Three-phase, full-wave, bridge rectifier.

tive. As the voltage of each of the three-phase lines goes positive, current flows through the load resistor to the neutral junction. This rectifier has a higher average voltage output and *less ripple* than a single-phase, full-wave rectifier.

The *three-phase, full-wave rectifier* is used in the alternator of an automobile, because it has less ripple than the half-wave rectifier. It does not require a center-tapped, wye-connected, three-phase transformer or alternator for operation. It needs only to be connected to three-phase power for operation. Therefore, power can be supplied by either a wye or delta system. Figure 36-16 shows a delta-wound automobile alternator bridge rectifier circuit. Six diodes are required for the bridge circuit. *Two* diodes must be conducting at all times in order to provide a complete path for the dc output. The diode with the anode voltage that is most positive will conduct. The diode with the cathode voltage that is most negative will also conduct. The three-phase, bridge-type rectifier also has a higher average dc voltage and less ripple than the three-phase, half-wave rectifier, because the bridge rectifier changes both the positive and negative halves of the ac voltage into dc.

36-5 Filter Circuits

The pulsating dc output of rectifier circuits is not smooth enough to properly operate most electronic devices. They do not produce pure dc. The power-supply output still has an ac component, which is called *ripple* (Figure 36-17). A *filter* is used to reduce the amount of ac ripple, thus providing a relatively pure form of dc.

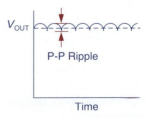

Figure 36-17   Ripple: ac component in a dc power supply.

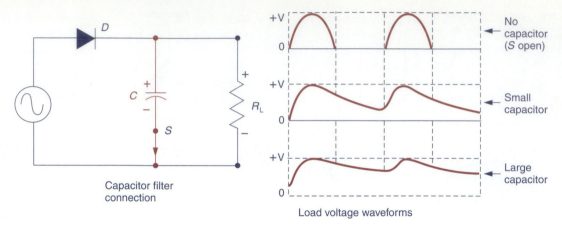

Figure 36-18 Capacitor filter.

The most common filter device is a *capacitor* connected in *parallel* with the output of the rectifier circuit. The filter capacitor is a large value electrolytic capacitor. It makes an excellent filter because of its ability to store electric charges. The schematic diagram for a simple capacitive filter is shown in Figure 36-18. It works by charging the capacitor when the diode conducts and discharging it when the diode does not conduct. When the rectifier circuit is conducting, the capacitor charges rapidly to approximately the *peak voltage* of the input wave. As the voltage from the rectifier drops, between the pulsations in the wave, the capacitor then discharges through the load. The capacitor, in effect, acts like a storage tank that accepts electrons at peak voltage and supplies them to the load when the rectifier output is low. The larger the load current, the larger the capacitor needed to provide adequate filtering. The capacitance value is determined from the value of the peak ripple voltage permitted.

The fact that a filter capacitor charges to the peak value of the ac waveform means that the capacitor must be rated for the *peak* voltage value. Most filter capacitors are of the polarized type and can *explode* if connected backwards. Always connect the negative (−) lead of the capacitor to the negative side of the circuit and the positive (+) lead to the positive side. Also, charged capacitors can present a shock hazard in high-voltage power-supply circuits even after the power is turned OFF. Always drain the charge from such capacitors before assuming that the circuit is dead.

If full-wave rectification is used, less capacitor storage is required. This is because the

full-wave output has little "OFF" time compared to the half-wave. The filtering action of the capacitor filter circuit can be improved by connecting a **choke** or *coil* in series with the load. This coil acts to resist any change in current. A rise in current through the coil produces a counter voltage, which opposes the rise; a decrease in current produces a counter voltage, which opposes the decrease. Used together the capacitor and choke provide a very effective filtering circuit (Figure 36-19).

The dc output voltage of a filtered power supply is *higher* than the output of a nonfiltered supply. This is due to the fact that the filter capacitors charge to the *peak* value of the wave-

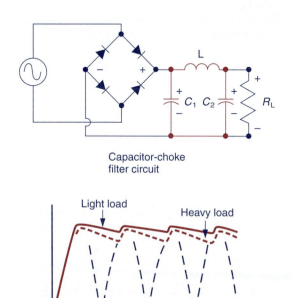

Figure 36-19 Capacitor-choke filter.

form. Therefore, when calculating the approximate dc output voltage, assume it is equal to the peak value of the ac input.

Most filtered power supplies will have less than 1 percent ripple. For example, a 10-V dc filtered supply should have less than 0.1-V peak-to-peak ac output.

Example 36-2

12-V ac is applied to the input of a bridge rectifier circuit, which is capacitively filtered. The dc output voltage is approximately the same as the peak voltage. Therefore,

$$V_{PEAK} = V_{rms} \times 1.414$$
$$V_{PEAK} = 12\,V \times 1.414$$
$$V_{PEAK} = 17\,V$$
$$V_{dc}\,Output = V_{PEAK}$$
$$= 17\,V$$

The amount of ac ripple is best measured using an oscilloscope. To measure ripple, place the dc/ac switch on the oscilloscope to *ac*. This blocks the dc, and the resultant peak-to-peak value is the ac ripple.

Example 36-3

A 20-V dc power supply shows 1 V of ac ripple when it is fully loaded. The percentage of ripple then is

$$Ripple = \frac{ac}{dc} \times 100\%$$
$$Ripple = \frac{1}{20} \times 100\%$$
$$Ripple = 5\%$$

36-6 Voltage Regulation

The output voltage of a power supply will usually decrease when a load is connected to it. The ability of a power supply to maintain a constant output voltage is indicated by finding its **percentage of voltage regulation.**

$$\%\ Voltage\ regulation = \frac{\left(\begin{array}{c}No\text{-}load\\volts\end{array}\right) - \left(\begin{array}{c}Full\text{-}load\\volts\end{array}\right)}{(Full\text{-}load\ volts)} \times 100$$

Example 36-4

Assume a power supply has a no-load voltage of 25-V dc. When loaded to its full load capacity, the output voltage drops to 21 V. The percentage of voltage regulation of the power supply is then:

$$\%\ Voltage\ regulation = \frac{\left(\begin{array}{c}No\text{-}load\\volts\end{array}\right) - \left(\begin{array}{c}Full\text{-}load\\volts\end{array}\right)}{(Full\text{-}load\ volts)} \times 100$$

$$= \frac{(25\ V) - (21\ V)}{(21\ V)} \times 100$$

$$\%\ Voltage\ regulation = 19\%$$

Unregulated dc *power-supply* circuits do not deliver pure dc. Some ac ripple voltage is always present in the dc output. Also, the dc output voltage changes with input line-voltage changes and changing load conditions.

A circuit that essentially eliminates variation in output power-supply voltage under conditions of changing input or changing load is called a *voltage regulator*. When a voltage regulator is connected to a rectifier circuit, the result is a **regulated power supply**. Many voltage regulators are designed to maintain voltage outputs within ±0.1 percent.

The schematic for a 5-V dc regulated power-supply circuit is shown in Figure 36-20. Transformer T_1 steps the 120-V, 60-Hz line voltage down to 8 V. The full-wave bridge rectifier, B_1, converts this ac voltage to a pulsating dc voltage. Capacitor C_1 smooths out these pulsations before this voltage is applied to the input lead of the regulator. The output lead of the regulator provides a fixed regulated voltage that is 5 V positive with respect to the common ground lead.

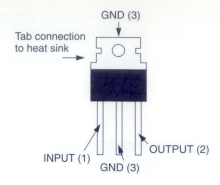

A. 7805 integrated-circuit voltage regulator

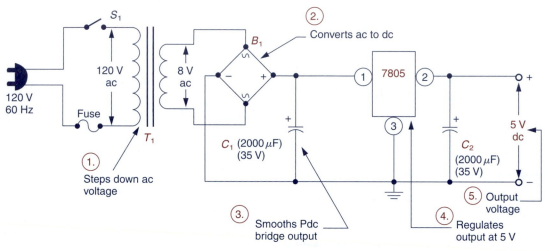

B. Circuit schematic

Figure 36-20 5-V regulated power supply circuit.

The 7805 integrated-circuit (IC) voltage regulator is designed to deliver a maximum output of about 1.5 A if the correct heat sink is provided. It is protected against excess load current by an internal thermal-shutdown circuit. If the chip becomes overheated due to an overload or short-circuit, the regulator output turns OFF. When the excessive-load current that caused the heat is removed, the chip cools down and turns itself back ON. This voltage regulator is ideal for operating digital logic circuits that require a 5-V regulated supply. The 7805 voltage regulator provides a regulated 5-V output for input voltages that vary from approximately 8 to 15 V.

DC-to-DC converters are used to change dc voltage levels and are usually well regulated. These devices are important where electronic equipment must be operated from a battery or

other dc source. Figure 36-21 shows an example of such a power supply used for automobile applications. With this type of power supply, damage can be done if power is not applied with proper polarity. For this reason, a protection diode is connected in the input circuit. If input polarity is applied in reverse of normal, the fuse will blow to open the circuit and protect the equipment.

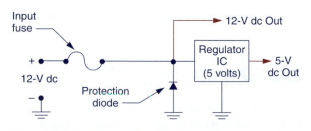

Figure 36-21 DC-to-DC power converter.

36-7 Voltage Multipliers

A *voltage-multiplier* circuit increases voltage without the extra cost involved in using a step-up transformer. These circuits are useful in producing high voltages for dc circuits that require low currents. The output voltage of these circuits depends upon charged capacitors and, as a result, there is very *poor voltage regulation*.

One popular type of voltage multiplier is a circuit called a *voltage doubler*. As the name implies, this circuit produces a maximum dc output voltage that is equal to twice the peak of the ac input voltage.

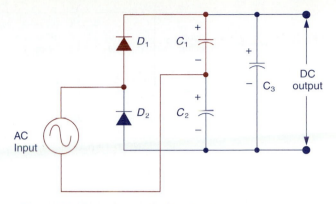

Figure 36-22   Voltage-doubler circuit.

Example 36-5

Assume a 12-V (rms) ac input is applied to the voltage-doubler circuit. The dc output voltage would be

Maximum dc output = 2 × Peak voltage
$$= 2 \times 12\,\text{V} \times 1.414$$

Maximum dc output = 34 V

The schematic of a full-wave voltage-doubler circuit is shown in Figure 36-22. During the first positive half-cycle, D_1 is forward biased and conducts charging C_1 to full-peak voltage. During the second half-cycle, the polarity of the ac line voltage reverses. As a result, D_2 becomes forward biased and conducts charging C_2 to full-peak voltage. The charges on C_1 and C_2 are in series, and C_3 charges to the sum of these two voltages. The output across C_3 is equal to two times the input peak voltage.

36-8 Switching Power Supplies

All power supply circuits discussed to this point are classified as conventional linear-regulated or linear supplies. The alternative to linear supplies is switch-regulated or switching supplies. Figure 36-23 shows a dc-to-dc *switching power supply* or converter. The applied dc voltage is switched on and off, producing a square wave across the secondary winding. The half-wave rectifier and capacitor filter then produce the dc output voltage.

Switching power supplies switch, or chop, the unregulated dc at a very high rate, then reconstruct the dc signal for use by the output load. This high radio frequency switching technique causes very little temperature rise in the supply so it can operate much cooler and because of the high frequency the ripple can be better filtered. Also, smaller components can be used in building a switching supply, so the supply can be manufactured in a smaller package. This is why switching power supplies are used in computers. Much less power is wasted in a

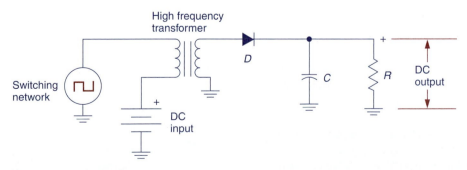

Figure 36-23   DC-to-dc switching power supply.

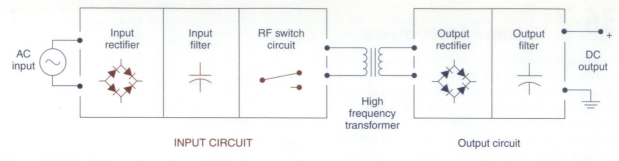

Figure 36-24   Major sections of a switching power supply.

switching supply than in a linear supply, so more output power can be developed for the same amount of input. Switching supplies can reach an efficiency of 85 percent compared to the 50 percent rate obtainable with linear supplies. The main disadvantage encountered with switching power supplies is the RF noise generated by the switching network.

Figure 36-24 shows the major components of a switching power supply. Like other power supplies it converts the ac line voltage into some value of dc. A high-frequency transformer connects the input and output components of the circuit. This transformer is especially designed to operate at the frequency of the RF switching circuit (20 kHz or higher). A conventional 60-Hz power transformer will not work at these high frequencies. The RF switching circuit rapidly chops the filtered dc from the input. This circuit is also responsible for providing the regulation necessary to keep the dc output constant.

36-9 Troubleshooting Power Supplies

Probably no skill area is more significant in establishing your value as a technician than your ability in getting equipment back into operation quickly after it has failed. Many times equipment failure is caused by troubles in the power supply. Troubleshooting any piece of equipment involves a systematic approach of observing the symptoms, analyzing the possible causes, and checking these failures by tests and measurements. Do not completely unload a switch mode power supply while troubleshooting, for this could cause the supply to self-destruct.

The multimeter and oscilloscope are two very useful pieces of test equipment used to troubleshoot power supply circuits. Listed below are some common problems associated with the rectifier circuit of Figure 36-25 that can be checked using these instruments.

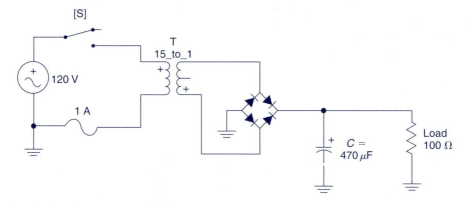

Figure 36-25   Troubleshooting power supply circuits.

Symptom	Possible Causes
Fuse blows as soon as power is switched on.	• Shorted diode in bridge rectifier. • Load is shorted. • Capacitor is shorted. • Transformer is shorted.
Zero dc output voltage.	• No ac input voltage. • Blown fuse. • Open line switch. • Open transformer winding. • Open in bridge rectifier.
Full-wave dc output signal.	• Capacitor open.
Low dc output voltage.	• Low ac input voltage. • Capacitor open.

HELP WANTED

Electronics Engineering Technician
Job Description
• Operate, test, and install electronic equipment.
• Assist engineers and technologists in research and development.
• Must be able to work from electrical circuit diagrams.
• Two-year college diploma required.

CHAPTER 36
Review and Applications

Related Formulas

$V_P \times I_P = V_S \times I_S$

$V_P = V_{rms} \times 1.414$

$V_{dc} = V_{PEAK} \times 0.637$

$\text{Ripple} = \dfrac{ac}{dc} \times 100\%$

$\% \text{ Voltage regulation} = \dfrac{\text{No-load volts} - \text{Full-load volts}}{\text{Full-load volts}} \times 100$

DC output (voltage-doubler) = $2 \times$ peak voltage

Review Questions

1. State the function of a power supply.
2. State five functions that a transformer may be required to perform in its role as part of a power-supply circuit.
3. (a) Explain what is meant by the term *rectification* as it applies to a power supply.
   (b) What electronic device is commonly used as a rectifier in power supplies?
4. Compare half-wave and full-wave rectifier circuits with regard to the value and amount of pulsation of the dc output voltage.
5. Compare a full-wave, center-tapped, bridge rectifier circuit with regard to:
   (a) the number of diode voltage drops along the path from the transformer to the load.
   (b) the amount of reverse voltage that the diodes must withstand.
6. (a) State two advantages of a three-phase, half-wave rectifier over a single-phase, full-wave rectifier.
   (b) How many diodes are required for a three-phase, full-wave bridge rectifier?
   (c) What type(s) of transformer or alternator system(s) can a three-phase bridge rectifier be operated from?
7. (a) Explain what is meant by the term *filtering* as it applies to a power supply.

(b) Name two electric devices commonly used as filters.
8. State three precautions that should be observed when selecting, connecting, and testing filter capacitors.
9. Explain the difference between a regulated and an unregulated power supply.
10. What special overcurrent protection feature is built into most IC voltage regulators?
11. What are dc-to-dc converters used for?
12. (a) What is the main advantage of using a voltage-multiplier circuit?
    (b) What is one limitation of the voltage-multiplier circuit?
13. Explain the principle of operation of a switching power supply.
14. (a) What are the two main advantages of switching power supplies over conventional linear power supplies?
    (b) What is the main disadvantage encountered with switching power supplies?

Problems

1. A step-down transformer has a turns ratio of 3:1 and 120 V rms applied to its primary.
   (a) What is the rms value of the secondary voltage?
   (b) What is the peak value of the secondary voltage?

2. (a) Calculate the approximate dc output voltage of a full-wave, bridge rectifier circuit that has 30-V ac applied to the input.
   (b) What would the value of dc output voltage be should one of the diodes burn open?

3. Calculate the approximate dc output voltage of a full-wave, bridge rectifier circuit that is capacitively filtered and has 24-V ac applied to its input. Assume a large size capacitor is used and a light load is connected.

4. Calculate the percentage of voltage regulation for a power supply with a no-load voltage of 12 V and a full-load voltage of 10 V.

5. A full-wave voltage doubler circuit has an input voltage of 120 V ac applied to it from an electrical wall outlet. Calculate the approximate value of the dc output voltage, assuming a light load is connected.

Critical Thinking

1. The input of an ac power supply is rated for 120 V ac, 60 Hz, and 100 W. The output is rated for 10 V dc. Is it correct to assume that a 100 W load can be safely connected to the output. Why?

Circuit Challenge

Using a Simulator

2. A bridge rectifier with a capacitor-choke filter circuit similar to that of Figure 36-19, is registering a low dc output voltage. For each of the possible faults listed, state which are possible and which can be eliminated as a possible cause. State the reason for each of your answers;
   (a) Fuse on the primary side of the transformer is blown.
   (b) ON/OFF switch on the primary side of the transformer is shorted.
   (c) One diode of the bridge rectifier is open.
   (d) Filter capacitor is open.
   (e) Heavier than normal load connected to the output.
   (f) An open in the choke coil.
   (g) Shorted windings on the transformer secondary.

Portfolio Project

• Obtain a discarded ac adapter, battery eliminator, battery charger, or power supply from a simple piece of electronic equipment. Identify the type of rectifier circuit used. Trace out the wiring to produce a schematic of the circuit. Carefully disassemble the parts of the power supply circuit. Display each part and explain its function in the power supply circuit.

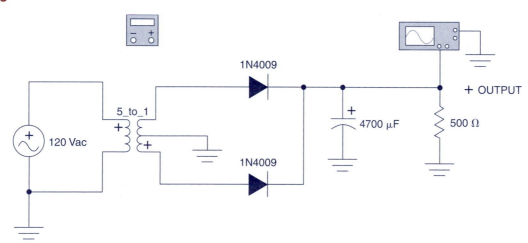

Procedure

(a) Construct the power supply circuit shown, using whatever simulation software package is available to you.

(b) Identify the type of rectifier circuit shown.
(c) Calculate the value of the ac voltage between the common ground and the

other two leads of the secondary of the transformer. Use the multimeter to verify your answer.

(d) Calculate the approximate value of the dc output voltage. Assume the capacitor filters the voltage to the peak value of the ac input.

(e) Use the multimeter to record the actual value of the dc output voltage.

(f) Examine the output voltage waveform on the scope and record the peak-to-peak value of the ac ripple voltage.

(g) Simulate an open capacitor fault, in the original circuit, by disconnecting the capacitor from the circuit. Record the value of the dc output voltage. Examine the output voltage waveform and identify the type of waveform displayed.

(h) Simulate an increase in load, in the original circuit, by changing the load resistance from 500 Ω to 50 Ω. Record the value of the dc output voltage. Examine the output voltage waveform on the scope and record the peak-to-peak value of the ac ripple voltage.

(i) Simulate an open diode, in the original circuit, by disconnecting one of the diodes from the circuit. Record the value of the dc output voltage. Examine the output voltage waveform on the scope and record the peak-to-peak value of the ac ripple voltage.

Transistors

The transistor's impact on electronics has been enormous. It is the basic component that is responsible for the operation of most complex electronic systems. The transistor has also led to all kinds of related inventions, such as integrated circuits and microprocessors. In this chapter, you will learn about the two main types of transistors: the bipolar junction transistor (BJT) and the field-effect transistor (FET).

Objectives

After studying this chapter, you should be able to:
1. Describe the ways in which transistors are used in circuits.
2. Compare the operation of bipolar and field-effect transistors.
3. Explain the operation of unijunction transistors and phototransistors.
4. Test a transistor.
5. Assemble and test experimental transistor circuits.

Key Terms

- transistor
- bipolar-junction transistor (BJT)
- field-effect transistor (FET)
- NPN transistor
- PNP transistor
- emitter
- collector
- base
- biasing
- DC ALPHA
- DC BETA
- current gain
- voltage gain
- power gain
- small-signal transistors
- large-power transistors
- Darlington pair
- source
- gate
- drain
- unipolar
- JFET
- transconductance
- MOSFET
- depletion-type MOSFET
- enhancement-type MOSFET
- unijunction transistor (UJT)
- phototransistor
- optocoupler (optoisolator)

- Keyence Corp. of America
- Newark Electronics
- Kelvin Electronics

37-1 Bipolar-Junction Transistors (BJTs)

Transistors, like diodes, are made using N-type and P-type semiconductor material. They have three leads, as opposed to the two leads of a diode. The transistor is basically a *current-control device*. A very small current input to one lead can control a much larger current that flows through the other two leads. Transistors can be used as *switches* to turn current ON and OFF, acting as *variable resistors* to vary current flow and as *amplifiers* to increase current or voltage.

There are two main families of transistors: *bipolar-junction transistors (BJTs)* and *field-effect transistors (FETs)*. The output current of the BJT is controlled by a small input current. The FET uses practically no input current. Instead, output current is controlled solely by a small input voltage. The basic operating characteristics of each type are quite different.

Basically, a *bipolar-junction transistor* (BJT) consists of a crystal containing three separate regions. A semiconductor may be *doped* (the deliberate adding of impure atoms to alter electronic conductivity) to form an *NPN* or a *PNP transistor* (Figure 37-1). (NPN and PNP take their designations from the N-type and P-type semiconductor material of which they are constructed.) The middle of the three regions in a transistor is known as the *base*. The two outer regions are known as the *emitter* and *collector*. The base region is much thinner and has fewer doping atoms than does the emitter and collector. The base is used to control current flow through the transistor. The schematic symbols for the NPN and PNP transistors are identical except for the arrow direction on the emitter lead. The direction in which this arrow points distinguishes one type from the other. Remember, for an NPN transistor, the arrow points *away* from the base; for a PNP transistor, the arrow always points *toward* the base.

The operation of a bipolar transistor can be compared to that of a water faucet. The emitter acts similar to the source of the water supply system. The base operates in much the same manner as the handle. The open end, where the water flows can be compared to the collector.

The bipolar transistor has two junctions that are similar to the junction of a diode. One is called the *emitter-base junction*; the other is called the *collector-base junction*. These junctions are drawn as two separate diodes on the equivalent circuit of a transistor (Figure 37-2).

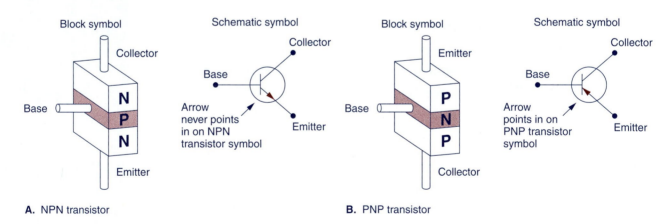

A. NPN transistor

B. PNP transistor

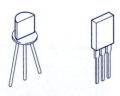

C. Actual appearance

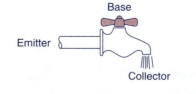

D. Water faucet analogy

Figure 37-1 Types of bipolar transistors.

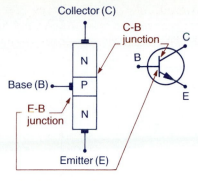

A. Actual circuit

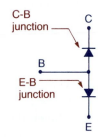

B. Equivalent circuit

Figure 37-2 NPN bipolar transistor equivalent circuit.

The PNP transistor is the complement of the NPN transistor. This means that *the two are electrically similar, except that opposite currents and voltages are involved.* For simplification, our discussion will just concentrate on the NPN type of transistor.

In the diode equivalent circuit of a NPN transistor, the base terminal is connected to the anode of the two diodes. The emitter is the cathode of one diode and the collector is the cathode of the other. However, two diodes connected back-to-back will not function as a transistor. The two PN junctions must be formed on a single wafer of silicon.

Biasing
The two junctions of the transistor must be correctly **biased** to ensure proper transistor action. The emitter-base junction is always connected in *forward bias* with voltage and resistance values selected to produce a small current flow (Figure 37-3). Actual conduction through this junction takes place in the same manner as in a forward-biased diode. As mentioned, transistors are current controllers, and it is this small input base current that is

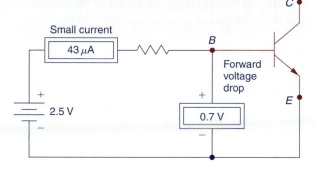

Figure 37-3 Emitter-base, forward-biased transistor.

used to control a much larger output-collector current.

The collector-base junction must always be connected in *reserve bias* to ensure proper transistor action. To accomplish this, the collector biasing voltage must be *higher* than that of the emitter-base forward-bias voltage. Due to the reverse-bias connection of this junction, little or no current flows in this circuit (Figure 37-4).

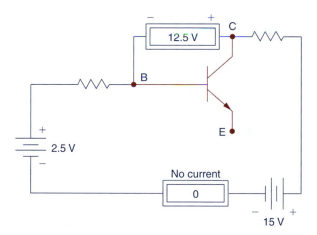

Figure 37-4 Collector-base, reverse-biased transistor.

Operation
The key to the operation of a bipolar-junction transistor is the *correct biasing* of the emitter and collector junctions, as well as the thinness and light doping of the base region. When both junctions are properly biased and connected into one circuit, the transistor is ready for action. Two voltage sources are used to forward bias the emitter-base junction and reverse bias the collector-base junction.

The *emitter* is very heavily doped and acts as the main *source* of the electron current. The *base* is very lightly doped and acts to *control* the flow of the current. In an NPN transistor, the *collector* is moderately doped and *receives* most of the electrons from the emitter (Figure 37-5).

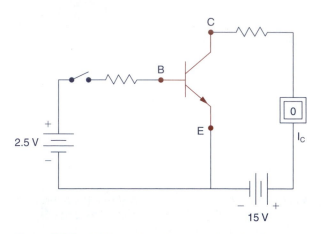

Figure 37-5 NPN transistor in a complete circuit.

Current in the base lead is called the *base current*, and the current in the collector lead is called the *collector current*. The amount of base current (I_B) determines the amount of collector current (I_C). A small increase in I_B results in a large increase in I_C, and thus, the base current acts to control the amount of collector current.

Figure 37-5 shows an NPN transistor in a complete circuit. All BJTs are normally off devices. When no current passes from emitter to collector, the base-to-emitter circuit is open or not activated.

If base current flows in a transistor, collector current will flow also. A forward-biased, emitter-base junction allows electrons to cross the emitter-base barrier. Because the base is very thin and has fewer doping atoms, a very small base current flows. Most of the current is attracted to the positive polarity of the collector. Therefore, the small base current can control a larger collector current (Figure 37-6). Because the base region is thin, excessive base current can destroy the transistor. Base voltages are kept low, and typical base currents range from a few microamperes to several milliamperes. On the other hand, collector-supply voltage must be larger in value to attract current carriers that are injected into the base region. Typical collector currents range from milliamperes to amperes. The emitter current is equal to the sum of the collector and base currents.

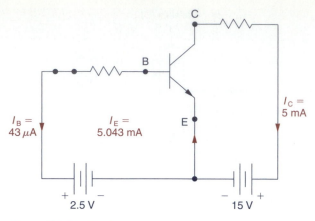

Figure 37-6 Bipolar transistor action.

More electrons leave the emitter than arrive at the collector. Therefore, the emitter current is always greater than the collector current. However, the base current is very small and can be ignored when calculating certain circuit values. When the value of the base current is ignored, the collector and emitter currents are said to be approximately equal to each other.

The *dc alpha* (symbolized α_{dc}) of a transistor is defined as the dc collector current divided by the dc emitter current. Since the collector current almost equals the emitter current, the dc alpha in a typical BJT is nearly 1 (usually between 0.95 and 0.99). The α_{dc} is a parameter set by the manufacturer and cannot be changed. This design ratio specifies how much collector current will flow for a given value of emitter current.

Current Gain   The *dc beta* (symbolized β_{dc}) of a transistor is defined as the dc collector current divided by the dc base current. Because the base current is very small compared to the collector, the dc beta of a BJT is relatively high. The dc beta of a transistor is also stated as the hFE on transistor data sheets. This parameter is also called the *current gain*, as it is a measure of the circuit capacity to amplify small base-current changes.

Typical betas or current gains for junction transistors range from a value as low as 10 to as high as several hundred.

$$\text{Current gain} = \frac{I_{\text{collector}}}{I_{\text{base}}}$$

Example 37-1

A current flow of $50\,\mu A$ (0.050 mA) through the base of a bipolar-junction transistor (BJT), produces a collector-current flow of 5 mA (Figure 37-7). The transistor has a current gain of

$$\text{Current gain} = \frac{I_{\text{collector}}}{I_{\text{base}}}$$

$$\text{Current gain} = \frac{5\,\text{mA}}{0.05\,\text{mA}}$$

$$\text{Current gain} = 100$$

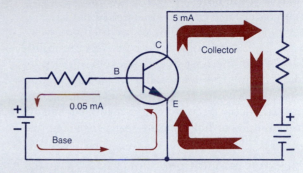

Figure 37-7

Example 37-2

The transistor of Figure 37-8 has a β_{dc} of 100. Calculate the base current, collector current, emitter current, and the voltage from collector-to-emitter. Assume a silicon transistor with a base-to-emitter forward voltage drop of 0.7 volts.

$$\text{Base current} = \frac{V_{\text{bb}} - \text{forward voltage drop (base-to-emitter)}}{\text{base resistor value}}$$

$$= \frac{2.5\,\text{V} - 0.7\,\text{V}}{39\,\text{k}\Omega}$$

$$= 46.2\,\mu A$$

$$\text{Collector current} = \beta_{\text{dc}} \times \text{base current}$$

$$= 100 \times 46.2\,\mu A$$

$$= 4.62\,\text{mA}$$

$$\text{Emitter current} = \text{base current} + \text{collector current}$$

$$= 46.2\,\mu A + 4.62\,\text{mA}$$

$$= 4.67\,\text{mA}$$

$$\text{Voltage collector-to-emitter} = V_{CC} (\text{collector current} \times \text{collector resistor value})$$

$$= 15\,\text{V} - (4.62\,\text{mA} \times 1\,\text{k}\Omega)$$

$$= 15\,\text{V} - 4.6\,\text{V}$$

$$= 10.4\,\text{V}$$

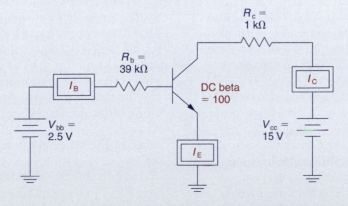

Figure 37-8

Gain is a term used to describe the ratio of increase. The most common types of gain are current gain, **voltage gain**, and **power gain**. Transistors are classified as *active* devices because they are capable of producing power gain. The BJT is basically a current amplifier, because an increase of collector current while maintaining the same collector voltage is an increase of power ($P = V \times I$). Depending on the transistor circuit configuration, the output can produce either current gain, voltage gain, or both.

Packages and Ratings   To protect the transistor and provide a means of attaching connecting leads, the transistor is mounted in a case, or package. **Small-signal transistors** are usually housed in a metal, molded epoxy, or plastic case (Figure 37-9). The transistor data sheet should be consulted to determine how each lead is identified.

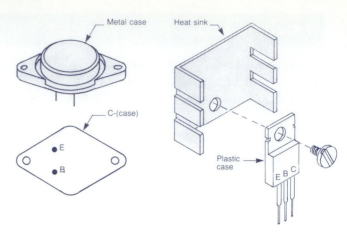

Figure 37-10   Large-power transistor packages.

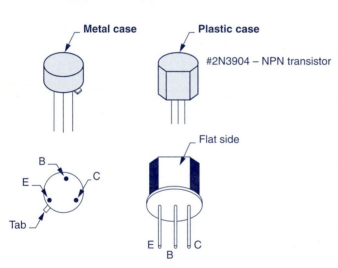

Figure 37-9   Small-signal transistor package.

Large-power transistors are designed to carry more current and, therefore, must dissipate more heat. To accomplish this, the semiconductor material and the case are made much larger (Figure 37-10). On this type of transistor, the connection to the collector often connects directly to the case. Silicon grease and washers are used to help heat flow away from power transistors. Power transistors are often mounted on *heat sinks* that help dissipate the heat from the transistor.

An important thing to keep in mind about BJTs is that they can be damaged by heat, high voltage, and reverse voltages. Avoiding transient voltages, using thermal grease to get rid of heat where required, and making sure that only the proper voltages are applied are measures that prevent damage to bipolar transistors.

Transistors are made with a wide variety of voltage, current, and power ratings. Large-power ratings can be obtained only if the transistors have been mounted on proper heat sinks and a thermal compound is used. Some transistors are designed to operate at a few thousand hertz, while others can operate at several hundred million hertz. Additionally, transistors can have:

- Voltage ratings from a few to several hundred volts.
- Current ratings from milliamperes to fifty or more amperes.
- Power ratings from milliwatts to several hundred watts.

Testing   Transistors, like *diodes*, can become defective as a result of excessive current or temperature. The two most common reasons for transistor failure are open and shorted junctions. The two junctions of a bipolar transistor can be tested using an ohmmeter or by using the diode function test of a digital multimeter. This is the same procedure used to test the single junction of a diode (Figure 37-11). If the junction is good, the ohmmeter will read a low resistance in one direction and a high resistance in the other. A digital multimeter that is set for the diode function test should read 700 mV in the forward-bias direction and overrange in the reverse-bias direction for a silicon-junction diode.

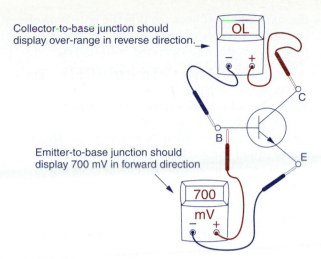

Collector-to-base junction should
display over-range in reverse direction.

OL

C

B

E

Emitter-to-base junction should
display 700 mV in forward direction

700
mV

Figure 37-11 Testing the emitter-base and collector-base junction of a bipolar transistor using the diode function test.

Sometimes a transistor can develop a *short circuit from the emitter to the collector.* Normally, the resistance between these two points is high in *both directions.* This is because at least one of the two transistor's junctions will be reverse biased regardless of the polarity of the applied voltage. To test for an emitter-to-collector short, connect an ohmmeter across these two points and measure the resistance in both directions (Figure 37-12). In both cases the measured value should be very high or infinite.

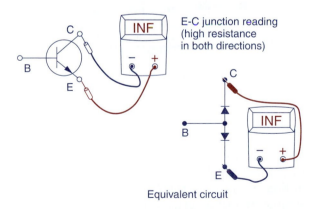

C

INF

B

E

E-C junction reading
(high resistance
in both directions)

C

INF

B

E

Equivalent circuit

Figure 37-12 Collector-to-emitter test of a transistor.

If a transistor passes the multimeter test of its junctions, then it is probably good. However, a transistor can pass the junction test and still be bad. This is because this check does not test

the beta of a transistor. For example, if the semiconductor material inside the transistor case is damaged due to excess heat, the transistor may still operate, but with different values of base and collector current. Consequently the beta of the transistor is altered, and the BJT no longer operates at its rated values. Transistor testers should be used to check the operating characteristics of a transistor.

When a transistor fails, there are big deviations in transistor currents and voltages. For this reason, in-circuit testing of transistors often involves checking for voltages at the transistor terminals that are different from the specification values.

Darlington Transistor Pair The ***Darlington transistor pair*** configuration is a way of hooking up two or more bipolar transistors so that a smaller transistor provides the base current for a larger transistor. This creates a transistor unit with a very high current-amplification factor. Two separate transistors may be used together, or—as is most often the case—the semiconductor manufacturer may put both transistors together in the same package (Figure 37-13).

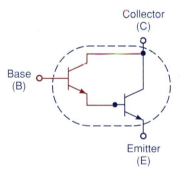

Collector
(C)

Base
(B)

Emitter
(E)

Figure 37-13 NPN Darlington transistor.

37-2 Field-Effect Transistors (FETs)

A BJT is a current-driven device. The current flow into or out of the base of the transistor controls the larger current that flows between the emitter and the collector. The *field-effect transistor* (FET) uses practically no input current. Instead, output current flow is controlled by a varying "electric field," which is created through the application of a voltage. This is the origin of the term *field-effect.*

FETs are used in the same way as bipolar transistors. FETs have three connections—*source*, *gate*, and *drain*. These terminals correspond to the emitter, base, and collector of the bipolar transistor, respectively. FETs are *unipolar*, referring to the fact that their working current flows through only *one* type of semiconductor material. This is in contrast to bipolar transistors, which have current flowing through both N-type and P-type regions (Figure 37-14).

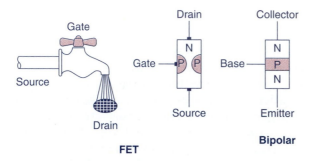

FET

Bipolar

Figure 37-14   The FET and bipolar transistor compared.

The field-effect transistor (FET) was designed to get around the two major disadvantages of the bipolar transistor: low switching speed and high drive power that are required because of base current. In addition, when the bipolar transistor is turned completely ON, there is a voltage loss, or drop, across the transistor due to its resistance. This loss generates heat in the transistor that must be disposed of to prevent damage. The loss also represents a waste of energy from the battery and a reduction in the power available to the load. FETs have less voltage loss when completely ON, thereby reducing the amount of power dissipation required and increasing the power available to the load.

In addition to having a higher input resistance than BJTs, the FET also provides better isolation between input and output circuits. Another advantage of the FET is that unlike the BJT, the FET is not subject to thermal runaway when the temperature of the device increases.

One area where BJTs are still better than FETs is by providing higher operating voltage and current. For example, the ignition module of an automobile uses a bipolar transistor to switch

the coil on and off because it is the only type of transistor capable of handling the higher currents and voltages associated with this circuit.

Junction Field-Effect Transistor (JFET)   The *junction field-effect transistor (JFET)* is basically a bar of doped semiconductor material with a junction formed near the center and across the surface of the bar (Figure 37-15). The bar is called the channel and may be of either P-type or N-type material. The devices are called P-channel and N-channel devices, respectively. The polarity of the junction material is opposite that of the channel material. Leads attached to either end of the channel are called the *drain* and *source* terminals. The junction terminal is called the *gate*.

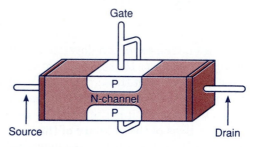

A.   N-channel type

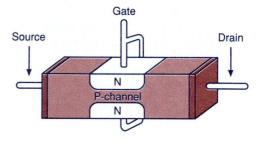

B.   P-channel type

Figure 37-15   Physical construction of a JFET.

The block symbol and schematic symbol for N-channel and P-channel JFETs are shown in Figure 37-16. In the N-channel schematic symbol, the thin vertical line represents the N-channel to which the source and drain connect. The gate and the channel form a PN diode; hence, the arrow on the gate lead points toward the channel. The schematic symbol for a P-channel JFET is similar to that for an N-channel JFET, except that the arrow points outward. We will only describe in detail the

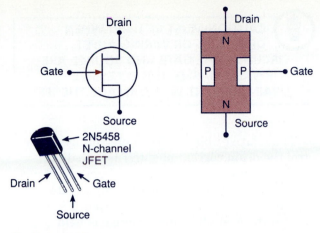

A. N-channel type

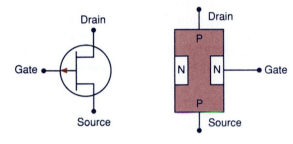

B. P-channel type

Figure 37-16   N-channel and P-channel JFETs.

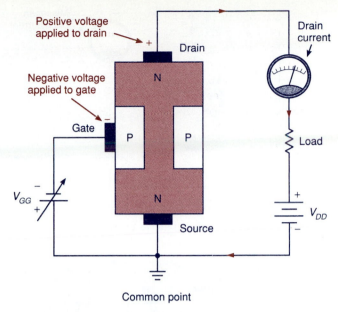

Figure 37-17   Biasing a JFET.

operation of an N-channel JFET. The action of a P-channel JFET is *complementary,* which means that it operates in exactly the same manner except all voltages and currents are reversed.

Figure 37-17 illustrates how current is conducted through an N-channel JFET. The N-channel of the JFET acts like a silicon resistor that conducts a current between the drain and source. The normal polarities for biasing the N-channel JFET are as indicated. A *positive* fixed V_{DD} supply voltage is connected between the source and drain. This sets up a current flow between the source and drain. A *negative* variable V_{GG} supply is connected between the gate and the source. This negative voltage at the gate increases the channel resistance and reduces the amount of current flow between the source and drain. Thus, the gate voltage controls the amount of drain current, and the JFET can be used as an amplifier, variable resistor, or switch. JFETs are *normally ON.* This means that JFETs allow full current to pass between the drain and the source when the gate circuit is *not activated.*

The gate of a JFET is always *reverse biased to prevent gate current flow.* This is standard for all JFET applications. As mentioned, the purpose of the gate is to vary the resistance of the channel and thereby control current flow between the source and drain. It does this as follows: The gate and channel form a PN-junction similar to a PN-junction diode. As with any PN-junction, a depletion region develops around the junction. The size of this region is controlled by the value of the reverse-bias voltage that is applied between the gate and source. With no negative bias voltage applied, the depletion region is minimal and the resistance of the channel is low, so maximum current flows between the source and drain (Figure 37-18A). As a small negative gate voltage is applied, the depletion region increases in size. This makes the conducting channel narrower, increasing its resistance and reducing the amount of current flow between the source and drain (Figure 37-18B). When the gate voltage is negative enough, the depletion layers touch or *pinch off,* and all current flow between source and drain is *cut off* (Figure 37-18C).

Since the gate-source junction of a JFET is operated in reverse-bias, ideally no current flows through it. As a result, all the current from the drain travels to the source. That is, drain current equals source current (Figure 37-19). Therefore, the only significant current

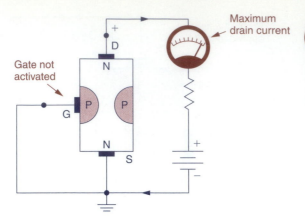

A. Full current passes from drain to source when gate is not activated.

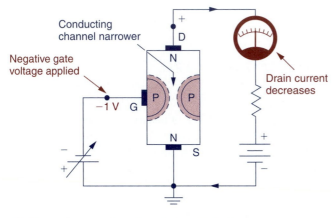

B. Channel narrows to reduce drain current when a negative voltage is applied to the gate.

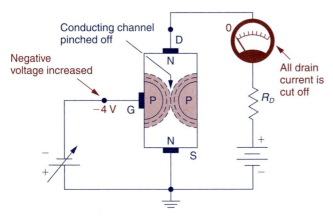

C. Increases the negative gate voltage and cuts off all drain current flow.

Figure 37-18   Gate control of drain current.

in a JFET is the source-to-drain current, which is usually just called the drain current and is designated as I_D.

> **CAUTION MUST BE USED WHEN DESIGNING OR WIRING A JFET CIRCUIT. IF THE GATE AND SOURCE ARE FORWARD BIASED, THE EXCESSIVE DRAIN CURRENT WILL DAMAGE THE JFET.**

The relative ability of the gate voltage to control the drain current of a JET is called **transconductance** (symbolized gm). In Figure 37-19, a change in the negative input gate voltage results in a change in the output drain current if the drain-to-source voltage is held constant. Ideally, a small change in the input voltage should produce a large change in output current. The amplifying ability or transconductance of a JFET is the ratio of the change in drain current to the change in gate-to-source voltage and is measured in siemens (S).

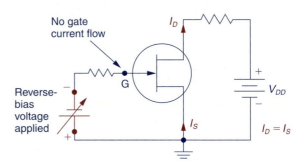

Figure 37-19   JFET currents.

The MOSFET   The *metal-oxide semiconductor FET (MOSFET)* is the most popular transistor in use today. The MOSFET is similar to the JFET because current flows through a channel region where the diameter changes to alter the resistance, and, consequently, the current flow through the transistor is altered. Unlike the JFET, the gate of a MOSFET has *no electric contact* with the source and drain. The input to a MOSFET acts as a capacitor. A glass-like insulation of silicon dioxide separates the gate's metal contact from the rest of the transistor. Because of this, gate current is extremely small whether the gate is *positive* or *negative*. The MOSFET is sometimes referred to as an *insulated-gate FET*.

A disadvantage of the JFET is the fact that whenever the gate becomes forward biased, a destructive current can flow from the source to the gate. Even if the current is not destructive, it will result in a highly distorted output signal because not all the current leaving the source arrives at the gate. The insulated gate of the MOSFET eliminates this problem. Except for this protection, the operation of the MOSFET is the same as that of the JET. The gate voltage controls the size of the depletion region and therefore controls the amount of drain current.

Like the JFETs, MOSFETs come in two varieties: N-channel and P-channel. The action of each is the same, but the polarities are reversed. There are also two different ways to make a MOS transistor: one is operated in the **depletion mode**, where, with no gate voltage, the transistor is *normally ON* (Figure 37-20). The other is operated in the **enhancement mode**, where, with no gate voltage, the transistor is *fully OFF* (Figure 37-21). Depletion-mode devices are often used in linear circuits, whereas enhancement devices are preferred for use in digital circuits.

The operation of the depletion-type (D-type) MOSFET is very similar to the JFET since both operate in the depletion mode. For D-type MOSFETs, the built-in channel allows conduction for gate-to-source voltages near zero. Figure 37-22 shows the correct biasing and typical characteristic curves for an N-channel D-type MOSFET. With zero gate bias voltage (V_{GS}) applied, drain current (I_D) passes through the channel between the source and drain. V_{GS} can swing positive or negative with reference to zero. When a positive voltage is applied to the gate, the channel becomes more conductive and the drain current flow increases. Similarly, when a negative voltage is applied to the gate, channel resistance increases and the drain current flow drops. If the negative voltage is high enough, no current will be conducted between the source and drain.

Enhancement-type (E-type) MOSFETs do not have a built-in channel, so the gate-to-source voltage must exceed the magnitude of the

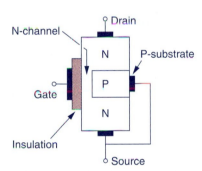

N-channel — Drain

N — P-substrate

Gate

P

N

Insulation — Source

The thin vertical line to the right of the gate represents the channel.

The channel is drawn as a solid line to signify that the circuit between the drain and source is normally complete and that the device is normally ON.

The schematic symbol for the P-channel type is the same except that the arrow points in the opposite direction.

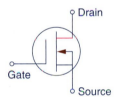

Drain

Gate

Source

Schematic symbol

Figure 37-20 N-channel depletion-type MOSFET.

Drain

N — P-substrate

Gate

P

N

Insulation — Source

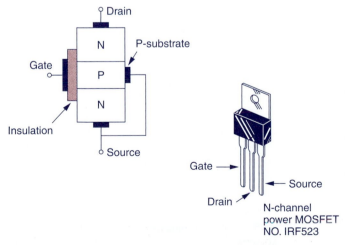

Gate →

← Source

Drain →

N-channel power MOSFET NO. IRF523

The channel is drawn as a broken line to signify that the circuit between the drain and the source is normally open and that the device is *normally OFF* until the proper gate voltage is applied.

The schematic symbol for the P-channel type is the same except that the arrow points in the opposite direction.

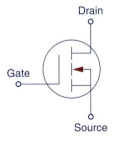

Drain

Gate

Source

Schematic symbol

Figure 37-21 N-channel enhancement-type MOSFET.

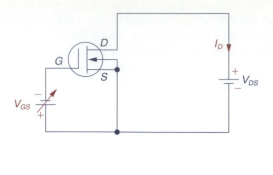

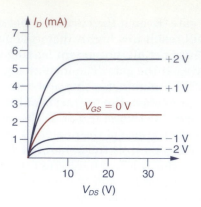

Biasing   Characteristic curves

Figure 37-22   D-type MOSFET operation.

threshold voltage for conduction to occur. Figure 37-23 shows the correct biasing and typical characteristic curves for an N-channel E-type MOSFET. If V_{GS} equals zero or any negative value, the device is off and no drain current flows. As the positive gate voltage is increased, a threshold voltage (typically 1 V) is reached, at which point the transistor's resistance begins to decrease and drain current flow increases to the saturation point.

Handling and Testing MOSFETs
As we have discussed, both depletion-type and enhancement-type MOSFETs use a thin layer of silicon dioxide to insulate the gate from the rest of the transistor. This insulating layer is kept as thin as possible to give the gate more control over drain current. Because the insulating layer is so thin, it is easily destroyed by too much gate voltage. These MOSFETs are *very sensitive to*

static charges (which can be very high voltage) that would not harm other devices. For this reason, MOSFETs are kept either in special *static protection packages* or with their leads shorted together until they are placed into a circuit. If a MOSFET is removed from or inserted into a circuit while the power is still ON, the transient inductive voltages produced can also destroy the MOSFET.

Some MOSFETs are protected by built-in zener diodes connected in parallel with the gate and source. The zener diode voltage rating is less than the maximum gate-to-source voltage rating of the device. Therefore, the zener diode conducts to shunt the current before any damage to the thin insulating layer occurs. The disadvantage of these internal zener diodes is that they reduce the MOSFETs' high input resistance.

Because of the fragile nature of the oxide, insulation testing with an analog ohmmeter is

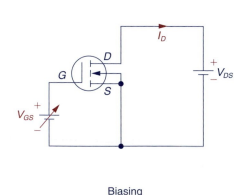

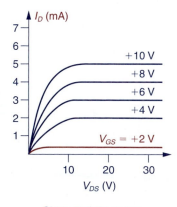

Biasing   Characteristic curves

Figure 37-23   E-type MOSFET operation.

548   **Chapter 37**   Transistors

absolutely *not* acceptable as a method of MOSFET testing. MOSFETs must be checked using only equipment with special provisions for MOSFET testing.

The ohmmeter can, as with bipolar devices, be used to determine whether or not a short or open exits within a unipolar device. Only ohmmeters with a low-ohm setting should be used. Devices such as MOSFETs are extremely sensitive to voltage and could be destroyed by ohmmeter tests. Field-effect transistors are often tested during in-circuit operation using a voltmeter or oscilloscope. The voltmeter or scope allows voltages and waveforms to be monitored and compared against those in the device service manual.

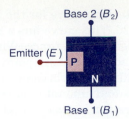

A. Structure

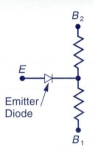

B. Equivalent circuit.

37-3 Unijunction Transistors

Unlike bipolar and field-effect transistors, a **unijunction transistor (UJT)** *does not amplify.* Instead, it works more like a voltage-controlled switch. The basic structure of a UJT is shown in Figure 37-24A. It contains one emitter and two bases. The emitter is heavily doped, whereas the N-region between the two bases is lightly doped. As a result, the resistance from Base 1 to Base 2 is relatively high when the emitter is open.

In the equivalent circuit of a UJT (Figure 37-24B), R_1 and R_2 represent the ohmic resistance of the silicon bar between B_1 and B_2 on either side of the PN-junction. The diode that is shown represents the PN-junction on the UJT. Supply voltage is applied between $B_2(+)$ and $B_1(-)$. The input signal is applied between the emitter and Base 1.

The uni (one) junction transistor gets its name from the fact that it has only one PN junction. Its schematic symbol (Figure 37-24C) is drawn with the arrow pointing in, indicating a positive emitter and a negative bar of N-type silicon. The slant on the arrow is used to distinguish the UJT from the N-channel JFET.

The basic operation and biasing of a UJT can be explained by referring to the circuit drawn in Figure 37-25. The voltage from B_1 to the emitter (VE) is monitored as the variable emitter supply voltage is increased gradually from zero. As the emitter supply voltage is increased the base-to-emitter voltage increases to some peak value and then decreases. When this peak

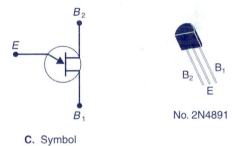

C. Symbol

Figure 37-24 Unijunction transistor (UJT).

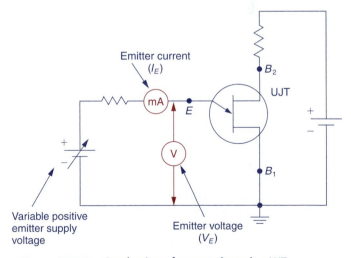

Figure 37-25 Conduction of current through a UJT.

value occurs, the B_1-to-emitter junction conducts a current.

The characteristic curve for the UJT is shown in Figure 37-26. This curve has a unique feature called *negative resistance*. When the emitter voltage reaches a certain point (V_P) on the curve, the emitter diode becomes forward-biased and a small amount of current begins to flow into the region between B_1 and the emitter. The additional charge carriers in this region increase the conductivity of the region. This reduction of resistance further increases the conductivity, which permits more current to flow, and so forth. As a result of this regenerative action, there is a sudden and dramatic increase in current accompanied by a degree in voltage. This results in a type of negative resistance, providing an increasing I for a decreasing V.

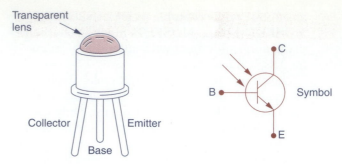

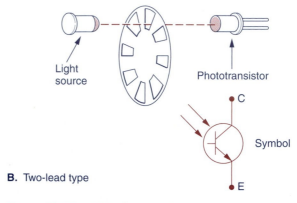

A. Three-lead type

B. Two-lead type

Figure 37-27 NPN phototransistor.

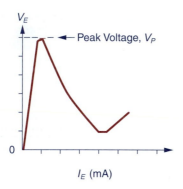

Figure 37-26 Characteristic curve of a UJT.

37-4 Phototransistors

The most common **phototransistor** is an NPN bipolar transistor with a light-sensitive, collector-base PN-junction (Figure 37-27). When this junction is exposed to light through a lens opening in the transistor package, it creates a control current flow that switches the transistor ON. This action is similar to that occurring with the base-emitter current of ordinary NPN transistors. A phototransistor can be either a two-lead or a three-lead device. The base lead may be brought out so that the device can be used as a conventional bipolar transistor, with or without the additional light-sensitivity feature.

Figure 37-28 shows an **optocoupler (also called an opto-isolator)** that uses the light from an LED to control a phototransistor. The primary function of the optocoupler is to provide electrical separation between input and output, especially in the presence of high voltages. Applying forward bias to the LED will cause it to produce infrared light and turn on the phototransistor and connect the load to the line. Since a light beam is used as the control medium, no voltage spikes or electrical noise produced on the load side can be transmitted to the control side.

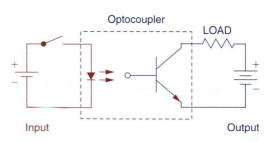

Figure 37-28 Optocoupler with LED and phototransistor.

In this device, the output of the phototransistor is used to control a second standard bipolar transistor. The result is a much higher current gain and collector current flow than a regular phototransistor. This arrangement is very light-sensitive, but is slower to respond than an ordinary NPN phototransistor.

A Photo-Darlington transistor consists of a phototransistor connected in a Darlington arrangement with a conventional transistor, all contained in the same package (Figure 37-29).

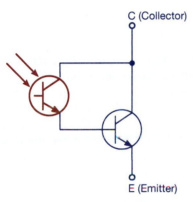

Figure 37-29   Photo-Darlington transistor.

CHAPTER 37
Review and Applications

Related Formulas

$$\text{DC beta (current gain)} = \frac{\text{dc collector current}}{\text{dc base current}}$$

$$\text{Current gain} = \frac{\text{change in collector current}}{\text{change in base current}}$$

Collector current = dc beta $\times$ base current

Emitter current = base current + collector current

Review Questions

1. State three ways in which transistors are used.
2. What are the two main families of transistors?
3. (a) In what two ways is the base region of a bipolar-junction transistor different from that of the emitter and collector?
   (b) For what purpose is the base region used?
4. Draw the schematic symbol and identify the leads of a PNP and an NPN bipolar-junction transistor (BJT).
5. What is the basic difference between the connection of an NPN and a PNP bipolar-junction transistor circuit?
6. (a) Name the two junctions found within a BJT.
   (b) State the type of biasing used for each junction in order to produce the required transistor action.
   (c) Compare the normal current flow through a forward-biased and reverse-biased junction.
7. With reference to a properly biased, complete bipolar-junction transistor circuit:
   (a) What determines the amount of collector current flow?
   (b) With no base current flow, how much collector current flows?
   (c) In what does a small increase in base current flow result?

(d) What is the operating current range of the base compared to that of the collector?
8. State two ways in which power transistor packages differ from the small-signal type.
9. A digital multimeter set for the diode function test is used to check the emitter-base junction of a bipolar-junction transistor. What readings should be expected if the junction is good?
10. An analog ohmmeter is used to check the collector-to-emitter circuit of a BJT. What should be the normal resistance reading between these two points?
11. (a) Explain the basic circuit configuration contained within a bipolar Darlington transistor.
    (b) What is the main advantage of this configuration?
12. Compare the control of output current of a bipolar-junction transistor (BJT) with that of a field-effect transistor (FET).
13. State three operating advantages of FETs over bipolar types of transistors.
14. Why are FETs said to be unipolar?
15. What are the two main types of FETs?
16. Draw the schematic symbol and identify the leads of an N-channel and a P-channel JFET.
17. The action of a P-channel JFET is complementary to that of an N-channel JFET. What does this mean?

18. What are the normal polarities of the drain and gate with respect to the source for an N-channel JFET?
19. Why are JFETs classified as normally ON devices?
20. In a JFET circuit, what gate-to-source biasing connection is normally used to prevent gate current flow?
21. Briefly explain how current flow through an N-channel JFET is controlled.
22. What is meant by the term *pinch-off* as it applies to a JFET circuit?
23. In a JFET circuit, why is the source current equal to the drain current?
24. What is the basic difference in construction between a JFET and a MOSFET?
25. (a) Draw the schematic symbol and identify the leads of an N-channel depletion-type MOSFET.
    (b) Why is the channel drawn as a solid line on this symbol?
    (c) Why can this MOSFET withstand a positive voltage at the gate without damage?
    (d) What happens to the drain current flow when a positive voltage is applied to the gate?
    (e) What happens to the drain current flow when a negative voltage is applied to the gate?
26. (a) Draw a schematic symbol and identify the leads of an N-channel enhancement type MOSFET.
    (b) Why is the channel drawn as a broken line on this symbol?
    (c) What happens to the drain current flow when a positive voltage is applied to the gate?
    (d) What happens to the drain current flow when a negative voltage is applied to the gate?
27. (a) Explain why MOSFETs can be damaged by static charges.
    (b) What special precaution is often taken for the packaging and shipping of MOSFETs?
    (c) Explain how built-in zener diodes can be used to protect MOSFETs against static charges.
    (d) Why should an analog ohmmeter not ordinarily be used for testing MOSFETs?

28. In what way is the unijunction transistor not like a true transistor?
29. State the basic function of a unijunction transistor.
30. (a) Draw the schematic symbol and identify the leads of a standard N-type unijunction transistor.
    (b) In what way is the UJT symbol different from that of a JFET symbol?
31. To what junction of the UJT is the input signal applied?
32. Assume a variable *positive* voltage is applied to the emitter of an N-type UJT. Describe what happens as this voltage is increased from zero.
33. Assume a variable *negative* voltage is applied to the emitter of an N-type UJT. Describe what happens as this voltage is increased from zero.
34. Briefly explain how a phototransistor operates to control current.
35. Draw the schematic symbol and identify the leads of a three-lead NPN phototransistor.
36. Why is the phototransistor more sensitive to changes in light intensity than the photodiode?

Problems

1. The base current of a bipolar transistor circuit is increased from $25\,\mu A$ to $60\,\mu A$. This change produces an increase in collector current from 3 mA to 7 mA. Calculate the current gain of the circuit.
2. Determine the emitter current in a transistor if the base current is 2 mA and the collector current is 160 mA.
3. What is the base current in a transistor if the emitter current is 58 mA and the collector current is 57 mA?
4. A BJT has a β of 200. Calculate the value of the collector current if the value of the base current is $300\,\mu A$.

Critical Thinking

1. When looking at a schematic diagram that shows NPN and PNP transistors how can you identify each type?

2. As part of a job interview you are asked how you would check a transistor in and out of circuit. What would your response be?

Portfolio Project

• Obtain an unmarked bipolar transistor. Use an ohmmeter to correctly identify the emitter, collector, and base leads. Determine if the transistor is a NPN or PNP type. Report on the procedures followed in obtaining this information.

Circuit Challenge

Using a Simulator

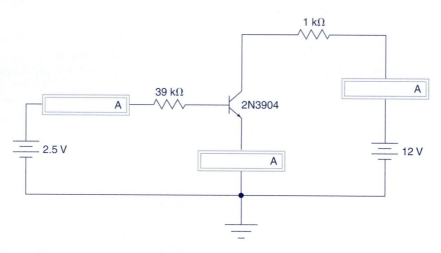

Procedure

(a) Construct the BJT circuit shown, using whatever simulation software package is available to you.

(b) Turn the simulation on and record the value of the emitter current, base current, and collector current.

(c) Using the values recorded in part (b), calculate the current gain of the circuit.

(d) Repeat parts (b) and (c) with the value of the base-emitter voltage source increased to 3 volts.

(e) Repeat parts (b) and (c) with the value of the base-emitter voltage source decreased to 2 volts.

Transistor Switching, Amplification, and Oscillation Circuits

Transistor switching and amplifier circuits are the most important and the most basic types of electronic circuits. These transistor circuits are connected in various ways for useful electronic functions. In this chapter, we will study some of these applications. We will also discuss the oscillator, which is a special type of amplification used in many communications circuits.

Objectives

After studying this chapter, you should be able to:
1. Discuss the use of transistors in switching applications.
2. Wire and test simple transistor-switching circuits.
3. Discuss the use of transistors in amplifying circuits.
4. Wire and test simple transistor-amplifying circuits.
5. Discuss the use of transistors in oscillator circuits.
6. Wire and test simple transistor-oscillator circuits.

Key Terms

- electronic switch
- saturation
- cutoff
- photoconductive (CdS) cell
- timed-switch
- dc amplifier
- ac amplifier
- voltage amplifier
- common-emitter amplifier
- amplifier coupling
- load resistor

- fixed-base bias
- quiescent operating point
- voltage-divider bias
- load line
- linear amplifier
- distortion
- cascaded amplifiers
- power amplifier
- bandwidth
- impedance matching
- class-A amplifier

- class-B amplifier
- push-pull amplifier
- audio amplifier
- decibels
- dB voltage gain
- tone control
- frequency response
- transistor oscillator
- resonant circuit
- multivibrator oscillator
- inverter

RESEARCH
INTERNET
SITES

- Circuit Specialists
- Heathkit
- Qkits

38-1 Transistor-Switching Circuits

Many electronic *digital* systems rely heavily on the transistor's ability to act as a switch. Used as a switch, the transistor has the advantages of having no moving parts, being able to operate ON and OFF at a very high rate of speed, and requiring very low driving voltages and currents in order to trigger the switching action.

When used as an **electronic switch**, a *BJT* normally operates in two states: either **cutoff** or **saturation**, but not in between. When a transistor is saturated, its emitter-base junction is forward-biased and base current is large enough to cause its collector current to reach its saturated value. Figure 38-1 illustrates the switching action of a BJT. The ON and OFF conditions of the BJT are shown with the mechanical switch equivalent circuit comparisons. When a transistor is saturated, it is like a closed switch from emitter-to-collector. When a transistor is cut-off, it is like an open switch from emitter-to-collector.

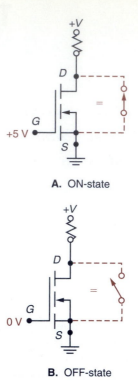

A. ON-state

B. OFF-state

Figure 38-2   Switching action of a MOSFET.

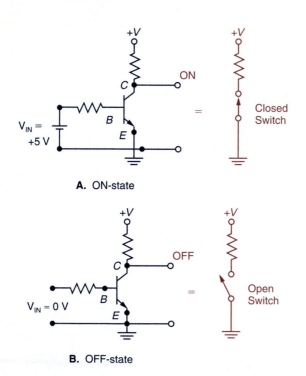

A. ON-state

B. OFF-state

Figure 38-1   Switching action of a BJT.

MOSFETs can be switched in the same fashion as BJTs. Figure 38-2 illustrates the switch-

ing action of a MOSFET. For the N-channel device shown, both the drain and gate are biased positive with respect to the source. The gate-to-source voltage V_{GS} is the input voltage, which is used to control the resistance between the drain and source (the channel resistance) and therefore determines whether the device is ON or OFF. When $V_{GS} = 0$ V, there is no conductive channel between source and drain, and the device is OFF. When $V_{GS} = +5$ V, the channel conducts maximum current and the device is switched ON. The channel resistance between the source and drain will switch from a very high resistance to a low resistance as the gate voltage switches from 0 to +5 volts.

Light-Controlled Transistor Switch

Electronic circuits are often required to be light-sensitive. One example is automatic control of house or street lights for night and day operation. Figure 38-3 shows the operation of an experimental *light-controlled* BJT switching circuit. A cadmium sulphate (**CdS**) *photoconductive cell* is used to sense different light levels. This CdS cell is effectively a resistor, which changes in resistance value depending on the

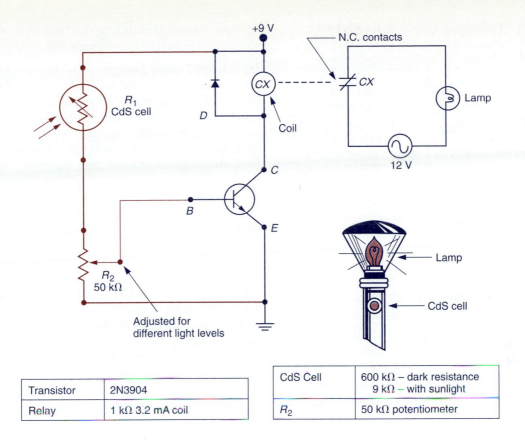

OPERATION

A. CdS CELL IN DARKNESS
- R_1 and R_2 form a voltage divider across the 9-V dc source.
- In darkness, the resistance of R_1 (CdS cell) is much *greater* than that of R_2.
- Almost all of the 9-V source voltage appears across R_1.
- The voltage across R_2 is too low to produce sufficient base current flow to switch the transistor ON.
- Relay coil is de-energized.
- The NC contacts are closed.
- Circuit to the lamp is completed and the lamp will be ON.

B. CdS CELL IN FULL SUNLIGHT
- In sunlight, the resistance of R_1 (CdS cell) is much *less* than that of R_2.
- Most of the 9-V source voltage appears across R_2.
- Increased voltage drop across R_2 produces sufficient base current flow to switch the transistor ON.
- Relay coil is energized.
- The NC contacts are open.
- Circuit to the lamp is opened and the lamp will be OFF.
- Potentiometer R_2 is adjusted to make the circuit operate at different levels of sunlight.

Figure 38-3 Light-controlled BJT switching circuit.

light energy present. In darkness, its resistance is very high, and in bright sunlight, it is lower.

The CdS-cell sensor does not have the voltage or current capacity to operate the lamp directly. Instead, it is used to vary the transistor base-to-emitter voltage and current in accordance with different sensed light levels. The *relay coil* acts as the collector load, energizing when the transistor is switched ON and de-energizing when the transistor is switched OFF. The *normally-closed (NC) relay contacts* open when the coil is energized and close when the coil is de-energized to switch the lamp OFF and ON.

The schematic of Figure 38-4 is that of a simple *phototransistor switching circuit*. The phototransistor Q_1 is used to drive the bipolar transistor Q_2. When the phototransistor is in darkness (PB_1 open), current flow through Q_1 is

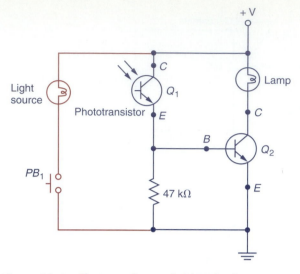

Figure 38-4 Phototransistor switching circuit.

insufficient to produce enough base current through Q_2 to switch it into conduction. As a result, no current is conducted between the collector and emitter of Q_2, and the lamp is OFF. When there is sufficient light present at the lens opening of Q_1 (PB_1 closed), it passes

enough current through to the base of Q_2 to switch the transistor and lamp ON.

Transistor-Timed Switch The transistor can also be used as a timed switch to turn a circuit ON or OFF for a preset period of time. A typical application of such a circuit is the lamp on an electronic garage door opener. When the door is first opened or closed, the lamp is turned ON for a short period of time then automatically turns OFF.

Figure 38-5 shows the operation of an experimental *transistor-timed switching circuit* that uses an NPN bipolar transistor. The length of the timing period is determined by the time it takes the capacitor to discharge after being initially charged.

Figure 38-6 shows the operation of an experimental N-channel MOSFET timer circuit. Unlike the base current drain of a bipolar-junction transistor, the gate current drain of a MOSFET is negligible. As a result, very long time delays of from minutes to hours are possible.

Another application for a timed-transistor switching circuit is the electronic-combination

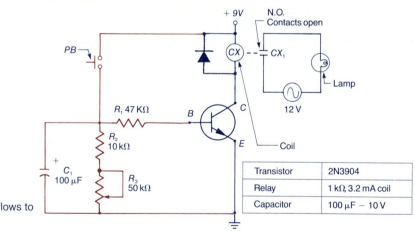

OPERATION

A. PUSH BUTTON CLOSED

- Sufficient base current flows to switch transistor ON.
- Relay coil energizes.
- NO relay contacts close to switch lamp ON.
- Capacitor charges to 9 V.
- The circuit remains in this state with the lamp ON as long as the pushbutton is closed.

B. PUSH BUTTON RELEASED

- Timing action begins.
- Base circuit to the 9-V dc source is opened.
- The capacitor begins to discharge its stored energy through the base circuit.
- Transistor continues to be switched ON until the charge on the capacitor is drained and then switches OFF.
- Relay coil de-energizes and relay contacts open to turn the lamp OFF.
- Fixed resistor R_2 and variable resistor R_3 form a second parallel-current discharge path across the capacitor.
- The discharge rate and, thus, the timing period can be adjusted by varying the resistance of R_3.

Figure 38-5 BJT timed-switching circuit.

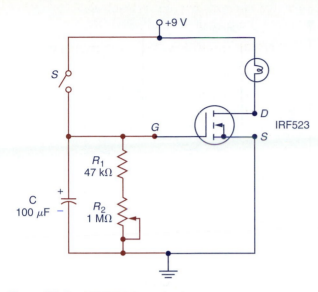

Figure 38-6 MOSFET timer circuit.

OPERATION

A. SWITCH CLOSED
- Positive voltage is applied to the gate.
- Transistor switches ON to switch lamp ON.
- Capacitor charges to 9 V.
- The circuit remains in this state with the lamp ON as long as the switch is closed.

B. SWITCH OPENED
- Timing action begins.
- Positive gate circuit to the 9-V source is opened.
- Positive charge on capacitor keeps the transistor switched ON.
- The capacitor begins to discharge its stored energy through R_1, and R_2 will still maintain a positive voltage at the gate.
- The transistor and lamp continue to conduct a current for as long as it takes the capacitor to discharge.
- Since the gate current flow is negligible, very long time delay periods are possible.
- The discharge rate and, thus, the timing period can be adjusted by varying the resistance of R_2.

lock. This circuit could be used in place of a key switch to manually operate an electronic garage door opener.

Figure 38-7 shows the operation of an experimental *electronic-combination lock* switching circuit. Basically, the first number is dialed and the button is pressed to charge the capacitor. Then the second number is dialed and the button is pressed to discharge the capacitor through the base circuit and switch the transistor into conduction. Dialing a wrong number and pressing the button causes any charge on the capacitor to immediately discharge.

38-2 Transistor Amplifiers

DC Amplifier   The transistor switching circuits are always operated at *saturation* or *cutoff*. Transistor *amplifiers* are operated in the region *between* these two points. In this region, the output signal varies in proportion to the input signal (Figure 38-8).

When operated as an amplifier, a transistor can act like a rheostat, or variable resistor, to *vary* (rather than switch ON and OFF) the amount of current delivered to a load. Transistors have many advantages over the rheostats they replace. They are smaller, lighter, and can control current much more efficiently than a rheostat.

An experimental *MOSFET lamp-dimmer* circuit is shown in Figure 38-9, p. 561. Basically, it is a *dc amplifier*. Adjustment of R_2 varies the amount of positive dc voltage applied to the gate. The MOSFET operates in the enhancement, or normally OFF, mode. Therefore, increasing the positive gate voltage lowers the channel resistance and increases drain current flow. Thus, changing the setting of R_2 changes the brightness of the lamp. This circuit illustrates how a transistor can be used as a variable resistor. The maximum load that can be safely connected into the circuit is determined by the power rating of the power MOSFET transistor.

AC Voltage Amplifier   Most pieces of electronic communications equipment make use of the transistor's ability to act as an *ac amplifier*. An ac amplifier is used to increase the level or strength of small ac signals. Basically, an ac amplifier is supplied with a dc operating voltage and small ac signal input voltage. The circuit produces an ac output signal that is an amplified version of the applied ac input signal (Figure 38-10, p. 561).

There are three basic configurations or methods of connecting transistors in circuits as ac amplifiers. These circuits are named according to which transistor element is *common* to both the input and output circuit. The three BJT connections are: common-emitter circuit,

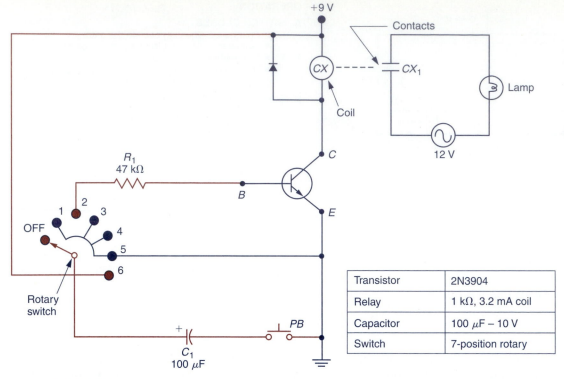

OPERATION

A. CHARGING CAPACITOR
- The number 6 is dialed on the rotary switch.
- The pushbutton is momentarily pressed closed.
- Capacitor charges to a full 9 V.

B. DISCHARGING THE CAPACITOR
- The number 2 is dialed on the rotary switch.
- The pushbutton is pressed closed.
- Capacitor discharges current through the emitter-base circuit to switch the transistor ON.
- Relay coil energizes to close the relay contacts and to switch the lamp ON.
- The unused rotary switch contacts are connected to ground to form a simple penalty circuit that ensures that the numbers must be selected in the correct sequence.

Figure 38-7 Electronic-combination lock circuit.

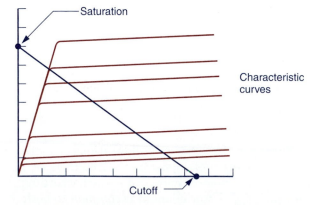

Figure 38-8 Transistor operating region—between saturation and cutoff.

common-base circuit, and common-collector circuit. Of the three, the **common-emitter** is the most widely used. Figure 38-11 shows the simplified circuit configurations for bipolar transistors and their corresponding equivalent field-effect transistors.

The common-emitter circuit, Figure 38-12, p. 562, can be used to illustrate how a BJT amplifier works. Basically, small base current changes are created by the microphone in accordance with the sound. These small base current changes are amplified by the collector circuit and are used to operate the headphones.

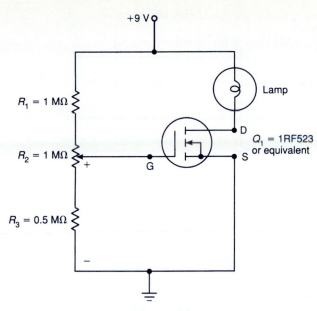

Figure 38-9 Power MOSFET lamp dimmer circuit.

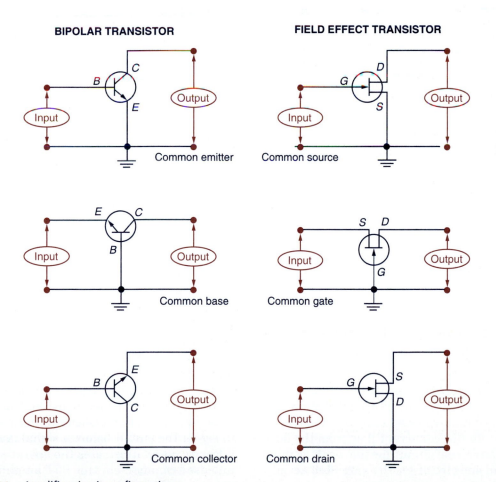

Figure 38-10 Basic transistor ac amplifier.

BIPOLAR TRANSISTOR

Common emitter

Common base

Common collector

FIELD EFFECT TRANSISTOR

Common source

Common gate

Common drain

Figure 38-11 Amplifier circuit configurations.

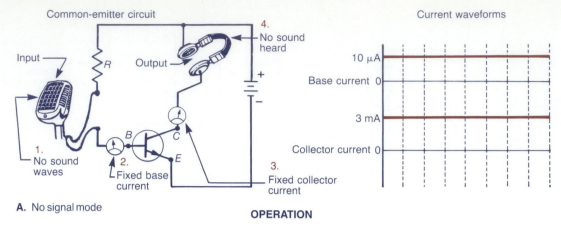

Common-emitter circuit

Input

R Output

No sound heard

4.

Current waveforms

10 μA
Base current 0

3 mA
Collector current 0

B
C
E

1.
No sound waves

2.
Fixed base current

3.
Fixed collector current

A. No signal mode

OPERATION

- A carbon microphone is used as the input transducer and a set of headphones is used as the output transducer.
- When no sound waves hit the microphone, the resistance of the microphone is fixed and a steady dc current is set up both in the base and collector circuits.

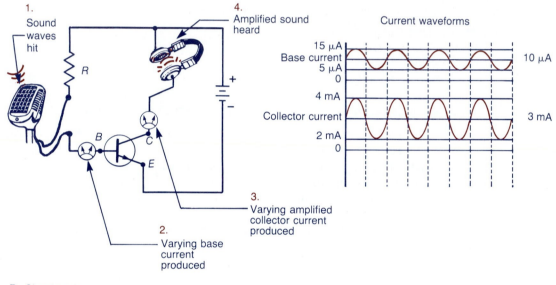

1.
Sound waves hit

4.
Amplified sound heard

Current waveforms

15 μA
Base current
5 μA
0

10 μA

4 mA
Collector current
2 mA
0

3 mA

R

B
C
E

3.
Varying amplified collector current produced

2.
Varying base current produced

B. Signal mode

OPERATION

- When sound waves hit the microphone the resistance of it varies in accordance with the sound.
- This produces a change in the base current corresponding to the sound waves that are present at the microphone.
- As a result of these small base current changes, a relatively larger collector current change takes place through the coils of the headphones.
- In this way the current flow through the headphones is controlled and amplified in exact proportion to the much smaller microphone current signal.

Figure 38-12 BJT common-emitter amplifier.

A transistor **voltage amplifier** is one that is designed to increase the voltage level of an ac signal. The circuit consists of three subcircuits: one circuit for direct current flow from the dc voltage source, one circuit for the weak ac input signal, and one circuit for the amplified ac output signal.

The dc circuit of the transistor amplifier is more commonly referred to as the *bias circuit*. Bias, as applied to transistor amplifiers, means to preset the circuit before a signal is applied. Figure 38-13 illustrates the operation of an unbiased common-emitter BJT amplifier. The *input* signal is applied to the *base* and the *out-*

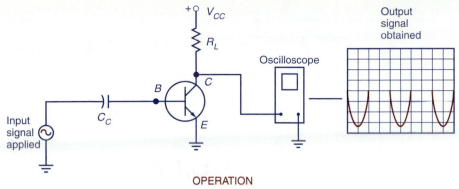

OPERATION

■ With no input signal applied, no base current flows so the transistor is turned completely OFF and the full V_{CC} voltage is dropped across the emitter collector.
■ As the input signal voltage begins to rise in a positive direction, base current begins to flow.
■ This lowers the emitter-collector resistance, causing collector current flow and a decrease in voltage across the emitter collector.
■ This decrease in voltage continues until the input sine-wave voltage reaches its peak value.
■ As the input signal voltage drops back to zero, the base current decreases, and the voltage across the transistor increases.
■ As the input voltage waveform increases in the negative direction, the base current remains at zero, since it already is completely OFF.
■ As a result, the amplified output waveform is inverse or opposite of the waveform applied to the base with one-half of it cut off.

Figure 38-13 Unbiased transistor amplifier.

put signal is obtained from the *collector* with the emitter being common to both. **Load resistor**, R_L, is used to limit the current through the emitter-collector circuit. **Coupling** *capacitor*, C_C, blocks any dc component from the signal source. This ensures that only an ac signal will be applied to the base of the transistor. Basically, this arrangement causes only one-half of the input signal to be amplified.

In order for the transistor to be able to reproduce the complete input signal, it must be biased. One of the simplest types of bias circuits is the **fixed-base bias** common-emitter circuit illustrated in Figure 38-14. With no input signal present, a dc current is applied to the base of the transistor and adjusted to a point that the transistor is turned *half* ON. This will permit the input signal to vary the collector current around a central point. With no ac input signal applied, the dc base and collector currents are such that they are operating at a steady value called the **quiescent value**. This will allow the applied signal to increase and decrease both the base and collector currents around this central operating point.

Heat has a significant influence on the operation of a BJT. The fixed-base bias circuit lacks the temperature stability required for many transistor-amplified circuits. With this type of

circuit, the operating point is dependent on base current, which can vary with changes in temperature. A better arrangement for biasing is the **voltage-divider bias** circuit. A BJT amplifier that uses a voltage-divider bias circuit is shown in Figure 38-15, p. 565. Resistors R_1 and R_2 form a series circuit or *voltage-divider* across the dc source. The voltage drop across R_2 is used to produce the emitter-base dc bias current. This circuit greatly increases the temperature stability of the amplifier.

A further step can be taken to improve temperature stability by connecting resistor R_4 in series with the emitter lead as shown in Figure 38-16, p. 565. The voltage across R_4 changes under different temperature conditions to adjust the base current up or down. This helps to counteract any base current changes that occur as a result of temperature changes. Used by itself, R_4 will reduce the overall voltage gain of the transistor amplifier stage. Used with bypass capacitor C_1 connected across it, R_4 maintains the temperature stabilizing effect as well as the voltage gain.

Figure 38-17, p. 565, shows a complete ac voltage amplifier circuit. The ac input signal to the transistor amplifier is applied through a coupling capacitor C_2 to the emitter-base circuit of the transistor. Use of a coupling capacitor

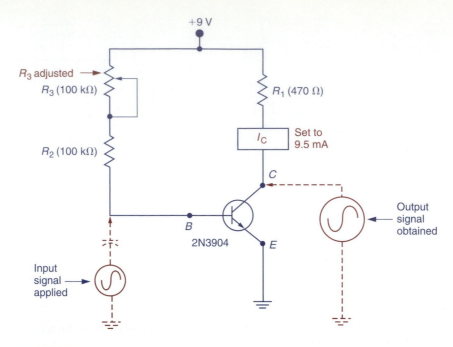

+9 V

R₃ adjusted →
R_3 (100 kΩ)

R_1 (470 Ω)

I_C Set to 9.5 mA

R_2 (100 kΩ)

C

Output signal obtained

B

2N3904

E

Input signal applied

OPERATION

- Variable resistor R_3 is adjusted to set the operating point of the circuit.
- In this circuit, R_3 is adjusted for a dc collector current of 9.5 mA. This will set the collector current at the midpoint between 0 and the maximum of 19 mA.
- The input signal to be amplified is applied through a capacitor to the base and emitter leads.
- The amplified signal appears across the collector-emitter terminals.

NOTE: The output signal is 180° out-of-phase with the input signal. When the input goes positive, the output goes negative and vice versa. This is called *phase inversion* and reflects the characteristics of a common-emitter amplifier.

Figure 38-14   BJT fixed-base bias circuit.

blocks the dc path to the input devices so that its resistance will not affect the dc bias currents. Similarly, the ac output circuit is taken from the collector to emitter through a second coupling capacitor C_3. Again, use of a capacitor blocks the dc path to the output load resistor R_5 so as not to change the bias current. Voltage gain of the amplifier is dependent on two factors. The first is the beta (I_C/I_B) of the transistor. For a given amplifier circuit, the greater the beta, the greater the voltage gain. Secondly, the value of the collector load resistor is an important factor. The higher the resistance, the greater the voltage change across it will be, given a change in the current.

The operating point of an amplifier can be demonstrated pictorially by plotting a **load line** on a set of characteristic curves. Characteristic curves are available for each type of transistor. Figure 38-18, p. 566 shows a load line drawn on a collector family of characteris-

tic curves for a BJT common-emitter amplifier. The load line is established in two simple steps. First, the transistor is considered to be an open circuit (*cutoff*). Under these conditions, the collector current will be zero, and the collector voltage will be V_{CC}. The second step is to consider the transistor at *saturation* (transistor short-circuited). Under the conditions, the collector voltage will be zero and the collector current is determined by using the equation:

$$I_C = \frac{V_{CC}}{R_L}$$

The best operating point is usually in the *center* of the load line.

Linear amplifiers have an output signal that is the same as the input signal, only it is larger. If the input is a sine wave, the output should be a sine wave (Figure 38-19A, p. 567). **Distortion** is any undesirable change in the original sig-

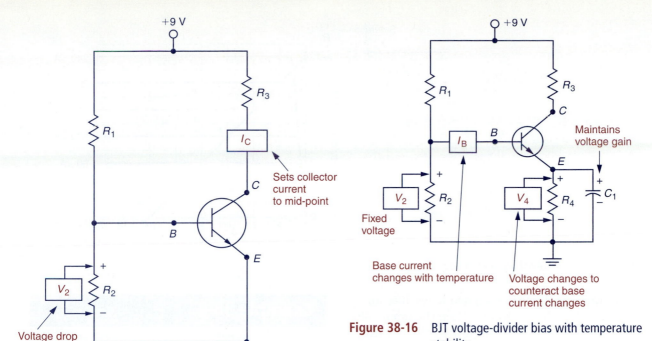

Figure 38-15 BJT voltage-divider bias circuit.

Figure 38-16 BJT voltage-divider bias with temperature stability.

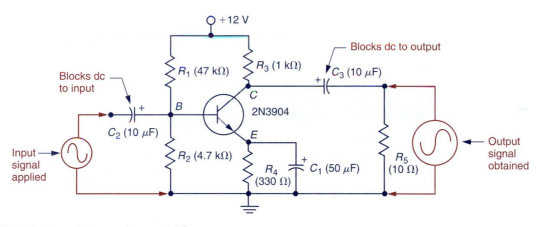

Figure 38-17 Complete ac voltage amplifier.

nal. For linear amplification, when selecting an operating point, the signal swing should never reach the extremes of saturation or cutoff. Figures 38-19B and 38-19C show signals might be distorted or clipped due to an operating point that is located too close to saturation or cutoff. When an input signal that is too large is applied to the amplifier, it is said to be **over-driven**. A good example of this is a volume control that is turned up too high. In this case, clipping occurs on both ends of the output waveform (Figure 38-19D, p. 567).

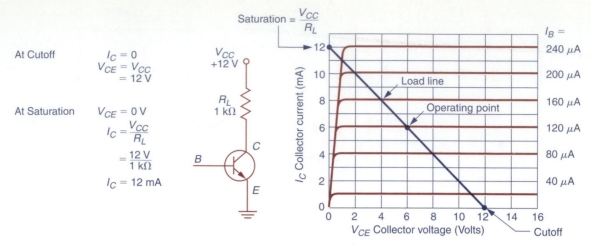

At Cutoff
$$I_C = 0$$
$$V_{CE} = V_{CC}$$
$$= 12\ V$$

At Saturation
$$V_{CE} = 0\ V$$
$$I_C = \frac{V_{CC}}{R_L}$$
$$= \frac{12\ V}{1\ k\Omega}$$
$$I_C = 12\ mA$$

Figure 38-18 Establishing a load line.

A voltage amplifier is designed to produce an output voltage that is greater than the input voltage applied. This type of amplifier will typically receive an input signal (V_{in}) measured in millivolts and produce an output signal (V_{out}) that is normally measured in volts. When a dc signal is applied, the amount of voltage amplification (A_V) is a ratio of the output signal voltage, to the input signal voltage. If an ac signal is applied, the voltage amplification is equal to the amount of change (Δ) in the input signal compared to the amount of change in the output signal.

Example 38-1

What is the voltage gain of a transistor amplifier when a dc voltage of 100 mV is applied to the input, and a dc voltage of 1 V (1000 mV) is obtained from the output?

$$A_V = V_{out}/V_{in}$$
$$= 1000mV/100mV$$
$$= 10$$

While one stage may have an amplification figure in the range of 100–300, stability and predictability of signal shape are sacrificed at this amplification level. In such a case, two or more amplifiers are used to obtain the gain required. Amplifiers connected in this manner are called *cascaded amplifiers*.

Figure 38-20, p. 568 shows the schematic for an experimental common source *JFET* voltage

Example 38-2

The value of the ac input signal applied to a voltage amplifier is measured and found to be 200 mV peak-to-peak. If the amplified output signal is 8 volts peak-to-peak, calculate the voltage gain.

$$A_V = V_{out}/V_{in}$$
$$= 8000mV/200mV$$
$$= 40$$

amplifier. In comparison to the BJT common emitter amplifier, the FET amplifier has much higher input impedance, but a lower voltage gain. The dc bias voltage is obtained from the voltage drop across R_S. Current flow through R_S produces a voltage drop that is positive at the source which reverse biases the source-to-gate junction. The gate is connected to common through R_G and, therefore, is negative with respect to the source. The ac input signal is applied across R_G and varies the negative gate voltage around the dc bias, causing a variation in drain current. Capacitors C_1 and C_2 function as coupling capacitors to block the flow of dc to the input and output circuits. C_S is used as a bypass capacitor to shunt ac signals around R_S and thus maintain a constant dc gate-to-source bias voltage with a signal applied.

The primary objective of the amplifier is to produce *gain*. Gain is a ratio of output to input and has *no unit of measure,* such as volts or amps, attached to it.

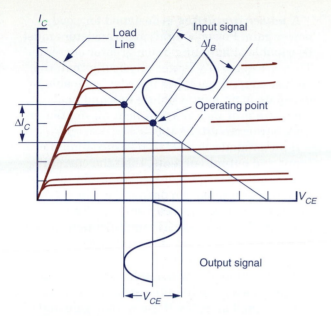

A. Linear amplifier

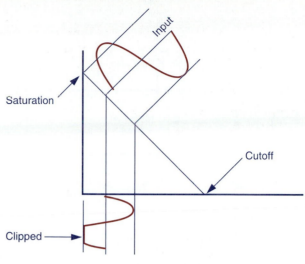

B. Clipping due to operating too close to saturation

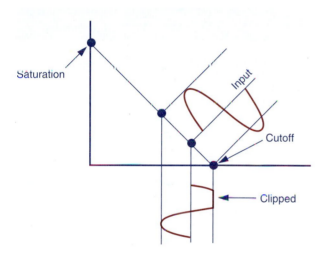

C. Clipping due to operating too close to cutoff

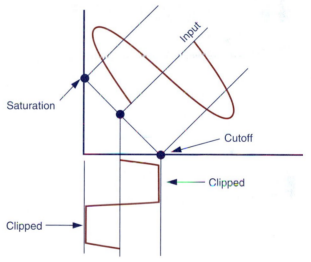

D. Clipping due to amplifier being overdriven

Figure 38-19   Linear and distorted amplifier outputs.

The cascade-amplifier configuration is obtained by simply connecting the output of one amplifier stage to the input of the next amplifier stage (Figure 38-21, p. 568). When dealing with a cascade amplifier, the final amplification figure is the *product* of each individual stage. For example, to calculate the final amplification of a two-stage cascade amplifier, if the voltage gain of one stage were 8 and the other were 9, the total gain would be 72.

The gain of an amplifier is not the same at all frequencies. Amplifiers are designed to operate within a given frequency range. If an amplifier is operated outside of this frequency range, the gain may decrease. The range of frequencies over which the gain of an amplifier is maximum and remains relatively constant is called its *bandwidth*.

Coupling refers to the method used to transfer the signal from one stage of an amplifier to the next. There are three basic types of coupling: capacitor, transformer, and direct. Capacitor coupling is useful in ac amplifiers because a capacitor will block direct current and allow the ac signal to be coupled. Transformer coupling can also be used in circuits

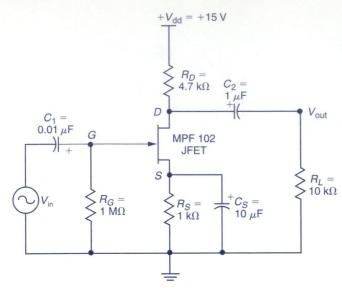

Figure 38-20 Common source JFET voltage amplifier.

A **power amplifier** is designed for good power gain. Power-amplifying transistors must be capable of handling much greater currents. They are generally larger physically than the voltage-amplifying types in order to be able to dissipate more heat. Cooling fins and heat sinks are provided to assist heat dissipation in high power circuits.

If an ammeter is used to measure amplifier input and output currents, then the current gain can be determined. A current amplifier is designed to produce an output current (I_{out}) greater than the input current (I_{in}). The amount of current gain or amplification can be calculated by the same gain ratio formula of output divided by input.

A power amplifier will amplify an input signal, typically measured in milliwatts, and produce an output signal in watts. If the voltage gain and

processing ac signals and has the added advantage of impedance matching. A direct coupled amplifier uses wire or some other dc path between stages. Remember that a Darlington circuit is an example of direct coupling.

AC Power Amplifier   All amplifiers are power amplifiers in the sense that they change the input dc power into output signal power. However, those dealing with small signals, where the most important function is voltage gain, are called ***voltage amplifiers***.

Example 38-3

Calculate the power gain of an amplifier that has a voltage gain of 10 and a current gain of 5.

$$P_{gain} = V_{gain} \times I_{gain}$$
$$= 10 \times 5$$
$$= 50$$

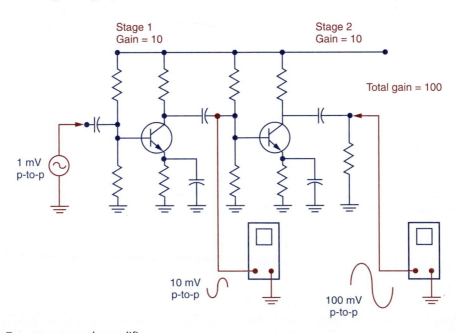

Figure 38-21 Two-stage cascade amplifier.

the current gain are both known, then the power gain can be determined by multiplying the two.

An experimental BJT power-amplifier circuit is shown in Figure 38-22. Basically, it has the same circuit requirements as the voltage amplifier except for the collector load resistor. An *output transformer* is used in place of a load resistor. The output transformer acts as an **impedance-matching** *device,* with the real load being the loudspeaker. Loudspeaker impedances are very low, from 4–16 Ω as a rule, whereas the output impedance of a transistor is likely to be 1000 times greater. To match these different impedances (which is necessary if maximum power is to be transferred) an output transformer is connected as a *step-down transformer.* The turns ratio is such that each circuit is presented with the impedance required to obtain the desired match.

An experimental MOSFET power-amplifier circuit is shown in Figure 38-23. The input signal to be amplified is applied from gate to source through coupling-capacitor C. A voltage-divider bias consisting of resistors R_1, R_2, and R_3 is used. The setting of R_2 controls the output volume. This circuit can be used to operate a speaker directly without the use of an output transformer.

The class of an amplifier indicates how it will be biased. A **class-A amplifier** is biased so that it conducts continuously. The bias is set so that the input varies the collector (or drain)

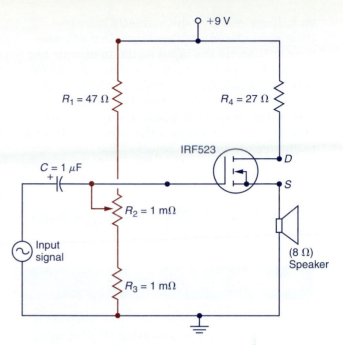

Figure 38-23   MOSFET power amplifier.

current to provide linear amplification. Usually we say that the class A amplifier conducts for *360° of an input sine wave.*

A **class-B amplifier** is biased at cutoff so that no current flows with zero input. The transistor conducts on only one-half of the sine wave input. In other words, it conducts for *180° of a sine wave input.* This means that only one-half of the sine wave is amplified.

Normally, two class-B amplifiers are connected in a **push-pull** arrangement so that both the positive and negative alterations of the input are amplified. Figure 38-24 shows a push-pull power amplifier. Each transistor operates

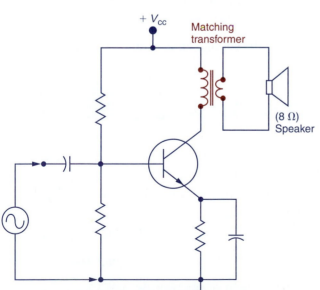

Figure 38-22   BJT power amplifier.

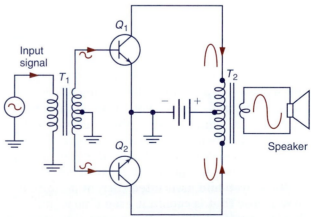

Figure 38-24   Class-B push-pull power amplifier.

on half-cycles only, which greatly increases their power handling capability. Driver transformer T_1 splits the input signal to provide two signals 180° out-of-phase that drive Q_1 and Q_2. Output transformer T_2 combines the two signals and supplies the output to the speaker.

It is important that a power amplifier has good efficiency. The job of the power amplifier is to change the input dc power of the source into output signal power. Its efficiency is determined by the equation:

$$\text{Efficiency} = \frac{\text{signal power output}}{\text{dc power input}} \times 100$$

Example 38-4

A class A power amplifier draws 5 A from a 12 V dc source while delivering 40 W of signal power. Determine the amount of dc input power output, efficiency of the amplifier and the amount of power dissipated in the amplifier itself.

$$\begin{aligned} \text{dc power input} &= V \times I \\ &= 12\,\text{V} \times 5\,\text{A} \\ &= 60\,\text{W} \end{aligned}$$

$$\begin{aligned} \text{Efficiency} &= \frac{\text{signal power output}}{\text{dc power input}} \times 100 \\ &= \frac{40\,\text{W}}{60\,\text{W}} \times 100 \\ &= 66.7\% \end{aligned}$$

Power dissipated in the amplifier = input supply power − signal output power

$$= 60\,\text{W} - 40\,\text{W}$$
$$= 20\,\text{W}$$

Class B amplifiers are much more efficient than class A types, since class B amplifiers are biased at cutoff. There is no drain on the power supply if no signal is being amplified. Class A type amplifiers operate near the center of the load line, which means about half of the supply voltage is dropped across the transistor. This voltage drop and associated current produce a power loss that is constant even if no signal is being amplified. The wattage needed to operate a class B push-pull amplifier at a given power

level is approximately one-fifth that needed for an equivalent class A type.

Audio Amplifier   An *audio amplifier* is designed to amplify audio-frequency signals (approximately 20 to 20,000 Hz). Since these frequencies are in the range of hearing of the human ear, audio amplifiers are associated with voice or other sound signals. For example, public address systems as well as radio and TV receivers always contain an audio amplifier.

The loudness response of human hearing is logarithmic. For this reason, logarithms are often used to describe the performance of audio systems. A scientific calculator can be used to find the logarithm of a number. The power gain of an audio amplifier can be measured in *decibels (dB)*, which are a logarithmic unit. Power gain in decibels can be calculated using the formula:

$$\text{dB power gain} = 10 \times \log\left(\frac{P_{out}}{P_{in}}\right)$$

One reason decibel power gain is often used is that smaller numbers result. For example, power gains of 100 and 100,000,000 can be expressed as 20 and 80 dB. When the circuit has gain, the decibel figure is positive. If the gain is less than 1, meaning there is an attenuation, the decibel figure is negative. To calculate the overall gain of several stages you simply add together the decibel gain of each stage.

The equation for finding *dB voltage gain* is slightly different from the one used for finding dB power gain.

$$\text{dB voltage gain} = 20 \times \log\left(\frac{V_{out}}{V_{in}}\right)$$

Example 38-5

An amplifier has an input of 3 mV and an output of 5 V. What is the dB voltage gain?

$$\begin{aligned} \text{dB voltage gain} &= 20 \times \log\left(\frac{V_{out}}{V_{in}}\right) \\ &= 20 \times \log\left(\frac{5}{0.003}\right) \\ &= 20 \times \log 1666.67 \\ &= 20 \times 3.22 \\ &= 64.4 \end{aligned}$$

The schematic for an experimental audio amplifier circuit is shown in Figure 38-25. It consists of a voltage-amplifying stage and a power-amplifying stage. The two stages are coupled, or connected together, by means of capacitor C_3. This coupling capacitor blocks the dc path between the two stages, while at the same time allowing the amplified ac signal to pass through.

The voltage-amplifier stage receives an audio-input signal from an input transducer such as a microphone. The purpose of the voltage amplifier is to increase the voltage level of the input signal to the point where it is sufficient to drive the power amplifier stage. This voltage increase is necessary, since the power amplifier requires a relatively large input voltage in order to develop a large current or power output. The amplified signal voltage at the output of the voltage amplifier is coupled to the power amplifier, which operates the speaker.

Additional features of an audio amplifier often include volume and tone controls. The **volume control** consists of a variable resistor that is manually adjusted to provide the desired listening output level from the speaker. The schematic for an experimental volume-control circuit is shown in Figure 38-26. It operates by setting the amplitude or voltage level of the signal input to the voltage-amplifier stage. The input-signal current flows through the volume-control resistor and develops a voltage across it. By adjustment of the sliding contact

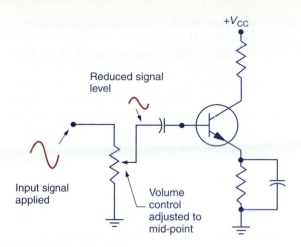

Figure 38-26 Amplifier volume control.

on the resistor, one can determine whatever percentage of the total signal voltage is to be applied to the input of the transistor. The larger the portion of resistance selected, the greater the voltage drop, and the larger the input signal to the stage.

The **tone control** of an audio amplifier varies the high- and low-frequency level of the amplified sound according to the listener's taste. By varying the **frequency response** of the audio amplifier the higher or lower frequencies within the audio range can be emphasized. Treble tone means a greater response for the high audio frequencies above 3000 Hz. Bass tone means

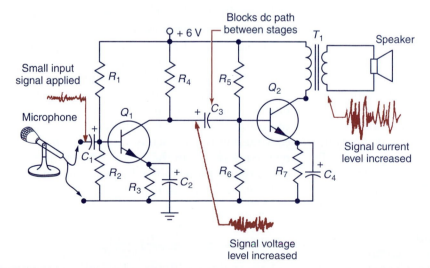

PARTS LIST	
R_1	47 kΩ 1 W
R_2, R_6	4.7 kΩ 1 W
R_3, R_7	330 Ω 1 W
R_4	1 kΩ 1 W
R_5	22 kΩ 1 W
C_1, C_3	10 μF
C_2, C_4	50 μF
T_1	1.5 kΩ : 8 Ω
Speaker	8 Ω
Q_1, Q_2	2N3904

Figure 38-25 Two-stage audio amplifier.

more response for the low audio frequencies from about 300 Hz down.

The schematic for an experimental tone control circuit is shown in Figure 38-27. It consists of a capacitor and a variable resistor that are connected across the primary coil of the output transformer. The tone is varied by adjusting the amount of resistance in series with the capacitor. Zero resistance results in more of the higher frequencies being shunted through the capacitor, giving more of a predominant bass, or low-frequency, response. Maximum resistance reduces the shunting effect so that the higher frequencies are attenuated less.

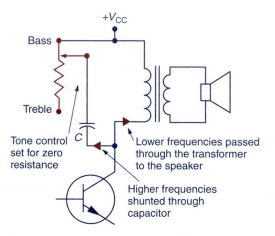

Figure 38-27   Amplifier tone control.

38-3 Transistor Oscillators

A *transistor oscillator* is a special transistor amplifier circuit that converts pure dc to either ac or to pulsating (varying) dc on its own with no external input signal being applied (Figure 38-28). As such, the *oscillator* can be viewed as being a type of signal generator. There are two major types of oscillators: sinusoidal oscillators and nonsinusoidal oscillators. **Sinusoidal oscillators** produce a *sine wave*-shaped output signal. **Nonsinusoidal oscillators** produce all other shaped waves, which include sawtooth, square wave, and pulses. Oscillators are used extensively in communications equipment, such as radio and TV circuits, as well as in computer and industrial electronic circuits.

A complete transistor oscillator circuit is made up of a **wave-producing circuit**, an **amplifying circuit**, and a positive **feedback**

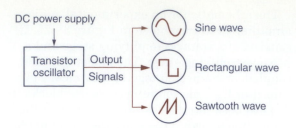

Figure 38-28   Transistor oscillator.

circuit (Figure 38-29). The wave-producing circuit generates the wave and determines the frequency of the oscillator. The weak output signal from the wave-producing circuit is fed to the input of the transistor amplifier for amplification and then on to the output. Part of the output from the amplifier is fed back to the original wave-producing circuit to keep this circuit operating.

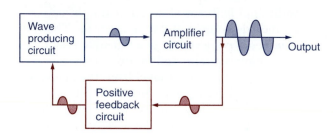

Figure 38-29   Transistor oscillator block diagram.

A *resonant circuit* (also known as a *tank circuit*) that consists of an inductor and a capacitor connected in parallel provides the simplest form of a wave-producing circuit to provide oscillations. Inductors and capacitors are both energy storage devices. Figure 38-30 illustrates the sequence of storing energy alternately in the electrostatic field of the capacitor and the magnetic field of the inductor, resulting in an oscillating current. The pushbutton is closed momentarily to charge the capacitor and start the oscillations. This action continues until the oscillations are damped out by the resistance in the circuit. The frequency of oscillations is determined by the values of the inductor and capacitor:

$$f_r = \frac{1}{2\pi\sqrt{LC}}$$

A transistor oscillator is basically a transistor amplifier in which a small amount of the out-

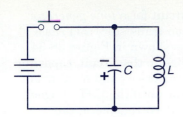

A. Button momentarily pressed closed to charge capacitor.

B. Capacitor discharges into the inductor. Magnetic field about the inductor builds up.

C. Magnetic field collapses. A voltage is induced in the inductor. Induced voltage causes current flow to continue and capacitor to be charged in the *opposite* polarity.

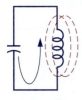

D. Capacitor discharges to transfer the energy to the magnetic field of the inductor.

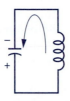

E. Magnetic field collapses again and the energy is returned to the capacitor. Capacitor is recharged to the *initial* polarity completing one cycle.

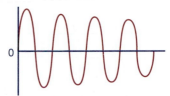

F. This action continues until oscillations are *dampened* out.

Figure 38-30   LC wave-producing circuit.

put voltage or current is fed back as an input in a direction which sustains oscillation.

An experimental ***code-practice oscillator circuit*** is shown in Figure 38-31. This circuit produces a nonsinusoidal-output wave within the audio frequency range that is used to drive a loudspeaker. Variable resistor R_1, fixed resistor R_2, and capacitor C_1 make up the wave-producing circuit. R_1 is varied to adjust the frequency of the wave-producing circuit for the

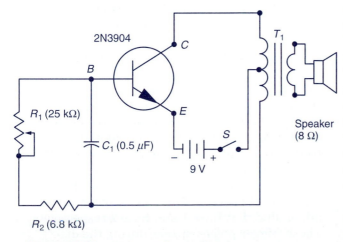

Figure 38-31   Code practice oscillator.

desired tone. The transistor amplifies the signal from this circuit and operates the speaker with its output through one-half of transformer T_1. The other half of the center-tapped transformer is used to provide a positive feedback to the wave-producing circuit. A key-switch is used to switch the dc voltage source ON and OFF to produce the desired audio code pulses.

An experimental circuit for an ***electronic siren oscillator*** is shown in Figure 38-32. This circuit uses an NPN and PNP transistor directly coupled together. The frequency of this circuit constantly changes to simulate a siren. When the pushbutton PB_1 is first closed, the speaker emits a tone that gradually increases in frequency from zero to maximum. When PB_1 is opened the speaker emits a tone that gradually decreases in frequency from maximum to zero. The rate of change in frequency is determined by the time it takes capacitor C_1 to charge and discharge. The charging rate of capacitor C_1 can be changed by increasing or decreasing the resistance of R_1.

An experimental circuit for a ***multivibrator oscillator*** is shown in Figure 38-33. This circuit has a square-wave output that is often found in computer circuits. In this particular

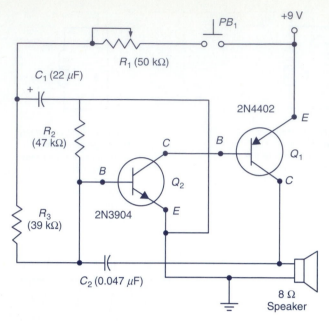

Figure 38-32 Electronic siren.

An experimental circuit for a *tone-generator* that uses a unijunction transistor (UJT) as a *relaxation* oscillator is shown in Figure 38-34. When voltage is applied to the circuit, capacitor C_1 charges through resistors R_1 and R_2. Current flows into the capacitor until the UJT's trigger voltage level is reached. At this point, the capacitor discharges through the emitter and Base-1 junction and speaker. The speaker conducts a current to emit sound until the capacitor is discharged. The charge-discharge cycle then repeats. The result is an audio frequency tone emitted by the speaker. Frequency rate is determined by the resistance setting of R_2. Reducing the resistance of R_2 increases tone frequency.

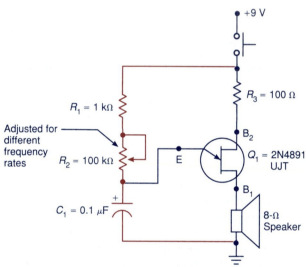

Figure 38-34 UJT tone generator circuit.

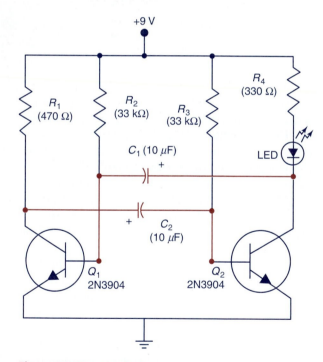

Figure 38-33 LED flasher.

application, the circuit is used as an LED *flasher*. Transistors Q_1 and Q_2 conduct alternately. When Q_2 conducts, the LED is flashed on. The rapid switching action is controlled by the values of the resistors R_2 and R_3 and capacitors C_1 and C_2. Increasing the values of C_1 and C_2 will slow the flash rate.

The UJT relaxation oscillator can be connected to provide signal waveforms used in computer logic circuits. An experimental **UJT signal-generator** circuit is shown in Figure 38-35. This circuit operates similar to that of the tone generator. Capacitor C_1 charges to the trigger voltage level and then discharges through the emitter. As a result, three useful output voltage waveforms are produced. A sawtooth waveform is produced at the emitter. The gradual increase in this sawtooth voltage is produced as a result of the charging of the capacitor. Positive and negative trigger pulses are produced at Base 1 and Base 2 respectively. These trigger pulses appear during the discharge of the capacitor because the UJT con-

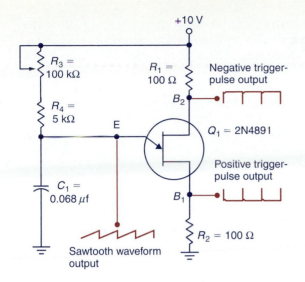

Figure 38-35 UJT signal generator circuit.

Figure 38-36 DC-to-ac power inverter.

ducts heavily at this time. Variable resistor R_3 can be adjusted for different frequency rates.

Power inverters are used to convert dc to ac. Some electric vehicles use an inverter to change the dc from the batteries into the ac needed for the vehicle's ac driven motors. People use them in remote areas to convert dc voltage from a 12-V battery into 120-V ac to power standard appliances.

A simple power inverter circuit is shown in Figure 38-36. The circuit is basically a free running multivibrator with transistors Q_1 and Q_2 conducting alternately. When Q_1 conducts, current flows in the upper half of the primary winding. When Q_2 conducts, current flows in the lower half of the primary winding. This action produces an ac voltage at the secondary of the transformer.

Review and Applications

Related Formulas

$$V_{gain} = \frac{V\ output}{V\ input}$$

V gain (cascaded amplifier) $=$ stage No. 1 gain $\times$ stage No. 2 gain . . .

$$P = V \times I$$

$$P_{gain} = V_{gain} \times I_{gain}$$

$$\%\ \text{Efficiency} = \frac{\text{signal power output}}{\text{dc power input}} \times 100$$

$$\text{dB power gain} = 10 \times \log\left(\frac{P_{out}}{P_{in}}\right)$$

$$\text{dB voltage gain} = 20 \times \log\left(\frac{V_{out}}{V_{in}}\right)$$

Review Questions

1. Outline three advantages of using a transistor as a switch.
2. (a) How does a photoconductive cell sense different light levels?
   (b) A photoconductive cell cannot operate a lamp directly. Why?
3. In a typical transistor-timer circuit, what component discharges to provide the timing action?
4. Compare the operating points of transistor switching and amplifying circuits.
5. When transistors are used like variable resistors, what advantages do they have over the rheostats that they replace?
6. What are ac amplifiers used for?
7. There are three basic configurations or methods of connecting transistors in circuits as amplifiers.
   (a) How are these circuits named?
   (b) Which circuit connection is most widely used with bipolar-junction transistors?
8. Name the three subcircuits of a transistor-voltage amplifier.
9. A bipolar-junction transistor (BJT) is to be biased in order to reproduce the complete ac input signal. With no input signal present, how are the dc current levels set?

10. What is the advantage of the voltage-divider bias circuit over the fixed-base bias circuit?
11. Why are coupling capacitors used on the input and output circuits of a transistor amplifier?
12. Explain how the value of (a) and (b) each affects the voltage gain of a given BJT amplifier.
    (a) the beta (I_C/I_B) of the transistor
    (b) the resistance value of the collector load resistor
13. (a) Explain the two simple steps involved to establish a load line on a collector family of curves for a BJT common-emitter amplifier.
    (b) What is usually the best-operating point along the load line?
14. Define what is meant by a linear amplifier.
15. List three ways in which an amplifier output can be clipped or distorted.
16. Compare voltage-amplifying and power-amplifying transistors.
17. Define the term *bandwidth* as it applies to amplifiers.
18. Explain the function of the output transformer as used in a BJT power amplifier circuit.
19. (a) What does a volume control consist of?

(b) Explain how it operates to set the volume.
20. How is tone control of an audio amplifier achieved?
21. Explain the function of a transistor-oscillator circuit.
22. State the purpose of each of the following subcircuits in a transistor oscillator circuit:
    (a) wave-producing circuit
    (b) amplifier circuit
    (c) feedback circuit
23. What causes the oscillating current between a charged capacitor and inductor connected in parallel?
24. What are power inverters used for?

Problems

1. Calculate the voltage gain of an amplifier if the output voltage is 4 V and the input signal is 50 mV.
2. If an amplifier has a voltage gain of 50, find the expected peak-to-peak value of the output signal when the input is 20 mV peak-to-peak.
3. Calculate the power gain of an amplifier that has a voltage gain of 5 and a current gain of 20.
4. Two amplifiers, one with a voltage gain of 10 and the other with a voltage gain of 5,

are cascaded. What is the total voltage gain of the circuit?
5. A 12 V power amplifier has an output rating of 70 W and an efficiency rating of 60 percent. Determine how much dc supply current this amplifier will require when delivering its rated output.

Critical Thinking

1. A transformer is used to couple two stages of an ac amplifier. Why will any dc component of the first stage not be passed on to the second stage.
2. In a two stage ac amplifier the collector of Q_1 is connected to the base of Q_2 through a coupling capacitor. What would happen if the coupling capacitor shorted?
3. What problem could occur when transistors used in push-pull amplifiers are not properly matched?
4. The output of an amplifier resembles a square-wave even though the input is a sine wave. What is the explanation?

Portfolio Project

• Construct one of the experimental circuits discussed in this chapter from scratch or using a kit. Complete a parts list for the circuit and a written explanation of how the circuit operates.

Circuit Challenge

Using a Simulator

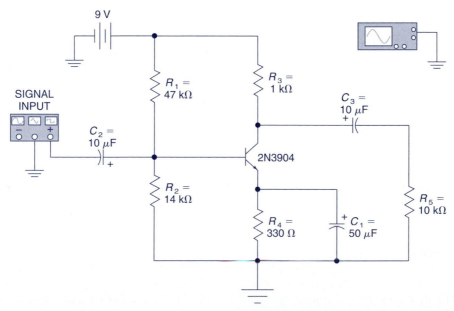

Procedure

(a) Construct the ac voltage amplifier circuit shown, using whatever simulation software package is available to you.

(b) Set the input voltage from the signal generator to a sine-wave voltage of 10 mV peak-to-peak at a frequency of 400 Hz. Use the scope to measure the peak-to-peak value of the output voltage across R_5.

(c) Calculate the voltage gain of the amplifier.

(d) Temporarily disconnect C_1 from the circuit and record the peak-to-peak value of the output voltage. Does this voltage remain the same, increase, or decrease?

(e) Temporarily increase the value of R_3 from $1\,k\Omega$ to $10\,k\Omega$ and report on the change in the shape and magnitude of the output waveform.

(f) Temporarily disconnect R_1 from the circuit. What happens to the output voltage?

(g) Temporarily increase the input voltage to 100 mV peak-to-peak and report on the change in shape of the output waveform.

Thyristors: The SCR and TRIAC

The term thyristor refers to a classification of solid-state devices that are used as electronic switches. Thyristors have the ability to control large amounts of power to a load while consuming very little activation power. When conduction is initiated, the device will latch or hold in its own state. The types of thyristors covered in this chapter are the silicon-controlled rectifier (SCR), the triac, and the diac.

Objectives

After studying this chapter, you should be able to:
1. Identify SCR, triac, and diac symbols.
2. Describe SCR, triac, and diac operating characteristics.
3. Use an ohmmeter to test an SCR.
4. Describe triac operating characteristics.
5. Discuss the use of phase shifting for SCR and triac circuits.
6. Describe diac operating characteristics.

Key Terms

- thyristor
- silicon-controlled rectifier (SCR)
- anode
- cathode
- gate
- forward-breakover voltage
- holding current
- gate trigger current
- commutation
- crowbar circuit
- triac
- MT_1
- MT_2
- diac
- lamp dimmer circuit

RESEARCH INTERNET SITES

- Sprint-Electronics
- Siemens
- ABB Power-Systems

39-1 Thyristors

Thyristors are a class of semiconductor devices that contain at least four layers of alternate P-type and N-type semiconductor materials. The *silicon-controlled rectifier (SCR)* and *triac* are the most frequently used of these devices. SCRs and triacs are the workhorses of industrial electronics. Common applications include: control of motors, robot control and manipulation, and heat and light control (Figure 39-1).

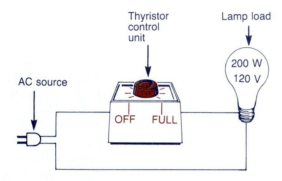

Figure 39-1 Thyristor lamp dimmer.

A thyristor is a semiconductor device that acts like a *switch*. Like a mechanical switch, it has two states: ON (conducting) and OFF (not conducting). There is *no* linear area in between these two states as there is in the transistor.

Although BJTs and FETs are capable of performing switching functions, these devices are limited in terms of their current-handling capabilities. With industrial electronic circuits, load currents can often be in the hundreds, or even thousands of amperes. Thyristors provide a highly efficient means of switching such heavy loads, and have excellent heat dissipation qualities.

39-2 Principle of SCR Operation

A *silicon-controlled rectifier*, commonly called an **SCR**, is a four-layer (PNPN) semiconductor device that makes use of three leads—the anode, cathode, and gate—for its operation (Figure 39-2). Unlike transistors, SCRs *cannot amplify* signals. They are used strictly as solid-state switches and are categorized according to the amount of current they can switch. Low-current SCRs can operate with an anode current of less than 1 A. High-current SCRs can handle load currents in the thousands of amperes. Most SCRs have provisions for some type of heat sink to dissipate internal heat.

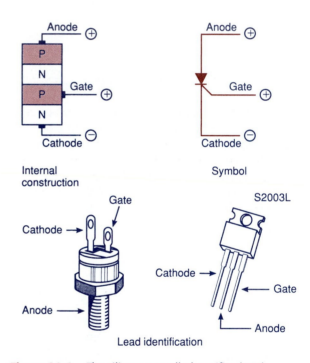

Figure 39-2 The silicon-controlled rectifier (SCR).

The schematic symbol for the SCR is very much like that of the diode rectifier. In fact, the SCR electrically resembles the diode, since it conducts in only one direction. In other words, the SCR must be forward biased from *anode* to *cathode* for current conduction. It is unlike the diode because of the presence of a **gate (G)** lead, which is used to turn the device ON.

SCR switching operation is best understood by visualizing its transistor equivalent circuit constructed using NPN and PNP transistors as shown in Figure 39-3. Current, trying to flow

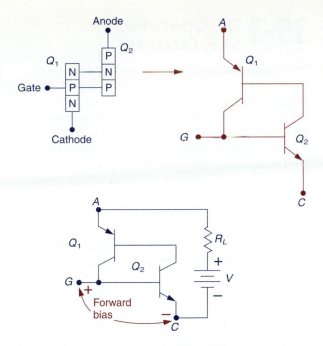

Figure 39-3 Transistor equivalent of a SCR.

current continues to flow into the base of the NPN, keeping it turned ON. The current continues to flow from cathode to anode until interrupted somewhere else in the circuit. This regenerative action is called ***latching*** because the SCR continues to conduct from cathode-to-anode even if the gate-to-cathode forward-bias voltage is removed.

An SCR can switch large amounts of power very efficiently. When it is in the OFF state, it does not draw much current. When it is ON, the voltage across it is around 1 V, regardless of the current through the device. This results in low power loss in the device. For instance, if the voltage drop across the device stays at 1 V with 20 A flowing through it, the power dissipated is only 20 W. All the rest of the power is transferred to the load.

Figure 39-4 shows the characteristic curve of a typical SCR. When reverse biased, the SCR operates as a regular diode; there will only be a very small leakage current until the breakdown voltage is reached. As in the diode, when the breakdown voltage is reached, avalanche conduction destroys the device. With forward bias applied, there is only a small forward-leakage current until the forward-breakover voltage is reached. At this point, the current increases rapidly and the resistance of the SCR is very small. The SCR then acts much like a closed

from cathode to anode, is blocked by both transistors. If forward current is pumped into the NPN base by way of the gate terminal, the NPN transistor turns ON. Its working current draws control current from the PNP base, which turns the PNP ON. Even after the original gate current has stopped, the PNP working

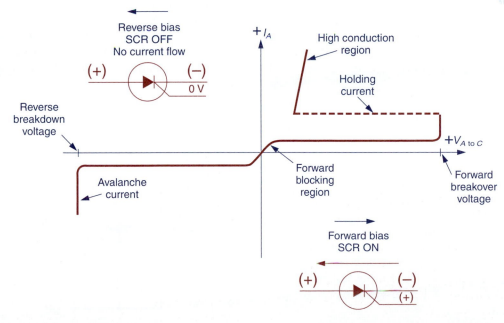

Figure 39-4 SCR characteristic curve.

switch and the current is limited only by the external load resistance.

Forward biasing the gate-cathode junction considerably lowers the **forward-breakover voltage** value. As a result, much less voltage between anode and cathode is required to turn the SCR ON. Once the device is in the ON state, the gate current can be shut OFF with no effect. The SCR will continue to conduct until the current reverses polarity or until it drops below the **holding current** (the amount of current required for the SCR to maintain conduction).

The SCR operates much like a mechanical switch. It is either ON or OFF. In normal operation, the anode-cathode circuit is forward biased and latched into the conducting state by applying suitable gate current. For the SCR to be most effective, this gate current must *not be* constantly applied. In most circuits a **pulse**, or sudden increase and decrease in voltage level, is used to trigger the SCR into conduction.

The SCR is a current operated device and requires a certain value of positive gate current to switch it into a conducting state. The amount of gate current required to switch ON an SCR is called the **gate trigger current** (I_{GT}) and it is required for only a brief instant (10 to $50\,\mu s$). To switch OFF an SCR, the holding current (I_H) must be reduced below its minimum value. The switching-off process is called **commutation**.

39-3 DC-Operated SCR Circuits

The amount of gate current needed to turn the SCR ON varies from one type to another. SCRs designed to control small amounts of power require a gate current of only a few microamps, while SCRs designed to control large amounts of power require a gate current of several hundred milliamps. However, the amount of gate current needed to fire the SCR is only a small fraction of the amount of current the device is designed to handle through the anode-cathode circuit.

In most dc circuits, the problem is not how to turn an SCR ON; the problem is how to turn it OFF. The only way to turn an SCR OFF is to reduce the anode current below the holding-current value. In dc circuits this is often accomplished by adding a manual reset switch to momentarily open the circuit.

The operation of an SCR is similar to that of a standard diode except that it requires a *momentary positive* voltage applied to the gate to switch it ON. The schematic of an SCR switching circuit that is operated from a dc source is shown in Figure 39-5. The anode is connected so that it is positive with respect to the cathode (*forward biased*). Momentarily closing pushbutton PB_1 applies a positive current-limited voltage to the gate of the SCR, which switches the anode-to-cathode circuit ON, or into conduction, thus, turning the lamp ON. Once the SCR is ON, it stays ON, even after the gate voltage is removed. The only way to turn the SCR OFF is to reduce the anode-cathode current to zero by removing the

Example 39-1

A data sheet lists the following specifications for an SCR:

$$V_{GT} = 0.75\,V$$
$$I_{GT} = 7\,mA$$
$$I_H = 6\,mA$$

What does this mean?

This means that the source driving the gate has to supply 7 mA at 0.75 V to latch the SCR and the anode current must not be lower than 6 mA for the SCR to maintain conduction.

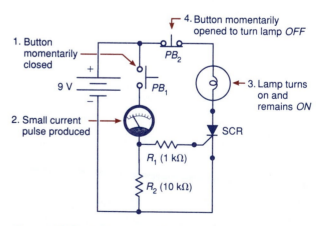

Figure 39-5   SCR operated from a dc source.

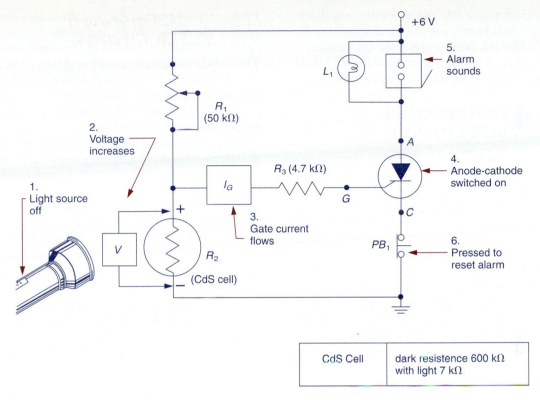

Figure 39-6 SCR light-operated alarm.

CdS Cell	dark resistence 600 kΩ with light 7 kΩ

source voltage from the anode-cathode circuit. This is accomplished by momentarily pressing pushbutton PB_2. It is important to note that the anode-cathode will switch ON in *only one direction*. This occurs only when the anode is positive with respect to the cathode and a positive voltage is applied to the gate.

One practical application for a dc-operated SCR is the light-operated alarm circuit shown in Figure 39-6. Variable resistor R_1 and CdS cell R_2 form a voltage divider across the 6-V dc source. With the light source shining on the CdS cell, its resistance and the resultant voltage drop across it will be low. As a result, it will not produce enough gate current to trigger the anode-cathode circuit into conduction. Therefore, the alarm is OFF when the light source is shining on the CdS cell. Momentarily blocking the light source to the CdS cell increases its resistance and the resultant voltage drop. Sufficient current now flows through the gate to trigger the anode-cathode circuit into conduction. The current path to the buzzer and light are completed—sounding the alarm. Once the anode-cathode circuit is switched ON it remains ON regardless of the CdS cell light

condition. Pushbutton PB_1 is momentarily opened to open the anode-cathode circuit and reset the alarm. Variable resistor R_1 is adjusted to vary the light level required to operate the circuit. The buzzer may be disconnected if a silent alarm is required.

The SCR **crowbar circuit** of Figure 39-7 is designed to deliberately short and blow the input fuse if the supply voltage exceeds a preset maximum voltage. This circuitry is used on some special equipment that could be damaged if the input voltage were to go too high. The SCR circuit is normally open at proper input voltages and has no effect on the operation. If the input

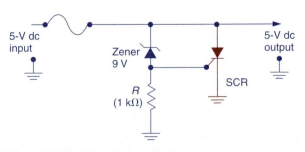

Figure 39-7 SCR crowbar circuit.

voltage rises above 9 V, the zener diode will conduct. This will produce a voltage drop across R that is sufficient to open the SCR into conduction. The fuse will blow, opening the supply line and thereby protecting the circuitry following.

39-4 AC-Operated SCR Circuits

The problem of SCR turnoff does not occur in ac circuits. The SCR is automatically shut OFF during each cycle when the ac voltage across the SCR approaches zero. As zero voltage is approached, anode current falls below the holding current value. The SCR stays OFF throughout the entire negative ac cycle, since it is *reverse bias*.

The SCR can be used for switching current to a load connected to an ac source. Because the SCR is a rectifier, it can conduct only one-half of the ac input wave, so the maximum output delivered to the load is 50 percent and is in the shape of a half-wave pulsating dc waveform. The schematic of an SCR switching circuit that is operated from an ac source is shown in Figure 39-8. The anode-cathode circuit can only be switched ON during the half-cycle when the anode is positive (forward biased). With pushbutton PB_1 open, no gate current flows so that the anode-cathode circuit is switched OFF.

Pressing pushbutton PB_1 continuously closed causes the gate-cathode and anode-cathode circuits to be forward biased at the same time. This produces a half-wave pulsating direct current through the lamp load. When pushbutton PB_1 is released, the anode-cathode current is automatically shut OFF when the ac voltage drops to zero on the sine wave.

The switching application of an SCR is primarily designed to take the place of a mechanical switch circuit. With the SCR static switch, it is possible to control large amounts of load current with a rather small gate current switch. A few milliamperes of gate current can be made to control several hundred amperes of load current. The SCR switch does not have contacts that spark and arc when the circuit is switched on and off.

The output to the load of an ac-operated SCR circuit can be doubled by the addition of a *bridge rectifier* to the circuit. The schematic of an SCR switching circuit that is operated from a bridge rectifier is shown in Figure 39-9. The ac input voltage is converted to a full-wave pulsating dc voltage by the bridge rectifier. Thus, a positive anode and gate voltage is produced for both halves of the ac input voltage. Since the anode voltage drops to zero during each half-cycle, the gate maintains a continuous control over the anode-cathode circuit. An output

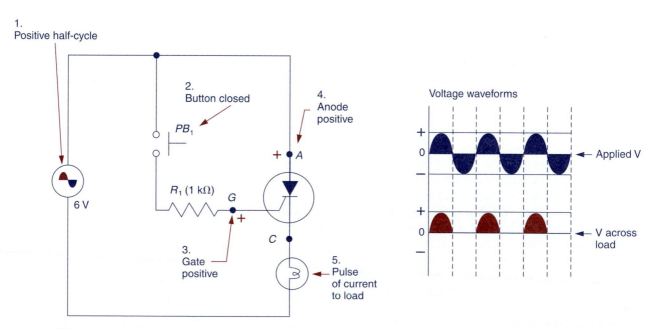

Figure 39-8 SCR operated from an ac source.

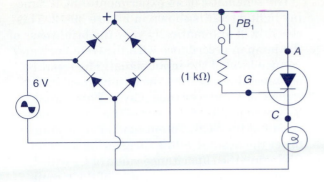

Voltage waveforms

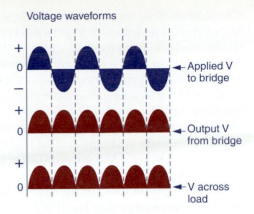

Figure 39-9 SCR operated from a bridge rectifier.

is obtained across the lamp as long as push-button PB_1 is closed. The output is full-wave pulsating dc.

Load control using SCR switching is limited to only on and off switching operations. When operated from an ac source the SCR also has the capability of varying the amount of power supplied to a load device. Variable power control permits the load to receive different values of load current.

This is the basis for its application in lamp dimmers and industrial motor-control circuits. Basically, it performs the same function as a rheostat, but is much more *efficient*.

Figure 39-10 illustrates the use of an SCR to both rectify and vary the amount of current supplied to load resistor R_L. Variable resistor R_2 is used to control the point in the sine wave where gate trigger current (I_{GT}) is reached. Diode (D_1) is used to ensure that only the positive voltage is applied to the gate. As the ac

supply voltage goes positive, the SCR is forward-biased, but will not conduct until the trigger current value I_{GT} is reached. If we assume a I_{GT} of 5 mA, R_2 can be adjusted so that the gate current will not reach the 5 mA level until the ac supply voltage reaches close to its peak value. If R_2 is decreased in value, the gate current can reach 5 mA sooner and the SCR will fire sooner during the cycle. Notice that the gate controls the firing of the SCR during only half of the positive cycle. As long as the voltage applied to the gate is locked in-phase with the voltage applied to the anode, the SCR cannot be fired after the ac supply voltage applied to the anode has reached 90° (peak position of ac waveform).

The main advantage of this system is that virtually no power is *wasted* by being converted to heat. Heat is generated when current is regulated, not when it is switched. The SCR is either fully ON or fully OFF. It never partially conducts current.

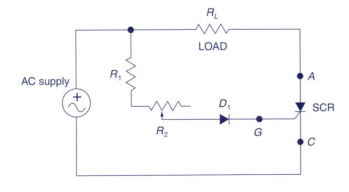

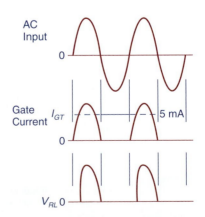

Figure 39-10 SCR variable output control.

The schematic of an experimental SCR lamp dimmer circuit is drawn in Figure 39-11. Variable resistor R_2 is used to vary the brightness of the lamp from 0 to about 30 percent of its normal visual light output. Resistors R_1, R_2, and diode D_1 form a dc voltage-divider circuit across the 12-V ac supply. The variable arm of resistor R_2 sets the value of positive voltage applied to the gate of the SCR. The higher this voltage is, the greater the output to the load. Output to the lamp load is pulsating dc. The maximum output of this circuit can be doubled by the addition of a bridge rectifier.

The schematic of an experimental SCR dancing light circuit is shown in Figure 39-12. This circuit will automatically vary the brightness of the lamp in accordance with the frequency and voltage level of the input signal. The input is obtained directly from the output speaker terminals of a radio or tape player. This signal is fed through a transformer to the gate-cathode circuit of the SCR. Potentiometer R_1 is adjusted to set the desired voltage-triggering level. Changing the capacitance value of C_1 will change the frequency response of the circuit.

To gain full control of the ac half cycle applied to the SCR, the voltage applied to the gate must be out-of-phase with the voltage applied to the anode. This type of control, called phase shift control, is popular in a variety of control systems. To shift the phase of the voltage applied to the gate, the gate must be separated from the voltage applied to the anode.

The schematic of a UJT phase-shifting circuit for an SCR is shown in Figure 39-13. The transformer and bridge rectifier are used to provide the low-voltage dc required to operate the UJT. The UJT is turned OFF until capacitor C_1 charges to a predetermined voltage level, at which time the UJT turns ON and discharges the capacitor through resistor R_2. This discharge produces a pulse of current through R_2, which triggers the gate of the SCR. The charge time of the capacitor and the pulse rate of the UJT are controlled by varying the resistance of R_1. The pulses produced by the UJT are

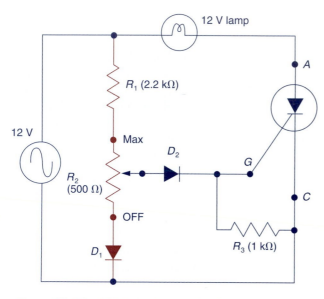

Figure 39-11 SCR lamp dimmer circuit.

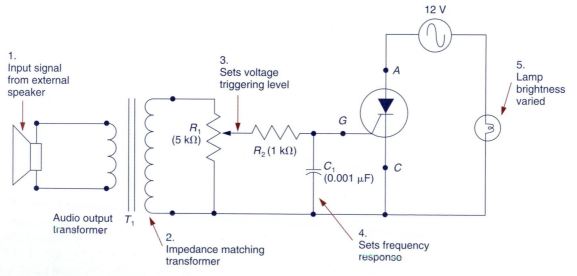

Figure 39-12 SCR dancing light circuit.

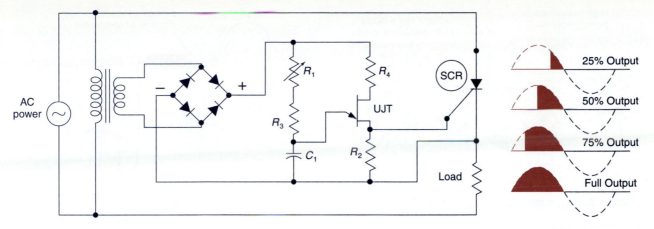

Figure 39-13 UJT phase shifting for an SCR.

entirely independent of the ac input power connected to the anode of the SCR. Since the UJT can be triggered at any time regardless of the ac waveform, the SCR can be switched ON at any point during the positive half-cycle of alternating current applied to it.

To achieve control for the full 360° of the ac supply cycle, it is necessary to use two SCRs connected anode-to-cathode as shown in Figure 39-14. By using two SCRs, it is possible to control the positive input voltage with one SCR and the negative input voltage with the other. This circuit is used for the "soft start" of a three-phase induction motor. Two SCRs are connected in reverse parallel to obtain full-wave control. In this connection scheme, one SCR controls the voltage when it is positive in the sine wave and the other when it is negative. Control of current and acceleration is achieved by triggering the firing of the SCR at different times within the half-cycle. If the gate pulse is applied early in the half-cycle, the output is high. If the gate pulse is applied late in the half-cycle, only a small part of the waveform is passed through and the output is low.

39-5 Testing the SCR

The most common SCR faults are either anode-cathode shorts or opens. Occasionally, the gate may short to either the anode or cathode, or become open. Most SCRs can be tested out of circuit using an ohmmeter. The resistance between the anode and cathode should measure high in both directions. High resistance also

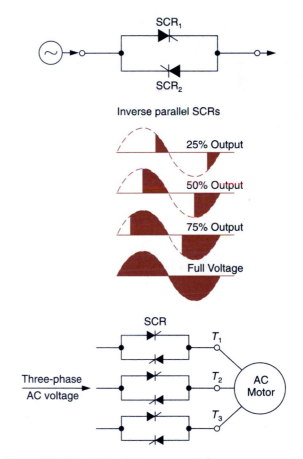

Figure 39-14 SCR soft start control of an ac motor.

should be measured from the anode to gate in both directions. However, the gate-to-cathode resistance should measure like a PN junction or diode; there should be low resistance in the forward-biased direction and high resistance in the reverse-biased direction (Figure 39-15).

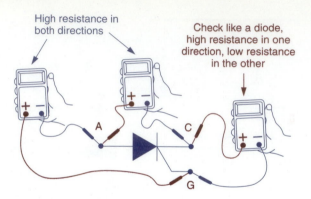

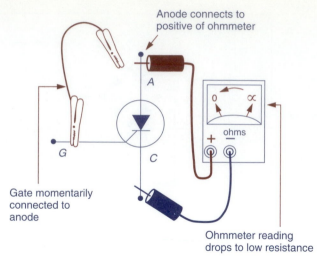

Figure 39-15 Checking an SCR with an ohmmeter.

An operational check of an SCR can be made using an analog ohmmeter connected as shown in Figure 39-16. With this test, the anode and cathode terminals are connected to the positive and negative leads of the ohmmeter. The ohmmeter reading should indicate a high resistance. Next, the gate lead is momentarily connected to the positive anode. The ohmmeter reading should show a drop to low resistance indicating the SCR is conducting. If the SCR is good, current continues to flow and the ohmmeter reads low resistance. This will only be true if the ohmmeter can supply the necessary holding current.

39-6 The Triac

The *triac* is a device very similar in operation to the SCR. When an SCR is connected into an ac circuit, the output voltage is rectified to direct current. The triac, however, is designed

Figure 39-16 SCR operational check.

to conduct on both halves of the ac waveform. Therefore, the output of the triac is *alternating current instead of direct current*. Triacs were developed to provide a means for improved control of ac power.

The triac operates as two SCRs in the same case. The equivalent triac circuit shown as two SCRs connected in reverse-parallel is illustrated in Figure 39-17. Note that the SCRs are inverted. As such, the triac is capable of conducting with either polarity of terminal voltage. It can also be triggered by *either polarity of gate signal*.

The schematic symbol and structure of a triac is shown in Figure 39-18. The symbol consists of two parallel diodes connected in opposite directions with a single gate lead attached. The main difference between the triac and the SCR is the triac will allow current flow regardless of

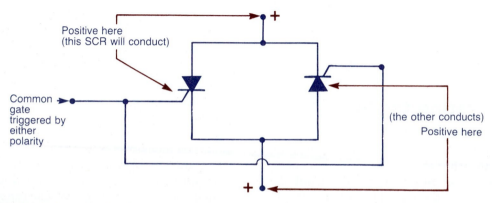

Figure 39-17 Equivalent triac circuit.

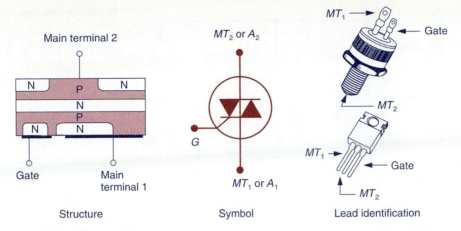

Main terminal 2

N   P   N
N
P
N   N

Gate   Main terminal 1

Structure

MT_2 or A_2

G

MT_1 or A_1

Symbol

MT_1 — Gate
MT_2

MT_1 — Gate
MT_2

Lead identification

Figure 39-18   Structure and symbol of the triac.

the biasing voltage and regardless of the polarity of the trigger pulses at the gate. Because there is no longer a specific anode or cathode, these leads are referred to as main terminals **MT_1** and **MT_2**. Main terminal MT_1 is used as the measurement reference terminal. Triacs are turned on by applying either a positive or negative voltage at the gate of the device. Current can flow between MT_1 and MT_2 and also between the gate and MT_1. Once the TRIAC has been latched into conduction, the gate has no further control over the device. The triac is commutated by lowering the holding current below its rated value.

The triac may be triggered into conduction in either direction by gate current moving either into or out of the gate. Once main terminal current flow is established in either direction, the triac has basically the same internal operating characteristic as the SCR. The triac has four possible triggering modes. These are, with respect to MT_1:

- MT_2 positive, gate positive
- MT_2 positive, gate negative
- MT_2 negative, gate positive
- MT_2 negative, gate negative

Two of these triggering modes are illustrated in the schematic diagrams shown in Figure 39-19.

Since the triac can conduct on both half-cycles, it is most useful for controlling loads that operate on ac. Full efficiency is gained by using both half-waves of the ac input voltage.

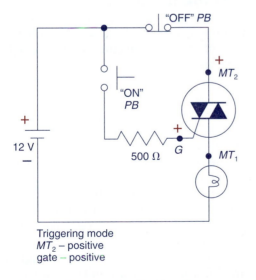

Triggering mode
MT_2 – positive
gate – positive

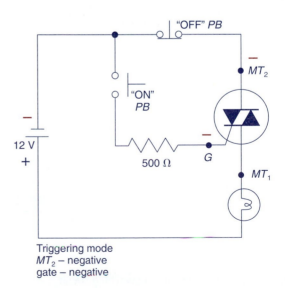

Triggering mode
MT_2 – negative
gate – negative

Figure 39-19   The triac triggering modes.

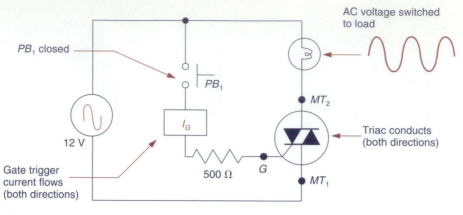

Figure 39-20   The triac ac switching circuit.

The schematic of a triac switching circuit operated from an ac source is shown in Figure 39-20. If the pushbutton PB_1 is held closed, a continuous trigger current is supplied to the gate. The triac conducts in both directions to switch all of the applied ac voltage to the load. If the pushbutton is opened, the triac turns OFF when the ac source voltage and holding current drop to zero, or reverse polarity. Note that unlike the similar SCR circuit, the output of this circuit is ac, not dc.

39-7 Triac Circuit Applications

One common application of a triac is switching ac current to an ac motor. The triac motor-switching circuit in Figure 39-21 illustrates the

ability of a triac to control a large amount of load current with a small amount of gate current. In this application, it operates like a solid-state relay. A 24-V step-down transformer is used to reduce the voltage in the thermostat circuit. The resistor limits the amount of current flow in the gate-MT_1 circuit when the thermostat contacts close to switch the triac and motor ON. The maximum current rating of the thermostat contacts is *much lower* than that of the triac and motor. If the same thermostat were wired in series with the motor to operate the motor directly, the contacts would be destroyed by the heavier current flow.

Triacs can be used to vary the average ac current going to an ac load as illustrated in Figure 39-22. The trigger circuit controls the point of the ac waveform at which the triac is switched ON. The resulting waveform is still alternating current, but the average current is changed. In a lighting circuit, varying the current to an incandescent lamp will vary the amount of light emitted by the lamp. Thus, the triac can be used as a light-dimming control. In some motor circuits, varying the current will change the speed of the motor.

The **diac** is a two terminal transistor-like device used primarily to control the triggering of SCRs and triacs (Figure 39-23). Unlike the transistor, the two junctions are heavily and equally doped. Notice from the symbol of the diac that it behaves like two diodes pointing in opposite directions. Current flows through the diac (in either direction) when the voltage

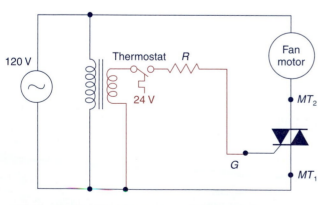

Figure 39-21   The triac motor-switching circuit.

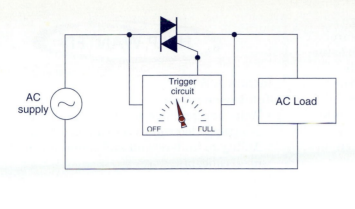

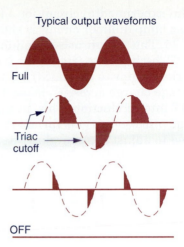
Typical output waveforms

Full

Triac cutoff

OFF

Figure 39-22 Triac used to vary current.

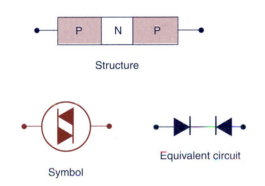

Structure

Symbol

Equivalent circuit

Figure 39-23 The diac.

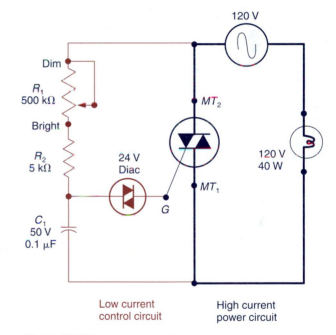

120 V

Dim

R_1
500 kΩ

Bright

R_2
5 kΩ

MT_2

24 V
Diac

120 V
40 W

MT_1

G

C_1
50 V
0.1 μF

Low current
control circuit

High current
power circuit

Figure 39-24 120-V triac/diac lamp dimmer.

across it reaches its rated breakover voltage. The current pulse produced when the diac changes from a nonconducting to a conducting state is used for SCR and triac gate triggering.

In operation, when the voltage applied to the diac exceeds the breakover voltage, the diac latches on, and remains in the conducting state until the voltage is removed.

An experimental triac/diac ***lamp dimmer circuit*** is shown in Figure 39-24. When the variable resistor R_1 is at its lowest value (Bright), capacitor C_1 charges rapidly at the beginning of each half-cycle of the ac voltage. When the voltage across C_1 reaches the breakover voltage of the diac, C_1 is discharged into the gate of the triac. Thus, the triac is ON early in each half-cycle and remains ON to the

end of each half-cycle. Therefore, current will flow through the lamp for most of each half-cycle and produce full brightness.

As the resistance R_1 increases, the time required to charge C_1 to the *breakover* voltage of the diac increases. This causes the triac to fire later in each half-cycle. Therefore, the length of time that current flows through the lamp is reduced and less light will be emitted.

An experimental circuit for a low-voltage triac lamp dimmer circuit is shown in Figure 39-25. This circuit uses a unijunction transistor (UJT) oscillator to trigger the triac gate. It works similar to the diac-trigger circuit. The voltage charge across C_1 is used to trigger the UJT into conduction. This, in turn, triggers the triac into conduction. Variable resistor R_1 is used to adjust the charging rate of C_1 which, in turn, controls the brightness of the lamp.

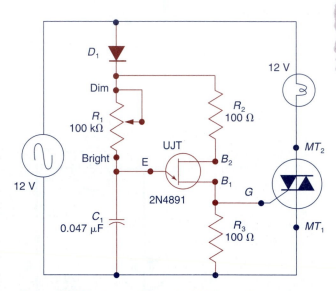

Figure 39-25 Low-voltage triac/UJT lamp dimmer circuit.

Review and Applications

Review Questions

1. State two practical applications for SCR and triac circuits.
2. In what way is the operation of a thyristor different from that of a transistor?
3. Compare the operation of an SCR with that of a conventional diode.
4. What does the regenerative latching action of an SCR refer to?
5. Why can an SCR switch large amounts of power very efficiently?
6. Define the term *holding current* as it applies to an SCR.
7. When an SCR switch circuit is operated from an ac source, what type of output is obtained?
8. The output to the load of an ac-operated SCR can be doubled by the addition of what type of circuit?
9. Explain the term *phase shift* as it applies to the control circuit of an SCR.
10. An SCR circuit, operated from an ac source, uses the pulses produced by a UJT oscillator to trigger the gate. Why are the pulses produced by the UJT considered to be "phase shifted"?
11. An ohmmeter is used to check the resistance between leads of an SCR. Assuming the SCR is not defective, what resistance reading should be indicated between the following (describe resistance in terms of high and low):
    (a) anode and cathode
    (b) gate and cathode
    (c) gate and anode
12. (a) When operated from an ac source, why is the triac more efficient than an SCR?
    (b) The SCR has only one triggering mode. What is it?
    (c) State the four possible triggering modes of a triac.
13. Under what conditions will a diac conduct a current?

Problems

1. State the polarity of the anode, cathode, and gate for a properly biased SCR.
2. An SCR has the following ratings: $I_H = 6\,\text{mA}$, $V_{GT} = 1\,\text{V}$, and $I_{GT} = 200\,\mu\text{A}$. What is the amount of current required to switch the SCR into conduction?
3. On a properly functioning SCR, what junction acts like a diode?
4. The triac and diac are both classified as bidirectional thyristors. What does this mean?

Critical Thinking

1. (a) With reference to the characteristic curve of an SCR, what does the forward breakover voltage point represent?
   (b) How is the value of this breakover voltage lowered?
2. When operated from an ac source, at what point on the sine wave does the SCR turn off automatically? Why?
3. Triacs are often in-circuit tested. Explain how they act when shorted and when open.

Portfolio Project

- Design and build an SCR alarm circuit, similar to the one discussed in this chapter. Use a switch, CdS cell, or thermistor for the sensor device. Include a written description of the circuit's operation and suggest several practical applications for the circuit.

Circuit Challenge

Using a Simulator

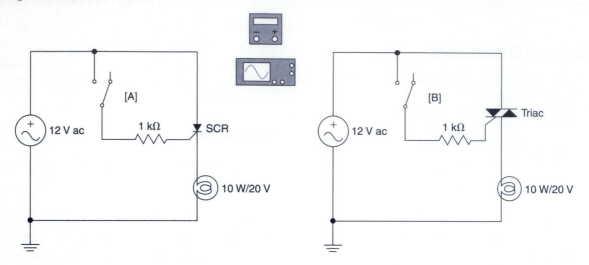

Procedure

(a) Construct the SCR and triac circuits shown, using whatever simulation software package is available to you.

(b) Connect the scope to view the output of the SCR circuit. Describe the type of waveform seen.

(c) Connect the scope to view the output of the triac circuit. Describe the type of waveform seen.

(d) Use the multimeter to record the value of the dc output voltage from the SCR circuit.

(e) Use the multimeter to record the value of the ac output voltage from the triac circuit.

Integrated Circuits (ICs)

The electronic components covered to this point are individually packaged devices that are connected with other individual devices to form a functional unit. These individually packaged devices are referred to as discrete components. In this chapter, we will introduce integrated circuits (ICs), in which many diodes, transistors, capacitors, and resistors are built into tiny chips and packaged in a single case to form an operational circuit.

Objectives

After studying this chapter, you should be able to:

1. Describe the basic construction and features of integrated circuits (ICs).
2. Compare the operation of digital and analog ICs and their applications.
3. Describe the characteristics of operational amplifiers (op-amp) and their applications.
4. Discuss the operation of a 555 Timer and its applications.
5. Assemble, test, and troubleshoot IC experimental circuits.

Key Terms

- integrated circuit (IC)
- discrete circuit
- dual in-line package (DIP)
- digital
- analog
- switch-type circuitry

- amplifying-type circuitry
- operational amplifier (op-amp)
- op-amp single supply
- op-amp dual voltage supply
- inverting input

- noninverting input
- differential input
- negative feedback
- inverting voltage amplifier
- noninverting voltage amplifier
- virtual ground

(continued)

RESEARCH
INTERNET
SITES

- **Texas Instruments**
- **Rockwell**
- **America Microsystems Inc. (AMIS)**

- voltage comparator
- common mode rejection
- summing amplifier
- difference amplifier
- 555 Timer IC
- monostable multivibrator
- free running multivibrator
- pulse-width modulation
- current sinking
- current sourcing

40-1 Integrated Circuit (IC) Construction

Integrated circuits (ICs) are probably the most important components in electronics today. An IC is a *complete* electronic circuit that is contained within a single chip of silicon. Often no larger than a transistor, it can contain as few as several to as many as hundreds of thousands of transistors, diodes, resistors, and capacitors, along with electric conductors processed and contained entirely within a single chip of silicon (Figure 40-1). ICs are often called **chips**, which are actually a component part of the IC.

Integrated circuits are made by using the same basic materials and techniques that are used in making transistors. The basic building block of this type of IC is a *doped silicon wafer*. This wafer is usually doped positive and called the *P substrate*. It is used as the base layer upon which the individual electronic components and conductors are formed. Components are formed by producing *PN junctions* at selected regions of the substrate. These compo-

About Electronics

AT&T currently has a chip that makes it possible to transmit 2.5 billion bits of information per second per channel as new developments continue to expand capacity. This chip, which is used for sorting and routing light at the receiving end of a fiberoptic transmission link, was developed by Lucent Technologies and uses the process of photolithography.

nents are interconnected to form a complete electronic circuit. Leads from the chip are connected to package pins to provide inputs and outputs to the circuit.

The simple three-component IC shown in Figure 40-2 will give you an idea of how an IC is produced. A single PN junction forms a diode, which is interconnected to the base of an NPN transistor. This transistor is formed using a pair of PN junctions. The transistor collector is interconnected to a resistor. This resistor is

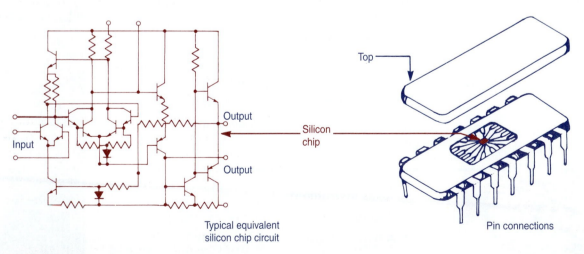

Input
Output
Output
Typical equivalent
silicon chip circuit

Top
Silicon
chip
Pin connections

Figure 40-1   IC chip.

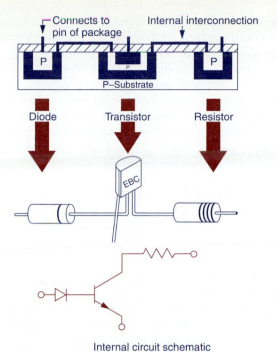

Connects to
pin of package   Internal interconnection

P-Substrate

Diode   Transistor   Resistor

EBC

Internal circuit schematic

Figure 40-2   Three-component integrated circuit.

formed from a small section of P-type silicon. The other lead from the diode, transistor-emitter, and resistor would connect to the pins of the package.

Discrete circuits use individual resistors, capacitors, diodes, transistors, and other devices to achieve the circuit function. These individual or discrete parts must be interconnected.

A *monolithic integrated circuit* is an electronic circuit that is constructed entirely on a single, small chip of silicon. All the components that make up the circuit—transistors, diodes, resistors, and capacitors—are an integral part of that single chip.

Hybrid integrated circuits combine two or more monolithic ICs in one package or combine monolithic ICs with discrete components. They are more expensive and larger than monolithic ICs, but smaller and cheaper than discrete circuits.

40-2 Advantages and Limitations of ICs

Small size is one major advantage of the integrated circuit. The electric circuitry within an IC chip is a miniature version of a circuit that

could just as well be constructed using individual or discrete parts. Miniature circuitry permits very complicated systems, such as computers, to have small physical dimensions.

Low cost is another advantage of integrated circuits. The cost of processing a single semiconductor chip, with its complex circuit of hundreds of transistors, is comparable in cost to that of a single transistor.

High reliability is perhaps the most important advantage of integrated circuits. This means that ICs usually operate for longer periods of time without a breakdown. They are much less likely to fail than the discrete versions of the same system. The major reason for this is that they require far fewer solder joints and mechanical connections. In many semiconductor systems, it is the failure of these interconnections that accounts for the breakdown.

The most important limitation of integrated circuits is that they cannot handle large currents or voltages. The current limitation results from the small size of the IC, and, hence, its inability to dissipate heat. High voltages can break down the insulation between the components in the IC because they are very close together. Therefore, most ICs are rated for low-operating currents (in the milliampere range) and low-operating voltages (5–20 V). Also, most ICs have a power dissipation rating of less than 1 W.

Integrated circuits are also limited in the size and type of components that can be formed on the chip. Inductors such as transformers are not suitable for IC packaging. High-value resistors and capacitors require too much space to be efficiently formed on the chip. Therefore, when required, these parts are added to the circuit as discrete components.

40-3 IC Symbols and Packages

Symbols for integrated circuits are very simple. Usually they are either a triangle or a rectangle. Because of the complexity of the internal chip circuit, no attempt is made to reproduce the internal circuit schematic on working diagrams (Figure 40-3).Manufacturers' data sheets should be consulted when additional data is required about a chip.

For protection purposes ICs are mounted in packages. Package designs for integrated

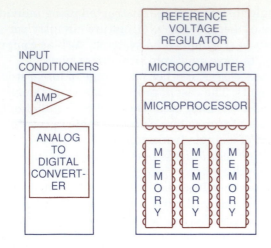

INPUT
CONDITIONERS

REFERENCE
VOLTAGE
REGULATOR

MICROCOMPUTER

AMP

MICROPROCESSOR

ANALOG
TO
DIGITAL
CONVERT-
ER

MEMORY   MEMORY   MEMORY

Figure 40-3 Typical symbols used to represent different IC packages.

circuits are usually round, square or rectangular. The most popular IC package is the *dual in-line package (DIP)*. A typical DIP is shown in Figure 40-4. This package is constructed from either plastic or ceramic materials. Notice that the package has two rows of mounting pins. These can either be plugged into matching sockets or soldered in place on a printed circuit board. Pin number one is often marked by a dot. All other pins are numbered in order in a counterclockwise direction.

When you replace an IC, make sure that you position it properly into the circuit and then you replace it with the same series number or a newer series. Inserting it backwards can cause an internal short, which can overheat and permanently damage the chip. Various symptoms can occur if one of the leads is bent as it is inserted into the IC socket. Also, when IC sockets are used, an oxide layer can sometimes

develop on the pins, causing the circuit to have an erratic output. If this should happen, remove the IC and clean the pins and the socket with contact cleaner to correct the problem.

Small-signal IC packages are designed for signal processing and generally are rated for low-operating currents (in the milliampere range) and require dc operating voltages in the 5–20-V range. *Power* IC packages are designed to handle higher power levels (over 2-W dissipation) and usually are mounted on an appropriate heat sink (Figure 40-5).

ICs fall into two broad categories, based on the major device used in the circuit fabrication. The *bipolar* types use the bipolar-junction transistor (NPN, PNP) as their principal circuit element. The *metal-oxide semiconductor (MOS)* types use *MOS field-effect transistors (MOSFETs)* as their principal circuit element. MOS integrated circuit packages are extremely vulnerable to damage from electrostatic discharge (ESD) and are sometimes marked with a warning label (Figure 40-6). When working with IC circuits, observe the following:

• Discharge the static charge on your body by touching a grounded metal surface before touching leads or pins.
• Use the special grounded-tip soldering irons that are available with MOSFET-type ICs to protect them from static damage.
• Do not connect or disconnect an IC from a circuit when power is applied.
• Do *not* apply voltage directly to an IC, unless the manufacturer's procedure calls for it.
• Store components in original antistatic-producing packaging.

As IC technology advances, the circuits tend to become more complex and more compact. This development leads to smaller chips with

8 7 6 5

741 ← Device number

Dot indicates pin 1

1 2 3 4

A. Typical dual in-line (DIP) IC package

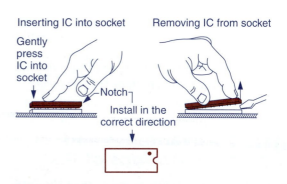

Inserting IC into socket   Removing IC from socket

Gently press IC into socket

Notch
Install in the correct direction

B. Removing and inserting IC chips

Figure 40-4 Typical DIP.

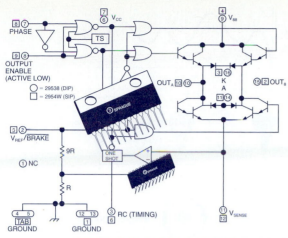

FUNCTIONAL BLOCK DIAGRAM

(Courtesy Sprague Semiconductor Group.)

Figure 40-5 Typical power IC package capable of delivering 100 W to a load.

(Courtesy International Rectifier Canada Ltd.)

Figure 40-6 Static-sensitive-warning label.

more inputs and outputs, thus requiring more pins on the package.

IC chips are capable of containing thousands or even millions of miniaturized transistors. Although ICs may consist of combinations of resistors, capacitors, diodes, and transistors, they are generally classified according to the number of transistors they contain. The terms small-scale integration (SSI), medium-scale integration (MSI), large-scale integration (LSI), and very large-scale integration (VLSI) have been used in the past to classify chips according to their complexity. The least complex are placed in the SSI category and usually have fewer than 200 transistors. VLSI circuits contain thousands (or hundreds of thousands) of components on a single chip. Ultra large-scale integration (ULSI) refers loosely to placing more than one million circuit elements on a sin-

gle chip. Nearly all modern microprocessor chips use the VLSI or ULSI architectures (Figure 40-7).

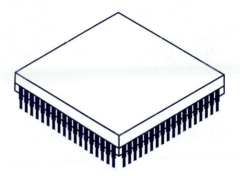

Figure 40-7 Pin grid array (PGA) microprocessor chip. The connecting pins for this type of package are located on the bottom in concentric squares.

Troubleshooting an IC package is relatively simple. An IC is basically a system within a system. Troubleshooting involves knowledge of how the system operates and what the inputs and outputs should be. Part of IC troubleshooting involves testing all the inputs to make sure they are correct and then testing all the outputs (Figure 40-8). If all inputs are good and some outputs are bad, it is likely the IC is at fault. Since ICs cannot be repaired, they must be replaced if they are defective.

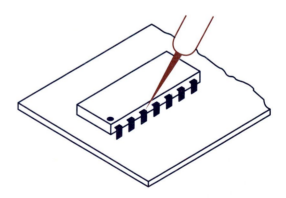

Figure 40-8 Checking IC inputs and outputs.

Unlike a diode or transistor, resistance measurements are not usually taken between the pins of an integrated circuit. The reason is that unlike diodes and transistors, not all IC pin connections go directly to a PN junction.

40-4 Analog and Digital ICs

Integrated circuits can be classified according to their application as being *digital* ICs or *analog* (linear) ICs. A digital IC is one containing **switch-type circuitry**. The input and output signals to a digital IC are designed to operate at two levels (zero and maximum), and the signals resemble a square wave (Figure 40-9). Zero voltage usually represents an OFF state and maximum voltage represents an ON state. Digital IC chips include binary logic gates, memories, and microprocessors.

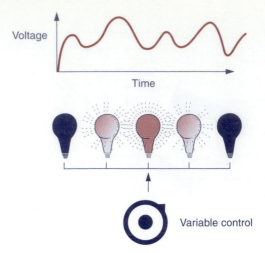

A. Typical analog waveform

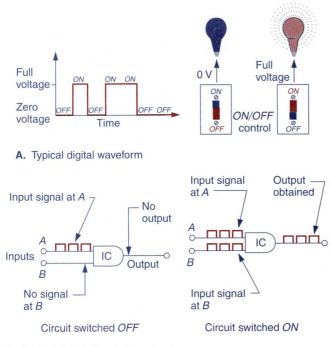

A. Typical digital waveform

B. Typical digital IC switching circuit

Figure 40-9 Digital ICs.

An *analog* (linear) IC is one that contains **amplifying-type circuitry**. The analog and digital processes can be seen in a simple comparison between a light dimmer and light switch. A light dimmer involves an analog process, which varies the intensity of light from fully OFF to fully ON over a range of brightnesses (Figure 40-10). The operation of a standard light switch, on the other hand, involves a *digital* process, meaning that the switch can be operated only to turn the light fully OFF or fully ON. Analog ICs are similar in operation to

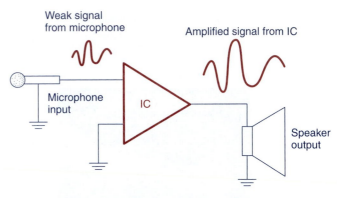

B. Typical analog IC amplifying circuit

Figure 40-10 Analog (linear) ICs.

transistor voltage or power *amplifiers*. They produce an output signal that is proportional to the input signal that is applied to the device. Analog IC chips include amplifiers, timers, oscillators, and voltage regulators. Some ICs combine analog and digital functions on a single chip.

40-5 Operational Amplifier ICs

Operational amplifier (*op-amp*) ICs are among the most widely used analog ICs. They take the place of amplifiers that formerly required many components. Some of their more important features can be summarized as follows:

- *High input impedance:* The inputs have very high impedance and draw essentially no current from the signal source.

- *High gain:* They are basically high gain amplifiers that are used to amplify weak ac or dc signals.
- *Low output impedance:* This feature allows them to deliver a signal directly to a low-impedance load without the need for impedance matching.
- *Common-mode rejection:* This feature gives them the ability to reduce hum and noise.

The schematic symbol for an op-amp is an arrowhead (Figure 40-11). This arrowhead symbolizes amplification and points from input to output. Op-amps have five basic terminals: two for supply voltage, two for input signals, and one for the output signal.

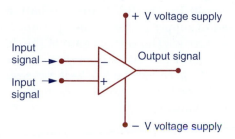

Figure 40-11 Operational amplifier (op-amp) symbol and terminals.

Most op-amp power supply terminals are labeled +V and −V. Operational amplifiers may be operated from a ***pair of supplies*** (positive and negative with respect to ground) or from a ***single supply*** (Figure 40-12). In either case, the op-amp will amplify only the difference between its two input lines. In some

instances, the supply connections are omitted from the symbol and assumed to be connected.

There is only *one* output terminal in an op-amp. The output is obtained between this output terminal and the common ground. There is a limit to the power that is available from the output terminal and this rating varies with the type of op-amp used. The output voltage limit is set by the value of the power supply voltages. Some op-amps, such as 741 and 380, have internal circuitry that automatically limits the current drawn from the output terminal in order to protect the device from damage due to output short circuits or overloads (Figure 40-13).

There are *two* input terminals on the op-amp which are labeled (−) ***inverting*** and (+) ***noninverting input*** (Figure 40-14). The polarity of a voltage applied to the inverting input is reversed at the output. The polarity of a voltage applied to the noninverting input is the same as the output. They are also called ***differential input*** terminals because of the effective input voltage to the op-amp depends on the difference in voltage between them.

40-6 Op-Amp Voltage Amplifiers

The op-amp is basically an amplifier that has a very high amplification level, or *gain* (a value of 500,000 or more). A very small change in voltage on either input will result in a very large change in output voltage. This makes the op-amp very unstable because the circuit is too

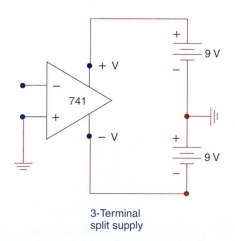

3-Terminal
split supply

Figure 40-12 Operational amplifier power supply connections.

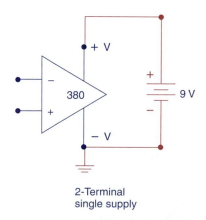

2-Terminal
single supply

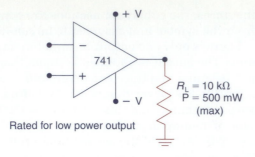

Figure 40-13 Operational amplifier output connection.

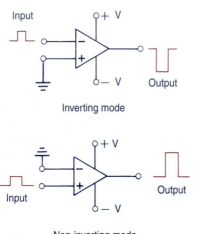

Figure 40-14 Operational amplifier input connections.

sensitive. As a result, the gain is usually reduced to a more practical level. The method used to accomplish this is to *feed back some of the output signal to the inverting* $(-)$ *input.* This *negative feedback* decreases the voltage

gain, but increases the bandwidth of the amplifier and makes it more stable.

The schematic diagram of a 741 op-amp ***inverting-voltage amplifier*** circuit is drawn in Figure 40-15. Two 9-V transistor batteries are used to supply power to the op-amp. The two resistors set the voltage gain of the amplifier. Resistor R_{in} is called the *input resistor.* Resistor R_{fb} is called the *feedback resistor.* The ratio of the resistance value of R_{fb} to that of the resistance value of R_{in} sets the voltage gain of the amplifier as follows:

$$\text{Op-amp gain} = \frac{R_{fb}}{R_{in}} \quad \begin{array}{l}\text{Inverting} \\ \text{amplifier}\end{array}$$

$$= \frac{500 \text{ k}\Omega}{50 \text{ k}\Omega}$$

$$\text{Op-amp gain} = 10$$

The ac input signal is applied through the input resistor R_{in} to the two input terminals of the op-amp. The ac output voltage is obtained across load resistor R_L. Part of the ac output

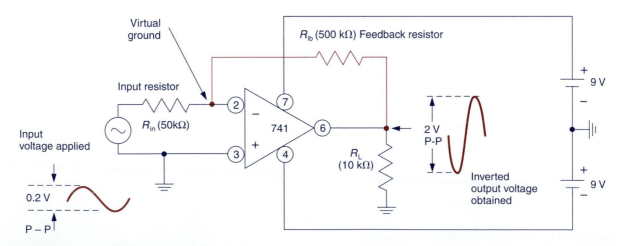

Figure 40-15 Op-amp inverting amplifier.

voltage is fed back to the inverting (−) input through the feedback resistor R_{fb}. According to the calculated gain (10), applying an ac signal of 0.2-V P-P to the input will produce an *inverted* amplified signal of 2-V P-P at the output. You can set the gain of the op-amp by simply using different values for R_{in} and R_{fb}.

An interesting fact about the inverting amplifier is that the feedback voltage tries to cancel out the input voltage. The end result is that the voltage at the inverting input of a good, working circuit is nearly zero volts. For this reason, in an inverting amplifier, the junction of R_{fb}, R_{in} and the inverting input is called the **virtual-ground** point. It is called virtual ground because the terminal is at ground potential, but is not physically connected to the system or chassis ground.

The schematic diagram of a 741 op-amp **non-inverting-voltage amplifier** circuit is drawn in Figure 40-16. In this circuit, the input signal does not become inverted in passing through the amplifier. Again, the feedback and input resistor on the inverting (−) input determines the voltage gain of the circuit. The output is sampled through a voltage-divider circuit and returned to the inverting input. This circuit will always have a gain of more than *one*. The gain formula is slightly different from that of an inverting op-amp and is calculated as follows:

$$\text{Op-amp gain} = \frac{R_{fb}}{R_{in}} + 1 \qquad \text{Noninverting amplifier}$$

$$= \frac{10\,\text{k}\,\Omega}{1\,\text{k}\,\Omega} + 1$$

$$\text{Op-amp gain} = 11$$

40-7 Op-Amp Voltage Comparator

When an op-amp uses negative feedback, the operation is called closed loop. If the op-amp is running wide open without negative feedback, the operation is known as open loop. A **voltage-comparator** circuit is used to compare two different voltage levels and to determine which is the larger of the two. The basic op-amp operated *without* a feedback circuit is ideal for this. A difference in one input voltage with respect to the other will produce a large change in the output voltage.

The schematic of a typical 741 voltage-comparator circuit is shown in Figure 40-17. In this circuit, the voltage at input A is compared with the voltage at input B. Whenever input A voltage is *greater* than input B voltage, the *LED* switches ON to indicate this condition. The operating mode can be reversed (that is, having the LED switch ON when input voltage B is greater than input voltage A) by exchanging the connection to the two inputs (pins 2 and 3).

The schematic of an op-amp bar-graph voltage indicator circuit is drawn in Figure 40-18. This circuit is basically a comparator circuit that causes the LEDs to glow in sequence as the input-sensor voltage rises. Variable resistor R_S is adjusted to control the sensitivity of the

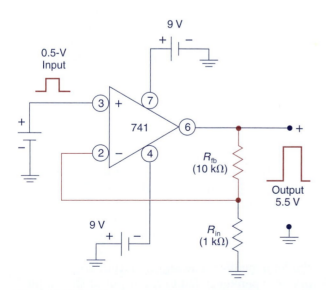

Figure 40-16 Op-amp noninverting amplifier.

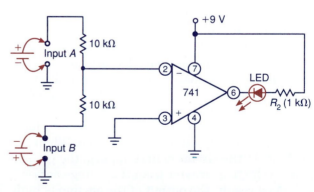

Figure 40-17 741 voltage-comparator circuit.

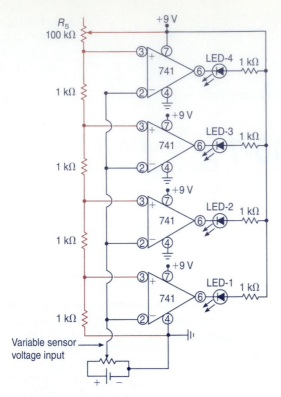

A. Schematic diagram

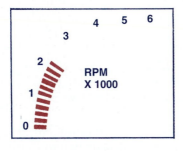

B. Typical bar-graph type gauge

Figure 40-18 Op-amp bar-graph voltage indicator.

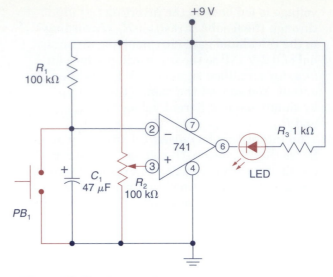

Figure 40-19 Op-amp timer.

circuit. In automobiles, bar-graph type displays are sometimes used to monitor fuel, pressure, temperature, voltage, and speed.

When operated as a comparator, the op-amp can also be used as a *timer*. The schematic for a simple op-amp timer circuit is shown in Figure 40-19. Resistor R_1 and *capacitor* C_1 form an RC charging circuit across the 9-V dc supply. Resistor R_2 sets the reference voltage level for the noninverting input of the op-amp. When voltage to the circuit is first applied, the reference voltage is greater than the voltage across C_1. As a result, the output of the op-amp is high and the LED is OFF. As the capacitor charges,

the voltage across it gradually increases. When the voltage on C_1 rises above the reference voltage, the output of the op-amp swings from high to low and the LED glows. Pushbutton PB_1 is pressed to reset the timing cycle. The time delay period can be changed by altering the values of R_1 and C_1 or the setting of R_2.

40-8 Linear-Power Amplifier

Power amplifiers are classified as those that can deliver more than 500 mW of output power. When used with a heat sink, some *linear* ICs are designed to operate in this higher power output range. They are often used in circuits that operate speaker outputs for tape recorders, record players, TV sound systems, AM-FM radios, and intercoms. The LM380 is one example of a linear-power amplifier IC. Used with the proper heat sink, it can safely deliver up to 2 W of output power.

The schematic for a simple intercom circuit that uses an LM380 IC is shown in Figure 40-20. Internally connected feedback resistors are used to set the gain of the amplifier at 50. The two speakers double as both loudspeakers and microphones. With the master switch in the *talk* position, the master speaker acts as the microphone and the remote speaker acts as the loudspeaker. The signal output from the master speaker is fed to the input of the amplifier through step-up transformer T_1. It is ampli-

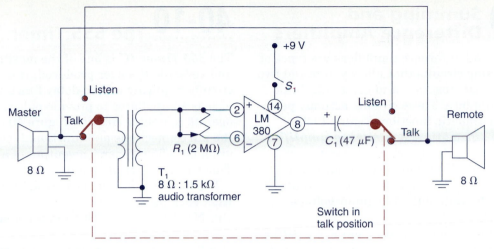

Figure 40-20 Intercom circuit.

fied and then fed on to the remote speaker through coupling capacitor C_1. Variable resistor R_1 provides the necessary volume control, while power to the circuit is controlled by switch S_1.

When the master switch is switched over to the *listen* position the input and output amplifier leads are reversed. The remote speaker now acts as a microphone, feeding a signal to the input of the amplifier. At the same time, the master speaker acts as a loudspeaker, receiving the amplified signal from the output of the amplifier.

As mentioned, op-amp circuits have the ability to reduce unwanted noise signals. The **common-mode rejection** circuit of Figure 40-21, illustrates this. Input signals A and B are unwanted noise signals of the same phase, frequency, and amplitude. Separately they produce outputs of opposite polarity. When both signals are applied they tend to cancel each

other resulting in near zero output. As a result, unwanted signals appearing with the same polarity on both input lines are essentially cancelled. This action is called common-mode rejection. Common-mode noise signals generally are the result of the pick-up of radiated energy from 60-Hz power lines or other sources.

The op-amp can also be configured as an oscillator. Figure 40-22 shows the schematic for an op-amp oscillator that produces a 1.5 kHz square wave. The output, fed to the speaker, produces an audible tone. Because of the op-amp's low output impedance, an impedance matching transformer is not required to operate the speaker. This circuit can be used for an operational check of an op-amp.

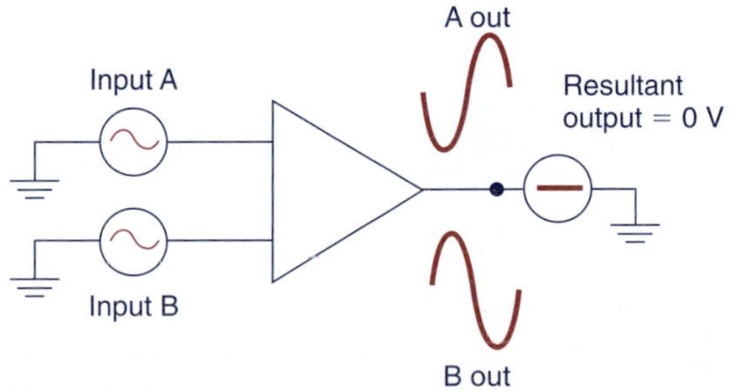

Figure 40-21 Common-mode rejection.

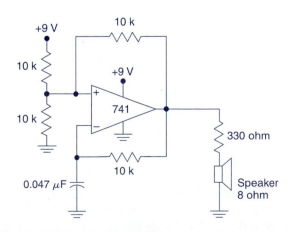

Figure 40-22 Op-amp oscillator.

40-9 Summing and Difference Amplifiers

Summing and difference amplifiers are types of linear op-amp circuits that allow you to add two or more signals together and get an algebraic sum or difference. These circuits actually perform an arithmetic function. Figure 40-23 shows the schematic of a **summing amplifier** with two inputs. In this case all resistors are equal in value. Therefore, each input has a closed loop voltage gain of 1 and the output voltage is the sum of the two input voltages:

$$V_{OUT} = -(V_{IN_1} + V_{IN_2})$$

Figure 40-24 shows the schematic of a **difference amplifier**. Again, all resistors are equal in value. Therefore, each input has a closed lamp voltage gain of 1. In this case, the output voltage is: $V_{OUT} = V_{IN_2} - V_{IN_1}$

40-10 The 555 Timer

The **555 Timer IC** is one of the most popular and versatile ICs ever produced. It is used in circuits requiring a time-delay function and also as an oscillator to provide pulses needed to operate computer circuits. Figure 40-25 shows the pin connections and functional block diagram for a 555 Timer. The output of a 555 Timer IC is a *digital* signal.

The following is a general explanation of what each pin does:

Pin No. 1 Ground   This pin is connected to ground the circuit.

Pin No. 2 Trigger   This pin must be connected to a voltage that is less than one-third V_{CC} (the applied voltage) to trigger the unit. This is generally done by connecting pin No. 2 to ground.

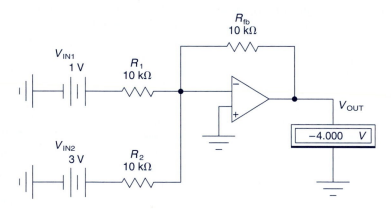

Figure 40-23   Summing amplifier.

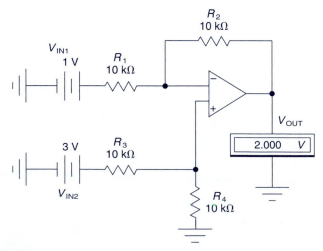

Figure 40-24   Difference amplifier.

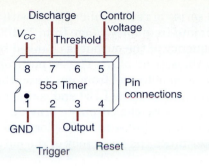

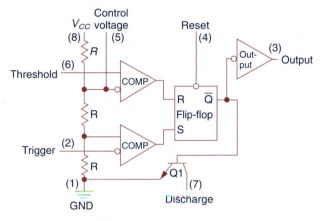

Figure 40-25 555 Timer.

Pin No. 3 Output The output turns ON when pin No. 2 is triggered and OFF when the discharge transistor is turned ON.

Pin No. 4 Reset When this pin is connected to V_{CC} it permits the unit to operate. When connected to ground it activates the discharge transistor and keeps the timer from operating.

Pin No. 5 Control Voltage If this pin is connected to V_{CC} through a variable resistor, the ON time becomes longer. The OFF time remains unaffected. If it is connected to ground through a variable resistor, the ON time becomes shorter. The OFF time is still not affected. If it is not used in the circuit, it is generally connected to ground through a small capacitor. This prevents circuit noise from affecting it.

Pin No. 6 Threshold When the voltage across the capacitor connected to pin No. 6 reaches two-thirds of the value of V_{CC}, the discharge transistors turns ON and the output turns OFF.

Pin No. 7 Discharge When pin No. 6 turns the discharge transistor ON, it discharges the capacitor connected to pin No. 6. The discharge remains ON until pin No. 2 retriggers the timer. It then remains OFF and the capacitor connected to pin No. 6 begins charging again.

Pin No. 8 V_{CC} This pin is connected to the positive applied voltage of the circuit which is known as V_{CC}. The 555 Timer operates on a wide range of voltages. Its operating range is between 3 and 16 volts dc.

Figure 40-26 shows the 555 Timer connected as a *one-shot timer* (also called **monostable multivibrator**). This simple circuit consists of only two timing components—R and C. During standby, the trigger input terminal is held

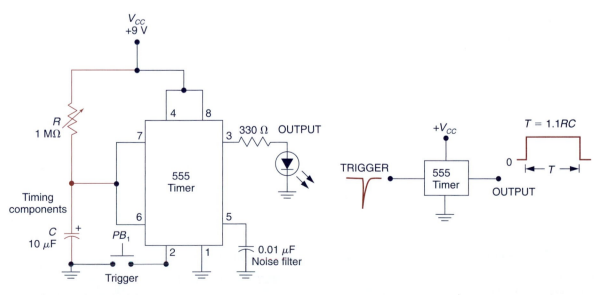

Figure 40-26 555 one-shot timer circuit.

higher than one-third V_{CC} and the output is low. When a trigger *pulse* appears with a level less than one-third V_{CC}, the timer is triggered and the timing cycle starts. The output rises to a high level near V_{CC}, and at the same time, C begins to charge toward V_{CC}. When the C voltage reaches two-thirds V_{CC}, the timing period ends with the output falling to zero, and the circuit is ready for another input trigger. The timing period is determined by values of R and C according to the formula

$$\text{Time (seconds)} = 1.1 \times R(\text{ohms}) \times C(\text{farads})$$
$$= 1.1(1\,\text{M}\Omega)(10\,\mu\text{F})$$
$$\text{Time} = 11\,\text{seconds}$$

In summary, the monostable 555 Timer IC produces a single pulse whose width is determined by the external R and C used. A 0.01-μF decoupling capacitor is used for noise immunity. Because of the internal-latching mechanism, the timer will always time out when triggered, regardless of any subsequent noise, such as bounce, on the trigger input. For this reason, the circuit can also be used as a bounceless switch by using a shorter RC time constant. A 100-kΩ resistor for R and a 1-μF capacitor for C would give a clean, 0.1-second output pulse when used as a bounceless switch.

The 555 Timer is used for many computer operations. One of its common functions is a *clock pulse generator* for timing control of the computing sequence of a computer. The schematic for a typical 555 pulse-generator circuit is shown in Figure 40-27. In this circuit, it operates as an *oscillator* (also called *astable* or **free-running multivibrator**). It produces a continuous string of pulses. No input signal is required to operate the circuit. The output signal is a series of rectangular pulses with an approximate frequency range from 1–100 Hz. The frequency of the output is changed by varying the setting of R_3. The output from this circuit can be obtained from pin 3 and the common terminal. The LED and series resistor R_5 connect across the output to provide a visual indication of the pulse rate. The formula that determines the frequency of the output pulses is:

$$f = \frac{1.44}{[R_1 + 2(R_2 + R_3)]C_1}$$

where:

$$f = \text{Frequency in hertz}$$
$$R = \text{Resistance in ohms}$$
$$C = \text{Capacitance in farads}$$

The output of this clock is not symmetrical; the high (ON) output lasts longer than the low (OFF) output state. The *duty cycle* of a pulse waveform is the percentage of time the output is high. It can be found by dividing the total period of the waveform into the time the output is high. Depending on the resistances of R_1, R_2, and R_3, the duty cycle is between 50–100 percent (Figure 40-28).

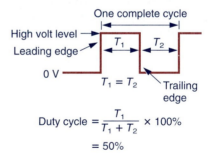

$$\text{Duty cycle} = \frac{T_1}{T_1 + T_2} \times 100\%$$
$$= 50\%$$

A. Square wave

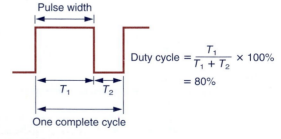

$$\text{Duty cycle} = \frac{T_1}{T_1 + T_2} \times 100\%$$
$$= 80\%$$

B. Rectangular wave

Figure 40-28 Pulse waveforms.

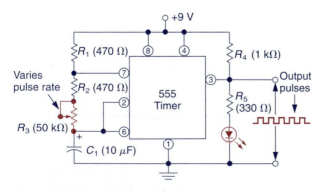

Figure 40-27 555 clock pulse generator.

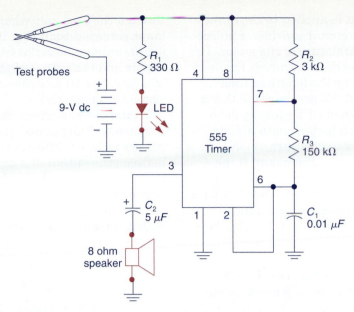

Figure 40-29   555 continuity tester.

Figure 40-29 shows how the 555 Timer can be used as a ***continuity tester*** that gives both a visual and an audible indication of when a complete circuit exists. The 555 Timer is connected to operate in the free-running or astable mode as a tone generator. When the test probes are connected together the circuit to the LED is completed and it lights to indicate a complete circuit. At the same time, power is provided to the timer and an audible tone is emitted from the speaker. The tone of the speaker can be adjusted by changing the value of resistor R_3 or capacitor C_1.

Applying a voltage to the control pin of the 555 Timer allows it to be used as a ***voltage-controlled oscillator*** (***VCO***) or as a ***pulse-width modulator*** (***PWM***). The external voltage overrides the internal voltage and, as a result, you can change the output frequency of the circuit by varying the control voltage.

The control voltage may come from a potentiometer, output of a transistor circuit, op-amp, or some other device. Figure 40-30 shows an experimental circuit for a VCO used to control the speed of a small permanent-magnet dc motor. The circuit works by changing the

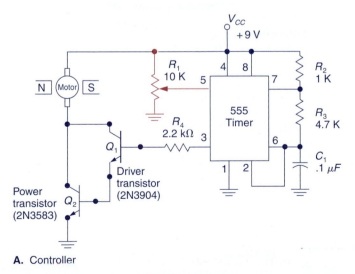

A. Controller

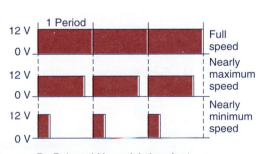

B. Pulse-width modulation chart

Figure 40-30   555 dc motor speed control.

amount of voltage that is applied to the armature of the motor. The circuit provides a pulsating dc voltage to the armature of the motor. The voltage applied across the armature is the *average* value determined by the length of time transistor Q_2 is turned ON as compared to the length of time it is turned OFF (duty cycle). Potentiometer R_1 controls the length of time the output of the timer will be turned ON, which controls the speed of the motor. If the wiper of R_1 is adjusted close to V_{CC} (higher positive voltage), the output will be turned ON for a long period of time as compared with the amount of time it will be turned OFF. This type of control is known as **pulse-width modulation**. Compared to conventional rheostat control, it offers a much higher degree of accuracy and control with a minimum of wasted energy.

40-11 Current Sinking and Sourcing of IC Outputs

In ICs, two types of output circuits are commonly used: current sourcing and current sinking. The names come from the position of the load in the output circuit. In **current sinking** (Figure 40-31), the load is connected between the output and V_{CC} supply. In this type of output circuit, the load is isolated from ground when the output is high. Current will flow

through the load only when the output is at a lower potential than V_{CC}. If the output goes low, a difference in potential will exist between the supply and the output of the IC, and current will flow. The IC switches or *sinks* the output current from ground.

In **current sourcing**, the load is connected between the output and ground. When the output goes high, a difference in potential exists between the output and ground. Current will then flow through the load (Figure 40-32). This circuit is called current sourcing because the IC provides a *source* of power to the load. An IC with a current-source output will normally be low, and go high when power is applied to the load.

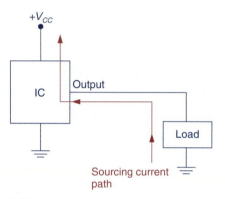

Figure 40-32   Current-sourcing IC output.

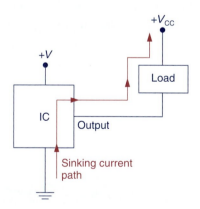

Figure 40-31   Current-sinking IC output.

Review and Applications

Related Formulas

Op-amp gain (inverting amplifier) $= R_{fb}/R_{in}$

Op-amp gain (noninverting amplifier) $= \dfrac{R_{fb}}{R_{in}} + 1$

V_{OUT} (summing amplifier) $= -(V_{IN_1} + V_{IN_2})$

V_{OUT} (difference amplifier) $= V_{IN_2} - V_{IN_1}$

Time (one-shot timer) $= 1.1 \times R \times C$

Frequency (pulse generator) $= \dfrac{1.44}{[R_1 + 2(R_2 + R_3)]C_1}$

% Duty cycle $= \dfrac{\text{pulse width}}{\text{period}} \times 100$

Review Questions

1. What is normally contained within a single IC chip?
2. List four common electronic component parts that can be formed on an IC chip.
3. Compare monolithic and hybrid integrated circuits (ICs).
4. List three major advantages that ICs have over conventional discrete-type circuits.
5. State two major limitations of ICs.
6. Why is no attempt usually made to reproduce the internal circuit schematic of an IC chip on diagrams?
7. What are the normal-operating voltages and current range for typical signal-processing ICs?
8. Compare the principal-circuit element used in bipolar-type and metal-oxide semiconductor-type ICs.
9. List four important things to observe when working with IC circuits.
10. Outline the basic steps involved in troubleshooting an IC package chip.
11. Compare the circuitry found in digital IC chips with that found in analog (linear) types.
12. What, basically, is an *op-amp?*
13. Draw the symbol and label the five basic terminals of an op-amp.
14. Explain the basic function of a comparator circuit.
15. What type of op-amp circuit makes an ideal comparator circuit?
16. Which IC chip is commonly used for a clock pulse generator?
17. Name the two edges of a clock pulse.
18. Under what condition would a string of pulses be classified as square-wave pulses?
19. Define the *duty cycle* of a pulse waveform.
20. Compare the output pulses produced by a 555 Timer IC when connected as a monostable and astable multivibrator.
21. Compare the position of the load to how it is switched ON for sourcing and sinking IC outputs.

Problems

1. (a) Calculate the voltage gain of the op-amp circuit shown in Figure 40-33.
   (b) If the value of the input voltage is 0.2 V, calculate the value of the output voltage.
2. With reference to the op-amp noninverting amplifier of Figure 40-16, determine the voltage gain and value of the output voltage if the value of R_{fb} is changed to $5\,k\Omega$.

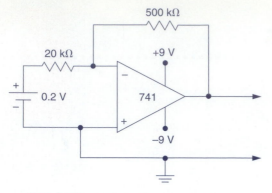

Figure 40-33   Circuit for Problem number 1.

3. With reference to the summing amplifier of Figure 40-23, determine the output voltage value when $V_{IN_1} = 4\,V$ and $V_{IN_2} = 2\,V$.
4. With reference to the difference amplifier of Figure 40-24, determine the output voltage value when $V_{IN_1} = 3\,V$ and $V_{IN_2} = 1\,V$.
5. With reference to the one-shot timer circuit of Figure 40-26, determine the timing period if variable resistor R is set to $0.5\,M\Omega$ and the value of the capacitor is changed to $15\,\mu F$.

6. With reference to the clock pulse generator circuit of Figure 40-27, determine the frequency of the output pulses when R_3 is set to $1\,k\Omega$.
7. Calculate the % duty cycle of a waveform that has a pulse width of $0.2\,\mu s$ and a period of $1.6\,\mu s$.

Critical Thinking

1. What is the best way to determine if any op-amp is faulty?
2. Assume the feedback resistor of the op-amp circuit shown in Figure 40-15 becomes open. What effect, if any, would this have on the shape of the output waveform? Why?

Portfolio Project

• Design and build a practical IC op-amp or timer circuit, similar to one studied in this chapter. Include a written description of the circuits operation and suggest several practical methods for troubleshooting the circuit.

Circuit Challenge

Using a Simulator

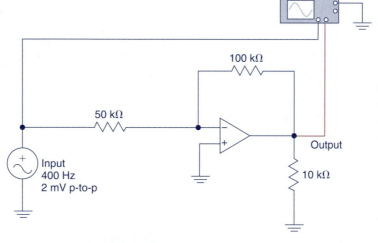

Procedure

(a) Construct the op-amp circuit shown, using whatever simulation software package is available to you.
(b) What is the phase relationship between the input and output?
(c) Based on the phase relationship between the two waveforms, what type of op-amp voltage amplifier is this?

(d) Calculate the voltage gain of the amplifier, based on the resistor values used.
(e) Record the p-to-p value of the output signal.
(f) Calculate the voltage gain based on the p-to-p value of the input and output signals.

Digital Fundamentals

*T*he world is moving from the industrial revolution to an information and communications revolution based on digital electronics. This chapter serves as an introduction to digital
technology. It focuses on the devices and circuits used to build
computers and other digital equipment.

Objectives

After studying this chapter, you should be able to:

1. Convert binary numbers to decimal and decimal numbers to binary.
2. Draw logic symbols and construct truth tables for fundamental logic
   gates.
3. Compare the operation of combinational and sequential logic circuits.
4. Explain the basic operating principles of comparators, adders, flip-
   flops, multiplexers, demultiplexers, encoders, and decoders.
5. Wire experimental digital circuits from given schematics.

Key Terms

- binary number
- bit
- byte
- word
- least significant bit (LSB)
- most significant bit (MSB)
- binary coded decimal (BCD)

- hexadecimal
- gate
- AND gate
- OR gate
- NOT (inverter) gate
- NAND gate
- NOR gate
- TTL
- CMOS

- combinational logic
- Boolean algebra
- binary addition
- binary subtraction
- encoder
- decoder
- multiplexer (MUX)
- serial transmission
- parallel transmission

(continued)

RESEARCH INTERNET SITES

- Micron Electronics
- U S Digital
- Compaq

- demultiplexer (DEMUX)
- sequential logic
- flip-flop
- clock

- storage register
- shift register
- counter
- analog-to-digital (A/D) converter

- digital-to-analog (D/A) converter
- digital logic probe

41-1 Digital Electronics

Digital electronics has had a great impact on the electronics industry. It has become the most important field within electronics. For years, applications of digital electronics were confined to computer systems. More recently, uses for these circuits have become widespread. In stores, cash registers read out digital displays and automatically control inventories. In industry, machine-like robots are controlled by digital circuits. In automobiles, digital circuits are being used to control and monitor engine functions. In the home, personal computers and many appliances are operated through the use of a digital control panel or a keyboard.

Digital circuits operate using *only **two-state signals***. These two states are normally represented within the circuitry by two different voltage levels: high and low. Because only two levels of voltage are permitted, it is convenient to associate a symbol with each level. Often the capital letters H and L represent the high and low levels, respectively. Another frequently

used technique is to let the digits 1 (ON) and 0 (OFF) stand for the two levels (Figure 41-1).

41-2 Binary Number System

In digital electronics, ***binary numbers*** serve as codes that represent decimal numbers, letters of the alphabet, and many other kinds of information. Binary coding of information is essential to the modern computer.

The binary number system is simply another way to represent numbers. It uses only two digits, 0 and 1. (*Bi* means 2; therefore, the *bi*nary system uses two digits). The *binary system*, with its two digits, is a *base-two system*. This system can be used with digital circuits because they process only high- and low-level digital signals. Each position of a binary number can only be a 0 or a 1, and then a 1 is carried to the next position to the left. Figure 41-2 shows binary numbers for decimal values 1 through 15.

The value of a decimal number depends on the digits that make up the number and the

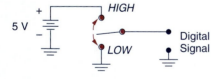

All voltages above this level will then be considered to represent an *ON* (1) signal, whereas all voltages below this will be considered to represent an *OFF* (0) signal.

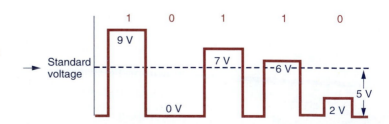

Figure 41-1 Typical digital signal waveform.

Count	Decimal number	Binary number
Zero	0	0
One	1	1
Two	2	10
Three	3	11
Four	4	100
Five	5	101
Six	6	110
Seven	7	111
Eight	8	1000
Nine	9	1001
Ten	10	1010
Eleven	11	1011
Twelve	12	1100
Thirteen	13	1101
Fourteen	14	1110
Fifteen	15	1111

Figure 41-2 Binary numbers representing decimal numbers 1-15.

place value of each digit. A *place value* (weight) is assigned to each position from right to left. In the decimal system, the first position, starting from the rightmost position, is 0; the second is 1; the third is 2; and so on up to the last position. The weighted value of each position can be expressed as the base (10 in this case) raised to the power of the position. For the decimal system, then, the position weights are 1, 10, 100, 1000, and so on. Figure 41-3 illustrates how the value of a decimal number can be calculated by multiplying each digit by the weight of its position and summing the results.

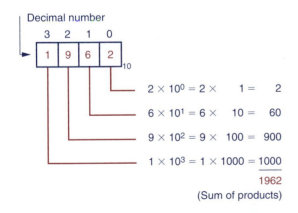

Figure 41-3 Weighted value in the decimal system.

The decimal equivalent of a binary number is calculated in a similar manner. In the binary system, the weighted values of the positions are

1, 2, 4, 8, 16, 32, 64, and so forth. The weighted value of each position is 2 raised to the power of the position. Figure 41-4 illustrates how the binary number 10101101 is converted to its decimal equivalent, 173.

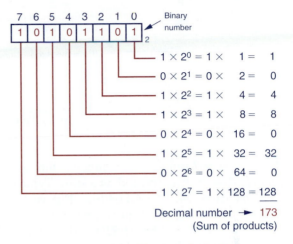

Figure 41-4 Converting binary to decimal.

The term *bit* refers to a single binary character, either 0 or 1. A pattern of 4 bits is called a **nibble**. A group of 8 bits is called a *byte*. A byte may represent a decimal number from 0 to 255.

Switches can be used to enter binary data into digital equipment. Each switch represents one bit of the binary number. The switches are set to either their binary 1 or binary 0 position to represent the desired number. If the switch is set to the UP position, a binary 1 is represented. If the switch is DOWN, a binary 0 is represented. Figure 41-5 shows a typical group of DIP switches set to represent the binary number 11000101, or decimal 197.

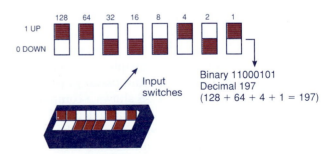

Figure 41-5 Using switches to enter binary data.

Indicator lights such as LEDs are sometimes used to read or display binary data in digital equipment. An ON light is a binary 1 and an OFF light is a binary 0. Figure 41-6 shows a group of LEDs displaying the binary number 01001101, or decimal 77.

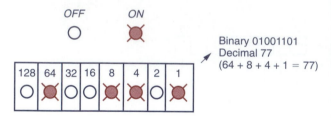

Figure 41-6   Using LEDs to read or display binary data.

The term *word* refers to a number of bits in sequence, that is, it is treated as a unit. Typical word sizes are 8, 16, 32, and 64 bits. Figure 41-7 illustrates a 16-bit word made up of 2 bytes. The *least significant bit (LSB)* is the digit that represents the least value and the *most significant bit (MSB)* is the digit that represents the greatest value.

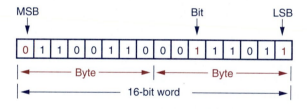

Figure 41-7   A 16-bit word made up of two bytes.

To express a number in the binary system requires many more digits than in the decimal system. A large number of binary digits can become cumbersome to read or write. To solve this problem, other number systems are used to make it easier and more efficient to communicate with digital circuits. These number systems have bases that are multiples of two and include octal, hexadecimal, and *binary coded decimal (BCD)*. Figure 41-8 shows a comparison of the hexadecimal, binary, and decimal numbering systems. *Hexadecimal* has base 16 and employs 16 digits (the numbers 0 through 9 and the letters A through F). It allows counting from 0 to 15 with single-character digits.

An 8-bit binary number can be written in hexadecimal by dividing it into two 4-bit numbers and substituting the hexadecimal value for

Hexadecimal	Binary	Decimal
0	0000	0
1	0001	1
2	0010	2
3	0011	3
4	0100	4
5	0101	5
6	0110	6
7	0111	7
8	1000	8
9	1001	9
A	1010	10
B	1011	11
C	1100	12
D	1101	13
E	1110	14
F	1111	15

Figure 41-8   Hexadecimal numbering system.

each 4-bit piece. Each 4-bit group represents a number between 0 and 15 (0000 and 1111). For example, the decimal number 47 in binary is 0010 1111, and in hexadecimal, it is 2F, where 2 = 0010 and F = 1111 (Figure 41-9). It is much easier to write a number with two hexadecimal characters (2F) than to write it as 00101111. It is also much easier to convert from binary to hexadecimal and hexadecimal to binary than it is to convert back and forth between binary and decimal.

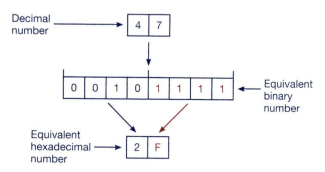

Figure 41-9   Hexadecimal representation of a decimal number.

41-3 Logic Gates

Two basic requirements must be met in order to process information using digital techniques. First, the data must be coded in a binary form. Second, the coded data must be controlled in a logical manner. The second requirement is sat-

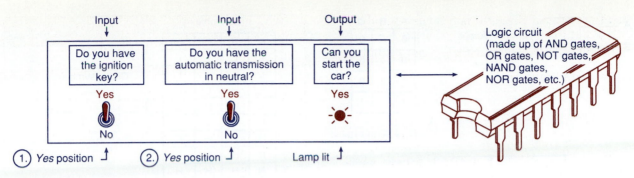

Figure 41-10   Using logic circuits to answer questions.

isfied using electric circuits known as **logic gates**.

Logic-gate circuits operate on the yes-no principle. For instance, a circuit can be designed to answer the question, "Can you start the car?" Suppose this circuit has been built inside a small box, as shown in Figure 41-10. On the outside of the box are two switches. Above the left-hand switch is the query, "Do you have the ignition key?" If you do, you place the switch in the *yes* position. Above the right-hand switch is the question, "Do you have the automatic transmission in neutral?" If you do, you place this switch in the *yes* position also. Under the question "Can you start the car?" the lamp lights to indicate the answer is *yes*. If either switch is in the *no* position, the lamp does not light.

Logic gates can be defined as devices that produce an output only when predetermined input conditions exist. They are referred to as *gates* because they open and close. That is, if all input conditions have been met, the output will be an open gate. If all input conditions are *not* met, the gate will remain closed. Although *logic* gates can be constructed using switches, relays, diodes, or discrete transistors, most modern digital logic circuits are constructed using ICs.

The AND Gate   The *AND gate* is a device with two or more inputs and one output. It is sometimes called the *all-or-nothing gate*. All inputs must be in the logic 1, or HIGH, state in order to obtain an output of logic 1. The diagram shown in Figure 41-11 illustrates logical AND using simple switches that are connected in series. Only when both switches A *and* B are closed, will the battery be able to turn the light ON.

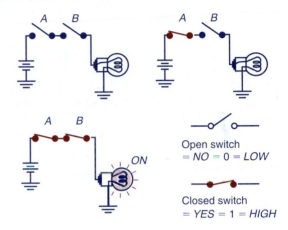

Figure 41-11   Logical AND.

Figure 41-12 shows a two-input AND gate constructed using a digital IC chip. Two SPDT (single-pole, double-throw) switches are used as the input devices. A low (0) input is produced when the switch is connected to the common negative supply, and a high (1) results when it is connected to the positive supply lead. Similarly, a low (0) output is indicated when the LED is not lit, and a high (1) is indicated when it is lit. As mentioned previously the output is high only when both inputs A and B are high. The **truth table** is used to show the resulting output for each possible input condition.

An AND gate can have more than two inputs. Additional inputs increase the decision-making power of a gate (Figure 41-13). To determine the number of combinations for a gate with any number of inputs use the formula:

$$N = 2^n$$

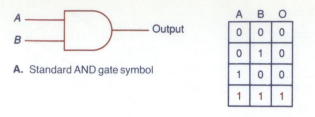

A. Standard AND gate symbol

A	B	O
0	0	0
0	1	0
1	0	0
1	1	1

B. AND truth table

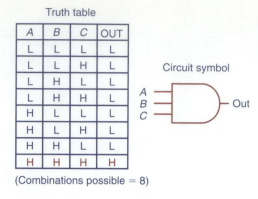

Truth table

A	B	C	OUT
L	L	L	L
L	L	H	L
L	H	L	L
L	H	H	L
H	L	L	L
H	L	H	L
H	H	L	L
H	H	H	H

(Combinations possible = 8)

Circuit symbol

Figure 41-13   Three-input AND gate.

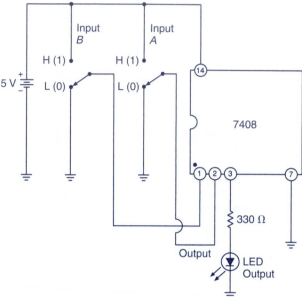

C. Circuit schematic

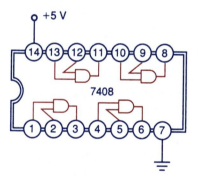

D. Typical quad two-input AND gate IC chip

Figure 41-12   Two-input AND gate.

where N is the total possible combinations and n is the number of inputs.

For *two* inputs $N = 2^2 = 4$

For *three* inputs $N = 2^3 = 8$

For *four* inputs $N = 2^4 = 16$

In most applications, the inputs to a gate are not constant levels of voltages, but digital pulses. In examining the pulsed operation of a gate, you must look at the inputs with respect to each other in order to determine the output level at any given time (Figure 41-14). A diagram of input and output waveforms that show time relationships is called a ***timing diagram***.

The OR Gate   The *OR gate* is also a decision-making circuit with two or more inputs and one output. It is sometimes called the *any-or-all gate*. Its output is logic 0, or LOW, unless any or all of its inputs are logic 1, or HIGH. Figure 41-15 illustrates logical OR using simple switches that are connected in parallel. Looking at the circuit, you can see that the output lamp will light when either A or B, or both, are closed. However, when both are open, no output is obtained.

Figure 41-16, p. 620, shows a two-input OR gate constructed using a digital IC chip. The OR gate output is 1, or HIGH, if one or more inputs is 1. The truth table shows the resulting output for each possible input condition.

The NOT (Inverter) Gate   Unlike the AND and OR gates, the ***NOT gate*** can have only *one* input. The NOT gate inverts, or complements, the logic state at its single input, so it is often called an *inverter*. The diagram of Figure 41-17, p. 621, illustrates the logical NOT using a single switch. The output lamp lights when the switch is open, or at logic 0. When the switch is closed, or at logic 1, the lamp is OFF, or at logic 0.

Figure 41-18, p. 621, shows a NOT gate, or INVERTER, constructed using a digital IC. It simply takes the input, reverses its state, and provides this as output. If each bit of a binary number were applied to the input of an inverter, the output number would have all of

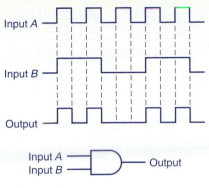

A. Timing diagram

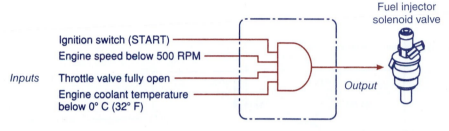

Fuel injector
solenoid valve

Inputs
- Ignition switch (START)
- Engine speed below 500 RPM
- Throttle valve fully open
- Engine coolant temperature below 0° C (32° F)

Output

(The amount of fuel injected is controlled by the length of time that the output is HIGH)

B. Typical automotive application

Figure 41-14 Pulsed gate operation.

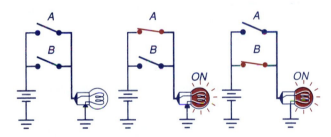

Figure 41-15 The logic OR.

its 1s changed to 0s and all of its 0s changed to 1s. Notice the use of the line above the output to show that *A* has been inverted, or complemented. Also, note that a "bubble" (°) is drawn on the output of the inverter symbol.

The NAND and NOR Gates   The AND and OR logical functions are basic to all digital systems. These functions are implemented with AND and OR gates, but a variation of these gates is even more widely used. These are NAND and NOR gates, which are a combination of either an AND gate or an OR gate and an INVERTER.

A *NAND gate* (NOT-AND) is a combination of an AND gate and an INVERTER (Figure 41-19, p. 621). The truth table for the NAND gate has output exactly *opposite* of the output of the AND gate. Notice that the symbol for the NAND gate looks almost like the one for the AND gate, except that a bubble or circle has been added near the output to indicate that the normal output of the AND has been inverted. The NAND gate is called a universal gate because it can be configured for any other type of logic function or gate.

A *NOR gate* (NOT-OR) is a combination of an OR gate and an INVERTER (Figure 41-20, p. 622). Again, notice that the output in the truth table is exactly opposite that of the OR gate.

Digital ICs, other than the basic gates we have discussed, include counters, converters, flip-flops, and microprocessor chips, to name a few. No special marking is used when a signal normally goes positive for an *active* or *required* input resulting in a desired output from the IC. If that pin must be negative, however, the input pin is designated with a small circle. In some circuits, letter designations for pins are used.

A. Standard OR gate symbol

A	B	OUT
0	0	0
0	1	1
1	0	1
1	1	1

B. OR truth table

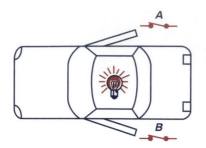

Car's interior light is activated
by either door being open.

C. Typical automotive application

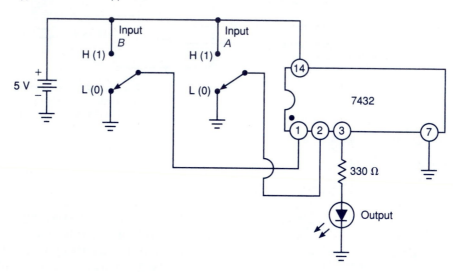

D. Circuit schematic

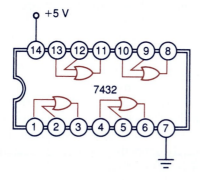

E. Typical quad two-input OR gate IC chip

F. Timing diagram

Figure 41-16 Two-input OR gate.

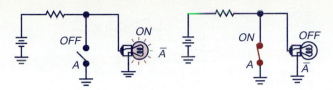

Figure 41-17 The logical NOT.

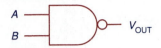

A. Standard NAND gate symbol

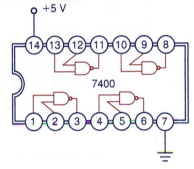

B. Equivalent NAND gate wired using an AND gate and INVERTER

A	B	V$_{OUT}$
0	0	1
0	1	1
1	0	1
1	1	0

C. NAND truth table

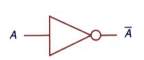

A	$\overline{A}$
0	1
1	0

A. Standard INVERTER symbol

B. INVERTER truth table

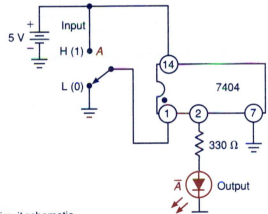

C. Circuit schematic

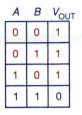

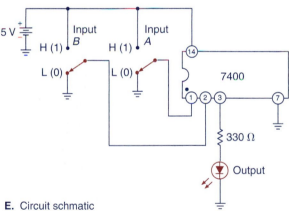

D. Typical quad two-input NAND gate IC chip

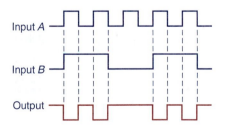

E. Circuit schmatic

D. Typical INVERTER IC chip

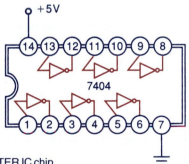

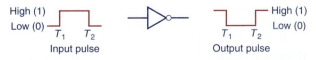

F. Timing device

E. Pulsed operation

Figure 41-19 Two-input NAND gate.

Figure 41-18 Inverter.

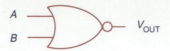

A. Standard NOR gate symbol

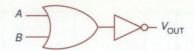

B. Equivalent NOR gate wired using an OR gate and INVERTER

A	B	V_{OUT}
0	0	1
0	1	0
1	0	0
1	1	0

C. NOR truth table

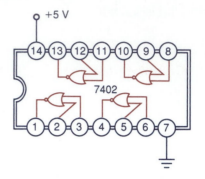

D. Typical quad two-input NOR gate IC chip

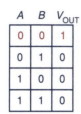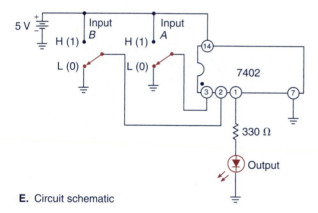

E. Circuit schematic

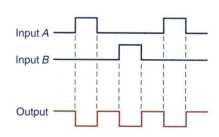

F. Timing diagram

Figure 41-20 Two-input NOR gate.

The letters have a bar over them if the pin must be negative (Figure 41-21).

Logic gates use diodes and transistors for their operation. The standard *TTL (transistor-transistor logic)* family is based on the use of multiple bipolar transistor circuits. Newer TTL families having a Schottky designation contain Schottky transistors and/or diodes and provide faster speed and lower power dissipation. Identification on standard TTL chips takes the form of 74XX. The XX represents numbers assigned to specific chip types. Chips identified as 74LXX are low-power Schottky TTL series. The difference between the various TTL series is indicated by their operating characteristics. The pin numbers remain the same. All TTL devices are electrically compatible, that is, they can be connected directly without any interfacing circuitry and operates at 5 V_{CC}.

The other type of logic gate is called *CMOS (Complimentary Metal Oxide Semiconductor)*. TTL and CMOS differ in that TTL uses bipolar transistors and CMOS uses FETs (Field Effect

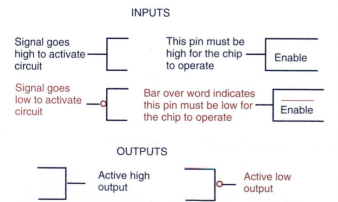

Figure 41-21 Digital signal coding.

Transistors). The logic circuits differ only in performance characteristics, not logic function. CMOS and TTL devices are not always compatible. CMOS voltage connections are labeled V_{dd}, and the ground is labeled V_{ss}. The operating voltage is 3 to 18 volts.

41-4 Combinational Logic

Individual logic gates produce many logic functions but they cannot perform all the operations required by a computer. *Combinational logic* circuits allow complex circuits to be made. They have no storage or memory and make decisions based on inputs. For every combination of signals at the input terminals, there is a definite combination of signals at the output terminals. A typical combinational logic circuit has multiple inputs and one or more outputs. The output is determined by the binary state of the inputs, the types of logic circuits used, and how they are interconnected.

The primary function of a combinational logic circuit is to *make decisions*. The combinational logic circuit looks at its inputs and then makes

a decision based on them. In many logic circuits, sensor switches act as the input devices. Figure 41-22 illustrates how multiple inputs can be connected to combinational logic circuits to obtain the desired logic. In analyzing a combinational logic circuit, we are concerned with determining in what way the output is dependent on the inputs; for example, for what combinations of inputs is the output a HIGH or a LOW?

An often-used combination of gates is that of the *exclusive-OR* (XOR) function (Figure 41-23). The output of this circuit is HIGH only when one input or the other is HIGH, but *not both*. The XOR gate is also available in an IC package, so it is not necessary for the designer to interconnect separate gates to build the function.

The exclusive-OR gate is commonly used for the *comparison* of two binary numbers. For example, binary numbers that represent temperature can be compared (bit by bit) with binary numbers that represent the voltage levels from a coolant sensor to determine the temperature of the coolant. The number representing the temperature is connected to one input of XOR gate, and the number representing

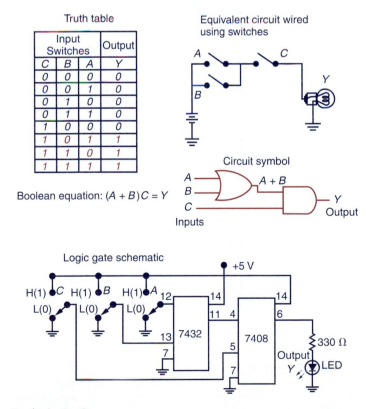

Truth table

Input Switches			Output
C	B	A	Y
0	0	0	0
0	0	1	0
0	1	0	0
0	1	1	0
1	0	0	0
1	0	1	1
1	1	0	1
1	1	1	1

Boolean equation: $(A + B)C = Y$

Figure 41-22   Combination logic circuits.

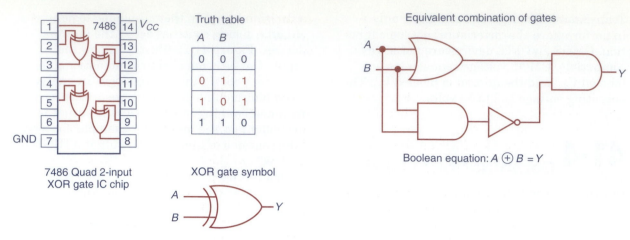

Figure 41-23   Exclusive-OR (XOR) function.

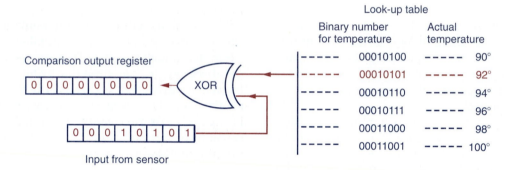

Figure 41-24   Comparing binary numbers.

coolant temperature is connected to the other input, bit by bit. If both inputs are 1 or both inputs are 0, then the output is 0. Therefore, when the codes are the same, each comparison yields an output code that has 0 in each bit position in the register, and the circuit knows it has a match. In the case illustrated in Figure 41-24, the number from the sensor matches the number in the look-up table for 92°.

Boolean algebra is a method used to express logical operations in mathematical form. The objective of effective digital design is the achievement of the desired system functions, or decision-making capability, with the minimum number of gates. Digital logic can be manipulated mathematically through the use of Boolean algebra. The solution of Boolean equations permits the necessary functions to be implemented with the minimum number of gates.

Figure 41-25 summarizes the basic operators of Boolean algebra as they are related to the

Logic symbol	Logic statement	Boolean equation
A B — AND — Y	Y is "1" if A and B are "1"	$Y = A \cdot B$ or $Y = AB$ or $Y = A \times B$
A B — OR — Y	Y is "1" if A or B is "1"	$Y = A + B$
A — NOT — Y	Y is "1" if A is "0" Y is "0" if A is "1"	$Y = \bar{A}$

Figure 41-25   Boolean algebra as related to AND, OR, and NOT functions.

basic AND, OR, and NOT functions. *Inputs* are represented by capital letters A, B, C, and so on. The output is represented by a *capital Y*. The multiplication sign ($\cdot$ or $\times$) represents the

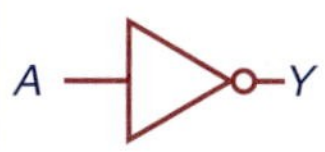

AND operation. Sometimes no sign is present (e.g., $Y = AB$). An addition sign ($+$) represents the OR operation; and a bar over the letter ($\overline{A}$) represents the NOT operation.

Combinational logic circuits may be designed using Boolean algebra. Circuit functions are represented by Boolean equations. Figure 41-26 illustrates how the basic AND, OR, and NOT functions are used to form Boolean equations.

Some laws of Boolean algebra are different from those of ordinary algebra. The following three basic laws illustrate the close comparison between Boolean algebra and ordinary algebra as well as one major difference between the two.

Commutative Laws:
The **commutative law of addition** for two variables is written algebraically as:

$$A + B = B + A$$

This equation states that the order in which the variables are ORed makes no difference.

The **commutative law of multiplication** for two variables is written algebraically as:

$$AB = BA$$

This equation states that the order in which the variables are ANDed makes no difference.

Associative Laws:
The **associative law of addition** for three variables is written algebraically as:

$$(A + B) + C = A + (B + C)$$

This equation states that in the ORing of several variables, the result is the same regardless of the grouping of the variables.

The **associative law of multiplication** for three variables is written algebraically as

$$(AB)C = A(BC)$$

This equation states that it makes no difference in what order the variables are grouped when ANDing several variables.

Distributive Law:
The **distributive law** for three variables is written algebraically as:

$$A(B + C) = AB + AC$$

This equation states that ORing several variables and ANDing the result with a single variable is equivalent to ANDing the single variable with each of the several variables and then ORing the products.

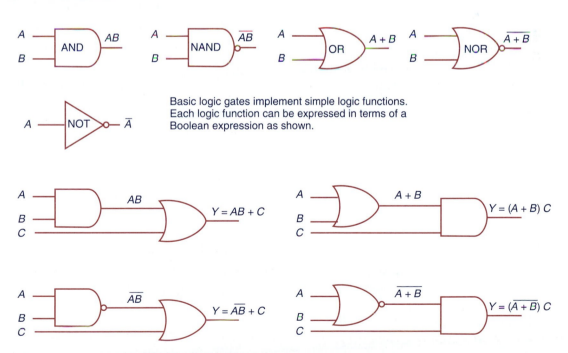

Basic logic gates implement simple logic functions. Each logic function can be expressed in terms of a Boolean expression as shown.

Any combination of control functions can be expressed as shown.

Figure 41-26 Examples of Boolean equations.

$$A + (BC) = (A + B)(A + C)$$

This equation is true only in Boolean algebra and states that ANDing several variables and ORing the result with a single variable is equivalent to ORing the single variable with each of the several variables and then ANDing the sums.

DeMorgan's Law is one of the most important results of Boolean algebra (Figure 41-27). It shows that any logical function can be implemented with AND gates and inverters or OR gates and inverters.

According to DeMorgan's law:

$$\overline{AB} = \overline{A} + \overline{B}$$

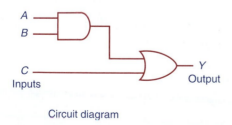

and

$$\overline{A + B} = \overline{A}\,\overline{B}$$

Figure 41-27 DeMorgan's law.

Figure 41-28 shows illustrated examples of the method used to develop a circuit from a Boolean expression. Figure 41-29 shows illustrated examples of the method used to write the Boolean equation for a given circuit.

Boolean expression: $Y = AB + C$

Gates required: (by inspection)
AND gate with input A and B
OR gate with input C and output from previous AND gate

A
B

C
Inputs

Y
Output

Circuit diagram

Since the output of a logic gate must supply current to other gates or to sink (absorb) the current from inputs, there is a limit to how many inputs that a single logic circuit can drive. The term *fan-out* is used in describing the number of inputs to other gates that a logic circuit can drive. The fan-out is simply a number that tells how many inputs can be connected to a logic circuit output. Typical fan-outs run from approximately 3–10 for most standard logic gates.

41-5 Binary Arithmetic

The heart of any computer is the **arithmetic-logic unit (ALU)**. The ALU performs all the mathematical operations that take place within the computer. Mathematical operations include addition, subtraction, multiplication, and division. In actuality, the only true mathematical operation that is performed by a computer is addition. Any other operation is just a variation of the additional process.

Combination logic can be designed to perform all the arithmetic functions. A widely used application of an exclusive-OR gate is as a binary adder. A binary adder is the heart of any digital arithmetic circuit.

Binary addition follows rules similar to decimal addition. When adding with binary numbers, there are only four conditions that can occur:

0	1	0	1
+0	+0	+1	+1
0	1	1	0 carry 1

Boolean expression: $Y = A(BC + D)$

Gates required: (by inspection)
AND gate with inputs B and C
OR gate with inputs $B \bullet C$ and D
AND gate wih inputs A and the output from the OR gate

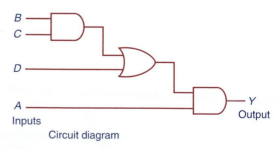

B
C

D

A
Inputs

Y
Output

Circuit diagram

Figure 41-28 Examples of circuit development from a Boolean expression.

Write the Boolean equation
for the following circuit:

Write the Boolean equation
for the following circuit:

Original circuit

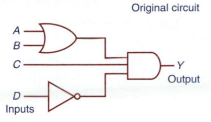

Original circuit

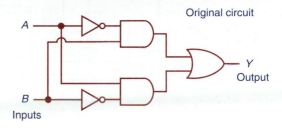

Circuit with Boolean terms

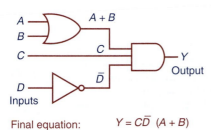

Final equation: $Y = C\bar{D}\,(A + B)$

Circuit with Boolean terms

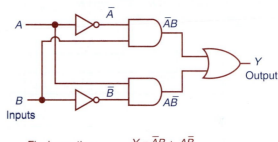

Final equation: $Y = \bar{A}B + A\bar{B}$

Figure 41-29   Examples of writing the Boolean equation for a given circuit.

Just as in the case of adding decimals, the first three conditions are easy, but the last condition is slightly different. In decimal, $1 + 1 = 2$. In binary, a 2 is written 10. Therefore, in binary, $1 + 1 = 0$, with a carry of 1 to the next most significant place value. Figure 41-30 shows the symbol for a half-adder. A half-adder has two inputs (A, B) and two outputs, a sum and a carry.

When adding larger binary numbers, the resulting 1s are carried into higher-order columns, as shown in the following examples:

Decimal   **Equivalent binary**

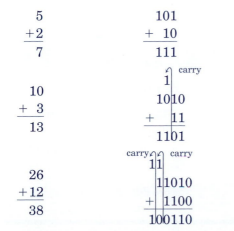

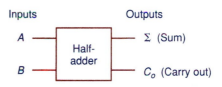

Figure 41-30   Half-adder symbol.

The half-adder can be used in only very limited applications, since it is restricted to receiving the two bits to be added as inputs and giving two outputs, a *sum* and a *carry*. In normal addition, we find that the basic addition requires *one more function,* the ability to add the carry from a previous addition. Figure 41-31 shows the symbol for a full-adder, which has that capability.

A four-bit adder consists of a string of four full adders connected in cascade so that two

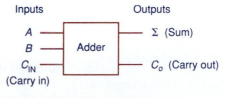

Figure 41-31   Full-adder symbol.

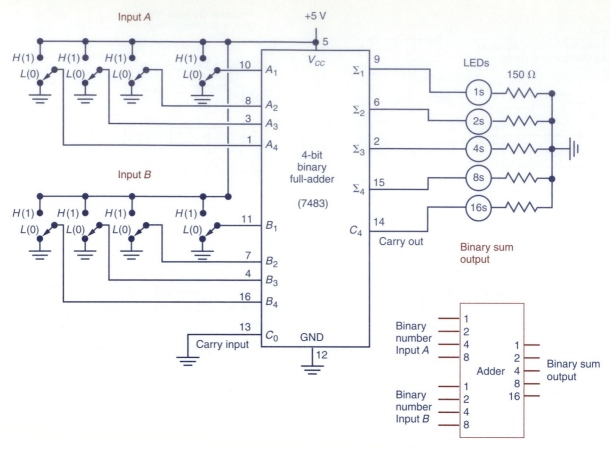

Figure 41-32 Four-bit adder circuit.

4-binary numbers can be added together. Figure 41-32 shows the 7483 full four-bit parallel adder that will add two 4-bit binary numbers and produce a full sum. A full sum is one that results from adding two binary numbers and a carry out does not result. Instead of a carry out, a fifth digit becomes the most significant bit (MSB).

Figure 41-33 shows the 7483 adder IC rewired for a four-bit parallel subtractor. The

---

Example 41-1

Problem Subtract decimal 6 from decimal 10 using 1s complement subtraction.

Step 1 Change the subtrahend to 1s complement.

Step 2 Add the two numbers.

Step 3 Remove the last carry and add it to the number (end-around carry).

Decimal		**Binary**	
10		1010	1010
− 6		−0110 $\xrightarrow{\text{1s complement}}$	+1001
4		100	10011
		End-around carry	→ +1
			100

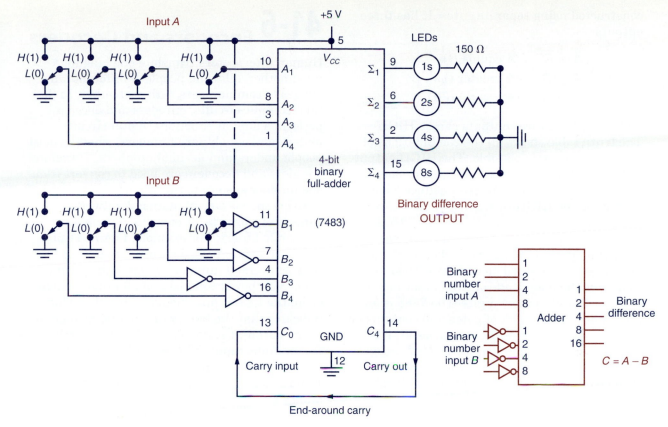

Figure 41-33 Four-bit parallel subtractor circuit.

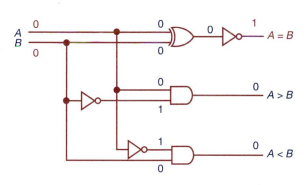

Figure 41-34 One-bit *A-B* comparator.

TRUTH TABLE

Number		Comparison		
A	B	A > B	A = B	A < B
0	0	0	1	0
1	0	1	0	0
0	1	0	0	1
1	1	0	1	0

operation is accomplished by using 1s complement and end-around carry subtraction. The 1s complement of a binary number is obtained by negating all its digits (change all 1s to 0s and all 0s to 1s) using inverters.

Since multiplication is simply a series of additions, it can be performed by using an adder in

conjunction with other circuits. Similarly, division is treated as repeated subtraction and can also be performed by using an adder in conjunction with other circuits.

The basic function of a *comparator* is to compare the relative magnitude of two quantities. Figure 41-34 shows a one-bit comparator that is constructed using separate gates. It has three outputs:

$$A = B \quad (A \text{ equals } B)$$
$$A > B \quad (A \text{ greater than } B)$$
$$A < B \quad (A \text{ less than } B)$$

Only one of the three outputs can be HIGH. The truth table shows all the combinations of inputs and outputs possible. The HIGH output tells you if the digital bits are equal or one is greater than the other. You can deduce which output will be HIGH by tracing signals through the gates.

Figure 41-35 shows an integrated circuit four-bit A-B magnitude comparator. The 7485 has all the gates contained on one chip. The circuit compares the magnitude of two binary numbers, each containing 4 digits. The 7485 is a combinational network of gates and contains no internal memory or storage. As a result, it is called an asynchronous device. *Asynchronous* means that it operates only when inputs are applied or changed and it is not controlled by a clock.

41-6 Encoders and Decoders

Human interpreters translate one language into another, allowing people of different languages to communicate with each other. In digital electronic circuits, encoders and decoders perform a similar function. Digital circuits understand only binary numbers, whereas most people understand decimal numbers. Therefore, encoders and decoders are used to convert from one number system to the other.

An *encoder* is a combinational network of gates that converts, or encodes, a nonbinary input into binary. For example, Figure 41-36 shows an experimental encoder circuit used to convert decimal numbers 0–9 to binary numbers. With power applied and all input switches open, the LED lights should all be OFF. This indicates that the binary number 0000 appears at the output. Closing decimal input-switch 3, for example, inputs the decimal number 3 to the encoder. This, in turn, causes the LEDs for 1 and 2 to come ON, indicating the equivalent binary number, 0011.

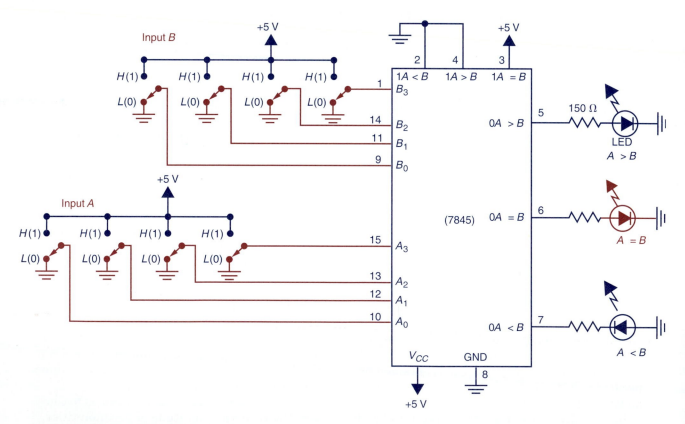

Figure 41-35 Four-bit A-B comparator.

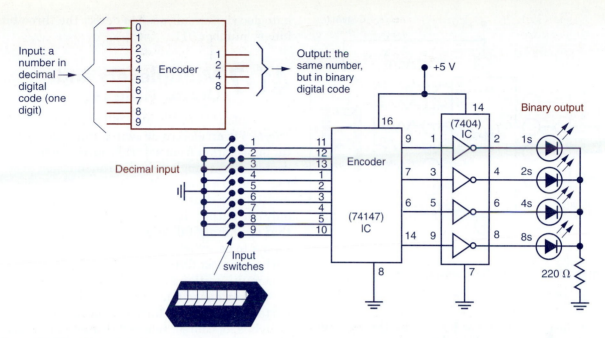

Figure 41-36 Decimal-to-binary encoder.

In digital electronics, it is often necessary to convert a binary number into some other format, and one common decoder application is the conversion of binary numbers into the format required to activate the appropriate segments in a *seven-segment decimal display*. These displays may be liquid crystal, fluorescent, or LED types. Figure 41-37 illustrates a seven-segment LED display. Each segment is identified by a letter: *A* through *G*. This pattern holds true for all seven-segment displays regardless of what type they are. When the correct voltages are applied to the LEDs, one or more of the segments will light up to form a number. Notice that all cathodes or all anodes are connected together depending on whether the display is a common cathode or common anode type.

Figure 41-38 shows a decoder circuit used to convert a four-bit binary code into the appropriate decimal digit. Binary-coded numbers are fed into the circuit by pressing the appropriate normally closed pushbutton. The decoder interprets the incoming binary number and produces an output that drives a seven-segment common-anode LED display. In order to obtain any digit from 0–9 on the LED display, you simply input the appropriate binary number code by means of the pushbuttons. When a binary number is present at the input of the decoder, the decoder puts out the right logic levels at its outputs to light up the correct LEDs to form the decimal equivalent of the binary number. The 330-ohm resistors connected between the outputs of the decoder and the display are used to

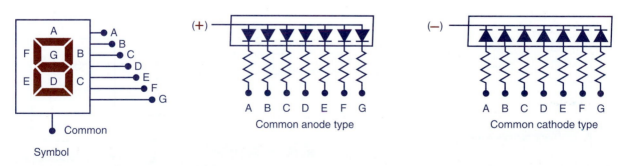

Figure 41-37 Seven-segment LED display.

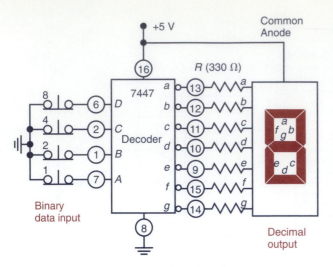

Figure 41-38 Binary-to-decimal decoder.

limit the current that will circulate through the LEDs of the display.

A *decoder* is a combinational logic circuit that recognizes the presence of a specific binary number or word. The main element in a decoder circuit is an AND gate. An AND gate generates a binary 1 output if all its inputs are binary 1. By connecting the bits of the binary number to be recognized to the AND gate, either directly or through inverters, the AND gate will be able to recognize the presence of a specific number. Figure 41-39 shows an AND

Figure 41-39 AND gate decoder.

gate decoder circuit used to detect the three-bit binary number 011.

41-7 Multiplexers and Demultiplexers

A *multiplexer* (*MUX*) (also called a *data selector*) is an electronic switch that allows digital information from several sources to be routed onto a single line for transmission over that line to a common destination. The basic multiplexer thus has several input lines but only *one* output. It also has select-control inputs that permit digital data on any one of the inputs to be switched to the output line.

The simplest form of a multiplexer is a single-pole multiposition switch, shown in Figure 41-40. Any one of four input signals can be connected to the output line. The mechanical switch is, of course, selected by hand to connect any one of the input lines *A–D* to output line *X*. The digital multiplexer performs the same operation with a binary code applied to the select-control inputs. The select-control inputs work as a decoder to allow the correct input line to be selected and passed through to the output.

Figure 41-41 shows an experimental 1 of 16 data-selector circuit that uses a 74150 IC. Data is transferred from any one of the 16 data inputs to the output depending on the state of the data select inputs. To operate the circuit load data (0 or 1) into the inputs and set the different digital counts on the data select inputs (0 through 15) to obtain the corresponding data on the output. The small circle at output W of 74150 IC means that the data will appear in an inverted form. The 7404 IC inverts the data so that the output equals the input data.

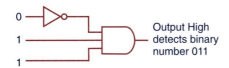

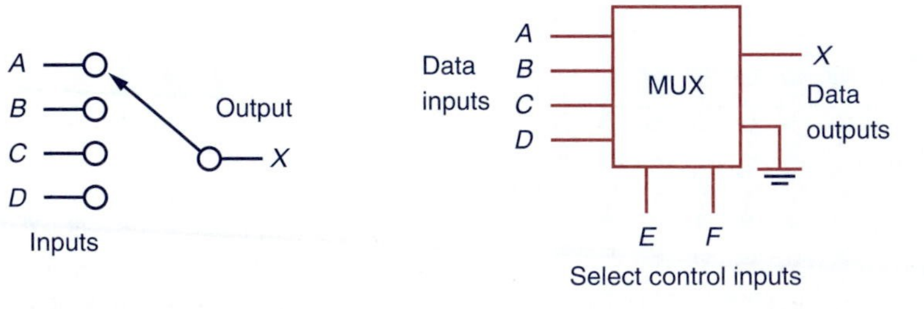

A. Mechanical multiplexer

Select control inputs		Input line directed to output *X*
E	*F*	
0	0	*A*
0	1	*B*
1	0	*C*
1	1	*D*

B. Digital multiplexer

Figure 41-40 Mechanical and digital multipliers.

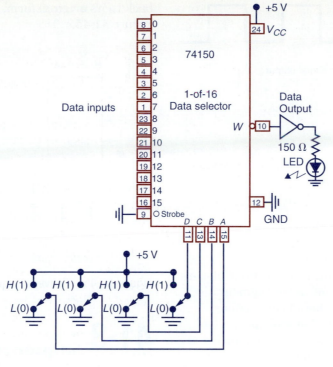

Figure 41-41   1 of 16 data selector IC.

Data can be transferred from one location to another within a digital system by either serial or parallel transmission. In **serial transmission**, data move from point to point on a single wire one bit at a time, as illustrated in Figure 41-42A. The rate at which the bits are transferred from A to B is called the *baud rate* and is expressed in bits per second. Thus serial systems are often referred to as operating at a certain baud rate. In **parallel transmission**, all bits in a given group are transferred simultaneously on separate lines, as illustrated in Figure 41-42B. As you can see, the advantage of serial transfer is that only one line is required. In parallel transfer multiple lines are required—1 for each bit.

Serial data transmission is the least expensive way to handle information, but it is also the slowest way, since each 0 and 1 in the sequence takes a certain amount of time. Parallel transmission is much faster, so it is used when the equipment being connected is close by (usually less than 25 feet). As the distance increases the cost and difficulty of parallel transmission of pulsed signals becomes a major limitation. For this reason, over long

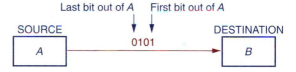

A. Serial data transmission

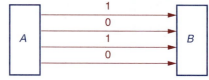

B. Parallel data transmission

Figure 41-42   Serial and parallel data transmission.

distances, serial transmission is preferred because it involves only one transmission line.

Figure 41-43 illustrates how a multiplexer can be used for parallel-to-serial data conversion. A four-bit binary number is applied to the four inputs of the one-of-four data-selector circuit. The A and B input-control lines are stepped one at a time through the 00, 01, 10, 11 binary sequence. Each set of control inputs

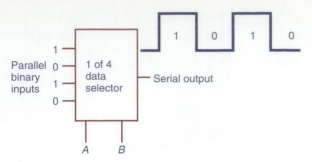

Figure 41-43 Multiplexer used for parallel-to-series data conversion.

appears for a fixed, but short, period. The result is a serial output version of the binary word applied to the input.

Another common application for multiplexers is controlling a number of seven-segment displays to reduce the number of connections required (Figure 41-44). The multiplexer lets all the digits in readout share a common set of terminals. It activates each digit or one segment in all the digits in rapid succession to fool the eye into thinking the display is continually illuminated. This basic method of display multiplexing can be extended to displays with any number of digits. The frequency must be high enough (about 30 Hz) to prevent visual flicker as the digit displays are multiplexed.

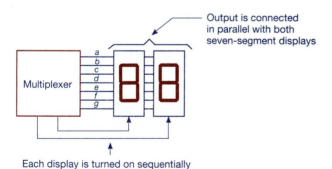

Figure 41-44 Multiplexing of displays.

A *demultiplexer* (*DEMUX*) (also called decoder) is opposite to a multiplexer in that it has a single input and multiple outputs. Like the multiplexer, the demultiplexer is also a data-routing circuit. Control inputs are required to determine to which output the input will be directed. Demultiplexers are used with multiplexers to convert multiplexed data

back to its original form, as illustrated in Figure 41-45.

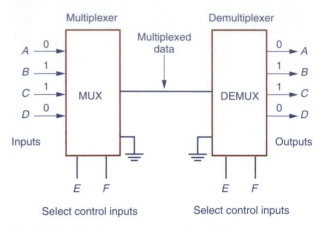

Figure 41-45 Converting multiplexed data.

41-8 Sequential Logic

Unlike combinational logic circuits, *sequential logic circuits* have *memory*. Their output(s) can reflect the effect of an input that occurred seconds, or even days, earlier. In the sequential system, the memory keeps track of what operations the system has performed in the immediate past, since future operations depend on that knowledge.

The main element in a sequential logic circuit is a *flip-flop*. A flip-flop is a circuit that changes state each time it receives a voltage pulse. The basic operation of a flip-flop can be explained by referring to the model of the circuit shown in Figure 41-46. This flip-flop has one input and two outputs. The outputs are called Q and $\overline{Q}$ (not Q). Only one of these outputs can be HIGH, or 1, at a time. When a single pulse is applied to the input of the flip-flop, it produces a switch in the output high between Q and $\overline{Q}$. Once the flip-flop is switched, it *stays switched* until the next pulse is received. In this way it acts as a simple electronic memory unit.

There are several different kinds of flip-flops, but all are capable of storing a single binary bit. The simplest flip-flop is the *reset-set* (*RS*) *flip-flop*. Figure 41-47 shows the logic symbol and truth table for an *RS* flip-flop as well as the circuit constructed using two NAND gates. Notice that the *RS* flip-flop has two inputs, labeled S and R, and two outputs, labeled Q and $\overline{Q}$

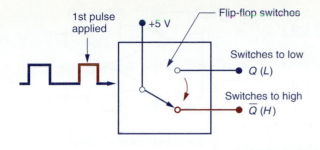

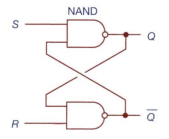

Figure 41-46   Basic flip-flop model.

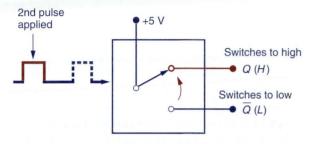

A. Logic symbol

S	R	Q	$\overline{Q}$
0	0	Not allowed	
0	1	1	0
1	0	0	1
1	1	No change	

B. Truth table

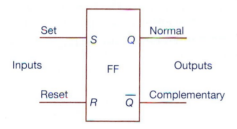

C. Wiring an RS flip-flop using NAND gates

Figure 41-47   *RS* flip-flop or latch.

(not Q). The two outputs are generally referred to as normal and complement. The outputs are always opposite, or complementary: If $Q = 1$, then $\overline{Q} = 0$. The letters S and R at the inputs are referred to as the *set* and *reset* inputs.

All flip-flops can assume one of two states. One of these is referred to as the *reset* state. When the flip-flop is reset, it is said to be storing a binary 0. The other condition of the flip-flop is the *set* state. When a flip-flop is set, it is said to be storing a binary 1.

The state of the flip-flop is determined by observing the normal output. If the flip-flop is reset, the normal output will be binary 0. If the flip-flop is set the normal output will be binary 1.

The state to which the flip-flop is set is determined by the signals applied to the inputs. The logic level required to put the flip-flop into one state or another can be either a binary 0 or binary 1, depending on how the flip-flop is constructed. Considering the RS flip-flop of Figure 41-47, applying a binary 0 to the S input causes the flip-flop to store a binary 1. Applying a binary 0 to the R input causes the flip-flop to store a binary 0. These inputs are applied only *momentarily* in order to put the flip-flop into the correct state. Otherwise, the set and reset inputs normally remain at the binary 1 level.

With both inputs at binary 1, the flip-flop is undisturbed and may be in either state. The condition where both inputs are binary 0 is an ambiguous state which is undesirable and usually avoided.

The basic RS flip-flop is *asynchronous;* it responds to inputs as soon as they occur. One way to synchronize the operation of the RS flip-flop with other logic circuits is to gate its inputs so they can respond only when activated by a logic 1 from a clock. A **clock** is a sequential circuit that produces a stream of alternating 0s and 1s. Figure 41-48 shows the logic symbol and truth table for a *clocked RS flip-flop,* as well as the circuit constructed using four NAND gates. Data is entered into this clocked latch on the occurrence of the clock pulse. Clocked digital circuits are referred to as *synchronous* because the digital circuitry operates in synchronism with the system's clock.

A D *flip-flop* is a further modification of a clocked RS flip-flop. A D flip-flop is formed from a clocked RS flip-flop by adding an inverter, as shown in Figure 41-49. It has only *one* data

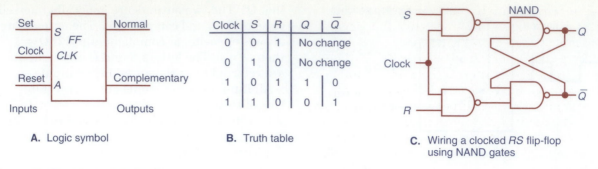

A. Logic symbol

Clock	S	R	Q	$\bar{Q}$	
0	0		1	No change	
0	1	0		No change	
1	0	1	1	0	
1	1	0	0	1	

B. Truth table

C. Wiring a clocked *RS* flip-flop using NAND gates

Figure 41-48 Clocked *RS* flip-flop.

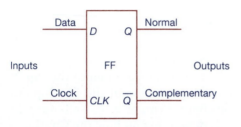

A. Logic symbol

D	CLK	Q
0	0	X
0	1	0
1	0	X
1	1	1

X = previous, or Last, state

B. Truth table

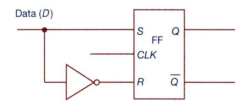

C. Wiring a D flip-flop using a clocked RS flip-flop and inverter

Notes:
— The prohibited, or unknown, state does not occur in this flip-flop.
— Because the inputs are kept complementary, the condition where both inputs are the same will never occur.
— When the clock is binary 0, the *Q* output can be either binary 0 or 1, depending on previous conditions.

Figure 41-49 *D* flip-flop.

input (*D*) and a clock input (*CLK*). To store a binary 0, a binary 0 is applied to the *D* input. To store a binary 1, a binary 1 is applied to the

D input. When the clock pulse occurs, the bit appearing at the *D* input is stored in the internal latch. The D flip-flop is often called a ***delay flip-flop***. The word *delay* describes what happens to the data, or information, at input *D*. The *Q* output is the same as the *D* input *one clock pulse later*. In this way, the flip-flop delays data from reading output *Q* by one clock pulse.

The ***JK flip-flop*** is a universal storage element that can perform the functions of both the *RS*- and *D*-type flip-flops. Figure 41-50 shows the logic symbol for *JK* flip-flops. Notice that it is similar to other flip-flop symbols that we have seen already in that it has the normal and complement outputs, which are used to determine the state of the flip-flop. Like the *D* flip-flop, it also has clock input. The *JK* flip-flop has all the attributes of the ideal memory device. It

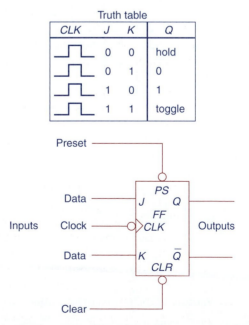

Truth table

CLK	J	K	Q
⎍	0	0	hold
⎍	0	1	0
⎍	1	0	1
⎍	1	1	toggle

Figure 41-50 *JK* flip-flop logic symbol.

has no indeterminate state, it can be set and reset, it has toggle and stop-action functions, and it is clocked-pulse controlled.

The preset and clear inputs of a JK flip-flop are similar in operation to the S and R inputs, respectively, or a basic latch. The preset and clear inputs are asynchronous in nature in that their change of state affects the JK flip-flop immediately. Ordinarily, the preset and clear inputs will be held in binary 1 position. In this state neither will affect the operation of the device. But to set or reset the flip-flop, the appropriate input is made a binary 0 for a short period of time.

The J and K inputs are only recognized at certain times during the occurrence of the clock signal. To store a binary 1, the J input must be binary 1 and the K input a binary 0 at the occurrence of the clock pulse. If the K input is binary 1 and the J input is binary 0, the flip-flop will reset at the occurrence of the clock pulse.

A JK flip-flop can be used to make a **toggle**, or **T, flip-flop**. The J and K inputs are tied together and called the T input. When a logic 1 is applied to T, the flip-flop changes state each time a clock pulse arrives. This OFF/ON action is like a toggle switch and is called *toggling*.

41-9 Registers

A **register** is a group of flip-flops used for the temporary storage of binary data. The number of flip-flops determines the amount of data per unit, often referred to as a *computer word, data word,* or just *word.* Registers are used in applications involving the *storage* and *transfer* of data in a digital system. There are two basic types of registers: *storage registers* and *shift registers.* As their names imply, a storage register holds binary data and a shift register processes it.

Figure 41-51 illustrates a typical **storage register**, which will temporarily store a four-bit binary word. This storage register is made up of four D flip-flops. If the word 0111 is presented as input data, it will be transferred through to the Q outputs when the clock pulse appears. If the clock pulse remains at 0 and the binary input data changes, it will have no effect on the output. The output word is latched, or stored in the register, until new data are clocked in.

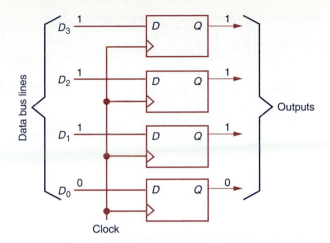

Figure 41-51 Typical four-bit storage register.

A storage register can be made far more useful by adding some combinational logic to *clear* the register as well as *preset* it. A typical logic circuit that provides for presetting and clearing is illustrated in Figure 41-52. A binary 1 at the preset input causes output Q to set at 1. A binary 1 at the clear input causes output Q to reset to 0. Pulses on these lines can ready the register for new data.

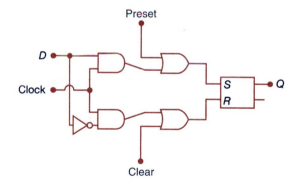

Figure 41-52 Adding register clear and preset inputs.

A **shift register** is considerably more versatile than a simple storage register. The storage elements in a shift register are connected in such a way that bits stored there can be moved, or shifted, from one element to another adjacent element. Figure 41-53 shows how a shift register operates. The shift register consists of three storage elements such as flip-flops. The binary word 101 is generated externally and is available to be input into the shift register serially. The data bits in the register are shifted

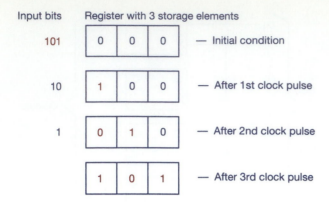

Input bits	Register with 3 storage elements	
101	0 0 0	— Initial condition
10	1 0 0	— After 1st clock pulse
1	0 1 0	— After 2nd clock pulse
	1 0 1	— After 3rd clock pulse

Figure 41-53 Shift register operation.

right one bit at a time by clock pulses to make room for the incoming bits. After three clock pulses, the three-bit number is shifted into the register.

Registers used in computer control systems usually can store either 8 or 16 bits. Input registers may be used to store information from various sensors. This information is later processed to control various operating conditions.

Universal shift registers that can accept and output data as serial bits or parallel words, as well as shift the data left or right, are available. The serial-load shift register shown in Figure 41-54 accepts data one bit at a time (serial input), while making available the contents of all its flip-flops simultaneously (parallel output). The data bits in the register are shifted right one bit at a time by clock pulses to make room for incoming bits.

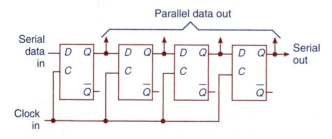

Figure 41-54 Serial-load shift register.

Automotive **scanners** or *scan tools* are electronic test instruments specifically designed to read information that is found in registers and sent along the data stream (Figure 41-55). They are able to read and interpret digital data and

(Courtesy Micro Processor System Inc. MPSI).

Figure 41-55 Typical scan tool.

thus provide useful information about the operation of the vehicle's computer control system.

41-10 Binary Counters

Most digital systems require the use of binary counters. A **counter** is a sequential logic circuit that is used to count the number of binary pulses applied to it. As input pulses are applied, the counter is incremented, and the binary number stored there changes to reflect the number of pulses that have occurred (Figure 41-56).

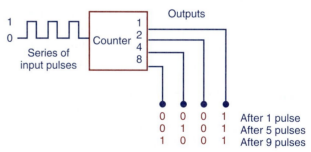

Figure 41-56 Binary counter.

Binary counters are usually constructed using flip-flops. Figure 41-57 illustrates how four flip-flop circuits can be connected to form a four-bit serial binary counter. When a pulse is applied to the input of the first flip-flop, it initiates a state of change that causes a HIGH, or 1, at output *A*. Arrival of the next pulse causes another change, and a LOW, or 0, now appears

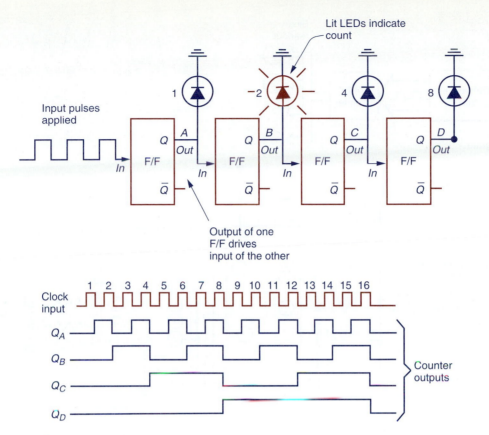

Figure 41-57 Four-bit binary serial counter.

at output *A*. As a result, two input pulses produce one output pulse, or a *divide-by-two* function. The four flip-flops are connected so that the output of one drives the input of the other to achieve a total count of up to 15. LEDs connected to the output of each flip-flop are used to indicate the count in binary numbers. The decimal-number value of each lighted LED is as indicated. To find the decimal value sum, add the decimal values of all lighted LEDs. The result is a 0000-to-1111 binary counter. The counter automatically recycles to 0000 after the sixteenth incoming pulse.

There are many different kinds of flip-flop counters. The *modulus* of a counter specifies the maximum count it reaches before recycling, and is determined by the number of flip-flop stages used. A modulus 10 counter is very popular because it recycles after the tenth input pulse and, therefore, provides a convenient way to counter in decimal.

Figure 41-58 shows the wiring diagram for a typical modulus 10, or decade, counter. The input pulses to be counted are produced by

manually operating the count button. Mechanical switches must be debounced prior to connecting them into logic circuits. If this conditioning of the switch input is not provided, the counter may receive several false make-and-break signals each time the button is closed. A special latch constructed using two inverters from a 7404 IC supplies the debouncing. Pulses are fed from the debouncing circuit into the 7490 decade counter via pin 14. The pulses cause the four flip-flops inside the 7490 to begin a counting sequence that recycles after ten counts. The binary count present in the decade counter is indicated by the LED display. Again, the decimal number value of each lighted LED is as indicated.

The counting string of input pulses that are used to operate a counter is often produced by an electronic oscillator called a *clock*. The schematic of a 7490 binary counter operated with a clock input is shown in Figure 41-59. A 555 Timer IC is wired as the variable clock input. The output from this clock is a square wave with an approximate frequency range of

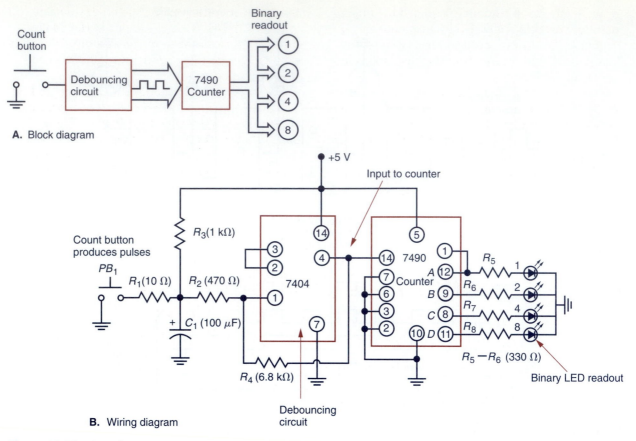

A. Block diagram

B. Wiring diagram

Figure 41-58 Decade counter.

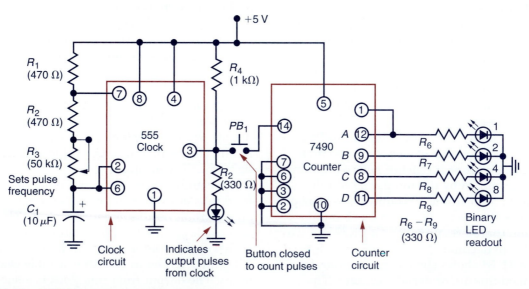

Figure 41-59 7490 binary counter with clock input.

from 1–100 Hz. Frequency is varied by adjustment of variable resistor R_3. The counting sequence can be observed by closing the count pushbutton PB_1 and setting the clock to its lowest output frequency.

The block and schematic diagrams of a complete digital counter with both a binary and a decimal readout are drawn in Figure 41-60. Clock pulses generated by the 555 IC are fed to the input of the 7490 IC counter. The output of the 7490 feeds four LED lamps, which indicate the count in binary code. The binary output of the 7490 also feeds the input of the 7447 decoder. The output from the decoder drives the seven-segment LED, which displays the decimal number count.

An **up counter** is one that is incremented by each input pulse. A **down counter** is one that is decremented by each input pulse. Figure 41-61 shows a 74192 IC up/down counter that can count up or down. To count up, the trigger pulses are connected to pin 5. To count down,

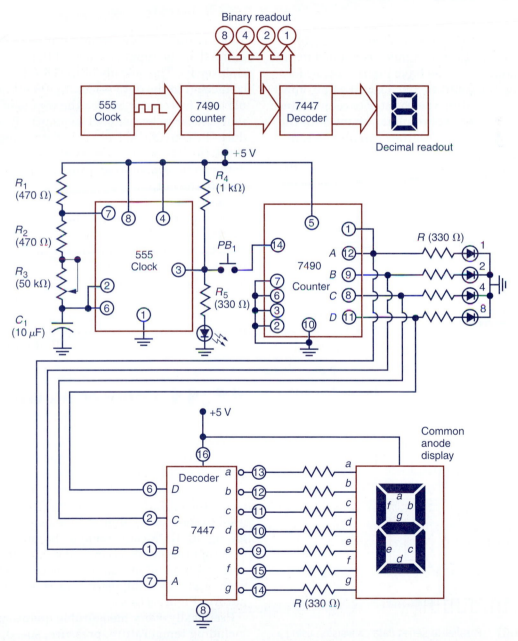

Figure 41-60 Binary counter with both binary and decimal readouts.

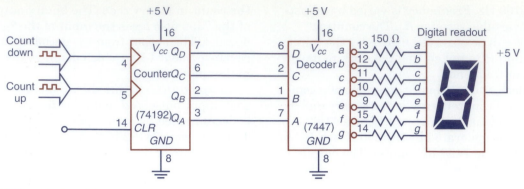

Figure 41-61   74192 up/down counter.

the trigger pulses are connected to pin 4. To clear the count to zero, a pulse is applied to pin 14. Counters may also have input controls for presetting the count to any desired value, and *enabling* (activating the counter to count). Since counters store the accumulated count until the next clock pulse arrives, they can also be considered storage registers.

Figure 41-62 shows an experimental parallel-to-serial data conversion circuit that uses a

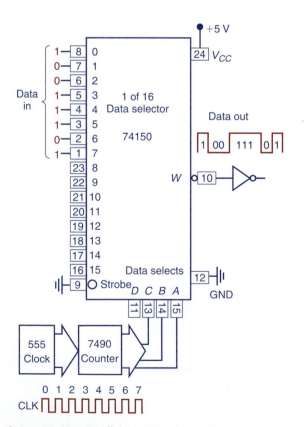

Figure 41-62   Parallel-to-series data conversion using a counter and multiplexer.

7490 counter IC. A word (eight bits long) is entered at the inputs (0 to 7) of the 74150 multiplexer IC. The outputs of the counter are connected to the data select inputs of the multiplexer. The counter causes the inputs of the multiplexer to be scanned sequentially to produce the conversion from parallel to serial data. If the clock is pulsed very fast, the parallel data can be transmitted quite quickly as serial data to the output.

Figure 41-63 shows an experimental serial data transmission system that uses a 74150 multiplexer and a 74154 demultiplexer along with a counter. The multiplexer first connects input 0 to the serial data transmission line. The bit is then transmitted to the demultiplexer, which places this bit of data at output 0. The multiplexer and demultiplexer proceed to transfer the data at input 1 to output 1, and so on. The bits are transmitted one bit at a time.

41-11 Connecting with Analog Devices

Analog signals must be coded into digital signals before they can be processed by digital circuits. The circuit that does this is called an *analog-to-digital (A/D) converter*. Its complement, the *digital-to-analog (D/A) converter*, is used to convert the digital code back to an analog signal. Data-conversion techniques allow us to accept commonly occurring analog inputs and outputs and utilize them most effectively with electronic equipment that is digital in nature (Figure 41-64).

Practically every measurable quantity, including temperature, pressure, speed, and time, is analog in nature. Basically, analog-to-

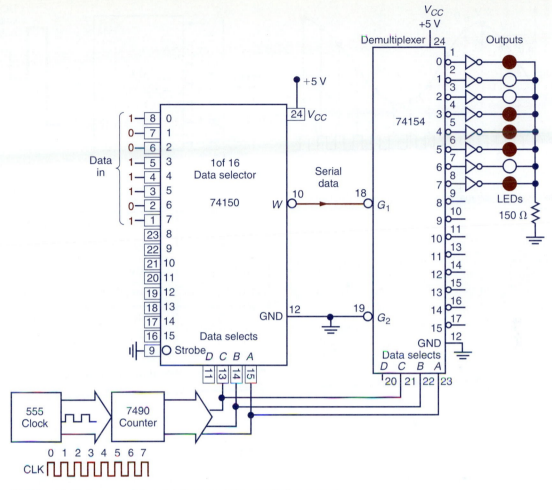

Figure 41-63 Multiplexer and demultiplexer serial transmission system.

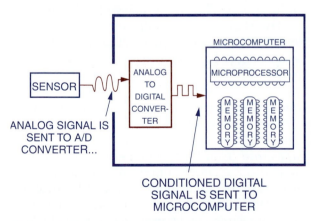

Figure 41-64 Using analog signals with digital equipment.

digital conversion is accomplished by assigning binary numerical values to represent the differ-

ent voltage levels of the analog signal. One method of A/D conversion is illustrated in Figure 41-65, where the analog signal from a sensor is ***sampled*** (or measured) at regular intervals. The equivalent binary number for the voltage level is then output as a digital signal. The *sampling rate* determines the accuracy with which the sequence of digital codes represents the analog input of the A/D converter.

Digital-to-analog converters are used in digital systems to convert digital signals into proportional voltages or currents to control analog devices. Typical analog output devices include small motors, valves, and analog meters. The input to a D/A converter is usually a parallel binary number. The output is a dc analog voltage that is proportional to the value of the binary input.

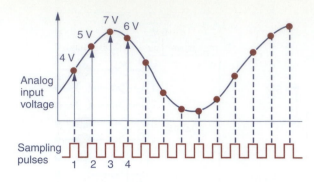

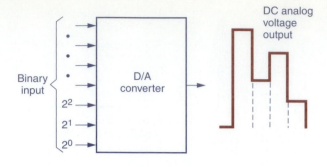

Sample	Analog input voltage	Binary output
1	4 V	0100
2	5 V	0101
3	7 V	0111
4	6 V	0110

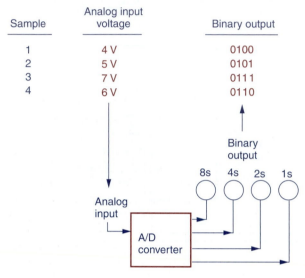

Figure 41-65　Sampling of values on an analog signal for conversion to digital.

Sample	Binary Input	DC analog voltage output
1	0000	0.0 V
2	0001	0.1 V
3	0010	0.2 V
4	0011	0.3 V
5	0100	0.4 V
6	0101	0.5 V

Figure 41-66　Digital-to-analog conversion.

Figure 41-66 shows a simplified illustration of a D/A converter. The D/A converter outputs a voltage level corresponding to the input binary code. During the sampling period, the voltage level remains constant, therefore, the output has stepped voltage levels. This causes the analog output voltage to have a staircase appearance.

41-12 Digital Logic Probe

The *digital logic probe* (Figure 41-67) is an inexpensive tool that can be used in signal-tracing digital signal pulses. It is especially designed to be used with digital circuits.

A slide switch on the digital logic probe is used to select the type of logic family under test, either TTL (bipolar transistors) or MOS (field-effect transistors). Typically, two leads

provide power to the logic probe. The red lead is connected to the positive (+) of the power supply, and the black lead of the logic probe is connected to the negative (−), or ground, of the power supply. After powering the logic probe, the needlelike probe is touched to a test point in the circuit. The logic probe will light either the HIGH or the LOW indicator. If neither indicator lights, it usually means that the voltage is somewhere between the HIGH and LOW values.

Each digital logic probe responds to slightly different voltage levels for logical HIGH and LOW indications. These voltage-detection levels are sometimes referred to as *voltage-threshold levels*. The voltage-threshold levels vary with the power-supply voltage value used to operate the probe and the type of IC logic family (TTL or MOS) selected. The test circuit of Figure 41-68 can be used to check the LOW and HIGH voltage thresholds of a logic probe. With the circuit wired as shown, the potentiometer is adjusted and the voltages required to turn the LOW and HIGH indicator lights ON (as indicated by the voltmeter) are recorded.

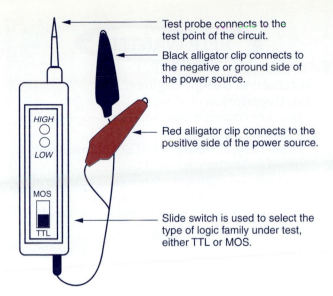

Test probe connects to the test point of the circuit.

Black alligator clip connects to the negative or ground side of the power source.

Red alligator clip connects to the positive side of the power source.

Slide switch is used to select the type of logic family under test, either TTL or MOS.

A. Typical digital logic probe

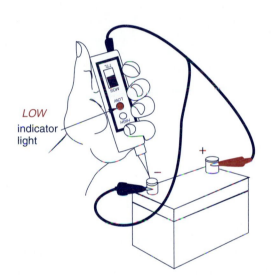

LOW indicator light

B. Logic probe power lead connections

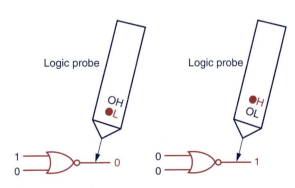

C. Using a logic probe to test a gate

Figure 41-67   Digital logic probe.

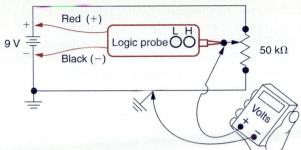

Figure 41-68   Test circuit used to check logic probe HIGH and LOW voltage thresholds.

Digital logic probes have some limitations when used to indicate voltage levels. Unlike digital voltmeters, they do not indicate exact voltage values. The most important application of a digital logic probe is its ability to detect the presence of digital signals. Digital signals that are switched between 0 and 5 V or 0 and 14 V are common in computer-controlled systems. The digital logic probe can be used to analyze the duty cycle of a pulse train (Figure 41-69). If the HIGH time is equal to the LOW time, the HIGH and LOW lights will be ON for the same length of time. If the HIGH time is longer than the LOW time, the HIGH light will be ON longer and appear brighter than the LOW light. If the LOW time is longer than the HIGH time, the LOW light will be ON longer and appear brighter than the HIGH light.

Most digital probes have a PULSE light indicator in addition to the HIGH and LOW light

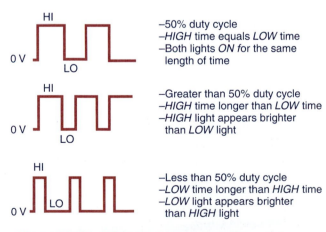

–50% duty cycle
–*HIGH* time equals *LOW* time
–Both lights *ON* for the same length of time

–Greater than 50% duty cycle
–*HIGH* time longer than *LOW* time
–*HIGH* light appears brighter than *LOW* light

–Less than 50% duty cycle
–*LOW* time longer than *HIGH* time
–*LOW* light appears brighter than *HIGH* light

Figure 41-69   Using a logic probe to analyze a duty cycle.

indicators. In the pulse mode, the PULSE light flashes ON momentarily every time there is a significant voltage transition or change at the probe tip. It can be used to observe the rhythm of the pulse train. It has no application when testing dc voltage levels.

Automation Design Engineer
Job Description
- Design of complex logic machines used to automate assembly processes.
- Experience in the design of microprocessor based products.
- Requires an engineer with a good balance of theoretical and practical design skills.
- Willingness to pay close attention to detail and accommodate change.

Review and Applications

Review Questions

1. Draw a typical digital signal waveform and properly identify the two operating levels.
2. What digits are used in the binary number system?
3. Define each of the following as it applies to a digital signal:
   - (a) bit
   - (b) nibble
   - (c) byte
   - (d) word
   - (e) least significant bit
   - (f) most significant bit
4. What is the advantage of the hexadecimal numbering system over the binary numbering system?
5. State two basic requirements that must be met in order to process information using digital techniques.
6. What is a logic gate?
7. How are most logic gates constructed?
8. Draw the logic symbol and construct a truth table for each of the following gates:
   - (a) Two-input AND gate
   - (b) Three-input OR gate
   - (c) NOT gate
9. What information about the logic gate is provided by its truth table?
10. Why is an AND gate called an *all-or-nothing gate?*
11. Why is an OR gate called an *any-or-all gate?*
12. Why is the NOT gate called an *inverter?*
13. (a) Draw the symbol for a NAND gate.
    - (b) What two basic gate functions are combined to form a NAND gate?
    - (c) How does the truth table for a NAND gate compare with that of an AND gate?
14. (a) Draw the symbol for a NOR gate.
    - (b) What two basic gate functions are combined to form a NOR gate?
    - (c) How does the truth table for a NOR gate compare with that of an OR gate?
15. What three things determine the output of a combinational logic circuit?
16. What is the primary function of a combinational logic circuit?
17. In many logic circuits, what device acts as the input?

18. Draw the logic symbol and construct a truth table for a two-input XOR gate.
19. Show the four conditions that can occur when adding two single-digit binary numbers.
20. What logic function is the heart of any digital arithmetic circuit?
21. Explain the basic function of a comparator circuit.
22. Explain the basic function of an encoder circuit.
23. Explain the basic function of a decoder circuit.
24. Explain how numbers are formed on a seven-segment LED display.
25. Explain the basic function of a multiplexer circuit.
26. How are data transferred by means of serial transmission?
27. How are data transferred by means of parallel transmission?
28. What is meant by the *baud rate* of data transmission?
29. Explain the basic function of a demultiplexer circuit.
30. In what way are sequential logic circuits different from combinational logic circuits?
31. What is the main logic element in a sequential logic circuit?
32. The Q and $\overline{Q}$ outputs of a flip-flop are said to be complementary. What does this mean?
33. In what way is the control of a clocked RS flip-flop different from that of a basic RS flip-flop?
34. How does the Q output state of a D flip-flop compare with that of the D input?
35. List four attributes of a JK flip-flop that make it an ideal memory device.
36. How does the toggle, or T, flip-flop react to a series of clock pulses?
37. What is a register?
38. In what types of applications are registers used?
39. (a) How is a storage register cleared?
    - (b) When the register is cleared, what happens to the Q outputs of the flip-flops?

40. Explain how the storage elements in a shift register control the data bits.
41. State one application for registers used in computer control systems.
42. (a) What is an automotive scanner, or scan tool?
    (b) Why is it useful as a diagnostic aid?
43. What is a binary counter?
44. What does the modulus of a counter specify?
45. Describe a typical automotive application for a binary counter.
46. Why are A/D and D/A converters of great importance in digital systems?
47. Explain how A/D conversion is accomplished.
48. Explain how the two leads and probe of a typical digital logic probe are connected to the circuit to be tested.
49. What is the most important application of a digital logic probe?

Problems

1. Convert each of the following decimal numbers to binary numbers:
   (a) 3   (b) 7   (c) 10   (d) 14
2. Convert each of the following binary numbers to decimal numbers:
   (a) 10   (c) 10101
   (b) 1001   (d) 11100011
3. Express the decimal number 197 (binary number 11000101) as a hexadecimal number.
4. (a) What is the sum of the binary numbers 10011 and 11010?
   (b) What are the decimal equivalents of each number and of the sum?
5. Subtract binary 0111 from binary 1001 using the 1s complement and end-round carry method of subtraction.

Critical Thinking

1. With reference to the accompanying combinational logic circuit (Figure 41-70), state whether output Y will be 1 or 0 for each of the following input conditions:
   (a) $A = 0; B = 0; C = 0$
   (b) $A = 1; B = 0; C = 0$
   (c) $A = 0; B = 1; C = 0$
   (d) $A = 1; B = 1; C = 0$
   (e) $A = 1; B = 1; C = 1$

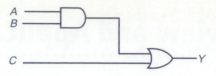

Figure 41-70

2. With reference to the accompanying combinational logic circuit (Figure 41-71), state whether output Y will be 1 or 0 for each of the following input conditions:
   (a) $A = 1; B = 0; C = 1; D = 0$
   (b) $A = 0; B = 1; C = 1; D = 0$
   (c) $A = 0; B = 0; C = 1; D = 1$
   (d) $A = 1; B = 1; C = 0; D = 0$

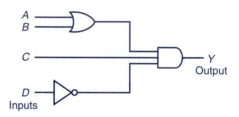

Figure 41-71

3. Develop a circuit for each of the following Boolean expressions using AND, OR, and NOT gates:
   (a) $Y = ABC + D$
   (b) $Y = AB + CD$
   (c) $Y = (A + B)(\overline{C} + D)$
   (d) $Y = \overline{A}(B + CD)$
   (e) $Y = AB + C$
   (f) $Y = (ABC + D)(E\overline{F})$
4. (a) Assume a pulse is applied to the set (S) input of an RS flip-flop. What will happen to the Q output as a result?
   (b) What happens to the Q output when the pulse is removed from the set input?
   (c) How is the state of the Q output changed?
   (d) Because of this action, what is an RS flip-flop often called?

Portfolio Project

• Build a binary counter with both binary and decimal readouts. Include a circuit wiring diagram and a written description of the circuit's operation.

Circuit Challenge

Using a Simulator

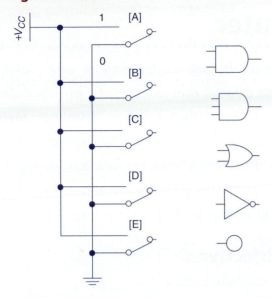

Procedure

(a) Using the switches and logic gates shown, design a logic circuit that will have the LED come on when all of the following conditions are met:
- Both switches A and B must be closed (logic 1).
- Either switch C or D or both must be closed (logic 1).
- Switch E must be open (logic 0).

(b) Construct the circuit shown, using whatever simulation software package is available to you.

The Microcomputer

Computers have become an indispensable tool in our society. We now find computers almost everywhere—in automobiles, industrial control circuits, banking machines, cameras, medical apparatus and the list goes on and on. The personal computer has both accelerated and simplified our work and increased our productivity. In this chapter, we will discuss some of the fundamentals related to the operation and application of computers.

Objectives

After studying this chapter, you should be able to:
1. Identify and explain the purpose of each of the major parts of a microcomputer system.
2. Identify types of computer input and output devices.
3. Explain how data is processed in a computer.
4. Discuss the role of computer software programs.
5. Discuss the operation of microprocessor-based control systems.
6. Explain the operating principles of a personal computer.
7. Do research on the Internet.

Key Terms

- **central processing unit (CPU)**
- **arithmetic-logic unit (ALU)**
- **microprocessor**
- **ROM memory**
- **RAM memory**
- **magnetic memory**
- **modem**
- **stepper motor**

- **programmable logic control (PLC)**
- **ladder-diagram language**
- **data bus**
- **operating system software**
- **application software**
- **I/O devices**
- **computerized engine control**

- **personal computer (PC)**
- **PC hardware devices**
- **PC ports**
- **DOS**
- **Windows**
- **Internet**
- **World Wide Web (WWW)**
- **Internet searches**

RESEARCH INTERNET SITES

- **Microsoft**
- **IBM**
- **Apple**
- **Intel**

42-1 Microcomputers

The computer has had a greater impact on modern technology than any other type of electronic device. A **computer** is basically a *digital* electronic system that can be programmed to perform various tasks, such as making calculations and decisions, at a very high rate of speed. In addition, it can store large amounts of data.

A computer system can be divided into four blocks as illustrated in Figure 42-1. These blocks are the input, memory, central processing unit (CPU), and output.

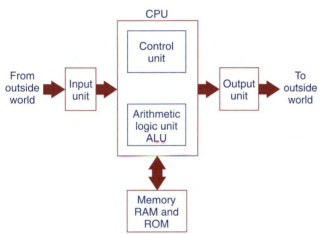

Figure 42-1 A basic computer system made up of input and output (I/O), central processing unit (CPU) and memory.

The **input** block consists of all the circuits needed to get programs and data into the computer. A particular input is selected by placing its *address* on the address bus by the CPU. The data is transferred to the CPU via the input data bus.

The **memory** block stores the program and data and consists of semiconductor RAM (random-access memory) or ROM (read/only memory) or magnetic storage or a combination of these. Data is stored in memory by a process called *write* and is retrieved from the memory by a process called *read*.

The *CPU (central processing unit)* is the "brain" of the system. It consists of a control unit and an **arithmetic-logic unit** (**ALU**). The control unit directs the operation of all other sections, telling them what to do and when to

do it. With the control unit telling it what to do and with memory feeding it data, the ALU provides answers to number and logic problems. The CPU contains many registers, counters, and combinational logic devices.

The *output* passes processed data from the CPU to the outside world. The output section often includes a video display to allow the user to see the processed data.

If the CPU is contained on *one* integrated circuit, this IC is called a *microprocessor*. The microprocessor is a VLSI (very large-scale integration) device that can perform arithmetic and logic operations and other functions in a prescribed sequence for the movement and processing of data.

A **microcomputer** is basically just a small computer whose CPU is a microprocessor. The typical microcomputer has three kinds of chips: microprocessor, memory, and input/output (I/O) (Figure 42-2). Solid-state technology has developed to the point where it is possible to put the CPU, memory, and input/output on one integrated circuit. These devices are called *single-chip microcomputers*.

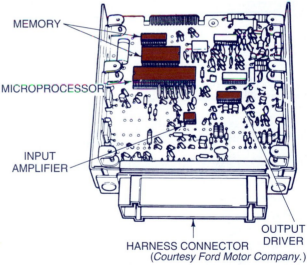

Figure 42-2 Internal components of a typical automobile engine control microcomputer.

42-2 Memory Units

A microprocessor does not normally store information, so storage devices called *memories* are added to a computer. Memories hold the program and data. A microprocessor can change the information in memory by *writing* new information into memory, or it can obtain information in memory by *reading* the information from memory.

Semiconductor memories used for primary memory fall into two main groups: ROM and RAM (Figure 42-3). *Read-only memory (ROM)* is a type of memory that is used to store information *permanently*. When the computer is built, the programs that control the microprocessor are stored in ROM. As the name implies, the microprocessor can read these instructions out of the ROM but cannot write in any new information. ROM is called a "nonvolatile" memory because the stored data is retained even when the power is turned OFF. That would classify ROM as permanent memory.

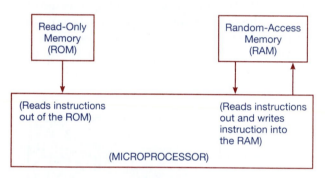

Figure 42-3 ROM and RAM.

Random-access memory (RAM) can be both written into and read from. RAMs act as the file cabinets for temporary storage of data. For example, the microprocessor may receive data it needs in order to make several different decisions. In such a case, it writes the data into RAM and then reads it out each time it is needed. RAMs are "volatile" memories because data is lost if power is lost for any reason. For this reason, RAM would be classified as temporary memory.

Magnetic memory storage includes hard disk drives, floppy disk drives, and magnetic tape. All of these are *nonvolatile* and often used for mass storage and backup.

The computing power of a computer is determined by the amount of digital bits of information it can store in its memory. In order for a computer to execute a large number of instruc-

tions, it must have a large memory capacity. Another important aspect of computer memory is speed. As memory gets larger it is important to be able to get data into or out of it quickly.

42-3 Input and Output

All microcomputer systems require means of inputting and outputting data. The input/output (I/O) provisions for a microprocessor-based system depend upon the application for which it is intended.

Figure 42-4 shows typical I/O provisions for a personal computer (PC). Typically, inputs come from keyboards, mouse or track-ball pointers, magnetic disks, CD-ROMs (compact disk read-only memory), microphones, telephone modems, scanners, cameras, and sending devices. Typically, outputs go to monitors (video displays), printers, magnetic disks, telephone modems, and speakers. In addition, outputs can be provided in order to facilitate data exchange with other microcomputers or with a modem.

The term *modem* is derived from *mo*dulator-*dem*odulator and is a device that converts serial data from the computer into tones that can be sent over telephone lines. This conversion process generates a certain tone frequency for a 1 and a different tone frequency for a 0. At the receiving end, the modem reverses the process and converts tones into zeros or ones.

Getting information into the microprocessor requires input interfacing. **Input interfacing** involves converting information in one form into digital electrical signals that can be used by the microprocessor.

Many microcomputer process control systems obtain information about a system's performance from input sensors. *Sensors* are devices that convert nonelectrical parameters (values) such as temperature, pressure, light position, and revolutions per minute, into electrical signals and send the data to the microcomputer. Usually, they do this by either generating a voltage or varying a reference voltage. Figure 42-5 shows typical microcomputer input sensors used in an automobile.

Figure 42-6, p. 654, illustrates the operation of an engine coolant temperature sensor. The sensor is basically a *thermistor* whose resistance varies with temperature. In this application the thermistor temperature-sensing unit

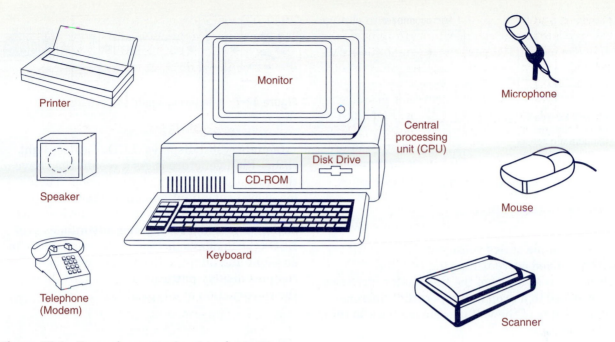

Figure 42-4 Personal computer inputs and outputs.

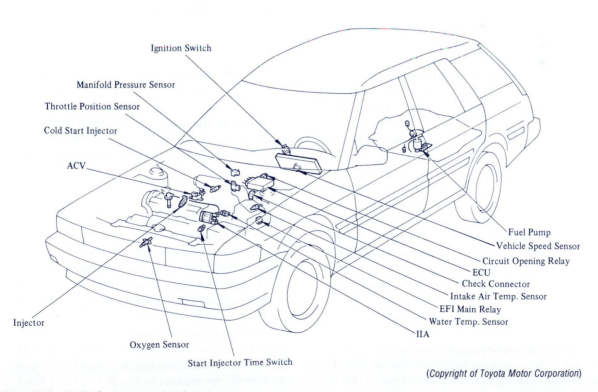

(Copyright of Toyota Motor Corporation)

Figure 42-5 Typical microcomputer input sensors.

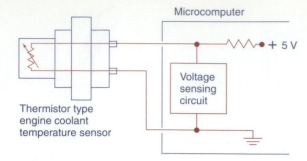

Figure 42-6 Engine coolant temperature sensor.

Figure 42-7 Interfacing inputs and outputs.

forms a simple series voltage-divider circuit with the fixed current-limiting resistor. Electrical resistance of the thermistor increases in response to a drop in temperature, increasing the value of the voltage relayed back to the microcomputer.

There are two principal types of signal voltages produced by input sensors: analog and digital. An *analog* signal is a continuously variable voltage signal, whereas a *digital* signal is one that has only two voltage stages: ON and OFF. A microprocessor uses only digital signals in a code that contains only 1s and 0s; therefore, all analog signals must be converted to digital signals. The job of converting an analog signal to a digital signal is performed by an A/D converter interface. Input interfacing is also used to amplify weak sensor signals, to pulse-shape digital voltage signals, and to electrically isolate the delicate electronics in the microprocessor from any high voltages applied to the inputs.

Output signals from the microprocessor must also be interfaced with the devices they operate. The output interface receives digital data from the microprocessor; the data is converted to a proper voltage or current to control the output device. Types of output devices include *video monitors* that display information and *actuators* that do work. Liquid-crystal displays (LCDs), light-emitting diodes (LEDs), vacuum-fluorescent displays (VFDs), indicator lights, and analog meters are used in automotive instrumentation to report measurement results. Figure 42-7 shows an example of how a microcomputer uses interfaces to handle the incoming information from the speed sensor and the outgoing information to the speedometer.

Liquid-crystal displays (LCDs) scatter light rather than generate it. As a result, they have the lowest power drain of any display. Their low-power consumption makes them ideally suited for use in battery-operated devices such as digital multimeters. Other advantages over LED-type displays include better readability in sunlight and adaptability for use with more complex display patterns. Figure 42-8 shows the construction of a typical LCD.

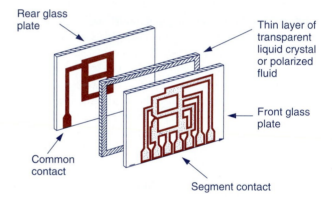

Figure 42-8 Construction of a liquid-crystal display.

The vacuum-fluorescent display (VFD), because of its durability and bright display qualities, is one of the most commonly used automotive displays. This device generates light in much the same way as the cathode-ray tube of an oscilloscope or television picture tube, by shooting electrons onto a phosphorus material. A simplified diagram of a VFD operation is shown in Figure 42-9. The display uses a filament that emits electrons when heated. These electrons are accelerated and directed to strike a phosphorus-coated anode by applying a positive voltage to the anode. The anode segments make up the visible portion of the display. They emit a blue-green light when struck by the electrons. Voltage is applied only to the anodes of the segments needed to form the character to be displayed. The brightness is

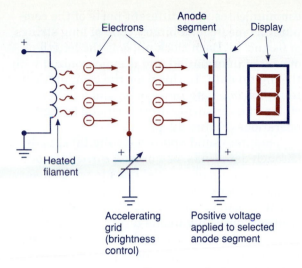

Figure 42-9   Operation of a vacuum-fluorescent display.

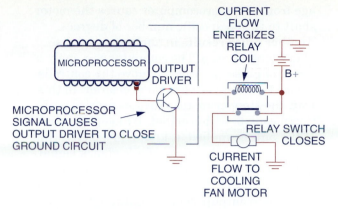

Figure 42-10   Relay control of cooling fan motor.

Control of the step sequence

Step	Switch position SA	SB	Counter-clockwise	Clock-wise
0	1	1		
1	2	1		
2	2	2		
3	1	2		
4	1	1		
etc.				

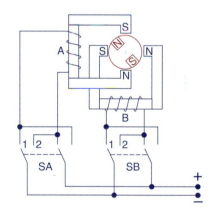

Stepper motor

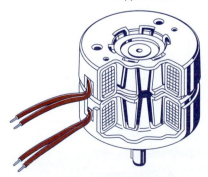

Figure 42-11   DC stepper motor.

controlled by varying the positive voltage on the accelerating grid.

An *actuator* is an output load device that delivers motion in response to an electrical signal. Actuators are usually operated by electrical output signals from the microcomputer to cause changes to system-operating conditions in response to decisions made by the microprocessor. Today's computers are capable of issuing millions of output commands per second. The principal types of actuators are solenoids, motors, and relays.

The output driver circuit from the microcomputer can be used to operate a *relay*, which, in turn, controls some secondary actuator. This application is often used for heavy current loads that normally draw too much current to be operated directly by the output driver's transistor. Such an application is illustrated in Figure 42-10. The microcomputer turns ON the appropriate driver transistor with a signal from the microprocessor. The transistor then completes the relay coil circuit path to ground to energize the relay coil and develop a magnetic field in the core. The magnetic field causes the relay switch contacts to close, completing the relay power circuit from the battery to the cooling fan motor, and the motor runs.

A type of motor that lends itself to digital control operation is the dc *stepper motor*. A typical stepper motor is constructed with a permanent magnet armature called a toothed rotor and two field coils (Figure 42-11). The motor is designed so that applying a single pulse of voltage from the microcomputer causes the motor shaft to turn a specific number of degrees. Another pulse results in another equal displacement, or step. The computer can apply a series of pulses in order to move the controlled device to whatever position is desired. In this way, the stepper motor provides precise positional control of movement. By keeping count of the pulses applied, it knows in exactly what position it is without use of a feedback signal.

Figure 42-12 shows the block diagram of a **programmable logic controller (PLC)** with inputs and outputs connected through *modules* to the central processing unit. Figure 42-12B shows the typical installation of the PLC. PLCs are used for control and operation of manufacturing process equipment, and machinery. PLCs are basically computers specifically designed to operate in the industrial environment.

42-4 Computer Communications

A clock pulse generator is used in the microcomputer for timing control of the computing sequence. Clock generators provide a continuous stream of square-wave pulses at a constant frequency. As mentioned, the microprocessor communicates with different parts of the computer by means of a binary code of long strings of 0s and 1s. Each clock pulse is the length of one bit of information and is synchronized with the digital data. In this way, the computer can tell when one data pulse ends and another starts, thus ensuring precise timing of digital operations (Figure 42-13).

Computers send and receive digital signals through data links, or buses. A **data bus**, or *link,* is a group of parallel lines to which a very large number of components with different functions can be connected. Some data links transmit data only in one direction, from one device to the other. Other data links can transmit both ways, from one device to the other and back again (Figure 42-14).

The binary data used by computer systems can be sent in two ways: serial and parallel. In *serial* form, the data is transmitted on a single line and appears as a stream of 0s and 1s, one after the other. Figure 42-15 shows one way to do this. A single switch is toggled ON and OFF sequentially for just the right amount of time to send the digital code 10100010. The serial data is synchronized with clock pulses so that the computer can distinguish between the different bits of data.

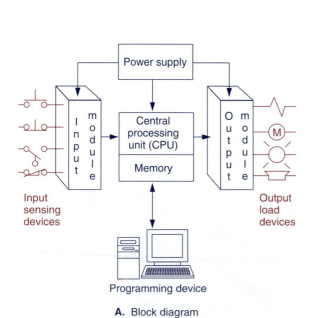

A. Block diagram

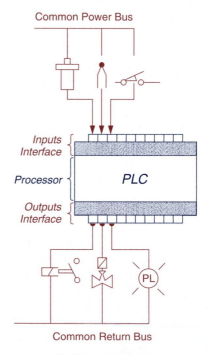

B. Typical installation

Figure 42-12 Programmable logic controller (PLC).

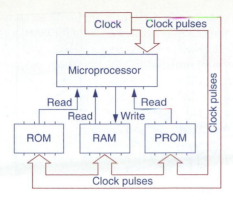

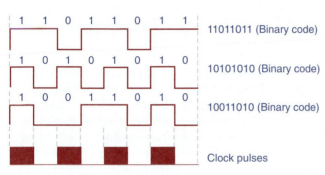

11011011 (Binary code)

10101010 (Binary code)

10011010 (Binary code)

Clock pulses

Figure 42-13   Computer clocks.

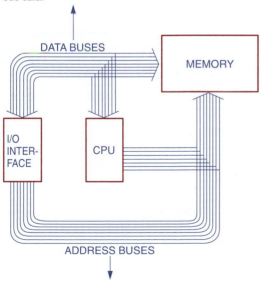

Over these buses flow the data to be written into
or read from the address specified by the address
bus data.

DATA BUSES

MEMORY

I/O
INTER-
FACE

CPU

ADDRESS BUSES

Over these buses flow sginals telling the memory
to/from which location (address) data should be
written/read.

Figure 42-14   Computer links, or buses.

Figure 42-15   Serial data transmission.

In *parallel* transmission of the same data,
each digit is sent at the same instant on its own
line, as shown in Figure 42-16. Eight switches
are toggled ON and OFF simultaneously along
eight parallel circuits to send the digital code
10100010.

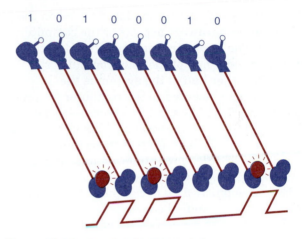

Figure 42-16   Parallel data transmission.

About ⬭ Electronics

With the latest on-board car electronics, you can
make hotel reservations, get sports scores, video
directions, or traffic reports. Perhaps, most
importantly, you can use these devices to notify
emergency service if you are having car trouble.

You can think of a computer bus as a highway
on which data travels. When used in reference
to personal computers, the term bus refers to
the internal bus that interconnects all the
internal components to the CPU and main
memory. There is also an expansion bus that
enables expansion boards to access the CPU
and memory. All buses consists of two parts; an
address bus and a data bus. The data bus
transfers actual data whereas the address bus
transfers information about where the data
should go. The size of a bus is determined by

the number of bits that control how much data can be transmitted at one time. For example, a 32-bit bus can transmit twice as much information at one time as a 16-bit bus. Every bus has a clock speed measured in MHz, which determines how fast applications will run. Most PCs also contain a local bus for data that requires especially fast transfer speeds, such as video data. The local bus is a high-speed pathway that connects directly to the processor.

42-5 Computer Software

Up to this point, discussion has focused on the physical, or hardware aspects of the computer system. The hardware components of the computer are only half of the total computer system. The other half of the system is the information presented to the computer for processing and the instructions about *how* that information is to be processed. This intelligence that controls the operation of the computer's hardware is called *programming* or *software*. Without good software to control its operation, the most sophisticated of computer hardware arrangements would be useless.

A *computer program* is a set of detailed instructions that a microcomputer follows when controlling a system. Programs are stored in memory and used by the microprocessor when processing information. The program breaks each computer task down to its most basic parts. For example, a program that tells a computer when to energize a solenoid can have dozens, or even hundreds of steps, depending on the complexity of the system. *The same computer can do many different types of jobs; all that is required is that the computer's program be changed* (Figure 42-17).

In automotive microcomputers, the memory chips permanently store all fixed programs, tables, and formulas used to perform calculations. These programs and data are permanently installed into the IC chip by the manufacturer and cannot normally be erased or reprogrammed. The program contains a detailed list of instructions used to direct the microprocessor through its processing operation. *Calibration tables* contain information about a specific vehicle, such as the number of cylinders and cylinder volume displacement. *Look-up tables* contain standard information

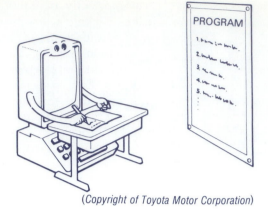

(Copyright of Toyota Motor Corporation)

Figure 42-17   The computer follows a program.

about how a vehicle should perform, such as the warm-up time of the engine for various engine temperatures (Figure 42-18).

Generally, personal computer software can be divided into two classes: operating systems and applications programs. **Operating systems software** consists of a group of special programs that control the computer's basic operation. They are involved in starting up the computer and allows the user to access and control the computer's memories, printers, and other peripheral devices. These programs also oversee the operation of applications programs. One common example of operating systems software used in personal computers is *MS-DOS* (*M*icrosoft *d*isk *o*perating *s*ystem).

Actual temp	Digital code for temp	Digital code for warm-up	Actual warm-up time
190°F	10111110	0000	0 min
180°F	10110100	0001	1 min
170°F	10101010	0010	2 min
160°F	10100000	0011	3 min
150°F	10010110	0100	4 min
140°F	10001100	0101	5 min

Figure 42-18   Typical automobile look-up table.

The second category, **applications software**, is made up of programs that perform specific tasks, such as word processing, accounting, and computer games. This type of software is available in two forms: (1) commercially available, user-oriented, applications packages, which can be bought and used directly, and (2) programming language packages, which allow specially trained users to "write" applications programs specifically for a desired task. Each programming language has unique features that make it useful for writing certain kinds of applications programs. Examples of programming languages include: Virtual Basic, C++, Java, and HTML.

42-6 Processing Information

As mentioned, the microprocessor processes information and makes decisions according to a program. In order to make informed decisions, the microprocessor receives data from many different sources. It is not always possible for the microprocessor to process data information immediately, so this information is initially stored in memory. To store information, the microprocessor writes the information into a specific RAM location called an *address*. To read information from RAM, the microprocessor specifies the address of the stored information and requests that the information contained at that address be retrieved (Figure 42-19).

In order to process information the microprocessor reads all inputs, takes these values and, according to the program, issues output commands. Normally all microprocessor address, data, read/write, and clock lines are active. Indeed they must be since the processor is continually fetching and executing instructions and data from memory and **I/O devices**. When observed on a scope, the patterns observed on the address, data, and read/write lines seem somewhat random. This apparent randomness is an illusion produced by the large number of bits in each repeat cycle. The processor is actually performing the same very long tasks over and over. If the processor's activity is limited to just a few tasks the repetition of the signals can be viewed on a scope.

The microprocessor can perform only a single function at a given instant, although it can perform operations very quickly. However, the microprocessor normally services several input and output devices by spending a little time on the task associated with each, then moving on to the next. It continually cycles through the

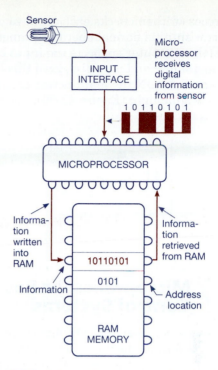

Figure 42-19   Information storage and retrieval.

servicing of all devices. Some input and output devices served by the microprocessor need constant attention, while others need only the processor's attention occasionally. The microprocessor repeats these cycles over and over, servicing one section at a time, yet the machine performs perfectly. This is possible because the microprocessor can execute a command every 5 microseconds.

Computers are often programmed to perform *self-diagnostic* tests of the system to verify that the hardware is working properly. If anything in the system is malfunctioning, these tests may cause an ERROR message to be displayed on the monitor's screen and/or a tone to be produced from the speaker.

Most computerized automotive control systems are equipped with built-in self-diagnostic systems that conduct a multitude of tests of the system. The basic microcomputer self-diagnosis system is designed to monitor the input and output signals of the sensors and actuators and to store any malfunctions in its memory as a *trouble code*. In general, checks are made for open and short circuits and illogical sensor readings. The computer is programmed so that it knows the voltage range within which each input should normally operate. The microprocessor then checks each input to make sure it is within its normal operating range. If it is not, the computer suspects the input circuit is malfunctioning and may disregard this information and turn ON the malfunction or indicator check engine light (Figure 42-20).

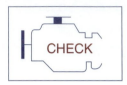

Figure 42-20   Malfunction indicator lamp.

42-7 Microprocessor-Based Control Systems

Most of us are familiar with the term "microprocessor" if for no other reason than the abundance of personal computers that are now in offices as well as homes. Because of this familiarity we tend to apply this same conceptual package to all microprocessor-based control systems as being complete with floppy disks, monitors, printers, and volatile memory. But such is not necessarily factual.

Microprocessors have replaced many discrete digital ICs in most consumer electronic products. In fact, a microprocessor can emulate almost any circuit constructed from discrete components. It is this feature that has made the microprocessor so popular with consumer product designers. A microprocessor usually allows the inclusion of more features in a product than would be economically feasible using discrete devices. Perhaps the best example of the impact a microprocessor can have on a product is its use to control the VCR. The microprocessor replaced an electromechanical control system that was expensive, bulky, and heavy. Its use reduced manufacturing costs and helped make the VCR affordable. The actual programs that are used in consumer products are written by the product designers and are not usually accessible to the user. Input to the program can normally be made through function buttons or a keypad (Figure 42-21).

Programmable Logic Controllers   In industry *programmable logic controllers* (*PLCs*) are used to perform logic functions previously accomplished by electromechanical relays. The programmable logic controller has eliminated much of the extensive wiring associated with conventional relay control circuits. In addition, today's programmable controllers are used to achieve factory automation, often interfacing with robots, numerical control equipment, CAD/CAM systems, as well as general-purpose computers.

Unlike personal computers PLCs can be programmed using a *ladder-diagram language*, which is similar to schematic diagrams used in the industry. The ladder-diagram language is basically a *symbolic* set of instructions used to create the controller program. These ladder instruction symbols are arranged to obtain the desired control logic that is to be entered into the memory of the PLC. Because the instruction set is composed of contact symbols, ladder-diagram language is also referred to as *contact symbology*. Representations of contacts and coils are the basic symbols of the logic ladder diagram instruction set.

To get an idea of how a PLC program operates, consider the simple process control problem illustrated in Figure 42-22. Here a mixer

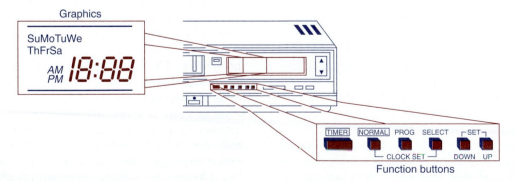

Figure 42-21   Inputs to a VCR program.

motor is to be used to automatically stir the liquid in a vat when the temperature and pressure reach preset values. In addition, direct manual operation of the motor is provided by means of a separate pushbutton station. The process is monitored with temperature and pressure sensor switches that close their respective contacts when conditions reach their preset values.

Now let us look at how a PLC might be used for this application. The same input field devices (pressure switch, temperature switch, and pushbutton) are used. These devices would be hard-wired to an appropriate input module according to the manufacturer's labeling scheme. Typical wiring connections for a 120-V ac input module are shown in Figure 42-24.

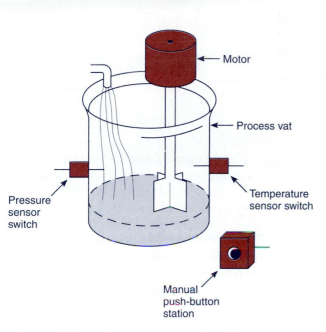

Figure 42-22 Mixer process control problem.

This control problem can be solved using the relay method for motor control shown in the relay ladder diagram of Figure 42-23. The motor starter coil (*M*) is energized when both the pressure and temperature switches are closed or when the manual pushbutton is pressed.

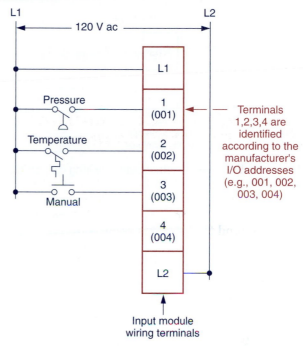

Figure 42-24 Typical input module wiring connection.

The same output field device (motor starter coil) would also be used. This device would be hard-wired to an appropriate output module according to the manufacturer's labeling scheme. Typical wiring connections for a 120-V ac output module are shown in Figure 42-25.

Next, the PLC ladder-logic diagram would be constructed and programmed into the memory of the CPU. A typical ladder-logic diagram for this process is shown in Figure 42-26. The format used is similar to the layout of the hard-wired relay ladder circuit. The individual symbols represent *instructions* while the numbers represent the instruction *addresses*. When programming the controller, these instructions are entered one by one into the processor memory from the operator terminal keyboard. Instructions are stored in the user program portion of the processor memory.

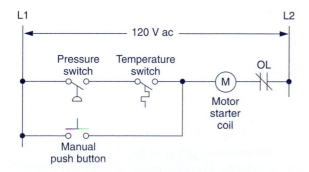

Figure 42-23 Process control relay ladder diagram.

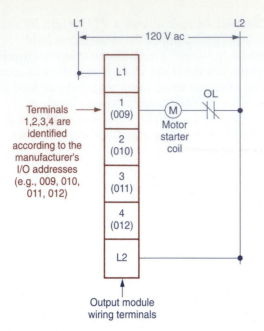

Figure 42-25 Typical output module wiring connection.

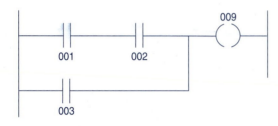

Note: Numbers 001, 002, 003, and 009 are identified with the pressure switch, temperature switch, manual push button, and motor starter coil, respectively.

Figure 42-26 Process control PLC ladder-logic diagram.

To operate the program, the controller is placed in the RUN mode, or operating cycle. During each operating cycle, the controller examines the status of input devices, executes the user program, and changes outputs accordingly. Each -||- can be thought of as a set of normally open (N.O.) contacts. The -()- can be considered to represent a coil that, when energized, will close a set of contacts. In the ladder-logic diagram of Figure 42-26, the coil 009 is energized when contacts 001 and 002 are closed, or when contact 003 is closed. Either of these conditions provides a continuous path from left to right across the rung that includes the coil.

The RUN operation of the controller can be described by the following sequence of events.

First, the inputs are examined and their status is recorded in the controller's memory (a closed contact is recorded as a signal that is called a logic 1 and an open contact by a signal that is called a logic 0). Then the ladder diagram is evaluated, with each internal contact given OPEN or CLOSED status according to the record. If these contacts provide a current path from left to right in the diagram, the output-coil memory location is given a logic 1 value and the output module interface contacts will close. If there is no conducting path on the program rung, the output-coil memory location is set to logic 0 and the output module interface contacts will be open. The completion of one cycle of this sequence by the controller is called a *scan*. The *scan time,* the time required for one full cycle, provides a measure of the speed of response of the PLC.

Computerized Engine Control Systems In today's automobiles the engine microcomputer and fuel injectors have replaced the conventional carburetor. A fuel injector (Figure 42-27) is a small, normally-closed precision solenoid valve. When the microcomputer outputs an injection signal to the injector, the coil built into the injector pulls the needle's valve back, and fuel is injected through the nozzle. The amount of fuel injected is controlled by the injection pulse duration, or the length of time that the coil remains energized.

Engines use a microcomputer to monitor and control the amount of fuel added to the air intake of the engine. Engines require a certain air-to-fuel ratio to operate efficiently. The air-to-fuel ratio is a measure of the weight of air in relation to the weight of fuel. The best air-to-fuel ratio is 14.7 parts of air to 1 part of fuel. This air-to-fuel ratio allows the catalytic converter to operate at maximum efficiency to remove pollutants from the engine exhaust gases while giving the best mileage and performance possible and protecting the engine. A *rich* mixture means there is too much fuel and not enough air. A *lean* mixture means there is too much air and not enough fuel.

Both open-loop and closed-loop control modes are used in automotive ***engine control systems***. In the closed-loop mode, the control loop is a complete cycle. A feedback-control *sensor* is used to measure the results of the microcomputer output. The operation of a typical automotive closed-loop engine-control system is

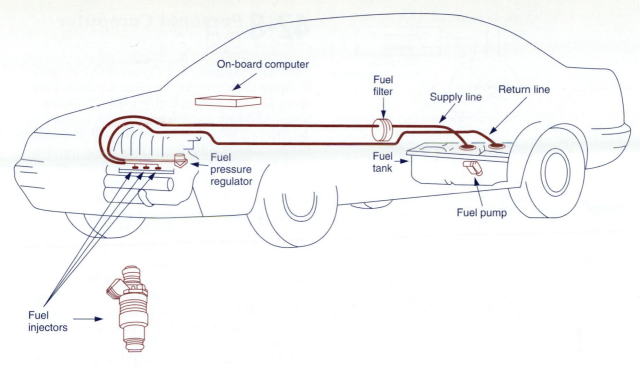

Figure 42-27 Basic electronic fuel injection system.

illustrated in Figure 42-28. Sensor inputs are sent to the microcomputer; the microcomputer processes them and sends a fuel injection pulse command to the fuel injector, which results in combustion of the air/fuel mixture. The exhaust gas oxygen (EGO) sensor provides feedback information about the amount of oxygen present in the exhaust manifold to the microcomputer. This allows the microcomputer to adjust the air/fuel ratio to achieve the proper level of oxygen.

When the engine is cold, most engine microcomputer control systems operate in the *open-loop mode*. In this mode, the control loop is *not* a complete cycle because it does not receive feedback information. Instead, the computer makes decisions based on preprogrammed information.

With a cold engine, the EGO sensor signal is not usable until the exhaust temperature reaches about 600°F (315°C). The open-loop mode can be activated by a signal from a temperature sensor (Figure 42-29). This signal tells the microprocessor to ignore the feedback signal from the EGO sensor until the engine has reached a preprogrammed temperature level. Systems with EGO sensors may also be programmed to go automatically into the open-loop

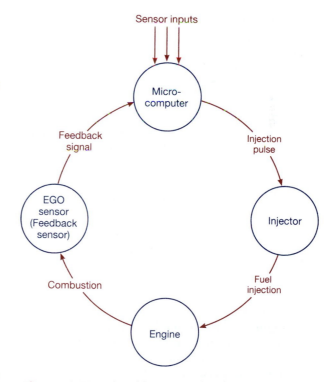

Figure 42-28 Closed-loop control cycle.

mode to maintain stabilized fuel combustion during deceleration; high-load, or high-speed,

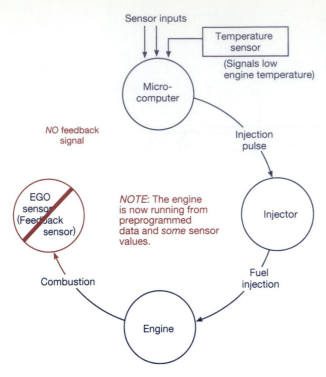

Sensor inputs

Temperature
sensor

(Signals low
engine temperature)

Micro-
computer

NO feedback
signal

Injection
pulse

EGO
sensor
(Feedback
sensor)

NOTE: The engine
is now running from
preprogrammed
data and *some* sensor
values.

Injector

Combustion

Fuel
injection

Engine

Figure 42-29   Open-loop control cycle.

operation; engine idling; malfunction of the EGO sensor or its circuit; and engine starting.

Although there is considerable variation among the configurations of computerized engine-control systems used by various manufacturers over various model years, the basic operating principles are the same. Earlier models of computer-controlled engine systems used separate subsystems for fuel control, emission control, and spark control. The current trend is toward a more integrated system, in which several subsystems are operated as separate functions of the *same* computer. Figure 42-30 provides an overview of an integrated computerized engine-control system.

The main objective of any computerized engine-control system is to allow the vehicle to give good fuel mileage and clean emission but not at the expense of good acceleration and smooth operation. Basically, the engine computer takes sensor readings and responds with signal outputs to control devices. Sensors can update the computer every 100 ms for general information and every 12.5 ms for critical emissions and drivability information. This precision control of the engine results in high engine performance at the lowest possible levels of emission.

***Personal computers* (*PCs*)** are either desktop or laptop units and are intended for single-user operation. You have probably used a PC many times. Although you don't need to understand in detail how a computer works to use it, it does help to have some familiarity with its basic operation.

All PCs consist of two basic components: hardware and software. The **computer hardware** is the physical component which makes up the computer system and includes:

Power supply. Converts the 120 V ac electricity from the line cord to low dc voltages that are needed by the computer circuits.

Disk drive. The disk drive allows you to read and save information on portable diskettes or floppy disks. You can use floppy disks to save information or load new software onto your computer.

CD-ROM drive. The CD-ROM drive reads data from a compact disk.

Hard drive. Allows information to be stored and read from nonremovable hard disk located inside your computer. The hard disk stores the computer's operating system, files, programs, and documents.

Motherboard. Is the circuit board that holds and electrically interconnects the major sections of the entire computer system. The CPU, ROM, RAM, power supply, and peripheral cards all plug into the motherboard (Figure 42-31, p. 666).

CPU. The **CPU**, or **processor**, interprets the instructions for the computer and performs the required process for each of these instructions.

Memory. This is the circuitry or device within the computer that holds information in an electrical or magnetic form. There are two kinds of memory: **read-only memory** (**ROM**) and **random-access memory** (**RAM**). You can write data into RAM and read data from RAM. This is in contrast to ROM, which permits you only to read data. Most RAM is volatile, which means that it requires a steady flow of electricity to maintain its contents. Computers always contain a small amount of read-only memory that stores instructions for computer start up.

Memory cache. A **memory cache** is a portion of memory made of high-speed static RAM (SRAM) instead of the slower and cheaper

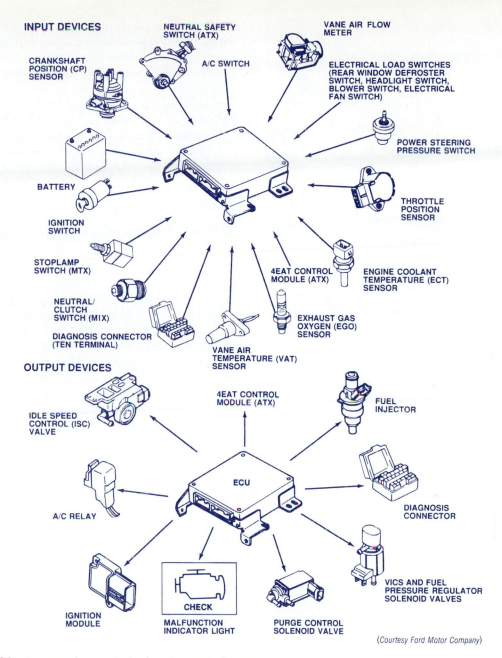

INPUT DEVICES

NEUTRAL SAFETY SWITCH (ATX)

VANE AIR FLOW METER

A/C SWITCH

CRANKSHAFT POSITION (CP) SENSOR

ELECTRICAL LOAD SWITCHES (REAR WINDOW DEFROSTER SWITCH, HEADLIGHT SWITCH, BLOWER SWITCH, ELECTRICAL FAN SWITCH)

POWER STEERING PRESSURE SWITCH

BATTERY

THROTTLE POSITION SENSOR

IGNITION SWITCH

4EAT CONTROL MODULE (ATX)

ENGINE COOLANT TEMPERATURE (ECT) SENSOR

STOPLAMP SWITCH (MTX)

NEUTRAL/ CLUTCH SWITCH (MIX)

EXHAUST GAS OXYGEN (EGO) SENSOR

DIAGNOSIS CONNECTOR (TEN TERMINAL)

VANE AIR TEMPERATURE (VAT) SENSOR

OUTPUT DEVICES

4EAT CONTROL MODULE (ATX)

FUEL INJECTOR

IDLE SPEED CONTROL (ISC) VALVE

ECU

DIAGNOSIS CONNECTOR

A/C RELAY

VICS AND FUEL PRESSURE REGULATOR SOLENOID VALVES

IGNITION MODULE

CHECK

MALFUNCTION INDICATOR LIGHT

PURGE CONTROL SOLENOID VALVE

(Courtesy Ford Motor Company)

Figure 42-30   Integrated computerized engine-control system.

dynamic RAM (DRAM) used for main memory. Memory caching is effective because most programs access the same data or instructions over and over. By keeping as much of this information as possible in SRAM, the computer avoids accessing the slower DRAM.

Peripheral cards. Allow accessory features and connects the computer to input and output devices such as printers, disk drives, display monitors, keyboards, and mice. For example, the sound card allows your

computer to reproduce music, sounds, and voices; while the video card is the part of the computer that sends the image to the monitor.

Expansion slots. Connectors used for the purpose of connecting other circuits to the motherboard.

Keyboard. Just like a typewriter keyboard, this device is the primary way of inputting data into programs.

Mouse. The ***mouse*** makes operating your computer easier. This handheld device is used for

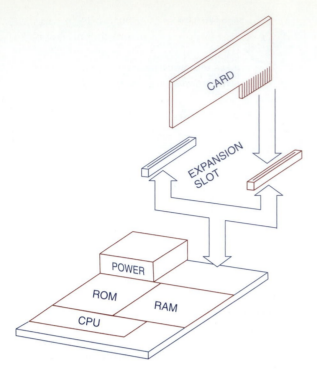

Figure 42-31 Motherboard is the circuit board that everything in the computer plugs into.

moving and pointing to objects on the screen and can replace many of the functions of the keyboard.

Monitor. The ***monitor*** allows you to see what you are doing. Color monitors are sometimes called RGB monitors because they accept three separate signals — red, green, and blue (Figure 42-32). Like televisions, monitor screen sizes are measured in diagonal inches, the distance from one corner to the opposite corner diagonally. The resolution of a monitor indicates how densely packed the pixels are. In general, the more pixels (often expressed in dots per inch), the sharper the image.

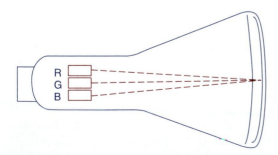

Figure 42-32 Color monitor uses red, green, and blue electron guns.

Modem. This mechanism connects a computer to a phone line so information can be sent from one computer to another, or the user can access online service or the Internet. Most modern ***modems*** are fax modems, which means that they can send and receive faxes.

Printer. The ***printer*** provides a hard copy of your work. There are four basic types of printers: dot matrix, inkjet, bubble jet, and laser. The dot matrix is the simplest. Most inkjets and bubble jets can print color and graphics, and a laser printer offers the best resolution at the highest speed.

Scanner. A ***scanner*** is a useful accessory when working with artwork or photos. This device can copy text, pictures, or photographs directly into your computer.

Software is the operating language for the computer. There are two major categories of PC software: operating system software and application software. The ***operating system*** is the most important program that runs on a computer. Every computer must have an operating system to run application programs. Operating systems perform basic tasks, such as recognizing input from the keyboard, sending output to the display screen, keeping track of files and directories on the disk, and controlling peripheral devices such as disk drives and printers. The application programs must be written to run in concert with a particular operating system. Therefore, your choice of operating system determines to a great extent the applications you can run.

Though there are many types of operating system, *MS-DOS* was the first widely installed operating system used in personal computers. MS refers to the manufacturer, Microsoft, and DOS to its ***Disk Operating System***. DOS created a common platform for all software used and allowed you to:

• Load programs.
• Read, write and edit files.
• Allocate memory.
• Organize a disk.
• Display the content of a disk.
• Control hardware.
• Manage files.
• View error messages.

A ***file*** is a collection of data stored under a single name. When information is stored on the disk, the file is saved or written to the disk. When the information is being used, the disk is being read or data is being loaded into the computer's memory. A computer can move, copy, rename, or delete the file at your command. Some important types of files include:

- *Executable file.* A list of instructions to the CPU.
- *Data file.* A list of information.
- *Text file.* A series of characters like a letter.
- *Graphics file.* A picture converted to digital code.

After a disk is formatted, writing or reading a file is a process that involves your software, DOS, the PC's *BIOS* (*Basic Input/Output System*), and the mechanism of the disk drive itself. The BIOS serves as a link between the computer hardware and the software programs. It is permanently stored in the computer's memory, which serves as a translator between the CPU and all the other peripherals.

When the computer is first turned on, the BIOS causes the computer to access the floppy disk drives to see whether there is a disk there containing DOS. If there is, the computer will load the DOS into itself. The process of loading DOS into your computer is called *booting*. If there are no floppy disks in your drives that contain DOS, the hard drive (C: drive) will be examined for DOS (Figure 42-33). If there is no DOS (or no hard drive), then an error message will be displayed.

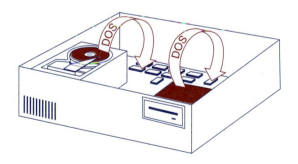

Figure 42-33   Loading DOS into memory.

DOS is simply a very large list of **commands** that the computer uses to control, manage, and run other programs. DOS is text oriented and not overly user friendly. Today's major operating systems provide a *graphical user interface* (*GUI*). The GUI familiar to most of us today is either the **Windows** or the **Macintosh** operating systems. Microsoft's Windows is a graphics-based interface for MS-DOS. It uses graphical

images to replace the text commands to run DOS. Instead of typing commands, you select graphic pictures (icons), using a mouse. The first Microsoft Windows operating system was really an application that operates with the MS-DOS operating system. Today, Windows operating systems continue to support DOS (or a DOS-like user interface) for special purposes by emulating the operating system. Typical elements of a GUI include such things as: windows, pull-down menus, buttons, scroll bars and iconic images (Figure 42-34).

Windows allows for **multitasking**. No computer can actually do two things at once, but it appears to do so by very rapidly shuffling back and forth from one program to another. Programs in Windows share the CPU resources, passing control back and forth at times when the computer is not being controlled by the user. Instead of having to develop an application that does everything, with Windows you can use multiple specialized applications that run concurrently and work together.

To work in Windows, you need to perform the following mouse actions:

- **Point.** Position the mouse so that the on-screen pointer touches an item.
- **Click.** Press and release the left mouse button once. Clicking an item usually selects it. Except when you're told to do otherwise (i.e., to right-click), you always use the left mouse button.
- **Double-click.** Press and release the left mouse button twice quickly. Double-clicking usually activates an item or opens a window, folder, or program.
- **Drag.** Place the mouse pointer over the element you want to move, press and hold down the left mouse button, then move the mouse to the new location. You might drag to move a window, dialog box, or file from one location to another.
- **Right-click.** Press and release the right mouse button once. Right-clicking usually displays a shortcut (or pop-up) menu from which you choose common commands.

The Windows screens contain a lot of different elements and controls. Most windows contain a title bar, a menu bar, scroll bars, and three control buttons. The **Minimize Button** collapses a window without closing its application. The minimized application can be made active again by clicking on its name in the

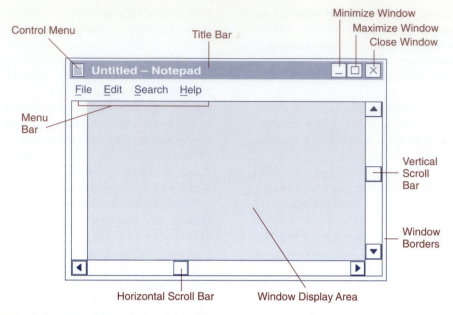

Control Menu   Title Bar   Minimize Window   Maximize Window   Close Window

Untitled – Notepad

File   Edit   Search   Help

Menu Bar

Vertical Scroll Bar

Window Borders

Horizontal Scroll Bar   Window Display Area

Figure 42-34   Typical elements of graphical user interface.

Taskbar along the bottom of the screen. The ***Maximize/Restore Button*** will expand a window to fill the screen. Clicking on it again will restore the window to its former size. Clicking on the ***Close Button*** will close a window and its application. A closed application will not appear on the taskbar; to reopen it you must launch the application again.

The Windows ***Desktop*** screen is much like your desk at school or at home. It is a flat area on which you will do your work. You may have several things on your desktop at a time. ***Icons*** are pictures that represent programs. Generally, your desktop will have icons to start programs that you use most often. The ***My Computer*** icon makes it easy to see what programs and data are on your computer, on diskettes, and on CD ROMs. the ***Recycle Bin*** icon acts like a trash can. You may drag files that you no longer need onto the Recycle Bin to dispose of them.

In all application windows, a ***menu bar*** is located along the top edge of the window under the title bar. Each word in the menu bar is a name for a menu (e.g. File, Edit, Search, and Help). Although the specific commands in the menu list vary from application to application, menus all work in the same way. For example, the ***Help*** menu displays help topics regarding the application you are running, while the ***Edit*** menu allows you to cut, copy, or clear data.

The ***Clipboard*** is a section in the computer's memory, which acts like a real clipboard. You can copy text or graphics from many Windows applications to the Clipboard, and then paste them into many other applications.

The ***Start*** menu can be thought of as the command center for Windows. It places all the icons, also known as ***Shortcuts***, into one organized location. The simplest way to start a program is as follows:

• Click the Start button.
• Point to Programs.
• Click the group that contains the program you want to start (such as Accessories).
• Click the program you want to start (such as Notepad).

A ***driver*** program is one that controls a device. Every device, whether it be a printer, disk drive, or keyboard, must have a driver program. Many drivers, such as the keyboard driver, come with the operating system. For other devices, you may need to load a new driver when you connect the device to your computer.

Plug-and-Play, or ***PnP***, is a specification built into the computer's operating system, which allows you to easily install peripheral devices built for it. All of the settings, jumpers, and drivers, are taken care of so that, theoretically, you can have your new hardware working for you in a couple of minutes. When you plug in a PnP device, Windows will detect the new hardware, automatically adjust the settings, and install the drivers.

Ports are the connecting devices that stick out the back end of the computer case (Figure 42-35). They are actually the back edge of an expansion card, providing the electrical door, or gateway, connecting the computer and a peripheral. Ports are usually a preset number of pins to allow the proper cable connection and are referred to as male or female connections (with pins or holes to fit the pins). Ports are usually referred to as *serial* (for a mouse connection) or *parallel* (for a printer connection). A serial port sends data one bit at a time over a single one-way wire; a parallel port can send several bits of data across eight parallel wires simultaneously. Because of its simplicity, the serial port has been used at one time or another to make a PC communicate with just about any device imaginable. The serial port is often referred to as an RS-232 port (Electronics Industries Association designation). The parallel port is often called a Centronics port.

Most computers now include *Universal Serial Bus* (*USB*) and *PS/2* ports. With a USB port, you can attach up to 127 different devices to each port, including scanners, cameras, mice, keyboards, and joysticks. The data rate for a USB port is very fast at 12 million bits per second. The PS/2 port is a type of port developed by IBM for connecting a mouse or keyboard to a PC. The PS/2 port supports a mini DIN plug containing just 6 pins. Most PCs have a PS/2 port so that the serial port can be used by another device, such as a modem.

42-9 The Internet

In the past, students had to rely almost exclusively on the resources that were physically on their campus to enhance their program of study. Their learning sphere was dictated by the knowledge of their instructor, the participation of the students in their classes, and the holdings of their campus library. Today the *Internet* extends any student's resources to schools and organizations around the world. The Internet offers vast stores of text, databases of information, electronic journals, bibliographies, software programs, and forums for information.

The Internet is an extensive network of interlinked yet independent computer networks. The Internet itself doesn't actually "do" anything. It simply transmits data from computers connected to its many branches. Routing packets of information around the world requires banks of hardware, but actual manipulation of data is left to the computers on either end of the connection. Some of the most commonly used Internet services are as follows:

- *E-mail. Electronic mail* (*e-mail*) is probably the most common Internet application. It allows you to send and receive messages from anyone else who has an Internet address. A

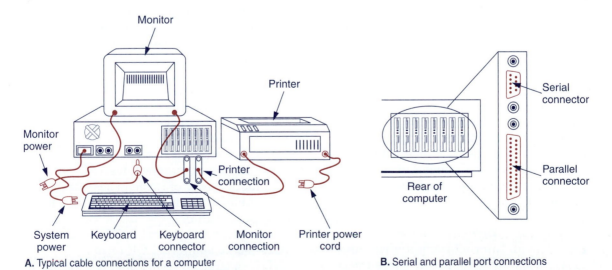

A. Typical cable connections for a computer

B. Serial and parallel port connections

Figure 42-35 Computer peripheral connections.

large number of e-mail programs are available to suit various needs. Generally they all: collect e-mail in a sort of "Inbox" or folder, store received messages, and let the user reply to or forward messages.

- **World Wide Web (WWW).** Many people use the terms "Internet" and "World Wide Web" interchangeably, as though they were the same thing. They are not. The **World Wide Web (WWW)** is a gateway to all the resources available on the Internet. The Web is an Internet application. What makes the Web different from other applications, such as e-mail, is that you do not send Web documents to anyone in particular. You just place them on a computer that is connected to the Internet. A special program called a **Web server** then makes the pages available to anyone who may be interested in browsing through them. To access the Web, you use a program called a *Web browser*. A browser is a program on your computer, which communicates with a Web server to retrieve the information you want.

- **File Transfer Protocol (FTP).** This process is used to transfer files across the Internet. With **FTP** you can transfer files between many different kinds of computers and have access to millions of files around the world.

- **Telnet. Telnet** is a way of connecting to a computer at a remote location and using that computer as if you were actually at that remote site. You might use Telnet to get to another site such as an online library catalog.

- **Usenet. Usenet**, short for Users Network, consists of thousands of discussion groups called newsgroups. Newsgroups are a great way to find information on any topic. With over 20,000 to choose from, there is surely one that will pique your interest.

- **Gopher. Gopher** is an Internet tool that allows you to go to hundreds of sites on the Internet using a menu structure. You maneuver your way through the Internet by selecting items listed on a screen menu. A Gopher menu organizes a vast number of options that help access all the Internet's features with a minimum of effort.

To put information on the Web, you must first have a Web site. A **Web site** is a directory of files stored on an Internet computer running Web server software. People can access those files with a **Web browser**. The two most well-known browsers are Netscape Navigator and Microsoft Internet Explorer. A Web site can have any number of pages. These pages are written using **HyperText Markup Language (HTML)**. Hypertext is a term that describes the system that allows documents to be cross-linked in such a way that the reader can explore related documents by clicking on a highlighted word or symbol. A page can simply be a block of text, but typically also includes graphics. Sound clips and full motion video might also be part of a page.

The home page is the first page a user sees when connecting and has a unique address or **URL (Uniform Resource Locator)**. You can think of a URL as a networked extension of the standard filename concept. All Internet addresses in the United States now end with one of six domain names: **.com** for commercial business, **.org** for nonprofit organizations, **.net** for networks, **.edu** for educational institutions, **.gov** for governmental bodies, or **.mil** for the military.

There are different types of **search engines**, which procure information and organize information in many different ways. Most of the search engines are unique in the way they do this. That is why there are so many different search engines. There is a single long data entry-line for each search engine where you can enter search words. In general, you should enter search words in all lower case letters separated by spaces. You should never use commas, dashes, or other delimiters. It doesn't hurt, and it may help, to enter capital letters for some formal names. Some search engines allow you to limit or narrow down your search in various ways which helps to make the process of searching easier and more efficient.

Two types of advanced searches are generally supported: Boolean searches and Phrase searches. **Boolean searching** refers to a form of logic applied to the search. When you perform a Boolean search, you search the computer for the keywords that best describe your topic. The unique power of Boolean searching is that you combine these keywords using the following logic operators:

AND—search on term1 AND term2
OR—search on term1 OR term 2
NOT—search on term1 but NOT term2

Phrase searching involves searching for a phrase or string of words. If you want a search engine to perform a phrase search you need to inform it that the words you are looking for need to be grouped together. This is done by enclosing the keywords within quotes.

As you use Internet sites, you are going to identify favorite places that you want to revisit frequently. Rather than type the long address each time you need to access the site, you may want to define a *bookmark* or *hot-list entry*. Your browser will provide a way to go directly to these documents without retyping a URL.

As you surf the Web, you will find places that let you download files or images. Before you start downloading files, you need to create a download directory. You can save all the files you download in this new directory. Some of these files may be in a compressed format to help make downloading faster. These files have to be uncompressed before they can be used.

HELP WANTED

Network Engineer
Job Description
- Design computer network equipment.
- Test and certify network designs.
- Provide technical support on installed networks.
- Act as project manager for network installations.
- Knowledge of Internet protocols and operation.
- B.S./M.S. in engineering, computer science, or related field.

Review and Applications

Review Questions

1. What is a computer?
2. Identify the four main sections of a computer and give a brief description of the function of each.
3. What is a microprocessor?
4. What is a microcomputer?
5. Compare read-only memory (ROM) with random-access memory (RAM).
6. Explain the function of input sensors that are used in conjunction with an automotive engine microcomputer.
7. Describe two types of outputs commonly operated by data received from the microprocessor.
8. What is a programmable logic controller (PLC)?
9. List three input devices and three output devices used with a personal computer.
10. What is a computer clock pulse generator used for?
11. What is a data bus?
12. Compare the way serial and parallel data is transmitted.
13. What is a microcomputer program?
14. Compare personal computer operating systems and applications programs software.
15. Explain two functions that can be performed by a computer's self-diagnostic program.
16. Why are microprocessors so widely used in consumer-oriented electronic products?
17. (a) What type of industrial circuits were PLCs specifically designed to replace?
    (b) What is ladder-diagram language similar to?
    (c) What are the two basic symbols of the logic ladder-diagram instruction set?
18. (a) What has replaced the conventional carburetor in today's automobiles?
    (b) Explain the basic operation of a typical closed-loop engine-control system.
    (c) What is the main objective of a computerized engine-control system?
19. State the PC hardware component used for each of the following functions:

(a) read data from a compact disk
(b) hold and electrically interconnect major sections of the entire computer system
(c) connect the computer to input and output devices
(d) connect the computer to a phone line
(e) hold information inside the computer
(f) convert the ac voltage from the line cord to low voltage dc
(g) access the same data over and over using high-speed static RAM
(h) read and save information on diskettes

20. List four different types of files along with a short description of each.
21. What is the major difference between DOS and GUI operating systems?

Problems

1. Refer to Figure 42-22. The PLC process control shown is to be changed so that the mixer motor will operate whenever the temperature sensor switch, pressure sensor switch, or manual push-button is closed.
   (a) What wiring changes, if any, would have to be made to the input module?
   (b) What wiring changes, if any, would have to be made to the output module?
   (c) Redraw the process control PLC ladder-logic to implement this new control process.
2. Refer to Figure 42-28. Assume that when the engine is operating in the closed-loop mode the EGO or oxygen sensor becomes defective and fails to generate any feedback signal. What do you think the response by the computer might be?

Critical Thinking

1. Assume a search by a search engine retrieves too many documents. If the search engine is Boolean-based, what Boolean operator would you use to narrow the scope of your search?

Portfolio Projects

- Prepare a written report on a microprocessor-based control system such as that used in automotive engine control, alarm systems, heating and air conditioning, or PLC control of a process. List all input and output devices and explain the purpose of each.

- Assuming cost is not a factor, write the specifications for what you would consider to be an ideal PC system. Include pricing and detailed information on things like CPU, hard disk drive, floppy drive, CD drive, cache, modem, bus, monitor, scanner, printer, sound card, speakers and ports.

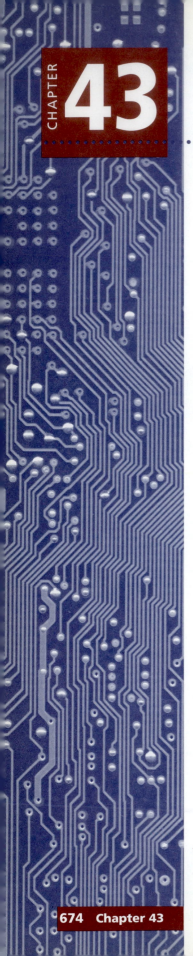

CHAPTER 43

Network Theorems

There are several "theorems" which allow you to calculate circuit parameters (volts, amps, ohms, etc.) in complex circuit networks. A network is a combination of interconnected components such as resistors connected in a way to produce a particular end result. However, networks generally need more than the rules of series and parallel circuits for analysis. In this chapter, you will learn how complex resistive networks can be reduced to a simpler form using the superposition, Thevenin and Norton theorems.

Objectives

After studying this chapter, you should be able to:
1. Compare voltage and current sources.
2. Apply the superposition theorem to a resistive network.
3. Apply the Thevenin theorem to a resistive network.
4. Apply the Norton theorem to a resistive network.

Key Terms

- constant voltage source
- internal resistance
- constant current source
- superposition theorem

- Thevenin's theorem
- Thevenin equivalent circuit
- Thevenin voltage (V_{TH})
- Thevenin resistance (R_{TH})

- Norton's theorem
- Norton equivalent circuit
- Norton resistance (R_N)
- Norton current (I_N)

RESEARCH INTERNET SITES

- Xycom
- Vickers
- Transcat

43-1 Voltage and Current Sources

A **constant-voltage source** (also known as an ideal voltage source) has zero internal resistance and supplies any amount of current with no change in its terminal voltage (Figure 43-1). While there are many situations that warrant such an assumption, there is in actuality, no such thing as a constant or ideal voltage source.

In reality, every voltage source has a series *internal resistance* (R_S) which causes some loss of voltage at its terminals when a current flows. A voltage source should have the smallest possible internal resistance, so that its output will remain constant regardless of whether a light load (i.e., large R_L, small I_L) or heavy load (i.e., small R_L, large I_L) is connected to its output terminals. When the internal resistance of a voltage source becomes a significant part of the total circuit resistance, you must take it into account when analyzing the circuit. Using the *voltage divider rule* (for a series circuit), the output voltage (or load voltage) can be determined in terms of R_S and R_L as seen in Example 43-1.

Up to this point, we have used voltage sources as a means of providing power for a circuit. We can also consider the source that powers a circuit to be a current source. A *constant-current source* (also known as an ideal current source) will supply the same *current* to any resistance connected across its terminals. Figure 43-3A shows the circuit of an ideal current source. A current source is represented by

Example 43-1

Problem   Assume a voltage source has a no-load voltage of 12 V and an internal resistance (R_S) of 1 Ω. Calculate the output voltage of Figure 43-2 if the load resistance (R_L) is equal to 10 Ω.

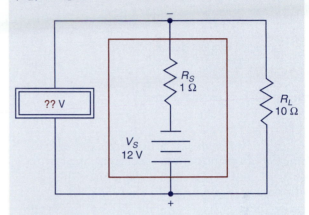

Figure 43-2

Solution
(Voltage-divider formula)

$$V_{out} = \frac{R_L}{R_L + R_S} \times V_S$$

$$= \frac{10\ \Omega}{10\ \Omega + 1\ \Omega} \times 12\ V$$

$$= 10.91\ V$$

an arrow in a circle rather than the battery or ac symbol. The arrow shows the direction of the electron flow (conventional) current produced by the source. Just as a voltage source has a voltage rating, the current source has a certain current rating (in this case 2 A).

Figure 43-3B shows the circuit of a constant current source, together with its source resistance R_S and a load resistance R_L. Note that the source resistance R_S is in parallel with the current source, not in series as is the case with a voltage source. Using the current divider rule (for a parallel circuit), the output current (or load current) can be determined in terms of R_S and R_L:

$$I_L = I_S \times \frac{R_S}{R_L + R_S}$$

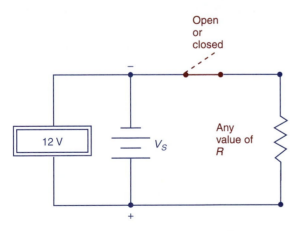

Figure 43-1   Constant voltage source supplies any amount of current with no change in its terminal voltage.

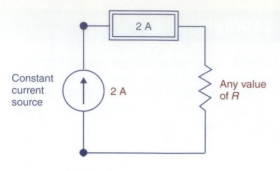

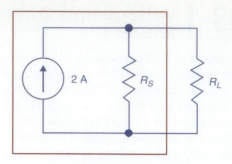

A. Ideal constant current source.

B. Constant current source together with its internal source resistance and a load.

Figure 43-3 Current sources.

Where the load resistance is *very much smaller* than the source resistance, a constant current source is assumed to have infinite parallel source resistance. In this case, all the source current is assumed to flow through the load.

Any voltage source with its series resistance can be converted to an equivalent current source with the *same value of resistance in parallel.* Just divide the source voltage V by its series resistance R, to calculate the current I for the equivalent current source of the same value (Figure 43-4). A similar procedure is followed when converting a current source to an equivalent voltage source. In this case, the source current I is multiplied by its parallel resistance R, to calculate voltage V for the equivalent voltage source in series with the same R.

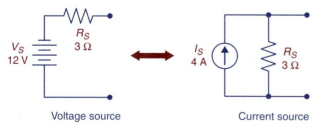

Figure 43-4 Voltage source and equivalent current source.

43-2 Superposition Theorem

The *superposition theorem* is very useful in solving resistive networks that have more than one source of voltage or current. This technique treats each source as an independent source in the network, then combines the individual results. The superposition theorem states the following: *In a resistor network with two or more sources, the current through or the voltage across any component is the algebraic sum of the effects due to each independent source.*

When applying the superposition theorem it is necessary to remove and "zero" all sources other than the one that is being examined. In order to zero a voltage source, we replace it with a short circuit, since the voltage across a short circuit is zero volts. In order to zero a current source, we replace it with an open circuit, since the current through an open circuit is zero amperes. When a voltage or current source contains an internal resistance, it must be taken into account when the source is removed.

The following steps are used applying the superposition theorem:

1. Zero all voltage or current sources but one.
2. Determine the current or voltage you need, along with its correct direction or polarity, just as if there were only one source in the circuit.
3. Repeat steps 1 and 2 for all the remaining sources in the circuit.
4. To find a specific current or voltage, algebraically combine the currents or voltages due to the individual sources. If the currents act in the same direction or the voltages are of the same polarity, add them. If the reverse is true, subtract them with the direction of the resultant current or voltage being the same as the *larger* of the original quantity.

Example 43-2

Problem Calculate the current through and the voltage across R_1 for the circuit of Figure 43-5, using the superposition theorem.

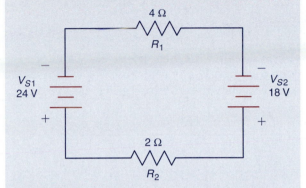

Figure 43-5

Step 1 Find the current through R_1 due to source V_{S1} by replacing V_{S2} with a short as shown in Figure 43-6.

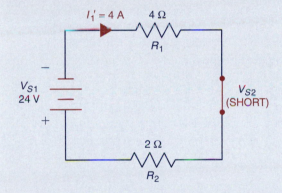

Figure 43-6

$$I_1' = \frac{V_{S1}}{R_1 + R_2} = \frac{24 \text{ V}}{6 \text{ }\Omega} = 4 \text{ A}$$

Step 2 Next find the current through R_1 due to source V_{S2} by replacing V_{S1} with a short as shown in Figure 43-7.

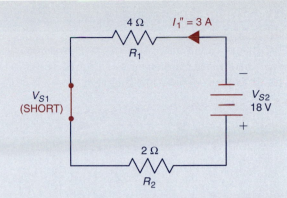

Figure 43-7

$$I_1'' = \frac{V_{S2}}{R_T} = \frac{18 \text{ V}}{6 \text{ }\Omega} = 3 \text{ A}$$

Step 3 Calculate the net current through R_1 subtracting the smaller current from the larger current since the two currents flow in opposite directions. The net current is in the same direction as the larger current as shown in Figure 43-8.

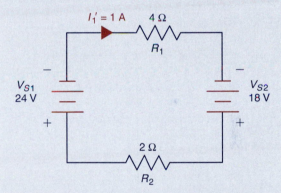

Figure 43-8

$$I_1 = I_1' - I_1'' = 4 \text{ A} - 3 \text{ A} = 1 \text{ A}$$

Step 4 The voltage across R_1 is then found using Ohm's law.

$$V_1 = I_1 \times R_1 = 1 \text{ A} \times 4 \text{ }\Omega = 4 \text{ V}$$

Example 43-3

Problem Calculate the value of current I_1, I_2 and I_3, for the circuit of Figure 43-9, using the superposition theorem.

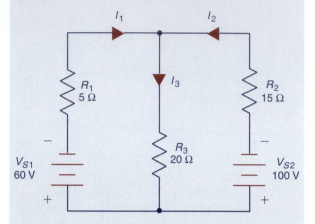

Figure 43-9

Step 1 Find the currents I_1', I_2' and I_3' due to source V_{S2} by replacing V_{S1} with a short as shown in Figure 43-10.

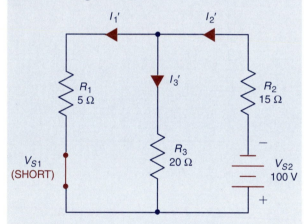

Figure 43-10

$$R_1||R_3 = \frac{R_1 \times R_3}{R_1 + R_3} = \frac{5\ \Omega \times 20\ \Omega}{5\ \Omega + 20\ \Omega} = 4\ \Omega$$

$$R_T = R_1||R_3 + R_2 = 4\ \Omega + 15\ \Omega = 19\ \Omega$$

$$I_T = V_{S2}/R_T = 100\ \text{V}/19\ \Omega = 5.263\ \text{A}$$

$$I_2' = I_T$$

$$V_1 = R_1||R_3 \times I_2' = 4\ \Omega \times 5.263\ \text{A}$$
$$= 21.052\ \text{V}$$

$$I_1' = V_1/R_1 = 21.052\ \text{V}/5\ \Omega$$
$$= -4.21\ \text{A (negative sign indicates current flow is in a direction opposite to that indicated in the original circuit)}$$

$$I_3' = I_2' - I_1' = 5.263\ \text{A} - 4.21\ \text{A}$$
$$= 1.053\ \text{A}$$

Step 2 Find the currents I_1'', I_2'', and I_3'' due to source V_{S1} by replacing V_{S2} with a short as shown in Figure 43-11.

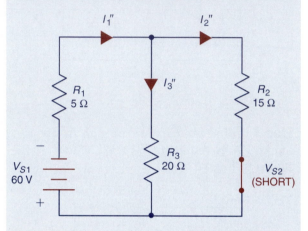

Figure 43-11

$$R_2||R_3 = \frac{R_2 \times R_3}{R_2 + R_3} = \frac{15\ \Omega \times 20\ \Omega}{15\ \Omega + 20\ \Omega}$$
$$= 8.571\ \Omega$$

$$R_T = R_2||R_3 + R_1 = 8.571\ \Omega + 5\ \Omega$$
$$= 13.571\ \Omega$$

$$I_T = V_{S1}/R_T = 60\ \text{V}/13.571\ \Omega$$
$$= 4.421\ \text{A}$$

$$I_1'' = I_T = 4.421\ \text{A}$$

$$V_2 = R_2||R_3 \times I_1'' = 8.571\ \Omega \times 4.421\ \text{A}$$
$$= 37.892\ \text{V}$$

$$I_2'' = V_2/R_2 = 37.892\ \text{V}/15\ \Omega$$
$$= -2.526\ \text{A (negative sign indicates current flow is in a direction opposite to that indicated in the original circuit)}$$

$$V_2 = V_3$$
$$I_3'' = I_1'' - I_2'' = 4.421 \text{ A} - 2.526 \text{ A}$$
$$= 1.895 \text{ A}$$

Step 3 Calculate the net current through each resistor. The actual current values and directions are as shown in Figure 43-12.

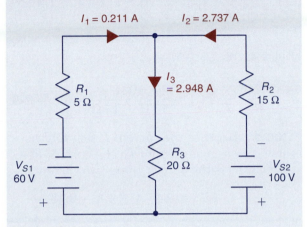

Figure 43-12

$$I_1 = I_1' + I_1'' = (-4.21 \text{ A}) + 4.421 \text{ A}$$
$$= 0.211 \text{ A}$$
$$I_2 = I_2' + I_2'' = 5.263 \text{ A} + (-2.526 \text{ A})$$
$$= 2.737 \text{ A}$$
$$I_3 = I_3' + I_3'' = 1.053 \text{ A} + 1.895 \text{ A} = 2.948 \text{ A}$$

Example 43-4

Problem

Calculate the value of current through resistor R_3, for the circuit of Figure 43-13, using the superposition theorem.

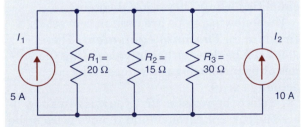

Figure 43-13

Step 1 Find the value of the current I_3' due to source I_1 by replacing I_2 with an open circuit, as shown in Figure 43-14.

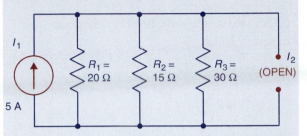

Figure 43-14

$$R_T = \frac{1}{1/R_1 + 1/R_2 + 1/R_3}$$
$$= \frac{1}{1/20 \ \Omega + 1/15 \ \Omega + 1/30 \ \Omega}$$
$$= 6.67 \ \Omega$$
$$V' = I_1 \times R_T = 5 \text{ A} \times 6.67 \ \Omega = 33.33 \text{ V}$$
$$I_3' = V'/R_3 = 33.33 \text{ V}/30 \ \Omega = 1.111 \text{ A}$$

Step 2 Find the value of the current I_3'' due to source I_2 by replacing I_1 with an open circuit, as shown in Figure 43-15.

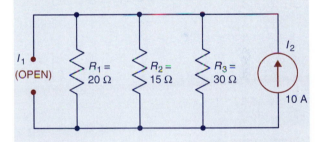

Figure 43-15

$$V'' = I_2 \times R_T = 10 \text{ A} \times 6.67 \ \Omega = 66.7 \text{ V}$$
$$I_3'' = V''/R_3 = 66.7 \text{ V}/30 \ \Omega = 2.223 \text{ A}$$

Step 3 The net current is then found by superimposing the two currents for I_3. Because the current I_3' acts in the same direction as I_3'' the net current is the sum of these two currents.

$$I_3 = I_3' + I_3'' = 1.111 \text{ A} + 2.223 \text{ A} = 3.334 \text{ A}$$

Example 43-5

Problem   Calculate the value of current through resistor R_2, for the circuit of Figure 43-16, using the superposition theorem.

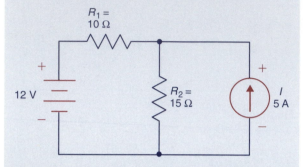

Figure 43-16

Step 1   Find the value of the current I' through R_2 due to the voltage source by replacing current source I with an open circuit, as shown in Figure 43-17.

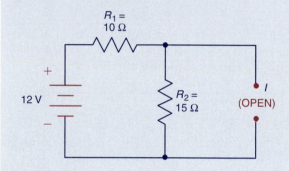

Figure 43-17

$$I' = \frac{V}{R_1 + R_2} = \frac{12\ \text{V}}{10\ \Omega + 15\ \Omega} = 0.48\ \text{A}$$

Step 2   Find the value of the current I'' through R_2 due to the current source by replacing voltage source V with a short circuit (Figure 43-18), and using the current-divider formula.

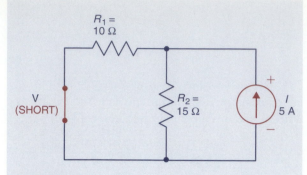

Figure 43-18

$$I'' = \frac{R_1}{R_1 + R_2} \times 1 = \frac{10\ \Omega}{10\ \Omega + 15\ \Omega} \times 5\ \text{A} = 2\ \text{A}$$

Step 3   Because the current I' flows in the same direction as current I'', the current flow through R_2 is the sum of these two currents.

$$I_{R2} = I' + I'' = 0.48\ \text{A} + 2\ \text{A} = 2.48\ \text{A}$$

43-3
Thevenin's Theorem

The most commonly used network theorem is *Thevenin's theorem*. It is particularly useful in situations where the circuit is complicated, but the interest is in the current through or the voltage across a particular resistor, which is generally referred to as the load. Thevenin's theorem states that: *Any network of voltage sources and resistors can be replaced by a single equivalent voltage source (V_{TH}) in series with a single equivalent resistance (R_{TH}) and the load resistor (R_L).* Figure 43-19 shows how a complex multiple resistor and source network is simplified using Thevenin's theorem.

The following steps are used in converting a circuit to its ***Thevenin equivalent***. These steps can be done either by calculating the values or by measuring the values if the circuit actually exists.

1. Select the resistor (R_L) you wish to know the current through and remove it from the circuit.
2. Measure or calculate the voltage across the points in the circuit where the resistor was connected. This is the ***Thevenin voltage*** (V_{TH}).

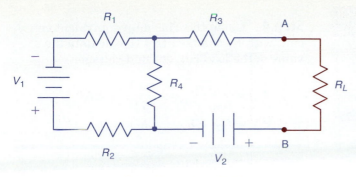

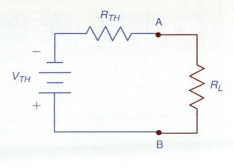

Original complex network.

Thevenin equivalent circuit.

Figure 43-19 Thevenin's theorem reduces any circuit to a single equivalent source in series with a single equivalent resistance.

3. Remove each voltage source and replace it with a short circuit. Be careful not to short across an active source in a working circuit.
4. Measure or calculate the resistance across the points in the circuit where the resistor was connected. This is the ***Thevenin resistance*** (R_{TH}).
5. Create a series circuit with the Thevenin voltage (V_{TH}) as the source and the Thevenin resistance (R_{TH}) placed in series with the resistor (R_L) that was removed. The polarity of V_{TH} must be such as to produce the same voltage polarity across R_{TH} as in the original circuit. This is the Thevenin circuit.
6. Finally, calculate the current in the Thevenin circuit using Ohm's law.

Figure 43-20

Step 1 Disconnect resistor R_L, and short the voltage source as shown in Figure 43-21. Looking back into terminals A-B, notice that R_1 and R_2 are in parallel. This becomes the Thevenin resistance R_{TH}.

Example 43-6

Problem   A voltage divider circuit is used to supply power to a load as shown in Figure 43-20. Determine the value of current through and the voltage across load resistor R_L, for the circuit values shown, using Thevenin's theorem. One advantage of using Thevenin equivalent circuit is that the effect of different values of R_L can be calculated easily. To illustrate this, determine the value of load current and voltage if R_L were changed to 4 Ω.

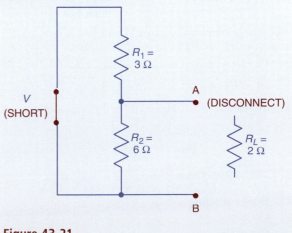

Figure 43-21

$$R_{\text{TH}} = \frac{R_1 \times R_2}{R_1 + R_2} = \frac{3\,\Omega \times 6\,\Omega}{3\,\Omega + 6\,\Omega} = 2\,\Omega$$

Step 2   Insert the voltage source back in the circuit as shown in Figure 43-22. Calculate the current flow in this circuit. Calculate the Thevenin voltage, which is the voltage across resistor R_2 as measured at terminals A-B.

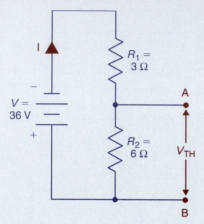

Figure 43-22

$$I = \frac{V}{R_1 + R_2} = \frac{36\,\text{V}}{3\,\Omega + 6\,\Omega} = 4\,\text{A}$$
$$V_{\text{TH}} = I \times R_2 = 4\,\text{A} \times 6\,\Omega = 24\,\text{V}$$

Step 3   The Thevenin equivalent circuit is as shown in Figure 43-23. Using this circuit calculate the current through and the voltage across R_L.

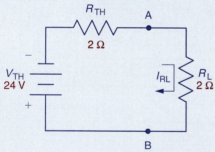

Figure 43-23

$$I_{\text{RL}} = \frac{V_{\text{TH}}}{R_{\text{TH}} + R_L} = \frac{24\,\text{V}}{4\,\Omega} = 6\,\text{A}$$
$$V_{\text{RL}} = I_{RL} \times R_L = 6\,\text{A} \times 2\,\Omega = 12\,\text{V}$$

Step 4   Using the Thevenin equivalent circuit, with R_L changed to 4 Ω, calculate the value of the load current and voltage.

$$I_{RL} = \frac{V_{\text{TH}}}{R_{\text{TH}} + R_L} = \frac{24\,\text{V}}{6\,\Omega} = 4\,\text{A}$$
$$V_{\text{RL}} = I_{RL} \times R_L = 4\,\text{A} \times 4\,\Omega = 16\,\text{V}$$

Example 43-7

Problem   Use Thevenin's theorem to determine the current through and the voltage across R_L, of the dual-voltage circuit shown in Figure 43-24.

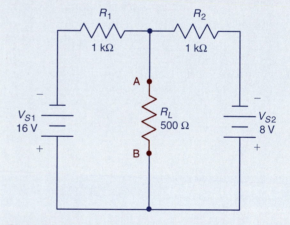

Figure 43-24

Step 1   Remove R_L and short voltage source V_{S2} as shown in Figure 43-25. Calculate the voltage drop at terminals A-B due to source V_{S1} (this will be the same as the voltage drop across R_2). Note the polarity of the voltage at terminals A-B.

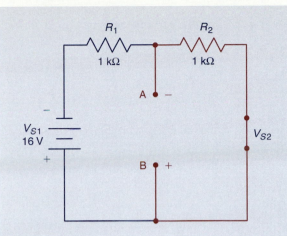

Figure 43-25

$$V'_{\text{A-B}} = \frac{R_2}{R_1 + R_2} \times V_{S1} = \frac{1000\ \Omega}{2000\ \Omega} \times 16\ \text{V} = 8\ \text{V}$$

Voltage at A is negative with respect to B.

Step 2 Short voltage source V_{S1} as shown in Figure 43-26. Calculate the voltage drop at terminals A-B due to source V_{S2} (this will be the same as the voltage drop across R_1). Note the polarity of the voltage at terminals A-B.

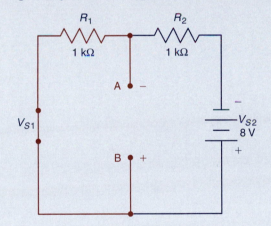

Figure 43-26

$$V''_{\text{A-B}} = \frac{R_1}{R_1 + R_2} \times V_{S2} = \frac{1000\ \Omega}{2000\ \Omega} \times 8\ \text{V} = 4\ \text{V}$$

Voltage at A is negative with respect to B.

Step 3 Calculate the net voltage at terminals A-B by adding the two voltages since they have the same polarity. This is the Thevenin voltage.

$$V_{\text{TH}} = V'_{\text{A-B}} + V''_{\text{A-B}} = 8\ \text{V} + 4\ \text{V} = 12\ \text{V}$$

Step 4 Short both voltage sources as shown in Figure 43-27. Looking back into the A-B terminals and you will see that the Thevenin resistance consists of R_1 and R_2 connected in parallel.

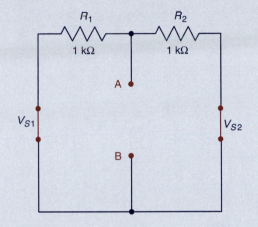

Figure 43-27

$$R_{\text{TH}} = R_1 || R_2 = \frac{1\ \text{k}\Omega \times 1\ \text{k}\Omega}{1\ \text{k}\Omega + 1\ \text{k}\Omega} = 500\ \Omega$$

Step 5 The Thevenin equivalent circuit is as shown in Figure 43-28. Using this circuit calculate the current through and the voltage across R_L.

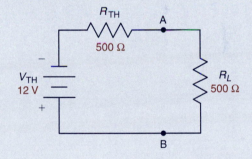

Figure 43-28

$$I_{\text{RL}} = \frac{V_{\text{TH}}}{R_{\text{TH}} + R_L} = \frac{12\ \text{V}}{1000\ \Omega} = 0.012\ \text{A} = 12\ \text{mA}$$

$$V_{\text{RL}} = I_{\text{RL}} \times R_L = 0.012\ \text{A} \times 500\ \Omega = 6\ \text{V}$$

Example 43-8

Problem Use Thevenin's theorem to determine the current through and the voltage across R_L, of the resistance bridge circuit shown in Figure 43-29.

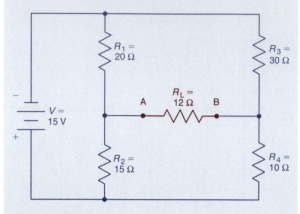

Figure 43-29

Step 1 Remove R_L and short voltage source V as shown in Figure 43-30. The circuit can then be simplified by re-drawing it as shown in Figure 43-31. The equivalent resistance, which is the Thevenin resistance, is then calculated.

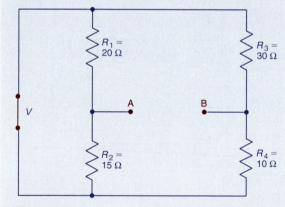

Figure 43-30

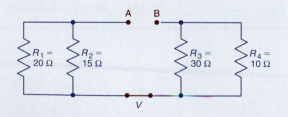

Figure 43-31

$$R_{\mathrm{TH}} = \frac{R_1 \times R_2}{R_1 + R_2} + \frac{R_3 \times R_4}{R_3 + R_4}$$

$$= \frac{20\ \Omega \times 15\ \Omega}{20\ \Omega + 15\ \Omega} + \frac{30\ \Omega \times 10\ \Omega}{30\ \Omega + 10\ \Omega}$$

$$= 8.57\ \Omega + 7.5\ \Omega = 16.07\ \Omega$$

Step 2 Insert the voltage source back into the circuit as shown in Figure 43-32. Use the voltage divider rule to find the value and polarity of the voltage across R_1 and R_3.

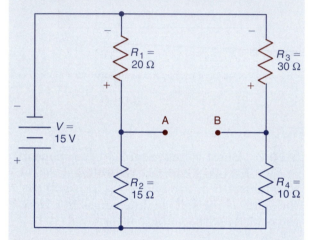

Figure 43-32

$$V_{\mathrm{R1}} = \frac{R_1}{R_1 + R_2} \times V = \frac{20\ \Omega}{20\ \Omega + 15\ \Omega} \times 15\ \mathrm{V}$$

$$= 8.57\ \mathrm{V}$$

$$V_{\mathrm{R3}} = \frac{R_3}{R_3 + R_4} \times V = \frac{30\ \Omega}{30\ \Omega + 10\ \Omega} \times 15\ \mathrm{V}$$

$$= 11.25\ \mathrm{V}$$

Step 3 The potential difference between terminals A and B is the difference in voltage between V_{R1} and V_{R3}. This is also the Thevenin voltage.

$$V_{\mathrm{TH}} = V_{\mathrm{A\text{-}B}} = V_{\mathrm{R1}} - V_{\mathrm{R3}} = 8.57\ \mathrm{V} - 11.25\ \mathrm{V}$$

$$= -2.68\ \mathrm{V}\ (\text{negative sign indicates that}$$
point A is negative with
respect to point B)

Step 4 The Thevenin equivalent circuit is as shown in Figure 43-33. Using this circuit calculate the current through and the voltage across R_L.

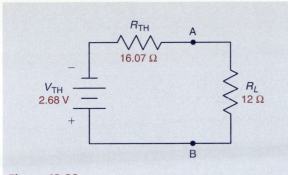

Figure 43-33

$$I_{RL} = \frac{V_{TH}}{R_{TH} + R_L} = \frac{2.68\text{ V}}{16.07\ \Omega + 12\ \Omega} = 95.48\text{ mA}$$

$$V_{RL} = I_{RL} \times R_L = 95.48\text{ mA} \times 12\ \Omega = 1.146\text{ V}$$

43-4 Norton's Theorem

Norton's theorem is used for simplifying a network in terms of currents instead of voltages. In certain cases, analyzing the division of currents may be easier than voltage analysis. Norton's theorem simplifes a resistive network and represents it with a **Norton equivalent current source (I_N)** in parallel with an equivalent **Norton resistance (R_N)** as shown in Figure 43-34. The basis of Norton's theorem is the use of a current source to supply a total load current that is divided among parallel branches.

As with any theorem, certain steps have to be performed to arrive at an equivalent circuit. The following steps are used to convert a resistive network into its Norton equivalent:

1. Calculate the Norton equivalent current source. This is equal to the current that would flow between terminals A-B if the

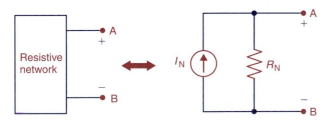

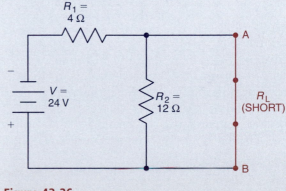

Figure 43-34 The network contained within the block can be reduced to the Norton equivalent parallel circuit.

load resistor was removed and replaced with a short-circuit.
2. Calculate the Norton equivalent resistance. This is equal to the resistance between terminals A-B when the voltage source is removed and replaced with a short-circuit.

Example 43-9

Problem Derive the Norton equivalent circuit for the small resistive network shown in Figure 43-35. Calculate the load current and voltage for the 6 Ω load. Repeat the calculations using a 3 Ω load.

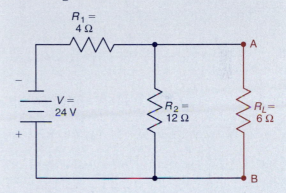

Figure 43-35

Step 1 Short-circuit the load terminals A-B as shown in Figure 43-36. Calculate the resulting current flow. This is the value of the Norton equivalent current source I_N. Note that the short circuit across terminals A-B short-circuits R_L and the parallel R_2. Then the only resistance in the circuit is R_1 in series with the voltage source.

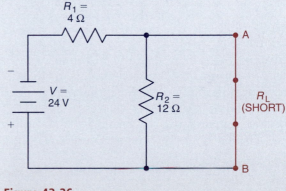

Figure 43-36

$$I_N = V/R_1 = 24\text{ V}/4\ \Omega = 6\text{ V}$$

Step 2 The Norton equivalent resistance (R_N) is equal to the resistance between terminals A-B with the load removed and the voltage source replaced with a short, as shown in Figure 43-37. The resistance seen looking back into terminals A-B is then R_1 in parallel with R_2.

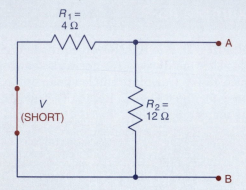

Figure 43-37

$$R_N = \frac{R_1 \times R_2}{R_1 + R_2} = \frac{4\ \Omega \times 12\ \Omega}{4\ \Omega + 12\ \Omega} = 3\ \Omega$$

Step 3 The resultant Norton equivalent circuit is shown in Figure 43-38. It consists of a 6 A source (I_N) shunted by a 3 Ω resistance (R_N).

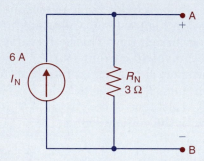

Figure 43-38

Step 4 To calculate I_L and V_L connect the load resistor to the equivalent Norton circuit as shown in Figure 43-39. The current source still delivers 6 A, but now that current is divided between the two branches of R_N and R_L. The load current can be calculated using the current divider rule and the load voltage can be calculated using Ohm's law.

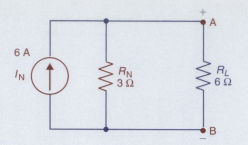

Figure 43-39

$$I_L = \frac{R_N}{R_N + R_L} \times I_N = \frac{3\ \Omega}{3\ \Omega + 6\ \Omega} \times 6\text{ A} = 2\text{ A}$$
$$V_L = I_L \times R_L = 2\text{ A} \times 6\ \Omega = 12\text{ V}$$

Step 5 When the value of R_L is changed the values of I_N and R_N remain the same. Therefore the load current and voltage for a 3 Ω load can be determined without recalculating the entire circuit.

$$I_L = \frac{R_N}{R_N + R_L} \times I_N = \frac{3\ \Omega}{3\ \Omega + 3\ \Omega} \times 6\text{ A} = 3\text{ A}$$
$$V_L = I_L \times R_L = 3\text{ A} \times 3\ \Omega = 9\text{ V}$$

Example 43-10

Problem Drive the Norton equivalent circuit for the resistive network shown in Figure 43-40.

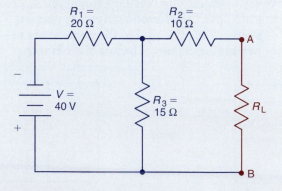

Figure 43-40

Step 1 Short circuit the load, as shown in Figure 43-41. Note that the short circuit current I_N in this example is a branch current not a main line current as in the previous example.

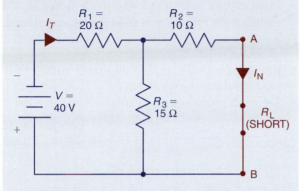

Figure 43-41

Resistors R_2 and R_3 are now in parallel so the total resistance seen by the voltage is:

$$R_T = \frac{R_2 \times R_3}{R_2 + R_3} + R_1$$

$$= \frac{10\ \Omega \times 15\ \Omega}{10\ \Omega + 15\ \Omega} + 20\ \Omega = 26\ \Omega$$

The total current is then:

$$I_T = V_T/R_T = 40\ \text{V}/26\ \Omega = 1.54\ \text{A}$$

The Norton equivalent current is then found by using the current divider rule.

$$I_N = \frac{R_3}{R_2 + R_3} \times I_T$$

$$= \frac{15\ \Omega}{10\ \Omega + 15\ \Omega} \times 1.54\ \text{A} = 0.924\ \text{A}$$

Step 2 Short circuit the voltage source, as shown in Figure 43-42, to find the Norton equivalent resistance of the circuit. R_1 and R_3 are in parallel, and R_2 is in series with this parallel connection.

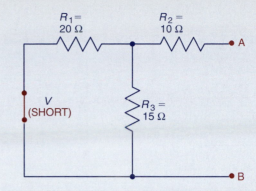

Figure 43-42

$$R_N = \frac{R_1 \times R_3}{R_1 + R_3} + R_2$$

$$= \frac{20\ \Omega \times 15\ \Omega}{20\ \Omega + 15\ \Omega} + 10\ \Omega = 18.57\ \Omega$$

Step 3 The Norton equivalent circuit is shown in Figure 43-43.

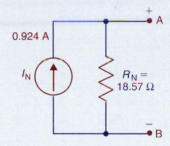

Figure 43-43

The Norton equivalent circuit may also be determined directly from the Thevenin equivalent circuit, and vice-versa, by using the source conversion technique developed at the beginning of this chapter. Figure 43-44 illustrates the relationship between the two circuits. The following formulas can be used to convert from one equivalent circuit to the other.

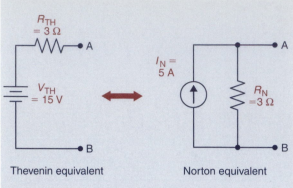

Thevenin equivalent   Norton equivalent

Figure 43-44 Norton-Thevenin conversions.

From Thevenin to Norton

$$R_N = R_{TH}$$
$$I_N = V_{TH}/R_{TH}$$

From Norton to Thevenin

$$R_{TH} = R_N$$
$$V_{TH} = I_N \times R_N$$

Review and Applications

···

Related Formulas

$$V_{\text{OUT}} = \frac{R_L}{R_L + R_S} \times V_S$$
(voltage-divider formula)

$$I_L = I \times \frac{R_S}{R_L + R_S}$$ (current-divider formula)

$$I_I' = \frac{V_{S1}}{R_T}$$ (Ohm's law)

$$I_1'' = \frac{V_{S2}}{R_T}$$

$$V = I \times R$$

$$R = V/R$$

$$R_T = \frac{R_1 \times R_2}{R_1 + R_2}$$ (two resistors in parallel)

$$R_T = \frac{1}{1/R_1 + 1/R_2 + 1/R_3}$$
(three or more resistors in parallel)

$$R_T = R_1 + R_2 + R_3$$ (resistors in series)

$$R_N = R_{TH}$$

$$I_N = V_{TH}/R_{TH}$$

$$V_{TH} = I_N \times R_N$$

Review Questions

1. Compare the operating characteristics of an ideal voltage source with that of a real voltage source.
2. A constant-current source has a rating of 10 A. Explain what this implies?
3. State the superposition theorem.
4. Outline how to calculate V_{TH} and R_{TH} in Thevenin equivalent circuits.
5. Compare the method used in simplifying circuits using the Thevenin and Norton theorems.
6. Outline how to calculate I_N and V_N in Norton equivalent circuits.

Problems

1. Refer to Figure 43-2. Calculate the value of the output of the voltage source with a 5 Ω load resistor connected.
2. Refer to Figure 43-3B. Assume the value of R_S is 10 Ω and the value of R_L is 40 Ω. Calculate the value of the current flow through each.
3. Refer to Figure 43-5. Assume R_2 is changed to 6 Ω. Calculate the current through and the voltage across R_2 using the superposition theorem.
4. Refer to Figure 43-16. Assume R_1 is changed to 5 Ω. Calculate the current

through and the voltage across R_1 using the superposition theorem.
5. Refer to Figure 43-20. Calculate the R_{TH} and the V_{TH} assuming the value of the voltage source is changed to 12 V.
6. Refer to Figure 43-29. Calculate the voltage across and the current through R_L assuming its resistance value is changed to 36 Ω.
7. Refer to Figure 43-35. Calculate the R_N and the I_N assuming the value of the voltage source is changed to 36 V and the value of R_1 is changed to 2 Ω.

Critical Thinking

1. A 10 Ω resistor is connected across a constant-current source of 2 A.
   (a) How much current would flow through the resistor?
   (b) How much current would flow through the resistor if its value were changed to 5 Ω? Why?
2. A voltage source of 48 V with an internal series resistance of 10 Ω is to be converted into an equivalent current source.
   (a) What would the current rating of the current source be?
   (b) What would be the value of the parallel connected R_S resistor?

3. Refer to Figure 43-9. Assume voltage source V_{S1} is changed to 100 V. Determine the value of current I_1, I_2, and I_3.
4. Refer to Figure 43-13. Determine the value of the current through resistor R_1.
5. Refer to Figure 43-24. Assume the polarity of voltage source V_{S2} is reversed.
   Determine:
   (a) V_{TH} (b) R_{TH} (c) V_{RL} (d) I_{RL}
6. Refer to Figure 43-40. Assume resistor R_3 becomes open. Determine the new value of I_N and R_N.

Portfolio Projects

- Design a dual-voltage resistive network circuit that can be used to verify the superposition theorem. Calculate the current and voltage for a key resistor. Breadboard the circuit and measure the actual voltage and current of the key resistor.
- Design a single-source resistive network circuit that can be used to verify the Thevenin theorem. Calculate the current and voltage point A-B of the load. Breadboard the circuit and measure the actual voltage and current of the load. Repeat using a different value of load resistor.

Circuit Challenge

Using a Simulator

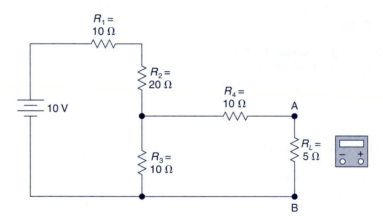

Procedure

(a) For the resistive network shown calculate V_{TH}, R_{TH}, V_{RL}, and I_{RL}.
(b) Construct the circuit shown, using whatever simulation software package is available to you. Record the measured value of V_{TH}, R_{TH}, V_{RL}, and I_{RL} for the simulated circuit.

Principles of AC Circuits

Alternating current (ac) is the most widely used type of electricity and plays a fundamental role in most applications. In this chapter, we will conduct a more indepth study of ac circuits. The focus will be on ac circuit analysis techniques, including ac versions of analysis used for dc circuits. The discussion of alternating current in this chapter is limited to sinusoidal waveforms.

Objectives

After studying this chapter, you should be able to:
1. Understand the relationship between resistance, reactance, and impedance.
2. Analyze the operation of series RLC circuits.
3. Analyze the operation of parallel RLC circuits.
4. Explain the function and operation of basic filter circuits.

Key Terms

- impedance (Z)
- inductance reactance (X_L)
- capacitive reactance (X_C)
- reactance (X)
- vector
- phasors
- lagging
- leading

- Ohm's law for ac circuits
- *RL* circuit
- *RC* circuit
- *RLC* circuit
- in-phase
- out-of-phase
- Sin
- Cos
- Tan

- resonant frequency
- apparent power (VA)
- true power (W)
- reactive power (VAR)
- power factor
- filter
- passive filter
- active filter
- low-pass filter
(*continued*)

RESEARCH
INTERNET
SITES

- Glencoe
- Amazon
- Borders

- high-pass filter
- band-pass filter
- band-stop filter
- cut-off frequency
- *RC* filter
- series resonant filter
- Q (quality)
- bandwidth
- attenuation
- Bode plot

44-1 Impedance in AC Series Circuits

The analysis of direct-current circuits is relatively simple because in these circuits the voltage and current are steady. The analysis of *alternating-current* (ac) circuits is more complex because of the fluctuating characteristics of alternating current. In previous chapters, ac amplitude variations of voltage and current were analyzed by use of effective or rms values. However, for more detailed analyses the time element introduced by inductance and capacitance must be considered. In ac circuits containing resistance, only the voltage and current were in phase (rising and falling at the same time). In purely inductive circuits, the voltage leads the current by 90 degrees and in purely capacitive circuits the current leads the voltage by 90 degrees.

In a direct-current circuit only the resistance of the circuit opposes current flow. In an alternating-current circuit, not only resistance, but also inductive reactance and capacitive reactance oppose current flow. The total opposition to current flow in an ac circuit is called *impedance* (Z).

The impedance Z of a circuit is expressed in ohms. The unit of resistance is the ohm, the unit of inductance is the henry, and the unit of capacitance is the farad. In order to make a simple conversion of the values of inductance and capacitance to ohms, it must be assumed that current and voltage are sinusoidal waveforms. Based on this assumption, the *inductive reactance,* (X_L) in ohms is:

$$X_L = 2\pi f L$$

where f is frequency in H_z and L is inductance in henrys. Likewise, capacitive reactance, (X_C) in ohms is:

$$X_C = \frac{1}{2\pi f C}$$

where f is the frequency in H_z and C is capacitance in farads.

Note that the ohmic values of these reactances depend upon the frequency of the ac. While a resistive element has a fixed resistance independent of the frequency, the same is not true of reactance. The ohmic values of the reactances vary as a function frequency.

In ac circuits that contain both reactance and resistance, the two combine to oppose current. For example, assume you want to find the impedance of a series combination consisting of a 40 Ω resistance and a 30 Ω inductive reactance as shown in Figure 44-1A. Recalling our study of Ohm's law for a dc series circuit, it initially appears feasible to determine the value of the impedance by simply adding R and X_L. This is incorrect because of the phase relationship between resistance and reactance. In an ac circuit that contains only resistance, the current and the voltage will be in phase, reaching their maximum values at the same time. In an ac circuit containing only inductive reactance, the current lags the voltage by $1/4$ of a cycle (90°). The effects of R and X_L are, therefore, 90° apart. The correct method of determining the impedance (Z) is to consider the effects of R and X_L to be 90° apart, or at right angles to each other. The combined impedance can be represented as the hypotenuse of the right-angle triangle whose sides are R and X_L (Figure 44-1B).

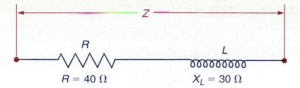

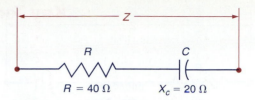

A. Resistor and inductor in series.

A. Resistor and capacitor in series.

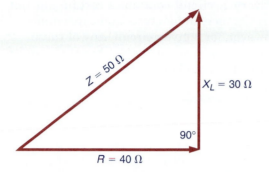

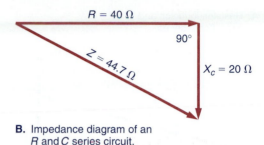

B. Impedance diagram of an R and L series circuit.

B. Impedance diagram of an R and C series circuit.

Figure 44-1 Impedance of a resistor and inductor in series.

Figure 44-2 Impedance of a resistor and capacitor in series.

This is also referred to as an impedance diagram. The impedance Z of this combination is then found by adding R and X_L vectorially as follows:

$$Z = \sqrt{R^2 + X_L^2}$$
$$= \sqrt{40^2 + 30^2}$$
$$= \sqrt{1600\ \Omega + 900\ \Omega}$$
$$= \sqrt{2500\ \Omega}$$
$$= 50\ \Omega$$

When dealing with *capacitive reactance* instead of inductive reactance a similar approach is taken. For example, assume you want to find the impedance of a series combination consisting of a 40 Ω resistance and a 20 Ω capacitive reactance as shown in Figure 44-2A. In this case, the current leads the voltage in the capacitor. It is customary to draw the line or vector representing the capacitive reactance in a downward direction as shown in Figure 44-2B. The line is drawn downward for X_C to indicate that it acts in a direction opposite to X_L. The impedance of this combination is:

$$Z = \sqrt{R^2 + X_C^2}$$
$$= \sqrt{40^2 + 20^2}$$
$$= \sqrt{1600\ \Omega + 400\ \Omega}$$

$$= \sqrt{2000\ \Omega}$$
$$= 44.7\ \Omega \text{ (approximately)}$$

Next, let us determine Z for a circuit consisting of all three elements R, L, and C connected in series as shown in Figure 44-3A. The effect of inductive reactance is to cause the current to lag the voltage, while that of capacitive reactance is to cause the current to lead the voltage. Since X_L and X_C are exactly opposite, they tend to cancel each other out, with the combined net effect then equal to the difference between their vector values (Figure 44-3B). This resultant, known as the equivalent *reactance*, is represented by the symbol X and expressed by the equation: $X = X_L - X_C$. The impedance of this combination:

$$Z = \sqrt{R^2 + (X_L - X_C)^2}$$
$$= \sqrt{40^2 + 10^2}$$
$$= \sqrt{1600\ \Omega + 100\ \Omega}$$
$$= \sqrt{1700\ \Omega}$$
$$= 41.2\ \Omega \text{ (approximately)}$$

Note that, as compared to the *RL* series circuit of Figure 44-1, where the impedance Z was equal to 50 Ω, the addition of C as in the *RLC* circuit has actually resulted in a reduction of Z to 41.2 Ω. Also, this *RCL* circuit is said to be inductive because the inductive reactance of

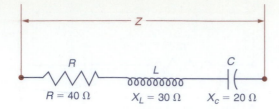

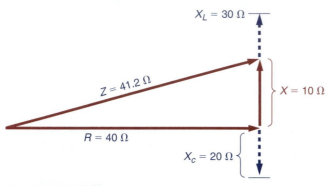

A. Resistor, inductor, and capacitor in series.

B. Impedance diagram.

Figure 44-3 Solving for Z in an RLC series circuit.

30 Ω is larger than the capacitive reactance of 20 Ω and therefore, cancels out the effects of the capacitive reactance. Since the circuit is inductive the voltage leads the current. The reverse is also true! When X_C is greater than X_L, the larger capacitive reactance cancels out the smaller inductive reactance and its effect. The circuit is then considered to be capacitive and the current leads the voltage.

Ohm's law for ac circuits may also be used to find the impedance Z for a series circuit (Figure 44-4). In applying Ohm's law to an ac

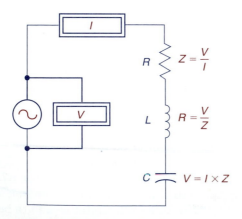

Figure 44-4 Ohm's law for ac circuits.

circuit, Z is substituted for R in the formulas as follows:

$$V = IZ$$
$$I = V/Z$$
$$Z = V/I$$

Every ac circuit contains a certain amount of resistance, inductance, and capacitance. However, for a given circuit, any of these factors may be so negligible that they can be disregarded, and only the significant factors considered.

44-2 Current and Voltage in AC Series Circuits

In an ac series circuit, as in a dc series circuit, there is only one path for current flow. This is true regardless of the type of ac circuit: *RL, RC,* or *RLC*. Since there is only one path the current flow is exactly the same in all parts of the circuit at any time. For this reason, the current is used as the common reference for ac series circuits (Figure 44-5). All phase angles are normally measured with respect to the circuit current (Figure 44-6). When the impedance and applied voltage of an ac series circuit are known, the circuit is calculated by dividing the voltage by the impedance.

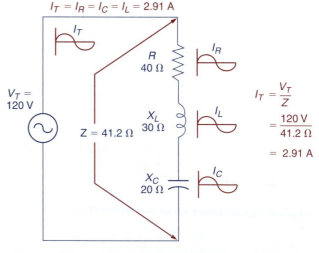

Figure 44-5 Finding the current when the impedance and voltage are known.

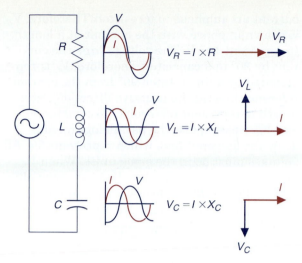

Figure 44-6 *RLC* series circuit voltage.

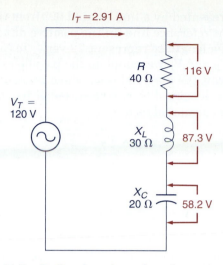

Figure 44-7 Finding the voltage drops in a series *RLC* circuit.

Once you know the value of current flowing through a series circuit, you can calculate the voltage across each element by multiplying the current by the ohmic value of the individual element. For the circuit of Figure 44-7 the voltages would be calculated as follows:

$$V_R = I \times R$$
$$= 2.91 \text{ A} \times \Omega$$
$$= 116 \text{ V}$$
$$V_L = I \times X_L$$
$$= 2.91 \text{ A} \times 30 \text{ }\Omega$$
$$= 87.3 \text{ V}$$
$$V_C = I \times X_C$$
$$= 2.91 \text{ A} \times 20 \text{ }\Omega$$
$$= 58.2 \text{ V}$$

Note that if you add the voltage drops of Figure 44-7, the total would be 262 volts, whereas V_T is only 120 volts. Conventional addition cannot be used to combine voltages or currents that are not in phase. Vectors, such as those previously used in the calculation of impedance, provide a simple means of adding out-of-phase voltages. This type of vector analysis is used extensively in studying the behavior of alternating current circuits.

A ***vector*** is a straight line used to represent the magnitude and direction of a given quantity. Vectors of electrical quantities are also known as ***phasors***. The length of the line drawn to scale represents magnitude. An arrow

at one end of the line indicates the direction. The angle of the vector or phasor from a reference line corresponds to the angle or phase of the voltage or current being represented. Vectors may be rotated like spokes of a wheel. Positive rotation is counterclockwise and generates positive angles.

As an example, let us combine the voltages of V_R and V_L in the ***RL*** series ***circuit*** of Figure 44-8. In a series circuit, current is used as the reference, so a horizontal reference line representing current is drawn first. To combine V_R and V_L, their vectors are drawn relative to the current reference line. The voltage and current are in phase in a resistor, therefore, V_R is drawn in phase with the current and of a length to represent 96 volts. The voltage ***leads*** the current by 90° in an inductor, therefore, V_L

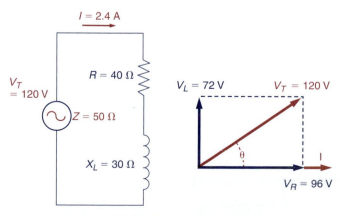

Figure 44-8 Combining voltages in a series *RL* circuit.

44-2 Current and Voltage in AC Series Circuits **695**

is represented by a line rotated 90° from the current reference line (in the positive direction), and of a length to represent 72 volts. To combine V_R and V_L we complete the parallelogram and draw the diagonal line, which represents V_T. The length of V_T can be scaled or its value can be calculated as follows:

$$V_T = \sqrt{V_R{}^2 + V_L{}^2}$$
$$= \sqrt{96^2 + 72^2}$$
$$= \sqrt{14400}$$
$$= 120 \text{ V}$$

Note that the answer of 120 V agrees with the total applied voltage of 120 V. The circuit phase angle θ between the total applied voltage V_T and circuit current can be measured from the scaled vector diagram using a protractor or calculated using the following trigonometric functions:

$$\theta = \textbf{Sin}^{-1} = \frac{\text{side opposite}}{\text{hypotenuse}} = \frac{V_L}{V_T} = \frac{72 \text{ V}}{120 \text{ V}}$$
$$= 0.6 = 36.9°$$

$$\theta = \textbf{Cos}^{-1} = \frac{\text{side adjacent}}{\text{hypotenuse}} = \frac{V_R}{V_T} = \frac{96 \text{ V}}{120 \text{ V}}$$
$$= 0.8 = 36.9°$$

$$\theta = \textbf{Tan}^{-1} = \frac{\text{side opposite}}{\text{side adjacent}} = \frac{V_L}{V_R} = \frac{72 \text{ V}}{96 \text{ V}}$$
$$= 0.75 = 36.9°$$

Figure 44-9 shows how voltages are combined in a series **RC circuit**. Since V_R and V_C are not in phase with each other they must be added vectorially. The horizontal reference line representing current is drawn first because current is the same in a series circuit. The voltage and

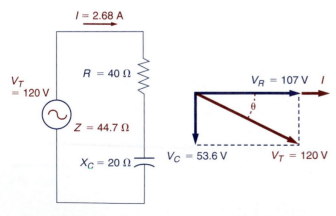

Figure 44-9 Combining voltages in a series *RC* circuit.

current are in phase in a resistor. Therefore, V_R is drawn in phase with the current at a length to represent 107 V. The voltage **lags** the current by 90° in a capacitor. Therefore, V_C is represented by a line rotated 90° from the current reference line (in the negative direction), at a length to represent 53.6 volts. To combine V_R and V_C, we complete the parallelogram and draw the diagonal line, which represents V_T. All voltages must be in the same units. When V_R and V_C are *rms* values, then V_T is an rms value. In calculating the value of V_T, first square V_R and V_C, then add and take the square root:

$$V_T = \sqrt{V_R{}^2 + V_C{}^2}$$
$$= \sqrt{107^2 + 53.6^2}$$
$$= \sqrt{14322}$$
$$= 120 \text{ V}$$

Note that the circuit phase angle between the applied voltage V_T and the current is negative, clockwise from the current reference line. This phase angle θ can then be calculated as follows:

$$\theta = \text{Tan}^{-1} = \frac{\text{side opposite}}{\text{side adjacent}} = \frac{V_C}{V_R} = \frac{53.6 \text{ V}}{107 \text{ V}}$$
$$= 0.501 = -26.6°$$

The negative sign means the angle is clockwise from zero, to indicate that V_T lags behind the leading *I*.

Next, let us combine the voltages V_R, V_L, and V_C, for the series *RLC* circuit shown in Figure 44-10. Once again, we start with the current reference line and draw V_R in phase with it at a length to represent 116 volts. V_L is represented by a line rotated 90° in the positive direction with its length proportional to 87.3 volts. V_C is represented by a line rotated 90° in the negative direction with its length proportional to 58.2 volts. Note that V_L and V_C are 180° out-of-phase with each other. For this reason, V_L and V_C subtract from each other, resulting in a net combined voltage V_X of plus 29.1 volts. To combine V_R and V_X, we complete the parallelogram and draw a diagonal line, which represents V_T. The applied voltage and phase angle can be calculated using the following equations:

$$V_T = \sqrt{V_R{}^2 + (V_L - V_C)^2}$$
$$= \sqrt{116^2 + (87.3 - 58.2)^2}$$
$$= \sqrt{116^2 + 29.1^2}$$

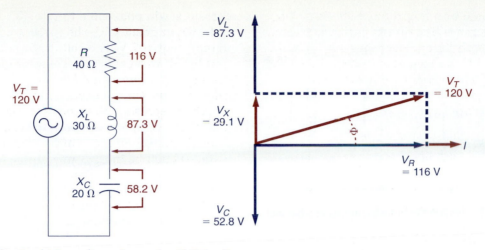

Figure 44-10 Combining voltages in a series *RLC* circuit.

$= 120 \text{ V}$

$$\theta = \text{Tan}^{-1} = \frac{\text{side opposite}}{\text{side adjacent}} = \frac{V_X}{V_R} = \frac{29.1 \text{ V}}{116 \text{ V}}$$

$= 0.251 = 14.1°$

A series ***RLC circuit*** will act inductively when the inductive reactance is greater than the capacitive reactance and capacitively when the reverse is true. The circuit will act purely resistive when the inductive reactance and capacitive reactance are equal. There is only one frequency at which X_L can be equal to X_C. You may recall that this is known as the *resonant frequency* and calculated using the *resonant frequency* equation:

$$f_r = \frac{1}{2\pi \sqrt{LC}}$$

In a series *RLC* circuit the voltage across the reactive elements can be much *higher* than the applied total voltage. This seemingly impossible condition is caused by the interaction between the capacitor and inductor. As an example, consider the series resonant *RLC* circuit of Figure 44-11. The voltage across the inductor and capacitor is 1200 V while the applied voltage is only 120 V. In this circuit, X_L cancels out X_C so the only limitation on the current is the 10-ohm resistance. The current is therefore 12 A, which flows in both reactive elements producing a 1200 V drop across each. The two reactive voltages cancel leaving the resistor voltage equal to the applied 120 V.

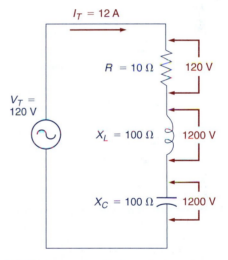

Figure 44-11 In a series *RLC* circuit the voltage across the reactive elements can be much higher than the voltage applied to the circuit.

44-3 Parallel AC Circuits

In a parallel dc circuit, the voltage across each of the parallel branches is equal. This is also true of the ac parallel circuit. The voltages across each parallel branch are the same value, equal in value to the total voltage V_T, and are all in phase with each other. Since voltage is the common quantity in parallel ac circuits, it is used as the reference for all phase relations.

As an example, let us analyze the parallel *RLC* circuit of Figure 44-12. Using Ohm's law,

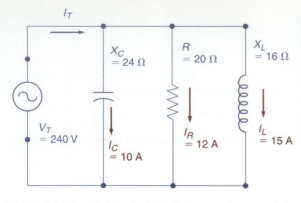

Figure 44-12 Finding the branch currents in a parallel *RLC* circuit.

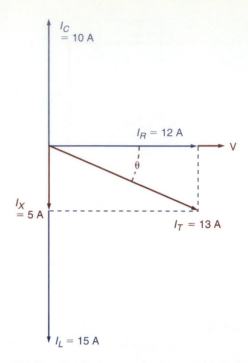

Figure 44-13 Combining currents in a parallel *RLC* circuit.

we can calculate the current through each branch as follows:

$$I_C = \frac{V_C}{X_C} = \frac{240 \text{ V}}{60 \text{ }\Omega} = 10 \text{ A}$$

$$I_R = \frac{V_R}{R} = \frac{240 \text{ V}}{40 \text{ }\Omega} = 12 \text{ A}$$

$$I_L = \frac{V_L}{X_L} = \frac{240 \text{ V}}{50 \text{ }\Omega} = 15 \text{ A}$$

The currents are all **out-of-phase** and, therefore, cannot be combined by direct addition. Current I_C leads the voltage by 90°, current I_R is in *phase* with the voltage, and current I_L lags the voltage by 90°. The vector diagram may be constructed in a manner similar to that used for the series *RLC*, except that the diagram now consists of current vectors. Figure 44-13 shows the relationship between them in the circuit. The reference line *V* is drawn with the current through the resistor I_R in phase with it. Because the inductive current lags the applied voltage, current I_L is drawn 90° in a negative clockwise direction from reference line *V*. Since the capacitive current leads the applied voltage, current I_C is drawn 90° in a positive counter-clockwise direction from reference line *V*. Note that I_C and I_L are 180° out-of-phase with each other. For this reason, I_C and I_L subtract from each other, resulting in a net combined current I_X of minus 5 A. To combine I_R and I_X, we complete the parallelogram and draw a diagonal line, which represents I_T. The total current and phase angle can be calculated using the following equations:

$$I_T = \sqrt{I_R{}^2 + I_X{}^2}$$

$$= \sqrt{12^2 + 5^2}$$

$$= \sqrt{169}$$

$$= 13 \text{ A}$$

$$\theta = \text{TAN}^{-1} = \frac{\text{side opposite}}{\text{side adjacent}} = \frac{I_X}{I_R} = \frac{5 \text{ A}}{12 \text{ A}}$$

$$= 0.417 = -22.6°$$

The impedance Z of this circuit can be calculated by applying Ohm's law for ac circuits as follows:

$$Z = \frac{V_T}{I_T} = \frac{240 \text{ V}}{13 \text{ A}} = 18.5 \text{ }\Omega$$

A parallel *RLC* circuit will act inductively when the inductor branch current is greater than the capacitor branch current, and capacitively when the reverse is true. The circuit will act purely resistive when the inductor and capacitor branch currents are equal. There is only one frequency at which I_L can be equal to I_C. This is known as the *resonant frequency* and is calculated using the resonant frequency equation.

You may recall from our study of power in ac circuits (Chapter 30) that the apparent power supplied to an ac circuit containing both

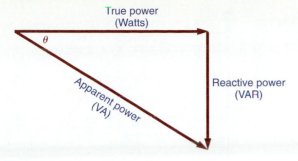

Figure 44-14 Power triangle.

resistance and reactance, consists of true power and reactive power. Figure 44-14 shows how the relationship between apparent power, true power, and reactive power may be displayed graphically using a right triangle.

The *apparent power* is the total power that the entire ac circuit is apparently using and is expressed in volt-ampere (**VA**) units. Thus the apparent power of the original *RLC* parallel circuit can be calculated as follows:

$$\text{Apparent power} = V_T \times I_T = 240 \text{ V} \times 13 \text{ A}$$
$$= 3120 \text{ VA}$$

Since the true power of pure inductive and capacitive circuits is zero, the only *true power* consumed by this circuit is that in the resistance branch. Thus the true power of this circuit can be calculated as follows:

$$\text{True power} = V_R \times I_R = 240 \text{ V} \times 12 \text{ A}$$
$$= 2880 \text{ W}$$

Since the net reactive current of this circuit is 5 amperes, the volt-amperes *reactive power* (*VAR*) can be calculated as follows:

$$\text{Reactive power} = V_X \times I_X = 240 \text{ V} \times 5 \text{ A}$$
$$= 1200 \text{ VAR}$$

The *power factor* is the ratio of true power to apparent power. Power factor is also equal to cosine of the phase angle θ. For this reason, the circuit phase angle is often referred to as the power factor angle. In an inductive circuit, where the current lags the voltage, the power factor is said to be a lagging power factor. If the current leads the voltage, the power factor is called a leading power factor. The value of the power factor depends on how much the current and voltage are out of phase with each other. The power factor for this parallel circuit is:

$$PF = \text{Cos } \theta = \text{Cos } 22.6° = 0.9232 \text{ lag PF}$$
$$= 92.32\% \text{ lag PF}$$

The power factor is an excellent indication of the relative amounts of resistance and reactance in a given circuit. In a purely resistive circuit, the true power and the apparent power are equal. Therefore, the power factor will be equal to 100%. The greater the power factor, the more resistive the circuit. More importantly, the power factor is an indication of the efficiency of a circuit. In any circuit, work is performed only by the current that flows through the resistive portion of the circuit. You may recall from our previous studies how the power factor of a circuit can be improved to make the circuit operate more efficiently. For example, the power factor of an inductive circuit can be corrected by connecting the proper value of capacitance in parallel with the original circuit.

Apparent power, reactive power, and true power are all products of voltage multiplied by current. The significance of this is that procedures used to calculate power factors are the same for both series and parallel circuits. Power factor is always the ratio of true power to apparent power, which is also the cosine of the phase angle.

44-4 Filter Circuits

A common use for reactive components in the field of electronics is filtering. A *filter* may either eliminate or pass certain frequencies. Filter circuits are used extensively in communication systems. They can separate desired signals from undesired signals, block interfering signals, enhance speech and video, and alter signals in other ways.

Filters are either passive or active. **Passive filters** use resistors, capacitors, and inductors, and do not contain an amplifying device. Passive filters never have an output equal to or greater than the input because the filter always has some power loss due to the resistance in the circuit. This is known as the insertion loss. An **active filter** uses some type of amplifying device, such as an operational amplifier in combination with resistors and capacitors to obtain the desired filtering effect with no insertion loss.

Filters can be classified into four main types depending on which frequency components of the input signal are passed on to the output signal. The four types of filters are low-pass, high-pass, band-pass, and band-stop:

- A *low-pass filter* is one that freely passes all frequencies from zero up to a certain cutoff value to pass through it and on to the load. Higher frequencies are blocked or attenuated.
- A *high-pass filter* is one that rejects all frequencies from zero up to a certain limit, but freely passes all higher frequencies.
- A *band-pass filter* is one that accepts all frequencies between two limits and blocks all frequencies above and below these limits.
- A *band-stop filter* is one that stops, or rejects, a specific band of frequencies from passing to a load. Also known as a band-rejection filter it blocks all frequencies between certain limits, but freely passes all frequencies above and below these limits.

Figure 44-15 illustrates the simplest form of a low-pass filter in an *RC* network. The input voltage is applied across the resistor and capacitor in series. The output voltage is taken across the capacitor. The voltage division ratio depends upon the value of resistance (R) and capacitive reactance (X_C). Assume that the input voltage has a fixed rms value, but that its frequency can be varied. When the frequency changes the value of R remains constant, but that of X_C changes (i.e., decreases as frequency increases). At low frequencies, X_C is greater than R, and most of the input signal will appear at the output. At very high frequencies, X_C is much less than R, and practically no voltage will appear across the output.

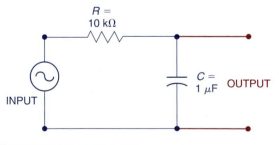

Figure 44-15 Low-pass filter.

Figure 44-16 illustrates the frequency response curve for the low-pass filter. This curve shows the amount of output voltage with

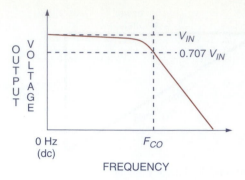

Figure 44-16 Low-pass filter frequency response.

respect to frequency. With a frequency of zero Hz or dc, the capacitor offers maximum opposition and the output voltage is equal to the input voltage. As the frequency increases, X_C begins to decrease, and the output voltage begins to drop off. At the *cut-off frequency* (f_{CO}), the output voltage is equal to approximately 70.7% of the input voltage. After the cut-off frequency is reached, the output voltage drops off at a constant rate. The cut-off frequency is a function of the resistor and the capacitor values. At the cut-off frequency, the capacitive reactance will equal the value of the resistance. The formula used to calculate the cut-off frequency is as follows:

$$f_{co} = \frac{1}{2\pi RC}$$
$$= \frac{1}{2\pi(10 \text{ k}\Omega \times 1 \text{ }\mu\text{F})}$$
$$= \frac{1}{2\pi(1 \times 10^4)(1 \times 10^{-6})}$$
$$= \frac{1}{6.28 \times 10^{-2}}$$
$$= 15.9 \text{ Hz}$$

This filter circuit will pass all frequencies below 15.9 Hz and will attenuate the frequencies above 15.9 Hz.

The placement of the capacitor in the circuit determines if the filter is a high-pass or low-pass. A simple *RC* high-pass *filter* is shown in Figure 44-17. Like the low-pass filter, it consists of a resistor and a capacitor connected to the input voltage. In the high-pass filter, however, the output voltage is taken across the resistor. At high input frequencies, the capacitive reactance will be very low. Thus the capaci-

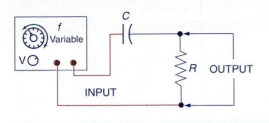

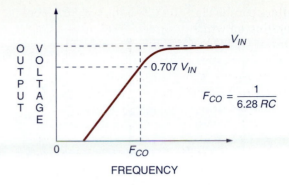

$$F_{CO} = \frac{1}{6.28\,RC}$$

Figure 44-17 High-pass filter.

tor acts like a short for higher frequencies to the resistor. Lower frequencies are blocked proportionally to the value of reactance in ohms. Note that at high frequencies, the output voltage is nearly equal to the input voltage. As the frequency decreases, the output voltage begins to decrease. At the cut-off frequency, the output voltage is approximately 70.7% of the input voltage. Below the cut-off frequency, **attenuation** increases and the output voltage drops accordingly.

Figure 44-18 illustrates a **series resonant band-pass filter**. The filter is the series resonant circuit composed of L and C. R_L is the load to which the output voltage is applied. You may recall from our previous study of the series resonance circuit (Chapter 30) that at the resonance frequency, the series resonant circuit has a very low impedance. Thus, it drops very little of the applied input voltage. Most of the voltage is developed across R_L. Therefore, at the resonant frequency, most of the input voltage will appear across R_L. Below the resonant frequency, the X_C of the capacitor is higher than the resistance of R_L and drops most of the input voltage. Above the resonant frequency, the X_L of

the inductor is higher than the resistance of R_L and drops most of the input voltage.

A band-pass filter is equivalent to combining a low-pass filter and a high-pass filter. The resulting response curve would appear as shown in Figure 44-19. The series resonant band-pass filter has two cut-off points f_1 and f_2. The bandwidth of a band-pass filter is the range of frequencies for which the output voltage is equal to or greater than 70.7% of its value at the resonant frequency. Band-pass filters are normally designed to have a very sharp, defined frequency response. The cut-off frequency of the high-pass section becomes the lower frequency limit (f_1) in the pass-band. The upper frequency limit in the pass-band (f_2) is the result of the cutoff frequency in the low-pass section. Other names for f_1 and f_2 are: -3dB frequencies, critical frequencies, band frequencies, and half-power frequencies. The bandwidth is the difference between f_2 and f_1.

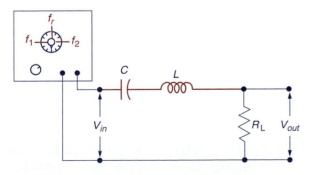

Figure 44-18 Series resonant band-pass filter.

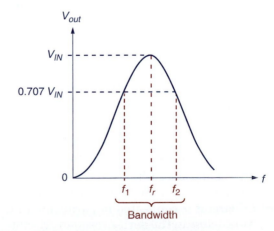

Figure 44-19 Typical response curve of a series resonant band-pass filter.

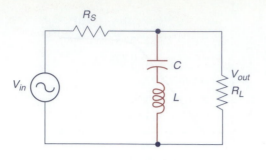

Figure 44-20 Band-stop filter.

At the resonant frequency (fr), $X_L = X_C$, and the output voltage is equal to the input signal.

The Q is the measure of **quality** of a series resonant band-pass filter. The higher the Q of a circuit, the sharper its frequency response curve, the greater its output voltage, and the narrower its bandwidth. Normally a high Q is desirable in band-pass filters. To obtain a high Q, the resistance R_L must be low in value and inductive reactance X_L must be high at the resonant frequency. The Q can be calculated in two different ways.

- Determining circuit Q with voltage ratio. The ratio of the voltage across the inductor is compared to that of the input, at resonance, as follows:

$$Q = \frac{V_L \text{ (at resonance)}}{V_{\text{INPUT}}}$$

- Determining circuit Q using X_L and X_C ratios. The ratio of the inductive reactance is compared to the resistance of the circuit as follows. If the circuit does not have a series resistor, then use the dc resistance of the inductor. An inductor will have some dc resistance due to its wire construction.

$$Q = \frac{X_L}{R_L}$$

The response of a band-stop filter is opposite that of the band-pass filter. Figure 44-20 shows a simple band-stop filter. Here a series resonant LC circuit is connected in parallel with the load. At *resonance*, the series resonant circuit offers very low impedance to current flow. This shorts or bypasses most of the current around

R_L and most of the input voltage is dropped across R_S. Above and below resonance, the impedance of the filter is much higher, and R_L is no longer bypassed. This type of filter is also referred to as a band-elimination filter, band-reject filter, or a notch filter.

Operational amplifier (op-amp) *active* filters can be used to implement all of the passive filter circuits discussed. RC networks are used in the input or feedback circuits. This makes the op-amp frequency sensitive. Some of the advantages of active filters include:

- High input resistance and low output resistance.
- No attenuation in the pass-band.
- Filtering is possible without using inductors.

Figure 44-21 shows an example of a simple low-pass active filter. Because there is no resistance in the feedback loop, the voltage gain of this particular filter is unity. The op-amp provides the high input impedance and low output impedance required for most filtering applications. As the input frequency increases above the cut-off frequency, the capacitive reactance decreases and reduces the noninverting input voltage. At very high frequencies, the capacitor becomes a short causing zero input voltage. The

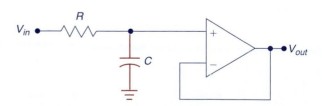

Figure 44-21 Low-pass active filter.

cut-off frequency for the low-pass active filter is found using the equation:

$$f_{CO} = \frac{1}{2\pi RC}$$

A **Bode plot** is a graph of a circuit's frequency response and is useful for analyzing filter circuits. Some simulation software programs include a virtual Bode plotter as part of their instrumentation set. Figure 44-22 is a typical circuit connection for the Bode plotter used in the Electronics Workbench simulation software package. The circuit shown is used to display the frequency response of a low-pass filter circuit. The Bode plotter generates a range of frequencies over a specified band and plots a graph of the circuit's frequency response.

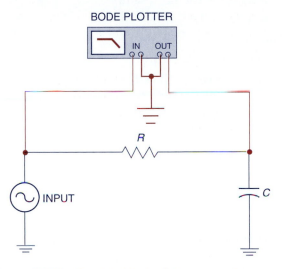

Figure 44-22   Simulated Bode plotter connection.

Review and Applications

Related Formulas

$X_L - 2\pi f L$

$X_C = \dfrac{1}{2\pi f C}$

$Z = \sqrt{R_2 + X_L^2}$

$Z = \sqrt{R_2 + X_C^2}$

$X = X_L - X_C$

$Z = \sqrt{R^2 + (X_L - X_C)^2}$

$V = IZ$

$I = V/Z$

$Z = V/I$

$V_R = I \times R$

$V_L = I \times X_L$

$V_C = I \times X_C$

$V_T = \sqrt{V_R^2 + V_L^2}$

$\theta = \text{Sin}^{-1} = \dfrac{\text{side opposite}}{\text{hypotenuse}}$

$\theta = \text{Cos}^{-1} = \dfrac{\text{side adjacent}}{\text{hypotenuse}}$

$\theta = \text{Tan}^{-1} = \dfrac{\text{side opposite}}{\text{side adjacent}}$

$V_T = \sqrt{V_R^2 + V_C^2}$

$V_T = \sqrt{V_R^2 + (V_L - V_C)^2}$

$f_r = \dfrac{1}{2\pi\sqrt{LC}}$

$I_C = \dfrac{V_C}{X_C}$

$I_R = \dfrac{V_R}{R}$

$I_L = \dfrac{V_L}{X_L}$

$I_T = \sqrt{I_R^2 + I_X^2}$

Apparent power $= V_T \times I_T$

True power $= V_R \times I_R$

Reactive power $= V_X \times I_X$

$\text{PF} = \text{Cos}\ \theta$

$\text{PF} = \dfrac{\text{True power}}{\text{Apparent power}}$

$f_{\text{CO}} = \dfrac{1}{2\pi RC}$

$Q = \dfrac{X_L}{R_L}$

Review Questions

1. List the three factors that can oppose the current flow in an ac circuit.
2. What is the total opposition to current flow in an ac circuit called?
3. State the ac phase relationship between voltage and current for each of the following load types: (a) resistive (b) pure inductive (c) pure capacitive.
4. In what way, if any, does an increase in frequency affect the:
   (a) R of a resistor?
   (b) X_L of an inductor?
   (c) X_C of a capacitor?

5. In a series circuit with R and X_L, what is the phase angle between:
   (a) I and V_R?
   (b) V_R and V_L?
6. How must out-of-phase voltages and currents be added?
7. How does the circuit phase angle for a series RL circuit differ from that of a series RC circuit?
8. Why can the V_L and V_C voltages in a series RLC circuit be directly subtracted from each other?
9. How do you determine if a parallel RLC circuit is acting inductively or capacitively?

10. How many frequencies are there at which a parallel *RLC* circuit will act purely resistive?
11. Under what condition does true power equal apparent power in a parallel *RLC* circuit?
12. State two methods of determining the power factor of a circuit.
13. Explain the basic function of a filter circuit.
14. What kind of filters use some type of amplifying device?
15. Describe the characteristics of a band-pass filter.
16. What characteristic of a capacitor makes it useful as a filter?
17. Explain what is meant by the bandwidth of a band-pass filter.
18. List three important output characteristics of a high *Q* series resonant band-pass filter.
19. List three advantages of active filters over passive filters.

Problems

1. A certain inductor has an X_L of 10 kΩ at 5 kHz. What will be its X_L at 10 kHz?
2. Find the impedance of a series combination consisting of a 50 Ω resistor and a 100 Ω inductive reactance.
3. Find the impedance of a series *RLC* circuit where $R = 20$ Ω, $X_L = 50$ Ω, and $X_C = 100$ Ω.
4. Find the current in a series *RLC* circuit where $Z = 200$ Ω and the applied voltage is 50 Vac.
5. Calculate the applied ac voltage to a series *RLC* circuit where $V_R = 70$ V, $V_L = 80$ V, and $V_C = 150$ V.
6. Calculate the total current flow of a parallel *RLC* circuit where the branch currents are as follows: $I_R = 100$ mA, $I_L = 50$ mA, and $I_C = 30$ mA.
7. Calculate the impedance of a parallel *RLC* circuit where the applied voltage is 12 Vac and the total circuit current is 3 mA.
8. In a given parallel *RLC* circuit $V_T = 120$ V, $I_T = 22.4$ A, $I_R = 10$ A, $I_L = 50$ A, and $I_C = 30$ A. Determine (a) apparent power (b) true power (c) reactive power (d) % power factor (lead or lag).
9. Calculate the cut-off frequency of an *RC* filter where $R = 1$ kΩ and $C = 1$ μf.

10. Calculate the *Q* of a series resonant band-pass filter that where $R_L = 2.5$ kΩ and $X_L - 250$ kΩ.

Critical Thinking

1. Refer to Figure 44-1. Assume an open circuit fault in the inductor. What would the value of the impedance be?
2. Refer to Figure 44-2. Assume a short fault in the capacitor. What would the value of the impedance be?
3. Refer to Figure 44-3. Assume the value of X_C to be 30 kΩ. What would the value of the impedance be?
4. Refer to Figure 44-4. Assume the voltmeter reads 250 V and the ammeter reads 3.33 A. What is the impedance of the circuit?
5. Refer to Figure 44-5. Assume the value of V_T to be 82.4 V. Determine the value of: (a) *I* (b) V_R (c) V_L (d) V_C.
6. Refer to Figure 44-8. Assume the value of *R* is increased to 80 Ω. Would the phase angle increase or decrease? Why?
7. Refer to Figure 44-12. Is this circuit acting inductively or capacitively? Why?
8. Refer to Figure 44-12. Assume the value of the branch current I_C was increased to 15 A. Determine the new value of: (a) I_X (b) I_T (c) the phase angle.
9. Refer to Figure 44-14. Assume the true power to be 480 W, the reactive power to be 640 VAR, and the apparent power to be 800 VA. Determine the (a) % PF (lead or lag) (b) phase angle.
10. Refer to Figure 44-15. Assume the value of *R* is decreased to 5 kΩ. Determine the value of the cut-off frequency.
11. Refer to Figure 44-18. In what way would you change the value of R_L to narrow the bandwidth? Why?
12. Refer to Figure 44-19. If $f_1 = 40$ kHz and $f_2 = 60$ kHz, what is the value of the bandwidth?

Portfolio Project

- Obtain copies of schematic diagrams for various types of communications equipment (radio, TV, computer, etc.). Identify the various forms of filter circuits found.

Circuit Challenge

Using a Simulator

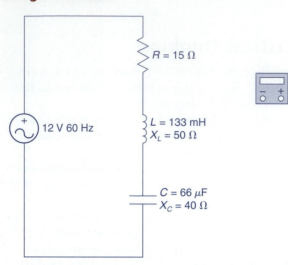

$R = 15\ \Omega$

12 V 60 Hz

$L = 133$ mH
$X_L = 50\ \Omega$

$C = 66\ \mu\text{F}$
$X_C = 40\ \Omega$

Procedure

(a) For the series RLC circuit shown, calculate X, Z, I, V_R, V_L, and V_C.

(b) Construct the circuit shown, using whatever simulation software package is available to you. Record the measured value of I, V_R, V_L, and V_C, for the simulated circuit.

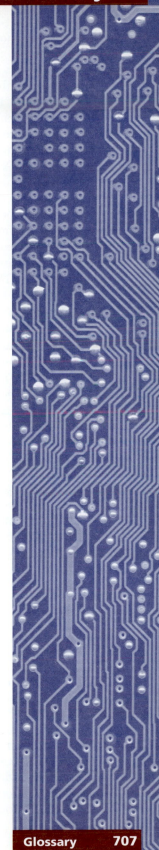

A

AC amplifier   An amplifier that boosts the voltage or current level of alternating-current signals.

Alternating current (ac)   Current from a power source that changes polarity with regular frequency.

Alternator   A device that supplies alternating current through mechanical, electrical, or electromechanical means.

American wire gauge (AWG)   Units of standardized wire sizes that range from the smallest, No. 40, to the largest, 4/0 (four aught).

Ammeter   An instrument used for measuring current.

Ampacity   The current-carrying capacity of conductors expressed in amperes.

Amperage (*A*)   The amount of electric current in amperes.

Ampere (A)   The basic unit of measurement for current.

Ampere-hour (Ah)   Unit of measurement for cell capacity. A cell or battery has a capacity of one ampere-hour if it can supply a current flow of one ampere for a period of one hour.

Amplifier   An electronic circuit that boosts the voltage or current level of a signal.

Amplitude modulation (AM)   A process of adding information to a radio wave by changing the amplitude of the wave.

Analog   Being continuous or having a continuous range of values as opposed to having discrete values. That branch of electronics dealing with infinitely varying quantities. Also referred to as linear electronics.

Analog-to-digital (A/D) converter   A circuit or device used to convert an analog signal or quantity to a digital form (usually binary).

AND logic   A principal logic operation characterized by having a true output only when all inputs are true.

Annunciator   Signal device using tabs or lights to indicate where a signal originated.

Anode   In a semiconductor diode, the anode is the P material that must be made positive with respect to the cathode in order to obtain current flow.

Apparent power   The product of the current times voltage in a circuit containing reactance and resistance; is measured in volt-amperes.

Armature   The moving part of an electromechanical device.

Audio amplifier   An amplifier designed to operate within the 20 Hz to 20,000 Hz range.

Autotransformer   A power transformer having one continuous winding that is tapped.

B

Ballast   A circuit element that provides starting voltage for certain types of lamps.

Bandwidth   The characteristic of certain electronic circuits that specifies the usable range of frequencies for which signals pass from input to output without significant reduction in amplitude.

Base   The element in a bipolar transistor that shares PN junctions with the emitter and the collector.

Battery   A direct-current power source composed of two or more chemical cells.

Bell transformer   Small transformer of special design, with built-in overload and short-circuit protection. Available in output voltage ranges, from 6 to 10 volts ac and 12 to 18 volts ac.

Bipolar-junction transistor (BJT)   A transistor that operates according to how a set of two PN junctions are handled: the two possible types are PNP and NPN.

Bit   A single binary digit (0 or 1).

Boolean algebra   A system of notation used to express logical operations such as AND, OR, and NOT in mathematical form.

Branch   One or more two-terminal elements connected in series.

Breadboarding   The assembly of a circuit without regard to the final locations of the components; used to prove the feasibility of a circuit.

Breakover   In an SCR, a transition into forward conduction caused by the application of excess anode voltage.

Bridge   An electrical instrument having four or more branches.

Bridge connection   A parallel connection that enables the frequent withdrawal of signal energy with no perceptible effect on the normal operation of the circuit.

Brush   A conductive block used to make sliding contact with an armature.

Busbar   A conductor used to carry a large amount of current or to make a connection between several circuits.

Byte   A 8-bit group that is commonly used to represent a number or code in digital electronics.

C
....

Cable   A stranded conductor or group of individual conductors insulated from each other.

Capacitance (C)   The ability of two conductors, separated by a dielectric, to store an electrical charge; measured in farads (F).

Capacitive reactance (X_C)   The opposition to current flow provided by capacitance; measured in ohms (Ω). It is inversely proportional to operating frequency and the capacitance value.

Capacitor   A device designed to provide a specified amount of capacitance.

Carrier   A radio-frequency oscillation that carries radio and television information.

Carrier frequency   The frequency of a carrier signal.

Cascade amplifier   An amplifier containing two or more stages arranged in the conventional series manner.

Cathode   In a semiconductor diode, the cathode is the N material that must be made negative with respect to the anode in order to obtain current flow.

Cathode-ray tube (CRT)   A vacuum tube that accelerates electrons toward a surface that emits visible light wherever the electrons strike it.

Cell   A device that transforms one form of energy into electrical energy.

Central processing unit (CPU)   The portion of a computer that performs the main logical, arithmetic, and data-transfer operations.

Circuit breaker   A circuit protection device which opens the circuit automatically when an overload or short circuit occurs.

Closed circuit   A circuit that provides a complete path for current flow.

CMOS   Complementary metal-oxide semiconductor. A type of IC fabrication which features extremely low power consumption. Uses opposite polarity field-effect transistors in its design.

Coil   A device consisting of a number of turns of wire that is used to introduce inductance into an electric circuit, to produce flux, or to react mechanically to a changing magnetic flux.

Cold resistance   The resistance of a heating element that is not conducting current.

Collector   The element in a BJT that accumulates the charges from the emitter and is controlled by the base.

Combination circuit   A circuit that includes components that are connected in both series and parallel configuration.

Combinational logic   Use of logic gates to produce desired output immediately. No memory or latching characteristics.

Commutator   A cylindrical device mounted on the armature shaft and consisting of a number of wedge-shaped copper segments arranged around the shaft.

Complete circuit   See Closed circuit.

Conduction   The movement of electrons through a conductor.

Conductor A low-resistance material that passes electric current.

Conventional current Term originally used for direction of current flow. Conventional current is thought of as running in the opposite direction as the actual electron flow.

Coulomb Base unit of electric charge equal to 6.25×10^{-25} electrons.

Counter emf A reverse voltage generated by magnetic lines of force that cut through the same conductor that carries the current that creates them.

Crystal A piezoelectric transducer used to change vibrations into electricity, control frequency, or filter frequencies. Also refers to the physical structure of semiconductors.

Current (I) The rate of flow of electrons through a conductor; measured in amperes (A).

Cycle One complete wave of alternating voltage or current.

D

Data bus In a computer system, parallel conductors used for communications between CPU, memories, and peripheral devices. Most systems have an address bus, data bus, and control bus.

DC amplifier A device that is capable of amplifying dc levels as well as ac signals.

Decibel (dB) A unit used to indicate the loss or gain of voltage or power between two points.

Decoder A logic device that translates from binary code to decimal.

Deflection The displacement of an electron beam from its path by an electrostatic or electromagnetic field.

Delta connection A connection in a three-phase system in which the individual phase elements are connected across pairs of the 3-phase power line.

Demodulator The stage in a receiver that converts changes in amplitude of the incoming signal into another form of signal.

Demultiplexer Combinational logic block that distributes data from a single input to one-of-X outputs. Can change serial data to parallel data.

Depletion In a semiconductor, the reduction of the charge-carrier density below the normal value for a given temperature and doping level.

Diac A silicon bilateral device used to gate other devices such as triacs.

Dielectric An insulator that is capable of concentrating electric fields.

Digital Related to digits or discrete quantities. Have a set of discrete values as opposed to a continuous range of values. That branch of electronics dealing with discrete signal levels. Most digital signals are binary: they are either high or low.

Digital IC An integrated circuit that is designed to work with the ON/OFF quality of binary operations.

Digital-to-analog (D/A) converter Conversion of a digital signal to its analog equivalent, such as a voltage.

Diode An electronic device that allows current to pass only from the cathode to the anode.

DIP Dual in-line package.

Direct current (dc) An electrical source with constant polarity.

Discharge The removal of electric charge from a battery, capacitor, or any other electric-energy storage device.

Doping The addition of impurities to a semiconductor to produce N- or P-type materials.

DOS Shorthand for MS-DOS (Microsoft disk operating system). Originally developed by Microsoft for IBM, MS-DOS was the standard operating system for IBM-compatible personal computers. A 16-bit operating system and does not support multiple users or multitasking.

Duty cycle A characteristic of a pulse waveform that indicates the percentage of time that a pulse is present during a cycle; the ratio of pulse width to period.

Dynamic braking A method in which a resistor is used to absorb kinetic energy to stop a motor quickly.

E

Eddy current Within the body of a conductor, an electric current induced by either movement through a nonuniform magnetic field or location in a region of changing magnetic flux.

Electric energy The energy carried by free electrons from a source to a load. Also, the potential energy of a stationary electric charge.

Electric power The rate at which energy is used in an electric circuit or in a load.

Electrolyte   The conducting solution used in primary and secondary cells.

Electrolytic capacitor   A capacitor with two electrodes separated by an electrolyte.

Electromagnet   A wire-wound steel or soft iron core that becomes magnetized when current flows through the coil.

Electromagnetic induction   The act of propelling free electrons into motion by means of a moving magnetic force cutting through a conductor.

Electromechanical relay   A relay that operates on the principle of electromagnetic attraction.

Electron   Negatively charged subatomic particles that make up electric current.

Electron flow   The flow of free electrons from the negative terminal of a power supply through the circuit to the positive terminal. Opposite to conventional current.

Electrostatic field   An electric field produced by stationary charges.

Emitter   That region of a bipolar junction transistor that sends the current carriers on to the collector.

Encoder   A logic device that translates from decimal to another code such as binary.

F
. . .

Farad (F)   The unit by which capacitance is measured.

Feedback   The return of output of a circuit to its input.

Field-effect transistor (FET)   A device that controls the flow of charged particles through a semiconductor material.

Filament   A metallic wire or ribbon used to produce heat by means of electron flow.

Filter   A circuit designed to separate one frequency, or a group of frequencies, from all other frequencies.

Flux (Θ)   Magnetic or electric lines of force.

Forward bias   Applying an external voltage to overcome the barrier potential of a PN-junction; this results in current flow through the junction. Also voltage applied across a semiconductor junction in order to permit current through the junction and the device.

Free electron   An electron that is not attached or held to any atom.

Frequency   The number of completed cycles of a periodic waveform during one second.

Frequency modulation (FM)   A process of adding information to a radio wave by changing the frequency of the wave.

Fuse   A protective device that interrupts current flow through a circuit when current exceeds a rated value for a period of time.

Fusible resistor   A protective device that limits current flow through resistance and protects against power surges.

G
. . . . .

Gain   The ratio of output to input by an amplifier, expressed in decibels.

Gate   Basic combinational logic device which performs a specific logic function (AND, OR, NOT, NAND, NOR).

Gauss (G)   The unit by which magnetic induction is measured.

Generator   A device that converts mechanical energy to electrical energy.

Ground   The common return path for electric current. A reference point connected to, or assumed to be at zero potential with respect to the earth.

Ground-Fault Circuit Interrupter (GFCI)   A special duplex receptacle or circuit breaker which provides extra protection against electric shock.

Grounding   The intentional connection to a reference conducting plane.

GUI   Graphical user interface. A program interface that takes advantage of the computer's graphics capabilities to make the program easier to use. The two most popular GUIs are Windows and Apple Macintosh operating systems.

H
. . . .

Hall effect   A transverse electric field developed in a current-carrying conductor that is placed in a magnetic field.

Hard drive   A stack of disks sealed inside a closed housing that is permanently mounted inside the computer.

Hard-wired   The physical interconnection of electric and electronic components with wire.

Heat sink   Any metal or device attached to a component, in order to absorb a portion of the heat.

Henry (H)   The unit by which inductance is measured.

Hertz (Hz)   The unit by which frequency is measured; in an ac waveform, the number of complete cycles per second.

Hum   An electrical disturbance occurring at the power supply frequency.

I

Impedance (Z)   The combination of resistance and reactance that creates opposition to current flow.

Impedance matching   Matching of the source impedance to the load impedance causes maximum power to be transferred.

Incandescent lamp   A lamp that produces visible light when a filament is heated.

Induced voltage   A voltage generated in a conductor or coil by a changing or moving magnetic force.

Inductance   Electrical property which opposes change in current.

Inductive reactance (X_L)   The opposition inductance offers to alternating current.

Inductor   A device that generates a counter emf that tends to oppose any change in current through the use of ac current flowing through a coil wire.

Insulation   A nonconducting material used to cover wires and components to remove shock hazards and to prevent short circuits.

Insulator   A high-resistance material that blocks the flow of electricity.

Integrated circuit (IC)   A silicon chip that has a number of diodes, transistors, and resistors etched upon it.

Interlock   A device actuated by the operation of another device with which it is directly associated. The interlock governs succeeding operations of the same or allied devices and may be either electrical or mechanical.

Internal resistance   Resistance contained within a power or energy source.

Internet   A global network connecting millions of computers.

Inverter   Basic logic function where the output is always opposite the input. Also called the NOT function. Also a device that changes dc to ac.

J

JFET   A field-effect transistor made up of a gate region diffused into a channel region. When a control voltage is applied to the gate, the channel is depleted or enhanced, and the current between source and drain is thereby controlled.

Jogging   A process by which a machine is controlled through the quickly repeated opening and closing of a circuit.

Joule   Base unit of energy.

Jumper   A short length of conduit used to make a connection between terminals around a break in a circuit.

K

Kilo (k)   Prefix for units of measurements, equal to a factor 10^3 or 1/1,000.

Kilohertz (kHz)   A measurement of frequency equal to 1000 Hz.

Kilowatthour   Unit of electric energy equal to 1,000 watt-hours or 3,600,000 watt-seconds (joules).

Kirchhoff's current law   The law used to describe the current characteristics of a parallel dc circuit with reference to the relationship between the total current flow in a parallel circuit and the current flow through each of its branches.

Kirchhoff's voltage law   The law used to describe the voltage characteristics of a series dc circuit by stating the relationship between the voltage drops and the applied voltage of a series circuit.

L

Lamp   Light- or heat-producing device usually consisting of a filament suspended in a vacuum.

Lampholder   A fixture to which a lamp is mounted.

Law of magnetic poles   Like poles repel, unlike poles attract.

LCD   Liquid crystal display. Nematic fluid in display changes reflectivity when energized changing display from silver to black characters.

Lenz's law The induced voltage always acts in a direction to oppose the current change that produced it.

Light-emitting diode (LED) A display device consisting of a semiconductor diode that emits light as current flows from the cathode to the anode across the PN junction.

Lightning arrester Circuit protection device which provides a ground path for atmospheric electricity.

Lightning rod Grounded metal rod attached to the highest point of tall buildings to protect them against lightning strokes.

Linear A circuit or component where the output is a straight line function of the input.

Load The power used by a machine or apparatus.

Logic A process of solving complex problems through the repeated use of simple functions that can be either true or false.

Logic probe A troubleshooting instrument which indicates logic levels and pulse streams in digital circuits.

M

Magnet An object which produces an invisible field of magnetic force in the surrounding space.

Magnetic field The region around a magnet or electromagnet in which the magnetic flux can be detected.

Magnetic flux See flux.

Magnetic materials Iron, nickel, cobalt and their alloys.

Magnetic pole One of the two points on any permanent or electromagnet at which the intensity of the magnetic force is strongest.

Magnetic saturation The condition in a magnetic material when all its magnetic atoms are aligned in the same direction or when it contains all the flux it can hold.

Magnetism The physical phenomena exhibited by magnets and electric currents that is characterized by fields of force.

Mega (M) Prefix for units of measurements, equal to a factor 10^6 or 1/1,000,000.

Memory The part of a programmable logic controller where data and instructions are stored either temporarily or semipermanently.

Metal-oxide semiconductor (MOS) A semiconductor device in which an electric field controls the conductance of a channel under a metal electrode called a gate.

Micro (Φ) Prefix for units of measurements, equal to a factor 10^{-6} or 1/1,000,000.

Microprocessor A CPU that is manufactured on a single integrated-circuit chip (or several chips) by utilizing large-scale integration technology.

Milli (m) Prefix for units of measurements, equal to a factor 10^{-3} or 1/1,000.

Modem Abbreviation for modulator-demodulator.

Modulator The section in a transmitter that combines the information to be broadcast with the carrier frequency.

Module An interchangeable, plug-in item containing electronic components.

MOSFET A field-effect transistor in which the insulating layer between the gate electrode and the channel is a metal-oxide layer.

Multimeter A multiple-range, multiple-function meter with either analog or digital display.

Multiplexer Combinational logic block that selects one-of-X inputs and directs the information to a single output. Also called a data selector. Can change parallel to series data.

N

NAND circuit A logic circuit that performs a NOT-AND inversion function.

National Electrical Code (NEC) A set of guidelines for the safe installation of electric wiring and apparatus.

Negative terminal That terminal of a voltage source which has an excess of electrons, and which therefore exerts a pushing force on the free electrons in a circuit connected to it.

Nonlinear A circuit or component that has a graph of current versus voltage that is not a straight line.

Nonvolatile memory A memory that retains data while its power supply is turned OFF.

NOR circuit A logic circuit that performs a NOT-OR inversion function.

North-seeking pole That pole of a permanent magnet or electromagnet which, when free to turn, points to the north magnetic pole of the earth.

Norton's theorem Method of reducing a complicated network to one current source with a shunt resistance.

NPN transistor   A BJT composed of a section of P-type material sandwiched between sections of N-type material.

Nuclear generating station   A nuclear powered thermal-electric generating station.

O
· · · · ·

Off-delay timer   An electromechanical relay that has contacts which change state at a predetermined time period after power is removed from its coil; upon re-energization of the coil, the contacts return to their shelf state immediately. A programmable logic controller instruction that emulates operation of the electromechanical off-delay relay.

Ohm (Ω) The unit by which resistance is measured.

Ohm's law   The amount of current flowing in a circuit is equal to the amount of applied voltage divided by the amount of resistance in the circuit.

Ohmmeter   An instrument that measures resistance.

On-delay timer   An electromechanical relay that has contacts which change state at a predetermined time period after the coil is energized; the contacts return to their shelf state immediately upon de-energization of the coil. A programmable logic controller instruction that emulates the operation of the electromechanical on-delay timer.

Operational amplifier   A highly stable dc amplifier that is resistant to oscillation; generally achieved by using a large negative feedback.

OR logic   An operation that yields a logic 1 output if one of any number of inputs is 1, and logic 0 if all inputs are 0.

Oscillator   A circuit that converts dc power into a waveform that is repeated at a constant frequency.

Oscilloscope (scope)   A test instrument that plots time against voltage, drawing a waveform on a screen.

Overload   A load in excess of that for which a circuit was designed.

P
· · · ·

Parallel circuit   A circuit in which there is more than one path for current flow.

Parallel transmission   The transmission of data in groups at the same time over multiple lines.

Period   The measurement of time in seconds required to complete one full cycle of a waveform.

Permanent magnet   A piece of hard steel or other alloy that has been strongly magnetized and will retain its magnetism indefinitely.

Permeability   The ability of a material to conduct lines of magnetic force.

Phase   A time relationship between two electrical quantities.

PN-junction   The boundary between N-type and p-type semiconductive materials.

PNP transistor   A BJT transistor composed of a section of N-type material sandwiched between sections of P-type material.

Polarity   The directional indication of electrical flow in a circuit. The indication of charge as either positive or negative, or the indication of a magnetic pole as either north or south.

Positive terminal   That terminal of a voltage source which has a lack of free electrons, and which therefore exerts a pulling force on the free electrons in the conductor connected to it.

Potentiometer   A variable-resistance-inducing switch that has terminals fixed at the extreme ends of the resistance and a third terminal connected to a wiper arm that is manually adjusted to select a desired amount of resistance.

Power factor (PF)   The ratio of the true power to the apparent power. It is equal to the cosine of the phase angle between the current and voltage in an ac circuit.

Power supply   A device used to convert an alternating current or direct current voltage of specific value to one or more direct current voltages of a specified value and current capacity.

Preamplifier   An amplifier that boosts a weak signal level before the signal is treated in any other fashion.

Primary cell   A device which converts chemical energy to electric energy by using up its ingredients, and which therefore cannot be recharged.

Primary winding   On a transformer, the winding that serves as the input connection.

Printed circuit board   A glass-epoxy card with copper foils for electric conductors and electronic components.

Pulse   A short change in the value of a voltage or current level characterized by a definite rise and fall time and a finite duration.

Pushbutton   A spring-loaded switch designed to operate by finger pressure.

R
....

Random-access memory (RAM)   A memory system that permits random accessing of any storage location for the purpose of either storing (writing) or retrieving (reading) information. Random-access memory systems allow the data to be retrieved and stored at speeds independent of the storage locations being accessed.

Reactance   Opposition to current which converts no energy (uses no power); a property of both inductance and capacitance.

Reactive power   Power which circulates between the source and the load as a result of reactance in the circuit; is measured in VARs (volt-amperes reactive).

Read-only memory (ROM)   A permanent memory structure where data are placed in the memory's storage locations at time of fabrication, or by the user at a speed much slower than it will be read. Information entered in a read-only memory is usually never changed once entered.

Rectifier   A solid-state device that converts alternating current to pulsed direct current.

Regenerative drive   A motor drive which regulates speed or stops a motor quickly by generating a voltage that opposes the supply voltage.

Relay   An electrically operated device that mechanically switches electric circuits.

Relay contacts   The contacts of a relay that are either opened and/or closed according to the condition of the relay coil. Relay contacts are designated as either normally open (NO) or normally closed (NC) in design.

Resistance (R)   The opposition to current flow, measured in ohms; an open circuit has infinite resistance.

Resistor   A component that offers a specified degree of resistance.

Resonance   A circuit condition in which $X_L = X_C$.

Resonant circuit   Circuit containing an inductor and capacitor tuned to resonate at a certain frequency.

Resonant frequency   The frequency at which a circuit or object will produce a maximum amplitude output.

Reverse bias   Applying an external voltage across a PN-junction to aid the barrier potential; the result is almost zero current flow through the junction. Also a polarity of current or voltage applied to a device that does not permit current flow.

***RF* power amplifier**   An amplifier that is designed to boost the power level of high-frequency signals.

Rheostat   Variable resistor with two terminals to vary current.

Ripple   At the output of a dc power supply, the small amount of ac voltage fluctuation that remains.

RMS   Root mean square. The value of an ac sine wave that indicates its heating effect, also known as the effective value. It is equal to 0.707 times the peak value.

S
....

Schematic diagram   A drawing of a circuit that shows connected components in the form of standardized symbols.

Secondary cell   A rechargeable storage cell, the basic unit of a storage battery.

Secondary winding   On a transformer, the winding that serves as the output connection.

Semiconductor   A doped, partly conductive material made from a highly purified insulator. Semiconductor materials can be either N type (wherein electrons carry the charges) or P type (wherein "holes" carry the charges).

Sensor   A device used to gather information by the conversion of a physical occurrence to an electric signal.

Sequential logic   A logic circuit whose logic state depends on asynchronous and synchronous inputs. Exhibit memory characteristics.

Serial transmission   The transmission of data one bit at a time.

Series circuit   A circuit in which the components or contact symbols are connected end to end, and all must be closed to permit current flow.

Shield   A barrier, usually conductive, that substantially reduces the effect of electric and/or magnetic fields.

Short circuit   An undesirable path of very low resistance in a circuit between two points.

Short-circuit protection   Any fuse, circuit breaker, or electronic hardware used to protect a circuit or device from severe overcurrent conditions or short circuits.

Shunt   A conductor that joins two points in an electric circuit and provides a parallel or alternative path through which a portion of current may pass.

Shunt winding   In a motor or generator, a winding that divides the armature current and leads a portion of it around the field-magnet coils.

Siemen (S)   Standard unit for conductance.

Silicon-controlled rectifier (SCR)   A semiconductor device that functions as an electronic switch.

Sine wave   A waveform whose amplitude at any time through a rotation of an angle from 0° to 360° is a function of the sine of an angle.

Software   The term applied to programs that control the processing of data in a computer, as contrasted to the physical equipment itself (hardware).

Solar cell   Also called a photovoltaic cell. A device which converts light energy into electricity.

Solder   A tin-lead alloy used to improve the conductivity and mechanical resilience of a connection or splice.

Solenoid   Electromagnet with a movable core or plunger that can perform mechanical operations.

Solid-state   The term that describes any circuit or component that uses semiconductors or semiconductor technology for operation.

Source   Alternate name for power supply or voltage source.

Specific gravity   The ratio of the density of electrolyte to that of pure water.

Static electricity   A stationary electric charge on a body.

Stator   The part of a motor or generator which does not rotate.

Step-down transformer   A transformer that reduces voltage at the output relative to the input.

Step-up transformer   A transformer that increases voltage at the output relative to the input.

Stepper motor   A type of motor that moves in discrete amounts with each electrical pulse.

Superposition theorem   Method of analyzing a network with multiple sources by using one source at a time and combining their effects.

Surge   A transient wave of current or power.

Switch   A device used to open and close circuits.

Synchronous speed   The rate at which the magnetic field in the stator of a motor rotates.

T

Terminal   A point used to make electrical connections.

Thermistor   A temperature sensor which exhibits a large change in resistance when subjected to a small change in temperature.

Thermocouple   A temperature-measuring device that utilizes two dissimilar metals for temperature measurement. As the junction of the two metals is heated, a proportional voltage difference is generated that can be measured.

Thevenin's theorem   Method of reducing a complicated network to one voltage source with a series resistance.

Three-phase   AC power consisting of three alternating currents of equal frequency and amplitude, each differing in phase from each other by one-third of a period.

Thyristor   A semiconductor switching device that can be bidirectional or unidirectional and have from two to four terminals.

Timer   In relay-panel hardware, an electromechanical device that can be wired and preset to control the operating interval of other devices.

Tolerance   The range above or below a specified value in terms of an allowable percentage or variance.

Torque   The measure of a motor's rotary force.

Transceiver   A device that both transmits and receives signals.

Transducer   A device used to convert physical parameters such as temperature or pressure into electric signals.

Transformer   A device that converts a circuit's electrical energy into a circuit or multiple circuits with output voltages and current ratings that differ from the input.

Transistor   A three-terminal active semiconductor device composed of silicon or germanium, which is capable of switching or amplifying an electric current.

Transmission line   A system of one or more electric conductors used to transmit electric signals or power from one place to another.

Transmitter　A device that places information onto a radio-frequency carrier signal that can be broadcast from an antenna.

Triode alternating current semiconductor switch (Triac)　A semiconductor switch that functions as an electrically controlled switch for alternating current loads.

True power　The product of the applied voltage and the current through a resistor; is measured in watts.

Truth table　A table that portrays information as either true or false.

TTL　Transistor-transistor logic. A type of digital IC fabrication using bipolar junction transistors.

U

Underwriters' Laboratories (UL)　The organization that tests and approves electrical devices and equipment for use in the United States.

V

Variable　A factor that can be altered, measured, or controlled.

Variable resistor　A device that can be adjusted across a specified range of resistance values.

Volt (V)　The unit by which voltage is measured.

Voltage (*V*)　In a complete circuit, the pressure that causes electrons to flow.

Voltage divider　A series circuit that divides the voltage.

Voltage drop　The difference in potential between two points caused by a current flow through a resistance or impedance.

Voltage ratio　The ratio describing primary-to-secondary voltage in a transformer.

Voltmeter　An instrument used to measure voltage.

W

Watt (W)　The unit of power equal to the work done by one joule per second or to the rate of work represented by a current of one ampere under a pressure of one volt. The unit by which wattage is measured.

Wattage　An amount of power expressed in watts.

Watt-hour (Wh)　A unit of energy equal to the power of one watt operating for one hour.

Wiring diagram　A graphic representation of the conducting path in a circuit or system of circuits.

Wye connection　A connection in a three-phase system in which one side of each of the 3-phases is connected to a common point or ground; the other side of each of the 3-phases is connected to the 3-phase power line.

Z

Zener diode　A semiconductor device that is designed to operate at a specified voltage level; used as a voltage regulating device.

A

Note: **Bold** page numbers indicate material in tables and figures.

AC motor, 373–379
  branch circuit and, **397**
  magnetic field in, 373
  SCR soft-start control of, **587**
AC power amplifier, 568–570
AC voltage amplifier, 559–568
Actuator, 655
A/D conversion, 96–97, 100, 642–644
Adjustable resistor, **126**, 126
Air-core inductor, 422
Alarm systems, **282**, 282–287
Alkaline cell, 212, 220
Allen keys, 21–22, **22**
Alnico, 189
Alternating current, 45, **46**, 260, 694–706
  current in series, 698–700
  as energy source, 76, **76**
  generator principle of, 261–262
  impedance in series, 695–697
  induction motor, 402–403
  power in, 436–438
  three-phase, 266–267
  voltage in series, 698–700
  waveforms of, 262–265, **373**
  (*See also* AC motor)
Alternator, 49, **49**
  automobile, 270
Aluminum, 117
Aluminum-leaf electroscope, 41–42, **42**
Ambient temperature, 118
American Wire Gauge table, 115, **116**
Ammeter, 92, **93**, 99–103
  clamp-on, 18, **18**, 102, **102**
  types of, **101**
Ampacity, 115–117
Ampere, 52
Ampere hour, 210
Ampere turn, 203
Amplifier tone control, 571–572
Amplifier volume control, 571
Amplitude, 494
Amplitude modulation, 499
Analog IC, 600, **600**
Analog meter, 92, **92**
  reading, 94–95
  scales on, **94**
  types of, 92
  typical movement of, **92**
Analog multimeter, 93, **93**, **95**
AND gate, 82, **82**, 617–618, 624–625, 626, 632
Annunciator, 276
  back-light, **276**
  circuit for, 276–277

Apparent power, 437, 445–446
Appliance grounding connection, 309–311, 317, **318**, 339–341
Appliance plug, 338
Arc tube, 359
Area protection, 282
Arithmetic-logic unit (ALU), 626, 651
Armature, 262
Arming device, 286
Armored cable, 298
Artificial respiration, 10–11, **11**
Associative law, 625
Atom, **31**, 31–33
Atomic mass, 32
Atomic number, 32
Atomic structure, 31–33
  Bohr model of, **31**, 31–33
  of conductors, 33–34, **34**
  of insulators, 34, **34**
Attenuator, 704
Audio amplifier, 570–572
Audio frequency generator, 497–498
Audio transformer, 452
Auto ranging, 95, 107
Automatic telephone dialer, 286
Automotive defogger, 124, **124**
Automotive look-up table, 658
Automotive scanner, 638
Autotransformer, 450–452, 522, **523**
Autotransformer starter, 404

B

Ballast, 354–355, **356**
Bandwidth, 567
Bar-code scanner, 470–471
Battery, **47**, 209–222
  in basic circuit, 56
  calculating internal resistance of, **216**
  carbon-zinc dry cell, **212**, 212
  discharge, 217, 221
  energy capacity of, 210
  lead-acid type, 216–219
    components of, 216–217
    testing, 218–219
  life of, 106
  nickel-cadmium type, 220
  nickel metal hydride (Ni-MH), 220
  parallel-connected, 214–215
  recharging, 216–219, 221–222
  series-connected, 213–214
  series-parallel connection, 215–216
  shelf life of, 211
  testing, 216
  voltage ratings of, 96, 210, **211**

Battery backup, 287
Battery charger, 221–222
Bell circuit, 205–206
Bell transformer, 274
Bias
  in bipolar-junction transistors, 539
  in diodes, 508–509
  fixed-base, 563
  forward, **107**, 508–509, 513, 539
  reverse, **107**, 509, 513, 514, 523
Bimetallic thermal-overload relay, **399**
Binary addition, 626–630
Binary arithmetic, 626–630
Binary counter, **638**, 638–642
Binary numbers, 614–616
  converting to decimal, **615**
  using switches to enter, **615**
Binary-to-decimal encoder, **632**
Bipolar-junction transistor (BJT), 538–543
  current gain in, 540–542
  Darlington pair of, 543
  operation of, 540–542
  rating of, 542
  switching action of, 556
  testing, 542–543
BJT common-emitter amplifier, 539–540, 559–560
BJT timed-switching circuit, 558–559
BJT voltage-divider biased circuit, 563, **564**
Black-box approach, 390–391
Bleeding, 8
Block diagram, 78, **80**, 356
Blower motor, **371**
Bode plot, 706
Bohr, Neils, 31
Bonding, 310
Bookmark, 671
Boolean algebra, 624–625
Boolean search, 670–671
Bound electron, 33
Branch, 163
Branch circuit, 292
  for motor, **397**
  residential, 315–332
    color-code chart for, **319**
    conductor size requirements in, 296–297
    distribution panel connections to, 294
    maximum outlets allowed, 267
Breadboard, 85–86, **87**
Bridge circuit, 500–502
Bridge rectifier, 525, 584–585
Bridging, 482
Brownout, 243–244
Brush, **371**
Burn, 8–10

Bus, 656
Byte, 615

C

Cable, 113
  coaxial, 70–72
  forms of, **113**
  nonmetallic sheathed, 297, 298–300
  protecting, 299–300, **300**
  support of, 298–300, 299
  two-conductor, **274**
Cable insulation stripper, 22, **300**
Cadmium sulphide photocell, 468
Capacitance, 427–434
Capacitive reactance, 432–434, 696
Capacitive sensor, 470
Capacitor, 427
  charging circuits for, **427**
  discharging circuits for, **427**
  microphone, 461–462
  ratings for, 428–430
  troubleshooting, 438–439
  types of, 430
Capacitor-choke filter, **528**
Capacitor filter, 528
Capacitor-start motor, **377**
Carbon-composition resistor, **125**, 125
Carbon microphone, 460–461
Carbon-zinc cell, 212, **212**
Card-access system, 278, **278**
Carriers, 506
Cartridge fuse, 229–230
Cascade amplifier, 566
Cathode, 354–355
Cathode-ray tube, 488–489, 654
CD-ROM, 465–467, 664
Cell
  parallel connection of, 214–215 (*See also* Parallel circuit)
  series connection of, 213–214 (*See also* Series circuit)
  series-parallel connection of, 215–216 (*See also* Series-parallel circuit)
Center punch, 21, **21**
Center-tapped transformer, 522, 524
Central processing unit (CPU), 651, 656–658, 664
Centrifugal switch, 376
Charge, measurement of, 40, 52
Chip resistor, 125
Chips, 596
Chisel, 23, **23**
Circuit
  AC-to-DC converters, 96
  analog-to-digital converters, 96–97, 100, 642–644
  automotive defoggers, 124, **124**
  basic electric, 56, **57**
  computer-simulated, **87**, 87–88

Circuit *(cont.)*
  duplex receptacle, switch, and lamp, 323–324
  energy limiting, 274
  filter, 527–529, 702–706
  four-way switch, 307, 326, **328**, **329**
  identifying quantities in, 128–129, **129**
  magnetic, 202–204
  parallel, 162–168, 173–174, 700–702
  pull-chain lampholder and duplex receptacle, 319
  series (*See* Series circuit)
  shorted, 104–105, 158, 168, **226**, 226–227, **227**, 341, 397–398
  simple, 80, **81**
  switch and lampholder, 319–322
  switched split-duplex receptacle, 325–326
  symbol for, 78
  three-way switch, 306–307, 326, **328**
  two lamps with one switch, 322–323
  voltage-conditioner, 96
  voltage-divider, 131
Circuit breaker, 56, 76–77, **77**, 232–235
  GFCI connection of, 342–343
  installation of, **296**, 296–297
  rating of, **228**
  resetting of, 232, 234–235
Circuit-control diode, 512–513
Circuit diagram, types of, 78, **80**
Circuit extender, **485**
Circuit symbol, **79**
Circular mil (CM) area, 115
Cladding, 475
Clamp, 299
Clamp-on ammeter, **18**, 18, **102**, 102
Clamping diode, 513
Class-A amplifier, 569
Class-B amplifier, 569
Clock, 635
Closed electric circuit, 56, **57**
Closed loop, 173, 409
CMOS (complimentary metal oxide semiconductor), 622–623
Coaxial cable, 70–72
Code-practice oscillator circuit, 573, **573**
Cogeneration, 242
Coil, 385
  left-hand rule for, 201
  magnetic field produced by live, 200–201
Cold-cranking amperes, 218
Cold-solder joint, 482
Collector-base junction, 538
Color code
  in branch circuits, **319**
  of five-band resistors, 128, **130**
  of four-band resistors, **128**, 128–129, **129**

Color code *(cont.)*
  safety, 13
  of wires, **115**
Combination circuit (*See* Series-parallel circuit)
Combination panelboard, **293**
Combination switch, 309
Combinational logic, 623–626
Common ground, 156
Common-mode rejection, 605
Common points, 85
Common terminal, 85
Commutation, 582
Commutative law, 625
Commutator, 267, 371
Compact disc player, 465–466, 664
Compact fluorescent bulb, 358–359
Comparator, 603–604, 629–630
Complimentary metal oxide semiconductor (CMOS), 622–623
Compound-type motor, 372
Compression connector, 64
Computer, 651
  cable connectors for, 69–72, **70**, **71**
  communication of, 656–658
  high-resistance connections in, 63
  processing information, 659–660
  simulated circuit in, **87**, 87–88
Computer-aided design, 478
Computer clock, 635
Condenser microphone, 461–462
Conductor, 33–34, 77, **77**
  ampacity rating, 115–117
  atomic structure of, 33–34, **34**
  in basic electronic circuit, 56
  forms of, 113–114
  insulation of, 114–115
  left-hand rule for, 199
  magnetic field around, 198–200
  measuring current flow in, **52**
  residential size requirements for, 296–297
  resistance, 117–118
  testing, **35**
Connection
  with butt-type connectors, 64, **65**
  cord, 337–339, 351
  with crimp-terminal lugs, 64, **65**
  electric, 351
  importance of correct, 63
  in light fixtures, 351
  preparing wire for, 63–64
  resistor, 129–134
  with screw-type terminals, 63, 64, **64**
  with set-screw connectors, 65, **67**
  by splicing, 64, **65**
  with twist-on connectors, 65, **66**
Connector
  crimp-on type, 64, **65**
  locking, **70**
  mechanical type, 65–66, **67**
Constant-current source, 675
Constant-voltage source, 675–676

Contact, 206–207, **386**
Continuity tester, 18–19, **19**, 34–35, **35**, 104, **105**
555 timers used as, 609
Continuous-current rating, 227
Control circuit, 386
Control device, 56, 77, **77**
parallel-connected, 83–84, **84**
series-connected, 80–82
Control panel, 284
Copper, 33, 113, **116**, 117, 118, 479
Cord, 114
connectors for, 337–339, 351
lamp, 336–337, 350–351
protecting from stress, 338–339
safety precautions for, 2, **3**
types of, 336–337
Corrosives, 13
Coulomb, 41, 52
Counter-emf, 400
Coupling, 563
Covalent bond, 506
Crimp-on connector, 64, **65**
Crimping tool, 64
Crowbar circuit, 583–584
Crystal cartridge, 462
Crystal microphone, 48–49, 461
Current, 8, 43–45, **46**, 52, 133, **138**, 139
effect of changes in resistance on, **140**
effect of changes in voltage on, **140**
effect on body, **9**
in electromechanical relay, 385–386
Kirchhoff's current law, 173–174
measurement of, 52, **53**, 99–103
Ohm's law in calculating, **141**, 141–142
relationship among voltage, resistance, and, 56–58, **140**
in series circuit, **130**, 150–151, 695–700
symbol for, 52
Current capacity, 226
Current limited voltage supply, 500
Current rating, of protective devices, 227–228, **228**, **312**
Current sinking, 610
Current sourcing, 610
Current transformer, 452
Cycle, 263

D

D flip-flop, 635–636
D-type connector, 72
DC amplifier, 559
DC motor, 369–372
electromagnetic type, 371–372
permanent-magnet type, 369–371, **370**
speed control of, **373**

DC stepper motor, 655
DC-to-DC power converter, 530
DC voltage supply, 500
batteries, 46–47
static electricity, 40–43
Decade counter, **640**
Decimal-to-binary encoder, **631**
Decoder, 471, 631–632
Delay flip-flop, 636
Delayed loop, 286
Delta-stator connection, 266–267
Demultiplexer, 634
Depletion mode, 547
Depletion region, 508
Desoldering, 482–483
Diac, 590–591
Dials, 251–253
Difference amplifiers, 606
Diffusion, 508
Digital display, 92, 97
Digital electronics, 614
Digital IC, 600
Digital logic probe, 644–646
Digital meter, 92, **92**, 92–94, 251–253
circuitry of, **92**
reading, **95**
types of, 92, 95–96
Digital multimeter, **18**, 18, 93–96
Digital-storage oscilloscopes (DSOs), 496
Digital-to-analog conversion, 642–644
Dimmer switch, 308–309
Diode, 268
bias in, 508–509
despiking, 513
operating characteristics of, 509
practical applications, 512–513
rating of, 509–510
special purpose, 514–516
testing of, 510–511
use of semiconductors in, 34, 505–507
Diode test, 107, **107**
Direct current, 45, **46**, 260, **260**
computing average DC value, 526
as energy source, 76, **76**
generation of, 267–268
produced by photovoltaic cells, 45–46, 242
sources of, 45–49
static charges produced by, 42
types of, **260**
(*See also* DC motor)
Disarming, 286
Discrete circuit, 597
Distortion, 564–565
Distribution law, 625–626
Distribution panel, 292
Door chime, circuit for, 274–276
Door latch mechanism, 276–278
Door-lock circuit, 276–278
Dopant, 506
Doped silicon wafer, 596
Double-insulated appliance, 341

Down counter, 641–642
Drill bits, types of, **23**
Drill chuck, 23
Drum switch, 417, **417**
Dry cell, 212
Dryer cable, 339
Dryer circuit, 318–319, **319**
Dryer receptacle, **339**
Dual in-line package, 597–598
Duplex receptacle connection, 303, **304**, 316, 323–324
DVD (digital versatile disk), 467
Dynamic breaking, 407–408
Dynamic microphone, 461

E

Eddy current, 445
Effective (rms) value, 265
Efficiency, 379–380
Electric braking, 408
Electric charge, law of, 40, **41**
Electric drill, 23, **23**, **341**
Electric energy, 244, 249–257
calculating, 250–251
costs, 253
management of, 253–257
Electric field of force, 40
Electric motor, 367–381
AC-type, 373–379, 402–403
DC-type, 369–372
developing torque in, **369**, 379–380
efficiency of, 379–380
induction, 374–375, **375**, 402–403
permanent-magnet type, 369–371, **370**
principle of, **268**, 368–369
right-hand rule for, 368–369
thermal overload protection in, 235
troubleshooting, 380, **380**
Electric power, 55, 239–246
in AC circuits, 436–438
calculating, 245–246
measuring, 246
rating of electrical devices, **55**
in series circuits, 151–152
Electric shock, 6–8, 10
Electrical interlock, 406–407
Electrical outlet
overloading of, 2
safety precautions for, 2, **3**
Electrical polarity, 45, 156, 189–191, **304**
Electrical receptacle, 303–305
Electricity, **33**, 33
high-voltage transmission of, 243–244
Electrodes, 46, 210
Electrolyte, 46, 210, 216–217
Electromagnet, 202–207
basic construction of, **202**
in record/play tape heads, 462–463

Electromagnet (cont.)
  self-shielding, 202
  strength factors of, **202**
  toroid-core, 202
  uses for, 204–207
Electromagnetism, 197–207, 371–372
Electromechanical relay, 385–388
  contact arrangements of, **386**
  controlling high currents with, 385–386
  controlling high voltages with, 385–386
  multiple-switching operation, 386
  operation of, 386–388
  rating of, 386–388
  symbol for, **385**
Electromotive force (emf), 43–45, 203
  counter-, 400
  generating
    chemical reaction, 46, **47**
    heat, **47**, 47–48
    light, 45–46, **46**
    mechanical-magnetic, 49
    piezoelectric, 48–49
  sources of, 45–49
Electron, 31
  bound, 33
  direction of flow, 58–59
  energy level of, 32–33
  magnetism and, 193–194
  shell arrangement of, **32**, 32–33
Electron tubes, 35–36
Electronic air cleaner, 42, **43**
Electronic combination-lock circuits, 559, **560**
Electronic devices, 35–36
Electronic mail (e-mail), 669–670
Electronic-overload devices, 399
Electronic sirens, 573
Electroscope, 41–42, **42**
Electrostatic discharge, 45, 548–549
Electrostatic field, 40, 427
Electrostatic lines of force, 40
Emitter-base junction, 538
Enamel insulation, 114–115
Encoder, 630
End-of-line resistor, 285
Energy, 250
  converting electric to mechanical, 368
  kinetic, 55–56
  mechanical, 49, 269–270, 368
Energy capacity, 210
Energy density, 211
Energy-efficient motors, 379–380
ENERGY GUIDE label, 256
Energy-limiting circuit, 274
Energy-limiting transformer, 452
Engine control systems, 662–664
Engine knock sensor, 49
Enhancement mode, 547
Environmental Protection Agency (EPA), 13
Equivalent resistance, 174

Etching, 478
Extension cord, 2, **3**, 338

F

Failure rate, 128
Fall time, 494
Fast-acting fuses, 228–229
Fastening devices, 25–27
Feedback circuit, 572
Ferrite-core inductor, 422, **422**
Ferrule-contact cartridge fuse, 229
Fiber optics, 72
  cable for, **72**, 516
Field coil, 262
Field-effect transistor (FET), 538, 543–549, 556
Filament, 349
File, 22, **22**, 666–667
File transfer protocol (FTP), 670
Film resistor, **125**, 125
Filter circuit, 527–529, 702–706
Fire
  classes of, **12**
  prevention of, 11–12
  procedure in event of electrical, 11
Fire extinguisher, 11, **12**
First aid, 8–10, **10**
Fish tape, 23, **23**
555 timer, 606–610
Fixed-base bias, 563
Fixed capacitor, 430
Fixed resistor, **126**, 126, 139
Floppy disk, 463–464
  drive for, **463**
Fluorescent lighting, 354–359
  instant-start type fixtures, 357–358
  preheat circuit in, 356–357
  preheat-type fixtures, **356**, 356–357
  rapid-start fixtures, 357
  troubleshooting chart for, **360**
Fluorescent tube, 354
Flux, 68, **68**, 480
  magnetic, 191–192, 202
Folding rules, **23**
42A block, 279
Forward bias, **107**, 508–509, 513, 539
Forward-breakover voltage, 582
Fossil fuel generating station, **241**
Free electron, 33, 53–55, 506
Free-running multivibrator, 608
Frequency, 264
Frequency counter, 500
Frequency modulation, 499
Frequency response, 571–572
Friction, 40, 42
Friction brake, 408
Function generator, 498, **498**
Fuse, 56, 77, **77**, 227–232
  connection of, **310**
  current rating of, **228**, 312

Fuse (cont.)
  fault indicators for, **229**
  ratings of, 227–228
  safety precautions for, 2
  testing, 232
  time-delay, 230–231, **231**, 399
  types, 229–232
Fusible link, **231**, 231–232
Fusible resistors, 125

G

Gallium arsenide, 515
Galvanometer, 92, 96, 444
Gate, 580
Gate trigger current, 582
General-purpose circuits, 296
Generating station, 240–241
  hydroelectric, 240, **240**
  solar-powered, 45–46, 242
  thermal-electric, 240, **241**
  wind-powered, 242
Generator, 49, **49**
  DC-type, 267–268
  magnetic field circuit for, 205
  prime mover for, 269–270
  self-excited, 268–269
  separately-excited, 268
  steam-powered, 240
Generator principle, 261–262
Germanium, 506
GFCI circuit breaker, 342
GFCI receptacle, 342
Glow-switch starter, **357**
Gopher, 670
Ground connection, in outlet box, **311**
Ground-fault circuit interrupter, 2, 317–319, 341–343, 400
Ground potential, 105
Grounded wire, 340
Grounding, 309–311
  of appliances, 309–311, 317, **318**, 339–341
  of electrical system, 309–310, **311**
  for protection, **310**

H

Hall-effect sensor, **102**, 103, 469–470
Halogen lamp, 347, **347**
Hammer, 20–21, **21**
Hand tools, 19–24
Hand-wired chassis, **475**, 475
Handset, 279
Hangar bars, 301–302
Hard disk, 464–465
Hardware, 664–666
Hazardous substances, 13–14
Hazardous waste, 13–14

Heat
creating with resistance, 124–125
as energy source, 47–48
in high-current circuits, **63**
and power dissipation, **164**
transformers and, 453–454
Heat shrink tubing, 67, **68**
Hcat-sink grease, 484
Heater-type cord, **336**, 338
Henry, 423
Hexadecimal number system, 616, **616**
High-intensity discharge lamp,
359–361
High-voltage transmission system, 243
Hold feature, 107
Holding current, 582
Hole, 507
Hollow-wall fasteners, 26–27, **27**
Hook-type probe, 484
Hook-up wire, 113
Horizontal sawtooth voltage, 488–489
Horseshoe magnet, **190**
in headphones, 460
uses of, 190
Hot wire, 293
HPD heater cord, **336**
HPN parallel lamp cord, **336**
Hydroelectric generating station, 240,
240
Hydrometer, 218, **219**
HyperText Markup Language
(HTML), 670
Hysteresis, 445

I

I/O devices, 659
Ignitable materials, 13
Ignition coil, 449–450
Impedance, 106, 434–436, 695–697
Impedance-matching device, 569
Impulse, transmission of, **52**
Incandescent lamp, 347–354, **349**
Incoming service, 292
overhead, 292
underground, 292, **293**
Induced rotor current, **375**
Inductance, 397, 422–427
Induction, 42
charging by, **43**
electromagnetic, 49, **49**
Induction motor, 374–375, **375**,
402–403
Inductive reactance, 424–426
Inductive sensor, 470
Inductor, 422
troubleshooting, 438–439
types of, 422
Infrared-emitting diode (IRED), 515
Infrared motion detector, 283–284,
362, **363**

Input block, 651
Input interfacing, 652
Instant loop, 286
Insulation
of conductors, 114–115
and dry skin, 2, **8**
removal of, 63–64
on repaired areas, 66–67
stripping, 63–64
Insulation removing devices, **22**, 22
Insulator
atomic structure of, **34**, 34
removing, **22**, 22
rupture of, 34
testing, **35**
Integrated-circuit (IC) chip, 107,
595–612
advantages and limitations of, 597
analog type, 600, **600**
construction of, 596–597
digital type, 600
with 555 timer, 609–610
hybrid, 597
linear, **600**, 604–605
monolithic, 597
op-amp type, 600–604
symbol for, 597–599
three-component, **597**
use of semiconductors in, 34–35
"Intelligent building," 257
Interlocking, 406–407
Internal resistance, 210–211, 216
Internet, 669–671
Interrupting-current rating, 227
Inverter, 242
Inverting-voltage amplifier, 602–603
Ion, 33, **33**
Ionization process, 33
Iron-core conductor, 422
Isolation, 522
Isolation transformer, 448–449

J

JFET (junction field-effect transistor),
544–546
JK flip-flop, 636–637
Jogging, 407
Joule, 55, 250
Junction field-effect transistor (JFET),
544–546

K

Keyboard, 665
Kilo-, 53
Kilowatt, 245
Kilowatthour, 56
energy meter, **56**, **251**, 251–253

Kirchhoff's current law, 173–174
Kirchhoff's voltage law, 173
Knife, 22, **22**, 63, **64**
Knife-blade cartridge fuse, **230**
Knockout, 300

L

Ladder-diagram language, 78,
660–662
Lagging current, 434–435
Laminate, 444–445, 475
Lamp
filament support for, 349
high-intensity discharge, 359–361
power ratings, **55**, 397–398
sodium, **361**
Lamp bases, 348
Lamp cord, 336–337, 350
Lamp dimmer circuit, 308–309, **591**,
591–592
Lamp socket, 350
Lampholder, 305
keyless connection of, 305
pull-chain type, 305, 319
wiring for switch and, 319–322
Large-power transistor, 542, **542**
Large-scale integration (LSI), 599
Laser diode, 516
Laser printer, 43
Latching, 581
Lead-acid battery
dry-charged, 218
maintenance-free, 218
recharging, 216–219
sealed, 218
Leads, 434, 698–699
Leakage current, 439
Least significant bit (LSB), 616
LED display, 106, 286, 515, 639, 654
LED flasher, 573–574
Left-hand coil rule, 201
Left-hand conductor rule, 199
Left-hand generator rule, 261–262
Level switch, 415, **416**
Light-emitting diode (LED) display,
106, 286, 515, 639, 654
Light energy, 45–46
Light fixtures
fluorescent, 356–358
installation of, 351–354
mounting, 351–352
recessed-type, 353
track-type, 353–354
wiring, **305**, 354
(*See also* Lamp)
Light sensors, 467–469
Lightening arrester, **235**, 236, **236**
Lightning rod, **235**, 235–236
Limit switch, 414–415
LINE, 343

Line voltage, 118–119, 452
Linear amplifier, 564–565
Linear IC, **600**, 604–605
Linear resistor, 145
Liquid-crystal (LCD) display, 106
Liquid-crystal display (LCD), 654
Lithium cell, 213
Lithium ion battery, 220
Live wire, 293
LOAD, 343
Load, 56, 77–78
　parallel-connected, 82–83, **83**
　series-connected, 80, **82**
Load device, 77–78, **78**
Load line, 564
Locked-rotor current, 400
Lockout, 6, **7**
Lodestone, 188
Logic gate, 616–623
Logic probe, 18, 19
Long-life bulbs, 348
Loop wiring systems, 298
Low-voltage signal systems, 273–287,
　400
Lumen, 348

M

Machine screw, 25–26
Magnetic cartridge, 462
Magnetic circuit, 202–204
　current of, 202–204
　reluctance in, 203
Magnetic circuit breaker, 233
Magnetic energy, 49
Magnetic field, 191–192, 422
　in AC motors, 373
　around live coils, 200–201
　around live conductors, 198–200
　patterns of, **192**
　shielding against, 192
　strength of, **198**
　using compass to trace, **198**
Magnetic flux, 191–192, 202
Magnetic lines of force, 191–192
Magnetic material, 188
　artificial, **188**
　natural, **188**, 189
　permeability of, 188, 202
Magnetic memory, 652
Magnetic-overload relay, 398, **398**
Magnetic polarity, 191
Magnetic pole, **189**, 189–191
　force between, 189–190
　law of, 189–191
Magnetic-reed relay, 389
Magnetic saturation, 193
Magnetic switch, 283
Magnetism, 187–195
　electron theory of, 193–194
　molecular theory of, 192–193
　retention of, **189**

Magnetite, 188
Magnetization process, 188
Magnetomotive force, **203**, 203
Manual starter, 402, **402**
Masonry fasteners, 26
Matter
　electron theory of, 31, 58–59
　phases of, 31–33
　structure of, 31–33
Measuring device, 18–19, 23
Mechanical energy, 49, 269–270, 368
Mechanical interlock, 406–407
Mechanical-lug cable connectors,
　65–66, **67**
Mechanically operated pilot switch,
　414–415
Medium-scale integration (MSI), 599
Mega-, 53
Megger, 114
Melting-alloy thermal-overload relay,
　399
Memory, 651–652
　battery, 220
　magnetic, 652
　random-access (RAM), 651, 652,
　　664
　read-only (ROM), 651, 652, 664
Memory block, 651
Memory cache, 664–665
Mercury cell, 213
Metal-oxide semiconductor
　(MOSFET), 546–549, 598
Metric prefixes, **138**, 138–139
Micro-, 53
Microammeter, 99
Microcomputer, 651, **653**
Microfarad, 428
Microphone, 48–49, 460–462
Microprocessor, 660–664
Microprocessor-based control panel,
　284
Microwave devices, 283
Milli-, 53, 95
Milliammeter, 99, 100, **100**
Millivoltmeter, 47
Modem, 652, 666
Modulation, 499
Molecule, 31
Monostable multivibrator, 607
MOSFET (metal-oxide semiconduc-
　tor), 546–549, 598
Most significant bit (MSB), 616, 628
Motherboard, 664
Motor (*See* Electric motor)
Motor contactor, 400–402, 406–407
Motor disconnect switch, 397
Motor drive, 408–412
Motor-generator (M-G) sets, 269–270
Motor overcurrent protection, 397–398
　external types of, 398–400
　internal types of, 398
Motor principle, 368–369
Motor starter, 400–405
Mouse, 665–666, 667–669

Multimeter, 18, 92–94
　analog-type, **95**
　digital-type, 18, **18**, 93–96
　special features of, 106–108
Multiple-switching circuit, 386
Multiplexer, **632**, 632–634
Multiplexing of display, **634**
Multiplier, 96, 531
Multispeed AC motor, 410
Mutual inductance, 423–424, 444

N

N-type material, 506–507, **507**
NAND gate, 619, **621**, 634, **635**
National Electrical Code, 117, 291,
　316–319, 336, 340
Needle-point probe, 484
Negative charge, 40, **40**
Negative electrode, 210
Negative terminal, 113
Neon test light, 18, **19**
Neoprene, 114
Net reactance, 435–436
Neutron, 31, **31**
Nibble, 615
Nickel-cadmium battery, 220
Nickel metal hydride battery, 220
Node, 174
Noninverting-voltage amplifier, 603
Nonlinear resistor, 145
Nonmetallic sheathed cable, 297,
　298–300
Nonsinusoidal oscillator, 572
NOR gate, 619
Normally closed (N.C.) contact, 385
Normally closed (N.C.) loop, 284–285
Normally open (N.O.) contact, 385
Normally open (N.O.) loop, 285
North-seeking pole, 191, 198
Norton's theorem, 687–691
NOT gate, 618–619, **621**, 624–625
Nuclear generating station, 240–241
Nuclear reactor, 240–241
Nucleus, 31, **31**
Null detector, 501
Number-sequence chart, 84–85, 319
Nut driver, 21, **22**

O

OFF-delay, 391, **392**
Off hook, telephone, 280
Ohm, 53, 424
Ohmmeter, 7, 18, 53, 92, **93**, 94
　measuring resistance with, 103–105
　reading, **95**
　using to test fuses, **232**
　using to test SCR, **588**
　using to test transformers, 452

Ohm's law, 58, 137–145
  for ac circuits, 697
  calculating current with, **141**,
    141–142
  calculating resistance with, 124,
    141, 142–143
  calculating voltage with, **141**, 142
  for electric circuits, 152
  formula triangle for, 143
  graphical form, 144–145
  for magnetic circuits, 204, **204**
  solving parallel circuits with,
    164–167
ON-delay, 391, **392**
On hook, telephone, 280
Open electric circuit, **57**, 157, 167–168
Open loop, 409
Open-loop control cycle, 663
Operating system, 666–667
Operational amplifier IC, 600–604, **601**
  output connection, **602**
  power-supply connection, **601**
  voltage amplifier type, 601–603
  voltage comparator type, 603–604
Optical disk drive, 465–467
Optocoupler, 550–551
Optoisolator, 550–551
OR gate, 84, **84**, 618, **620**, 624–625,
  626
Oscilloscope, 19, **19**, 488–497
  displaying waveforms and, **497**
  dual-trace type, 495–496
  front-panel controls, 489–490, **491**
  measuring voltage with, 490–493
  probe connection of, **496**
OSHA, 13
Outdoors
  lighting and, 363–364
  safety and, 2, **4**
Outlet box, 300–303
  ganged, 300–301
  ground connection in, **311**
  nonmetallic-type, 301
Output block, 651
Overcurrent-protection device, 227,
  311–312
Overhead incoming service, 292
Overload, 226
Overloaded circuit, **227**

P
••••

P substrate, 596
P-type material, **507**, 507–508
Parallel circuit, 162–168, 173–174,
  700–702
  current in, 163, **163**, 173–174
  power dissipation in, **164**, 164
  resistance in, 163–164, **164**
  solving, 164–167
  tracing currents in, **174**

Parallel circuit (*cont.*)
  troubleshooting, 167–168
  voltage in, 163, **163**
Parallel components, 177–179
Parallel conductor, 199–200
Parallel-connected resistor, 132–133
  as current dividers, 133
  total resistance of, 132
Parallel connection, 97
  of batteries, 214–215
  of control devices, 83–84, **84**
  of lamps, 319–322
  of load, 82–83, **83**
Parallel data transmission, 633, 657
Part-winding starter, 405
Peak-to-peak value, 264–265
Peak value, 264
Pencil-lead resistor, **124**
Penstock, 240
Pentavalent, 507
Perimeter protection, 282
Period, 264, 494
Peripheral cards, 665
Permanent-capacitor motor, **377**
Permanent magnet, 188–189
  ceramic-type, 189
  DC-motor type, 369–371, **370**
  uses of, 194–195
Permanent-magnet loudspeaker,
  459–460
Personal computer, 664–669
Phase angle, 494–495
Phase shift, **587**
Phasors, 698
Phonograph cartridge, 462
Phonograph pickup, 49
Photoconductive cell, 468–469,
  556–558
Photocopier, 42–43, **44**
Photodiode, 516
Photoelectric control unit, **361**, 362
Phototransistor, 550–551
Phototransistor switching circuit,
  556–558
Photovoltaic cell, 45, 242, 467–469
Phrase search, 671
Pictorial diagram, 78, **80**
Piezoelectricity, 48–49
Pigtail splice, **65**
Pilot devices, 412
Pilot-duty load, 414
Plastic tape, 66–67, **67**
Plated through hole (PTH), 475
Pliers, 20, **20**
Plug-and-play, 668–669
Plug cap, 337–338
  dead-front, 337
  polarized, 338
  three-wire, 338
  two-prong, 337–338
Plug fuse, 229
Plugging, 407
PN junction, 596
PN-junction diode, 508–509

Polarity
  electrical, 45, 156, 189–191, **304**
  magnetic, 191
  relative, 156
Polarized capacitor, 430
Pole, 374
Ports, 669
Positive charge, 40, **40**
Positive electrode, 210
Positive terminal, 113
Potential barrier, 508
Potentiometer, **127**
  taper of, 128
  trimmer-type, 127–128
  uses of, 127
Power amplifier, 568
Power dissipation, 143–144, 164
Power factor, 437–438, **438**, 702
Power formula, 143–144, 245
Power gain, 542
Power inverter, 575
Power loss, 119
Power rating, of electric devices, **55**,
  244, **255–256**
Power resistor, 125
Power supply, 520–533, 664
Power tools, 23–24
Power transformer, 443–455
  action of, 444–445
  circuits, 521–522
  laminations in, 444–445
  power transfer in, **445**
  terminal marking of, 449
  testing, 452–453
  types of, 450–452, **451**
Precision resistor, 126
Pressure switch, 415, **416**
Primary cell, 210, 213, 216
Primary coil, 444
Primary-resistance starter, 404
Primary winding, 448, 450
Prime mover, 269–270
Printed circuit board, **63**, 114, **114**,
  474–485
  component assembly and soldering
    on, 479–482
  planning layout for, 476–478
  printing, 478–479
  servicing, 482–485
  single-sided, **476**
  and surface-mounted resistors, 125
  unprocessed, **475**
Printer, 43, 666
Product-over-the-sum formula, 426
Programmable logic controller (PLC),
  656, 660
Programmable timer, 362
Protection device, 56, 225–236
Protector, 278–279
Protons, 31, **31**
Proximity sensor, 470
Proximity switch, 415–417, **416**
PS/2 port, 669
Pulse, 582

Pulse dialing, 280–281
Pulse generator waveform, 498
Pulse-width modulator, 410–411, **411**, 609–610
Punch, 21
Push-pull arrangement, 569–570
Pushbuttons, 84
PVC conduit, 297

Q

Quiescent value, 563

R

Radio frequency (RF) generator, 498–499
Radio frequency transformer, 452
RAM, 651, 652, 664
Random-access memory, 651, 652, 664
Range cable, 339
Range circuit, 318–319
Range receptacle, **339**
RC circuit, 699, 703
RC-time constant, 430–432
Reactive materials, 13
Reactive power, 436–437, 702
Read-only memory, 651, 652, 664
Reader, 277
Real power, 436
Receiver, 279
Receptacle, 303–305
Rechargeable battery
    alkaline, 220
    lead-acid type, 216–219
    lithium ion, 220
    nickel-cadmium type, 220
    nickel metal hydride (Ni-MH), 220
Rectification, 512
Rectifier, 96
    bridge, 525, 584–585
    full-wave circuits for, 523–526
    half-wave circuits for, 523
    three-phase type, 526–527
Rectifier diode, 512
Reduced-voltage starter, 403–405
Reflection, 448
Regenerative drives, 412
Register, 637–638
Regulated voltage supply, 500
Regulator, 410
Relative polarity, 156
Relay, 400–402, 655
Relay coil, **557**
Reluctance, 203
Remote-control lighting, 362–363, 386, **387**
Remote module, 362
Renewable fuse, 231–232
Reserve-capacity rating, 218

Residential service
    size requirements for conductors, 296–297
    three-wire distribution system, **294**
    wiring diagram, **295**, 319
    (*See also* Branch circuit)
Resist, 478
Resistance, 7–8, 53–55, **138**, 139
    of conductors, 117–118
    effect of changes on current, **140**
    in electric circuits, **76**
    improper connections and, 63
    internal, 210–211, **216**
    measurement of, **8**, 53, **54**, 103–105
    Ohm's law in calculating, 124, **141**, 142–143
    relationship among current, voltage, and, 56–58, **140**
    in series circuits, 128–131, 151
    and wet skin, 2, **8**
Resistance wire, 124
Resistor, 123–134
    color coding, **128**, 128–130, **129**, **130**
    five-band, 128, **130**
    four-band, **128**, 128–129, **130**
    identification of, 128–129
    nonlinear, 145
    overheated, **152**
    parallel connection, 132–133
    series connection, 129–131
    series-parallel connection, 133–134
    type of, 125–127
Resistor network, **126**, 126
Resistor rating, 124–125
Resolution, 106
Resonant circuit, 572
Resonant frequency, 436
Response time, 107, 228
Retentivity, 188
Reverse bias, **107**, 509, 513, 514, 523
Rheostat, 127, **127**
    used with electromagnets, 205
Right-hand motor rule, 368–369
Rigid conduit, 297, **297**
Ring magnet, 190–191
Ring wire, 279
Ripple, 523, 527
Rise time, 494, 497
RL series circuits, 698–699
RLC circuit, 700–702
RMS value, 107, **108**, 265
ROM, 651, 652, 664
Rotating magnetic field, 373
Rotor, 371
RS flip-flop, 634–635

S

Safe clothing, 5, **6**
Safety
    color codes, 13

Safety (*cont.*)
    common hazards, 2, **3**
    fire prevention and, 11, **12**
    in the home, 2, **3**
    meter, 105–106
    minimum requirements, 291–292
    in the outdoors, 2, **4**
    in the school laboratory, 2–5
    in the workplace, 5–6, **7**
Sampling rate, 496, 643
Saw, 21, **21**
Scanner, 470–471, 638, 666
Schematic diagram, 78, **80**, 319
    ladder-type, 78, 660–662
    using, 84–85
SCR operational check, **588**
Screw-base fuse, 229
Screw fasteners, 25–26
Screwdriver, 19–20
    proper and improper use of, **20**
    types of, 19–20
Search engine, 670
Secondary cell, 210
Secondary coil, 444
Secondary winding, 446–448, 450
Security lighting, 362–364
Self-inductance, 423
Self-tapping screw, 26
Semiconductor, **34**, 36
    in diodes, 34, 505–507
    metal-oxide semiconductor (MOSFET), 546–549
Semiconductor memory, 651, 652
Sensing devices, 45–46
Sensors, 467–471, 662–664
Sequence control, 417–418
Sequential logic, 634–637
Serial data transmission, 633
Series circuit, 80–82, **82**, 129–131, 149–158
    current in, **130**, 150–151, 695–700
    identifying quantities in, 150
    impedance in, 695–697
    power dissipation in, 151, **152**
    resistance in, 128–131, 151
    solving, 152–156
    tracing voltage in, **173**
    troubleshooting, 157–158
    voltage drop in, 151, 156
Series components, 177–179
Series-connected battery, 213–214
Series-connected control device, 80–82
Series-connected load, 80, **82**
Series-connected resistor, 129–131
Series-parallel circuit
    solving, 174–181
    troubleshooting, 182–183
Series-parallel-connected battery, 215–216
Series-parallel connected resistor, 133–134
Series resonant filter, 704
Series-type motor, 371–372
Service box, 292

Service mast, 292
Set-screw connectors, 65, **67**
Shaded pole motor, **378**
Shift register, 637–638
Short circuit, 104–105, 158, 168, **226**, 226–227, **227**, 341, 397–398
Shunt, 99–100
Shunt generator, 269
Shunt motor, 372, **372**
Signal generator, 497–499
Silicon, 506
Silicon-controlled rectifier, 580–588
    AC-operated, 584–587
    bridge rectifiers and, 584–585
    crowbar circuits in, 583–584
    DC-operated, 582–584, **583**
    principle of operations, 580–582
    testing, 587–588
Sine wave, 262–266
    current values and, 265
    cycles and, 263
    frequency and, 264
    period and, 264
Sine-wave voltage, measurement of, 492
Single-phase motor, 373
Single-phase reluctance motor, **379**
Sinusoidal oscillator, 572
SJ junior hard service cord, **337**
Slip, 375
Slip ring, 262
Slow-blow fuses, 228–229
Small-appliance circuits, 296–297
Small-scale integration (SSI), 599
Small-signal transistor, 542, **542**
"Smart building," 257
Soft start, 405
Software, 658–659
Solar cell, 45–46
Solar-powered generating station, 242
Solder, 68
Soldering equipment, 24
    cleaning of, **481**
    removing insulation with, 64
    temperature-controlled type, 480, **481**
Soldering gun, **24**, 68–69
Soldering pencil, **24**
Soldering process, 479–482
Solenoid, 205, 386
    in automobile starter motor circuits, **388**
    in door-chime circuits, 274
Solid-state relay
    optically coupled, 390
    symbol for, **390**
Solid-state relay (SSR), 389–391
Specific gravity, 218
Splicing, 64, **65**
Split-link receptacles, 317
Split-phase motor, **376**
SPT parallel lamp cord, **336**
Staple, 298
Starter motor circuit, **388**

Starting winding, 376
Static electricity, 40–43, 548–549
    DC voltage and, 42
    discharge of, 45
    photocopying and, 42–43, **44**
    propagation of, 40–41
    testing for, 41–42
Stator, 262, 371
Steel tape, **23**
Step-down transformer, 446, **447**, 521
Step-up transformer, 446, 522
Storage register, 637
Strain relief, 338–339
Strap, 298
Stray capacitance, 428
Subatomic particle, 31, **31**, 33
Summing amplifiers, 606
Superposition theorem, 676–681
Surface-mount technology, 125, 475–476, 480
SVT vacuum cleaner cord, **337**
Switch, 56
    double-pole, double-switch (DPDS), 369–370
    double-pole, single-throw type (DPST), 306
    four-way type, 307, 326, **328**, **329**
    single-pole, single-throw type (SPST), 306
    three-way type, 306–307, 326, **328**
Switching power supply, 531–532
Symbol
    for accident prevention, **5**
    in circuit diagrams, **79**
    for current, 52
    for electromechanical relays, **385**
    for integrated circuit chips, 597–599
    for solid-state relays, **390**
    for timing relays, **392**
    of triacs, 588–589
    (*See also* Color code)
Synchronous motor, 376
Synchronous speed formula, 374, 409–410

T

T flip-flop, 637
Table lamp
    basic parts and assembly of, **298**
    repairing, 349–350
Tagout, 6, **7**
Tape recorder head, 462–463
Taper, 128
Telephone circuit, 278–282
Telephone connection, **279**
Telephone dialing, 280–282, 286
Telephone systems, **279**
Telnet, 670
Temporary magnet, 188–189
Terminal, connecting wires to, 64–66, **70**

Thermal-overload protection, 233–235
Thermal-overload relay, 398–399, **399**
Thermistor, 469, 652
Thermocouple, 47
Thermopile, 47–48
Thermoplastic, 114
Thermostat, 415
Thevenin's theorem, 681–686
Three-phase alternating current, 266–267
Three-phase motor, **373**
    reversing, 406–407
    squirrel-cage induction type, **375**
Three-phase synchronous motor, **378**
Three-phase transformer systems, 453–455
Three-wire control, 412, **413**
Three-wire distribution systems, 292–293
    live wires in, 293
    residential, **294**
Thyristor, 506, 580–588
Time-current characteristics, 228
Time-delay fuse, 230–231, **231**, 399
Timing diagram, 618
Timing relay, 391–393
    pneumatic type, **391**, 391
    solid-state type, 391, **391**
    symbol for, **392**
Tinning, 68–69, **69**, 481
Tip wire, 279
Tone control, 571–572
Tone dialing, 281–282
Tool case, 24, **25**
Tool pouch, 24, **24**
Tools, 17–27
    common types of, 19–23
    grounding of power, 5, **7**
    measuring devices, 18–19, 23
    organization and use of, 24
Toroidal transformer, 452
Torque, **369**, 379–380
Toxic materials, 13
Transconductance, 546
Transducer, 459
    electric-input and -output, **459**
    types of, 459–467
Transformer, 205, 243, 274, 452–454, 521–522
Transformer action, 444–445
Transformer circuit, 205–207, **206**
Transistor, 389, 537–551
    bipolar-junction types, 538–543
        current gain in, 540–542
        operation of, 540–542
        rating of, 542
        switching action of, **557**
    unijunction types, 549–550
    use of semiconductors in, 34
Transistor-controlled relay, **388**
Transistor oscillator, 572–575
Transmission network, 243
Transmitter, 279

Triac, 389, 588–592
   circuit applications for, 590–591
   structure and symbol of, 588–589
   triggering modes for, **589**
Trickle charger, 222
Trilight lamp, 348–349
"Tripped" circuit breakers, 234
Trivalent, 507–508
Trouble code, 659
True RMS, 107, **108**
Truth table, 82, **82**, 617
TTL (transistor-transistor logic), 622
Tungsten filament, 347
Turbine, 240
Twist-on connectors, 65, **66**
Two-wire control, 412, **413**

U

UJT phase shifting for SCR, 586–587
UJT signal generator circuits, 574–575
Ultra large-scale integration (ULSI),
   599
Ultrasonic devices, 283
Ultrasonic sensor, 470
Underground incoming service, 292,
   293
Underwriters' Laboratories, 291
Universal motor, 372–373, **373**
Universal Serial Bus (USB), 669
Up counter, 641–642
URL (Uniform Resource Locator), 670
Usenet, 670

V

Vacuum fluorescent display (VFD),
   654
Valence electron, 33
Valence shell, 33–34, 506
Varactor diode, 514
Variable capacitor, 430
Variable frequency drive, 410
Variable resistor, **126**, 126–127
Varicap diode, 514
Vector, 698

Vibration-service filament, 349
Video monitor, 654, 666
Virtual ground, 603
Virtual meters, 109
Volt, 52–53
Volt-amperes reactive (VAR), 437
Volt-Ohm-Milliammeter (VOM), 93,
   94
Voltage, 8, 52–53, **53**, **76**, **138**, 139
   of batteries, 210
   effect of changes on current, **140**
   in electromechanical relay, 385–386
   Kirchhoff's voltage law, 173
   measurement of, 53, **53**, **54**, 96–99
   Ohm's law in calculating, **141**, 142
   relationship among current, resis-
      tance, and, 56–58, **140**
   series-aiding, 156–157
   series circuit, 698–700
   series-opposing, 157
   (*See also* Electromotive force)
Voltage amplifier, 568
Voltage-controlled oscillator, 609
Voltage drop
   polarity of, 156
   in series circuits, 151, 156, 173
   testing for excessive, 98, **99**
   zero-line, 118–119
Voltage gain, 542
Voltage multiplier, 531
Voltage rating, 96, 210, **211**, 227, 347
Voltage ratio, 446
Voltage regulation, 529–530
Voltage tester, 18, **18**, **100**
Voltage threshold levels, 644
Voltaic cell, 45–46, **210**, 210 (*See also*
   Battery)
Voltmeter, 18, 53, 92, **93**, 96–99
   using to test fuses, 232, **233**
   using to test transformers, 452–453

W

Watt, 55, 124–125, 244, 245
Wattage, 143–144
Wattage rating, 347–348
Watthour, 55–56
Wattmeter, 246

Weatherproof fixture, 343, **343**
Web browser, 670
Web server, 670
Web site, 670
Weber, 202
Western Union splice, **65**
Wet cell, 212, 213
Wheatstone bridge circuit, 501–502
Wind-powered generator, 242
Wire gauge, 115, **117**
Wire marker, 64, **64**
Wire size, 115
Wire splice
   methods of, 64, **65**, **69**
   soldering, 69, **69**
Wire splicing, 64, **65**
Wire stripper, 22, **22**, **63**, 63–64, **64**
Wire-wound resistor, 125, **125**
Wire-wrapping, **72**
Wireless, 283
Wiring diagram, 78, **80**, **85**
Wiring installation, general require-
   ments for, 294–298
Wiring projects, steps for planning, **85**,
   326–332
Wood-screw, 26
World Wide Web (WWW), 670
Wound-rotor motor controller,
   411–412
Wrench, 21, **22**
Wye-connected alternator, 266–267
Wye-delta starter, 404–405
Wye-stator connection, 266–267

X

XOR gate, 623–624

Z

Zener diode, 514
Zero-line voltage drop, 118–119
Zero-speed switch, 407
Zinc-chloride cell, 212
Zone, 286